# Flora Australiensis

## A Description of the Plants of the Australian Territory

### VOLUME 1: RANUNCULACEAE TO ANACARDIACEAE

GEORGE BENTHAM
FERDINAND VON MUELLER

CAMBRIDGE
UNIVERSITY PRESS

CAMBRIDGE UNIVERSITY PRESS

Cambridge, New York, Melbourne, Madrid, Cape Town,
Singapore, São Paolo, Delhi, Tokyo, Mexico City

Published in the United States of America by Cambridge University Press, New York

www.cambridge.org
Information on this title: www.cambridge.org/9781108037389

© in this compilation Cambridge University Press 2011

This edition first published 1863
This digitally printed version 2011

ISBN 978-1-108-03738-9 Paperback

# CAMBRIDGE LIBRARY COLLECTION

*Books of enduring scholarly value*

## Life Sciences

Until the nineteenth century, the various subjects now known as the life sciences were regarded either as arcane studies which had little impact on ordinary daily life, or as a genteel hobby for the leisured classes. The increasing academic rigour and systematisation brought to the study of botany, zoology and other disciplines, and their adoption in university curricula, are reflected in the books reissued in this series.

## Flora Australiensis

George Bentham (1800–84) was one of Britain's most influential botanists, whose own collection of plant specimens numbered more than 100,000. Although he donated his herbarium to the Royal Botanic Gardens, Kew in 1854, he continued to make significant contributions to the field, including this exhaustive, seven-volume work detailing the plant life of Australia, which was published from 1863 to 1878. It was part of a series of works commissioned by the British government to document the flora in its colonies. Using the extensive numbers of specimens at Kew – and with the help of Ferdinand Mueller (1825–96), a German botanist in Australia – Bentham was able to compile descriptions of more than 8,000 species of Australian plants, making these volumes the first completed compendium of the flora of any large continental area. Volume 1, published in 1863, introduces the project and describes 39 orders of the dicotyledon class of flora.

Cambridge University Press has long been a pioneer in the reissuing of out-of-print titles from its own backlist, producing digital reprints of books that are still sought after by scholars and students but could not be reprinted economically using traditional technology. The Cambridge Library Collection extends this activity to a wider range of books which are still of importance to researchers and professionals, either for the source material they contain, or as landmarks in the history of their academic discipline.

Drawing from the world-renowned collections in the Cambridge University Library, and guided by the advice of experts in each subject area, Cambridge University Press is using state-of-the-art scanning machines in its own Printing House to capture the content of each book selected for inclusion. The files are processed to give a consistently clear, crisp image, and the books finished to the high quality standard for which the Press is recognised around the world. The latest print-on-demand technology ensures that the books will remain available indefinitely, and that orders for single or multiple copies can quickly be supplied.

The Cambridge Library Collection will bring back to life books of enduring scholarly value (including out-of-copyright works originally issued by other publishers) across a wide range of disciplines in the humanities and social sciences and in science and technology.

# FLORA AUSTRALIENSIS.

# FLORA AUSTRALIENSIS:

## A DESCRIPTION

OF THE

# PLANTS OF THE AUSTRALIAN TERRITORY.

BY

## GEORGE BENTHAM, F.R.S., P.L.S.,

ASSISTED BY

## FERDINAND MUELLER, M.D., F.R.S. & L.S.,

GOVERNMENT BOTANIST, MELBOURNE VICTORIA.

VOL. I.

*RANUNCULACEÆ TO ANACARDIACEÆ.*

PUBLISHED UNDER THE AUTHORITY OF THE SEVERAL GOVERNMENTS
OF THE AUSTRALIAN COLONIES.

LONDON:
LOVELL REEVE AND CO., 5, HENRIETTA STREET, COVENT GARDEN.
1863.

·

JOHN EDWARD TAYLOR, PRINTER,
LITTLE QUEEN STREET, LINCOLN'S INN FIELDS.

# PREFACE.

For a general view of the progress of botanical discovery in Australia, and an enumeration of the Botanists, Navigators, Travellers, Collectors, or Residents who have supplied the materials for describing its Flora, or have published more or less of their descriptions, I must for the present refer to the valuable Essay on the Flora of Australia, prefixed by Dr. J. D. Hooker to his 'Flora of Tasmania.' Should life be spared to me to bring the present work to a conclusion, I purpose, with the last volume, to give a sketch of the labours of all those who, to my knowledge, have contributed to the investigation of the vegetation of Australia. But, in the meantime, I would mention in a few words, the principal sources from which I am now enabled to draw materials for the present Flora.

The chief foundation of the work may be said to be the vast herbarium of Sir William J. Hooker, with a few smaller collections under his charge at Kew. I need not here repeat the detail of the rich stores of Australian plants it contains, enumerated in Dr. Hooker's Essay, but I cannot forbear thus early expressing my acknowledgment of the liberality of the arrangements sanctioned by Sir William for the admission of botanists to these collections, for which he has made so many sacrifices, and amongst which I have been enabled to work as if they were my own, with the free use of one of the most extensive practical botanical libraries. Here also I have had the benefit of continual friendly assistance from Dr. J. D. Hooker, Assistant Director of the Royal Gardens, and from Professor D. Oliver, Librarian, who have invariably allowed me to consult them upon all points of difficulty which have arisen; from Mr. A. Black, the intelligent and zealous Curator, whose activity, combined with a very great knowledge of plants, has brought the herbarium into such a state of order that few of the additions which are continually arriving remain many months without being laid into their places; and from Mr. W. Hemsley, a young but able assistant, who has carefully checked my proofs with the herbarium as they have issued from the printer's hands. The value of this herbarium for a work like the present, is also greatly increased by the notes and determinations it contains from the hands of various botanists who have worked in it, and especially of Dr. Planchon, who had examined and corrected the determination of a large portion of the specimens it contained during several years that he had the charge of it. But the importance of this herbarium, will be best appreciated by the consideration that it contains specimens of almost every species described in the present work.* The very few exceptions will be found to be specially noted by a reference to the herbarium in which I have seen them, given in a parenthesis after the habitat, or by an indication of the sources whence the description has been derived.

To my friend Mr. J. J. Bennett, the Head of the Botanical Department of the British Museum, I am indebted for the important and essential aid derived from the inspection of the Australian herbarium of the late Robert Brown. This extraordinary collection, the main foundation of our knowledge of Australian vegetation, would be alone sufficient to show the powers of observation, the sagacity, the zeal, and industry of that eminent man, dwelt upon by Dr. Hooker, in the above-mentioned Essay. He seems during his short visits often almost to have exhausted the Flora of the points he touched at; his specimens are gathered with great judgment, and there still remain in his herbarium, in most cases, several of each species in an excellent state of preservation, and detailed descriptive notes on them all were made at the time. These specimens, now the property of Mr. Bennett, have been kindly brought by him successively to the British Museum for my use, where I have also been allowed to consult Mr. Brown's notes. Two or three small parcels have been unfortunately mislaid, but of those I have in some cases found specimens in a duplicate set laid out for the Banksian herbarium.

In the Banksian herbarium I have verified several species of which the types are there deposited, and inspected several of the original specimens of Banks and Solander, of which some, gathered above ninety years back, have never yet been published. Whilst at the British Museum, I should also gladly have availed myself of the valuable Australian collections there hoarded,—and certainly nothing can exceed the obliging

* All the specimens examined for the present work (often very numerous) are marked in the Hookerian herbarium in red ink.

readiness with which Mr. Bennett gives every assistance to those who come to visit the Botanical Department, and to myself in particular,—but the system now so long pursued by the managing trustees is one which interferes much with the use of those collections which, like Herbaria, are made for the purposes of science, not for the public gaze. It would appear as if the whole object were to accumulate stores, without caring to make them available for use. The rich herbaria collected at the public expense by the late A. Cunningham in his various expeditions under Captain King and others, by the Officers of the 'Beagle' under Captain Wickham and Captain Stokes, and many others either presented to the Museum or purchased out of the annual grants, have been stored away, many of them from a quarter to half a century, unarranged in their original parcels, without any thought of providing the staff and funds necessary to render them of use to scientific botanists. No system of separating duplicates for making exchanges has, I believe, been adopted. And for those who wish to work in the Botanical Department, notwithstanding the readiness of the officers to afford them every assistance, the want of a practical botanical library in the department, the regulations preventing the use of any apparatus for heating water, and the defective construction of the room as to light, are serious drawbacks.

With regard to the late A. CUNNINGHAM's plants, however,—a collection second only to R. Brown's in the influence it has had, by its variety and extent, on our knowledge of Australian Botany,—I have, I believe, been able to examine the whole of them. Besides the nearly complete set deposited in the Hookerian herbarium, MR. R. HEWARD, to whom Mr. Cunningham's private herbarium, containing the set he had reserved for himself, had been left, on hearing that I was engaged in the preparation of the present work, most generously presented the whole of his plants to the Kew herbarium, in order that I might there have the free use of them.

Another herbarium of which I have always had the free use, is that of my friend DR. LINDLEY, who, for the last thirty-five years, has ever been ready to afford me every assistance in my botanical works. I had already received from him, at the time, nearly complete sets of the plants of the late SIR WILLIAM MITCHELL's various expeditions; and I have now examined, in Dr. Lindley's own herbarium, the very few types of these or of other Australian plants published by him, which may have been wanting in the Hookerian herbarium or in my own, now part of the national collection at Kew.

I have found in the herbarium of the late SIR JAMES E. SMITH, now the property of the Linnean Society, the types of the Australian species described by him, chiefly in Rees's Cyclopædia.

With the few Australian species described from the herbarium of the late A. B. LAMBERT, I have had much difficulty. At his death the preparation of his collections for sale was so ill-managed, that it is very difficult to ascertain where any particular portions of it may now be deposited. A few have found their way to the Kew herbaria, many were purchased for Berlin and St. Petersburg, and other distant Continental towns; some were, I believe, bought by the British Museum, and are still lying among their unarranged collections; and some others, but, as I understand, not the Australian portion, are in the Fielding herbarium at Oxford. I have, therefore, in most instances been obliged to rely chiefly on circumstantial evidence for the identification of such of these plants as are only known by the brief diagnoses of G. Don and others.

Of the important and extensive West Australian collections of MR. JAMES DRUMMOND I have had for examination complete sets of excellent specimens in the Kew herbaria, and in the majority of instances I have seen them in different sets so as to check the one with the other. I have thus been enabled to identify nearly the whole of the species published by TURCZANINOW in the ' Bulletin de la Société Impériale des Naturalistes de Moscou.' As these collections are very generally distributed, I have quoted the numbers attached to the specimens where I could do it with any certainty. Unfortunately there is much confusion in some of these numbers, Mr. Drummond having recommenced a fresh series with each of the five collections he sent over, besides one or two supplementary sets. The first collection, of which many were published by Lindley and others, were not originally numbered, but numbers were afterwards added in a few additional sets sent home. In the Hookerian herbarium, owing to the belief at the time that these numbers were not certain enough for quotation, they were often not preserved; in most instances where they are kept there is no indication of which series they belong to, and in other herbaria I have often found them referred to a wrong series. These numbers cannot therefore be relied on absolutely for identification without checking them by descriptions.

To DR. O. W. SONDER, of Hamburg, Dr. Harvey's able collaborator in the ' Flora Capensis,' I have to offer my best thanks for the liberality with which he transmits to me for examination the whole of his Australian herbarium,—an invaluable aid, inasmuch as it comprises a nearly complete series of typical specimens of the PLANTÆ PREISSIANÆ. As many portions of that rich collection were confided for publication to such botanists as the late Dr. Steudel, it would have been impossible to identify them without such an inspection of authentic specimens. This herbarium contains also several authentic specimens of Labillardière and some other French botanists, and often also several of the plants sent over by Dr. F. Mueller, of which he himself had kept fragments only or nothing at all. I find also specimens authentically named by Steetz, Bartling, Schlechtendal, and other German botanists.

Thanks to the liberality with which the late P. B. WEBB distributed his duplicates, I have seen in various herbaria the majority of LABILLARDIÈRE'S plants; but as there were several others, described in the first volume of De Candolle's 'Prodromus' and other works, from the herbarium of the Jardin des Plantes, about which I had some doubts, I paid a visit, in January last, to Paris, where I met, as usual, with every attention on the part of the gentlemen connected with the establishment. I there verified these doubtful species up to the end of *Rutaceæ*, which I had then completed, and since then, my friend M. A. BRONGNIART, as the head of the botanical department of the museum, has most obligingly transmitted to me notes and flowers for examination of a few species belonging to the subsequent Orders.

With regard to the originals of the species described in BARON HUEGEL's 'Enumeratio Plantarum' and other works, published at Vienna, I was enabled to bring over with me specimens of several, especially of those which I had myself described, and I have identified many others by means of specimens compared with the Vienna types. Those published from F. BAUER's collections occur necessarily also in R. Brown's herbarium; and when I have had any doubts as to any of the remaining ones, they have been cleared up by full notes communicated to me by my friend DR. FENZL, Director of the Imperial Garden and Herbarium.

There remains for me to mention the very essential assistance received from the distinguished Government Botanist of Victoria, DR. FERDINAND MUELLER. His extensive journeys and important labours during the first ten years of his residence in Australia, have been adverted to by Dr. Hooker in the above-mentioned Essay. Since that time, his botanical explorations have been chiefly in the Victorian mountains and in the neighbourhood of Twofold Bay and Cape Otway, whilst his zeal, talent, and indefatigable industry have been still more fully exemplified in the various publications which have issued from the Melbourne press. Not to mention minor papers or reports on expeditions, we have a first volume of an elaborate illustrated quarto Flora of Victoria, under the title of 'The Plants indigenous to the Colony of Victoria,' and three octavo volumes, all but complete, of 'Fragmenta Phytographiæ Australiæ,'— comprising above a thousand detailed descriptions of plants, whose general accuracy will bear the test of a very close examination.  When indeed it was first contemplated to bring out a general Flora of Australia under Government sanction, Dr. Mueller was naturally looked to as the botanist the best qualified for undertaking the task of preparing it.; and in the hope that it would be entrusted to him, he had devoted his utmost energies to collecting the necessary materials.  But there was one indispensable step, the examination of European herbaria where the published types were deposited, which he was unable to take ; and it is a signal proof of the generosity of his disposition and the absence of all selfishness, that when it was proposed to him that the preparation of the Flora should be confided to me, on account of the facilities which my position here gave me for the examination of the Australian collections I have mentioned above, he not only gave up his long-cherished projects in my favour, but promised to do all in his power to assist me, a promise which he has fulfilled with the most perfect faith.  A joint work was at first thought of, but, independently of the ordinary drawbacks attending on joint works, the distance which separates us, requiring four months to obtain an answer to every trivial doubt or query, put this quite out of the question.  I alone am therefore responsible for the details of this work, for the limitation given to genera and species, for their characters and description.  But important observations have been frequently suggested by the published works of Dr. Mueller, or by his manuscript notes, which he has freely communicated; and a still more essential and generous contribution to the work has been the loan of the very rich herbarium he had amassed for the Australian Flora, which he remits to me in instalments.  One beneficial result to science of the course he has thus pursued is that there will be for future reference duplicate authentic specimens here and in Australia of the great majority of Australian species.

This herbarium comprises chiefly :—

1. The specimens collected by Dr. Mueller himself in the course of his extensive land-journeys in Australia (upwards of 20,000 miles), as well as during his residence in Victoria.  Of one important portion of these plants, the North Australian collection, the set in the Hookerian herbarium is better and more complete than his own.  Dr. Mueller at that time did not contemplate the publications he has since undertaken, and with his usual generosity he wrote to Sir W. J. Hooker, in 1857,

" You receive always the whole of the specimens of every rare kind, nothing of many species having been retained at all, or I satisfied myself with a solitary leaf, or flower, or fruit in many cases; . . . the plants being so much more useful at Kew than in Australia. All my wishes are concentrated upon the point to discharge my duties faithfully and to the satisfaction of the Government." (Hook. Kew Journ. ix. 195.) So also of several of those which he had in early days collected in the north as well as in Victoria and in South Australia, he sent the best specimens to Dr. Sonder for description and publication in Germany, and unfortunately, a great proportion of the principal botanical treasures of the northern expedition were destroyed by damp in the 'Messenger.' But of the results of Dr. Mueller's subsequent herborizations his herbarium contains good, instructive, and well-preserved specimens.

2. The collections made during various exploring expeditions in the interior of Australia, and entrusted to Dr. Mueller for determination or publication. These are necessarily, from the difficulties attending these expeditions, although highly interesting as to species, often fragmentary or unsatisfactory as specimens. Among the most important of them are those of MR. BABBAGE's expedition to the north-west interior of S. Australia, of MR. AUGUSTUS GREGORY's expedition to Cooper's Creek, and of MR. E. FITZALAN, in LIEUT. SMITH's expedition to the estuary of the Burdekin, all specially reported on by Dr. Mueller; of MR. J. M'DOUALL STUART, who, notwithstanding the obstacles opposed by the arduous nature of his journey, appears never to have neglected Natural History; and the collections made by MR. PEMBERTON WALCOTT and MR. MAITLAND BROWN, in MR. FRANCIS GREGORY's expedition to the north-west. As I have not been able always to make out from the labels which of these two gentlemen actually gathered the specimens, I have generally quoted them as the results of Mr. Gregory's expedition. The herbarium also contains some specimens from MR. LANDSBOROUGH's expeditions, and to this class I should perhaps add a large number of the late DR. LEICHHARDT's plants, entrusted to Dr. Mueller on loan by the trustees of the Sydney Museum on the proposition of SIR WILLIAM DENISON. These were chiefly collected in the back country from Moreton Bay during two years previous to his celebrated expedition, together with a few saved from the general wreck of the plants of that expedition. I have also seen a few of Dr. Leichhardt's specimens in the herbarium of the Paris Museum.

3. Collections made by gentlemen more or less employed as collectors for the botanical department at Melbourne, among whom, those who

have most contributed to the herbarium are :—Dr. H. BECKLER, who first collected for himself in the country to the back of Moreton Bay, and afterwards for Melbourne in the jungle-forest about the Hastings, Richmond, Macleay, and Clarence rivers, and, still more recently, between the Darling and the Barrier range, as botanist and surgeon to Burke's unfortunate expedition; his specimens are remarkably good and well selected. MR. J. DALLACHY, whose principal journey was one to the Darling desert. MR. G. MAXWELL, from whom there are numerous species from W. Australia, chiefly from the southern districts. MR. C. STUART, who collected in Tasmania, and afterwards more largely in New England, in the neighbourhood of Tenterfield. A considerable set of the latter has also been presented to the Kew herbarium by SIR STUART DONALDSON; MR. F. WATERHOUSE, who made large collections for the Government of S. Australia, chiefly in Kangaroo Island; and MR. AUGUSTUS OLDFIELD, an acute observer as well as an intelligent collector, who, besides the Tasmanian contributions mentioned in Dr. Hooker's Flora, made large additions to the West Australian plants previously known; in the first instance from the neighbourhood of Murchison river, and afterwards from the south-western districts. Mr. Oldfield is now in this country, and has most generously offered the use of his own Australian herbarium to the Kew Museum, as a contribution towards the present Flora.

4. Collections presented to Dr. Mueller by friends chiefly resident in Australia. These, owing to the greater facilities for drying and preserving enjoyed by stationary collectors, are usually the most satisfactory to the working botanists. The first of them in importance are those of MR. C. MOORE, Superintendent of the Botanic Garden at Sydney, and of MR. W. HILL, Superintendent of the Botanic Garden at Brisbane; the former chiefly from the northern districts of New South Wales, and the latter from the vicinity of Moreton Bay. Amongst the numerous amateur contributions, I notice those of MR. W. ALLITT from Portland, of MISS LOUISA ATKINSON from the Blue Mountains, of DR. H. BEHR (now in California) from South Australia, of MR. E. BOWMAN from Queensland, of MR. J. NERNST (unfortunately, from a misreading of the labels, spelt *Vernet* in the first sheets of this volume) from Ipswich, of MR. A. THOZET from Queensland, of MR. W. VERNON from Sydney, of the REV. W. WHAN from Shipton, of MR. C. WILHELMI from Port Lincoln, of the REV. S. E. WOODS from the Tattiara country, and of MR. W. WOOLLS from Paramatta.

Besides the above-mentioned names and those enumerated in Dr.

Hooker's Essay, some others may be found quoted in the present work in connection with species they have collected.  To supply any omissions I may have inadvertently made, and in the hope of doing full justice to all who may have directly or indirectly contributed to the investigation of the Australian Flora, it is my purpose, with the last volume to give a general alphabetical list, with a sketch of their labours, of all those whose collections are deposited in the public or private herbaria to which I have access.

With regard to the form and language adopted in the present work, they are those which, after much consideration, were adopted and sanctioned by Sir W. J. Hooker for colonial Floras in general, and exemplified in the 'Flora Hongkongensis.'  I may therefore here repeat what I then stated, that it has been my endeavour to follow out the principles laid down in the " Outlines of Botany " prefixed to each of these Floras, so as to facilitate as much as possible the finding out the name of any plant gathered in the territory, by the comparison of specimens with the descriptions given.  For this purpose, although I cannot yet give an analytical key to the Orders, until at least the *Polypetalæ* shall have been gone through, the genera of each Order, and the species of each genus, are universally preceded by analytical tables, in which their more prominent characters are contrasted.  These tables may be considered as another form for the short diagnoses of Linnæus and his immediate followers, or for the italicized portions of many modern diagnoses, and can refer only to the differentiation of known species.  It is the vain attempt to introduce characters which might absolutely distinguish a species from all others to be hereafter discovered,—to contrast the known with the unknown,—that has occasioned those long and tedious diagnoses, which render many modern descriptive works almost unmanageable.  A long description in the ablative absolute, supposed to contain the essential characters only, and another in the nominative with the accessory ones, often fail in their purpose, for some of the most striking features, such as stature, dimensions, colour, etc., because they are less absolute than the others, are conventionally considered as accessory ; and the descriptions containing them are usually first glanced over by the botanist seeking to name a plant, before he wades through the confused mass of ablatives in which he is to find the important characters.  In my descriptions, therefore, which I have been obliged to shorten as much as consistent with their practical use, I have endeavoured to select the characters most important to observe for their identification.  Many of these descriptions are, I am aware, as yet very imperfect, and some may be in some respect erroneous, especially with regard to stature, colour, and dimensions, owing to the insufficiency of specimens and the want of reliable memoranda by those who have seen the plants in a living state. Travellers, therefore, making use of this work in the country, will have to guard against attaching much importance to discrepancies in characters which dried specimens cannot show, when the descriptions apply well to the plant they are examining as to form and structure. With regard to dimensions, especially, it must be borne in mind that those here given are the average limits between which the organs vary in their full-grown normal state. Starvation, inordinate luxuriance, the imperfect development of the first- or last-formed organs of each kind, and other similar circumstances, may reduce or extend the dimensions beyond the limits assigned, but the general aspect of the specimens, if tolerably good, will generally indicate whether the organs are or not in any such abnormal conditions.

With regard to the synonymy, I have endeavoured to give a complete reference to all published names of endemic Australian plants, as well as to all names which have been specially given with reference to Australian specimens. But in the case of well-known extra-Australian species extending into our Flora, I have thought it unnecessary to repeat the whole of the synonyms, already given in the general works I have quoted, adding only such new ones as my researches for the identification of Australian species have enabled me to verify.

In order to facilitate the use of this work as a separate Flora of each of the colonial territories whose Governments have supported it by separate grants, I have thought it right to indicate by a prominent typographical arrangement the particular colonies in which each species is to be found. For this purpose I have considered Queensland as extending (as indicated in our most recent maps) to Cape York, and have designated under the general name of North Australia the whole of the unsettled territory to the westward within the tropics. Sharks Bay and its neighbourhood are considered as belonging to West Australia; and I have taken as the northern limits of South Australia, the 26th parallel S. latitude, as I find it marked in our maps.

In giving the various stations at which each species has been found, it has been my plan to enumerate all those I find in R. Brown's herbarium, all Cunningham's except the Tasmanian ones, and generally all others that I find authentically recorded on labels accompanying the specimens, excepting where many collectors have gathered the same plant at such well-known localities as Port Jackson, King George's

Sound, etc., in which case I have mentioned only R. Brown, or some others of those who first collected it, and excepting also Tasmania and Victoria. For the two latter colonies, I have usually extracted or abridged the stations (always verified on the specimens) given in the elaborate Floras of J. D. Hooker and F. Mueller.

Many of the varieties which I have indicated will be considered as distinct species by a large number of general botanists; on the other hand, there are many forms which I have adopted as species which Dr Mueller is disposed to reduce. In some cases I have yielded to his opinion, rather against the conclusion I should have come to from the examination of dried specimens, because, for Victoria plants especially, he has the great advantage of observing them living in their native stations. Having had myself much experience in describing plants both with and without this aid, and of testing descriptions made with and without it, I can fully appreciate the great use that can be made of it, provided due caution be observed, for it often acts as a snare. It rarely occurs that many species of a genus are found together so as to admit of comparison in a growing state, and we are too apt in regard to them to trust to recollections of general impressions. I do not consider it safe therefore to unite forms usually regarded as distinct and appearing so in a large number of specimens from a great variety of stations, on account of generally observed variations unconfirmed by specimens, nor even on account of single apparently intermediate specimens, unless the history of such abnormal specimens is ascertained. Little as we know, for instance, of the influence of natural hybridizing in Europe, it has been still less, if ever, observed in Australia; and many other causes may have produced apparent passages between species really distinct. I have, therefore, wherever there is a difference of opinion between Dr. Mueller and myself, adopted the conclusion which has appeared to me the most probable, and mentioned the objection to it for the consideration and, if possible, the decision of future botanists.

At the moment of sending these pages to press, several additional collections have arrived at Kew from Dr. Mueller, from Mr. Oldfield, and from Mr. B. Lowrie. Were I to delay the publication of this volume for the purpose of inserting any additions they might supply, it is probable that others again might come to hand, and to such delays there would be no limit. As it is probable, also, that the first use of this volume may be the means of detecting many errors or inaccuracies, I think it better to reserve all "Addenda and Corrigenda" for a Supplement, to be issued with the second volume.

I should here have added an introductory sketch of the geography of Australian vegetation and of the history of its botany ; but the need for it is for the present obviated by the elaborate review contained in Dr. Hooker's above-mentioned Essay. It is true that recent discoveries as well as a more careful examination of the Australian species previously deposited in our herbaria, may require some corrections in the statistical details given, or slight modifications, as to the proportions in which the Australian Flora is connected with those of other countries ; but the general features of its geographical distribution, so ably sketched out by Dr. Hooker, are only confirmed as further research renders them more definite, and the minor corrections may be much more satisfactorily given with the close of the work, when the whole Flora shall have been gone through.

# CONTENTS.

CONTENTS.

# INTRODUCTION.

## OUTLINES OF BOTANY, WITH SPECIAL REFERENCE TO LOCAL FLORAS.

### CHAP. I. DEFINITIONS AND DESCRIPTIVE BOTANY.

1. The principal object of a **Flora** of a country, is to afford the means of *determining* (*i. e.* ascertaining the name of) any plant growing in it, whether for the purpose of ulterior study or of intellectual exercise.

2. With this view, a Flora consists of descriptions of all the wild or native plants contained in the country in question, so drawn up and arranged that the student may identify with the corresponding description any individual specimen which he may gather.

3. These descriptions should be *clear, concise, accurate,* and *characteristic,* so as that each one should be readily adapted to the plant it relates to, and to no other one; they should be as nearly as possible arranged under *natural* (184) divisions, so as to facilitate the comparison of each plant with those nearest allied to it ; and they should be accompanied by an *artificial key* or index, by means of which the student may be guided step by step in the observation of such peculiarities or *characters* in his plant, as may lead him, with the least delay, to the individual description belonging to it.

4. For descriptions to be clear and readily intelligible, they should be expressed as much as possible in ordinary well-established language. But, for the purpose of accuracy, it is necessary not only to give a more precise technical meaning to many terms used more or less vaguely in common conversation, but also to introduce purely technical names for such parts of plants or forms as are of little importance except to the botanist. In the present chapter it is proposed to define such *technical* or *technically limited* terms as are made use of in these Floras.

5. At the same time mathematical accuracy must not be expected. The forms and appearances assumed by plants and their parts are infinite. Names cannot be invented for all ; those even that have been proposed are too numerous for ordinary memories. Many are derived from supposed resemblances to well-known forms or objects. These resemblances are differently appreciated by different persons, and the same term is not only differently applied by two different botanists, but it frequently happens that the same writer is led on different occasions to give somewhat different meanings to the same word. The botanist's endeavours should always be, on the one hand, to make as near an approach to precision as circumstances will allow, and on the other hand to avoid that prolixity of detail and overloading with technical terms which tends rather to confusion than clearness. In this he will be more or less successful. The aptness of a botanical description, like the beauty of a work of imagination, will always vary with the style and genius of the author.

## § 1. *The Plant in General.*

6. The **Plant,** in its botanical sense, includes every being which has *vegetable life,* from the loftiest tree which adorns our landscapes, to the humblest moss which grows on its stem, to the mould or fungus which attacks our provisions, or the green scum that floats on our ponds.

7. Every portion of a plant which has a distinct part or *function* to perform in the operations or phenomena of vegetable life is called an **Organ.**

8. What constitutes *vegetable life,* and what are the functions of each organ, belong to *Vegetable Physiology;* the microscopical structure of the tissues composing the organs, to *Vegetable Anatomy;* the composition of the substances of which they are formed, to *Vegetable Chemistry;* under *Descriptive and Systematic Botany* we have chiefly to consider the forms of organs, that is, their *Morphology,* in the proper sense of the term, and their general structure so far as it affects classification and specific resemblances and differences. The terms we shall now define belong chiefly to the latter branch of Botany, as being that which is essential for the investigation of the Flora of a country. We shall add, however, a short chapter on Vegetable Anatomy and Physiology, as a general knowledge of both imparts an additional interest to and facilitates the comparison of the characters and affinities of the plants examined.

9. In the more perfect plants, their organs are comprised in the general terms **Root, Stem, Leaves, Flowers,** and **Fruit.** Of these the three first, whose function is to assist in the growth of the plant, are *Organs of Vegetation;* the flower and fruit, whose office is the formation of the seed, are the *Organs of Reproduction.*

10. All these organs exist, in one shape or another, at some period of the life of most, if not all, *flowering plants,* technically called *phænogamous* or *phanerogamous plants;* which all bear some kind of flower and fruit in the botanical sense of the term. In the lower classes, the ferns, mosses, fungi, moulds or mildews, seaweeds, etc., called by botanists *cryptogamous plants,* the flowers, the fruit, and not unfrequently one or more of the organs of vegetation, are either wanting, or replaced by organs so different as to be hardly capable of bearing the same name.

11. The observations comprised in the following pages refer exclusively to the flowering or phænogamous plants. The study of the cryptogamous classes has now become so complicated as to form almost a separate science. They are therefore not included in these introductory observations, nor, with the exception of ferns, in the present Flora.

12. **Plants** are

*Monocarpic,* if they die after one flowering-season. These include *Annuals,* which flower in the same year in which they are raised from seed; and *Biennials,* which only flower in the year following that in which they are sown.

*Caulocarpic,* if, after flowering, the whole or part of the plant lives through the winter and produces fresh flowers another season. These include *Herbaceous perennials,* in which the greater part of the plant dies after flowering, leaving only a small perennial portion called the Stock or Caudex, close to or within the earth; *Undershrubs, suffruticose* or *suffrutescent* plants, in which the flowering branches, forming a considerable portion of the plant, die down after flowering, but leave a more or less prominent perennial and woody base; *Shrubs (frutescent* or *fruticose plants),* in which the perennial woody part forms the greater part of the plant, but branches near the base, and does not much exceed a man's height; and *Trees (arboreous* or *arborescent plants)* when the height is greater and forms a woody *trunk,* scarcely branching from the base. *Bushes* are low, much branched shrubs.

13. The terms *Monocarpic* and *Caulocarpic* are but little used, but the other distinctions enumerated above are universally attended to, although more useful to the gardener than to the botanist, who cannot always assign to them any precise character. Monocarpic plants, which require more than two or three years to produce their flowers, will often, under certain circumstances, become herbaceous perennials, and are generally confounded with them. Truly perennial herbs will often commence flowering the first year, and have then all the appearance of annuals. Many tall shrubs

and trees lose annually their flowering branches like undershrubs. And the same botanical species may be an annual or a perennial, a herbaceous perennial or an undershrub, an undershrub or a shrub, a shrub or a tree, according to climate, treatment, or variety.

14. Plants are usually *terrestrial*, that is, growing on earth, or *aquatic*, *i.e.* growing in water ; but sometimes they may be found attached by their roots to other plants, in which case they are *epiphytes* when simply growing upon other plants without penetrating into their issue, *parasites* when their roots penetrate into and derive more or less nutriment from the plant to which they are attached.

15. The simplest form of the perfect plant, the annual, consists of—

(1) The **Root,** or descending axis, which grows downwards from the stem, divides and spreads in the earth or water, and absorbs food for the plant through the extremities of its branches.

(2) The **Stem,** or ascending axis, which grows upwards from the root, branches and bears first one or more leaves in succession, then one or more flowers, and finally one or more fruits. It contains the tissues or other channels (217) by which the nutriment absorbed by the roots is conveyed in the form of *sap* (192) to the leaves or other points of the surface of the plant, to be *elaborated* or *digested* (218), and afterwards redistributed over different parts of the plant for its support and growth.

(3) The **Leaves,** usually flat, green, and horizontal, are variously arranged on the stem and its branches. They *elaborate* or *digest* (218) the nutriment brought to them through the stem, absorb carbonic acid gas from the air, exhaling the superfluous oxygen, and returning the assimilated sap to the stem.

(4) The **Flowers,** usually placed at or towards the extremities of the branches. They are destined to form the future seed. When perfect and complete they consist : 1st, of a *pistil* in the centre, consisting of one or more *carpels,* each containing the germ of one or more seeds ; 2nd, of one or more *stamens* outside the pistil, whose action is necessary to *fertilize* the pistil or enable it to ripen its seed ; 3rd, of a *perianth* or *floral envelope,* which usually encloses the stamens and pistil when young, and expands and exposes them to view when fully formed. This complete perianth is double ; the outer one, called *Calyx,* is usually more green and leaf-like ; the inner one, called the *Corolla,* more conspicuous, and variously coloured. It is the perianth, and especially the corolla, as the most showy part, that is generally called the flower in popular language.

(5) The **Fruit,** consisting of the pistil or its lower portion, which persists or remains attached to the plant after the remainder of the flower has withered and fallen off. It enlarges and alters more or less in shape or consistence, becomes a *seed-vessel,* enclosing the seed until it is ripe, when it either opens to discharge the seed or falls to the ground with the seed. In popular language the term *fruit* is often limited to such seed-vessels as are or look juicy and eatable. Botanists give that name to all seed-vessels.

16. The herbaceous perennial resembles the annual during the first year of its growth ; but it also forms (usually towards the close of the season), on its *stock* (the portion of the stem and root which does not die), one or more *buds,* either exposed, and then popularly called *eyes,* or concealed among leaves. These buds, called *leaf-buds,* to distinguish them from *flower-buds* or unopened flowers, are future branches as yet undeveloped ; they remain dormant through the winter, and the following spring grow out into new stems bearing leaves and flowers like those of the preceding year, whilst the lower part of the stock emits fresh roots to replace those which had perished at the same time as the stems.

17. Shrubs and trees form similar leaf-buds either at the extremity of their branches, or along the branches of the year. In the latter case these buds are usually *axillary,* that is, they appear in the *axil* of each leaf, *i. e.* in the angle formed by the leaf and the branch. When they appear at any other part of the plant they are called *adventitious.* If these buds by producing roots (19) become distinct plants before separating from the parent, or if adventitious leaf-buds are produced in the place of flowers or seeds, the plant is said to be *viviparous* or *proliferous.*

## § 2. *The Root.*

18. **Roots** ordinarily produce neither buds, leaves, nor flowers. Their branches, called *fibres* when slender and long, proceed irregularly from any part of their surface.

19. Although roots proceed usually from the base of the stem or stock, they may also be produced from the base of any bud, especially if the bud lie along the ground, or is otherwise placed by nature or art in circumstances favourable for their development, or indeed occasionally from almost any part of the plant. They are then often distinguished as *adventitious*, but this term is by some applied to all roots which are not in prolongation of the original radicle.

20. **Roots** are

*fibrous*, when they consist chiefly of slender fibres.

*tuberous*, when either the main root or its branches are thickened into one or more short fleshy or woody masses called *tubers* (25).

*taproots*, when the main root descends perpendicularly into the earth, emitting only very small fibrous branches.

21. The stock of a herbaceous perennial, or the lower part of the stem of an annual or perennial, or the lowest branches of a plant, are sometimes underground and assume the appearance of a root. They then take the name of *rhizome*. The rhizome may always be distinguished from the true root by the presence or production of one or more buds, or leaves, or scales.

## § 3. *The Stock.*

22. The **Stock** of a herbaceous perennial, in its most complete state, includes a small portion of the summits of the previous year's roots, as well as of the base of the previous year's stems. Such stocks will increase yearly, so as at length to form dense tufts. They will often preserve through the winter a few leaves, amongst which are placed the buds which grow out into stems the following year, whilst the under side of the stock emits new roots from or amongst the remains of the old ones. These perennial stocks only differ from the permanent base of an undershrub in the shortness of the perennial part of the stems and in their texture usually less woody.

23. In some perennials, however, the stock consists merely of a branch, which proceeds in autumn from the base of the stem either aboveground or underground, and produces one or more buds. This branch, or a portion of it, alone survives the winter. In the following year its buds produce the new stem and roots, whilst the rest of the plant, even the branch on which these buds were formed, has died away. These *annual stocks*, called sometimes *hybernacula*, *offsets*, or *stolons*, keep up the communication between the annual stem and root of one year and those of the following year, thus forming altogether a perennial plant.

24. The stock, whether annual or perennial, is often entirely underground or rootlike. This is the *rootstock*, to which some botanists limit the meaning of the term *rhizome*. When the stock is entirely root-like, it is popularly called the *crown* of the root.

25. The term *tuber* is applied to a short, thick, more or less succulent rootstock or rhizome, as well as to a root of that shape (20), although some botanists propose to restrict its meaning to the one or to the other. An Orchis tuber, called by some a *knob*, is an annual tuberous rootstock with one bud at the top. A potato is an annual tuberous rootstock with several buds.

26. A *bulb* is a stock of a shape approaching to globular, usually rather conical above and flattened underneath, in which the bud or buds are concealed, or nearly so, under *scales*. These scales are the more or less thickened bases of the decayed leaves of the preceding year, or of the undeveloped leaves of the future year, or of both. Bulbs are annual or perennial, usually underground or close to the ground, but occasionally buds in the axils of the upper leaves become transformed into bulbs. Bulbs are said to be *scaly* when their scales are thick and loosely imbricated, *tunicated* when the scales are thinner, broader, and closely rolled round each other in concentric layers.

27. A *corm* is a tuberous rootstock, usually annual, shaped like a bulb, but in which the bud or buds are not covered by scales, or of which the scales are very thin and membranous.

## § 4. *The Stem.*

**28. Stems** are

*erect*, when they ascend perpendicularly from the root or stock; *twiggy* or *virgate*, when at the same time they are slender, stiff, and scarcely branched.

*sarmentose*, when the branches of a woody stem are long and weak, although scarcely climbing.

*decumbent* or *ascending*, when they spread horizontally, or nearly so, at the base, and then turn upwards and become erect.

*procumbent*, when they spread along the ground the whole or the greater portion of their length; *diffuse*, when at the same time very much and rather loosely branched.

*prostrate*, when they lie still closer to the ground.

*creeping*, when they emit roots at their nodes. This term is also frequently applied to any rhizomes or roots which spread horizontally.

*tufted* or *cæspitose*, when very short, close, and many together from the same stock.

29. Weak climbing stems are said to *twine*, when they support themselves by winding spirally round any object; such stems are also called *voluble*. When they simply climb without twining, they support themselves by their leaves, or by special clasping organs called *tendrils* (169), or sometimes, like the Ivy, by small root-like excrescences.

30. *Suckers* are young plants formed at the end of creeping, underground rootstocks. *Scions*, *runners*, and *stolons*, or *stoles*, are names given to young plants formed at the end or at the nodes (31) of branches or stocks creeping wholly or partially aboveground, or sometimes to the creeping stocks themselves.

31. A *node* is a point of the stem or its branches at which one or more leaves, branches, or leaf-buds (16) are given off. An *internode* is the portion of the stem comprised between two nodes.

**32. Branches** or **leaves** are

*opposite*, when two proceed from the same node on opposite sides of the stem.

*whorled* or *verticillate* (in a *whorl* or *verticil*), when several proceed from the same node, arranged regularly round the stem; *geminate*, *ternate*, *fascicled*, or *fasciculate*, when two, three, or more proceed from the same node on the same side of the stem. A tuft of fasciculate leaves is usually in fact an axillary leafy branch, so short that the leaves appear to proceed all from the same point.

*alternate*, when one only proceeds from each node, one on one side and the next above or below on the opposite side of the stem.

*decussate*, when opposite, but each pair placed at right-angles to the next pair above or below it; *distichous*, when regularly arranged one above another in two opposite rows, one on each side of the stem; *tristichous*, when in three rows, etc. (92).

*scattered*, when irregularly arranged round the stem; frequently, however, botanists apply the term *alternate* to all branches or leaves that are neither opposite nor whorled.

*secund*, when all start from or are turned to one side of the stem.

33. **Branches** are *dichotomous*, when several times forked, the two branches of each fork being nearly equal; *trichotomous*, when there are three nearly equal branches at each division instead of two; but when the middle branch is evidently the principal one, the stem is usually said to have two opposite branches; *umbellate*, when divided in the same manner into several nearly equal branches proceeding from the same point. If however the central branch is larger than the two or more lateral ones, the stem is said to have opposite or whorled branches, as the case may be.

34. A *culm* is a name sometimes given to the stem of Grasses, Sedges, and some other Monocotyledonous plants.

## § 5. *The Leaves.*

35. The ordinary or perfect **Leaf** consists of a flat *blade* or *lamina*, usually green, and more or less horizontal, attached to the stem by a stalk called a *footstalk* or *petiole*. When the form or dimensions of a leaf are spoken of, it is generally the blade that is meant, without the petiole or stalk.

36. The end by which a leaf, a part of the flower, a seed, or any other organ, is

attached to the stem or other organ, is called its *base*, the opposite end is its *apex* or summit, excepting sometimes in the case of anther-cells (115).

37. **Leaves** are

*sessile*, when the blade rests on the stem without the intervention of a petiole.

*amplexicaul* or *stem-clasping*, when the sessile base of the blade clasps the stem horizontally.

*perfoliate*, when the base of the blade not only clasps the stem, but closes round it on the opposite side, so that the stem appears to pierce through the blade.

*decurrent*, when the edges of the leaf are continued down the stem so as to form raised lines or narrow appendages, called *wings*.

*sheathing*, when the base of the blade, or of the more or less expanded petiole, forms a vertical sheath round the stem for some distance above the node.

38. Leaves and flowers are called *radical*, when inserted on a rhizome or stock, or so close to the base of the stem as to appear to proceed from the root, rhizome, or stock; *cauline*, when inserted on a distinct stem. Radical leaves are *rosulate* when they spread in a circle on the ground.

39. **Leaves** are

*simple* and *entire*, when the blade consists of a single piece, with the margin nowhere indented, *simple* being used in opposition to *compound*, *entire* in opposition to *dentate*, *lobed*, or *divided*.

*ciliate*, when bordered with thick hairs or fine hair-like teeth.

*dentate* or *toothed*, when the margin is only cut a little way in, into what have been compared to teeth. Such leaves are *serrate*, when the teeth are regular and pointed like the teeth of a saw; *crenate*, when regular and blunt or rounded (compared to the battlements of a tower); *serrulate* and *crenulate*, when the serratures or crenatures are small; *sinuate*, when the teeth are broad, not deep, and irregular (compared to bays of the coast); *wavy* or *undulate*, when the edges are not flat, but bent up and down (compared to the waves of the sea).

*lobed* or *cleft*, when more deeply indented or divided, but so that the incisions do not reach the midrib or petiole. The portions thus divided take the name of *lobes*. When the lobes are narrow and very irregular, the leaves are said to be *laciniate*. The spaces between the teeth or lobes are called *sinuses*.

*divided* or *dissected*, when the incisions reach the midrib or petiole, but the parts so divided off, called *segments*, do not separate from the petiole, even when the leaf falls, without tearing.

*compound*, when divided to the midrib or petiole, and the parts so divided off, called *leaflets*, separate, at least at the fall of the leaf, from the petiole, as the whole leaf does from the stem, without tearing. The common stalk upon which the leaflets are inserted is called the *common petiole* or the *rhachis*; the separate stalk of each leaflet is a *petiolule*.

40. Leaves are more or less marked by *veins*, which, starting from the stalk, diverge or branch as the blade widens, and spread all over it more or less visibly. The principal ones, when prominent, are often called *ribs* or *nerves*, the smaller branches only then retaining the name of *veins*, or the latter are termed *veinlets*. The smaller veins are often connected together like the meshes of a net, they are then said to *anastomose*, and the leaf is said to be *reticulate* or *net-veined*. When one principal vein runs direct from the stalk towards the summit of the leaf, it is called the *midrib*. When several start from the stalk, diverge slightly without branching, and converge again towards the summit, they are said to be *parallel*, although not mathematically so. When 3 or 5 or more ribs or nerves diverge *from the base*, the leaf is said to be 3-*nerved*, 5-*nerved*, etc., but if the lateral ones diverge from the midrib a little above the base, the leaf is *triplinerved*, *quintuplinerved*, etc. The arrangement of the veins of a leaf is called their *venation*.

41. The **Leaflets, Segments, Lobes**, or **Veins** of leaves are

*pinnate* (feathered), when there are several succeeding each other on each side of the midrib or petiole, compared to the branches of a feather. A pinnately lobed or divided leaf is called *lyrate* when the terminal lobe or segment is much larger and broader than the lateral ones, compared, by a stretch of imagination, to a lyre; *run-*

*cinate*, when the lateral lobes are curved backwards towards the base of the leaf; *pectinate*, when the lateral lobes are numerous, narrow, and regular, like the teeth of a comb.

*palmate* or *digitate*, when several diverge from the same point, compared to the fingers of the hand.

*ternate*, when three only start from the same point, in which case the distinction between the palmate and pinnate arrangement often ceases, or can only be determined by analogy with allied plants. A leaf with ternate lobes is called *trifid*. A leaf with three leaflets is sometimes improperly called a ternate leaf: it is the leaflets that are ternate; the whole leaf is *trifoliolate*. Ternate leaves are leaves growing three together.

*pedate*, when the division is at first ternate, but the two outer branches are forked, the outer ones of each fork again forked, and so on, and all the branches are near together at the base, compared vaguely to the foot of a bird.

42. Leaves with pinnate, palmate, pedate, etc., leaflets, are usually for shortness called *pinnate, palmate, pedate*, etc., *leaves*. If they are so cut into segments only, they are usually said to be *pinnatisect, palmatisect, pedatisect*, etc., although the distinction between segments and leaflets is often unheeded in descriptions, and cannot indeed always be ascertained. If the leaves are so cut only into lobes, they are said to be *pinnatifid, palmatifid, pedatifid*, etc.

43. The teeth, lobes, segments, or leaflets, may be again toothed, lobed, divided, or compounded. Some leaves are even three or more times divided or compounded. In the latter case they are termed *decompound*. When twice or thrice pinnate (*bipinnate* or *tripinnate*), each primary or secondary division, with the leaflets it comprises, is called a *pinna*. When the pinna of a leaf or the leaflets of a pinna are in pairs, without an odd terminal pinna or leaflet, the leaf or pinna so divided is said to be *abruptly pinnate*; if there is an odd terminal pinna or leaflet, the leaf or pinna is *unequally pinnate* (*imparipinnatum*).

44. The number of leaves or their parts is expressed adjectively by the following numerals, derived from the Latin:—

| uni-, | bi-, | tri-, | quadri-, | quinque-, | sex-, | septem-, | octo-, | novem-, | decem-, | multi- |
|-------|------|-------|----------|-----------|-------|----------|--------|---------|---------|--------|
| 1-, | 2-, | 3-, | 4-, | 5-, | 6-, | 7-, | 8-, | 9-, | 10-, | many- |

prefixed to a termination, indicating the particular kind of part referred to. Thus—

*unidentate, bidentate, multidentate*, mean one-toothed, two-toothed, many-toothed, etc.

*bifid, trifid, multifid*, mean two-lobed, three-lobed, many-lobed, etc.

*unifoliolate, bifoliolate, multifoliolate*, mean having one leaflet, two leaflets, many leaflets, etc.

*unifoliate, bifoliate, multifoliate*, mean having one leaf, two leaves, many leaves, etc.

*biternate* and *triternate*, mean twice or thrice ternately divided.

*unijugate, bijugate, multijugate*, etc., pinnæ or leaflets, mean that they are in one, two, many, etc., pairs (*juga*).

45. **Leaves** or their parts, when **flat**, or any other flat organs in plants, are

*linear*, when long and narrow, at least four or five times as long as broad, falsely compared to a mathematical line, for a linear leaf has always a perceptible breadth.

*lanceolate*, when about three or more times as long as broad, broadest below the middle, and tapering towards the summit, compared to the head of a lance.

*cuneate*, when broadest above the middle, and tapering towards the base, compared to a wedge with the point downwards; when very broadly cuneate and rounded at the top, it is often called *flabelliform* or *fan-shaped*.

*spathulate*, when the broad part near the top is short, and the narrow tapering part long, compared to a spatula or flat ladle.

*ovate*, when scarcely twice as long as broad, and rather broader below the middle, compared to the longitudinal section of an egg; *obovate* is the same form, with the broadest part above the middle.

*orbicular, oval, oblong, elliptical, rhomboidal*, etc., when compared to the corresponding mathematical figures.

*transversely oblong*, or *oblate*, when conspicuously broader than long.

*falcate*, when curved like the blade of a scythe.

46. Intermediate forms between any two of the above are expressed by combining two terms. Thus, a *linear-lanceolate* leaf is long and narrow, yet broader below the middle, and tapering to a point; a *linear-oblong* one is scarcely narrow enough to be called linear, yet too narrow to be strictly oblong, and does not conspicuously taper either towards the summit or towards the base.

47. The *apex* or *summit* of a leaf is

*acute* or *pointed*, when it forms an acute angle or tapers to a point.

*obtuse* or *blunt*, when it forms a very obtuse angle, or more generally when it is more or less rounded at the top.

*acuminate* or *cuspidate*, when suddenly narrowed at the top, and then more or less prolonged into an *acumen* or *point*, which may be acute or obtuse, linear or tapering. Some botanists make a slight difference between the *acuminate* and *cuspidate* apex, the acumen being more distinct from the rest of the leaf in the latter case than in the former; but in general the two terms are used in the same sense, some preferring the one and some the other.

*truncate*, when the end is cut off square.

*retuse*, when very obtuse or truncate, and slightly indented.

*emarginate* or *notched*, when more decidedly indented at the end of the midrib; *obcordate*, if at the same time approaching the shape of a heart with its point downwards.

*mucronate*, when the midrib is produced beyond the apex in the form of a small point.

*aristate*, when the point is fine like a hair.

48. The base of the leaf is liable to the same variations of form as the apex, but the terms more commonly used are *tapering* or *narrowed* for acute and acuminate, *rounded* for obtuse, and *cordate* for emarginate. In all cases the petiole or point of attachment prevent any such absolute termination at the base as at the apex.

49. A leaf may be *cordate* at the base whatever be its length or breadth, or whatever the shape of the two lateral lobes, called *auricles* (or *little ears*), formed by the indenture or notch, but the term *cordiform* or *heart-shaped* leaf is restricted to an ovate and acute leaf, cordate at the base, with rounded auricles. The word auricles is more particularly used as applied to sessile and stem-clasping leaves.

50. If the auricles are pointed, the leaf is more particularly called *auriculate*; it is moreover said to be *sagittate*, when the points are directed downwards, compared to an arrow-head; *hastate*, when the points diverge horizontally, compared to a halbert.

51. A *reniform* leaf is broader than long, slightly but broadly cordate at the base, with rounded auricles, compared to a kidney.

52. In a *peltate* leaf, the stalk, instead of proceeding from the lower edge of the blade, is attached to the under surface, usually near the lower edge, but sometimes in the very centre of the blade. The peltate leaf has usually several principal nerves radiating from the point of attachment, being, in fact, a cordate leaf, with the auricles united.

53. All these modifications of division and form in the leaf pass so gradually one into the other that it is often difficult to say which term is the most applicable— whether the leaf be toothed or lobed, divided or compound, oblong or lanceolate, obtuse or acute, etc. The choice of the most apt expression will depend on the skill of the describer.

54. **Leaves,** when **solid, Stems, Fruits, Tubers,** and other parts of plants, when not flattened like ordinary leaves, are

*setaceous* or *capillary*, when very slender like bristles or hairs.

*acicular*, when very slender, but stiff and pointed like needles.

*subulate*, when rather thicker and firmer like awls.

*linear*, when at least four times as long as thick; *oblong*, when from about two to about four times as long as thick, the terms having the same sense as when applied to flat surfaces.

*ovoid*, when egg-shaped, with the broad end downwards, *obovoid* if the broad end is upwards; these terms corresponding to *ovate* and *obovate* shapes in flat surfaces.

*globular* or *spherical,* when corresponding to *orbicular* in a flat surface. *Round* applies to both.

*turbinate,* when shaped like a top.

*conical,* when tapering upwards : *obconical,* when tapering downwards, if in both cases a transverse section shows a circle.

*pyramidal,* when tapering upwards ; *obpyramidal,* when tapering downwards, if in both cases a transverse section shows a triangle or polygon.

*fusiform,* or spindle-shaped, when tapering at both ends ; *cylindrical,* when not tapering at either end, if in both cases the transverse section shows a circle, or sometimes irrespective of the transverse shape.

*terete,* when the transverse section is not angular ; *trigonous, triquetrous,* if the transverse section shows a triangle, irrespective in both cases of longitudinal form.

*compressed,* when more or less flattened laterally ; *depressed,* when more or less flattened vertically, or at any rate at the top ; *obcompressed* (in the achenes of *Compositæ*), when flattened from front to back.

*articulate* or *jointed,* if at any period of their growth (usually when fully formed and approaching their decay, or in the case of fruits when quite ripe) they separate, without tearing, into two or more pieces placed end to end. The joints where they separate are called *articulations,* each separate piece an *article.* The name of *joint* is, in common language, given both to the articulation and the article, but more especially to the former. Some modern botanists, however, propose to restrict it to the article, giving the name of *joining* to the articulation.

*didymous,* when slightly two-lobed, with rounded obtuse lobes.

*moniliform,* or beaded, when much contracted at regular intervals, but not separating spontaneously into articles.

55. In their consistence **Leaves** or other organs are

*fleshy,* when thick and soft ; *succulent* is generally used in the same sense, but implies the presence of more juice.

*coriaceous,* when firm and stiff, or very tough, of the consistence of leather.

*crustaceous,* when firm and brittle.

*membranous,* when thin and not stiff.

*scarious* or *scariose,* when very thin, more or less transparent and not green, yet rather stiff.

56. The terms applied botanically to the consistence of solids are those in general use in common language.

57. The mode in which unexpanded leaves are disposed in the leaf-bud is called their *vernation* or *præfoliation ;* it varies considerably, and technical terms have been proposed to express some of its varieties, but it has been hitherto rarely noticeḑ in descriptive botany.

## § 6. *Scales, Bracts, and Stipules.*

58. **Scales** (*Squamæ*) are leaves very much reduced in size, usually sessile, seldom green or capable of performing the respiratory functions of leaves. In other words, they are organs resembling leaves in their position on the plant, but differing in size, colour, texture, and functions. They are most frequent on the stock of perennial plants, or at the base of annual branches, especially on the buds of future shoots, when they serve apparently to protect the dormant living germ from the rigour of winter. In the latter case they are usually short, broad, close together, and more or less *imbricated,* that is, overlapping each other like the tiles of a roof. It is this arrangement as well as their usual shape that has suggested the name of *scales,* borrowed from the scales of a fish. Imbricated scales, bracts, or leaves, are said to be *squarrose,* when their tips are pointed and very spreading or recurved.

59. Sometimes, however, most or all the leaves of the plant are reduced to small scales, in which case they do not appear to perform any particular function. The name of *scales* is also given to any small broad scale-like appendages or reduced organs, whether in the flower or any other part of the plant.

60. **Bracts** (*Bracteæ*) are the upper leaves of a plant in flower (either all those of the flowering branches, or only one or two immediately under the flower), when differ-

ent from the stem-leaves in size, shape, colour, or arrangement. They are generally much smaller and more sessile. They often partake of the colour of the flower, although they very frequently also retain the green colour of the leaves. When small they are often called *scales*.

61. *Floral leaves* or *leafy bracts* are generally the lower bracts on the upper leaves at the base of the flowering branches, intermediate in size, shape, or arrangement, between the stem-leaves and the upper bracts.

62. *Bracteoles* are the one or two last bracts under each flower, when they differ materially in size, shape, or arrangement from the other bracts.

63. **Stipules** are leaf-like or scale-like appendages at the base of the leaf-stalk, or on the node of the stem. When present there are generally two, one on each side of the leaf, and they sometimes appear to protect the young leaf before it is developed. They are however exceedingly variable in size and appearance, sometimes exactly like the true leaves except that they have no buds in their axils, or looking like the leaflets of a compound leaf, sometimes apparently the only leaves of the plant; generally small and narrow, sometimes reduced to minute scales, spots or scars, sometimes united into one opposite the leaf, or more or less united with, or *adnate* to the petiole, or quite detached from the leaf, and forming a ring or sheath round the stem in the axil of the leaf. In a great number of plants they are entirely wanting.

64. *Stipellæ*, or secondary stipules, are similar organs, sometimes found on compound leaves at the points where the leaflets are inserted.

65. When scales, bracts, or stipules, or almost any part of the plant besides leaves and flowers are stalked, they are said to be *stipitate*, from *stipes*, a stalk.

### § 7. *Inflorescence and its Bracts.*

66. The **Inflorescence** of a plant is the arrangement of the flowering branches, and of the flowers upon them. *An Inflorescence* is a flowering branch, or the flowering summit of a plant above the last stem-leaves, with its branches, bracts, and flowers.

67. A single flower, or an inflorescence, is *terminal* when at the summit of a stem or leafy branch, *axillary* when in the axil of a stem-leaf, *leaf-opposed* when opposite to a stem-leaf. The inflorescence of a plant is said to be *terminal* or *determinate* when the main stem and principal branches end in a flower or inflorescence (not in a leaf-bud), *axillary* or *indeterminate* when all the flowers or inflorescences are axillary, the stem or branches ending in leaf-buds.

68. A *Peduncle* is the stalk of a solitary flower, or of an inflorescence; that is to say, the portion of the flowering branch from the last stem-leaf to the flower, or to the first ramification of the inflorescence, or even up to its last ramifications; but the portion extending from the first to the last ramifications or the axis of inflorescence is often distinguished under the name of *rhachis*.

69. A *Scape* or *radical Peduncle* is a leafless peduncle proceeding from the stock, or from near the base of the stem, or apparently from the root itself.

70. A *Pedicel* is the last branch of an inflorescence, supporting a single flower.

71. The branches of inflorescences may be, like those of stems, opposite, alternate, etc. (32, 33), but very often their arrangement is different from that of the leafy branches of the same plant.

72. **Inflorescence** is

*centrifugal*, when the terminal flower opens first, and those on the lateral branches are successively developed.

*centripetal*, when the lowest flowers open first, and the main stem continues to elongate, developing fresh flowers.

73. Determinate inflorescence is usually centrifugal. Indeterminate inflorescence is always centripetal. Both inflorescences may be combined on one plant, for it often happens that the main branches of an inflorescence are centripetal, whilst the flowers on the lateral branches are centrifugal; or *vice versâ*.

74. An **Inflorescence** is

a *Spike*, or *spicate*, when the flowers are sessile along a simple undivided axis or rhachis.

a *Raceme*, or *racemose*, when the flowers are borne on pedicels along a single undivided axis or rhachis.

a *Panicle*, or *paniculate*, when the axis is divided into branches bearing two or more flowers.

a *Head*, or *capitate*, when several sessile or nearly sessile flowers are collected into a compact head-like cluster. The short, flat, convex or conical axis on which the flowers are seated, is called the *receptacle*, a term also used for the torus of a single flower (135). The very compact flower-heads of *Compositæ* are often termed *compound flowers*.

an *Umbel*, or *umbellate*, when several branches or pedicels appear to start from the same point and are nearly of the same length. It differs from the head, like the raceme from the spike, in that the flowers are not sessile. An umbel is said to be *simple*, when each of its branches or *rays* bears a single flower; *compound*, when each ray bears a *partial umbel* or *umbellule*.

a *Corymb*, or *corymbose*, when the branches and pedicels, although starting from different points, all attain the same level, the lower ones being much longer than the upper. It is a flat-topped or *fastigiate* panicle.

a *Cyme*, or *cymose*, when branched and centrifugal. It is a centrifugal panicle, and is often corymbose. The central flower opens first. The lateral branches successively developed are usually forked or opposite (dichotomous or trichotomous), but sometimes after the first forking the branches are no longer divided, but produce a succession of pedicels on their upper side forming apparently unilateral centripetal racemes; whereas if attentively examined, it will be found that each pedicel is at first terminal, but becomes lateral by the development of one outer branch only, immediately under the pedicel. Such branches, when in bud, are generally rolled back at the top, like the tail of a scorpion, and are thence called *scorpioid*.

a *Thyrsus*, or *thyrsoid*, when cymes, usually opposite, are arranged in a narrow pyramidal panicle.

75. There are numerous cases where inflorescences are intermediate between some two of the above, and are called by different botanists by one or the other name, according as they are guided by apparent or by theoretical similarity. A spike-like panicle, where the axis is divided into very short branches forming a cylindrical compact inflorescence, is called sometimes a spike, sometimes a panicle. If the flowers are in distinct clusters along a simple axis, the inflorescence is described as an *interrupted* spike or raceme, according as the flowers are nearly sessile or distinctly pedicellate; although when closely examined the flowers will be found to be inserted not on the main axis, but on a very short branch, thus, strictly speaking, constituting a panicle.

76. The *catkins* (*amenta*) of *Amentaceæ*, the *spadices* of several Monocotyledons, the *ears* and *spikelets* of Grasses are forms of the spike.

77. **Bracts** are generally placed singly under each branch of the inflorescence, and under each pedicel; bracteoles are usually two, one on each side, on the pedicel or close under the flower, or even upon the calyx itself; but bracts are also frequently scattered along the branches without axillary pedicels; and when the differences between the bracts and bracteoles are trifling or immaterial, they are usually all called bracts.

78. When three bracts appear to proceed from the same point, they will, on examination, be found to be really either one bract and two stipules, or one bract with two bracteoles in its axil. When two bracts appear to proceed from the same point, they will usually be found to be the stipules of an undeveloped bract, unless the branches of the inflorescence are opposite, when the bracts will of course be opposite also.

79. When several bracts are collected in a whorl, or are so close together as to appear whorled, or are closely imbricated round the base of a head or umbel, they are collectively called an *Involucre*. The bracts composing an involucre are described under the names of *leaves, leaflets, bracts*, or *scales*, according to their appearance. *Phyllaries* is a useless term, lately introduced for the bracts or scales of the involucre of *Compositæ*. An *Involucel* is the involucre of a partial umbel.

80. When several very small bracts are placed round the base of a calyx or of an

involucre, they have been termed a *calycule,* and the calyx or involucre said to be *calyculate,* but these terms are now falling into disuse, as conveying a false impression.

81. A *Spatha* is a bract or floral leaf enclosing the inflorescence of some Monocotyledons.

82. *Paleæ, Pales,* or *Chaff,* are the inner bracts or scales in *Compositæ, Gramineæ,* and some other plants, when of a thin yet stiff consistence, usually narrow and of a pale colour.

83. *Glumes* are the bracts enclosing the flowers of *Cyperaceæ* and *Gramineæ.*

## § 8. *The Flower in General.*

84. A *complete* **Flower** (15) is one in which the calyx, corolla, stamens, and pistils are all present; a *perfect* flower, one in which all these organs, or such of them as are present, are capable of performing their several functions. Therefore, properly speaking, an *incomplete* flower is one in which any one or more of these organs is wanting; and an *imperfect* flower, one in which any one or more of these organs is so altered as to be incapable of properly performing its functions. These imperfect organs are said to be *abortive* if much reduced in size or efficiency, *rudimentary* if so much so as to be scarcely perceptible. But, in many works, the term *incomplete* is specially applied to those flowers in which the perianth is simple or wanting, and *imperfect* to those in which either the stamens or pistil are imperfect or wanting.

85. A **Flower** is

*dichlamydeous,* when the perianth is double, both calyx and corolla being present and distinct.

*monochlamydeous,* when the perianth is single, whether by the union of the calyx and corolla, or the deficiency of either.

*asepalous,* when there is no calyx.

*apetalous,* when there is no corolla.

*naked,* when there is no perianth at all.

*hermaphrodite* or *bisexual,* when both stamens and pistil are present and perfect.

*male* or *staminate,* when there are one or more stamens, but either no pistil at all or an imperfect one.

*female* or *pistillate,* when there is a pistil, but either no stamens at all, or only imperfect ones.

*neuter,* when both stamens and pistil are imperfect or wanting.

*barren* or *sterile,* when from any cause it produces no seed.

*fertile,* when it does produce seed. In some works the terms *barren, fertile,* and *perfect* are also used respectively as synonyms of *male, female,* and *hermaphrodite.*

86. The flowers of a plant or species are said collectively to be *unisexual* or *diclinous* when the flowers are all either male or female.

*monœcious,* when the male and female flowers are distinct, but on the same plant.

*diœcious,* when the male and female flowers are on distinct plants.

*polygamous,* when there are male, female, and hermaphrodite flowers on the same or on distinct plants.

87. A head of flowers is *heterogamous* when male, female, hermaphrodite, and neuter flowers, or any two or three of them, are included in one head; *homogamous,* when all the flowers included in one head are alike in this respect. A spike or head of flowers is *androgynous* when male and female flowers are mixed in it. These terms are only used in the case of very few Natural Orders.

88. As the scales of buds are leaves undeveloped or reduced in size and altered in shape and consistence, and bracts are leaves likewise reduced in size, and occasionally altered in colour; so the parts of the flower are considered as leaves still further altered in shape, colour, and arrangement round the axis, and often more or less combined with each other. The details of this theory constitute the comparatively modern branch of botany called *Vegetable Metamorphosis,* or *Homology,* sometimes improperly termed *Morphology* (8).

89. To understand the arrangement of the floral parts, let us take a *complete* flower, in which moreover all the parts are *free* from each other, *definite* in number, *i. e.* always the same in the same species, and *symmetrical* or *isomerous, i. e.* when each whorl consists of the same number of parts.

90. Such a complete symmetrical flower consists usually of either four or five whorls of altered leaves (88), placed immediately one within the other.

The **Calyx** forms the outer whorl. Its parts are called *sepals*.

The **Corolla** forms the next whorl. Its parts, called *petals*, usually *alternate* with the sepals; that is to say, the centre of each petal is immediately over or within the interval between two sepals.

The **Stamens** form one or two whorls within the petals. If two, those of the outer whorl (the *outer stamens*) alternate with the petals, and are consequently opposite to, or over the centre of the sepals; those of the inner whorl (the *inner stamens*) alternate with the outer ones, and are therefore opposite to the petals. If there is only one whorl of stamens, they most frequently alternate with the petals; but sometimes they are opposite the petals and alternate with the sepals.

The **Pistil** forms the inner whorl; its carpels usually alternate with the inner row of stamens.

91. In an axillary or lateral flower the *upper* parts of each whorl (sepals, petals, stamens, or carpels) are those which are next to the main axis of the stems or branch, the *lower* parts those which are furthest from it; the intermediate ones are said to be *lateral*. The words *anterior* (front) and *posterior* (back) are often used for lower and upper respectively, but their meaning is sometimes reversed if the writer supposes himself in the centre of the flower instead of outside of it.

92. The number of parts in each whorl of a flower is expressed adjectively by the following numerals derived from the Greek :—

| mono-, | di-, | tri-, | tetra-, | penta-, | hexa-, | hepta, | octo-, | ennea-, | deca-, etc., | poly- |
|--------|------|-------|---------|---------|--------|--------|--------|---------|--------------|-------|
| 1-, | 2-, | 3-, | 4-, | 5-, | 6-, | 7-, | 8-, | 9-, | 10-, | *many-* |

prefixed to a termination indicating the whorl referred to.

93. Thus, a **Flower** is

*disepalous, trisepalous, tetrasepalous, polysepalous,* etc., according as there are 2, 3, 4, or many (or an indefinite number of) sepals.

*dipetalous, tripetalous, polypetalous,* etc., according as there are 2, 3, or many petals.

*diandrous, triandrous, polyandrous,* etc., according as there are 2, 3, or many stamens.

*digynous, trigynous, polygynous,* etc., according as there are 2, 3, or many carpels.

And generally (if symmetrical), *dimerous, trimerous, polymerous,* etc., according as there are 2, 3, or many (or an indefinite number of) parts to each whorl.

94. Flowers are *unsymmetrical* or *anisomerous*, strictly speaking, when any one of the whorls has a different number of parts from any other; but when the pistils alone are reduced in number, the flower is still frequently called symmetrical or isomerous, if the calyx, corolla, and staminal whorls have all the same number of parts.

95. Flowers are *irregular* when the parts of any one of the whorls are unequal in size, dissimilar in shape, or do not spread regularly round the axis at equal distances. It is however more especially irregularity of the corolla that is referred to in descriptions. A slight inequality in size or direction in the other whorls does not prevent the flower being classed as *regular*, if the corolla or perianth is conspicuous and regular.

### § 9. *The Calyx and Corolla, or Perianth.*

96. The **Calyx** (90) is usually green, and smaller than the corolla; sometimes very minute, rudimentary, or wanting, sometimes very indistinctly whorled, or not whorled at all, or in two whorls, or composed of a large number of sepals, of which the outer ones pass gradually into bracts, and the inner ones into petals.

97. The **Corolla** (90) is usually coloured, and of a more delicate texture than the calyx, and, in popular language, is often more specially meant by the *flower*. Its petals are more rarely in two whorls, or indefinite in number, and the whorl more rarely broken than in the case of the calyx, at least when the plant is in a natural state. *Double flowers* are in most cases an accidental deformity or monster in which the ordinary number of petals is multiplied by the conversion of stamens, sepals, or even carpels into petals, by the division of ordinary petals, or simply by the addition of supernumerary ones. Petals are also sometimes very small, rudimentary, or entirely deficient.

98. In very many cases, a so-called *simple perianth* (15) (of which the parts are usually called *leaves* or *segments*) is one in which the sepals and petals are similar in form and texture, and present apparently a single whorl. But if examined in the young bud, one half of the parts will generally be found to be placed outside the other half, and there will frequently be some slight difference in texture, size, and colour, indicating to the close observer the presence of both calyx and corolla. Hence much discrepancy in descriptive works. Where one botanist describes a simple perianth of six segments, another will speak of a double perianth of three sepals and three petals.

99. The following terms and prefixes, expressive of the modifications of form and arrangement of the corolla and its petals, are equally applicable to the calyx and its sepals, and to the simple perianth and its segments.

100. The Corolla is said to be *monopetalous* when the petals are united, either entirely or at the base only, into a cup, tube, or ring; *polypetalous* when they are all free from the base. These expressions, established by a long usage, are not strictly correct, for *monopetalous* (consisting of a single petal) should apply rather to a corolla really reduced to a single petal, which would then be on one side of the axis; and *polypetalous* is sometimes used more appropriately for a corolla with an indefinite number of petals. Some modern botanists have therefore proposed the term *gamopetalous* for the corolla with united petals, and *dialypetalous* for that with free petals; but the old-established expressions are still the most generally used.

101. When the petals are partially united, the lower entire portion of the corolla is called the *tube*, whatever be its shape, and the free portions of the petals are called the *teeth*, *lobes*, or *segments* (39), according as they are short or long in proportion to the whole length of the corolla. When the tube is excessively short, the petals appear at first sight free, but their slight union at the base must be carefully attended to, being of importance in classification.

102. The **Æstivation** of a corolla, is the arrangement of the petals, or of such portion of them as is free, in the unexpanded bud. It is

*valvate*, when they are strictly whorled in their whole length, their edges being placed against each other without overlapping. If the edges are much inflexed, the æstivation is at the same time *induplicate; involute*, if the margins are rolled inward; *reduplicate*, if the margins project outwards into salient angles; *revolute*, if the margins are rolled outwards; *plicate*, if the petals are folded in longitudinal plaits.

*imbricate*, when the whorl is more or less broken by some of the petals being outside the others, or by their overlapping each other at least at the top. Five-petaled imbricate corollas are *quincuncially* imbricate when one petal is outside, and an adjoining one wholly inside, the three others intermediate and overlapping on one side; *bilabiate*, when two adjoining ones are inside or outside the three others. Imbricate petals are described as *crumpled* (*corrugate*) when puckered irregularly in the bud.

*twisted*, *contorted*, or *convolute*, when each petal overlaps an adjoining one on one side, and is overlapped by the other adjoining one on the other side. Some botanists include the twisted æstivation in the general term *imbricate;* others carefully distinguish the one from the other.

103. In a few cases the overlapping is so slight that the three æstivations cannot easily be distinguished one from the other; in a few others the æstivation is variable, even in the same species, but, in general, it supplies a constant character in species, in genera, or even in Natural Orders.

104. In general shape the **Corolla** is

*tubular*, when the whole or the greater part of it is in the form of a tube or cylinder.

*campanulate*, when approaching in some measure the shape of a cup or bell.

*urceolate*, when the tube is swollen or nearly globular, contracted at the top, and slightly expanded again in a narrow rim.

*rotate* or *stellate*, when the petals or lobes are spread out horizontally from the base, or nearly so, like a wheel or star.

*hypocrateriform* or *salver-shaped*, when the lower part is cylindrical and the upper portion expanded horizontally. In this case the name of *tube* is restricted to the cylindrical part, and the horizontal portion is called the *limb*, whether it be divided to the base or not. The orifice of the tube is called its *mouth* or *throat*.

*infundibuliform* or *funnel-shaped*, when the tube is cylindrical at the base, but enlarged at the top into a more or less campanulate limb, of which the lobes often spread horizontally. In this case the campanulate part, up to the commencement of the lobes, is sometimes considered as a portion of the tube, sometimes as a portion of the limb, and by some botanists again described as independent of either, under the name of *throat* (*fauces*). Generally speaking, however, in campanulate, infundibuliform, or other corollas, where the lower entire part passes gradually into the upper divided and more spreading part, the distinction between the *tube* and the *limb* is drawn either at the point where the lobes separate, or at the part where the corolla first expands, according to which is the most marked.

105. Irregular corollas have received various names according to the more familiar forms they have been compared to. Some of the most important are the

*bilabiate* or *two-lipped* corolla, when, in a four- or five-lobed corolla, the two or three upper lobes stand obviously apart, like an *upper lip*, from the two or three lower ones or *under lip*. In *Orchideæ* and some other families the name of lip, or *labellum*, is given to one of the divisions or lobes of the perianth.

*personate*, when two-lipped, and the orifice of the tube closed by a projection from the base of the upper or lower lip, called a *palate*.

*ringent*, when very strongly two-lipped, and the orifice of the tube very open.

*spurred*, when the tube or the lower part of the petal has a conical hollow projection, compared to the spur of a cock; *saccate*, when the spur is short and round like a little bag; *gibbous*, when projecting at any part into a slight swelling; *foveolate*, when marked in any part with a slight glandular or thickened cavity.

*resupinate* or *reversed*, when a lip, spur, etc., which in allied species is usually lowest, lies uppermost, and *vice versâ*.

106. The above terms are mostly applied to the forms of monopetalous corollas, but several are also applicable to those of polypetalous ones. Terms descriptive of the special forms of corolla in certain Natural Orders, will be explained under those Orders respectively.

107. Most of the terms used for describing the forms of leaves (39, 45) are also applicable to those of individual petals; but the flat expanded portion of a petal, corresponding to the blade of the leaf, is called its *lamina*, and the stalk, corresponding to the petiole, its *claw* (*unguis*). The stalked petal is said to be *unguiculate*.

## § 10. *The Stamens.*

108. Although in a few cases the outer stamens may gradually pass into petals, yet, in general, **Stamens** are very different in shape and aspect from leaves, sepals, or petals. It is only in a theoretical point of view (not the less important in the study of the physiological economy of·the plant) that they can be called altered leaves.

109. This usual form is a stalk, called the *filament*, bearing at the top an *anther* divided into two pouches or *cells*. These anther-cells are filled with *pollen*, consisting of minute grains, usually forming a yellow dust, which, when the flower expands, is scattered from an opening in each cell. When the two cells are not closely contiguous, the portion of the anther that unites them is called the *connectivum*.

110. The filament is often wanting, and the anther sessile, yet still the stamen is perfect; but if the anther, which is the essential part of the stamen, is wanting, or does not contain pollen, the stamen is imperfect, and is then said to be *barren* or *sterile* (without pollen), *abortive*, or *rudimentary* (84), according to the degree to which the imperfection is carried. Imperfect stamens are often called *staminodia*.

111. In unsymmetrical flowers, the stamens of each whorl are sometimes reduced in number below that of the petals, even to a single one, and in several Natural Orders they are multiplied indefinitely.

112. The terms *monandrous* and *polyandrous* are restricted to flowers which have really but one stamen, or an indefinite number respectively. Where several stamens are united into one, the flower is said to be *synandrous*.

113. **Stamens** are

*monadelphous*, when united by their filaments into one cluster. This cluster either forms a tube round the pistil, or, if the pistil is wanting, occupies the centre of the flower.

*diadelphous,* when so united into two clusters. The term is more especially ap plied to certain *Leguminosæ,* in which nine stamens are united in a tube slit open on the upper side, and a tenth, placed in the slit, is free. In some other plants the sta mens are equally distributed in the two clusters.

*triadelphous, pentadelphous, polyadelphous,* when so united into three, five, or many clusters.

*syngenesious,* when united by their anthers in a ring round the pistil, the filaments usually remaining free.

*didynamous,* when (usually in a bilabiate flower) there are four stamens in two pairs, those of one pair longer than those of the other.

*tetradynamous,* when (in *Cruciferæ*) there are six, four of them longer than the two others.

*exserted,* when longer than the corolla, or even when longer than its tube, if the limb be very spreading.

114. An **Anther** (109) is

*adnate,* when continuous with the filament, the anther-cells appearing to lie their whole length along the upper part of the filament.

*innate,* when firmly attached by their base to the filament. This is like an adnate anther, but rather more distinct from the filament.

*versatile,* when attached by their back to the very point of the filament, so as to swing loosely.

115. Anther-cells may be *parallel* or *diverging* at a less or greater angle; or *divaricate,* when placed end to end so as to form one straight line. The end of each anther-cell placed nearest to the other cell is generally called its *apex* or *summit,* and the other end its *base* (36); but some botanists reverse the sense of these terms.

116. Anthers have often, on their connectivum or cells, appendages termed *bristles* (setæ), *spurs, crests, points, glands,* etc., according to their appearance.

117. Anthers have occasionally only one cell: this may take place either by the disappearance of the partition between two closely contiguous cells, when these cells are said to be *confluent ;* or by the abortion or total deficiency of one of the cells, when the anther is said to be *dimidiate.*

118. Anthers will open or *dehisce* to let out the pollen, like capsules, in *valves, pores,* or *slits.* Their dehiscence is *introrse,* when the opening faces the pistil; *extrorse,* when towards the circumference of the flower.

119. Pollen (109) is not always in the form of dust. It is sometimes collected in each cell into one or two little wax-like masses. Special terms used in describing these masses or other modifications of the pollen will be explained under the Orders where they occur.

## § 11. *The Pistil.*

120. The carpels (91) of the **Pistil,** although they may occasionally assume, rather more than stamens, the appearance and colour of leaves, are still more different in shape and structure. They are usually sessile; if stalked, their stalk is called a *podocarp.* This stalk, upon which each separate carpel is supported above the receptacle, must not be confounded with the *gynobasis* (143), upon which the whole pistil is sometimes raised.

121. Each carpel consists of three parts :

1. The **Ovary,** or enlarged base, which includes one or more cavities or *cells,* containing one or more small bodies called *ovules.* These are the earliest condition of the future seeds.

2. the **Style,** proceeding from the summit of the ovary, and supporting—

3. the **Stigma,** which is sometimes a point (or *punctiform* stigma) or small head (a *capitate* stigma) at the top of the style or ovary, sometimes a portion of its surface more or less lateral and variously shaped, distinguished by a looser texture, and covered with minute protuberances called *papillæ.*

122. The style is often wanting, and the stigma is then sessile on the ovary, but in

the perfect pistil there is always at least one ovule in the ovary, and some portion of stigmatic surface. Without these the pistil is imperfect, and said to be *barren* (not setting seed), *abortive*, or *rudimentary* (84), according to the degree of imperfection.

123. The ovary being the essential part of the pistil, most of the terms relating to the number, arrangement, etc., of the carpels, apply specially to their ovaries. In some works each separate carpel is called a pistil, all those of a flower constituting together the *gynœcium;* but this term is in little use, and the word *pistil* is more generally applied in a collective sense. When the ovaries are at all united, they are commonly termed collectively a compound ovary.

124. The number of carpels or ovaries in a flower is frequently reduced below that of the parts of the other floral whorls, even in flowers otherwise symmetrical. In a very few genera, however, the ovaries are more numerous than the petals, or indefinite. They are in that case either arranged in a single whorl, or form a head or spike in the centre of the flower.

125. The terms *monogynous, digynous, polygynous*, etc. (with a pistil of one, two, or more parts), are vaguely used, applying sometimes to the whole pistil, sometimes to the ovaries alone, or to the styles or stigmas only. Where a more precise nomenclature is adopted, the flower is

*monocarpellary*, when the pistil consists of a single simple carpel.

*bi-, tri-*, etc., to *poly-carpellary*, when the pistil consists of two, three, or an indefinite number of carpels, whether separate or united.

*syncarpous*, when the carpels or their ovaries are more or less united into one compound ovary.

*apocarpous*, when the carpels or ovaries are all free and distinct.

126. A *compound* ovary is

*unilocular* or *one-celled*, when there are no partitions between the ovules, or when these partitions do not meet in the centre so as to divide the cavity into several cells.

*plurilocular* or *several-celled*, when completely divided into two or more cells by partitions called *dissepiments (septa)*, usually vertical and radiating from the centre or axis of the ovary to its circumference.

*bi-, tri-*, etc., to *multi-locular*, according to the number of these cells, two, three, etc., or many.

127. In general the number of cells or of dissepiments, complete or partial, or of rows of ovules, corresponds with that of the carpels, of which the pistil is composed. But sometimes each carpel is divided completely or partially into two cells, or has two rows of ovules, so that the number of carpels appears double what it really is. Sometimes again the carpels are so completely combined and reduced as to form a single cell, with a single ovule, although it really consist of several carpels. But in these cases the ovary is usually described as it appears, as well as such as it is theoretically supposed to be.

128. In apocarpous pistils the styles are usually free, each bearing its own stigma. Very rarely the greater part of the styles, or the stigmas alone, are united, whilst the ovaries remain distinct.

129. Syncarpous flowers are said to have

*several styles*, when the styles are free from the base.

*one style, with several branches*, when the styles are connected at the base, but separate below the point where the stigmas or stigmatic surfaces commence.

*one simple style, with several stigmas*, when united up to the point where the stigmas or stigmatic surfaces commence, and then separating.

*one simple style, with a branched, lobed, toothed, notched, or entire stigma* (as the case may be), when the stigmas also are more or less united. In many works, however, this precise nomenclature is not strictly adhered to, and considerable confusion is often the result.

130. In general the number of styles, or branches of the style or stigma, is the same as that of the carpels, but sometimes that number is doubled, especially in the stigmas, and sometimes the stigmas are dichotomously or pinnately branched, or *penicillate*, that is, divided into a tuft of hair-like branches. All these variations sometimes make it a difficult task to determine the number of carpels forming a compound ovary, but the point is of considerable importance in fixing the affinities of plants, and, by careful

consideration, the real as well as the apparent number has now in most cases been agreed upon.

131. The *Placenta* is the part of the inside of the ovary to which the ovules are attached, sometimes a mere point or line on the inner surface, often more or less thickened or raised. *Placentation* is therefore the indication of the part of the ovary to which the ovules are attached.

132. Placentas are

*axile*, when the ovules are attached to the axis or centre, that is, in plurilocular ovaries, when they are attached to the inner angle of each cell ; in unilocular simple ovaries, which have almost always an excentrical style or stigma, when the ovules are attached to the side of the ovary nearest to the style ; in unilocular compound ovaries, when the ovules are attached to a central protuberance, column, or axis rising up from the base of the cavity. If this column does not reach the top of the cavity, the placenta is said to be *free* and *central*.

*parietal*, when the ovules are attached to the inner surface of the cavity of a one-celled compound ovary. Parietal placentas are usually slightly thickened or raised lines, sometimes broad surfaces nearly covering the inner surface of the cavity, sometimes projecting far into the cavity, and constituting partial dissepiments, or even meeting in the centre, but without cohering there. In the latter case the distinction between the one-celled and the several-celled ovary sometimes almost disappears.

133. Each **Ovule** (121), when fully formed, usually consists of a central mass or *nucleus* enclosed in two bag-like *coats*, the outer one called *primine*, the inner one *secundine*. The *chalaza* is the point of the ovule at which the base of the nucleus is confluent with the coats. The *foramen* is a minute aperture in the coats over the *apex* of the nucleus.

134. **Ovules** are

*orthotropous* or *straight*, when the chalaza coincides with the base (36) of the ovule, and the foramen is at the opposite extremity, the axis of the ovule being straight.

*campylotropous* or *incurved*, when the chalaza still coinciding with the base of the ovule, the axis of the ovule is curved, bringing the foramen down more or less towards that base.

*anatropous* or *inverted*, when the chalaza is at the apex of the ovule, and the foramen next to its base, the axis remaining straight. In this, one of the most frequent forms of the ovule, the chalaza is connected with the base by a cord, called the *raphe*, adhering to one side of the ovule, and becoming more or less incorporated with its coats, as the ovule enlarges into a seed.

*amphitropous* or *half-inverted*, when the ovule being as it were attached laterally, the chalaza and foramen at opposite ends of its straight or curved axis are about equally distant from the base or point of attachment.

§ 12. *The Receptacle and Relative Attachment of the Floral Whorls.*

135. The **Receptacle** or *torus* is the extremity of the peduncle (above the calyx), upon which the corolla, stamens, and ovary are inserted. It is sometimes little more than a mere point or minute hemisphere, but it is often also more or less elongated, thickened, or otherwise enlarged. It must not be confounded with the receptacle of inflorescence (74).

136. A *Disk*, or *disc*, is a circular enlargement of the receptacle, usually in the form of a cup (*cupular*), of a flat disk or quoit, or of a cushion (*pulvinate*). It is either immediately at the base of the ovary within the stamens, or between the petals and stamens, or bears the petals or stamens or both on its margin, or is quite at the extremity of the receptacle, with the ovaries arranged in a ring round it or under it.

137. The disk may be *entire*, or *toothed*, or *lobed*, or *divided* into a number of parts, usually equal to or twice that of the stamens or carpels. When the parts of the disk are quite separate and short, they are often called glands.

138. *Nectaries*, are either the disk, or small deformed petals, or abortive stamens, or appendages at the base of petals or stamens, or any small bodies within the flower which do not look like petals, stamens, or ovaries. They were formerly supposed to

supply bees with their honey, and the term is frequently to be met with in the older Floras, but is now deservedly going out of use.

139. When the disk bears the petals and stamens, it is frequently adherent to, and apparently forms part of, the tube of the calyx, or it is adherent to, and apparently forms part of, the ovary, or of both calyx-tube and ovary. Hence the three following important distinctions in the relative insertion of the floral whorls.

140. Petals, or as it is frequently expressed, flowers, are

*hypogynous* (*i. e.* under the ovary), when they or the disk that bears them are entirely free both from the calyx and ovary. The ovary is then described as *free* or *superior*, the calyx as *free* or *inferior*, the petals as being *inserted on the receptacle*.

*perigynous* (*i. e.* round the ovary), when the disk bearing the petals is quite free from the ovary, but is more or less combined with the base of the calyx-tube. The ovary is then still described as *free* or *superior*, even though the combined disk and calyx-tube may form a deep cup with the ovary lying in the bottom ; the calyx is said to be *free* or *inferior*, and the petals are described as *inserted on the calyx.*

*epigynous* (*i. e.* upon the ovary), when the disk bearing the petals is combined both with the base of the calyx-tube and the base outside of the ovary ; either closing over the ovary so as only to leave a passage for the style, or leaving more or less of the top of the ovary free, but always adhering to it above the level of the insertion of the lowest ovule (except in a very few cases where the ovules are absolutely suspended from the top of the cell). In epigynous flowers the ovary is described as *adherent* or *inferior*, the calyx as *adherent* or *superior*, the petals as *inserted on* or *above the ovary.* In some works, however, most epigynous flowers are included in the perigynous ones, and a very different meaning is given to the term *epigynous* (144), and there are a few cases where no positive distinction can be drawn between the epigynous and perigynous flowers, or again between the perigynous and hypogynous flowers.

141. When there are no petals, it is the insertion of the stamens that determines the difference between the hypogynous, perigynous, and epigynous flowers.

142. When there are both petals and stamens,

in hypogynous flowers, the petals and stamens are usually free from each other, but sometimes they are combined at the base. In that case, if the petals are distinct from each other, and the stamens are monadelphous, the petals are often said to be *inserted on* or *combined with the staminal tube ;* if the corolla is gamopetalous and the stamens distinct from each other, the latter are said to be *inserted in the tube of the corolla.*

in perigynous flowers, the stamens are usually inserted immediately within the petals, or alternating with them on the edge of the disk, but occasionally much lower down within the disk, or even on the unenlarged part of the receptacle.

in epigynous flowers, when the petals are distinct, the stamens are usually inserted as in perigynous flowers ; when the corolla is gamopetalous, the stamens are either free and hypogynous, or combined at the base with (inserted in) the tube of the corolla.

143. When the receptacle is distinctly elongated below the ovary, it is often called a *gynobasis, gynophore,* or *stalk of the ovary.* If the elongation takes place below the stamens or below the petals, these stamens or petals are then said to be *inserted on the stalk of the ovary,* and are occasionally, but falsely, described as *epigynous.* Really epigynous stamens (*i. e.* when the filaments are combined with the ovary) are very rare, unless the rest of the flower is epigynous.

144. An *epigynous disk* is a name given either to the thickened summit of the ovary in epigynous flowers, or very rarely to a real disk or enlargement of the receptacle closing over the ovary.

145. In the relative position of any two or more parts of the flower, whether in the same or in different whorls, they are

*connivent,* when nearer together at the summit than at the base.

*divergent,* when further apart at the summit than at the base.

*coherent,* when united together, but so slightly that they can be separated with little or no laceration ; and one of the two cohering parts (usually the smallest or least important) is said to be *adherent* to the other. Grammatically speaking, these two terms convey nearly the same meaning, but require a different form of phrase ; practically however it has been found more convenient to restrict *cohesion* to the union of parts of the same whorl, and *adhesion* to the union of parts of different whorls.

*connate*, when so closely united that they cannot be separated without laceration. Each of the two connate parts, and especially that one which is considered the smaller or of the least importance, is said to be *adnate* to the other.

*free*, when neither coherent nor connate.

*distinct* is also used in the same sense, but is also applied to parts distinctly visible or distinctly limited.

### § 13. *The Fruit.*

146. The **Fruit** (15) consists of the ovary and whatever other parts of the flower are *persistent* (*i. e.* persist at the time the seed is ripe), usually enlarged, and more or less altered in shape and consistence. It encloses or covers the seed or seeds till the period of maturity, when it either opens for the seed to escape, or falls to the ground with the seed. When stalked, its stalk has been termed a *carpophore*.

147. Fruits are, in elementary works, said to be *simple* when the result of a single flower, *compound* when they proceed from several flowers closely packed or combined in a head. But as a fruit resulting from a single flower, with several distinct carpels, is compound in the sense in which that term is applied to the ovary, the terms *single* and *aggregate*, proposed for the fruit resulting from one or several flowers, may be more appropriately adopted. In descriptive botany a fruit is always supposed to result from a single flower unless the contrary be stated. It may, like the pistil, be syncarpous or apocarpous (125); and as in many cases carpels united in the flower may become separate as they ripen, an apocarpous fruit may result from a syncarpous pistil.

148. The involucre or bracts often persist and form part of aggregate fruits, but very seldom so in single ones.

149. The receptacle becomes occasionally enlarged and succulent; if when ripe it falls off with the fruit, it is considered as forming part of it.

150. The adherent part of the calyx of epigynous flowers always persists and forms part of the fruit; the free part of the calyx of epigynous flowers or the calyx of perigynous flowers, either persists entirely at the top of or round the fruit, or the lobes alone fall off, or the lobes fall off with whatever part of the calyx is above the insertion of the petals, or the whole of what is free from the ovary falls off, including the disk bearing the petals. The calyx of hypogynous flowers usually falls off entirely or persists entirely. In general a calyx is called *deciduous* if any part falls off. When it persists it is either enlarged round or under the fruit, or it withers and dries up.

151. The corolla usually falls off entirely; when it persists it is usually withered and dry (*marcescent*), or very seldom enlarges round the fruit.

152. The stamens either fall off, or more or less of their filaments persists, usually withered and dry.

153. The style sometimes falls off or dries up and disappears; sometimes persists, forming a point to the fruit, or becomes enlarged into a wing or other appendage to the fruit.

154. The *Pericarp* is the portion of the fruit formed of the ovary, and whatever adheres to it exclusive of and outside of the seed or seeds, exclusive also of the persistent receptacle, or of whatever portion of the calyx persists round the ovary without adhering to it.

155. Fruits have often external appendages called *wings* (alæ), *beaks, crests, awns*, etc., according to their appearance. They are either formed by persistent parts of the flower more or less altered, or grow out of the ovary or the persistent part of the calyx. If the appendage be a ring of hairs or scales round the top of the fruit, it is called a *pappus*.

156. Fruits are generally divided into *succulent* (including *fleshy, pulpy,* and *juicy* fruits) and *dry*. They are *dehiscent* when they open at maturity to let out the seeds, *indehiscent* when they do not open spontaneously but fall off with the seeds. Succulent fruits are usually indehiscent.

157. The principal kinds of succulent fruits are

the *Berry*, in which the whole substance of the pericarp is fleshy or pulpy, with the exception of the outer skin or rind, called the *Epicarp*. The seeds themselves are usually immersed in the pulp ; but in some berries, the seeds are separated from the pulp by the walls of the cavity or cells of the ovary, which forms as it were a thin inner skin or rind, called the *Endocarp*.

the *Drupe*, in which the pericarp, when ripe, consists of two distinct portions, an outer succulent one called the *Sarcocarp* (covered like the berry by a skin or epicarp), and an inner dry endocarp called the *Putamen*, which is either *cartilaginous* (of the consistence of parchment) or hard and woody. In the latter case it is commonly called a *stone*, and the drupe a *stone-fruit*. When the putamen consists of several distinct stones or nuts, each enclosing a seed, they are called *pyrenes*, or sometimes *kernels*.

158. The principal kinds of dry fruits are

the *Capsule* or *Pod*,* which is dehiscent. When ripe the pericarp usually splits longitudinally into as many or twice as many pieces, called *valves*, as it contains cells or placentas. If these valves separate at the line of junction of the carpels, that is, along the line of the placentas or dissepiments, either splitting them or leaving them attached to the axis, the dehiscence is termed *septicidal* ; if the valves separate between the placentas or dissepiment, the dehiscence is *loculicidal*, and the valves either bear the placentas or dissepiments along their middle line, or leave them attached to the axis. Sometimes also the capsule discharges its seeds by *slits, chinks*, or *pores*, more or less regularly arranged, or bursts irregularly, or separates into two parts by a horizontal line ; in the latter case it is said to be *circumsciss*.

the *Nut* or *Achene*, which is indehiscent and contains but a single seed. When the pericarp is thin in proportion to the seed it encloses, the whole fruit (or each of its lobes) has the appearance of a single seed, and is so called in popular language. If the pericarp is thin and rather loose, it is often called an *Utricle*. A *Samara* is a nut with a wing at its upper end.

159. Where the carpels of the pistil are distinct (125) they may severally become as many distinct berries, drupes, capsules, or achenes. Separate carpels are usually more or less compressed laterally, with more or less prominent inner and outer edges, called *sutures*, and, if dehiscent, the carpel usually opens at these sutures. A *Follicle* is a carpel opening at the inner suture only. In some cases where the carpels are united in the pistil they will separate when ripe ; they are then called *Cocci* if one-seeded.

160. The peculiar fruits of some of the large Orders have received special names, which will be explained under each Order. Such are the *siliqua* and *silicule* of Cruci-feræ, the *legume* of Leguminosæ, the *pome* of *Pyrus* and its allies, the *pepo* of Cucur-bitaceæ, the *cone* of Coniferæ, the *grain* or *caryopsis* of Gramineæ, etc.

### § 14. *The Seed.*

161. The **Seed** is enclosed in the pericarp in the great majority of flowering plants, called therefore *Angiosperms*, or *angiospermous plants*. In *Coniferæ* and a very few allied genera, called *Gymnosperms*, or *gymnospermous plants*, the seed is naked, without any real pericarp. These truly gymnospermous plants must not be confounded with *Labiatæ*, *Boragineæ*, etc., which have also been falsely called gymnospermous, their small nuts having the appearance of seeds (158).

162. The seed when ripe contains an *embryo* or young plant, either filling or nearly filling the cavity, but not attached to the outer skin or the seed, or more or less immersed in a mealy, oily, fleshy, or horn-like substance, called the *albumen*, or *peri-sperm*. The presence or absence of this albumen, that is, the distinction between *albu-minous* and *exalbuminous* seeds, is one of great importance. The embryo or albumen can often only be found or distinguished when the seed is quite ripe, or sometimes only when it begins to germinate.

163. The shell of the seed consists usually of two separable *coats*. The outer coat, called the *testa*, is usually the principal one, and in most cases the only one attended to in descriptions. It may be hard and *crustaceous*, woody or bony, or thin and *mem-*

---

* In English descriptions, *pod* is more frequently used when it is long and narrow; *capsule*, or sometimes *pouch*, when it is short and thick or broad.

*branous* (skin-like), dry, or rarely succulent. It is sometimes expanded into *wings*, or bears a tuft of hair, cotton, or wool, called a *coma*. The inner coat is called the *tegmen*.

164. The *funicle* is the stalk by which the seed is attached to the placenta. It is occasionally enlarged into a membranous, pulpy, or fleshy appendage, sometimes spreading over a considerable part of the seed, or nearly enclosing it, called an *aril*. A *strophiole* or *caruncle* is a similar appendage proceeding from the testa by the side of or near the funicle.

165. The *hilum* is the scar left on the seed where it separates from the funicle. The *micropyle* is a mark indicating the position of the foramen of the ovule (133).

166. The **Embryo** (162) consists of the *Radicle* or base of the future root, one or two *Cotyledons* or future seed-leaves, and the *Plumule* or future bud within the base of the cotyledons. In some seeds, especially where there is no albumen, these several parts are very conspicuous, in others they are very difficult to distinguish until the seed begins to germinate. Their observation, however, is of the greatest importance, for it is chiefly upon the distinction between the embryo with one or with two cotyledons that are founded the two great classes of phænogamous plants, *Monocotyledons* and *Dicotyledons*.

167. Although the embryo lies loose (unattached) within the seed, it is generally in some determinate position with respect to the seed or to the whole fruit. This position is described by stating the direction of the radicle next to or more or less remote from the *hilum*, or it is said to be *superior* if pointing towards the summit of the *fruit*, *inferior* if pointing towards the base of the *fruit*.

## § 15. *Accessory Organs.*

168. Under this name are included, in many elementary works, various external parts of plants which do not appear to act any essential part either in the vegetation or reproduction of the plant. They may be classed under four heads : *Tendrils* and *Hooks*, *Thorns* and *Prickles*, *Hairs*, and *Glands*.

169. **Tendrils** (*cirrhi*) are usually abortive petioles, or abortive peduncles, or some-times abortive ends of branches. They are simple or more or less branched, flexible, and coil more or less firmly round any objects within their reach, in order to support the plant to which they belong. *Hooks* are similar holdfasts, but of a firmer consistence, not branched, and less coiled.

170. **Thorns** and **Prickles** have been fancifully called the weapons of plants. A *Thorn* or *Spine* is the strongly pointed extremity of a branch, or abortive petiole, or abortive peduncle. A *Prickle* is a sharply pointed excrescence from the epidermis, and is usually produced on a branch, on the petiole or veins of a leaf, or on a peduncle, or even on the calyx or corolla. When the teeth of a leaf are pungent, they are also called *prickles*, not *thorns*. A plant is *spinous* if it has thorns, *aculeate* if it has prickles.

171. **Hairs,** in the general sense, or the *indumentum* (or clothing) of a plant, include all those productions of the epidermis which have, by a more or less appropriate comparison, been termed *bristles, hairs, down, cotton,* or *wool.*

172. Hairs are often branched. They are said to be *attached by the centre*, if parted from the base, and the forks spread along the surface in opposite directions ; *plumose,* if the branches are arranged along a common axis, as in a feather ; *stellate,* if several branches radiate horizontally. These stellate hairs have sometimes their rays connected together at the base, forming little flat circular disks attached by the centre, and are then called *scales,* and the surface is said to be *scaly* or *lepidote.*

173. The *Epidermis,* or outer skin, of an organ, as to its surface and indumentum, is
  *smooth,* when without any protuberance whatever.
  *glabrous,* when without hairs of any kind.
  *striate,* when marked with parallel longitudinal lines, either slightly raised or merely discoloured.
  *furrowed* (*sulcate*) or *ribbed* (*costate*) when the parallel lines are more distinctly raised.

*rugose,* when wrinkled or marked with irregular raised or depressed lines.

*umbilicate,* when marked with a small round depression.

*umbonate,* when bearing a small boss like that of a shield.

*viscous, viscid,* or *glutinous,* when covered with a sticky or clammy exudation.

*scabrous,* when rough to the touch.

*tuberculate* or *warted,* when covered with small, obtuse, wart-like protuberances.

*muricate,* when the protuberances are more raised and pointed but yet short and hard.

*echinate,* when the protuberances are longer and sharper, almost prickly.

*setose* or *bristly,* when bearing very stiff erect straight hairs.

*glandular-setose,* when the setæ or bristles terminate in a minute resinous head or drop. In some works, especially in the case of *Roses* and *Rubus,* the meaning of *setæ* has been restricted to such as are glandular.

*glochidiate,* when the setæ are hooked at the top.

*pilose,* when the surface is thinly sprinkled with rather long simple hairs.

*hispid,* when more thickly covered with rather stiff hairs.

*hirsute,* when the hairs are dense and not so stiff.

*downy* or *pubescent,* when the hairs are short and soft ; *puberulent,* when slightly pubescent.

*strigose,* when the hairs are rather short and stiff, and lie close along the surface all in the same direction ; *strigillose,* when slightly strigose.

*tomentose* or *cottony,* when the hairs are very short and soft, rather dense and more or less intricate, and usually white or whitish.

*woolly (lanate),* when the hairs are long and loosely intricate, like wool. The wool or tomentum is said to be *floccose* when closely intricate and readily detached, like fleece.

*mealy (farinose),* when the hairs are excessively short, intricate and white, and come off readily, having the appearance of meal or dust.

*canescent* or *hoary,* when the hairs are so short as not readily to be distinguished by the naked eye, and yet give a general whitish hue to the epidermis.

*glaucous,* when of a pale bluish-green, often covered with a fine bloom.

174. The meanings here attached to the above terms are such as appear to have been most generally adopted, but there is much vagueness in the use practically made of many of them by different botanists. This is especially the case with the terms *pilose, hispid, hirsute, pubescent,* and *tomentose.*

175. The name of **Glands** is given to several different productions, and principally to the four following :—

1. Small wart-like or shield-like bodies, either sessile or sometimes stalked, of a fungous or somewhat fleshy consistence, occasionally secreting a small quantity of oily or resinous matter, but more frequently dry. They are generally few in number, often definite in their position and form, and occur chiefly on the petiole or principal veins of leaves, on the branches of inflorescences, or on the stalks or principal veins of bracts, sepals, or petals.

2. Minute raised dots, usually black, red, or dark-coloured, of a resinous or oily nature, always superficial, and apparently exudations from the epidermis. They are often numerous on leaves, bracts, sepals, and green branches, and occur even on petals and stamens, more rarely on pistils. When raised upon slender stalks they are called *pedicellate* (or *stipitate*) *glands,* or *glandular hairs,* according to the thickness of the stalk.

3. Small, globular, oblong or even linear vesicles, filled with oil, imbedded in the substance itself of leaves, bracts, floral organs, or fruits. They are often very numerous, like transparent dots, sometimes few and determinate in form and position. In the pericarp of *Umbelliferæ* they are remarkably regular and conspicuous, and take the name of *vittæ.*

4. Lobes of the disk (137), or other small fleshy excrescences within the flower, whether from the receptacle, calyx, corolla, stamens, or pistil.

## Chap. II. Classification, or Systematic Botany.

176. It has already been observed (3) that descriptions of plants should, as nearly as possible, be arranged under natural divisions, so as to facilitate the comparison of each plant with those most nearly allied to it. The descriptions of plants here alluded to are descriptions of *species;* the natural divisions of the Flora refer to natural *groups of species.*

177. A **Species** comprises all the individual plants which resemble each other suffi-; ciently to make us conclude that they are all, or *may have been* all, descended from a common parent. These individuals may often differ from each other in many striking particulars, such as the colour of the flower, size of the leaf, etc., but these particulars are such as experience teaches us are liable to vary in the seedlings raised from one individual.

178. When a large number of the individuals of a species differ from the others in any striking particular they constitute a **Variety.** If the variety generally comes true from seed, it is often called a *Race.*

179. A *Variety* can only be propagated with certainty by grafts, cuttings, bulbs, tubers, or any other method which produces a new plant by the development of one or more buds taken from the old one. A *Race* may with care be propagated by seed, although seedlings will always be liable, under certain circumstances, to lose those particulars which distinguish it from the rest of the species. A real *Species* will always come true from seed.

180. The known species of plants (now near 100,000) are far too numerous for the human mind to study without classification, or even to give distinct single names to. To facilitate these objects, an admirable system, invented by Linnæus, has been universally adopted, viz. one common substantive name is given to a number of species which resemble each other more than they do any other species; the species so collected under one name are collectively called a **Genus**, the common name being the *generic* name. Each species is then distinguished from the others of the same genus by the addition of an adjective epithet or *specific name.* Every species has thus a botanical name of two words. In Latin, the language usually used for the purpose, the first word is a substantive and designates the genus; the second, an adjective, indicates the species.

181. The genera thus formed being still too numerous (above 6,000) for study without further arrangement, they have been classed upon the same principles; viz. genera which resemble each other more than they do any other genera, have been collected together into groups of a higher degree called **Families** or **Natural Orders,** to each of which a common name has been given. This name is in Latin an adjective plural, usually taken from the name of some one *typical* genus, generally the best known, the first discovered, or the most marked (e. g. *Ranunculaceæ* from *Ranunculus*). This is however for the purpose of study and comparison. To speak of a species, to refer to it and identify it, all that is necessary is to give the generic and specific names.

182. Natural Orders themselves (of which we reckon near 200) are often in the same manner collected into **Classes ;** and where Orders contain a large number of genera, or genera a large number of species, they require further classification. The genera of an Order are then collected into minor groups called *Tribes,* the species of a genus into *Sections,* and in a few cases this intermediate classification is carried still further. The names of these several groups the most generally adopted are as follows, beginning with the most comprehensive or highest :—

| | |
|---|---|
| Classes. | Genera. |
| *Subclasses* or *Alliances.* | *Subgenera.* |
| Natural Orders or Families. | Sections. |
| *Suborders.* | *Subsections.* |
| Tribes. | Species. |
| *Subtribes.* | Varieties. |
| *Divisions.* | |
| *Subdivisions.* | |

183. The characters (3) by which a species is distinguished from all other species of

the same genus are collectively called the *specific character* of the plant ; those by which its genus is distinguished from other genera of the Order, or its Order from other Orders, are respectively called the *generic* or *ordinal* character, as the case may be. The *habit* of a plant, of a species, a genus, etc., consists of such general characters as strike the eye at first sight, such as size, colour, ramification, arrangement of the leaves, inflorescence, etc., and are chiefly derived from the organs of vegetation.

184. Classes, Orders, Genera, and their several subdivisions, are called *natural* when, in forming them, all resemblances and differences are taken into account, valuing them according to their evident or presumed importance ; *artificial*, when resemblances and differences in some one or very few particulars only are taken into account independently of all others.

185. The number of species included in a genus, or the number of genera in an Order, is very variable. Sometimes two or three or even a single species may be so different from all others as to constitute the entire genus ; in others, several hundred species may resemble each other so much as to be all included in one genus ; and there is the same discrepancy in the number of genera to a Family. There is moreover, unfortunately, in a number of instances, great difference of opinion as to whether certain plants differing from each other in certain particulars are varieties of one species or belong to distinct species ; and again, whether two or more groups of species should constitute as many sections of one genus, or distinct genera, or tribes of one Order, or even distinct Natural Orders. In the former case, as a species is supposed to have a real existence in nature, the question is susceptible of argument, and sometimes of absolute proof. But the place a group should occupy in the scale of degree is very arbitrary, being often a mere question of convenience. The more subdivisions upon correct principles are multiplied, the more they facilitate the study of plants, provided always the main resting-points for constant use, the Order and the Genus, are comprehensive and distinct. But if every group into which a genus can be divided be erected into a distinct genus, with a substantive name to be remembered whenever a species is spoken of, all the advantages derived from the beautiful simplicity of the Linnæan nomenclature are gone.

CHAP. III. VEGETABLE ANATOMY AND PHYSIOLOGY.

§ 1. *Structure and Growth of the Elementary Tissues.*

186. If a very thin slice of any part of a plant be placed under a microscope of high magnifying power, it will be found to be made up of variously shaped and arranged ultimate parts, forming a sort of honeycombed structure. These ultimate parts are called *cells*, and form by their combination the *elementary tissues* of which the entire plant is composed.

187. A cell in its simplest state is a closed membranous sac, formed of a substance permeable by fluids, though usually destitute of visible pores. Each cell is a distinct individual, separately formed and separately acting, though cohering with the cells with which it is in contact, and partaking of the common life and action of the tissue of which it forms a part. The membranes separating or enclosing the cells are also called their *walls*.

188. Botanists usually distinguish the following tissues :—

(1) *Cellular tissue*, or *parenchyma*, consists usually of thin-walled cells, more or less round in form, or with their length not much exceeding their breadth, and not tapering at the ends. All the soft parts of the leaves, the pith of stems, the pulp of fruits, and all young growing parts, are formed of it. It is the first tissue produced, and continues to be formed while growth continues, and when it ceases to be active the plant dies.

(2) *Woody tissue*, or *prosenchyma*, differs in having its cells considerably longer than broad, usually tapering at each end into points and overlapping each other. The cells are commonly thick-walled ; the tissue is firm, tenacious, and elastic, and constitutes

the principal part of wood, of the inner bark, and of the nerves and veins of leaves, forming, in short, the framework of the plant.

(3) *Vascular tissue*, or the *vessels* or *ducts* of plants, so called from the mistaken notion that their functions are analogous to those of the vessels (veins and arteries) of animals. A *vessel* in plants consists of a vertical row of cells, which have their transverse partition-walls obliterated, so as to form a continuous tube. All phænogamous plants, as well as ferns and a few other cryptogamous plants, have vessels, and are therefore called *vascular plants;* so the majority of cryptogams having only cellular tissue are termed *cellular plants.* Vessels have their sides very variously marked; some, called *spiral vessels*, have a spiral fibre coiled up their inside, which unrolls when the vessel is broken; others are marked with longitudinal slits, cross bars, minute dots or pits, or with transverse rings. The size of vessels is also very variable in different plants; in some they are of considerable size and visible to the naked eye in cross sections of the stem, in others they are almost absent or can only be traced under a strong magnifier.

189. Various modifications of the above tissues are distinguished by vegetable anatomists under names which need not be enumerated here as not being in general practical use. *Air-vessels, cysts, turpentine-vessels, oil-reservoirs,* etc., are either cavities left between the cells, or large cells filled with peculiar secretions.

190. When tissues are once formed, they increase, not by the general enlargement of the whole of the cells already formed, but by *cell-division*, that is, by the division of young and vitally active cells, and the enlargement of their portions. In the formation of the embryo, the first cell of the new plant is formed, not by division, but around a segregate portion of the contents of a previously existing cell, the embryo-sac. This is termed *free cell-formation*, in contradistinction to cell-division.

191. A young and vitally active cell consists of the *outer wall*, formed of a more or less transparent substance called *cellulose*, permeable by fluids, and of ternary chemical composition (carbon, hydrogen. and oxygen); and of the *cell-contents*, usually viscid or mucilaginous, consisting of *protoplasm*, a substance of quaternary chemical composition (carbon, hydrogen, oxygen, and nitrogen), which fills an important part in cell-division and growth. Within the cell (either in the centre or excentrical) is usually a minute, soft, subgelatinous body called the *nucleus*, whose functions appear to be intimately connected with the first formation of the new cell. As this cell increases in size, and its walls in thickness, the protoplasm and watery cell-sap become absorbed or dried up, the firm cellulose wall alone remaining as a permanent fabric, either empty or filled with various organized substances produced or secreted within it.

192. The principal organized contents of cells are

*sap*, the first product of the digestion of the food of plants; it contains the elements of vegetable growth in a dissolved condition.

*sugar*, of which there are two kinds, called *cane-sugar* and *grape-sugar.* It usually exists dissolved in the sap. It is found abundantly in growing parts, in fruits, and in germinating seeds.

*dextrine,* or vegetable mucilage, a gummy substance, between mucilage and starch.

*starch* or *fecula*, one of the most universal and conspicuous of cell-contents, and often so abundant in farinaceous roots and seeds as to fill the cell-cavity. It consists of minute grains called *starch-granules*, which vary in size and are marked with more or less conspicuous concentric lines of growth. The chemical constitution of starch is the same as that of cellulose; it is unaffected by cold water, but forms a jelly with boiling water, and turns blue when tested by iodine. When fully dissolved it is no longer starch, but dextrine.

*chlorophyll*, very minute granules, containing nitrogen, and coloured green under the action of sunlight. These granules are most abundant in the layers of cells immediately below the surface or epidermis of leaves and young bark. The green colouring matter is soluble in alcohol, and may thus be removed from the granules.

*chromule*, a name given to a similar colouring matter when not green.

*wax, oils, camphor,* and *resinous* matter, are common in cells or in cavities in the tissues between the cells, also various mineral substances, either in an amorphous state or as microscopic crystals, when they are called *Raphides.*

§ 2. *Arrangement of the Elementary Tissues, or Structure of the Organs of Plants.*

193. Leaves, young stems, and branches, and most parts of phænogamous plants, during the first year of their existence consist anatomically of

1, a *cellular system*, or continuous mass of cellular tissue, which is developed both vertically as the stem or other parts increase in length, and horizontally or laterally as they increase in thickness or breadth. It surrounds or is intermixed with the fibro-vascular system, or it may exist alone in some parts of phænogamous plants, as well as in cryptogamous ones.

2, a *fibro-vascular system*, or continuous mass of woody and vascular tissue, which is gradually introduced vertically into, and serves to bind together, the cellular system. It is continued from the stem into the petioles and veins of the leaves, and into the pedicels and parts of the flowers, and is never wholly wanting in any phænogamous plant.

3, an *epidermis*, or outer skin, formed of one or more layers of flattened (horizontal), firmly coherent, and usually empty cells, with either thin and transparent or thick and opaque walls. It covers almost all parts of plants exposed to the outward air, protecting their tissues from its immediate action, but is wanting in those parts of aquatic plants which are constantly submerged.

194. The epidermis is frequently pierced by minute spaces between the cells, called *Stomates.* They are oval or mouth-shaped, bordered by *lips*, formed of two or more elastic cells so disposed as to cause the stomate to open in a moist, and to close up in a dry state of the atmosphere. They communicate with intercellular cavities, and are obviously designed to regulate evaporation and respiration. They are chiefly found upon leaves, especially on the under surface.

195. When a phænogamous plant has outlived the first season of its growth, the anatomical structure of its stem or other perennial parts becomes more complicated and very different in the two great classes of phænogamous plants called *Exogens* and *Endogens*, which correspond with very few exceptions to the two classes Dicotyledons and Monocotyledons (167), founded on the structure of the embryo. In Exogens (Dicotyledons) the woody system is placed in concentric layers between a central *pith* (198, 1), and an external separable *bark* (198, 5). In Endogens (Monocotyledons) the woody system is in separate small bundles or fibres running through the cellular system without apparent order, and there is usually no distinct central pith, nor outer separable bark.

196. The anatomical structure is also somewhat different in the different organs of plants. In the **Root**, although it is constructed generally on the same plan as the stem, yet the regular organization, and the difference between Exogens and Endogens, is often disguised or obliterated by irregularities of growth, or by the production of large quantities of cellular tissue filled with starch or other substances (192). There is seldom, if ever, any distinct pith, the concentric circles of fibro-vascular tissue in Exogens are often very indistinct or have no relation to seasons of growth, and the epidermis has no stomates.

197. In the **Stem** or branches, during the first year or season of their growth, the difference between Exogens and Endogens is not always very conspicuous. In both there is a tendency to a circular arrangement of the fibro-vascular system, leaving the centre either vacant or filled with cellular tissue (pith) only, and a more or less distinct outer rind is observable even in several Endogens. More frequently, however, the distinction is already very apparent the first season, especially towards its close. The fibro-vascular bundles in Endogens usually anastomose but little, passing continuously into the branches and leaves. In Exogens the circle of fibro-vascular bundles forms a more continuous cylinder of network emitting lateral offsets into the branches and leaves.

198. The Exogenous stem, after the first year of its growth, consists of

1, the *pith*, a cylinder of cellular tissue, occupying the centre or longitudinal axis of the stem. It is active only in young stems or branches, becomes dried up and compressed as the wood hardens, and often finally disappears, or is scarcely distinguishable in old trees.

2, the *medullary sheath*, which surrounds and encases the pith. It abounds in spiral vessels (188, 3), and is in direct connection, when young, with the leaf-buds and branches, with the petioles and veins of leaves, and other ramifications of the system. Like the pith, it gradually disappears in old wood.

3, the *wood*, which lies immediately outside the medullary sheath. It is formed of woody tissue (188, 2), through which, in most cases, vessels (188, 3) variously dispersed. It is arranged in annual concentric circles (211), which usually remain active during several years, but in older stems the central and older layers become hard, dense, comparatively inactive, and usually deeper coloured, forming what is called *heart-wood* or *duramen*, the outer, younger, and usually paler-coloured living layers constituting the *sapwood* or *alburnum*.

4, the *medullary rays*, which form vertical plates, originating in the pith, and, radiating from thence, traverse the wood and terminate in the bark. They are formed of cellular tissue, keeping up a communication between the living portion of the centre of the stem and its outer surface. As the heart-wood is formed, the inner portion of the medullary rays ceases to be active, but they usually may still be seen in old wood, forming what carpenters call the *silver grain*.

5, the *bark*, which lies outside the wood, within the epidermis. It is, like the wood, arranged in annual concentric circles (211), of which the outer older ones become dry and hard, forming the *corky layer* or *outer bark*, which, as it is distended by the thickening of the stem, either cracks or is cast off with the epidermis, which is no longer distinguishable. Within the corky layer is the *cellular*, or *green*, or *middle bark*, formed of loose thin-walled pulpy cells containing chlorophyll (192); and which is usually the layer of the preceding season. The innermost and youngest circle, next the young wood, is the *liber* or *inner bark*, formed of long tough woody tissue called *bast-cells*.

199. The Endogenous stem, as it grows old, is not marked by the concentric circles of Exogens. The wood consists of a *matrix* of cellular tissue irregularly traversed by vertical cords or bundles of woody and vascular tissue, which are in connection with the leaves. These vascular bundles change in structure and direction as they pass down the stem, losing their vessels, they retain only their bast- or long wood-cells, usually curving outwards towards the rind. The old wood becomes more compact and harder towards the circumference than in the centre. The epidermis or rind either hardens so as to prevent any increase of diameter in the stem, or it distends, without increasing in thickness or splitting or casting off any outer layers.

200. In the **Leaf,** the structure of the petioles and principal ribs or veins is the same as that of the young branches of which they are ramifications. In the expanded portion of the leaf the fibro-vascular system becomes usually very much ramified, forming the smaller veins. These are surrounded and the interstices filled up by a copious and very active cellular tissue. The majority of leaves are horizontal, having a differently constructed upper and under surface. The cellular stratum forming the upper surface consists of closely set cells, placed vertically, with their smallest ends next the surface, and with few or no stomates in the epidermis. In the stratum forming the under surface, the cells are more or less horizontal, more loosely placed, and have generally empty spaces between them, with stomates in the epidermis communicating with these intercellular spaces. In vertical leaves (as in a large number of Australian plants) the two surfaces are nearly similar in structure.

201. When leaves are reduced to scales, acting only as protectors of young buds, or without taking any apparent part in the economy of vegetable life, their structure, though still on the same plan, is more simple; their fibro-vascular system is less ramified, their cellular system more uniform, and there are few or no stomates.

202. Bracts and floral envelopes, when green and much developed, resemble leaves in their anatomical structure, but in proportion as they are reduced to scales or transformed into petals, they lose their stomates, and their systems, both fibro-vascular and cellular, become more simple and uniform, or more slender and delicate.

203. In the stamens and pistils the structure is still nearly the same. The fibro-vascular system, surrounded by and intermixed with the cellular tissue, is usually simple in the filaments and style, more or less ramified in the flattened or expanded parts, such as the anther-cases, the walls of the ovary, or carpellary leaves, etc. The pollen consists of granular cells variously shaped, marked, or combined, peculiar forms being constant in the same species, or often in large genera, or even Orders. The stigmatic portion of the pistil is a mass of loosely cellular substance, destitute of epidermis, and

usually is in communication with the ovary by a channel running down the centre of the style.

204. Tubers, fleshy thickenings of the stem or other parts of the plant, succulent leaves or branches, the fleshy, woody, or bony parts of fruits, the albumen, and the thick fleshy parts of embryos, consist chiefly of largely developed cellular tissue, replete with starch or other substances (192), deposited apparently in most cases for the eventual future use of the plant or its parts when recalled into activity at the approach of a new season.

205. Hairs (171) are usually expansions or processes of the epidermis, and consist of one or more cells placed end to end. When thick or hardened into prickles, they still consist usually of cellular tissue only. Thorns (170) contain more or less of a fibro-vascular system, according to their degree of development.

206. Glands, in the primary sense of the word (175, 1), consist usually of a rather loose cellular tissue without epidermis, and often replete with resinous or other substances.

### § 3. *Growth of the Organs.*

207. Roots grow in length constantly and regularly at the extremities only of their fibres, in proportion as they find the requisite nutriment. They form no buds containing the germ of future branches, but their fibres proceed irregularly from any part of their surface without previous indication, and when their growth has been stopped for a time, either wholly by the close of the season, or partially by a deficiency of nutriment at any particular spot, it will, on the return of favourable circumstances, be resumed at the same point, if the growing extremities be uninjured. If during the dead season, or at any other time, the growing extremity is cut off, dried up, or otherwise injured, or stopped by a rock or other obstacle opposing its progress, lateral fibres will be formed on the still living portion ; thus enabling the root as a whole to diverge in any direction, and travel far and wide when lured on by appropriate nutriment.

208. This growth is not however by the successive formation of terminal cells attaining at once their full size. The cells first formed on a fibre commencing or renewing its growth, will often dry up and form a kind of terminal cap, which is pushed on as cells are formed immediately under it ; and the new cells, constituting a greater or lesser portion of the ends of the fibres, remain some time in a growing state before they have attained their full size.

209. The roots of Exogens, when perennial, increase in thickness like stems by the addition of concentric layers, but these are usually much less distinctly marked ; and in a large number of perennial Exogens and most Endogens the roots are annual, perishing at the close of the season, fresh adventitious roots springing from the stock when vegetation commences the following season.

210. The Stem, including its branches and appendages (leaves, floral organs, etc.), grows in length by additions to its extremity, but a much greater proportion of the extremity and branches remains in a growing and expanding state for a much longer time than in the case of the root. At the close of one season, leaf-buds or seeds are formed, each containing the germ of a branch or young plant to be produced the following season. At a very early stage of the development of these buds or seeds, a commencement may be found of many of the leaves it is to bear ; and before a leaf unfolds, every leaflet of which it is to consist, every lobe or tooth which is to mark its margin, may often be traced in miniature, and thenceforth till it attains its full size, the branch grows and expands in every part. In some cases however the lower part of a branch and more rarely (*e. g.* in some *Meliaceæ*) the lower part of a compound leaf attains its full size before the young leaves or leaflets of the extremity are yet formed.

211. The perennial stem, if exogenous (198), grows in thickness by the addition every season of a new layer or ring of wood between the outermost preceding layer and the inner surface of the bark, and by the formation of a new layer or ring of bark within the innermost preceding layer and outside the new ring of wood, thus forming a succession of concentric circles. The sap elaborated by the leaves finds its way, in a manner not as yet absolutely ascertained, into the *cambium-region,* a zone of tender thin-walled cells connecting the wood with the bark, by the division and enlargement of which new

cells (190) are formed. These cells separate in layers, the inner ones constituting the new ring of wood, and the outer ones the new bark or liber. In most exogenous trees, in temperate climates, the seasons of growth correspond with the years, and the rings of wood remain sufficiently distinct to indicate the age of the tree; but in many tropical and some evergreen trees, two or more rings of wood are formed in one year.

212. In endogenous perennial stems (199), the new wood or woody fibre is formed towards the centre of the stem, or irregularly mingled with the old. The stem consequently either only becomes more dense without increasing in thickness, or only increases by gradual distention, which is never very considerable. It affords therefore no certain criterion for judging of the age of the tree.

213. Flowers have generally all their parts formed, or indicated by protuberances or growing cells at a very early stage of the bud. These parts are then usually more regularly placed than in the fully developed flower. Parts which afterwards unite are then distinct, many are present in this rudimentary state which are never further developed, and parts which are afterwards very unequal or dissimilar are perfectly alike at this early period. On this account flowers in this very early stage are supposed by some modern botanists to be more *normal*, that is, more in conformity to a supposed type; and the study of the early formation and growth of the floral organs, called *Organogenesis*, has been considered essential for the correct appreciation of the affinities of plants. In some cases, however, it would appear that modifications of development, not to be detected in the very young bud, are yet of great importance in the distinction of large groups of plants, and that Organogenesis, although it may often assist in clearing up a doubtful point of affinity, cannot nevertheless be exclusively relied on in estimating the real value of peculiarities of structure.

214. The flower is considered as a *bud* (*flower-bud, alabastrum*) until the perianth expands, the *period of flowering* (*anthesis*) is that which elapses from the first expanding of the perianth, till the pistil is set or begins to enlarge, or, when it does not set, until the stamens and pistil wither or fall. After that, the enlarged ovary takes the name of *young fruit.*

215. At the close of the season of growth, at the same time as the leaf-buds or seeds are formed containing the germ of future branches or plants, many plants form also, at or near the bud or seed, large deposits, chiefly of starch. In many cases,—such as the tubers of a potato or other root-stock, the scales or thickened base of a bulb, the albumen or the thick cotyledons of a seed,—this deposit appears to be a store of nutriment, which is partially absorbed by the young branch or plant during its first stage of growth, before the roots are sufficiently developed to supply it from without. In some cases, however, such as the fleshy thickening of some stems or peduncles, the pericarps of fruits which perish long before *germination* (the first growth of the seed), neither the use nor the cause of these deposits has as yet been clearly explained.

## § 4. *Functions of the Organs.*

216. The functions of the Root are,—1. To fix the plant in or to the soil or other substance on which it grows. 2. To absorb nourishment from the soil, water, or air, into which the fibres have penetrated (or from other plants in the case of parasites), and to transmit it rapidly to the stem. The absorption takes place through the young growing extremities of the fibres, and through a peculiar kind of hairs or absorbing organs which are formed at or near those growing extremities. The transmission to the stem is through the tissues of the root itself. The nutriment absorbed consists chiefly of carbonic acid and nitrogen or nitrogenous compounds dissolved in water. 3. In some cases roots secrete or exude small quantities of matter in a manner and with a purpose not satisfactorily ascertained.

217. The Stem and its branches support the leaves, flowers, and fruit, transmit the crude sap, or nutriment absorbed by the roots and mixed with previously organized matter, to the leaves, and re-transmit the assimilated or elaborated sap from the leaves to the growing parts of the plant, to be there used up, or to form deposits for future use (204). The transmission of the ascending crude sap appears to take place chiefly through the elongated cells associated with the vascular tissues, passing from one cell to another by a process but little understood, but known by the name of *endosmose.*

218. Leaves are functionally the most active of the organs of vegetation. In them is chiefly conducted digestion or *Assimilation*, a name given to the process which accomplishes the following results :—1. The chemical decomposition of the oxygenated matter of the sap, the absorption of carbonic acid, and the liberation of pure oxygen at the ordinary temperature of the air. 2. A counter-operation by which oxygen is absorbed from the atmosphere and carbonic acid is exhaled. 3. The transformation of the residue of the crude sap into the organized substances which enter into the composition of the plant. The exhalation of oxygen appears to take place under the influence of solar heat and light, chiefly from the under surface of the leaf, and to be in some measure regulated by the stomates; the absorption of oxygen goes on always in the dark, and in the daytime also in certain cases. The transformation of the sap is effected within the tissues of the leaf, and continues probably more or less throughout the active parts of the whole plant.

219. The Floral Organs seldom contribute to the growth of the plant on which they are produced ; their functions are wholly concentrated on the formation of the seed with the germ of a future plant.

220. The Perianth (calyx and corolla) acts in the first instance in protecting the stamens and pistils during the early stages of their development. When expanded, the use of the brilliant colours which they often display, of the sweet or strong odours they emit, has not been adequately explained. Perhaps they may have great influence in attracting those insects whose concurrence has been shown in many cases to be necessary for the due transmission of the pollen from the anther to the stigma.

221. The pistil, when stimulated by the action of the pollen, forms and nourishes the young seed. The varied and complicated contrivances by which the pollen is conveyed to the stigma, whether by elastic action of the organs themselves, or with the assistance of wind, of insects, or other extraneous agents, have been the subject of numerous observations and experiments of the most distinguished naturalists, and are yet far from being fully investigated. Their details, however, as far as known, would be far too long for the present outline.

222. The fruit nourishes and protects the seed until its maturity, and then often promotes its dispersion by a great variety of contrivances or apparently collateral circumstances, *e. g.* by an elastic dehiscence which casts the seed off to a distance ; by the development of a pappus, wings, hooked or other appendages, which allows them to be carried off by winds, or by animals, etc., to which they may adhere; by their small specific gravity, which enables them to float down streams ; by their attractions to birds, etc., who taking them for food drop them often at great distances, etc. Appendages to the seeds themselves also often promote dispersion.

223. Hairs have various functions. The ordinary indumentum (171) of stems and leaves indeed seems to take little part in the economy of the plant besides perhaps some occasional protection against injurious atmospheric influences, but the root-hairs (216) are active absorbents, the hairs on styles and other parts of flowers appear often materially to assist the transmission of pollen, and the exudations of glandular hairs (175, 2) are often too copious not to exercise some influence on the phenomena of vegetation. The whole question, however, of vegetable exudations and their influence on the economy of vegetable life, is as yet but imperfectly understood.

---

CHAP. IV. COLLECTION, PRESERVATION, AND DETERMINATION OF PLANTS.

224. Plants can undoubtedly be most easily and satisfactorily examined when freshly gathered. But time will rarely admit of this being done, and it is moreover desirable to compare them with other plants previously observed or collected. *Specimens* must, therefore, be selected for leisurely observation at home, and preserved for future reference. A collection of such specimens constitutes a *Herbarium*.

225. A botanical **Specimen,** to be perfect, should have *root, stem, leaves, flowers* (both open and in the bud), and *fruit* (both young and mature). It is not, however, always possible to gather such complete specimens, but the collector should aim at

completeness. Fragments, such as leaves without flowers, or flowers without leaves, are of little or no use.

226. If the plant is small (not exceeding 15 in.) or can be reduced to that length by folding, the specimen should consist of the whole plant, including the principal part of the root. If it be too large to preserve the whole, a good flowering-branch should be selected, with the foliage as low down as can be gathered with it; and one or two of the lower stem-leaves or radical leaves, if any, should be added, so as to preserve as much as possible of the peculiar aspect of the plant.

227. The specimens should be taken from healthy uninjured plants of a medium size. Or if a specimen be gathered because it looks a little different from the majority of those around it, apparently belonging to the same species, a specimen of the more prevalent form should be taken from the same locality for comparison.

228. For bringing the specimens home, a light portfolio of pasteboard, covered with calico or leather, furnished with straps and buckles for closing, and another for slinging on the shoulder, and containing a few sheets of stout coarse paper, is better than the old-fashioned tin box (except, perhaps, for stiff prickly plants and a few others). The specimens as gathered are placed between the leaves of paper, and may be crowded together if not left long without sorting.

229. If the specimen brought home be not immediately determined when fresh, but dried for future examination, a note should be taken of the time, place, and situation in which it was gathered; of the stature, habit, and other particulars relating to any tree, shrub, or herb of which the specimen is only a portion; of the kind of root it has; of the colour of the flower; or of any other particulars which the specimen itself cannot supply, or which may be lost in the process of drying. These memoranda, whether taken down in the field, or from the living specimen when brought home, should be written on a label attached to the specimen or preserved with it.

230. To dry specimens, they are laid flat between several sheets of bibulous paper, and subjected to pressure. The paper is subsequently changed at intervals, until they are dry.

231. In laying out the specimen, care should be taken to preserve the natural position of the parts as far as consistent with the laying flat. In general, if the specimen is fresh and not very slender, it may be simply laid on the lower sheet, holding it by the stalk and drawing it slightly downwards; then, as the upper sheet is laid over, if it be slightly drawn downwards as it is pressed down, it will be found, after a few trials, that the specimen will have retained a natural form with very little trouble. If the specimen has been gathered long enough to have become flaccid, it will require more care in laying the leaves flat and giving the parts their proper direction. Specimens kept in tin boxes, will also often have taken unnatural bends which will require to be corrected.

232. If the specimen is very bushy, some branches must be thinned out, but always so as to show where they have been. If any part, such as the head of a thistle, the stem of an *Orobanche*, or the bulb of a Lily, be very thick, a portion of what is to be the under side of the specimen may be sliced off. Some thick specimens may be split from top to bottom before drying.

233. If the specimen be succulent or tenacious of life, such as a *Sedum* or an *Orchis*, it may be dipped in boiling water *all but the flowers*. This will kill the plant at once, and enable it to be dried rapidly, losing less of its colour or foliage than would otherwise be the case. Dipping in boiling water is also useful in the case of Heaths and other plants which are apt to shed their leaves during the process of drying.

234. Plants with very delicate corollas may be placed between single leaves of very thin unglazed tissue-paper. In shifting these plants into dry paper the tissue-paper is not to be removed, but lifted with its contents on to the dry paper.

235. The number of sheets of paper to be placed between each specimen or sheet of specimens, will depend, on the one hand, on the thickness and humidity of the specimens; on the other hand, on the quantity and quality of the paper one has at command. The more and the better the paper, the less frequently will it be necessary to change

it, and the sooner the plants will dry. The paper ought to be coarse, stout, and unsized. Common blotting-paper is much too tender.

236. Care must be taken that the paper used is well dried. If it be likewise hot, all the better; but it must then be very dry; and wet plants put into hot paper will require changing very soon, to prevent their turning black, for hot damp without ventilation produces fermentation, and spoils the specimens.

237. For pressing plants, various more or less complicated and costly presses are made. None is better than a pair of boards the size of the paper, and a stone or other heavy weight upon them if at home, or a pair of strong leather straps round them if travelling. Each of these boards should be double, that is, made of two layers of thin boards, the opposite way of the grain, and joined together by a row of clenched brads round the edge, without glue. Such boards, in deal, rather less than half an inch thick (each layer about 2¼ lines) will be found light and durable.

238. It is useful also to have extra boards or pasteboards the size of the paper, to separate thick plants from thin ones, wet ones from those nearly dry, etc. Open wooden frames with cross-bars, or frames of strong wire-work lattice, are still better than boards for this purpose, as accelerating the drying by promoting ventilation.

239. The more frequently the plants are shifted into dry paper the better. Excepting for very stiff or woody plants, the first pressure should be light, and the first shifting, if possible, after a few hours. Then, or at the second shifting, when the specimens will have lost their elasticity, will be the time for putting right any part of a specimen which may have taken a wrong fold or a bad direction. After this the pressure may be gradually increased, and the plants left from one to several days without shifting. The exact amount of pressure to be given will depend on the consistence of the specimens and the amount of paper. It must only be borne in mind that too much pressure crushes the delicate parts, too little allows them to shrivel, in both cases interfering with their future examination.

240. The most convenient specimens will be made, if the drying-paper is the same size as that of the herbarium in which they are to be kept. That of writing-demy, rather more than 16 inches by 10½ inches, is a common and very convenient size. A small size reduces the specimens too much, a large size is both costly and inconvenient for use.

241. When the specimens are quite dry and stiff, they may be packed up in bundles with a single sheet of paper between each layer, and this paper need not be bibulous. The specimens may be placed very closely on the sheets, but not in more than one layer on each sheet, and care must be taken to protect the bundles by sufficient covering from the effects of external moisture or the attacks of insects.

242. In laying the specimens into the herbarium, no more than one species should ever be fastened on one sheet of paper, although several specimens of the same species may be laid side by side. And throughout the process of drying, packing, and laying in, great care must be taken that the labels be not separated from the specimens they belong to.

243. To examine or dissect flowers or fruits in dried specimens it is necessary to soften them. If the parts are very delicate, this is best done by gradually moistening them in cold water; in most cases, steeping them in boiling water or in steam is much quicker. Very hard fruits and seeds will require boiling to be able to dissect them easily.

244. For dissecting and examining flowers in the field, all that is necessary is a penknife and a pocket-lens of two or three glasses from 1 to 2 inches focus. At home it is more convenient to have a mounted lens or simple microscope, with a stage holding a glass plate, upon which the flowers may be laid; and a pair of dissectors, one of which should be narrow and pointed, or a mere point, like a thick needle, in a handle; the other should have a pointed blade, with a sharp edge, to make clean sections across the ovary. A compound microscope is rarely necessary, except in cryptogamic botany and vegetable anatomy. For the simple microscope, lenses of ¼, ½, 1, and 1½ inches focus are sufficient.

245. To assist the student in *determining* or ascertaining the name of a plant belonging to a Flora, analytical tables should be prefixed to the Orders, Genera, and

Species. These tables should be so constructed as to contain, under each bracket, or equally indented, two (rarely three or more) alternatives as nearly as possible contradictory or incompatible with each other, each alternative referring to another bracket, or having under it another pair of alternatives further indented. The student having a plant to determine, will first take the general table of Natural Orders, and examining his plant at each step to see which alternative agrees with it, will be led on to the Order to which it belongs ; he will then compare it with the detailed character of the Order given in the text. If it agrees, he will follow the same course with the table of the genera of that Order, and again with the table of species of the genus. But in each case, if he finds that his plant does not agree with the detailed description of the genus or species to which he has thus been referred, he must revert to the beginning and carefully go through every step of the investigation before he can be satisfied. A fresh examination of his specimen, or of others of the same plant, a critical consideration of the meaning of every expression in the characters given, may lead him to detect some minute point overlooked or mistaken, and put him into the right way. Species vary within limits which it is often very difficult to express in words, and it proves often impossible, in framing these analytical tables, so to divide the genera and species, that those which come under one alternative should absolutely exclude the others. In such doubtful cases both alternatives must be tried before the student can come to the conclusion that his plant is not contained in the Flora, or that it is erroneously described.

246. In those Floras where analytical tables are not given, the student is usually guided to the most important or prominent characters of each genus or species, either by a general summary prefixed to the genera of an Order or to the species of the genus, for all such genera or species ; or by a special summary immediately preceding the detailed description of each genus or species. In the latter case this summary is called a *diagnosis*. Or sometimes the important characters are only indicated by italicizing them in the detailed description.

247. It may also happen that the specimen gathered may present some occasional or accidental anomalies peculiar to that single one, or to a very few individuals, which may prevent the species from being at one recognized by its technical characters. It may be useful here to point out a few of these anomalies which the botanist may be most likely to meet with. For this purpose we may divide them into two classes, viz. :

1. *Aberrations from the ordinary type or appearance of a species for which some general cause may be assigned.*

A bright, light, and open situation, particularly at considerable elevations above the sea, or at high latitudes, without too much wet or drought, tends to increase the size and heighten the colour of flowers, in proportion to the stature and foliage of the plant.

Shade, on the contrary, especially if accompanied by richness of soil and sufficient moisture, tends to increase the foliage and draw up the stem, but to diminish the number, size, and colour of the flowers.

A hot climate and dry situation tend to increase the hairs, prickles, and other productions of the epidermis, to shorten and stiffen the branches, rendering thorny plants yet more spinous. Moisture in a rich soil has a contrary effect.

The neighbourhood of the sea, or a saline soil or atmosphere, imparts a thicker and more succulent consistence to the foliage and almost every part of the plant, and appears not unfrequently to enable plants usually annual to live through the winter. Flowers in a maritime variety are often much fewer, but not smaller.

The luxuriance of plants growing in a rich soil, and the dwarf stunted character of those crowded in poor soils, are too well known to need particularizing. It is also an everyday observation how gradually the specimens of a species become dwarf and stunted as we advance into the cold damp regions of the summits of high mountain-ranges, or into high northern latitudes ; and yet it is frequently from the want of attention to these circumstances that numbers of false species have been added to our Enumerations and Floras. Luxuriance entails not only increase of size to the whole plant, or of particular parts, but increase of number in branches, in leaves, or leaflets of a compound leaf ; or it may diminish the hairiness of the plant, induce thorns to grow out into branches, etc.

Capsules which, while growing, lie close upon the ground, will often become larger, more succulent, and less readily dehiscent, than those which are not so exposed to the moisture of the soil.

Herbs eaten down by sheep or cattle, or crushed underfoot, or otherwise checked in their growth, or trees or shrubs cut down to the ground, if then exposed to favourable circumstances of soil and climate, will send up luxuriant side-shoots, often so different in the form of their leaves, in their ramification and inflorescence, as to be scarcely recognizable for the same species.

Annuals which have germinated in spring, and flowered without check, will often be very different in aspect from individuals of the same species, which, having germinated later, are stopped by summer droughts or the approach of winter, and only flower the following season upon a second growth. The latter have often been mistaken for perennials.

Hybrids, or crosses between two distinct species, come under the same category of anomalous specimens from a known cause. Frequent as they are in gardens, where they are artificially produced, they are probably rare in nature, although on this subject there is much diversity of opinion, some believing them to be very frequent, others almost denying their existence. Absolute proof of the origin of a plant found wild, is of course impossible; but it is pretty generally agreed that the following particulars must always co-exist in a *wild hybrid*. It partakes of the characters of its two parents; it is to be found isolated, or almost isolated, in places where the two parents are abundant; if there are two or three, they will generally be dissimilar from each other, one partaking more of one parent, another of the other; it seldom ripens good seed; it will never be found where one of the parents grows alone.

Where two supposed species grow together, intermixed with numerous intermediates bearing good seed, and passing more or less gradually from the one to the other, it may generally be concluded that the whole are mere varieties of one species. The beginner, however, must be very cautious not to set down a specimen as intermediate between two species, because it appears to be so in some, even the most striking characters, such as stature and foliage. Extreme varieties of one species are connected together by transitions in all their characters, but these transitions are not all observable in the same specimens. The observation of a single intermediate is therefore of little value, unless it be one link in a long series of intermediate forms, and, when met with, should lead to the search for the other connecting links.

2. *Accidental aberrations from the ordinary type, that is, those of which the cause is unknown.*

These require the more attention, as they may sometimes lead the beginner far astray in his search for the genus, whilst the aberrations above-mentioned as reducible more or less to general laws, affect chiefly the distinction of species.

Almost all species with coloured flowers are liable to occur occasionally with them all white.

Many may be found even in a wild state with double flowers, that is, with a multiplication of petals.

Plants which have usually conspicuous petals will occasionally appear without any at all, either to the flowers produced at particular seasons, or to all the flowers of individual plants, or the petals may be reduced to narrow slips.

Flowers usually very irregular, may, on certain individuals, lose more or less of their irregularity, or appear in some very different shape. Spurs, for instance, may disappear, or be produced on all instead of one only of the petals.

One part may be occasionally added to, or subtracted from, the usual number of parts in each floral whorl, more especially in regular polypetalous flowers.

Plants usually monœcious or diœcious may become occasionally hermaphrodite, or hermaphrodite plants may produce occasionally unisexual flowers by the abortion of the stamens or of the pistils.

Leaves cut or divided where they are usually entire, variegated or spotted where they are usually of one colour, or the reverse, must also be classed amongst those accidental aberrations which the botanist must always be on his guard against mistaking for specific distinctions.

# INDEX OF TERMS, OR GLOSSARY.

*The Figures refer to the Paragraphs of the Outlines.*

# FLORA AUSTRALIENSIS.

———◆———

## Class I. DICOTYLEDONS.

Stem, when perennial, consisting of a pith in the centre, of one or more concentric circles of woody tissue, and of the bark on the outside. Embryo with two cotyledons, the young stem in germination proceeding from between the two lobes of the embryo or from a notch at its summit.

The above characters are the most constant to separate *Dicotyledons* from *Monocotyledons*; these two great classes have, however, each a peculiar habit, which in most cases is easily recognized. All Australian trees and shrubs, except *Palms*, a few *Ferns*, and *Bamboos*, and a few others with linear grass-like leaves, are *Dicotyledons*; so also are almost all plants with opposite, or whorled, or netted-veined leaves, or with the parts of the flower in fours, fives, or eights, or with indefinite stamens, all these characters being very rare in *Monocotyledons*.

(The following list of Orders contained in this first volume is intended to show the arrangement adopted. The characters given are not absolute, nor without exception, and are inserted for the purpose of calling attention to one or two of the most striking or most important features of each Order. In some cases, where an Order is represented in Australia only by some anomalous genus, its exceptional character is placed in a parenthesis. An analytical key to the Orders will be given at the close of the work.)

## SUBCLASS I. POLYPETALÆ.

Petals several, distinct (wanting in a few genera, very rarely united).

SERIES I. THALAMIFLORÆ.—Torus small or elongated, rarely expanded in a disk. Ovary superior. Stamens definite or more frequently indefinite.

**Alliance** (Cohors) **I. Ranales.**—*Stamens indefinite, or if definite, opposite the petals. Carpels distinct or united at the base only, superior, or rarely enclosed in a fleshy torus. Embryo small, in a fleshy albumen.*
(Carpels united in *Eupomatia* and *Nymphæa*. Embryo large, without albumen in some *Menispermaceæ* and in *Nelumbium*.)

I. RANUNCULACEÆ. Herbs with radical or alternate leaves, or climbers with opposite leaves. No stipules. Sepals usually coloured and deciduous. Petals in a single series or none. Stamens indefinite. No arillus.
II. DILLENIACEÆ. Shrubs or undershrubs with alternate leaves. No stipules. Sepals usually herbaceous and persistent. Petals in a single series. Stamens usually indefinite. Seeds with an arillus or strophiola.

2 DICOTYLEDONS.

III. MAGNOLIACEÆ. Shrubs or trees, with alternate leaves. Petals indefinite. Stamens indefinite. No arillus. (Calyx entire in the bud, irregularly split.)

IV. ANONACEÆ. Shrubs, trees, or woody climbers, with alternate leaves. No stipules. Sepals 3. Petals in 2 series of 3 each (excepting *Eupomatia*, where sepals and petals are combined in a mass). Stameus indefinite. Carpels indefinite. Albumen ruminate.

V. MENISPERMACEÆ. Twiners, with alternate leaves. No stipules. Flowers small, diœcious. Sepals in 2 or more series of 3 or 2 each. Petals smaller than the inner sepals, or none. Stamens definite, opposite the petals. Carpels 6 or fewer.

VI. NYMPHÆACEÆ. Aquatic herbs. Leaves usually peltate. Sepals or petals indefinite, or rarely in threes. Stamens indefinite. Carpels free or united, the ovules not in the inner angle.

**Alliance II. Parietales.**—*Stamens definite or indefinite. Ovary syncarpous, with 2 or more parietal placentas, either 1-celled, or incompletely divided by the placentas protruding in the cavity, or divided by false dissepiments connecting the placentas. Ovules usually several to each placenta, rarely solitary.*

VII. PAPAVERACEÆ. Herbs, with alternate leaves. No stipules. Sepals 2. Petals 4. Flowers regular, with indefinite stamens, or irregular, with diadelphous definite stamens. Albumen copious. Embryo small.

VIII. CRUCIFERÆ. Herbs, with alternate leaves. No stipules. Sepals 4. Petals 4. Stamens 6, tetradynamous or rarely 4. Placentas 2, connected by a false dissepiment. No albumen. · Embryo curved.

IX. CAPPARIDEÆ. Herbs, shrubs, or trees. Stipules often prickly. Sepals 4 (2 outer ones sometimes united). Petals 4 (rarely more, or none, or united). Stamens indefinite, or if few, not tetradynamous. Placentas 2 or more. No albumen. Embryo curved.

X. VIOLARIEÆ. Herbs or shrubs. Stipules herbaceous or small. Sepals 5. Petals 5 (often irregular). Anthers 5, on short filaments, connivent or connected in a ring round the pistil. Placentas usually 3. Albumen fleshy. Embryo rather large.

XI. BIXINEÆ. Trees or shrubs. Stipules none. Sepals 5 or fewer. Petals various, often none. Stamens indefinite. Placentas 2, 3, or more (meeting in the axis in *Cochlospermum*). Albumen fleshy. Embryo rather large.

**Alliance III. Polygalineæ.**—*Sepals and petals 5 each, rarely fewer. Stamens the same number or twice as many, or fewer when the flowers are irregular. Ovary usually 2-merous (although in most genera occasionally 3–5-merous), partially or completely divided into as many cells. Ovules indefinite, or solitary with a superior micropyle. Albumen fleshy.*

XII. PITTOSPOREÆ. Trees, shrubs, undershrubs, or twiners, with alternate leaves. No stipules. Flowers regular or oblique. Stameus as many as petals. Embryo minute.

XIII. TREMANDREÆ. Shrubs often heath-like, with alternate or whorled or opposite leaves. No stipules. Flowers regular. Stamens twice as many as petals. Embryo small or minute.

XIV. POLYGALEÆ. Herbs, undershrubs, or shrubs, with alternate leaves. No stipules. Flowers irregular. Stamens monadelphous. Embryo rather large, sometimes almost or quite without albumen.

**Alliance IV. Caryophyllineæ.**—*Sepals or calyx-lobes 5 or fewer. Petals 5 or fewer. Stamens as many or twice as many, or indefinite. Ovary 1-celled, with central placentas (except Frankenia). Albumen mealy. Embryo curved, or rarely straight when the albumen is scanty.*

(Ovary half-inferior in *Portulaca*.)

XV. FRANKENIACEÆ. Small or prostrate undershrubs, or herbs, with small opposite leaves. No stipules. Calyx angular, toothed. Petals isomerous with the calyx. Stamons definite. Placentas parietal.

XVI. CARYOPHYLLEÆ. Herbs, rarely undershrubs, with opposite entire leaves. Stipules none or scarious. Calyx toothed or sepals free. Petals isomerous with the calyx. Stamens definite. Placentas central.

XVII. PORTULACEÆ. Herbs, often succulent, with alternate or opposite leaves. Stipules scarious or changed into hairs. Sepals 2. Petals more numerous than the sepals. Stamens indefinite or rarely definite. Placentas central.

**Alliance V. Guttiferales.**—*Sepals imbricate. Petals as many as sepals, or rarely more. Stamens indefinite (except* Elatineæ). *Ovary divided into cells, with axile placentas.*

XVIII. ELATINEÆ. Herbs or undershrubs, with small opposite leaves. Stipules small. Flowers hermaphrodite. Stamens definite.

XIX. HYPERICINEÆ. Herbs or shrubs, with opposite leaves. No stipules. Flowers hermaphrodite. Stamens indefinite.

XX. GUTTIFERÆ. Trees or shrubs, with opposite leaves. No stipules. Flowers polygamous or unisexual. Stamens indefinite.

**Alliance VI. Malvales.**—*Sepals valvate (except* Echinocarpus). *Petals as many as sepals, or none. Stamens indefinite or monadelphous (except* Lasiopetaleæ). *Ovary divided into cells with axile placentas.*

XXI. MALVACEÆ. Herbs, shrubs, or trees, with alternate leaves. Stipules usually present. Stamens monadelphous. Anthers 1-celled.

XXII. STERCULIACEÆ. Herbs, shrubs, or trees, with alternate leaves. Stipules usually present. Stamens monadelphous, or, if free, definite and alternating with the petals. Anthers 2-celled.

XXIII. TILIACEÆ. Trees or shrubs, rarely herbs, with alternate leaves. Stipules usually present. Stamens indefinite, free, or scarcely united at the base. Anthers 2-celled.

SERIES II. DISCIFLORÆ.—Torus usually thickened or expanded into a disk, either free or adnate to the ovary, or to the calyx, or to both, rarely reduced to glands, or wanting. Stamens as many or twice as many as petals, or fewer. Ovary superior, or partially immersed in the disk, divided into cells with axile placentas, or the carpels distinct.

(Stamens indefinite in a very few exceptional species. Ovary inferior or enclosed in the calyx-tube in most *Rhamneæ*; 1-celled in some *Olacineæ*.)

**Alliance VII. Geraniales.**—*Disk within the stamens, or confluent with the staminal tube, or reduced to glands, or obsolete. Gynœcium lobed or apocarpous, or sometimes entire. Ovules usually 1 or 2 in each cell, 1 or both pendulous with a ventral raphe.*

XXIV. LINEÆ. Herbs or shrubs, with undivided alternate leaves. Stipules often present. Disk small, glandular, or none. Ovary entire. Ovules usually 2 in each cell. Albumen fleshy, rarely wanting.

XXV. MALPIGHIACEÆ. Woody climbers (rarely trees or shrubs), with opposite (rarely alternate) leaves. Stipules present. Two glands on the outside of some or all the calyx-lobes (wanting in the Australian genera). Disk not large. Gynœcium lobed or apocarpous. Ovules solitary in each cell. No albumen.

XXVI. ZYGOPHYLLEÆ. Herbs or shrubs, usually articulate or succulent, without glandular dots. Leaves 2-foliolate or pinnate, rarely simple. Stipules present. Disk fleshy. Ovary angular or lobed. Ovules 2 or more in each cell. Albumen fleshy or none.

XXVII. GERANIACEÆ. Herbs or shrubs, articulate or not, with toothed, divided, or compound leaves without glandular dots. Stipules usually present. Disk reduced to 5 glands or obsolete. Ovary angular or lobed. Ovules 1, 2, or rarely more in each cell. Albumen none or rarely fleshy.

XXVIII. RUTACEÆ. Trees or shrubs, very rarely herbs, with compound or rarely simple leaves, always marked with pellucid glandular dots. No stipules. Disk within the stamens. Ovary rarely entire, usually lobed or the carpels distinct, with the styles connate or gynœcium entirely apocarpous. Ovules 2 in each cell. Albumen fleshy or none.

B 2

XXIX. SIMARUBEÆ. Characters of *Rutaceæ*, except that the leaves are not dotted and the ovules are usually solitary in each cell. Taste generally bitter.

XXX. BURSERACEÆ. Trees or shrubs, not dotted, but with a balsamic juice. Leaves pinnately or ternately compound. No stipules. Disk free or adnate to the calyx-tube. Ovary entire. Ovules usually 2 in each cell. Albumen none. Cotyledons much folded or rarely thick and fleshy.

XXXI. MELIACEÆ. Trees or shrubs, with compound or rarely simple leaves. No stipules. Stamens monadelphous. Anthers sessile or rarely stipitate within or on the top of the staminal tube. Ovary entire. Ovules 2 in each cell. Albumen none or fleshy.

**Alliance VIII. Olacales.**—*Disk various or none. Ovary entire. Ovules 1 to 3 in a solitary cell, or 1 in each cell, pendulous with a dorsal raphe, the integuments not distinct from the nucleus. Seeds solitary in the fruit or in the cells. Albumen copious.*

XXXII. OLACINEÆ. Trees or shrubs, rarely undershrubs or climbers. No stipules. Petals or corolla-lobes valvate (except *Villaresia*). Ovary 1-celled or incompletely 3- to 5-celled. Fruit 1-seeded.

XXXIII. ILICINEÆ. Trees or shrubs. No stipules. Petals or corolla-lobes imbricate. Ovary 3- or more celled.

**Alliance IX. Celastrales.**—*Disk thick and fleshy or adnate to the calyx, the stamens outside or upon it. Ovary entire (except Stackhousia). Ovules 1 or 2 in each cell, erect with a ventral raphe.*

XXXIV. CELASTRINEÆ. Trees or shrubs, with simple leaves. Stipules none, or minute and deciduous. Calyx-lobes imbricate. Petals spreading. Stamens alternating with the petals or fewer. Ovary entire.

XXXV. STACKHOUSIEÆ. Herbs or undershrubs, with simple leaves. Calyx-lobes imbricate. Petals erect, usually connate. Stamens alternating with the petals. Ovary lobed.

XXXVI. RHAMNEÆ. Trees or shrubs, with simple leaves. Stipules usually present. Calyx-lobes valvate. Petals small, concave (or none). Stamens opposite the petals. Ovary entire, often inferior.

XXXVII. AMPELIDEÆ. Climbers, with simple or compound leaves, the petiole usually expanded into a stipule. Calyx-lobes imbricate. Petals valvate. Stamens opposite the petals. Ovary entire. Albumen cartilaginous. Embryo small.

**Alliance X. Sapindales.**—*Disk fleshy or adnate to the calyx, within or under or outside the stamens. Gynœcium entire, lobed or apocarpous. Ovules 1 or 2 in each cell, ascending with a ventral raphe, or reversed, or suspended from an erect funiculus, or pendulous with an inferior micropyle.*

XXXVIII. SAPINDACEÆ. Trees, shrubs, or climbers, with compound or simple leaves. Stamens anisomerous with the petals, or twice as many as petals or of the same number, often (but not always) within the disk. Style 1. Ovules ascending.

XXXIX. ANACARDIACEÆ. Trees or shrubs, with compound or simple leaves. Stamens as many or twice as many as petals, never within the disk. Ovules suspended from an erect funicle or from the top or side of the cell with an inferior micropyle.

## ORDER I. RANUNCULACEÆ.

Sepals 3 or more, most frequently 5, usually petal-like and deciduous. Petals of the same number or more, or sometimes none, or very small and deformed. Stamens indefinite, hypogynous, free. Anthers innate. Gynœcium of several carpels, usually free; ovules anatropous, either solitary and ascending, with a ventral raphe, or pendulous with a dorsal raphe, or several. Fruit of one or more indehiscent achenes or berries, or follicular capsules, the distinct styles usually persistent as short points, or lengthened into long, often bearded tails. Seeds without any arillus. Embryo very small, near the base of a copious albumen.—Herbs either annual or with a perennial rootstock, or creeping stolons, with radical or alternate leaves, or climbers with opposite leaves. Leaves entire, or palmately or pinnately lobed or divided, the petiole often dilated and sheathing at the base, or rarely accompanied by stipular appendages. Hairs, when present, simple. Flowers regular (or in a few genera, not Australian, irregular), terminal or leaf-opposed, rarely axillary, solitary paniculate or racemose.

The Order is chiefly numerous in the temperate regions of the northern hemisphere, rare within the tropics, and not represented by many species in the southern hemisphere. The Australian ones are all extratropical, and belong to genera more numerously represented in the north.

TRIBE I. **Clematideæ.**—*Sepals valvate. Carpels indehiscent, with 1 pendulous ovule or seed in each. Stems often climbing. Leaves opposite.*

Petals none . . . . . . . . . . . . . . . . . . . . 1. CLEMATIS.

TRIBE II. **Anemoneæ.**—*Sepals imbricate. Carpels indehiscent, with 1 pendulous ovule or seed in each. Herbs. Leaves radical or alternate or forming an involucre below the flower.*

Petals none. Involucre below the flower. Achenes in a short head . . 2. ANEMONE.
Petals minute, narrow. No involucre. Achenes very numerous, in a
long, close, slender spike . . . . . . . . . . . . . 3. MYOSURUS.

TRIBE III. **Ranunculeæ.**—*Sepals imbricate. Carpels indehiscent, with 1 ascending ovule or seed in each. Herbs. Leaves radical or alternate.*

Sepals deciduous. Petals 3, 5, or more . . . . . . . . . 4. RANUNCULUS.

TRIBE IV. **Helleboreæ.**—*Sepals imbricate. Carpels usually opening along the inner edge, containing several ovules or seeds. Herbs. Leaves radical or alternate.*

Petals none . . . . . . . . . . . . . . . . . . . . 5. CALTHA.

## 1. CLEMATIS, Linn.

Sepals 4, or rarely 5 to 8, petal-like, valvate in the bud. Petals none, or smaller than the sepals, and passing gradually into the stamens. Carpels many, with one pendulous ovule in each. Achenes capitate, sessile, or scarcely stipitate, terminating in a plumose or simple tail, formed by the persistent and enlarged style.—Stem woody and climbing, or rarely dwarf or prostrate. Leaves opposite, pinnately or ternately divided into three or more petiolulate segments, or rarely simple, the petiole often twisted or twining. Flowers axillary or terminal, solitary, or in panicles, which are shortened branches with the leaves reduced to small bracts, and often polygamous or dioecious.

A large genus, dispersed over the temperate regions both of the New and the Old World, rare within the tropics. The Australian species are all endemic, although one is closely connected with a South Pacific one. They have all simple or once- or twice-ternately divided leaves, dioecious, apetalous, white or cream-coloured flowers, the males usually without any ovaries, the females with a few imperfect stamens, and the carpels of all have plumose tails.

Anthers linear or oblong, tipped by a subulate or oblong appendage.
Woody climbers. Leaflets mostly once or twice ternate.
Anther-points slender. Leaflets almost coriaceous, when
large usually toothed, when small twice ternate . . . 1. *C. aristata.*
Anther-points very short. Leaflets usually 3, rather large,
thin, and entire . . . . . . . . . . . . . 3. *C. glycinoides.*

Stem prostrate, creeping, or shortly erect. Leaves simple or
with 3 leaflets. Flowers large, usually solitary. Anther-
tips very short . . . . . . . . . . . . . . 2. *C. gentianoides.*
Anthers short, without any appendage.
Leaflets ternate, rather large, loosely pubescent underneath . 3. *C. glycinoides,*
var. *submutica.*
Leaflets mostly twice ternate, small or narrow, glabrous or
closely pubescent . . . . . . . . . . . . . 4. *C. microphylla.*

1. **C. aristata,** *R. Br. in DC. Syst. Veg.* i. 147. A woody climber,
trailing over rocks and bushes, or ascending into tall trees, glabrous, or softly
pubescent, especially on the inflorescence. Leaves mostly on long petioles,
and divided into 3 petiolulate segments or leaflets, varying from ovate-cordate
to narrow-lanceolate, obtuse or acute, 1 to 2 or even 3 in. long, usually
irregularly toothed when large, entire when small, and of a firm consistence
when full grown, but some of the leaves near the base of the flowering
branches are occasionally simple, and others have often twice ternate leaflets.
Flowers white or yellowish, usually in short panicles or clusters in the upper
axils. Sepals 4, or very rarely 5, oblong or linear-lanceolate, usually ¾ to
1 in. long when fully out, glabrous or pubescent. Anthers oblong-linear,
tipped by a subulate appendage, often as long as the cells, usually rather
shorter, but seldom so short as in the two following species, the outer anthers
on long filaments, the inner ones almost sessile. Achenes numerous, ovate
or lanceolate, pubescent or glabrous, with a plumose tail often attaining 1½ in.
—F. Muell. Pl. Vict. i. 3; Bot. Reg. t. 238.

**N. S. Wales.** Port Jackson, *R. Brown, Sieber, n.* 273, and others, and southward
to Illawara, *Backhouse* and others; Twofold Bay, *F. Mueller.*
**Victoria.** Moist forest localities, chiefly along banks of rivers and rivulets as far west
as the Grampians, *F. Mueller.*
**Tasmania.** Abundant throughout the island, *J. D. Hooker.*
**W. Australia.** Swan River, *Huegel, Drummond, Preiss, n.* 1344, 1345, *and* 1346,
and others; from King George's Sound to the northern parts of the colony, *Herb. F.
Mueller.*
The different forms assumed by the numerous specimens we have of this species may be
classed under the following principal varieties:—
*a. coriacea.* Leaflets large, usually once ternate. Flowers often pubescent or villous.
Carpels pubescent.—*C. coriacea,* DC. Syst. Veg. i. 146; Hook. f. Fl. Tasm. i. 2.—From
Port Jackson to Tasmania.
*b. blanda.* Leaflets usually small and often twice ternate (sometimes incompletely so, the
leaves appearing at first sight simply pinnate with 5 leaflets). Flowers and carpels glabrous.
*C. clitorioides,* DC. Syst. Veg. i. 158; *C. blanda,* Hook. Journ. Bot. i. 241; Hook. f. Fl.
Tasm. i. 3.—South coast of Victoria and Tasmania.
*c. occidentalis.* Like *a,* but usually more pubescent, with narrower sepals and shorter
appendages to the anthers; some western specimens cannot however be distinguished from
some of the Port Jackson ones.—*C. pubescens,* Hueg. Enum. 1; *C. elliptica,* Endl. in
Hueg. l. c.; *C. indivisa,* Steud. in Pl. Preiss. ii. 262, not Willd.; *C. discolor,* Steud. l. c.
*C. cognata,* Steud. l. c. 263; *C. Gilbertiana,* Turcz. in Bull. Mosc. 1854, ii. 273.—West
Australia.

2. **C. gentianoides,** *DC. Syst. Veg.* i. 159. Believed by F. Mueller
to be a variety of *C. aristata,* but, if so, it is so strongly marked a one as to
have all the appearance of a distinct species. The stem creeps underground,
throwing up short tufts of flowering branches, or lies prostrate on the ground,
to the length of 3 or 4 feet at most. Leaves usually simple or with 3 seg-

ments, large, ovate-lanceolate or lanceolate, and firm.   Flowers large, usually glabrous, solitary, or few in loose clusters.  Anther-appendages short.  Achenes villous, narrow.—Deless. Ic. Sel. i. t. 5 ; Hook. f. Fl. Tasm. i. t. 3.

**Tasmania.**  Not so common as *C. aristata*, but found in various parts of the colony, always in poor soil, *J. D. Hooker.*

3. **C. glycinoides,** *DC. Syst. Veg.* i. 145.   A woody climber, very near to those forms of *C. aristata* which have simply ternate rather large ovate-lanceolate or cordate leaflets, but these leaflets are usually of a thinner consistence, often broader, and quite entire or rarely with a single tooth near the base.  Flowers usually smaller, the sepals narrow, from ½ to ¾ in., pubescent or rarely glabrous.  Anthers rather shorter, with a very short obtuse and almost gland-like appendage.  Achenes glabrous or pubescent, usually narrower than in *C. aristata*, with tails of about 2 in.—*C. stenosepala*, DC. Syst. Veg. i. 147.

**Queensland.**  Keppel Bay, *R. Brown* (a form with 3 large broad segments).
**N. S. Wales.**  Port Jackson and Port Macquarie, *R. Brown* and others ; Lord Howe Island.  From the latter station we have a small specimen, gathered by *Milne*, with the foliage of Brown's specimen from Keppel Bay.  Another female specimen, gathered in Lord Howe Island by *M'Gillivray*, who states it to be very abundant there, has several of the leaves large, simple, and orbicular-cordate, with 7 to 9 nerves.  This connects it very closely with *C. cocculifolia*, A. Cunn. in Ann. Nat. Hist. ser. 1. iv. 260, from Norfolk Island, which has most of the leaves simple and orbicular, and with *C. Pickeringii*, A. Gray, in Bot. Amer. Expl. Exped. i. 1, from the Fiji Islands, which has three large leaflets.  All these plants have similar floral characters, and may not unlikely prove to be varieties of one species.
Var. *? submutica.*  Leaf-segments loosely pubescent underneath, sepals shorter, broader, and more villous than in the other forms, anthers short, tipped by a minute gland or entirely without appendage, as in *C. microphylla.*—Clarence river and Brisbane river, *Herb. F. Mueller,* upon whose authority I insert it as a variety of *C. glycinoides,* the specimens being as yet insufficient to determine whether it may not really be a distinct species.

4. **C. microphylla,** *DC. Syst. Veg.* i. 147.   A tall woody climber, with the habit of the smaller-leaved varieties of *C. aristata.*  Leaflets mostly twice ternate, narrow, from ovate-lanceolate or oblong to nearly linear, ½ to 1 in. long, but sometimes simply ternate and larger and broader, or three times ternate and much smaller.  Flowers rather smaller than in *C. aristata,* usually numerous in short panicles.  Sepals cream-coloured, from oblong-lanceolate to narrow-linear, mostly about ½ in. rarely near 1 in. long, glabrous or pubescent.  Stamens with unequal filaments as in *C. aristata,* but the anthers are always very shortly oblong or ovate and very obtuse, without any terminal appendage.  Achenes of *C. aristata,* but usually with thicker, often wrinkled or warted margins and longer tails.—F. Muell. Pl. Vict. i. 4 ; *C. linearifolia*, Steud. ; Hook. f. Fl. Tasm. i. 4, *t.* 1 ; *C. stenophylla,* Fras. ; Hook. in Mitch. Trop. Aust. 368.

**Queensland.**  On the Maranoa, *Mitchell ;* Moreton Bay, *Herb. F. Mueller.*
**N. S. Wales.**  Frequent in the western interior, *A. Cunningham, Fraser,* and others.
**Victoria.**  South coast, *R. Brown;* not rare along the coast and on the banks of rivers near the sea, much less frequent inland, *F. Mueller.*
**Tasmania.**  Sandhills, George Town and Flinders Island, *Gunn.*
**S. Australia.**  Banks of the Torrens, *Whittaker,* and other points along the coast, *F. Mueller.*

**W. Australia.** King George's Sound, *Collie;* Swan River, *Drummond; Preiss, n.* 1343.

Var. *occidentalis.* Carpels narrower and seldom wrinkled, with tails often of 3 to 4 inches. Sepals usually long and narrow.—*C. linearifolia,* Steud. in Pl. Preiss. ii. 262. Apparently the usual form in West Australia.

Var. *leptophylla,* F. Muell. Leaf-segments very small and narrow. Trailing over granite rocks on the Snowy River and Mitta Mitta, *F. Mueller.*

## 2. ANEMONE, Linn.

Involucre of 3 or more leaves or lobes either close to the flower or on the peduncle below it. Sepals 4 to 20, petal-like. Petals none. Carpels indefinite, with 1 pendulous ovule in each. Achenes in a globular or oblong head, glabrous or woolly, pointed by the persistent style, which is sometimes lengthened into a bearded tail.—Herbs, with a perennial rootstock. Leaves radical, cut or lobed. Scapes radical, leafless except the involucre. Flowers terminal, variously coloured, but not bright yellow. Stamens shorter than the sepals.

A large genus, chiefly dispersed over the temperate or mountainous regions of the northern hemisphere. A few species are found in South America and southern Africa, but they are further removed even than some of the northern ones from the Australian one, which is strictly endemic.

1. **A. crassifolia,** *Hook. Ic. Pl. t.* 257. Radical leaves on rather long petioles; segments 3, distinct but sessile, obovate or almost orbicular, from ½ to ¾ in. long or rarely 1 inch, more or less deeply divided into 3 or more broad obtuse lobes, thick and almost succulent or coriaceous, glabrous or sprinkled with rigid appressed hairs. Scape 6 to 8 in. high, clothed with appressed hairs, especially in the upper part. Involucre rather above the middle, irregularly divided into 2 or 3 sessile lobed segments. Sepals usually 6 or 7, white, ovate or obovate, ½ to ¾ in. long. Achenes in a globular head, glabrous, rather inflated, terminating in a glabrous point about two lines long, hooked at the extremity.—Hook. f. Fl. Tasm. i. 4.

**Tasmania.** Mountains of the Black Bluff range and west of Cape St. Clair, at an elevation of 4000 to 5000 feet, *Gunn, Milligan.*

## 3. MYOSURUS, Linn.

Sepals usually 5, produced below their insertion into a small spur. Petals 5, small and very narrow, almost tubular at the top, often wanting. Carpels numerous, with one pendulous ovule in each. Achenes closely packed in a long slender spike, flat on the back, or with a raised nerve ending in the short persistent style.—Small annuals with linear radical entire leaves. Flowers very small, on leafless scapes.

A genus comprising, besides the following, only one other species, *M. aristatus,* Geyer, distinguished by the more prominent and spreading points of the achenes, which although originally described from North America and from Chili, has also been found in New Zealand, and may not improbably appear in Australia.

1. **M. minimus,** *Linn.; DC. Prod.* i. 25. Leaves sometimes not an inch long, sometimes attaining 2 or even 3 inches, including their long petiole. Scapes shorter or longer than the leaves. Sepals yellowish or pale green, very small; petals rarely longer than the calyx, and in the Australian

specimens often deficient. Stamens usually 4 or 5, and seldom above 10. Achenes sometimes near 300, the head lengthening into a spike of 1 to 2 inches, which has been compared to a mouse's tail.—F. Muell. Pl. Vict. i. 4 ; A. Gray, Gen. Ill. t. 8 ; *M. australis,* F. Muell. in Trans. Phil. Soc. Vict. i. 6.

**Victoria.** Moist places near permanent waters, or open places where rain-water lodges from time to time, *F. Mueller.*

The species is widely spread over Europe, temperate Asia, northern and western America, and may possibly have been introduced into Australia.

## 4. RANUNCULUS, Linn.

Sepals usually 5, deciduous. Petals as many or more, usually marked with a small nectariferous pit, or a minute scale near the base. Carpels several, with a single ascending ovule in each. Achenes in a globular or ovoid head or oblong spike, tipped or beaked by the persistent hooked or straight style. —Herbs either annual or with a perennial rootstock, and tufted entire or variously cut radical leaves. Flowering stems either a leafless scape, or several-flowered, bearing few leaves and chiefly at the base of the peduncles. Flowers yellow, white, or red.

A large genus abounding in the temperate and colder regions of both the northern and southern hemispheres, but more especially in the former, and almost confined in the tropics to the higher mountain ranges. The Australian species have no peculiar character, but belong to the three principal sections of the genus, and two at least are specifically identical with widely-spread northern species.

Sect. 1. **Batrachium.** — *Carpels transversely wrinkled. Water-plants with their leaves when submerged finely divided into segments. Flowers white.* 1. *R. aquatilis.*

Sect. 2. **Hecatonia.**—*Carpels smooth. Perennials (in Australia) with a tufted rootstock, or creeping or floating stolons. Flowers white or yellow.*

Radical leaves pinnate, with narrow-linear, entire or divided, rather distant segments.
Rootstock a cluster of short thick fibres. Stems mostly 2-flowered, longer than the leaves. (Fl. yellow?) . . . . . . . . 2. *R. Robertsoni.*
Rootstock tufted with long fibres. Scapes 1-flowered, shorter than the leaves. Fl. white. . . . . . . . . . . . . . 3. *R. Millani.*
Radical leaves orbicular, with numerous overlapping lobes. Stem-leaves similar but sessile. Flowers large, white . . . . . 4. *R. anemoneus.*
Radical leaves with numerous narrow-linear segments, pinnate but crowded at the top of the petiole. Flowers yellow.
Carpels numerous, tapering into a beak either straight or slightly hooked. Petals narrow, often more than 6.
Sepals from ¾ to nearly as long as the petals . . . . . . 5. *R. Gunnianus.*
Sepals not half so long as the petals . . . . . . . . . 6. *R. dissectifolius.*
Carpels with a much recurved point. Leaf-segments less crowded. Petals usually 5, obovate . . . . . . . . . . . 7. *R. lappaceus,* var.
Radical leaves pinnate, with flat segments or digitate. Flowers yellow. Stems tufted or erect or decumbent, without stolons. Petals usually 5.
Calyx appressed or spreading, not reflexed.
Carpels with a much recurved point. Plant hispid, or silky hairy, or nearly glabrous. Leaves pinnatisect, or 3- to 5-lobed, or entire . . . . . . . . . . . . . . 7. *R. lappaceus.*
Carpels numerous, tapering into a straight or slightly hooked beak. Leaves thick, entire or 3-lobed, silky underneath, with long tubercular hairs above . . . . . . . . 8. *R. Muelleri.*

Calyx reflexed.    Stem weak, hirsute.    Leaves not pinnate.
Flowers small . . . . . . . . . . . . . 9. *R. plebeius.*
Stems creeping, floating, or stoloniferous. Plant glabrous or nearly
so. Leaves digitate. Petals usually 6 to 10 . . . . . 10. *R. rivularis.*

SECT. 3. **Echinella.** — *Carpels tuberculate or muricate or hispid on the sides.*
*Annuals. Flowers yellow.*

Flowers lateral, sessile, or on peduncles shorter than the leaves.
Hairy plant, with very small flowers, often sessile. Carpels usually
about 1 line long, with a small recurved point . . . . . . 11. *R. parviflorus.*
Glabrous plant. Flowers all pedunculate. Carpels much muricate,
2 lines long or more, with a stout beak . . . . . . . . *R. muricatus.* (p. 15.)
Flowers terminal, pedunculate . . . . . . . . . . . *R. philonotis.* (p. 15.)

1. **R. aquatilis,** *Linn.; DC. Prod.* i. 26. A most variable species, easily known by its stem either floating in water or creeping in half-dried mud, by its white flowers and very small ovoid carpels marked with transverse wrinkles. It is always glabrous, excepting sometimes the carpels and their receptacle. In the Australian specimens the leaves are all submerged and divided into numerous very fine linear segments; in northern ones, there are frequently also a few upper leaves spreading on the surface of the water, which are rounded and more or less cut into 3 or 5 wedge-shaped, obovate, or rounded lobes. Peduncles axillary and 1-flowered. Petals 5 or sometimes more, white, without any scale or spot at the base; in most Australian specimens they are scarcely longer than the calyx, and the stamens are very few, but sometimes the petals are fully twice as long, and the stamens numerous. —Hook. f. Fl. Tasm. i. 5.; F. Muell. Pl. Vict. i. 5.

**Victoria.** Bacchus Marsh, Murray river, Mitta-Mitta river, etc., *F. Mueller.*
**Tasmania.** Lake river, near Grindelwald and Formosa, *Gunn;* South Esk river and near Evandale, *C. Stuart.*
**S. Australia.** Near Adelaide, on the Lower Murray river, etc., *Behr, F. Mueller.*
The species is abundant in the waters of the northern hemisphere.

2. **R. Robertsoni,** *Benth.* Allied to *R. Millani,* but distinguished from all Australian species, and in some measure connected with some of the European ones by its rootstock consisting of a cluster of short thick fibres. Radical leaves usually 2 or 3 in. long, pinnately divided in their upper portion into a few rather distant narrow linear segments, which are often again divided into 2 to 5 lobes, not unlike those of *R. Millani,* glabrous or with a few silky appressed hairs. Flower-stems often 2-flowered, 3 to 8 in. high, with 1 or 2 narrow and not much cut leaves. Flowers rather large, appearing yellowish in the dried specimens, but possibly white. Sepals not half so long as the petals. Petals 5, obovate, with a small glandular pit. Achenes in an ovoid head on a slender glabrous receptable, glabrous and smooth, tapering into a long and slightly hooked beak.

**Victoria.** Forest land near the Glenelg, and in Nangela Vale, *Robertson.*

3. **R. Millani,** *F. Muell. in Hook. Kew Journ.* vii. 358, and *Pl. Vict.* i. 6. A dwarf tufted perennial, with long clustered fibres, occasionally emitting a short stolon terminating in another tuft. Leaves all radical, 1 to 2 in. long, pinnately divided in their upper portion into a few narrow-linear segments either entire or again divided, most of them terminating in a small gland, glabrous or hispid, with a few long hairs. Scapes 1-flowered, leafless,

shorter than the leaves and often very short.  Flowers white, although sometimes appearing yellowish when dry.  Sepals not above half as long as the petals.  Petals 5 to 10, obovate or oblong-cuneate, the glandular pit very small.  Achenes in a globular head with a short recurved style ; receptacle hairy, very short.

**Victoria.**  Gravelly places on most of the summits of the Australian Alps, *F. Mueller.*

4. **R. anemoneus,** *F. Muell. in Trans. Phil. Soc. Vict.* i. 97, *and Pl. Vict.* i. 7. *t.* 1.  A rather stout perennial, hirsute with long soft hairs, or glabrous.  Rootstock thick, with long clustered fibres, and bearing several broad thin scales at the base of the leaves and stems.  Radical leaves on long petioles of 5 to 10 in., nearly orbicular, 2 to 4 in. diameter, deeply divided into 3 or 5 segments, which are again digitately cut and lobed, the segments overlapping each other so as to make the leaf appear peltate, the ultimate lobes short and lanceolate.  Stem 9 in. to 1 ft. high, 1- to 3-flowered, with a sessile, deeply-lobed, nearly orbicular leaf at the base of each peduncle.  Flowers large and white.  Sepals 5 to 7, rarely more than half the length of the petals.  Petals usually numerous, oblong-cuneate, often ¾ in. long, the glandular pit rather large.  Carpels numerous, in a globular head, tapering into a straight or scarcely hooked beak.

**Victoria.**  Along springs near the summits of the Munyang mountains, *F. Mueller.* A very distinct species, allied in some respects to *R. nivicola*, from New Zealand, but readily known by the sessile stem-leaves.

5. **R. Gunnianus,** *Hook. Journ. Bot.* i. 244. *t.* 133.  Rootstock thick, sometimes horizontal or shortly creeping, with long fibres.  Leaves all radical and glabrous, or with a few long hairs, the petioles varying from 2 to 6 in., pinnately divided at the top into crowded linear or linear-lanceolate segments, most of them again once or twice divided, all thicker and firmer than in *R. Millani*, mostly tipped by a small gland.  Scapes leafless and 1-flowered, usually longer than the leaves, silky hairy, at least at the summit.  Flowers rather large, yellow, but often, especially the sepals, purple outside.  Sepals nearly as long as the petals, glabrous.  Petals 5, 6, or rarely more, cuneate-oblong, 6 to 9 lines long, usually with three glandular pits, the central one rather longer than the other, but sometimes only 1 and occasionally 5 pits to each petal.  Carpels numerous, in a globular head, with a conical triquetrous or flattened beak, not hooked at the point.—Hook. f. Fl. Tasm. i. 5 ; F. Muell. Pl. Vict. i. 9.

**Victoria.**  Grassy places throughout the greater portion of the Australian Alps at an elevation of from 4500 to 7000 ft., *F. Mueller.*

**Tasmania.**  Hampshire hills, Western mountains, Ben Lomond, and as far north as Mount Lapeyrouse, etc., at about 4000 ft. elevation, *Lawrence, Gunn.*

The large loose grains of the albumen mentioned by Hooker, do not appear to be in their normal state ; for I find the albumen of apparently quite ripe seeds, dense and fleshy as in other *Ranunculi.*

6. **R. dissectifolius,** *F. Muell. Herb.*  Considered by F. Mueller as a variety of *R. lappaceus*, but it appears to me to be more nearly allied to *R. Gunnianus*, and although intermediate, as it were, between the two species, yet separated from both by characters not to be neglected.  Leaves divided into numerous linear lobes and segments, crowded at the top of the petiole,

and often tipped with a gland, especially when very narrow, and achenes numerous, with straight or scarcely hooked beaks, as in *R. Gunnianus.* Hairs usually copious and spreading, and sepals not half so long as the petals, as in *R. lappaceus.* Scapes usually 1-flowered and leafless, or with a single leaf. Petals more than 5, usually 8 to 10, narrow, the glandular pit usually very faint and sometimes quite imperceptible.

**Victoria.** Wet alpine meadows of the Munyang mountains, at an elevation of 5000 to 6000 ft., *F. Mueller.*

7. **R. lappaceus,** *Sm.; DC. Prod.* i. 39. A perennial, more or less clothed with soft spreading or rarely silky and appressed hairs. Rootstock short, with long fibres and no stolons. Leaves chiefly radical, on long petioles, usually divided into 3 or 5 deep lobes or segments, ovate or rhomboid-cuneate, either pinnately distinct or, if confluent, almost palmate, although the middle lobe is generally longer than the lateral ones, each lobe or segment is often again lobed or toothed and sometimes much cut into narrow lobes, more rarely the leaves are all entire or shortly 3-lobed. Flowering stems either a leafless 1-flowered scape or branching and erect or decumbent, bearing several flowers and a few leaves, smaller and less divided than the radical ones. Flowers of a rich yellow. Sepals hairy or rarely glabrous, usually much shorter than the petals, appressed or open, but not closely reflexed. Petals usually 5, broadly obovate and rather large, with a small glandular pit near the base. Carpels in a globular head, compressed or rarely turgid, glabrous and smooth, with a recurved style, usually short, but longer and slender in some western specimens.—Hook. f. Fl. Tasm. i. 6; F. Muell. Pl. Vict. i. 7; *R. colonorum,* Endl. in Hueg. Enum. 1; *R. discolor,* Steud. in Pl. Preiss. i. 263 (calyx certainly not reflexed).

**N. S. Wales.** Port Jackson and in the interior, apparently common, *R. Brown* and others.

**Victoria.** Grassy places, from the lowlands to the limits of eternal snow; here and there also in boggy and swampy localities, *F. Mueller.*

**Tasmania.** Very common all over the island up to the highest summits, *J. D. Hooker, Gunn.*

**S. Australia.** In the pasture lands, *Behr.*

**W. Australia.** In sandy shady woods not far from the sea, *Preiss, n.* 1347. Blackwood river, *Oldfield.*

The following forms, all united by F. Mueller with *R. lappaceus,* and certainly appearing sometimes to pass into the common one by intermediate gradations, are nevertheless sufficiently well characterized to be considered at least as marked varieties :—

Var. *pimpinellifolius.* A small plant, with spreading hairs. Leaves all radical, distinctly pinnate, with usually 5 short, broad, 3- or 5-lobed segments. Scapes 1-flowered, leafless or with one small bract. Pit of the petals usually distant from the base. *R. pimpinellifolius,* Hook. Journ. Bot. i. 243, and Ic. Pl. t. 260. *R. hirtus,* Hook. f. Fl. Tasm. i. 6, but scarcely of Banks and Solander, which has the reflexed calyx and narrow petals of *R. plebeius.*—Australian Alps, *F. Mueller.* Tasmania, in moist places chiefly in the mountains, *Gunn,* including an alpine form, with much smaller petals.

Var. *scapigerus.* Very villous. Leaves all radical, short and broad, deeply 3- or 5-lobed, with obovate cuneate lobes, the middle one scarcely longer than the lateral ones. Scapes 1-flowered and leafless, or few-flowered with small leaves. Flowers small. Calyx almost reflexed.—*R. scapigerus,* Hook. Journ. Bot. i. 244; Hook. f. Fl. Tasm. i. 7.—Australian Alps, *F. Mueller.* Tasmania, mountains, *Gunn.* This form seems to pass almost into *R. plebeius* as to technical characters, but the habit is very different.

Var. *subsericeus.* Hairs all appressed and silky. Leaves usually narrow, entire, 3-lobed or pinnately divided into 3 or 5 entire segments. Scapes 1-flowered.—Summits of the Australian Alps, *F. Mueller.* Tasmania, in the Hampshire hills and Western Mountains, *Gunn.*

Var. *nanus.* Dwarf and nearly glabrous. Leaves all radical, usually 3-lobed or of 3 segments. Flowers small, on short scapes.—*R. nanus,* Hook. Journ. Bot. i. 242 ; Hook. f. Fl. Tasm. 1, 7 ; *R. cuneatus,* Hook. Journ. Bot. i. 242 ; Hook. f. Fl. Tasm. 1, 8.—Australian Alps, *F. Mueller.* Tasmania, alpine districts, summits of the Western Mountains, Arthur's Lakes, etc., *Gunn.*

8. **R. Muelleri,** *Benth.* Allied to *R. lappaceus,* var. *subsericeus,* but the achenes are too different to admit of its being united in the same species, at least until better known. Leaves all radical, undivided, entire or coarsely 3-toothed, oblong or cuneate, ½ to 1 in. long, very thick, covered on the upper surface with long hairs proceeding from tubercles, and underneath with appressed short silky hairs. Scapes 1-flowered. Flowers nearly of *R. lappaceus.* Sepals very obtuse, not half so long as the petals. Petals 5, narrow-obovate. Achenes numerous, in a dense globular head, narrower than in *R. lappaceus,* and attenuated into a rigid, straight, or scarcely hooked point.

**Victoria.** Summits of the Munyang mountains, *F. Mueller.*

9. **R. plebeius,** *R. Br. in DC. Syst. Veg.* i. 288. Hirsute with spreading or rarely nearly appressed hairs. Radical leaves on long petioles, digitately divided into 3 deeply lobed and toothed cuneate or rhomboid segments. Stems weak, decumbent or erect, often above a foot long and branched, with a few leaves, the lower ones more divided than the radical ones, with the primary segments petiolate, the others smaller, more sessile, and less cut. Flowers several, small, on long peduncles. Calyx reflexed, shorter than the petals, very deciduous. Petals obovate or oblong, seldom above 2 lines long. Achenes few or numerous, more or less compressed, rather small, with a hooked or recurved slender style.—Steud. in Pl. Preiss. i. 263 ; *R. hirtus,* Banks and Sol. in DC. Syst. Veg. i. 289 ; F. Muell. Pl. Vict. i. 8.

**N. S. Wales.** Port Jackson, *R. Brown,* and northward to the Hastings river.

**Victoria.** Moe Swamp and Snowy River, Narracan river and Baw-baw mountains, *F. Mueller.*

**W. Australia.** In the interior, *Preiss, n.* 1348.

The New Zealand *R. hirtus,* Banks and Sol., appears to be a slight variety of this species. A closely allied South African one has a rather different foliage, and the carpels often tuberculate or muricate, which never occurs in Australian specimens ; it passes under the name of *R. pinnatus,* Poir., which was originally given to an East Indian plant, very near to and perhaps identical with the Cape species, and that again almost passes into some European ones ; but I do not think that any except the New Zealand *R. hirtus* can be absolutely identified with *R. plebeius.*

10. **R. rivularis,** *Banks and Sol. in DC. Syst. Veg.* i. 270. Stems creeping or stoloniferous, producing at every node tufts of radical leaves and erect scapes, or weak slightly branched flowering stems, rarely forming short thick rhizomes. Leaves on long petioles, digitately divided into 3, 5, or 7 segments, varying from cuneate to narrow-linear, rarely entire, usually 3-lobed, and sometimes much cut, but never pinnate, either quite glabrous, as well as the whole plant, or rarely with a very few appressed hairs. Flowers yellow,

usually small, the sepals not reflexed.   Petals 6 to 10, about twice as long as
the sepals, or 5 only in small-flowered varieties, narrow-oblong.   Achenes
rather small and broad, with a firm or slender recurved or rarely nearly
straight point, not tubercled or muricate.—F. Muell. Pl. Vict. i. 8.

**Queensland.**   Moreton Bay, *W. Hill.*
**N. S. Wales.**   Abundant about Port Jackson, *Herb. Hooker.*
**Victoria.**   In swamps, rivulets, marshes, or inundated places from the coast to the
higher Alps, as well in brackish as in fresh water, *F. Mueller.*
**Tasmania.**   Abundant in wet places, sometimes growing in deep water, *J. D. Hooker,
Gunn.*
**S. Australia.**   In swampy lands, *Behr.*; extending to the Darling and St. Vincent's
Gulf, but rare in the Colony, *F. Mueller.*

This very variable species is recognizable in perfect specimens by its creeping or floating
stolons; where these are wanting, the glabrous digitate leaves and narrow petals are the
best marks of distinction from the *R. lappaceus.*   The following are the most marked forms
it assumes.

Var. *major.*   Tufts erect.   Leaf-segments ½ to 1 in. long or more, often very narrow and
much cut, on petioles of 2 to 6 inches.   Flowers rather large.—*R. inundatus,* R. Br. in
DC. Syst. Veg. 1, 269.   *R. glabrifolius,* Hook. Journ. Bot. i. 243; Hook. f. Fl. Tasm. i. 9.
*R. incisus,* Hook. f. Fl. Nov. Zeal. 1, 10. t. 4.

Var. *subfluitans.*   Very slender and creeping, or half floating in large masses, with small
leaves, not much divided, and small flowers and achenes.—*R. rivularis,* Banks and Sol. in
DC. Syst. Veg. i. 270.   *R. inundatus,* Hook. f. Fl. Tasm. i. 8.

Var. *inconspicuus.*   Still smaller, with very small flowers.—*R. inconspicuus,* Hook. f. Fl.
Tasm. i. 9. t. 2 B; Gunn, n. 1018, 1019.—An alpine form, which in the dried state might
be confounded with some of the minute specimens of *R. lappaceus nanus.*

The New Zealand specimens appear identical with the Australian ones.   The nearest approach to it in other countries is the Antarctic-American *R. biternatus,* Sm.; but that
has biternate petiolate leaf-segments, and thick broad, almost reniform achenes, very
different from those of any Australian specimen I have seen.   *R. acaulis,* Banks, from
New Zealand and from Auckland Islands, referred to *R. rivularis* by F. Mueller, comes certainly near to the var. *inconspicuus,* but appears to me to be distinct, although perhaps a
reduced form of *R. biternatus.*   The New Zealand *R. macropus,* Hook., is also supposed by
F. Mueller to be a variety of *R. rivularis,* but is too different in several points to be admitted
without having seen connecting specimens.

11.  **R. parviflorus,** *Linn.; DC. Prod.* i. 42 : *var. australis.*   A
slender hairy annual, either with tufted erect stems of a few inches, or weak,
procumbent, and lengthening to a foot or even more.   Leaves small, orbicular, the lower ones often only 3- or 5-lobed, but mostly divided into
three segments, either entire or 3-lobed, or again cut into narrow segments.
Flowers small, leaf-opposed, sessile, or on short slender peduncles.   Sepals
rarely above 1 line long and very deciduous.   Petals 5 or fewer, seldom much
longer than the calyx.   Achenes in a small globular head, much compressed,
with a smooth margin, seldom much exceeding a line in breadth in Australian
specimens, the sides covered with short hairs, or tubercles, or short hooked
bristles, the style forming usually a very short recurved point, more rarely
rigid and dilated at the base.—F. Muell. Pl. Vict. i. 9; *R. sessiliflorus,* R. Br.
in DC. Syst. Veg. i. 302; Hook. f. Fl. Tasm. i. 9; *R. collinus,* R. Br. l. c.
i. 271; *R. pumilio,* R. Br. l. c. i. 271; Hook. f. Fl. Tasm. i. 10; *R. leptocau-
lis,* Hook. Journ. Bot. i. 244; *R. pilulifer,* Hook. Ic. Pl. t. 600.

**Queensland.**   In water-holes on the tops of the ranges in the interior, *Mitchell.*
**N. S. Wales.**   Moist pastures and banks of rivers and lagoons, *R. Brown* and others.
**Victoria.**   Common in similar stations, *F. Mueller.*

**Tasmania.** *R. Brown*, common, *J. D. Hooker, Gunn.*
**W. Australia.** *Drummond.*
The Australian variety above described, which occurs also in New Zealand, has smaller flowers and achenes, and they are more frequently sessile than in the usual typical form, which is widely spread over Europe.

**R. muricatus,** *Linn.; DC. Prod.* i. 42.—A densely-tufted annual, much larger and coarser than *R. parviflorus;* leaves much longer and usually less divided; flowers larger, yellow, on leaf-opposed peduncles; carpels flat, much muricated, fully 2 lines long, with a flat, stout, recurved beak: a common weed in southern Europe and many parts of Asia, has now become wild about Melbourne.

**R. philonotis,** *Retz; DC. Prod.* i. 41. An annual, with 3-lobed or divided leaves like some of those of *R. parviflorus,* but larger and less hairy, and with much larger yellow flowers on terminal peduncles, with a closely-reflexed calyx: a common European species, has been found near the seacoast at Southport, in Tasmania, by C. Stuart.

## 5. CALTHA, Linn.

Sepals 5 or more, coloured and petal-like. Petals none. Carpels several, sessile, distinct, bearing several ovules in a double row along their inner angle, opening into follicles when ripe. Seeds obovoid; testa crustaceous, smooth, the raphe usually very prominent.—Glabrous, tufted, or stoloniferous herbs. Leaves mostly radical, entire or crenate, with palmate nerves, cordate at the base, or sagittate with the auricles or basal lobes turned upwards over their face. Scapes 1-flowered and leafless, or few-flowered with a small leaf at the base of each peduncle. Flowers yellow or rarely white.

The genus is confined to the temperate and cold regions of both the northern and southern hemispheres. The southern ones are almost always distinguished by the turned-up basal lobes of the leaves. The only Australian species is endemic, unless it prove a variety of the New Zealand one.

1. **C. introloba,** *F. Muell. in Trans. Phil. Soc. Vict.* i. 98, *and Pl. Vict.* i. 10. A dwarf, glabrous, somewhat succulent perennial. Rootstock thick, often elongated, producing numerous stoutish fibres. Leaves all radical, the petioles ½ to 3 in. long, with broad, sheathing, membranous bases, forming a stem-like sheath, reaching to half their length, the blade hastate-ovate or ovate-lanceolate, ½ to 1 in. or rather more in length, the 2 basal lobes turned over the upper surface, often reaching above half its length. Scapes 1-flowered, sometimes scarcely exceeding the leaf-sheaths, sometimes 6 to 8 in. high. Sepals 5 to 8, linear-lanceolate, 4 to 5 lines long. Stamens usually few. Carpels sometimes 5 or 6, sometimes above 20, ovate-falcate or shortly oblong, 2 to 3 lines long, and the outer ones almost horizontal when ripe, tipped by the persistent and usually straight style, containing 3 to 5 seeds.—Hook. f. Fl. Tasm. ii. 355.

**Victoria.** In gravelly places irrigated by the melting snows in the Australian Alps, *F. Mueller.*
**Tasmania.** Western Mountains, *Archer.*
Very closely allied to the *C. Novæ-Zelandiæ,* Hook. f., from New Zealand, which indeed appears only to differ in its broader and shorter leaves and recurved styles. It has also yellow flowers, whilst the Australian one has them white, perhaps only when fading; but the same difference in the colour of the flowers occurs in different plants of *C. palustris* in the Himalayas.

ORDER II. **DILLENIACEÆ.**

Sepals usually 5, persistent, imbricate in the bud. Petals 5 or rarely fewer, deciduous, imbricate in the bud. Stamens hypogynous, indefinite, few or numerous, or rarely definitely 10, free or rarely united in clusters. Anthers innate or adnate. Gynœcium of carpels several, free and distinct or cohering at the base, or rarely single and excentrical, 1-celled, with 1 or more ovules in each. Styles quite distinct and diverging. Fruit-carpels either indehiscent and succulent, or opening along the inner edge, or in two valves. Seeds furnished with an arillus; testa crustaceous. Embryo very small, at the base of a fleshy albumen.—Trees, shrubs, climbers, or herbs. Leaves alternate or very rarely opposite. Stipules minute or none. Flowers usually yellow or white.

A considerable Order, of which rather the larger portion, with regularly pinnate veins prominent on the under side of the leaves, is entirely tropical, and represented in Australia by a single species of *Wormia*. The remainder of the Order, forming the tribe *Hibbertieæ*, with the midrib of the leaf alone prominent, or rarely with reticulate veins, is almost entirely Australian, there being besides only one species known from New Caledonia and two from Madagascar.

Anthers elongated, opening in two pores at the top. Trees with large
    leaves, with raised parallel veins underneath . . . . . . . 1. WORMIA.
Anthers opening longitudinally. Undershrubs, shrubs, or rarely climbers.
    Leaves with a prominent midrib and obscure or reticulate veins.
    Perfect stamens free or nearly so, more than 10, or, if fewer, on one
        side of the pistil . . . . . . . . . . . . . . . 2. HIBBERTIA.
    Stamens united in 5 clusters, or in 3 clusters with two separate stamens 3. CANDOLLEA.
    Perfect stamens 10 or fewer, in a complete ring round the pistil.
      No staminodia within the perfect stamens . . . . . . . . 4. ADRASTÆA.
      Two staminodia within the perfect stamens. Branches leafless . . 5. PACHYNEMA.

1. **WORMIA,** Rottb.

Sepals 5, spreading. Petals 5. Stamens numerous, with erect linear anthers opening at the summit in two pores, the inner ones often longer and recurved. Carpels 5 to 10, scarcely cohering, with several ovules in each, dehiscent when ripe. Seeds with an arillus.—Trees often very lofty. Leaves large, with raised parallel veins diverging from the midrib, the petioles often bordered with narrow deciduous wings. Flowers large, in loose terminal panicles.

A tropical genus, extending over tropical Asia and the Indian Archipelago, with one Madagascar species. The only Australian one is endemic.

1. **W. alata,** *R. Br. in DC. Syst. Veg.* i. 434. Glabrous, or the young parts very slightly hoary. Leaves oval or nearly orbicular, rounded at both ends, 4 to 8 in. long, entire or slightly sinuate, rather rough to the touch, with about 9 prominent veins on each side of the midrib and transversely reticulate veinlets, the petiole 1 in. long or more, with longitudinal wings about 1 line broad, which fall off in the greater part of their length. Peduncles terminal, not usually exceeding the leaves, bearing 2 or 3 large flowers on pedicels of nearly 1 in. Sepals 6 to 8 lines long, ovate, concave,

ciliate.  Petals obovate, 1½ in. long, narrowed at the base.  Stamens very numerous, the inner ones long and recurved, the others shorter, and the outermost sometimes small and barren.  Gynœcium of 5 to 8 glabrous carpels, tapering into long recurved styles.  Ovules 6 to 8 in each carpel.

**Queensland.**  Endeavour river, *Banks, A. Cunningham*; Cape York, *M'Gillivray.*

## 2. HIBBERTIA, Andr.

(Hemistemma, Pleurandra, *and* Hibbertia, *DC.*; Ochrolasia, *Turcz.*; Hemistephus, *Drummond.*)

Sepals 5, spreading, sometimes shortly united at the base.  Petals 5.  Stamens indefinite, rarely fewer than 12, and then usually all on one side of the carpels, either all perfect or some of them reduced to staminodia, all free or the filaments shortly and irregularly united at the base; anthers erect, oblong, or rarely ovate or orbicular, opening in longitudinal slits.  Carpels usually 2 to 5, rarely solitary or more than 5, free or shortly cohering on their inner edge, with 2 to 6 or rarely only 1 or more than 6 ovules in each.  Styles filiform, diverging, terminal or almost dorsal.  Fruit-carpels usually dehiscent at the top.  Seeds reniform or nearly globular, with an entire or divided arillus.—Shrubs or undershrubs, usually much branched and low, erect or procumbent, sometimes almost herbaceous or climbing, rarely 5 or 6 feet high.  Leaves usually small, alternate in all the Australian species, with a midrib prominent underneath, the lateral veins reticulate and rarely prominent.  Flowers yellow or white, solitary and terminal, or (owing to the shortness or abortion of the flowering shoot) apparently axillary sessile in a tuft of floral leaves or pedunculate.

Besides the Australian species, there are only two known, both from Madagascar, belonging to the section *Hemistemma*, but with opposite leaves.  The species of the first three of the following sections are usually distributed into two separate genera, *Hemistemma* and *Pleurandra*, the *Hemipleurándras* being referred sometimes to the one, sometimes to the other; but their characters appear to be much less important and less conformable to habit than was originally supposed, and I have followed Mueller in uniting them with *Hibbertia* as sections only.

Sect. I. **Hemistemma.**—*Perfect stamens and staminodia all on one side of the carpels, the staminodia outside.  Peduncles mostly 2- or more-flowered, except in* H. verrucosa.—All tropical species except *H. verrucosa.*

Leaves oblong or lanceolate, flat or the margins slightly recurved.
  Leaves obtuse.
    Leaves with recurved margins, narrowed into a petiole, rusty-brown underneath.  Sepals obtuse . . . . . . . . 1. *H. Banksii.*
    Leaves flat, closely sessile with a rounded base, white underneath.  Sepals acute . . . . . . . . . . . . 2. *H. Brownei.*
  Leaves acute or mucronate, white underneath.
    Spikes terminal, several-flowered . . . . . . . . 3. *H. dealbata.*
    Peduncles lateral, 2- or 3-, rarely 1-flowered . . . . . 4. *H. candicans.*
Leaves narrow-oblong or linear, the margins revolute.
  Leaves oblong-linear, thick, about ½ in. long.
    Leaves and calyx glabrous or scabrous with stiff stellate hairs.  Peduncles 1-flowered . . . . . . . . . . . . 8. *H. verrucosa.*
    Leaves tomentose underneath.  Sepals densely and softly villous.  Peduncles mostly 2- or 3-flowered . . . . . . 7. *H. ledifolia.*

Leaves narrow linear, about 1 in. long.
    Softly hairy . . . . . . . . . . . . . . . . 6. *H. Muelleri.*
    Glabrous.  Leaves white underneath . . . . . . . . 5. *H. angustifolia.*
(*Hemistemma? Leschenaultii*, DC. Syst. Veg. i. 414, is a species of *Beyeria*.)

SECT. II. **Hemipleurandra.**—*Perfect stamens all on one side of the carpels ; staminodia 2 or 3 on each side of them, or more numerous and continued round the carpels, very rarely any outside the perfect stamens.*—All western species.
Peduncles bearing 2 or more sessile flowers in a one-sided spike.
    Leaves glabrous.  Staminodia completing the ring of stamens . 9. *H. spicata.*
    Leaves or sepals hirsute.  Staminodia few . . . . . . . 10. *H. polystachya.*
Peduncles 1-flowered.
    Leaves oblong or linear, very obtuse, stellate-tomentose or hoary
        underneath.
        Leaves mostly above 1 in., the margins scarcely recurved.
            Ovules 4 . . . . . . . . . . . . . . . . . 11. *H. furfuracea.*
        Leaves mostly ½ in., the margins much revolute.  Ovules 2 . 12. *H. hypericoides.*
    Leaves rigid, glabrous.
        Leaves short, convex, reflexed . . . . . . . . . . 13. *H. microphylla.*
        Leaves narrow-linear, the margins very closely revolute.
            Leaves 2 to 4 lines, whitish, obtuse or recurved at the end 14. *H. recurvifolia.*
            Leaves mostly ½ in., straight, obtuse . . . . . . . 15. *H. lineata.*
            Leaves very pointed . . . . . . . . . . . 16. *H. acerosa.*
Flowers sessile.
    Plant glabrous or nearly so.  Leaves mostly ½ in. Sepals shining 17. *H. aurea.*
    Leaves very obtuse, 2 to 3 lines long, hoary.  Sepals pubescent 18. *H. crassifolia.*

SECT. III. **Pleurandra.**—*Stamens all on one side of the carpels without any staminodia.  Peduncle 1-flowered or none.*—Species all southern and eastern except *H. pedunculata* and *H. mucronata*, which are western.
Leaves obtuse or with a callous point, oblong or linear.
    Flowers sessile.
        Leaves with flat or slightly recurved margins, glabrous or
            slightly hairy.
            Calyx glabrous . . . . . . . . . . . . . 19. *H. nitida.*
            Calyx very villous . . . . . . . . . . . . 20. *H. bracteata.*
        Leaves with their margins much revolute.
            Leaves softly pubescent or villous, oblong or linear.
                Sepals 3 to 5 lines.  Floral leaves usually as long or
                    longer.  Petals broadly obcordate . . . . . . . 21. *H. sericea.*
                Sepals 2 lines.  Floral leaves small.  Petals narrow . 22. *H. hirsuta.*
            Leaves narrow-linear, rigid, glabrous or scabrous . . . 23. *H. stricta.*
    Flowers pedunculate.
        Ovules 4 or more in each carpel.
            Leaves obovate, oblong, or shortly linear . . . . . 25. *H. Billardieri.*
            Leaves narrow-linear.
                Stems virgate, or with numerous ascending branches, or
                    divaricately branched.  Calyx glabrous, stellate-tomentose, or, if hirsute, pedicels very short . . . . 23. *H. stricta.*
                Stems prostrate.  Calyx hirsute, on rather long pedicels 24. *H. humifusa.*
        Ovules 2.  Peduncles slender.
            Leaves obovate, oblong, or shortly linear. Peduncles usually
                short . . . . . . . . . . . . . . 25. *H. Billardieri.*
            Leaves narrow-linear.  Peduncles slender, ⅓ to 1 inch.
                Stems diffuse . . . . . . . . . . . . . 26. *H. gracilipes.*
Leaves narrow-linear, very acute, mostly pungent.
    Flowers on slender peduncles.  Stems procumbent or diffuse.
        Leaves much revolute or nearly terete, slightly pointed . . 26. *H. gracilipes.*

Leaves nearly flat, rigidly pungent . . . . . . . . . 27. *H. acicularis.*
Flowers sessile or shortly peduncled. Leaves loose, channelled
    underneath. Sepals rather obtuse . . . . . . . . . . 23. *H. stricta.*
Flowers sessile. Leaves crowded, convex underneath. Outer
    sepals mucronate or aristate . . . . . . . . . . . 28. *H. mucronata.*

(*Pleurandra reticulata,* Hook. Journ. Bot. i. 245, described from a single specimen in leaf only, is probably some *Pultenæa.*)

SECT. IV. **Euhibbertia.**—*Stamens placed all round the carpels, with occasionally small staminodia outside.*

§ 1. *Tomentosæ.*—Carpels usually tomentose or scaly and 2-ovulate. Stamens numerous, without any or rarely with small staminodia outside. Leaves flat or the margins slightly revolute, usually stellately tomentose or scaly. Flowers pedunculate, axillary.

Leaves oval, oblong, or cuneate.
    Tomentum rigid, stellate, mixed with simple hairs. Leaves cu-
      neate, ½ to ¾ in. . . . . . . . . . . . . . . . . . 29. *H. hermanniæfolia.*
    Tomentum soft and velvety. Leaves oblong, 1 to 2 in. . . 30. *H. velutina.*
    Tomentum close and whitish, stellate with a scale-like base.
      Leaves ¾ to 1 in. long, with an intramarginal vein underneath.
      Peduncles 1 to 2 lines long . . . . . . . . . . . 31. *H. oblongata.*
      Leaves ¼ to ½ in., without intramarginal veins. Peduncles 1
      to 2 lines long . . . . . . . . . . . . . 32. *H. tomentosa.*
      Leaves ½ to 1½ in., without intramarginal veins. Peduncles
      1 to 1½ in. long . . . . . . . . . . . . . . 33. *H. cistifolia.*
      Leaves scabrous with scattered stellate hairs. Sepals very
      scaly. Peduncles 1 to 2 lines . . . . . . . . . 34. *H. echiifolia.*
Leaves narrow-linear.
    Tomentum stellate. Peduncles ¾ to 1¼ in. . . . . . . . 35. *H. scabra.*
    Tomentum of peltate scales. Peduncles 1 to 3 lines . . . . 36. *H. lepidota.*

§ 2. *Vestitæ.*—Carpels (usually 3) villous, 4–6-ovulate. Stamens with or without staminodia outside. Leaves small, narrow, with revolute margins.

Flowers sessile, or peduncles not exceeding the leaves.
    Stamens above 30, with several staminodia . . . . . . . 37. *H. vestita.*
    Stamens under 15, without staminodia . . . . . . . . 38. *H. serpyllifolia.*
Peduncles longer than the leaves. Stamens 15 to 25 . . . . 39. *H. pedunculata.*

§ 3. *Ochrolasiæ.*—Carpels glabrous, 6–8-ovulate. No staminodia.
Leaves with revolute margins. Bracts small . . . . . . 40. *H. ochrolasia.*

§ 4. *Fasciculatæ.*—Carpels glabrous, 2–6-ovulate. No staminodia. Leaves very narrow, convex underneath, the margins not revolute. Bracts small. Flowers sessile.

Ovules 6 in each carpel. Plant glabrous, procumbent . . . . 41. *H. procumbens.*
Ovules 2, or rarely 3 or 4 in each carpel. Leaves usually fine,
    much clustered, often hirsute or pubescent . . . . . . . 42. *H. fasciculata.*

§ 5. *Bracteatæ.*—Carpels glabrous, 1–2-ovulate. No staminodia. Leaves flat or convex underneath. Flowers (except in *H. rostellata*) closely sessile within broad brown shining bracts, like those of some of the *Hemihibbertiæ.*

Leaves very narrow, convex underneath.
    Leaves obtuse.
      Glabrous and green. Leaves not dilated at the top . . . 43. *H. virgata.*
      More or less hoary. Leaves mostly slightly cuneate . . . 44. *H. inclusa.*
    Leaves recurved and mucronate at the top . . . . . . 45. *H. rostellata.*
Leaves flat, mostly oblong.
    Glabrous. Leaves seldom above ½ in. . . . . . . . . 46. *H. glomerata.*
    Densely silvery-tomentose. Leaves ½ in. or more . . . . 47. *H. argentea.*

Loosely pilose or pubescent.  Leaves mostly above ½ in.
  Sepals very densely silky-hairy.  Brown bracts very con-
    spicuous . . . . . . . . . . . . . . . . . . 49. *H. montana.*
  Sepals loosely hairy.  Brown bracts short and thin  . . . 48. *H. pilosa.*
  Sepals glabrous.  Staminodia several . . . . . . . . 61. *H. Mylnei.*

§ 6. *Subsessiles.*—Carpels glabrous.  Stamens usually numerous, without staminodia.
Leaves flat or the margins slightly recurved.  Bracts small or passing into the sepals.
Flowers sessile or nearly so.

Carpels 1–2-ovulate.  Stems erect or diffuse.
  Leaves mostly under 1 in. long.
    Leaves linear-oblong or scarcely enlarged above the middle.
      Stems usually erect or ascending . . . . . . . . . 50. *H. linearis.*
    Leaves obovate or cuneate.  Stems usually diffuse or prostrate 51. *H. diffusa.*
  Leaves 1 to 3 in. long.  Plant softly hairy.
    Leaves obovate-oblong, obtuse . . . . . . . . . . 50. *H. linearis,* var.
    Leaves lanceolate . . . . . . . . . . . . . . 52. *H. saligna.*
Carpels 6–8-ovulate.  Stems twining or trailing.  Leaves large . 53. *H. volubilis.*

§ 7. *Hemihibbertiæ.*—Carpels glabrous or rarely villous.  Stamens very numerous, with
several small, subulate or clavate staminodia outside.  Leaves flat.  Flowers pedunculate,
except in *H. Mylnei.*

Leaves distinctly petiolate, ovate, or oblong, mostly toothed. . .
  Carpels 10 or more, villous, 2-ovulate . . . . . . . . 54. *H. grossulariæfolia.*
  Carpels 3, glabrous, 6- to 8-ovulate . . . . . . . . . 55. *H. dentata.*
Leaves stem-clasping or tapering near the base and again dilated,
  glabrous.
  Leaves ovate or oblong.
    Leaves all perfoliate, the auricles combined.  Sepals lanceolate 56. *H. perfoliata.*
    Auricles rounded, shortly decurrent . . . . . . . . 57. *H. bracteosa.*
    Auricles of most of the leaves distinct, angular, projecting be-
      yond the stem.  Sepals ovate-lanceolate . . . . . . 58. *H. amplexicaulis.*
  Leaves linear, mostly auricled . . . . . . . . . . . 59. *H. Cunninghami.*
  Leaves oblong-lanceolate, tapering at the base, and half stem-
    clasping . . . . . . . . . . . . . . . . . 60. *H. glaberrima.*
Leaves sessile, oblong, very hairy.  Bracts at the base of the pe-
  duncle broad and brown, as in the *Bracteatæ.*
  Sepals glabrous.  Carpels 3.  Flowers sessile . . . . . 61. *H. Mylnei.*
  Sepals very silky-hairy.  Carpels 5.
    Larger leaves obovate-oblong, toothed.  Carpels villous . 62. *H. lasiopus.*
    Larger leaves narrow-oblong, entire.  Carpels glabrous . 63. *H. potentillæflora.*

§ 8. *Brachyantheræ.*—Carpels glabrous.  Stamens about 15 to 20, without staminodia.
Anthers (except in *H. pungens*) ovate or orbicular, flattened, with introrse cells.  Leaves
narrow-linear.  Flowers pedunculate.

Leaves rigid, pungent.  Sepals about 2 lines.  Anthers oblong . 64. *H. pungens.*
Leaves rigid, recurved at the top.  Sepals 5 to 6 lines.  Anthers ovate 65. *H. nutans.*
Leaves slender, but stiff and almost cylindrical.  Sepals not 2 lines.
  Anthers orbicular . . . . . . . . . . . . . . . 66. *H. leptopus.*
Leaves thin, flat.  Sepals about 2 lines.  Anthers broader than long 67. *H. stellaris.*

SECTION 1. HEMISTEMMA, *R. Br. in DC. Syst. Veg.* i. 412 (as a distinct
genus).—Stamens usually numerous, all inserted on one side of the pistil,
with smaller imperfect ones or staminodia outside of them ; filaments short,
anthers linear-oblong.  Carpels 2, villous, with 2 or 3 ovules in each.

1. **H. Banksii,** *Benth.*  Young branches and under side of the leaves

densely clothed with a short, soft, rusty tomentum. Leaves oblong, obtuse, 2 to 3 in. long, ¼ to near 1 in. broad, the margins more or less recurved, narrowed into a short petiole, ·glabrous above and somewhat shining when old, the pinnate and anastomosing veins prominent underneath. Spikes terminal, 1-sided, rusty-villous, about 1 in. long, the flowers closely sessile. Sepals about 4 lines long. Petals longer. Stamens about 20, obtuse, with half as many staminodia outside, about one-third shorter.—*Hemistemma Banksii*, R. Br. in DC. Syst. Veg. i. 414.

**Queensland.** Endeavour river, *Banks.*

2. **H. Brownei,** *Benth.* Young branches clothed with a short rusty down. Leaves oblong-lanceolate, obtuse or scarcely pointed, 2 to 3 in. long, closely sessile and very obtuse or rounded at the base, the margins flat, glabrous, and at length almost shining above, white underneath, with the midrib alone prominent and rust-coloured. Spikes terminal, 1-sided, silky-villous. Sepals scarcely 4 lines long, acute. Stamens nearly as in *H. Banksii.*

**N. Australia?** *R. Brown.* (*Hb. R. Br.*)

3. **H. dealbata,** *Benth.* Young branches minutely rusty-downy. Leaves oblong or oblong-lanceolate, obtuse with a small callous point, or rarely acute, 2 to 3 in. long, ½ to ¾ in. broad, narrowed at the base, but sessile or very shortly stalked, the margins flat, glabrous above, white underneath, with a very close tomentum, the anastomosing veins rust-coloured. Spikes terminal, 1-sided, simple or forked, 1 to 2 in. long, rusty-tomentose or silky. Flowers closely sessile within lanceolate bracts. Stamens as in *H. Banksii.*— *Hemistemma dealbatum*, R. Br. in DC. Syst. Veg. i. 413; Deless. Ic. Sel. i. t. 76.

**N. Australia.** Arnhem's Land, *R. Brown;* Port Essington, *Armstrong, A. Cunningham, Leichhardt.*

4. **H. candicans,** *Benth.* Like *H. dealbata* in the white tomentum that covers the under side of the leaves, but it is rather more silky or rusty on the peduncles and calyx, the leaves are rather narrower, and the inflorescence is very different; peduncles all axillary, ½ to 1 in. long, bearing at their extremity 1 to 3 sessile flowers, and bracts and sepals usually broader. Stamens and carpels the same as in *H. Banksii.*—*Hemistemma candicans*, Hook. f. in Kew Journ. Bot. ix. 48, t. 2.

**Queensland.** Cape York, *M'Gillivray;* Albany Island, *F. Mueller.*

5. **H. angustifolia,** *Benth.* Branches very slender, with a very minute rusty down. Leaves very narrow-linear, obtuse or acute, 1 to 2 in. long, the margins revolute, glabrous and shining above, white or slightly ferruginous underneath, with a prominent rusty midrib. Spikes on slender terminal peduncles, consisting of 2 to 5 sessile flowers. Sepals about 3 lines long, densely and softly villous.—*Hemistemma angustifolium*, R. Br. in DC Syst. Veg. i. 414; Deless. Ic. Sel. i. t. 77.

**N. Australia.** Arnhem's Land, *R. Brown.* (*Hb. R. Br.*)

6. **H. Muelleri,** *Benth.* Branches slender, as in *H. angustifolia*, but loosely villous with soft spreading hairs, intermixed with a closer tomentum. Leaves narrow-linear as in that species, and about 1 line long, nearly glabrous

above, white-cottony and hairy on the under surface, which is however almost concealed by the revolute margins. Spikes terminal or lateral, about 3-flowered. Sepals softly hairy, about 4 lines long. Stamens and carpels as in *H. Banksii* and *dealbata*.

**N. Australia.** Barren places at the mouth of the Victoria, Providence Hill, etc., *F. Mueller.*

7. **H. ledifolia,** *Benth.* Branches rigid, the young ones as well as the under side of the leaves densely covered with a rusty or whitish down. Leaves oblong-linear, about ½ in. long, obtuse, rather thick, with the margins revolute, hoary above when young, but soon glabrous. Peduncles short, terminal, 1- to 3-flowered. Sepals ovate, about 5 lines long, thick and densely villous as well as the bracts. Petals scarcely longer. Stamens about 20, with about 15 shorter staminodia outside. Carpels very villous, with usually 3 ovules in each.—*Hemistemma ledifolium,* A. Cunn. Herb.

**N. Australia.** York Sound, *A. Cunningham.*

8. **H. verrucosa,** *Benth.* Much branched, the young shoots and leaves very scabrous, with tubercles forming the base of stellate hairs. Leaves linear-oblong, obtuse, ¼ to ½ in. long, the margins very revolute. Peduncles all 1-flowered, very short, or seldom 4 or 5 lines long. Calyx about 3 lines, sometimes nearly glabrous, more frequently more or less covered with stellate hairs, which are sometimes stipitate, the outer sepals always acute, the inner more obtuse. Petals obovate, slightly obcordate. Stamens often under 10, with at least as many smaller staminodia outside. Carpels as in the allied species, 2, hairy and biovulate.—*Pleurandra verrucosa,* Turcz. in Bull. Mosc. 1852, ii. 139.

**W. Australia.** Cape Riche?, *Drummond, 5th Coll. n.* 289 ; Bald Island and Mount Monypeak, *Maxwell.*—In habit and inflorescence this species resembles *H. hypericoides,* but the acute sepals, and especially the stamens, readily distinguish it.

SECTION II. HEMIPLEURANDRA.—Stamens rarely more than 12, all on one side of the pistil ; staminodia small, usually subulate or club-shaped, either 2 or 3 on each side of the fertile ones, or continued round to the opposite side of the pistil, with very rarely any outside the fertile ones. Peduncles in two species bearing a 1-sided spike of several flowers, in all the others 1-flowered. Carpels 2, villous, with 2 or rarely 4 ovules in each. The species are all West Australian.

9. **H. spicata,** *F. Muell. Fragm.* ii. 1. Glabrous or very slightly and minutely pubescent. Leaves linear, usually obtuse, ¼ to 1 in. long, the margins much revolute. Peduncles lateral, usually longer than the leaves, bearing a 1-sided spike of 4 to 8 flowers. Sepals about 3 lines long, pubescent or shortly hairy. Petals deeply obcordate. Stamens usually 8 to 10 on one side of the carpels, with a ring of short, subulate or spathulate staminodia continued all round the carpels, and a few even behind the fertile ones.—*Hemistephus linearis,* J. Drumm. and Harv. in Kew Hook. Journ. vii. 52.

**W. Australia.** Flinders' Bay, *Collie;* Port Gregory, *Walcott* and *Oldfield;* northern districts, *Drummond.*

10. **H. polystachya,** *Benth.* Procumbent and much branched, with

spreading hairs, or at length scabrous only or nearly glabrous. Leaves narrow-linear, obtuse, 3 to 5 lines, or in some specimens ½ in. long, the margins much revolute. Peduncles lateral, usually above 1 in. long, bearing a 1-sided spike of 2 to 4 flowers. Sepals broader and more scarious than in *H. spicata,* from which this species differs chiefly in its hairs, and in the staminodia, which although continued from the fertile stamens round the rest of the torus, yet are usually entirely wanting, or there is only a single one behind the perfect stamens. The 2 ovules in this and the last species do not appear to be really superposed, although one is usually borne on a much longer funiculus than the other.

**W. Australia.** Swan River, *Drummond;* Blackwood river, *Oldfield.*

11. **H. furfuracea,** *Benth.* Rather coarse and erect, 2 to 4 ft. high, the branches thickly clothed with rust-coloured, loosely stellate hairs. Leaves narrow-oblong or linear, very obtuse, 1 to 2 in. long, the margins revolute, but leaving the under surface open, villous above when young, scabrous when old, closely tomentose and white or hoary underneath. Peduncles mostly axillary, 1-flowered, ¼ to 1 in. long. Outer sepals ovate or ovate-lanceolate, sometimes near 5 lines long, inner ones shorter and rounder. Petals 2-lobed. Stamens 8 to 12, with numerous small staminodia on each side, and on the opposite side of the carpels. Carpels 2, globose, villous, 4-ovulate. Arillus very short.—*Pleurandra furfuracea,* R. Br. in DC. Syst. Veg. i. 417; Deless. Ic. Sel. i. t. 80; *Hibbertia astrophylla,* Steud. in Pl. Preiss. i. 270; *Hemistemma asperifolium,* F. Muell. Fragm. i. 161.

**W. Australia.** Rocky hills, from King George's Sound to the Stirling range, *R. Brown, A. Cunningham, Drummond,* and others; rocks on the western side of Mount Clarence, *Preiss, n.* 2167.

12. **H. hypericoides,** *Benth.* Branches spreading, the young ones as well as the leaves hoary, with a short stellate down. Leaves linear-oblong, very obtuse, ½ in. long or rather more, those of the smaller branches half as long, the thick margins much revolute. Peduncles mostly terminal, 1-flowered, ¼ to ½ in. long. Sepals broad, very concave and obtuse, shorter than in *H. furfuracea,* hoary outside. Petals 2-lobed. Stamens 12 to 15, with rather numerous (or rarely very few) small spathulate or clavate staminodia on each side or on the opposite side of the carpels. Carpels connate at the base, globular, 2-ovulate.—*Pleurandra hypericoides,* DC. Syst. Veg. i. 421; Deless. Ic. Sel. i. t. 81; *Hibbertia trachyphylla,* Steud. in Pl. Preiss. i. 271; *H. aspera,* Steud. l. c. i. 270; *H. proxima* and *H. cinerascens,* Steud. l. c. i. 271.

**W. Australia.** Common about Perth, *Preiss, n.* 2132 and 2136 *a, Drummond* and others; Cape Leeuwin, *Collie;* Port Gregory and Blackwood river, *Oldfield;* Darling range, *Preiss, n.* 2147; Cataract Valley, *Preiss, n.* 2140; between Perth and King George's Sound, *Harvey;* Stokes' Inlet, *Maxwell.*

13. **H. microphylla,** *Steud. in Pl. Preiss.* i. 273. Branches erect and rigid, or sometimes slender and decumbent or diffuse, minutely pubescent or glabrous. Leaves usually 1 to 1½ line long, ovate and very convex, sometimes more linear and 2 lines long, always very convex and very patent or closely reflexed on the stem, glabrous or rough, with a minute pubescence. Peduncles 1-flowered, slender, often ½ to ¾ in. long, arranged in the upper

axils so as to form a kind of leafy raceme towards the ends of the branches. Sepals 2 to near 3 lines long, glabrous or stellate-pubescent. Stamens 8 to 10 on one side of the pistil, with 1, 2, or 3 small spathulate staminodia on each side. Carpels 2-ovulate. Arillus very short.—*H. lepidophylla*, F. Muell. Fragm. i. 217 ; *Hemistemma revolutum*, Turcz. in Bull. Mosc. 1849, ii. 4.

**W. Australia.** King George's Sound, *Menzies, R. Brown ;* and thence to the Stirling range, *Drummond, Preiss, n.* 2154 and 2180, *Oldfield,* and others.

14, **H. recurvifolia,** *Benth.* A shrub with the foliage nearly of *H. rostellata* or of *Candollea uncinata,* but with the flowers of a *Hemipleurandra.* Leaves narrow-linear, rigid, obtuse and hooked or recurved at the extremity, 2 to 4 lines long, convex underneath, but furrowed by the closely recurved margins, whitish on both sides but glabrous, or with a minute tuft of short hairs at the tip. Peduncles 3 to 5 lines long, nearly glabrous. Sepals whitish, about 2 lines long, the outer ones keeled and acute, surrounded by 2 or 3 small bracts. Stamens about 8 on one side of the pistil, with a few small staminodia on each side or behind them. Carpels villous, 2-ovulate.— *Pleurandra recurvifolia,* Steud. in Pl. Preiss. i. 264.

**W. Australia.** Gravelly places at the foot of the Konkoberup hills, *Preiss, n.* 2170; Phillips river, *Maxwell.*

Var. *virens.* Leaves rather longer, the margins more prominently revolute, green but rough with small tubercles or a short stellate pubescence.—Point Henry, *Oldfield.*

15. **H. lineata,** *Steud. in Pl. Preiss.* i. 272. Intermediate as it were between *H. hypericoides, H. recurvifolia,* and *H. acerosa,* differing from the first by its leaves much narrower, with the margins closely revolute so as to appear 2- or 3-grooved on the under side, either glabrous or rough, with scattered tubercles or a few spreading hairs ; from *H. recurvifolia,* by the leaves nearly twice as long, not hoary, quite straight or scarcely perceptibly recurved at the tip ; and from *H. acerosa* by the leaves not pungent, either obtuse or with a minute recurved point. The flowers in Preiss's original specimens are rather larger than in *H. acerosa,* of which species this plant may prove to be a variety.

**W. Australia.** Shady woods on the north side of Mount Wuljenup, *Preiss, n.* 2151 ; Mount Monypeak river, *Maxwell.*

Var. *parviflora.* Flowers small, as in *H. acerosa,* midrib of the leaves less prominent underneath.—*Pleurandra diamesogenos,* Steud. in Pl. Preiss. i. 265.—Boggy woods, Sussex district, *Preiss, n.* 2141. This variety approaches *H. gracilipes* in aspect, but is readily distinguished by the presence of staminodia.

16. **H. acerosa,** *Benth.* Usually low and very much branched, but sometimes throwing up ascending stems of nearly 1 ft. from a thick base, glabrous or rough with short spreading hairs. Leaves linear-subulate or broader at the base, very pointed and usually pungent, 4 lines to 1 in. long, erect or spreading, the margins closely revolute, but much narrower than the broad prominent midrib. Peduncles 1-flowered, slender, ¼ to 1 in. long. Flowers nearly those of *H. acicularis,* except that there are always 1, 2, or 3 small club-shaped or spathulate staminodia on each side of the fertile stamens. Carpels 2-ovulate.—*Pleurandra acerosa,* R. Br. in DC. Syst. Veg. i. 422 ; *P. cognata,* Steud. in Pl. Preiss. i. 265 ; *P. juniperina,* Turcz. in Bull. Mosc. 1849, ii. 6.

**W. Australia.** King George's. Sound, *R. Brown, Fraser,* and others ; Swan River, *Drummond,* 1st *Coll. and* 1845, *n.* 2 ; Mount Melville, *Preiss, n.* 2156 ; Champion Bay, *Oldfield.*

Var. *ulicifolia.* Leaves stouter and not so long. King George's Sound, *Baxter.*

17. **H. aurea,** *Steud. in Pl. Preiss.* i. 272. Rigid, and somewhat virgate, perfectly glabrous, or the leaves slightly scabrous, and sometimes shortly ciliate. Leaves narrow-linear and stiff, shortly pointed, the lower ones ½ to ¾ in., those near the flowers about half as long, the margins much revolute. Flowers terminal, sessile, with 2 or 3 small sepal-like bracts at their base. Outer sepals fully 3 lines long, stiffly coriaceous and almost shining, with a prominent keel projecting into a sharp point, inner ones less pointed, broader and thinner. Petals broad. Stamens about 10, one-sided, with 2 to 4 small staminodia on each side of them. Carpels 2-ovulate.—*H. pallida,* Steud. in Pl. Preiss. i. 272.

**W. Australia.** Swan River, *Drummond ;* in gravelly places at the foot of Darling range, *Preiss, n.* 2152 *a* and 2152 *b.*

Var. *obtusa.* Leaves obtuse, sepals scarcely keeled or pointed.—*Pleurandra glaucophylla,* Steud. in Pl. Preiss. i. 262 ? The fragments I have seen without flowers agree with this variety ; but Steudel describes the ovaries as glabrous, which I have not observed in any *Hemipleurandra.* He does not describe the stamens, but I know of no other western groups to which his specimen could be referred. Swan River, *Drummond ;* sandy places near Avon Dale, York District, *Preiss, n.* 2159.

18. **H. crassifolia,** *Benth.* Erect, with the habit of some of the hoary varieties of *H. stricta.* Leaves linear-oblong, very obtuse, 2 to 3 lines long, the margins much rolled back, rather thick, hoary or rough with very short stellate hairs, the floral ones ovate-lanceolate passing into the bracts. Flowers closely sessile, solitary, and terminal. Sepals ovate, brown, slightly hoary, nearly 3 lines long, surrounded by several bracts. Stamens about 12, one-sided, with 3 or 4 spathulate staminodia on each side of them, and not half so long. Carpels 2-ovulate.—*Pleurandra crassifolia,* Turcz. in Bull. Mosc. 1849, ii. 5.

**W. Australia.** *Drummond,* 4th *Coll. n.* 120.

SECTION III. PLEURANDRA.—Stamens often very few, and rarely more than 15, all on one side of the pistil, and often more or less united at the base, without any staminodia. Peduncles 1-flowered, or flowers sessile, solitary, or in terminal heads. Carpels 2, villous or tomentose, or very rarely glabrous, with 2, 4, or more ovules in each.

19. **H. nitida,** *Benth.* Erect, much branched and glabrous. Leaves crowded, especially under the flowers, oblong, obtuse, or with a short point, ½ to ¾ in. long, narrowed at the base, the margins flat or slightly recurved, somewhat coriaceous and shining. Flowers sessile within the last leaves, and surrounded by a few short bracts. Sepals lanceolate or oblong, very pointed and quite glabrous, 3 to 5 lines long. Petals broad and notched. Stamens about 11. Carpels hairy, 4-ovulate. Arillus slightly sinuate.—*Pleurandra nitida,* R. Br. in DC. Syst. Veg. i. 416 ; *P. Cneorum,* DC. l. c. i. 416.

**N. S. Wales.** About Port Jackson, *R. Brown, Sieber, n.* 141 *and Fl. Mixt. n.* 508, and others.

20. **H. bracteata,** *Benth.* Erect and much branched, with the aspect

of *Pultenæa daphnoides,* and resembles also *H. nitida,* but is not so glabrous. Leaves narrow-oblong, mostly obtuse, with a short callous point, ½ to ¾ in. long, narrowed at the base, the margins slightly recurved, somewhat rusty, with a minute tomentum underneath, glabrous and shining or scabrous above, or occasionally bearing a few long hairs. Flowers terminal, or on very short axillary branches, sessile within a tuft of floral leaves, which are mostly longer than the flowers, except a few of the innermost, which are much shorter and more hairy. Sepals oblong-lanceolate, fully 5 lines long, densely clothed with long silky hairs. Petals broad, notched. Stamens about 16. Carpels hairy, with 4 to 6 ovules in each.—*Pleurandra bracteata,* R. Br. in DC. Syst. Veg. i. 415; Deless. Ic. Sel. i. t. 78.

**N. S. Wales.** Port Jackson to the Blue Mountains, *R. Brown* and others; Emu Plains, *A. Cunningham.*

**21. H. sericea,** *Benth.* A variable species which sometimes scarcely differs from *H. bracteata,* except in being much more hairy and the leaves more revolute on the margin, but is usually more diffuse or procumbent, softly villous all over, with the floral leaves not much longer than the others. Leaves rarely much above ½ in. long, and in some varieties much shorter, obtuse, with the margins much revolute, clothed with stellate down, especially underneath, with longer hairs on the upper surface. Flowers sessile among crowded floral leaves, as in the last two species. Sepals rather shorter and broader, villous. Stamens usually 10 to 12. Carpels tomentose or villous, with 4 to 6 ovules in each.—*Pleurandra sericea,* R. Br. in DC. Syst. Veg. i. 416; Deless. Ic. Sel. i. t. 79; Hook. f. Fl. Tasm. i. 16; *H. densiflora,* F. Muell. Pl. Vict. i. 15. *Pleurandra cinerea,* R. Br. in DC. l. c. i. 417, is a slight variety with shorter pubescence, and shorter, more oblong leaves, the flowers often very shortly pedicellate.

**Victoria.** Port Phillip, *R. Brown;* sandy heathy places on barren scrubby ridges, and occasionally on rocky ranges from the Glenelg to the Murray rivers, and thence to Port Phillip, *F. Mueller* and others.

**Tasmania.** Common on sandy soil, on the coast only, all round the island, *J. D. Hooker.*

**S. Australia.** Near Adelaide, *Macarthur, F. Mueller.*

Var. *densiflora.* More villous. Leaves, especially the floral ones, shorter. Stems usually more procumbent.—*Pleurandra densiflora,* Hook. f. in Journ. Bot. i. 245. The Tasmanian specimens belong chiefly, but not entirely, to this variety, and a few of the Victorian ones are referrible to it.

**22. H. hirsuta,** *Benth.* A low, prostrate, densely branched species, with much smaller leaves and flowers than in any of the same section, resembling some forms of *H. fasciculata,* and shortly hirsute all over. Leaves linear-oblong, obtuse, 1½ to 2, or seldom 3 lines long, with revolute margins. Flowers axillary or terminal, sessile within leaves often as long as the calyx, the innermost of which are however much smaller. Sepals ovate, villous, scarcely 2 lines long. Petals narrow and entire or very slightly obcordate. Stamens very few. Ovaries 2, pubescent, with 4, or very rarely only 2 ovules in each.—*Pleurandra hirsuta,* Hook. Comp. Bot. Mag. i. 273; Hook. f. Fl. Tasm. i. 17.

**Tasmania.** Among stones in basaltic soil, George Town and Hobart Town, *J. D. Hooker, Gunn,* and others.

23. **H. stricta,** *R. Br. Herb.; F. Muell. Pl. Vict.* i. 15. Erect, spreading, or diffuse, but scarcely prostrate, sometimes throwing up almost simple stems of 6 in. from a thick rhizome, sometimes attaining several feet in height, more or less hoary or scabrous, with a minute stellate tomentum, although sometimes appearing glabrous at first sight. Leaves narrow-linear, erect or spreading, rather obtuse, mostly ¼ to ½ in. long, the closely revolute margins disclosing little more than the midrib underneath. Flowers nearly sessile, or on pedicels of 2 or 3 lines in length. Sepals usually about 3 lines long, oblong, lanceolate, or the inner ones ovate. Stamens usually 8 to 12. Carpels tomentose, or very rarely glabrous, with 4 to 6, or very rarely more ovules in each. Arillus usually very small.—*Pleurandra stricta,* R. Br. in DC. Syst. Veg. i. 422; *P. riparia,* R. Br. in DC. l. c. i. 419; *P. ericifolia,* DC. l. c. i. 420; Hook. f. Fl. Tasm. i. 17; *P. cistiflora,* Sieb. in Spreng. Syst. Cur. Post. 191; Reichb. Icon. Exot. t. 79.

**Queensland.** Port Curtis, *M'Gillivray;* Moreton Bay, *F. Mueller,* and inland to the ranges on the Burritt river, *D. Moore,* and Maranoa river, *Mitchell.*

**N. S. Wales.** Port Jackson, *R. Brown* and others, and apparently throughout the colony.

**Victoria.** In sandy, rocky, or heathy localities of the lowlands and hills, not rare, *F. Mueller.*

**Tasmania.** Abundant throughout the island, *J. D. Hooker.*

**S. Australia.** From the Murray to Streaky Bay, *Whitaker, F. Mueller, Warburton,* and others.

**W. Australia.** Only at the extreme eastern limits on the south coast, *Maxwell.*

This is a very variable species, with the flowers seldom so closely sessile as in the preceding ones, nor borne on peduncles so long as in most of the following ones. There are a few specimens, however, which come near to the narrow-leaved forms of *H. Billardieri,* and others which are very close upon *H. humifusa.* The following are the most striking forms :—

*a. glabriuscula.* Glabrous or nearly so, procumbent or erect. Flowers nearly sessile. Calyx not hoary. Carpels tomentose. Ovules 4 to 6. The commonest form in N. S. Wales, southern Victoria, and Tasmania, including Sieber's n. 150 (*P. riparia*), 151 (*P. stricta*), and 147 (*P. fumana*), the latter a straggling variety approaching *H. Billardieri* in habit. No. 148 (*P. cistiflora*) is the same, with longer, more acute, sometimes almost pungent leaves, from the Blue Mountains; and a form with very short obtuse leaves appears to be common about Lake Hindmarsh, in Victoria.

*b. leiocarpa.* Procumbent and perfectly glabrous, even the carpels. Ovules 4. From the south coast of W. Australia, east of Stokes Inlet, *Maxwell.*

*c. canescens.* Leaves and calyx more or less hoary with stellate hairs. Flowers pedunculate or more rarely nearly sessile. Ovules usually 4.—*Pleurandra incana,* Lindl. in Mitch. Three Exped. ii. 156. Apparently common in Victoria, extending also over N. S. Wales into Queensland and westward to Spencer's Gulf. In this I should include *P. microphylla,* Sieb. Pl. Exs. n. 143; Spreng. Syst. Cur. Post. 191, a small-flowered and small-leaved form from the Blue Mountains and from Tasmania, *Gunn. n.* 1020; and *P. cistoidea,* Hook. in Mitch. Trop. Austr. 363, from New England, *C. Stuart,* and Queensland, *Mitchell.*

*d. calycina.* Leaves narrow and acute or almost pungent. Calyx hirsute, almost as in the var. *hirtiflora.*—*Pleurandra calycina,* DC. Syst. Veg. i. 422 (judging from a specimen of Caley's named *P. pilosa* in Herb. Brown, but which quite agrees with De Candolle's description of *P. calycina*). N. S. Wales, *Caley ;* Avon Ranges, Gipps' Land, *F. Mueller.*

*e. hirtiflora.* Leaves nearly as in the var. *canescens.* Calyx usually large, more sessile, and hirsute with spreading hairs. Ovules usually 6 to 8 or more.—*P. calycina,* A. Cunn. in Field N. S. Wales, 338. On the Maranoa river, *Mitchell ;* Moreton Bay, *F. Mueller ;* New England Ranges, *C. Stuart ;* near Bathurst, *A. Cunningham ;* and almost the same form from Spencer's Gulf and Streaky Bay, *Herb. Mueller.*

24. **H. humifusa,** *F. Muell. Pl. Vict.* i. 16, *t. Suppl.* 1. Prostrate,

much branched, hoary, and more or less hirsute, like the *H. hirsuta*, with linear obtuse leaves, the margins much revolute, but these leaves are usually longer and the flowers much larger, always borne on a pedicel of from ¼ to ½ in. From some specimens of *H. stricta*, var. *hirtiflora*, it differs chiefly in its low, prostrate habit, in being more hairy, and the peduncles much longer. Sepals 4 to 5 lines long, and very hairy. Petals, stamens, and carpels of *H: stricta.* Ovules usually 6.

**Victoria.** Barren scrubby plains near Mount Zero, *F. Mueller.*

25. **H. Billardieri,** *F. Muell. Pl. Vict.* i. 14. Stems weak, sometimes short and erect, but more frequently trailing to the length of two or three feet or more over other shrubs, the branches clothed with stellate hairs, often mixed with long spreading ones. Leaves from obovate, ovate or oval-oblong to oblong-cuneate or narrow-oblong, the larger ones ½ to 1 in. long, but in the commoner slender varieties not half that size, the margins recurved, more or less stellately pubescent, especially underneath, and scabrous above, but becoming glabrous with age. Pedicels terminating short, leafy shoots, or apparently axillary, slender, and recurved, ¼ to ½ in. long. Sepals 2 to 3 lines long, or in some varieties rather shorter or longer, the outer ones usually pointed, the inner broader and more obtuse, glabrous, or nearly so. Petals broad. Stamens usually 10 to 12. Carpels downy or villous, with 2 to 4 ovules. Arillus sometimes almost enveloping the seed, sometimes very short.—*Pleurandra ovata*, Labill. Pl. Nov. Holl. ii. 5, t. 143; Hook. f. Fl. Tasm. i. 16.

**Queensland.** Glasshouse Mountains, *F. Mueller.*
**N. S. Wales.** Port Jackson, *R. Brown, Sieber, n.* 144, and others; Hastings river, *Beckler.*
**Victoria.** Scattered over the southern part of the colony, *F. Mueller.*
**Tasmania.** Sandy soils on the coast in various places, *J. D. Hooker, Gunn.*
**S. Australia.** Spencer's Gulf, *F. Mueller.*
Although apparently not so common as *H. stricta*, this species appears to be more variable, and the following forms have in general the appearance of distinct species, but are always too much connected by intermediate specimens to admit of their being characterized as such.

*a. monadelpha*, F. Muell. mss. Leaves large, obovate or oblong. Flowers large. Ovules 4. Sealers' Cove, *F. Mueller;* Flinders Island, *Gunn.*

*b. obovata.* Leaves and flowers of *a*, but ovules only 2.—*Pleurandra obovata*, R. Br. Herb., from Port Dalrymple; Hastings river, *Beckler;* West Head, Tasmania, *Gunn.*

*c. ovata.* Leaves and flowers small, ovate or oblong. Ovules 2. The most common Tasmanian and N. S. Wales form.

*d. scabra.* Leaves narrow, seldom (except a few of the lower ones) above 4 lines long, and usually much revolute on the margin.—*Pleurandra scabra*, R. Br. in DC. Syst. Veg. i. 418; *P. empetrifolia*, DC. l. c. i. 420; *P. asterotricha*, Sieb. in Spreng. Syst. Cur. Post. 191; Pl. Exs. n. 149, and Fl. Mixt. n. 505 (n. 139, *P. cinerea*, is a rather more canescent form). Common about Sydney.

*e. parviflora.* Slender and much branched. Leaves 2 to 4 lines long, from obovate to linear-oblong, flat or much revolute. Sepals under 2 lines long. Ovules 2, or rarely 4.—*Pleurandra parviflora*, R. Br. in DC. Syst. Veg. i. 418; *Hibbertia aspera*, DC. Syst. Veg. i. 430. Port Jackson, *R. Brown ; Sieber, n.* 144, and *Fl. Mixt. n.* 504, and others.

26. **H. gracilipes,** *Benth.* Nearly glabrous, diffuse or prostrate, and much branched, with much of the appearance of *H. acicularis*, but the leaves are usually broader and not pungent. They are narrow-linear, usually very

obtuse, 2 to 4, or even 5 lines long, with the margins revolute, and often slightly scabrous. Peduncles slender, ½ to 1 in. long, thickened under the flowers. Sepals 2 to nearly 3 lines long, membranous, obtuse. Stamens usually about 10. Carpels glabrous or downy, 2-ovulate.—*Pleurandra pedunculata*, R. Br. in DC. Syst. Veg. i. 419.

**W. Australia.** South coast ?, *Drummond, n.* 16, 9, 4; Lucky Bay, *R. Brown ;* King George's Sound and Gordon river, *Oldfield.*

27. **H. acicularis,** *F. Muell. Pl. Vict.* i. 17. Nearly or quite glabrous, procumbent or diffuse, with a thick woody stock, and numerous branches, shoit and intricate, or lengthened to a foot. Leaves narrow-linear, rigid, with a stiff, often pungent point, about 3 to 6 lines long, the margins recurved. Pedicels terminal or axillary, often on very short shoots, with a few leaves at the base sometimes reduced to minute bracts, recurved, ¼ to ½. in. long. Sepals glabrous, or very slightly downy, about 2 lines long. Stamens usually 8, or fewer. Carpels downy, or rarely glabrous, with 2, or very rarely 4 ovules.—*Pleurandra acicularis*, Labill. Pl. Nov. Holl. ii. 6, t. 144 ; Hook. f. Fl. Tasm. i. 15.

**Queensland.** Moreton Island, *F. Mueller.*
**N. S. Wales.** Port Jackson, *R. Brown* and others ; sterile bushy hills in Wellington Valley, and westward to Croker's range, *A. Cunningham ;* New England, *C. Stuart.* The Port Jackson specimens include a variety with more rigid leaves and larger flowers, and another with glabrous ovaries.
**Victoria.** Heathy ground, particularly in moist localities near the coast, *F. Mueller.* Some Port Adelaide specimens are the only ones I have seen with 4 ovules to each carpel.
**Tasmania.** Sandy land at George Town, sea-coast E. of Port Dalrymple, and islands of Bass's Straits, *J. D. Hooker, Gunn.*
*Pleurandra triandra*, Turcz. in Bull. Mosc. 1854, ii. 280, described from a specimen said to have been gathered by Gunn " near Sydney in Tasmania," may possibly belong to this species.

28. **H. mucronata,** *Benth.* Erect and rigid, the young branches shortly villous. Leaves crowded, erect, rigid, linear, and very pungent, mostly 4 to 6 lines long, semiterete, but marked with a furrow on each side of the midrib indicating the revolute margins, glabrous, or the young ones bearing a few spreading, silky hairs. Flowers sessile, the leaves of the very short floral shoots passing into 2 or 3 subulate bracts. Sepals 3 to 4 lines long, loosely villous, the outer ones with long pungent points, the inner ones shorter and less pointed. Petals broadly 2-lobed. Stamens about 5. Carpels very villous, 2-ovulate.—*Pleurandra mucronata*, Turcz. in Bull. Mosc. 1849, ii. 139.

**W. Australia.** Between Swan River and Cape Riche, *Drummond, 5th Coll. n.* 290 ; King George's Sound, *R. Brown ;* W. Mount Barren, *Maxwell.*

SECTION IV. EUHIBBERTIA.—Stamens usually numerous, and rarely fewer than 12, arranged all around the pistil, although sometimes more numerous on one side than on the other, either without any staminodia, or with few or many small subulate or clavate staminodia outside the perfect stamens.—*Hibbertia proper*, as limited by De Candolle, and most authors.

§ 1. *Tomentosæ.*—Carpels 2 (or very rarely and exceptionally 3), tomentose, or covered with peltate scales, with 2, or very rarely 1 or 3 ovules in each. Stamens numerous, without any, or rarely with small staminodia outside. Leaves ovate, obovate, cuneate, oblong, or linear, flat, or with the margins slightly revolute, usually covered with stellate hairs or peltate scales. Flowers axillary, pedunculate, with a small bract under the sepals, those at the base of the peduncle minute or wanting. The species are all tropical or subtropical.

29. **H. hermanniæfolia,** *DC. Syst. Veg.* i. 431. Resembles in general aspect *H. furfuracea*, but very different in the stamens. Whole plant covered with a rather rigid stellate down, mixed, especially on the upper side of the leaves, with simple hairs. Leaves from obovate-oblong to cuneate, very obtuse or retuse, $\frac{1}{2}$ to $\frac{3}{4}$ in. long, the margins not recurved. Peduncles axillary, mostly about $\frac{1}{2}$ in. long. Sepals about 4 lines, rather obtuse, membranous, pubescent. Stamens about 15. Carpels 2, villous, with 2 (or perhaps sometimes 4?) ovules in each.

**N. S. Wales?** "Dovedale," *Caley.* I have been unable to find the locality in any of our maps. (*Hb. Brit. Mus.*)

30. **H. velutina,** *R. Br. Herb.* Whole plant clothed with a soft, velvety tomentum. Leaves oval or oval-oblong, sometimes slightly cuneate, obtuse, 1 to 2 in. long, the margins scarcely recurved, and very soft. Peduncles axillary, $\frac{1}{4}$ to $\frac{1}{2}$ in. long. Sepals about 3 lines long, softly tomentose. Petals broadly obovate. Stamens numerous. Carpels 2?, tomentose.

**Queensland.** N. E. Coast, *R. Brown.* (*Hb. R. Br.*)

31. **H. oblongata,** *R. Br. in DC. Syst. Veg.* i. 431. Branches rather slender and elongated, covered as well as the leaves with a close whitish tomentum consisting of stellate hairs more or less united into a scale at their base. Leaves narrow-oblong, obtuse or with a very short slightly recurved point, mostly $\frac{3}{4}$ to 1 in. long, the margins flat, the lateral veinlets converging on the under side into an intramarginal vein. Peduncles axillary, seldom above 2 lines long. Inner sepals about 3 lines long, obtuse, the outer shorter and more acute. Petals 2-lobed. Stamens above 20, all perfect or rarely one or two on the side where there are fewest reduced to small staminodia. Carpels 2, scaly-tomentose, 2-ovulate.

**N. Australia.** Gulf of Carpentaria, *R. Brown;* rocky situations, Sims' Island, *A. Cunningham;* sandstone ravines on the table-land and rocks on the Fitzmaurice river, *F. Mueller.*

Var. *brevifolia.* Leaves mostly 3 to 4 lines long.—Upper Victoria river, *F. Mueller.*

32. **H. tomentosa,** *R. Br. in DC. Syst. Veg.* i. 432. Allied to *H. oblongata*, but more slender and much more branched. Leaves oblong-linear, 3 to 4 lines long or very seldom $\frac{1}{2}$ in., hoary on both sides, with a minute close tomentum, and without the intramarginal vein of *H. oblongata.* Flowers smaller, with the sepals more prominently keeled.

**N. Australia.** Gulf of Carpentaria, *R. Brown.* (*Hb. R. Br.*) This and some other species of the present group may possibly, when better known, be reduced to varieties.

33. **H. cistifolia,** *R. Br. in DC. Syst. Veg.* i. 431. Resembles *H. oblongata* in the whitish tomentum, consisting of stellate hairs proceeding from a scale-like base, which covers every part, but the branches appear to be diffuse or shortly trailing from a woody rhizome, the leaves are broader, from

obovate to oblong, $\frac{1}{2}$ to $1\frac{1}{2}$ in. long, and without the intramarginal nerve, and above all, the flowers are borne on peduncles of 1 to $1\frac{1}{2}$ in. long. They are also larger, and have above 50 stamens without any staminodia. Carpels 2, very scaly, 2-ovulate.

**N. Australia.** Gulf of Carpentaria, *R. Brown;* Port Essington, *Armstrong.*

34. **H. echiifolia,** *R. Br. Herb.* Branches diffuse, flexuose, hoary with a minute scabrous tomentum, with prominent angles decurrent from the base. Leaves oblong or ovate-oblong, very obtuse, mostly about $\frac{1}{2}$ in., but the larger ones often above an inch long, rigid, not hoary but very rough with minute stellate scales. Peduncles very short, rarely 2 lines long, axillary, or more frequently terminating short leafy branches. Sepals broad, concave, rigid, about 3 lines long, densely covered with peltate scales. Stamens numerous. Carpels 3 or 4, scaly (2-ovulate?).

**N. Australia.** N. coast, *R. Brown.* (*Hb. R. Br.*)

35. **H. scabra,** *R. Br. Herb.* Branches slender, scabrous as well as the upper side of the leaves with minute stellate hairs. Leaves like those of *H. angustifolia,* narrow-linear, $\frac{3}{4}$ to $1\frac{1}{2}$ in. long, acute or scarcely obtuse, the margins slightly revolute, very closely and minutely tomentose underneath. Peduncles axillary, $\frac{3}{4}$ to $1\frac{1}{4}$ in. long. Sepals about 3 lines, acute, tomentose outside. Petals obovate. Stamens numerous. Carpels 2 or 3, tomentose, 2-ovulate (according to R. Brown's notes).

**N. Australia.** N. coast, *R. Brown.* (*Hb. R. Br.*)

36. **H. lepidota,** *R. Br. in DC. Syst. Veg.* i. 432. Branches stiff but slender, covered as well as the leaves and sepals with a close silvery or slightly rusty tomentum, consisting of minute peltate scales with scarious edges. Leaves linear, rather acute, mostly $\frac{1}{2}$ to $\frac{3}{4}$ in. long, concave, the margins not revolute. Flowers rather small, on pedicels of 1 to 3 lines, solitary or 2 or 3 together in the axils. Sepals broad, very obtuse, about 2 lines long, or 3 when in fruit, the 2 outer rather shorter. Stamens about 12, mostly, but not all, on one side of the carpels, with several small staminodia outside. Carpels 2, scaly-tomentose, 2-ovulate.

**N. Australia.** Gulf of Carpentaria, *R. Brown, A. Cunningham;* rocky barren sandstone table-land at the sources of Roper river, at the head of Macarthur river, Upper Victoria river, and near M'Adam range, *F. Mueller.*

§ 2. *Vestitæ.*—Carpels usually 3, villous, with 4 to 6 ovules in each. Stamens rather numerous, with small staminodia outside, or fewer without staminodia. Leaves small, narrow, with revolute margins. Bracts small. Flowers sessile or pedunculate.

37. **H. vestita,** *A. Cunn. Herb.* Branches elongated, decumbent or erect, clothed as well as the young leaves with short spreading hairs. Leaves narrow-linear, obtuse, 3 to 4 lines long, rigid with recurved margins, often glabrous when full grown. Flowers nearly sessile, in clusters of floral leaves shorter than them, the inner ones passing into small linear bracts. Sepals ovate-lanceolate, obtuse, or the outer ones scarcely acute, 3 or even 4 lines long, with rather silky hairs outside. Petals obovate, deeply emarginate. Stamens above 30, with several short filiform or clavate staminodia outside.

Carpels 3, villous, 6-ovulate. The general aspect is sometimes that of *H. serpyllifolia*, but it is readily known by the stamens.

**Queensland.** Open forest-land near Moreton Bay, *A. Cunningham;* Stradbrooke Island, *Fraser;* Glasshouse mountains, *F. Mueller;* swamps towards Durval, *Leichhardt.*
**N. S. Wales.** Clarence river, *Beckler.*
Var. *thymifolia.* Leaves shorter, often recurved at the end.—Near Moreton Bay, *A. Cunningham.*

38. **H. serpyllifolia,** *R. Br. in DC. Syst. Veg.* i. 430. Decumbent or prostrate, much branched, and either glabrous or the branches and young parts clothed with short spreading hairs. Leaves (like those of *H. vestita*) narrow-linear, obtuse, 2 to 4 lines long, rigid with recurved margins. Peduncles very short, rarely attaining 2 or 3 lines, with 2 or 3 small bracts at their base. Sepals about 2 lines long, acute or the inner ones obtuse, glabrous or hairy. Stamens about 12, without staminodia. Carpels 3, villous, 4-ovulate.—*H. ericifolia,* Hook. f. Fl. Tasm. i. 14. t. 3; F. Muell. Pl. Vict. i. 17.

**N. S. Wales.** Port Dalrymple, *Caley;* Shoalwater Bay and Passage, *R. Brown.*
**Victoria.** Stony mountains, particularly in the highlands; also on subalpine meadows, *F. Mueller.*
**Tasmania.** Common on the serpentine formation, Asbestos hills; also Launceston and George Town, *Gunn.*
Var.? *minutifolia.* Leaves 1 to 2 lines long. Mount Aberdeen, *F. Mueller.* These specimens may possibly belong to the small-leaved variety of *H. pedunculata,* but the shortness of the peduncle and general aspect bring them nearer to *H. serpyllifolia.*

39. **H. pedunculata,** *R. Br. in DC. Syst. Veg.* i. 430. Stems diffuse, prostrate, or rarely erect, much branched, glabrous or clothed as well as the leaves with a few very short spreading hairs. Leaves narrow-linear, rigid, obtuse, usually 2 to 3 lines long, the margins revolute, numerous but not clustered. Peduncles ¼ to ½ in. long or sometimes more, the bracts at the base inconspicuous or wanting. Sepals 2 to nearly 3 lines long, ovate, very obtuse, usually minutely pubescent outside. Petals obovate, slightly emarginate. Stamens 15 to 25, accompanied usually by one or two small staminodia outside. Carpels 3, villous (or rarely glabrous?), with 4 or 6 ovules in each.—*Pleurandra intermedia,* DC. Syst. Veg. i. 420 (according to an unnamed specimen of Caley's, in Herb. R. Br.).

**N. S. Wales.** Port Jackson, *R. Brown;* to the Blue Mountains, *A. Cunningham.* In the mountains and Paramatta, *Caley;* and southward to the lower part of the Australian Alps, *F. Mueller.* These specimens, with elongated, divaricate branches, about 15 stamens and 4 ovules, occur in some herbaria under the name of *H. minutifolia,* F. Muell., as well as those of a var. of *H. serpyllifolia.*
Var. *corifolia.* Stems short, diffuse or prostrate. Stamens about 20. Ovules usually 6.
—*H. corifolia,* Bot. Mag. t. 2672; *H. pedunculata,* Bot. Reg. t. 1001. The carpels are described in the Botanical Magazine as glabrous, but in the Register, where the same garden-plant is represented, they are said to be silky, as I have always found them.

§ 3. *Ochrolasiæ.*—Carpels glabrous, with 6 to 8 ovules. No staminodia. Leaves narrow, with revolute margins, as in the *Vestitæ.* Flowers sessile, without the broad brown bracts of the *Bracteatæ.*

40. **H. ochrolasia,** *Benth.* Branches rigid, divaricate, glabrous. Leaves linear, obtuse, 2 to 3 lines long, the margins much revolute, rather thick and

rigid, whitish, but without hairs or asperities.　Flowers solitary, or 2 or 3 together at the ends of the branches, nearly sessile, surrounded by a few bracts like the sepals, but smaller.　Sepals 3 to 4 lines long, densely clothed with long golden hairs.　Petals broad.　Stamens 15 to 20.　Carpels 2.—*Ochrolasia Drummondi*, Turcz. in Bull. Mosc. 1849, ii. 4.

**W. Australia.**　*Drummond, 4th Coll. n.* 119.

§ 4. *Fasciculatæ.*—Carpels glabrous.　Ovules 2 to 6.　No staminodia. Leaves narrow-linear, convex below, the margins not recurved.　Flowers sessile or nearly so, but without the broad brown bracts of the *Bracteatæ.*

41. **H. procumbens,** *DC. Syst. Veg.* i. 427.　Diffuse or prostrate and much branched, resembling in habit some of the varieties of *H. fasciculata*, with which F. Mueller unites it ; but the leaves are broader, the larger ones above ½ in. long and 1 line broad, glabrous or rarely hairy, the flowers much larger, the sepals 4 to 5 lines long, broadly membranous, the stamens at least 20, and the carpels 4 or 5, with almost always 6 ovules in each.—*Dillenia procumbens*, Labill. Pl. Nov. Holl. ii. 16, t. 156 ; *H. angustifolia*, Salisb. Parad. Lond. under n. 73.

**Victoria.**　Albert river, Gipps' Land, *F. Mueller.*
**Tasmania.**　*R. Brown ;* abundant in open heathy places, *J. D. Hooker.*

42. **H. fasciculata,** *R. Br. in DC. Syst. Veg.* i. 428.　Stems erect, procumbent or prostrate.　Leaves very narrow-linear, clustered and crowded, 2 to 3 lines or rarely ½ in. long, hirsute with soft rather spreading hairs, or at length glabrous, obtuse, or scarcely pointed, the margins never revolute or recurved, but rather turned upwards so as to leave the under surface convex with the prominent midrib.　Flowers sessile, on very short leafy shoots along the branches, with 2 or 3 small sepal-like bracts at their base.　Sepals 2 to 3 lines long, broadly ovate, membranous at the edge, the outer ones narrower and less obtuse.　Petals obcordate.　Stamens usually 8 to 12, without staminodia.　Carpels usually 3, glabrous, with 2 erect ovules in each.—Hook. f. Fl. Tasm. i. 13 ; *H. angustifolia* (partly), F. Muell. Pl. Vict. i. 18 ; *H. virgata*, Hook. Ic. Pl. t. 267, not R. Br.; *H. prostrata*, Hook. Journ. Bot. i. 246 ; *Pleurandra camforosma*, Sieb. in Spreng. Syst. Cur. Post. 191 ; *H. camphorosma*, A. Gray, Bot. Amer. Expl. Exped. i. 21.

**N. S. Wales.**　Port Jackson, *R. Brown, Sieber, n.* 146, and *Fl. Mixt. n.* 506, and others.
**Victoria.**　Port Phillip, *R. Brown ;* sand ridges, heathy ground, and dry, barren places throughout the colony, *F. Mueller.*
**Tasmania.**　Abundant throughout the colony, ascending to 2000 or 3000 ft., *J. D. Hooker.*
**S. Australia.**　Extending as far as Spencer's Gulf, *F. Mueller* and others.
　Var. *crassifolia.*　Stems prostrate, the habit sometimes nearly that of *H. linearis*, but the margins of the leaves involute not revolute, sometimes very pubescent like the following variety.—*H. glandulosa*, Schlecht. Linnæa, xx. 626.　Chiefly in S. Australia.
　Var. *pubigera.*　Very hoary all over with very short, stiff hairs.　Leaves 3 to 6 lines, thicker and less clustered than in the ordinary form.　Flowers terminating loosely-leaved branches, but scarcely pedunculate above the last leaf.　Flowers as in the common form, except that the sepals are more hairy and the carpels usually 4-ovulate.　S. Australia, *Atherstone.*
　The species is said, in Pl. Preiss. ii. 236, to have been found in York district, W. Australia.

I have not seen Preiss's specimen referred to, n. 2171, but should think it very probable that *Candollea teretifolia* may have been mistaken for it.

§ 5. *Bracteatæ.*—Carpels glabrous. Ovules 1 or 2, erect or ascending. Stamens usually under 20 in the first five species, more numerous in the following ones, without any staminodia. Leaves flat, or when narrow, convex underneath, the margins not prominently revolute. Flowers closely sessile within broad brown shining bracts (except in *H. rostellata*).

43. **H. virgata,** *R. Br. in DC. Syst. Veg.* i. 428. Diffuse or erect, glabrous, with numerous thin but stiff and often wiry branches. Leaves narrow-linear, obtuse or scarcely acute, mostly about ½ in. long, but sometimes much longer, stiff and rather thick, the margins not revolute, and sometimes almost terete. Flowers sessile, surrounded by 2 or 3 very broad scarious pale brown bracts, fully half as long as the calyx. Sepals about 4 lines long, obtuse or more frequently acute, or with a short sharp point, glabrous and more scarious than in any other species. Petals broadly obovate, scarcely emarginate. Stamens 10 to 15, without staminodia. Carpels 3, glabrous, 2-ovulate. —Hook. f. Fl. Tasm. i. 14 ; *H. angustifolia,* var., F. Muell. Pl. Vict. i. 19.

**N. S. Wales.** Port Jackson, *R. Brown.*

**Victoria.** Murray river, and near Mount William and Port Phillip, *F. Mueller ;* Mount Lockhart, *Moreton.*

**Tasmania..** Sandy soil on the road from George Town to Currie's River, *Gunn.*

44. **H. inclusa,** *Benth.* Allied to *H. virgata,* but much more rigid, the leaves and young branches more or less hoary, and always hirsute, with short white hairs about the floral leaves. Leaves narrow-linear or slightly cuneate, obtuse, ¼ to ½ in. long, rather thick, convex underneath, the floral ones clustered. Flowers closely sessile within them, surrounded by short broad brown scarious bracts. Sepals glabrous, about 3 lines long. Petals obovate, entire. Stamens 12 to 15, without staminodia. Carpels 2 or 3, glabrous, 1-ovulate.

**W. Australia.** Swan River, *Drummond, n.* 13.

45 ? **H. rostellata,** *Turcz. in Bull. Mosc.* 1849, ii. 8. Branches rigid and glabrous. Leaves rigid, thick, narrow-linear, 3 to 4 lines long, hooked at the extremity, with a short recurved sharp point, convex underneath or nearly terete, but marked laterally with a slight furrow indicating the recurved margins which however are not prominent. Flowers nearly sessile. Bracts much smaller and narrower than in any of this group. Sepals glabrous, obtuse, rather above 2 lines long. Stamens 15 to 20, without staminodia. Carpels 5, glabrous, 2-ovulate.

**W. Australia.** *Drummond, 4th Coll. n.* 121. The position of this species is somewhat doubtful ; the foliage is nearly that of *H. recurvifolia* or of *Candollea uncinata,* from both of which it widely differs in the stamens. It has not the broad brown bracts of the *Bracteatæ,* but in other respects comes nearer to them than to any other group.

46. **H. glomerata,** *Benth.* Rather rigid, much branched and often tortuous, quite glabrous and often rather glaucous, or rarely with a very minute pubescence on the young parts. Leaves from linear-cuneate to oblong or cuneate, obtuse truncate or retuse, usually ¼ to ½ in. long, flat or with the edges slightly recurved, and the midrib prominent underneath, the floral ones

shorter and clustered, sometimes nearly ovate. Flowers rather small, sessile in the tufts of floral leaves, and surrounded by short broad brown scarious bracts. Sepals lanceolate, usually acute, stiffly membranous, quite glabrous, nearly 3 lines long. Petals broadly obcordate. Stamens 10 to 15, or rarely above 20, without staminodia. Carpels 3, glabrous, 1- or 2-ovulate.

**W. Australia.** Swan River, *Drummond,* 1*st Coll. n.* 8 of 1843.

Var. ? *canescens.* Leaves hoary, with a minute appressed pubescence. Sepals larger but glabrous. Gordon river, *Oldfield ;* rock at Oolingarran, *Herb. Mueller.* The specimens are insufficient for accurate determination.

47. **H. argentea,** *Steud. in Pl. Preiss.* i. 268. Allied to *H. montana,* but the whole plant is silvery-white, with densely appressed silky hairs. Leaves narrow-oblong, ½ to ¾ in. long, obtuse or with a minute point, slightly contracted at the base. Flowers closely sessile in tufts of floral leaves, and surrounded by broad short bracts, brown on the edges, but more or less silky-hairy on the back, and not so obtuse as in *H. montana.* Flowers smaller. Sepals 3 to 4 lines long, lanceolate, acute, very silky-hairy. Petals broad, emarginate, almost 2-lobed. Stamens above 40, without staminodia. Carpels 3, glabrous, 2-ovulate. Arillus very short.

**W. Australia.** *Drummond ;* Cape Riche, *Preiss, n.* 2144.

Var. *diffusa.* Dwarf, with obovate-oblong leaves of 1 in. or rather more. Flowers large. —Stoney hills, Tone river, *Oldfield.*

48. **H. pilosa,** *Steud. in Pl. Preiss.* i. 272. Branches slender, weak, loosely pubescent or hairy. Leaves narrow-oblong or oblong-oval, above 1 in. long, the margins slightly recurved, nearly glabrous, scabrous, or loosely hairy. Flowers closely sessile, surrounded by broad brown scarious bracts, usually mucronate, and shorter and thinner than in *H. montana.* Sepals hairy, with loose spreading not silky hairs, acute, about 3 lines long. Stamens and carpels of *H. montana,* of which this plant may possibly hereafter prove to be a variety only.

**W. Australia.** Dense shady places, Darling's Range, *Preiss, n.* 2130 (*Hb. Sonder.*).

49. **H. montana,** *Steud. in Pl. Preiss.* i. 270. Stems usually erect, from a thick rhizome, 1 ft. high or rather more, pubescent. Leaves in the normal form linear-oblong, obtuse, with a minute point, ½ to 1 in. long, the margins slightly recurved, narrowed at the base, usually glabrous above, silky-hairy underneath. Flowers closely sessile, and surrounded by 2 or 3 orbicular shining brown bracts. Sepals very densely clothed with long silky hairs, the outer ones acuminate, and often above 5 lines long. Petals obovate, emarginate. Stamens very numerous, without staminodia. Carpels 3, glabrous, 2-ovulate.—*H. discolor* and *H. commutata,* Steud. in Pl. Preiss. i. 267.

**W. Australia.** Hills of Swan River and Canning river, and Darling Range, *Collie, Drummond, Preiss, n.* 2135, 2136, and 2137, and others.

Var. *confertifolia.* Leaves and flowers smaller.—*H. confertifolia,* Steud. in Pl. Preiss. i. 267. King George's Sound and neighbouring districts, *Oldfield, Preiss, n.* 2143, and others.

Var. *major.* Larger and more branched and often more or less hirsute, with long spreading hairs. Leaves usually larger, on luxuriant shoots often above 1½ or 2 in. long, broad and coarsely toothed, almost all less contracted at the base than in the normal form, and closely sessile.—*H. ovata,* Steud. in Pl. Preiss. i. 270.—Swan River, *Drummond ;* Darling Range, *Preiss, n.* 2134. Some specimens of this variety look so different from *H. montana,* with their coarse habit, long spreading hairs, and broad-toothed leaves, that I had at first retained

D 2

them as a distinct species; but they pass into the smaller forms through so many intermediates, that I have been quite unable to draw any definite limits between them.

§ 6. *Subsessiles.*—Carpels glabrous, usually 3, with 1 or 2 ovules in each, but in one species 5 or more, with 6 or more ovules in each. Stamens usually numerous, without staminodia. Leaves flat or the margins slightly recurved. Bracts small or passing into the sepals. Flowers sessile or nearly so.

50. **H. linearis,** *R. Br. in DC. Syst. Veg.* i. 428. Much branched, erect or divaricate, or rarely decumbent, glabrous in all its parts, or with a very minute pubescence on the young shoots. Leaves in the normal forms linear, rather acute or obtuse, with a short recurved point, 4 to 8 lines long, or nearly 1 in. when luxuriant, the margins flat or slightly recurved, and not convex underneath. Flowers on very short peduncles, and usually surrounded by rather longer floral leaves, with small acuminate brown bracts at the base of the peduncle, and one or two at the summit passing into the sepals. Sepals all or the inner ones only obtuse, glabrous with thin margins, 2½ to 3 lines long. Petals obovate, scarcely notched. Stamens 15 to 20, without staminodia. Carpels usually 3, rarely 2 or 1, glabrous, 2-ovulate.

**Queensland.** Moreton Island, *M'Gillivray, F. Mueller.*

**N. S. Wales.** Port Jackson, *R. Brown, Sieber, n.* 138, and *Fl. Mixt. n.* 503, and others; and northward to New England, *C. Stuart.*

Var. *floribunda.* Sepals more acute and rather hairy. Stamens more numerous.—Peel's Island, *A. Cunningham.*

Var. *grandiflora.* Sepals above 4 lines long. Stamens about 50.—New England, *C. Stuart.*

Var. ? *obtusifolia.* More rigid than the normal form, more frequently erect, and more or less hairy, with a minute crisped or shortly stellate tomentum, sometimes densely and softly pubescent, and very rarely glabrous. Leaves from linear to broadly oblong-spathulate, very obtuse or truncate, in some southern specimens above 1½ in. long, and mostly narrowed into a short petiole. Flowers rather larger than in the normal variety, with numerous stamens.—*H. obtusifolia,* DC. Syst. Veg. i. 429 ; *H. canescens,* Sieb. in Spreng. Syst. Cur. Post. 211.

**Queensland.** Brisbane and Burnett rivers, *F. Mueller.*

**N. S. Wales.** Port Jackson, *Sieber, n.* 140 ; Twofold Bay, *F. Mueller ;* and other places south of Sydney, *A. Cunningham.*

**Victoria.** Goulburn river, towards the Dandenong ranges, and on the northern slopes of the Australian Alps, *F. Mueller ;* also in *Mitchell's* collections. The majority of specimens of this variety have a very different aspect from those of the typical *H. linearis ;* but as there are certainly numerous intermediates, I feel compelled to follow F. Mueller in uniting them as varieties. He also includes in the same species the following *H. diffusa,* which, however, appears to me to be rather more constant in its characters. The specimens described by De Candolle were from Port Jackson, not from Van Diemen's Land.

51. **H. diffusa,** *R. Br. in DC. Syst. Veg.* i. 429. Stems low, usually diffuse or prostrate, with numerous short ascending branches, pubescent or at length glabrous. Leaves from obovate to linear-cuneate, very obtuse or truncate, seldom above ¼ in. long, and then often 2- or 3-toothed. Peduncles very short. Sepals broadly oblong, obtuse, about 4 lines long, the outer ones rather shorter and narrower. Petals obovate, entire. Stamens about 20 to 25, without staminodia. Carpels usually 3, or rarely 2 or 4, glabrous, 2-ovulate.

**N. S. Wales.** Port Jackson, *R. Brown, Sieber, n.* 145, and *Fl. Mixt. n.* 501, and others.

Var. *dilatata.* More erect and very much branched. Leaves small, broadly spathulate, and much contracted at the base, with a petiole often longer than the blade. Carpels 1, 2, or 3.—*H. monogyna,* R. Br. in DC. Syst. Veg. i. 429 ; *H. dilatata,* A. Cunn. Herb.— Port Jackson, *R. Brown* and others ; and southward to Yowaka river, *F. Mueller.*

**52. H. saligna,** *R. Br. in DC. Syst. Veg.* i. 427. Branches elongated, flexuose, apparently diffuse or half trailing, softly pubescent when young. Leaves oblong-linear or lanceolate, usually shortly pointed, 1½ to 3 in. long, narrowed below, with a broader stem-clasping base, leaving a raised ring on the branch, glabrous or nearly so above, loosely villous underneath. Flowers sessile in a cluster of floral leaves. Sepals oval-oblong, 6 to 8 lines long, the inner ones obtuse, the outer ones more lanceolate and pointed, very silky-hairy outside. Petals broadly obovate, scarcely notched. Stamens 20 to 30, without staminodia. Carpels 3, glabrous, 2-ovulate.

**N. S. Wales.** Port Jackson, *R. Brown* and others ; to the Blue Mountains, *A. Cunningham, Miss Atkinson,* and others.

**53. H. volubilis,** *Andr. Bot. Rep. t.* 126. Stems woody, short and trailing, or twining and climbing to the height of 2 to 4 ft., the young parts more or less clothed with silky hairs. Leaves from obovate to lanceolate, obtuse or acute, 1½ to 3 in. long, narrowed below, but slightly enlarged and stem-clasping at the base, leaving a raised ring on the stem, as in most *Candolleas,* glabrous above, silky-hairy underneath. Flowers the largest of the genus, nearly sessile, the upper leaves passing into sepal-like bracts. Sepals 8 lines to 1 in. long, ovate-acuminate, very silky-hairy outside. Petals obovate, entire. Stamens very numerous, without staminodia. Carpels usually 5, but sometimes up to 8, glabrous, 6- to 8-ovulate.—*Dillenia scandens,* Willd. Spec. ii. 1251 ; *Dillenia volubilis,* Vent. Choix. t. 11 ; *D. speciosa,* Bot. Mag. t. 449, not of Thunb.

**Queensland.** Loose sand and sides of rocks near the sea, Moreton Island, *M'Gillivray, F. Mueller.*

**N. S. Wales.** N. shore, Port Jackson, *R. Brown* and others ; Kiama, *Harvey ;* Hastings river, *Beckler ;* Paramatta, *Woolls.*

§ 7. *Hemihibbertiæ.*—Carpels glabrous, except in *H. grossulariæfolia* and *H. lasiopus.* Stamens very numerous, with several, often numerous, small subulate or clavate staminodia round the outside. Leaves flat. Flowers pedunculate.

**54. H. grossulariæfolia,** *Salisb. Parad. Lond. t.* 73 (*Burtonia* on the plate). Stems weak and prostrate or trailing, loosely pubescent. Leaves distinctly petiolate, ovate or oval-oblong, obtuse, 1 to 1½ in. long, undulate and coarsely toothed, prominently pinnate-veined underneath, glabrous or scabrous above, more or less pubescent or hairy underneath. Flowers rather small, on filiform peduncles of 1 in. or more, with 2 or 3 narrow bracts at their base. Sepals ovate or lanceolate, acuminate, about 3 lines long, silky-hairy. Petals obovate, entire or nearly so. Stamens numerous, with several filiform or clavate staminodia outside ; anthers short but oblong. Carpels 10 to 15, villous, 2-ovulate.—Bot. Mag. t. 1218 ; DC. Syst. Veg. i. 425 ; Reichb. Ic. et Descr. Pl. t. 74 ; *H. crenata,* Andr. Bot. Rep. t. 472 ; *H. latifolia,* Steud. in Pl. Preiss. i. 269 ; *Warburtonia potentillina,* F. Muell. Fragm. i. 230. t. 9 ; ii. 182.

**W. Australia.** Sandy and rocky places near the sea, King George's Sound, *R. Brown, Menzies;* Swan River, *Collie, Drummond, Preiss, n.* 2126; Cape Naturaliste, *Oldfield.*

**55. H. dentata,** *R. Br. in DC. Syst. Veg.* i. 426. Stems woody at the base only, trailing or twining, glabrous or the young branches pubescent. Leaves distinctly petiolate, oblong, obtuse or acute, 1½ to 2½ in. long, flat, marked with a few distant callous teeth, or slightly sinuate, rounded at the base, glabrous or pubescent when young. Flowers rather large, on short peduncles, with 1 or 2 small bracts at their base. Sepals ovate, ½ in. long, the inner ones obtuse, the outer rather shorter and more acute, rarely all acuminate, pubescent or silky-hairy. Petals obovate, entire or scarcely notched. Stamens very numerous with slender filaments, the anthers short, although not so broad as in the *Brachyantheræ*, and a considerable number of filiform or clavate staminodia outside. Carpels 3, glabrous, 6- to 8-ovulate.—F. Muell. Pl. Vict. i. 217 ; Bot. Reg. t. 282 ; Bot. Mag. t. 2338.

**N. S. Wales.** Woods and stony places near the sea, Port Jackson, *R. Brown, Caley,* and others; northward to Hastings and Clarence rivers, *Beckler;* and southward to Illawara, *A. Cunningham;* and Twofold Bay, *F. Mueller.*
**Victoria.** Stony forest declivities, near the Genoa river, Genoa Peak, and other localities at the S. E. limit of Gipps' Land, *F. Mueller.*

**56. H. perfoliata,** *Endl. in Hueg. Enum.* 3. Stems weak, procumbent, ascending or shortly erect, or sometimes shortly trailing, quite glabrous as well as the whole plant. Leaves ovate, acute, 1 to 2 in. long, often edged with minute distant teeth, perfoliate near the base, the auricles quite united behind the stem. Peduncles 1 to 2 in. long. Sepals lanceolate, acute or acuminate, 4 to 5 lines long. Petals obovate, entire. Stamens numerous, with a few short filiform staminodia outside. Carpels 3, 4, or 5, glabrous. —Bot. Reg. 1843, t. 64.

**W. Australia.** Marshes, Swan River, *Huegel;* Freemantle, *Collie;* shady boggy places about Perth, *Preiss, n.* 2127; Vasse river, *Oldfield;* King George's Sound, *A. Cunningham.*

**57. H. bracteosa,** *Turcz. in Bull. Mosc.* 1852, ii. 140. Stems erect, somewhat compressed, with 2 prominent angles, 1 to 1½ ft. high, glabrous like the whole plant. Leaves broadly obovate, very obtuse, 1 to 2 in. long, closely clasping the stem at their base, the auricles slightly decurrent or projecting beyond the stem. Peduncles leaf-opposed or axillary, 1 in. long or more. Flowers large. Sepals ovate, 5 to 6 lines long, the inner ones obtuse, the outer more acute. Petals very broadly obcordate. Stamens very numerous, with a few filiform staminodia outside. Carpels 5, glabrous, 3- or 4-ovulate.

**W. Australia.** *Drummond, n.* 286 ; Plantagenet, Stirling, Perongerup ranges, *Maxwell.*

**58. H. amplexicaulis,** *Steud. in Pl. Preiss.* i. 266. Perfectly glabrous like the last two, with ascending or perhaps half-trailing stems of 1 to 2 ft. Leaves broadly lanceolate or oblong, acute, 2 to 3 in. long, embracing the stem by two ovate auricles, quite free or occasionally united beyond the stem. Peduncles flexuose, 1 to 2 in. long. Flowers rather large. Sepals fully 6 lines, ovate-lanceolate, and very acute in the original specimens, broader and very obtuse in many others. Petals broadly obovate, entire or slightly notched.  Stamens very numerous, with a few filiform staminodia outside. Carpels 4 or 5, glabrous, 4-ovulate.

**W. Australia.**  King George's Sound, *Menzies;* and thence to Vasse and Swan rivers, *Drummond, Preiss, n.* 2129, *Oldfield,* and others.

Some specimens have the auricles of the lower leaves more or less united, thus showing an approach to *H. perfoliata,* and have been described as species under the names of *H. bupleurifolia,* Lehm. Nov. Hort. Hamb. and Linnæa, xxv. 307, and of *H. disticha,* Lehm. l. c. 309.  They may be readily distinguished from *H. perfoliata,* by the thicker rigid pedicels, larger broader sepals, etc.  On the other hand, narrow-leaved branches appear almost to pass into *H. Cunninghamii.*

59. **H. Cunninghamii,** *Hook Bot. Mag t.* 3183.  Perfectly glabrous, with slender branches apparently tending to climb.  Leaves linear, mostly pointed, 1 to 1½ or rarely 2 in. long, the edges scarcely recurved, narrowed below the middle, but expanded again into a stem-clasping or sagittate base. Peduncles axillary, ½ to ¾ in. long, with a few small leafy bracts at their base. Sepals thin, about 3 lines long, broadly ovate, the outer ones more acute. Petals slightly notched.  Stamens numerous, with numerous short filiform staminodia outside.  Carpels 5, glabrous, 3- or 4-ovulate.—*Candollea Cunninghamii,* Benth. in Maund. Bot. ii. t. 83 ; *Hibbertia lactucæfolia,* Steud. in Pl. Preiss. i. 267.

**W. Australia.**  King George's Sound, *R. Brown, A. Cunningham,* and others; Cape Riche, *Harvey ;* shady places, Sussex and Plantagenet districts, *Preiss, n.* 2161 and 2173 ; Stirling range, *Maxwell ;* Cape Naturaliste, *Oldfield.*

Var. *hastata.*  Leaves rather broader, the broadest nearly 3 lines, and carpels, according to Steudel, 2 only.  I have only seen fragments.—*H. hastata,* Steud. in Pl. Preiss. i. 266.— S. W. Australia, *Preiss, n.* 2128.

60. **H. glaberrima,** *F. Muell. Fragm.* iii. 1.  Perfectly glabrous. Leaves (the upper ones only known) oblong-lanceolate, obtuse with a short glandular point, 1 to 1½ in. long, quite entire, tapering below the middle almost into a petiole, and slightly expanded so as to half-clasp the branch. Peduncles axillary or terminal, about 1½ in. long.  Innermost sepals fully 6 to 7 lines long, and very broad, the others gradually diminishing to the outermost, which is lanceolate and about 3 lines.  Petals not much longer than the calyx.  Stamens very numerous (200 to 300), with numerous (2 or 3 dozen) short clavate staminodia outside.  Carpels 3, glabrous, with about 8 ovules in each.

**S. Australia.**  In the interior at Brinkley's Bluff, near Macdonnell's Range, *M'Douall Stuart.*  Evidently nearly allied to *H. amplexicaulis,* but without the basal auricles of the leaf.

61. **H. Mylnei,** *Benth.* Resembles, at first sight, some of the hairy varieties of *H. montana,* but the flowers are different.  Stems in our specimens short and erect from a thick rhizome, hispid as well as the leaves with long spreading or reflexed hairs.  Leaves oblong, obtuse, or shortly pointed, mostly about 1 in. long, slightly contracted, and half stem-clasping at the base, the margins scarcely recurved.  Flowers closely sessile in a cluster of smaller floral leaves, and surrounded by brown scarious bracts as in *H. montana,* but the sepals (5 or 6 lines long) are glabrous, the petals almost 2-lobed, and the numerous stamens, with slender filaments and short anthers, are surrounded by small, filiform or slightly clavate staminodia.  Carpels 3, glabrous, 2-ovulate.

**W. Australia.**   Swan River, *Mylne.*

62. **H. lasiopus,** *Benth.*   Stems usually rather short, with a short pubescence, mixed with long spreading hairs, in our specimens nearly simple and erect from a thick rhizome. Leaves from obovate to oblong, 1 to 2 in. long, or rather more, the larger ones often coarsely toothed and more or less hairy, the younger ones often deeply toothed, narrowed but half-stem-clasping at the base. Flowers on very hairy peduncles of ½ to 1½ in., surrounded at the base by broad brown scarious bracts. Sepals very densely silky-hairy, ½ in. long, acuminate. Petals broadly obovate, deeply notched. Stamens very numerous, with a ring of filiform or clavate staminodia outside. Carpels 5, very villous, 2-ovulate.

**W. Australia.**   Swan River, *Drummond, Mylne.*

63. **H. potentillæflora,** *F. Muell. Herb.*   Stems either nearly simple, erect, from a thick rhizome, and ½ to 1 foot high, or longer, and branched, hoary, with a short, close, somewhat silky pubescence. Leaves oblong-linear or lanceolate, usually obtuse, 1 to 2 in. long, the margins flat or slightly recurved, silky-hairy on both sides when young, but nearly glabrous above when old, narrowed below, and scarcely stem-clasping. Peduncles clustered, or rarely solitary, silky-hairy, 1 to 1½ in. long, surrounded at the base by broad brown scarious bracts. Sepals silky-hairy, ovate, rather acute, about 5 lines long, with membranous edges. Petals obovate, retuse, stamens very numerous, more or less clustered between the carpels, but free, with a considerable number of subulate staminodia outside. Anthers oval-oblong, opening laterally. Carpels 5, glabrous, 2-ovulate.

**W. Australia.**   Swan River, *Drummond,* 1st *Coll.* ; Murchison River, *Oldfield.*

§ 8. *Brachyantheræ.*—Carpels glabrous. Stamens about 15 to 20, without staminodia. Anthers (except in *H. pungens*) ovate or orbicular, flattened, with the cells opening on the inner face. Leaves narrow-linear, glabrous. Flowers pedunculate.

64. **H. pungens,** *Benth.*   Glabrous and rigid with the pungent leaves of *H. acicularis* and *H. acerosa,* but very different stamens. Leaves narrow-linear, or linear-subulate, often fasciculate, the longest about ½ in. long, very rigid, with a fine pungent point. Peduncles shorter than the leaves, recurved. Sepals about 2 lines long, broad, obtuse, or the outer ones with a short, fine point, quite glabrous. Carpels 5, glabrous, 2-ovulate. Stamens about 15, without staminodia. Anthers oblong.

**W. Australia.**   E. Mount Barren and Phillip's River, *Maxwell (Hb. F. Muell.).*

65. **H. nutans,** *Benth.*   Branches rigid, rather wiry, and erect from a thick rhizome, the young ones ash-coloured, but glabrous. Leaves rigid, linear, with a short recurved point, mostly about ½ in. long, the margins slightly recurved, the midrib underneath very thick, whitish, but glabrous. Peduncles recurved, about ½ in. long. Sepals 5 to 6 lines, glabrous, the inner ones with membranous edges. Petals not seen. Stamens about 20, without staminodia. Anthers ovate, flat, opening inwards, the connective ending in an obtuse, prominent point. Carpels 5, glabrous, 2-ovulate.

**W. Australia.**   Swan River, *Drummond, Coll.* 1843, *n.* 10.

66. **H. leptopus,** *Benth.* Glabrous and slender, like *H. stellaris,* but stiffer and less branched, and the branches usually ashy-white. Leaves narrow-linear, obtuse, or nearly so, $\frac{1}{2}$ to $1\frac{1}{2}$ in. long, the edges so revolute as to make them nearly terete. Pedicels very slender, usually about $\frac{1}{2}$ in. long. Flowers of *H. stellaris,* but smaller, the sepals more herbaceous. Anthers nearly orbicular, and very concave on the inner face. Carpels of *H. stellaris.*

**W. Australia.** Swan River, *Drummond, n.* 11.

67. **H. stellaris,** *Endl. in Hueg. Enum.* 3. Glabrous, with numerous slender branches. Leaves linear, flat, acute, and somewhat falcate, mostly about 1 in. long, narrowed below the middle, the floral ones often slightly enlarged and sheathing, or stem-clasping at the base. Flowers numerous, on slender peduncles of $\frac{1}{2}$ to $\frac{3}{4}$ in. Sepals orbicular, membranous, very obtuse, about 2 lines long. Petals nearly twice as long, broad, deeply notched and more persistent than in most species. Stamens about 15, without staminodia, the anthers short, broad, and flattened, turned over the ovaries, and opening on the inner face. Carpels 3, very truncate, glabrous, 1- or 2-ovulate.—*H. tenuiramea,* Steud. in Pl. Preiss. i. 268.

**W. Australia.** Sandy places, Swan River, *Hugel, Preiss, n.* 2145; from Geographer Bay and Gordon river to Murchison river, *Maxwell, Oldfield,* and others.

### 3. CANDOLLEA, Labill.

Sepals 5. Petals 5. Stamens united to the middle or higher up, into five bundles, each bearing 2 to 6 anthers, and alternating with the carpels when there are five carpels, or when the carpels are reduced to 3 or 2, 2 or 3 of the bundles are often reduced to a single stamen, and in some species there is a free stamen within each bundle. No staminodia. Carpels usually 3 or 5, very rarely reduced to 2, always glabrous, with 1, 2, or very rarely 3 ovules in each. Styles and fruit of *Hibbertia.*—Shrubs or undershrubs with the habit of *Hibbertia.*

All the known species are from West Australia.

Flowers sessile within the floral leaves.
  Leaves with flat, or slightly recurved, not revolute margins.
    Leaves obovate or oblong. Carpels 5, 2- or 3-ovulate.
      Leaves obovate or shortly obovate-cuneate. Petals slightly
        exceeding the calyx . . . . . . . . . . . . . 1. *C. cuneiformis.*
      Leaves narrow-oblong, 1 to 2 in. Petals much longer than
        the calyx, deeply notched . . . . . . . . . 2. *C. tetrandra.*
    Leaves linear or subulate. Carpels 3 to 5, 1-ovulate.
      Leaves linear-cuneate, enlarged at the base into a broad
        sheath . . . . . . . . . . . . . . . . 10. *C. glaberrima.*
      Leaves linear, slightly dilated at the base, obtuse or trun-
        cate, $\frac{3}{4}$ to 1 in. Carpels 5, rarely 3 . . . . . . . 3. *C. glomerosa.*
      Leaves heathlike, clustered, mostly 2 to 4 lines. Carpels 3 4. *C. teretifolia.*
  Leaves linear, with revolute margins
    Leaves heathlike, glabrous, mostly 2 to 4 lines. Flowers
      small. Sepals glabrous . . . . . . . . . . . . 4. *C. teretifolia.*
    Leaves clustered, mostly $\frac{1}{2}$ in., the floral ones and sepals hairy.
      Carpels 3.
        Stem shrubby. Leaves rigid, the floral ones long, gla-
          brous at the tips . . . . . . . . . . . . 5. *C. desmophylla.*

Stem half herbaceous. Leaves very hirsute, the floral
    ones not exceeding the flowers . . . . . . . . 6. *C. helianthemoides.*
Carpels 5. Stem half herbaceous . . . . . . . . 7. *C. fasciculata.*
Leaves mostly 1 to 2 in. long and scarcely clustered.
    Glabrous. Leaves rigid, mostly acute. Staminal bundles of
    about 5 . . . . . . . . . . . . . . . . 8. *C. Huegelii.*
    Silky-hairy. Leaves less rigid, more obtuse. Staminal
    bundles of 2 or 3 each . . . . . . . . . . . 9. *C. pachyrrhiza.*
Flowers pedunculate.
    Peduncles shorter than the enlarged sheaths of the floral leaves.
    Leaves flat, obtuse, or truncate.
    Blade of the floral leaves longer than their sheaths . . . . 10. *C. glaberrima.*
    Sheaths of the floral leaves ½ in., with the blade reduced to a
    short point . . . . . . . . . . . . . . . 11. *C. vaginata.*
    Peduncles longer than the sheaths of the floral leaves. Leaves
    flat or the margins scarcely recurved, obtuse or truncate.
    Plant very glaucous. Leaves thick, broadly linear, mostly
    above 1 in. Peduncles tomentose, scarcely longer than the
    flowers . . . . . . . . . . . . . . . . 12. *C. Preissiana.*
    Plant slightly glaucous. Leaves narrow, ½ to 1 in. Peduncles
    long, slender, glabrous . . . . . . . . . . . 13. *C. pedunculata.*
    Peduncles short. Leaves narrow-linear, rigid, thick, without
    sheaths.
    Leaves with a straight pungent point . . . . . . . . 14. *C. exasperata.*
    Leaves recurved at the top . . . . . . . . . . . 15. *C. uncinata.*

1. **C. cuneiformis,** *Labill. Pl. Nov. Holl.* ii. 34, *t.* 176. An erect shrub,
attaining sometimes above a man's height, but often much lower, with numerous
short, crowded branches, the young ones slightly hairy. Leaves from oblong-
cuneate to obovate, obtuse, truncate, or with a few teeth at the top, seldom
above 1 in. long, flat, narrowed into a short stem-clasping petiole, leaving a
prominent ring on the branch. Flowers sessile among the crowded floral
leaves. Sepals ovate-oblong, the 2 outer ones thick, about ½ in. long, the
inner shorter, thinner, and broader. Petals rather longer, broad, and deeply
notched. Stamens in 5 bunches of 3 to 5 each, with one free one within
each bunch. Carpels 5, glabrous, 2-ovulate. Arillus more than half as long
as the seed.—Bot. Mag. t. 2711 ; *Hibbertia obcuneata,* Salisb. Parad. Lond.
under n. 73.

**W. Australia.** King George's Sound, *R. Brown* and others ; Point Possession, *Collie ;*
Champion Bay, *Bower ;* Geographer Bay and Bald Island, *Oldfield.*

2. **C. tetrandra,** *Lindl. Bot. Reg.* 1842, *Misc.* 39, *and* 1843, *t.* 50.
Branches elongated, angular, shortly pubescent. Leaves from narrow-oblong
to oblong-obovate, obtuse, or shortly acuminate, but not truncate, 1 to 2¼ in.
long, the larger ones obscurely or coarsely toothed, narrowed at the base, and
stem-clasping, as in *C. cuneiformis.* Flowers as in that species, but larger,
the outer almost acute sepals often 8 lines, and the petals fully 1 in. Stamens
of *C. cuneiformis.* Carpels 5, glabrous, with 2 or rarely 3 ovules in each.
Ripe carpels black, and somewhat fleshy. Seeds more or less enveloped in
an orange-coloured lobed arillus.—*C. latifolia,* Steud. in Pl. Preiss. i. 273.

**W. Australia.** Swan River, *Drummond, Coll.* 1843, *n.* 6 ; shady places, Port Leschenault, *Preiss, n.* 2162.

*C. calycina,* Steud. in Pl. Preiss. i. 274, from Port Leschenault and Sussex district,
*Preiss, n.* 2131, appears to be the same species, although the petals are said to be smaller.

The specimens I have seen are bad, and the petals shrivelled or fallen off, the carpels nearly ripe.

3. **C. glomerosa,** *Benth.* Stems virgate, usually glabrous, except about the floral leaves. Leaves linear, obtuse, or truncate, mostly ¾ to 1 in. long, glabrous, the margins flat or recurved, but not revolute, narrowed below the middle, and slightly enlarged and stem-clasping at the base. Flowers nearly or quite sessile, usually surrounded by 2 or 3 ovate glabrous bracts, sometimes passing into the sepals. Calyx clothed with long, silky, or woolly hairs, or sometimes quite glabrous, the outer sepals ovate-lanceolate, acute, 3 to 4 lines long, the inner broad and more obtuse. Petals broad, notched. Stamens in 5 bundles of 4 to 6 each, often with a free one inside. Carpels 5, glabrous, 1-ovulate. Seeds brown, with a short, entire, or lobed arillus.

**W. Australia.** Swan River, *Drummond;* Port Gregory, *Oldfield.*
Var. *subsericea.* More silky; stamens fewer, two of the clusters reduced to single stamens, and carpels 3 only.—Swan River, *Drummond.*

4. **C. teretifolia,** *Turcz. in Bull. Mosc.* 1849, ii. 7. Perfectly glabrous. Branches slender, erect, virgate· Leaves heath-like, often clustered, linear, semiterete, slender, and rather acute, usually 2 or 3 lines, but in some specimens ½ in. long, the margins scarcely or not at all revolute. Flowers small, sessile in the clusters of leaves. Sepals ovate, membranous, coloured, scarcely 2 lines long, with 2 or 3 short orbicular bracts. Petals broadly obovate, entire. Stamens in 3 clusters of about 3 each, often less united than in most *Candolleas,* and 2 single stamens. Carpels 3, glabrous, 1-ovulate. The general aspect is very much that of the small glabrous-leaved specimens of *Hibbertia fasciculata,* but the stamens and ovaries are very different.—*Pleurandra enervia,* DC. Syst. Veg. i. 421?, Steud. in Pl. Preiss. i. 264; *P. hemignosta* and *P. hibbertioides,* Steud. l. c. i. 265.

**W. Australia.** King George's Sound, *Harvey, Oldfield;* ironstone gravel of the Darling Hills, *Drummond, 1st Coll., also 4th Coll. n.* 124; sandy places, Plantagenet district, and along places on the N. side of Mount Bakewell, *Preiss, n.* 2155, 2163, 2164, and 2172; and eastward to Phillips river, *Maxwell.*—I have been unable to find authentic specimens of the plant described by De Candolle in the Lambertian Herbarium, now dispersed.
In one specimen from the East River flats, Stokes' Inlet, *Maxwell,* the leaves are not so slender, very obtuse or recurved at the top, and grooved underneath by the slightly recurved margins, but the flowers are precisely the same.

5. **C. desmophylla,** *Benth.* Stems rigid, divaricately branched, glabrous, or the young ones loosely pubescent. Leaves densely clustered, linear, obtuse, mostly about ½ in. long, the margins closely revolute, rather dilated at the base, clothed with long, loose, spreading hairs, to about the middle, glabrous, smooth, and almost terete above. Flowers sessile in the clusters, much shorter than all except the innermost leaves, and immediately surrounded by a few imbricate membranous bracts, with brown tips, passing into similar but longer sepals, of which the innermost are 2½ lines long and scarious, without the brown tips. Petals obovate, obtuse. Stamens in 3 bundles of 3 or 4 each, and 2 single ones. Carpels 3, glabrous, 1-ovulate.

**W. Australia.** *Drummond;* Murchison river, *Oldfield.*

6. **C. helianthemoides,** *Turcz. in Bull. Mosc.* 1849, ii. 8. Stem

erect or procumbent, rather slender, and apparently half herbaceous, about 1 foot long, the branches clustered or dichotomous, the young ones as well as the leaves softly hairy.    Leaves usually clustered, linear or linear-lanceolate, obtuse, 4 to 8 lines long, the margins rather thick and revolute.    Flowers sessile within the clusters of leaves, the bracts at their base small, or none. Sepals oblong, obtuse, about 2½ lines long, membranous and coloured. Petals broadly 2-lobed, narrowed into a claw.    Stamens in 5 bundles, of which usually 3 have 3 or 4 each, and 2 have only 2 each.    Carpels 3, glabrous, 1-ovulate.

**W. Australia.**  *Drummond, 4th Coll. n.* 118.

7. **C. fasciculata,** *R. Br. in DC. Syst. Veg.* i. 424.    Stems procumbent, half herbaceous, loosely clothed as well as the leaves with silky or almost woolly hairs, which wear off with age.    Leaves clustered below the branches and about the flowers, distant on the branches, linear, obtuse, ½ to 1 in. long, or much shorter on the smaller branches, all with the margins revolute.    Flowers sessile in the clusters of leaves, which are all longer than them, except a few of the innermost.    Sepals membranous, about 3 lines long, slightly hairy, the outer ones acute, the inner ones less so.    Stamens in 5 bundles, usually of 3 each, without free inner ones.    Carpels 5, glabrous, 1-ovulate.—*Hibbertia depressa,* Steud. in Pl. Preiss. i. 268 ; *C. kochioides,* Turcz. in Bull. Mosc. 1849, ii. 7 (from the description given).

**W. Australia.**  King George's Sound, *R. Brown* and others ; in woody places, *Mylne ;* sandy hills near Albany, *Preiss, n.* 2153.

8. **C. Huegelii,** *Endl. in Hueg. Enum.* 2.    Branches stiff, but often elongated, glabrous and shining, or shortly villous about the floral leaves. Leaves narrow-linear, with the margins so closely revolute as to appear almost terete, acute, but frequently broken off at the ends so as to appear truncate, 1 to 2 in. long, or even more on vigorous shoots, the floral ones dilated and stem-clasping at the base.    Flowers nearly sessile in clusters of floral leaves, with small lanceolate acuminate bracts at their base.    Sepals fully ½ in. long, ovate-acuminate, usually pubescent outside.    Petals narrow-obovate, entire, or nearly so.    Stamens in 5 bundles of about 5 each, with one free one inside each bundle.    Carpels 5, or very rarely 4, glabrous, 1-ovulate.—*C. striata,* Steud. in Pl. Preiss. i. 275.

**W. Australia.**  Swan River, *Drummond* and others; in sandy places near Perth, *Preiss, n.* 2148; between Perth and King George's Sound, *Harvey.*—I have not seen Huegel's original specimen, but have no doubt of the identity of the species.

9. **C. pachyrrhiza,** *Benth.*    Nearly allied to *C. Huegelii,* and possibly a variety only, the stems are more erect, apparently arising from a thick rhizome, and more or less silky-hairy, as well as the leaves.    Leaves usually shorter and more obtuse, yet still exceeding 1 in. and nearly terete.    Flowers similar to those of *C. Huegelii,* but smaller, and with fewer stamens, there being usually only 2 or 3 to each bundle, and the inner free ones often deficient.—*Hibbertia pachyrrhiza,* Steud. in Pl. Preiss. i. 269 ; *H. basitricha,* Steud. l. c. 268.

**W. Australia.**  Swan River, *Drummond ;* between Perth and King George's Sound, *Harvey*  sandy and stony places, Darling Range, *Preiss, n.* 2149 *and* 2165.

10. **C. glaberrima,** *Steud. in Pl. Preiss.* i. 274. Apparently procumbent, much branched and somewhat glaucous, either quite glabrous or slightly pubescent on the smaller shoots. Leaves linear or linear-cuneate, obtuse with a small point, $\frac{1}{2}$ to 1 in. long, or rather more, suddenly enlarged at the base into a stem-clasping sheath 2 to 3 lines long, leaving a ring round the stem when they fall off. Pedicels included in the sheath, with 2 or 3 lanceolate bracts at their base. Sepals lanceolate, acute, 4 to 5 lines long, more distinctly united than in most species into a short tube at the base, quite glabrous, keeled, membranous on the edges. Petals narrow-obovate, entire. Stamens in 3 bundles of 2 or 3 each, and 2 single ones. Carpels 3, glabrous, 1-ovulate.—*C. subvaginata,* Steud. in Pl. Preiss. i. 275 ; *C. rupestris,* Steud. l. c. (sheaths of the floral leaves rather shorter).

**W. Australia.** Swan River, *Drummond;* sandy, shrubby, and woody places, ˙Perth district, *Preiss, n.* 2157 ; Hay district, *n.* 2160 ; and clefts of rocks of Darling Range, n. 2158.

11. **C. vaginata,** *Benth.* Stems numerous, erect from a thick rhizome, and but little branched, the whole plant glaucous and glabrous, except a slight pubescence on the flowering shoots. Lower leaves linear or linear-lanceolate, acute, 1 to 2 in. long, narrowed below the middle, and scarcely enlarged at the base, the floral ones very much enlarged and sheathing below, the upper ones reduced to broad loose acute sheaths of about $\frac{1}{2}$ in. Pedicels very short and included in the sheaths, bearing a few minute bracts, and a larger one under the flower. Sepals glabrous, ovate or ovate-lanceolate, about 3 lines long. Petals obovate, etuse. Stamens in 3 or rarely 2 bundles of 2 or 3 each, and 2 or rarely 3 single ones. Carpels 3, glabrous, 1-ovulate.

**W. Australia.** Swan River, *Drummond.*

12. **C. Preissiana,** *Steud. in Pl. Preiss.* i. 274. Much branched, and more or less glaucous and glabrous, or with a slight down or woolly hairs at the base of the floral leaves. Leaves linear-oblong or linear-cuneate, obtuse with a short point, or more frequently truncate or 3-toothed, $\frac{3}{4}$ to $1\frac{1}{2}$ in. long, and mostly $1\frac{1}{2}$ to 2 lines broad, rather thick, flat, narrowed below the middle, but mostly, especially the floral ones, again dilated and stem-clasping at the base, leaving a prominent ring. Flowers irregularly clustered in the upper axils, on pedicels of 2 to 5 lines. Sepals 3 to 4 lines long, thin and yellow especially on the edges, the outer ones acute, the inner obtuse and petal-like. Petals narrow-obovate, slightly notched. Stamens in 3 or 2 bundles of about 3 each, and 2 or 3 single ones. Carpels 3, glabrous, 1-ovulate.

**W. Australia.** *Burges;* maritime rocks, Perth district, *Preiss, n.* 2159 *b* ; Port Gregory, *Oldfield.* This may probably prove to be a variety of *C. pedunculata.*

13. **C. pedunculata,** *R. Br. in DC. Syst. Veg.* i. 424. Stems usually rather weak, branching, erect or ascending from a thick rhizome to about a foot, but sometimes more rigid with short branches ; glabrous, except a few hairs about the floral leaves. Leaves linear or linear-cuneate, obtuse, truncate or emarginate, $\frac{1}{2}$ to $1\frac{1}{2}$ in. long, the margins recurved, narrowed below, with a broader stem-clasping or sheathing base, leaving a raised ring round the stem, glabrous and in the larger specimens somewhat glaucous. Peduncles usually clustered with small leaves in the upper axils, slender, $\frac{1}{4}$ to $\frac{1}{2}$ in.

long, forming a kind of leafy raceme. Sepals about 2 lines long, obtuse, or the outer ones acute, glabrous, membranous on the edge. Petals clawed, obovate-oblong, entire. Stamens in 3 or 4 bundles of 3 or 4, with 2 or 1 single. Carpels 3 or 4, glabrous, 1-ovulate.—*C. racemosa*, Endl. in Hueg. Enum. 2; *C. tridentata*, Turcz. in Bull. Mosc. 1849, ii. 140; *C. assimilis*, Steud. in Pl. Preiss. i. 273; *C. parviflora*, Steud. l. c. i. 276; *Hibbertia subexcisa*, Steud. in Pl. Preiss. i. 269.

**W. Australia.** King George's Sound, *R. Brown* and others; Swan River, *Drummond, 5th Coll. n. 288, Oldfield;* sands near Perth, *Preiss, n.* 2133 *b,* 2146, and 2150; and northwards to Murchison river, *Oldfield.*

14. **C. exasperata,** *Steud. in Pl. Preiss.* i. 276. Rigid, much branched and glabrous. Leaves narrow-linear, thick and rigid, pointed and almost pungent, about ½ in. long, slightly hoary or scabrous, but glabrous, the recurved margins slightly indicated by two striæ underneath. Peduncles 1 to 2 lines long, erect, with small bracts at their base, and a large sepal-like one under the calyx. Sepals broad, obtuse, stiff, and dry, the inner ones nearly 4 lines, the outer shorter and often slightly hoary on the bud. Petals obovate, rather narrow, notched. Stamens scarcely united above the middle in 5 bundles of 3 or sometimes 2 each, without single ones. Carpels 5, glabrous, 2-ovulate.—*Hibbertia squamosa*, Turcz. in Bull. Mosc. 1849, ii. 9.

**W. Australia.** Swan River, *Drummond, 4th Coll. n.* 122; *Roe ;* gravelly places, Quanyen Plains, Victoria district, *Preiss, n.* 2175. The foliage is nearly that of *Hibbertia mucronata.*

15. **C. uncinata,** *Benth.* Rigid, much branched and glabrous. Leaves narrow-linear, rigid, recurved upwards and obtuse, or with a minute reflexed point, 2 to 4 lines long, the margins closely revolute, smooth or marked with slight asperities. Pedicels 1 to 3 lines long, with a few narrow pointed bracts at their base, but none under the flower. Sepals broad, concave, very obtuse, glabrous, about 2 lines long. Petals broadly obovate, retuse. Stamens in 5 bundles of usually 3 each, without any free ones. Carpels 5, glabrous, 2-ovulate.

**W. Australia.** *Drummond.* The foliage resembles that of *Hibbertia recurvifolia* and *H. rostellata.*

*C. cygnorum*, Steud. in Pl. Preiss. i. 275, is unknown to me. It is described as having leaf-opposed peduncles, bracteate in the middle, which is so unlike the inflorescence of any *Dilleniacea*, that I cannot but suspect it is some very different plant incorrectly described.

## 4. ADRASTÆA, DC.

Sepals 5. Petals 5. Stamens 10, or occasionally fewer, in a single series, filaments dilated and regularly cohering in a short tube round the pistil. Carpels and fruit of *Hibbertia.*

The genus consists of only one species, with the habit of a *Hibbertia* or *Candollea.*

1. **A. salicifolia,** *DC. Syst. Veg.* i. 424. Branches rather slender, apparently erect, the young ones silky-hairy. Leaves linear or linear-oblong, mostly with a minute fine point, ¾ to 1½ in. long, often bordered by a few remote and minute callous teeth, glabrous above when old, more or less silky underneath. Flowers small, sessile in clusters of small leaves in the older axils. Sepals lanceolate, very acute, nearly 3 lines long. Petals scarcely

longer, obovate-oblong, obtuse.   Anthers oblong, longer than the filaments.
Carpels 2, glabrous, 1-ovulate.—*Hibbertia salicifolia*, F. Muell. Fragm. i. 161.

**Queensland.**   Freshwater swamps and rushy peat bogs about Moreton Bay and
Moreton and Peel's Islands, *A. Cunningham, M'Gillivray, F. Mueller.*

**N. S. Wales.**   Port Jackson, *R. Brown ;* margins of bogs, *A. Cunningham.*

## 5. PACHYNEMA, R. Br.

### (Huttia, *Drumm. and Harv.*)

Sepals 5.   Petals 5, rarely reduced to 4 or 3.   Stamens usually 10, outer
ones in a single series all round the carpels, either all perfect, or 2 or 3 of
them reduced to small staminodia ; filaments either thickened and ovoid, or
flat, short, and broad ; anthers erect ; two inner staminodia alternating with
the carpels, and similar to the perfect stamens, except that the anthers are
small and empty or wanting.   Carpels 2, 2-ovulate.   Styles and fruit of
*Hibbertia.*—Perennial herbs or undershrubs, with erect, branching, rush-like
or flattened stems, apparently leafless, the leaves being all reduced to minute
scales, except sometimes a few at the base of the stem.   Flowers small, on
very short recurved lateral peduncles.   Bracts minute.

A small genus, entirely Australian.   The three species of one section all tropical, the
fourth western.

SECT. 1. **Huttia.**—*Filaments flat, very short.   Anthers long.*   1. *P. conspicuum.*

SECT. 2. **Pachynema.**—*Filaments thick, ovoid.   Anthers small, the cells somewhat
diverging.*

Stem and branches terete and rush-like . . . . . . . . . . 2. *P. junceum.*
Stem and branches flat.
  Branches 1 to 2 lines broad, not glaucous . . . . . . . . 3. *P. complanatum.*
  Branches ¼ to ½ in. broad or more, very glaucous . . . . . 4. *P. dilatatum.*

SECTION I. HUTTIA.—Filaments flat and very short.   Anthers long.—
*Huttia* (genus), Drumm. and Harv.

1. **P. conspicuum,** *Benth.*   Stems erect, from a thick rhizome, 1 to 1½
ft. high, branching, terete and rush-like, glabrous or slightly hirsute at the
base.   Leaves few and small at the base of the stem, narrow and mostly 3-
lobed, the upper ones all reduced to minute distant scales.   Peduncles few
towards the top of the branches, 2 to 4 lines long, rather thick and recurved,
each bearing 1 flower, much larger than in the other species.   Sepals fully 4
lines long, the outer ones lanceolate and acute, the inner broader, more obtuse
and membranous on one side.   Petals obovate or orbicular, entire.   Stamens
of the outer row usually 7 only, the anthers oblong-linear, with the cells opening laterally, the three others reduced to minute staminodia ; the 2 inner
staminodia like the perfect stamens, except that the anthers are lanceolate
and petal-like, their cells empty with the inner valve smaller than the outer
one.—*Huttia conspicua*, Drumm. and Harv. in Hook. Kew Journ. vii. 51.

**W. Australia.**   Between Moore and Murchison rivers, *Drummond ;* Murchison river,
*Oldfield.*

SECTION II. PACHYNEMA.—Filaments ovoid, tapering at the top, with
short terminal anthers.

2. **P. junceum,** *Benth.*   Stems erect, branching, 1 to 1½ ft. high,

terete and rush-like, or very slightly compressed, but scarcely angular, finely striate. Leaves all reduced to minute distant scales. Peduncles usually solitary, slender, recurved, 1 to 3 lines long, or terminating the branches. Sepals orbicular, about 2 lines long, the outer ones rather smaller. Petals obovate-orbicular, entire, about the same size as the sepals. Stamens of the outer row usually 7 or 8, perfect, the filaments thick, fleshy and ovoid at the base, tapering at the top, where they bear 2 small innate diverging cells, the 3 or 2 other outer stamens reduced to minute staminodia, the 2 inner staminodia like the perfect stamens, but without anthers. Carpels 2, glabrous, tapering into pointed styles so as very much to resemble the stamens in shape. Ovules 2 in each ovary.

**N. Australia.** N. coast, *R. Brown ;* Victoria river, *Bynoe.*

3. **P. complanatum,** *R. Br. in DC. Syst. Veg.* i. 412. Erect, leafless and glabrous, like the last species, and the lower part of the stem at length terete, but the branches are all flattened with thin edges, more or less thickened in the middle, and seldom above 2 lines broad. Scales minute and distant. Peduncles exceedingly short, usually several together in a little cluster or short raceme. Flowers as in *P. junceum.* In the one I opened there were 8 perfect outer stamens, and I could not find the 2 minute abortive ones to complete the ring. The inner staminodia and carpels precisely as in *P. junceum.*—Deless. Ic. Sel. i. t. 73.

**N. Australia.** N. coast, *R. Brown ;* Melville Island, *Fraser ;* Port Essington, *A. Cunningham, Leichhardt.*

4. **P. dilatatum,** *Benth.* Allied to *P. complanatum,* but apparently taller and more robust, of a very glaucous hue, and the branches thick and angular, dilated upwards to the breadth of from ½ to 1 in., and 2 to 3 lines broad even on the smallest branches. Peduncles on the edges of the branches or in the forks. Flowers as in the last two species. In one of those I examined I found all 10 of the outer stamens perfect.

**N. Australia.** Macadam range, *F. Mueller.*

ORDER III. **MAGNOLIACEÆ.**

Sepals and petals several, imbricate, and often passing gradually from the one to the other, deciduous; or in the Australian genus the calyx exceptionally 2- or 3-cleft. Stamens indefinite, hypogynous; filaments often thickened or dilated, anthers adnate. Carpels indefinite, rarely solitary, free or partially cohering. Ovules 2 or more, attached to the inner angle of the cavity, or rarely ascending from the base. Stigma sessile. Ripe carpels opening in 2 valves or indehiscent. Seeds with a crustaceous testa, often succulent externally; albumen copious, oily. Embryo minute, near the hilum, with divaricate cotyledons.—Trees or shrubs, often aromatic. Leaves alternate, undivided, reticulately penninerved, entire or toothed, with or without stipules. Flowers axillary or terminal, solitary or fasciculate, often large.

An Order chiefly distributed over tropical and eastern temperate Asia and North America, and only represented by one somewhat anomalous genus in the southern hemisphere.

# 1. DRIMYS, Forst.

### (Tasmannia, *R. Br.*)

Sepals 2 or 3, membranous, united in the bud in a globular calyx, irregularly split or separating when open. Petals usually few. Filaments thick, the anther-cells parallel or divergent. Carpels various in number, mostly solitary in the Australian species, containing several ovules. Berries indehiscent.—Glabrous and aromatic trees or shrubs. Leaves marked with pellucid dots. Peduncles (in the Australian species 1-flowered) arising from the axils of deciduous scales at the base of the new shoots, but as these shoots are rarely developed till the fruit has ripened, the flowers appear to be in terminal umbels with a central bud. Flowers of a greenish-yellow or white, or in some species (not Australian) pink.

Besides the two Australian species, there are one in New Zealand, one or more in New Caledonia, one in Borneo, and one in South America.

Leaves tapering into a short petiole. Berries small, globular . . . 1. *D. aromatica.*
Leaves narrowed below, but obtuse or 2-auriculate at the very base.
Berries ovoid, about ½ in. long . . . . . . . . . . . . 2. *D. dipetala.*

1. **D. aromatica,** *F. Muell. Pl. Vict.* i. 20. A bushy shrub or small tree, rarely attaining the height of 30 ft., and very dwarf in alpine stations. Leaves from elliptic-oblong and scarcely 1 in. long in alpine forms, to oblong-lanceolate, and fully 3 in. long in luxuriant specimens, obtuse or acute, always tapering at the base into a short petiole. Flowers polygamous, apparently in terminal umbels, on pedicels rarely exceeding ½ in., the scaly bracts very small. Sepals usually 2, 1½ to 2 lines long. Petals 2 to 8, nearly twice as long. Carpels solitary, or rarely 2 or 3. Stigma linear, terminal at first, but soon becoming lateral by the unequal growth of the carpel. Berries globular, about the size of a pea.—*Tasmannia aromatica*, R. Br. in DC. Syst. Veg. i. 445; Deless. Ic. Sel. i. t. 84; Bot. Reg. 1845, t. 43; Hook. f. Fl. Tasm. i. 11.

**Victoria.** Humid forest-ranges from Mount Disappointment and the Dandenong mountains to the Australian Alps, ascending to at least 5000 ft., *F. Mueller.*

**Tasmania.** *R. Brown;* abundant in many parts of the island, from the level of the sea to the height of 4000 ft. on the mountains, *J. D. Hooker.*

2. **D. dipetala,** *F. Muell. Pl. Vict.* i. 21. A tall shrub. Leaves oblong-lanceolate or rarely oval-oblong, acute or acuminate, usually 3 to 5 in. long, narrowed towards the base, but all (except sometimes a few of the smaller leaves of lateral shoots) abruptly obtuse or minutely biauriculate at the very base, on an exceedingly short broad petiole, or almost sessile. Peduncles longer than in *D. aromatica*, and flowers rather larger. Sepals and petals usually 2 each. Carpels often 2 or 3, but one only usually enlarges. Stigma short or linear, more or less unilateral. Berry ovoid, fully ½ in. long, and more succulent than in *D. aromatica.*—*Tasmannia insipida*, R. Br. in DC. Syst. Veg. i. 445; *T. dipetala*, R. Br. ms. ex DC. Prod. i. 78; *T. monticola*, A. Rich. Sert. Astrolab. 50, t. 19.

**N. S. Wales.** Port Jackson, *Brown;* and in the interior, extending northward to Mount Lindsay, *W. Hill;* and Clarence and Hastings rivers, *Beckler;* southward to Illawarra, *A. Cunningham, Macarthur,* who gives it as the *Pepper shrub* of the colonists.

# Order IV. ANONACEÆ.

Sepals usually 3, distinct, or more or less united in a 3-lobed or 3-toothed calyx (in *Eupomatia* united in one mass with the petals). Petals usually 6, hypogynous, in 2 rows, 3 outer ones alternating with the sepals, 3 inner ones alternating with the outer, sometimes all united in a ring at the base, those of each row valvate or imbricate in the bud. Stamens indefinite, usually very numerous, closely packed on the thickened torus, round or under the carpels, linear or wedge-shaped, with 2 adnate anther-cells on the back or edges, often concealed by the more or less dilated summit of the connectivum. Gynœcium of several, often very many carpels, distinct (except in *Eupomatia*), closely packed on the centre of the torus, terminating each in a capitate stigma, or in a thick oblong or rarely more slender style, stigmatic on the top or inner side. Ovules in each carpel either 1 or 2, ascending from the base , or 2 or more attached to the inner angle of the cavity, anatropous. Fruit either of several distinct carpels sessile or stalked, indehiscent and fleshy or pulpy, sometimes opening along the inner edge, or the carpels more or less united in a single mass. Seeds with or without an arillus. Albumen copious, always ruminate. Embryo very small, near the hilum.—Trees, shrubs, or woody climbers. Leaves alternate, simple, and quite entire, without stipules. Fowers sessile, or on 1-flowered pedicels, solitary, or few together, terminal, lateral, or axillary, usually of a greenish-yellow or purple colour.

A large Order, widely distributed over the New World as well as the Old, but chiefly confined to the tropics. Of the 6 Australian genera, 5 are more numerously represented in tropical Asia or Africa, the sixth is endemic. None are American.

Petals 6, nearly equal.
  Petals spreading.
    Petals broad, imbricate in the bud. Ovules or seeds several in
      each carpel . . . . . . . . . . . . . . . . 1. Uvaria.
    Petals narrow, valvate in the very young bud, but soon spreading.
      Ovules 1 or 2, erect in each carpel . . . . . . . . . 2. Polyalthia.
    Petals concave, not spreading, valvate.
      Ovules 1 in each carpel, erect. (Flowers 3 to 4 lines diameter) . 3. Popowia.
      Ovules several in each carpel. (Flowers about 6 lines diameter) 4. Melodorum.
Petals, 3 outer like the sepals, 3 inner large, erect, very concave . . 5. Saccopetalum.
Petals and sepals united in a conical mass, which falls off entire . . 6. Eupomatia.

## 1. UVARIA, Linn.

Sepals broad. Petals 6, imbricate in the bud in each row, spreading. Stamens numerous and closely packed, rather flat, the connective produced into a shortly ovoid, or truncate appendage, concealing the cells in the normal species. Receptacle slightly raised. Carpels numerous, with a short truncate style, and several ovules in 2 rows along the inner angle. Berries distinct, sessile, or stalked, usually with several seeds.—Stems climbing or trailing. Flowers usually rather large, leaf-opposed or axillary.

A considerable genus, chiefly Asiatic, with a few African species. The following Australian ones are both endemic, and one of them a doubtful congener.

Petals all broad. Anthers dilated at the top, concealing the lateral cells 1. *U. membranacea.*

Inner petals narrow.  Anthers shortly dilated at the top, showing the
dorsal parallel cells . . . . . . . . . . . . . . .  2. *U. heteropetala.*

1. **U. membranacea,** *Benth.*  A long woody trailer, quite glabrous,
except a slight tomentum on the petioles and buds.  Leaves on short stalks,
oval-oblong, obtuse, or with a very short, broad point, 5 to 6 in. long, 3 to
3½ in. broad, oblique, and somewhat cordate at the base, thin and membranous,
with distant primary veins branching into the reticulate smaller venation.
Flowers large, solitary, on peduncles of about ½ in.  Petals obovate, very
obtuse, fully 1 in. long, narrowed, and slightly united at the base.  Connective
truncate and dilated above the anther-cells.  Carpels very numerous, but not
seen in fruit.

**N. Australia.**  Scrub at Cape York, *M'Gillivray.*

2. **U. (?) heteropetala,** *F. Muell. Fragm.* iii. 1.  A scrubby shrub of 8
to 10 ft., the young branches densely pubescent.  Leaves on very short
petioles, broadly ovate, obtuse, or shortly acuminate, 2 to 4 in. long, not
coriaceous, glabrous above, loosely pubescent underneath.  Flowers dark
purple, solitary, on very short recurved terminal or lateral pedicels.  Sepals
ovate-lanceolate, villous, 3 to 4 lines long.  Petals imbricate in each series,
the outer ones broadly ovate, attaining at least 7 lines, and probably longer
when full grown, silky-villous outside, glabrous inside, the inner ones narrower and perhaps longer.  Stamens numerous, the short triangular terminal
appendage not dilated, showing the rather large dorsal parallel cells.  Carpels
numerous, densely hirsute ; stigma small.  Ovules 6 to 8 in each carpel, in
2 series.  Fruit unknown.

**Queensland.**  Port Denison, *Fitzalan.*  This plant differs from *Uvaria* in the stamens,
which are those of *Saccopetalum.*  The habit and foliage are also more those of the latter
genus than of *Uvaria,* but the petals certainly appear to be imbricate in each row, and the
outer ones are much more developed than is usual in *Saccopetalum.*  The flowers in the
specimens seen are however still young, and insufficient for fixing the precise affinities of
the species.

### 2. POLYALTHIA, Blume.

Sepals broad.  Petals 6, valvate in the very young bud, in two rows, but
spreading or open long before they have attained their full size, nearly equal
and flat, usually narrow.  Stamens numerous, narrow-wedge-shaped, the connective flattened at the top, concealing the cells.  Torus slightly raised.
Carpels several, with a short, oblong, or capitate style, and 1 or 2 erect ovules.
Berries stalked, globular or ovoid.—Trees or shrubs.  Flowers solitary or
clustered, axillary or leaf-opposed.

A considerable genus, chiefly Asiatic, with one African species.  The following Australian
one extends to New Caledonia.

1. **P. nitidissima,** *Benth.*  A tree of 15 to 50 or 60 ft., glabrous in
all its parts.  Leaves elliptical, or the upper ones almost lanceolate, obtuse
or obtusely acuminate, 2 to 3 in. long, narrowed into a petiole varying from
2 to 5 lines, smooth and shining, the veins fine and reticulate, but not
numerous.  Peduncles solitary, axillary, 3 to 6 lines long, or more when in
fruit, with 2 or 3 small bracts near the base.  Sepals short and broad.  Petals
linear, rather thick, 5 or 6 lines long when fully out, but spreading very early.

E 2

Stamens very short, and closely packed. Carpels 10 to 20 in the flower, much fewer in the fruit, and then globular or shortly ovoid, 1-seeded, shortly stalked.—*Unona nitidissima*, Dun. Anon. 109, t. 23; *Unona fulgens*, Labill. Sert. Austr. Caled. 57, t. 56; *Unona nitens*, F. Muell. Fragm. iii. 2.

**Queensland.** In brushes on islands in Moreton Bay, *A. Cunningham;* Port Denison, *Fitzalan.* Also found in New Caledonia.

In some specimens the torus, after flowering, becomes thick and woody, enclosing several cavities, probably a deformity occasioned by the puncture of some insect. Labillardière describes and figures the carpels as having several ovules, but this is a mistake; his own specimens, quite similar to the Australian ones, have but one erect ovule in each.


### 3. POPOWIA, Endl.

Sepals ovate. Petals 6, valvate in the bud in 2 rows, short, broad, concave, those of the 2 rows nearly equal, but the outer ones rather more open. Stamens numerous, closely packed, wedge-shaped, the connective flattened at the top, concealing the cells. Torus but little raised. Carpels indefinite (sometimes few), with a short obovate or capitate style and 1 or 2 erect ovules. Berries stalked, globular or ovoid.—Trees or shrubs. Flowers small, axillary or leaf-opposed, on short pedicels.

A small genus, scattered over tropical Africa and Asia, with one species endemic in Australia. As a genus it is scarcely sufficiently distinct from *Polyalthia.*

1. **P. australis,** *Benth.* Probably a shrub. Leaves ovate-lanceolate or oblong, 3 to 5 in. long, obtuse, rounded at the base with a very short broad petiole, glabrous on both sides, the primary veins prominent underneath. Pedicels solitary or 2 or 3 together in the axils of the older leaves, longer than in most species of the genus, attaining near 1 in. Expanded flowers 3 or 4 lines diameter. Petals broadly ovate, rather thick, pubescent and strictly valvate in each row. Carpels numerous, hairy. Ovule solitary, erect.

**N. Australia.** Barrow Bay, Port Essington, *Armstrong.*


### 4. MELODORUM, Dun.

Sepals small, united at the base. Petals 6, valvate in the bud in 2 rows, the outer ones broad, thick, concave, connivent or scarcely open, the inner ones smaller. Stamens numerous, the connective ovate or truncate, concealing the cells. Torus convex or conical. Carpels several, with an oblong thick style and 2 or more ovules in each, attached to the inner angle. Berries distinct, sessile or stalked.—Stems woody, usually climbing. Primary veins of the leaves prominent underneath. Flowers terminal or leaf-opposed.

The genus comprises several species dispersed over tropical Asia and the Indian Archipelago, the Australian one endemic.

1. **M. Leichhardtii,** *Benth.* A shrub or tree, with flexuose (or somewhat climbing?) branches, the younger ones slightly rusty-tomentose. Leaves much like those of *M. elegans*, Hook. f. and Thoms., but with very much shorter petioles, oblong, obtuse or obtusely acuminate, about 3 in. long, coriaceous, glabrous and shining, sprinkled on the under side with a few minute, almost microscopic, fringed scales or stellate hairs, the veins much less prominent than in most

species.    Peduncles ¼ to ¾ in. long, rusty-tomentose.    Flowers nearly ½ in.
in diameter.    Sepals 3 lines long, spreading.    Outer petals about 6 lines,
slightly tomentose, very obtuse, concave and connivent, inner ones thicker
and rather shorter.    Stamens very.numerous.    Berries stipitate, either depressed-globose, 4 or 5 lines diameter and 1-seeded, or somewhat oblong,
2-seeded with a slight transverse furrow between the seeds, or moniliform,
consisting of 2 depressed-globose 1-seeded or oblong 2-seeded portions.—
*Unona Leichhardtii,* F. Muell. Fragm. iii. 41.

**Queensland.**  Wide-Bay, *Bidwill;* Mount Torampa and woods at M‘Connell's Brush,
*Leichhardt;* near Ipswich, *J. Vernet;* Rockhampton, *Thozet;* Brisbane river, *A. Cunningham, F. Mueller.*

**N. S. Wales.**  Clarence river, *Beckler.*

## 5. SACCOPETALUM, Benth.

Sepals small.  Petals 6, valvate in 2 rows, the outer ones small and resembling
the sepals, the inner large, erect, and very concave    Stamens numerous but
loosely imbricate, showing the anther-cells on their back just below the short
tips.    Torus nearly globular.    Carpels several, with an ovoid or oblong thick
style, and 6 or more ovules in each attached to the inner angle.    Berries
globular.—Trees or shrubs, with deciduous leaves.    Flowers usually appearing on the young shoots before or with the young leaves.

A small genus, dispersed over India and the Archipelago; the Australian species endemic.

1. **S. Bidwilli,** *Benth.*  Apparently a shrub, with rather weak branches,
densely hirsute with short rusty hairs.    Leaves very shortly stalked, oblong
or obovate-oblong, obtuse or very shortly acuminate, 3 to 4 in. long, rounded
at the base, glabrous above, hairy underneath.    Flowers lateral, solitary or 2
together, on very short pedicels.    Sepals thin, lanceolate, hairy, about 2 lines
long.    Outer petals similar, but twice as long.    Inner petals when fully developed 1½ in. long, not saccate at the base only, as in most other species of
the genus, but hollowed into a broad boat-shape all the way up, with the
upper end turned inwards, thin, and very hairy both inside and out.    Stamens
numerous, the anther-cells contiguous and conspicuous, terminated by the
small flat tip of the connectivum.    Carpels very hairy in the flower, when
ripe nearly sessile, oblong, 6 to 8 lines long, thick and hard, covered with
rusty hairs, containing 3 to 6 flattened seeds.

**Queensland.**  Wide Bay, *Bidwill.*

## 6. EUPOMATIA, R. Br.

Sepals and petals completely consolidated into one mass, the upper part
falling off in a conical lid, leaving the lower campanulate tube (or enlarged
peduncle) filled with the thick flat-topped torus.    Stamens inserted on the
margin of the torus, the inner ones in many rows, converted into petal-like
obovate staminodia, the outer ones in fewer rows, perfect, linear-lanceolate,
curved, with acuminate tips and longitudinal dorsal anther-cells.    Carpels
many, immersed in the torus, appearing like the cells of a single inferior
ovary, the stigmas adnate on the flat areolate surface; ovules several in each
carpel or cell.    Fruit several-celled, formed of the enlarged perianth-tube
more or less enclosing the carpels, becoming turbinate or urceolate and suc-

culent. Seeds 1 or 2 in each cell, irregularly angular; albumen ruminate, and embryo precisely as in the more normal *Anonaceæ.*—Shrubs or under-shrubs, quite glabrous. Leaves alternate, entire, shortly petiolate. Peduncles short, 1-flowered, terminal or lateral.

The genus is confined to Australia.

Petioles shortly decurrent. Flowers terminal. Outer staminodia spreading and longer than the stamens. Fruit turbinate . . . . . . 1. *E. Bennettii.*
Petioles not decurrent. Flowers lateral. Staminodia all connivent,
shorter than the stamens. Fruit urceolate . . . . . . . . . 2. *E. laurina.*

1. **E. Bennettii,** *F. Muell. Fragm.* i. 45. A shrub or undershrub, 1 to 2 ft. high and quite glabrous. Leaves oblong-lanceolate, acuminate or acute, 3 to 5 in. long, narrowed at the base into a short petiole, which is again en-larged at the base and shortly decurrent on the stem, leaving oblique raised lines when they fall off. Flowers solitary, terminal, on a short peduncle above the last leaf, when fully expanded rather more than 1 in. diameter. Petal-like staminodia very numerous, yellow, the outer ones stained with orange or blood-red, beset with stipitate glands and bordered with stellate hairs spreading and completely concealing the perfect stamens, which are re-flexed on the peduncle, the inner staminodia shorter and connivent. Fruit turbi-nate, about $\frac{3}{4}$ in. diameter, the pericarp thin, the top convex, with the tips of the carpels distinctly prominent, the base of the perianth scarcely projecting as a slight ring round the edge.—*E. laurina,* Hook. in Bot. Mag. t. 4848.

**Queensland.** Brisbane river, *Herb. Mueller.*

2. **E. laurina,** *R. Br. in Flind. Voy.* ii. 597, *t.* 2. An erect gla-brous shrub with weak branches. Leaves evergreen, oblong or almost el-liptical, shortly acuminate, 3, 4, or sometimes 5 in. long, narrowed into a short petiole which is not decurrent on the branch. Flowers solitary, on short lateral or nearly axillary peduncles, the buds at first oblong, becoming nearly globular and about $\frac{1}{2}$ in. diameter before opening; when the bud has fallen the stamens expand to about 1 in. diameter. Petal-like staminodia connivent or the outer ones scarcely open, glabrous or with a very few stipi-tate glands; perfect stamens longer, erect or spreading, the linear anthers tipped by a short fine point, the filaments dilated. Fruit urceolate-globular, nearly $\frac{3}{4}$ in. diameter, the persistent base of the perianth forming a narrow rim projecting above the nearly flat top.—*F. Muell. Fragm.* i. 45.

**Queensland.** Brisbane river, *F. Mueller;* Pine river, *Fitzalan.*
**N. S. Wales.** Woods and thickets in the colony of Port Jackson, especially in the mountainous districts, and on the banks of the principal rivers, *R. Brown,* and apparently along the whole coast from Clarence river, *Beckler,* to Twofold Bay, *F. Mueller.*

## Order V. MENISPERMACEÆ.

Flowers diœcious. Sepals usually 6 in 2 series, rarely 9 or 12 in 3 or 4 series, or very rarely 5 or fewer, imbricate or very rarely valvate in each series, the inner ones the largest. Petals usually 6, smaller than the sepals (except in *Sarcopetalum*), nearly equal but imbricate in 2 series in the bud, rarely fewer or none. Male fl.: Stamens usually 6, free and opposite the petals, or united in a central column, rarely 9 or more or only 3. Female fl.: Staminodia usually 6, free. Carpels distinct, usually 3, sometimes 6 or more or only 1, containing 1 or very rarely 2 amphitropous ovules peltately attached to the inner angle. Style terminal, usually recurved, and often expanding into a short sessile stigma. Fruit-carpels drupaceous, nearly straight, or more frequently curved, so that the remains of the style are near the base, the putamen then becoming more or less horseshoe-shaped, with an inner projection of the endocarp bearing the placentæ. Seed taking the shape of the cavity, with a thin membranous testa. Albumen sometimes fleshy, entire or ruminate, sometimes thin or none. Embryo nearly as long as the albumen or occupying the whole seed, the radicle pointing to the remains of the style.—Climbers, usually woody, or in a very few non-Australian species erect herbs or shrubs. Leaves alternate, without stipules, entire or rarely palmately lobed, usually with 3 or more palmate ribs at the base. Flowers small, in axillary panicles, racemes, or cymes.

A considerable tropical Order, both in the New and the Old World, a very few species extending into more temperate regions in North America, eastern Asia, or southern Africa. Of the 7 Australian genera 3 are endemic, the others Asiatic or African.

Sepals imbricate. Petals 6. Stamens 6, free. Carpels 3.
  Flowers in simple racemes.
    Inner sepals broad and thin. Carpels of the fruit ovoid, the
      style at the top. Seed albuminous, nearly straight . . . . 1. TINOSPORA.
    Inner sepals narrow-ovate. Carpels of the fruit broad, the style
      near the base. Seed without albumen . . . . . . . 5. PACHYGONE.
    Flowers in much-branched cymes. Carpels of the fruit broad, the
      style near the base. Seed albuminous . . . . . . . . 2. PERICAMPYLUS.
Sepals imbricate or open. Petals usually 3 to 5. Stamens united in
  a central column. Carpels broad, the style near the base. Seed
  albuminous.
    Sepals very small. Petals thick and fleshy, almost globular. An-
      thers 2 or 3. Carpels 3 to 6. Flowers racemose . . . . . 3. SARCOPETALUM.
    Petals smaller than the sepals, concave. Anthers 4 or 5. Carpels
      solitary. Flowers umbellate . . . . . . . . . . . 4. STEPHANIA.
Inner sepals valvate. Petals 6. Stamens 3. Carpels about 6, when
  in fruit broad, the style near the base. No albumen . . . . . 6. PLEOGYNE.
Petals imbricate. Petals 3. Stamens 9 to 12. Carpels 3, 2-ovulate 7. ADELIOPSIS.

## 1. TINOSPORA, Miers.

Sepals 6, in 2 series, the inner ones large. Petals 6, smaller than the sepals, nearly flat. Male fl.: Stamens 6, free, thickened towards the top, the anther-cells lateral. Female fl.: Staminodia 6. Carpels 3, stigmas jagged. Drupes ovoid, the remains of the style nearly terminal. Putamen slightly concave on the inner face, the internal projection hemispherical and hollow, forming an empty cell. Seed disk-shaped, albuminous. Cotyledons ovate, spreading laterally.—Leaves cordate or truncate at the base. Flowers usually clustered in long simple racemes.

A small genus, chiefly Asiatic, but extending also to tropical Africa. The Australian species endemic.

Leaves ovate-cordate, entire. . . . . . . . . . . . . . . 1. *T. smilacina.*
Leaves broad, obtusely 3-lobed, much veined . . . . . . . . 2. *T. Walcottii.*

1. **T. smilacina**, *Benth. in Journ. Linn. Soc. v. Suppl. 52.* A glabrous

twiner, the branches somewhat succulent. Leaves ovate, deeply and broadly cordate at the base, or almost hastate with rounded auricles, obtuse or scarcely acuminate, 3 or 4 in. long, 5-nerved, the smaller pinnate veins scarcely prominent, on petioles of about 1 in. Flowers green, the male racemes 2 or 3 in., the females about 1 in. long; pedicels about 1 line. Sepals, 3 outer ones very small and triangular, 3 inner ones about 1 line long, ovate, thin, spreading. Petals about half as long as the inner sepals, obovate. Anthers terminal, ovoid, almost globular, the cells almost parallel. Drupes oblong, about 3 lines long.

**N. Australia.** Islands of the Gulf of Carpentaria, *R. Brown;* common in many parts of Arnhem's Land and thence to the Burdekin, *F. Mueller.*—Nearly allied to the Asiatic *T. crispa,* but the leaves are rather differently shaped and the fruits much smaller.

2. **T. Walcottii,** *F. Muell. Herb.* Of this I have only seen fragments of a fruiting specimen with the drupes not quite ripe, but sufficiently so to show the peculiar form of *Tinospora,* with the somewhat succulent branches and with the racemes of *T. smilacina,* but the leaves appear to be as broad as long, obscurely 3-lobed, cuneate and not cordate at the base, of a thinly coriaceous texture, with prominent reticulate veins.

**N. Australia.** Nichol Bay, *Walcott.*


## 2. PERICAMPYLUS, Miers.

Sepals 6 in 2 series, the inner ones larger. Petals 6, smaller than the sepals, the edges embracing the stamens. Male fl. : Stamens 6, free, the anther-cells lateral. Female fl. : Staminodia 6. Carpels 3, the styles 2-cleft. Drupes globular, somewhat flattened, the remains of the style near the base. Putamen horseshoe-shaped, crested on the back, the sides concave. Seed horseshoe-shaped. Embryo in the axis of the albumen, with narrow cotyledons closed against each other.—Leaves broad. Cymes dichotomously branched.

The genus is limited to the following species.

1. **P. incanus,** *Miers; Hook. and Thoms. Fl. Ind.* i. 194. Achenium with the younger branches shortly tomentose or at length glabrous. Leaves nearly orbicular, sometimes slightly peltate, 2 to 4 in. or sometimes above 5 in. diameter, glabrous above, usually hoary underneath, on petioles of 1 to 2 in. Flowers very small, in axillary dichotomous cymes, shorter than the leaves. Sepals hairy on the back. Drupes red.—*Cocculus Moorei,* F. Muell. Fragm. i. 162.

**Queensland.** Woody valleys, Moreton Bay and Wide Bay, *C. Moore, W. Hill, F. Mueller.*
**N. S. Wales.** *R. Brown;* Illawarra, Port Macquarie, Pooral on the Karuak river, and Port Stephens, *Backhouse.*—Common in eastern India and the Malayan Archipelago, extending northward to S. China.


## 3. SARCOPETALUM, F. Muell.

Sepals 2 to 5, small. Petals 3 to 6, thickly fleshy, nearly globular. Male fl. : Stamens united in a column, divided at the top into 2 or 3 short horizontal lobes, each bearing a 2-celled anther. Female fl. : Carpels 3 to 6, with recurved lobed stigmas. Drupes flattened, the remains of the style near the base. Putamen horseshoe-shaped, the sides concave. Seed horseshoe-shaped.

Embryo curved, linear, in rather copious albumen; cotyledons closed.—Racemes simple.

The genus is limited to the following species.

1. **S. Harveyanum,** *F. Muell. Pl. Vict.* i. 27 *and* 221, *t. suppl.* 3.  A tall woody climber, with thick terete stems.  Leaves broadly ovate or orbicular, acuminate or rarely obtuse, and sometimes angular or lobed, attaining 4 to 6 in. in breadth, deeply cordate at the base or sometimes slightly peltate, 7- to 9-nerved, quite glabrous, on a petiole of 1 to 3 in.  Racemes simple, axillary or mostly lateral below the leaves, solitary or clustered, 1 to 3 in. long.  Bracts small.  Pedicels about 1 line long.  Flowers reddish-yellow, scarcely 2 lines diameter, the sepals usually shorter than the thick almost gland-like petals.  Drupes 3 or 4 lines diameter, almost pear-shaped.

**Queensland.**  Moreton Bay, *W. Hill.*

**N. S. Wales.**  Port Jackson to the Blue Mountains, *R. Brown* and others; southward of the colony, *A. Cunningham*, to Twofold Bay, *F. Mueller.*

**Victoria.**  Forests near the mouth of Snowy river, *F. Mueller.*

## 4. STEPHANIA, Lour.

(Clypea, *Blume.*)

Male fl.: Sepals 6, 8, or 10, in 2 series.  Petals 3, 4, or 5, shorter than the sepals, obovate.  Stamens united in a column bearing a flat disk, with the sessile anthers confluent into a single ring round the margin.  Female fl.: Sepals 3, 4, or 5.  Petals as many.  Carpel 1, with a divided stigma.  Drupe compressed, the scar of the style not far from the base.  Putamen horseshoe-shaped, with an open concavity on each side.  Seed curved, with little albumen.  Embryo linear, with closed cotyledons.—Leaves mostly peltate.  Flowers in simple or compound umbels.

A small genus, extending over tropical or subtropical Africa and Asia.  The Australian species common over the whole range.

1. **S. hernandiæfolia,** *Walp.; Hook. and Thoms. Fl. Ind.* i. 196.  A glabrous or more or less pubescent climber.  Leaves broadly ovate, orbicular, or nearly triangular, usually more or less peltate at the base, the larger ones 3 or 4 in. long, on a petiole of 2 or 3 in., but often much smaller, glabrous or pubescent underneath.  Peduncles axillary, shorter than or rather longer than the petioles, bearing an umbel of about 5 rays, each ray terminated by a head or partial umbel of 8 to 12 small sessile or shortly pedicellate flowers, or the partial umbel again compound.—F. Muell. Pl. Vict. i. 220; *Clypea hernandifolia*, W. and Arn. Prod. i. 14 ; Wight, Ic. t. 939.

**N. Australia.**  N. coast, *R. Brown;* rocky declivities and cataracts of Fitzroy and Stokes' Range, *F. Mueller.*

**Queensland.**  Keppel Bay, *R. Brown;* tropical districts, *A. Cunningham;* Moreton Bay, Taylor's Range, and Burnett river, *F. Mueller.*

**N. S. Wales.**  Near Sydney, *American Exploring Expedition, Harvey,* and others, northward to Clarence river, *Beckler,* and southward to Illawara and Twofold Bay, *F. Mueller,* but rare in the latter locality.

**Victoria.**  Forest glens, S. E. extremity of Gipps' Land, *F. Mueller.*

The glabrous form, *S. australis,* Miers ; A. Gray, in Bot. U. S. Expl. Exped. i. 38, and the pubescent one, *S. Gaudichaudi,* A. Gray, in Bot. U. S. Expl. Exped. i. 37, have been distinguished as species: but they almost always grow together, and pass gradually from the one to the other.

The species extends from eastern Africa almost all over India and the Archipelago, and northward to China.

## 5. PACHYGONE, Miers.

Sepals 6 or 9, in 2 or 3 series, the inner ones larger, imbricate. Petals 6, shorter than the sepals, embracing the stamens at the base. Male fl.: Stamens 6, free, incurved at the top, anthers small, globose-didymous. Female fl.: Staminodia 6. Carpels 3, with thick horizontal stigmas. Drupes reniform, the scar of the style near the base; putamen slightly excavated, with an internal process. Seed horseshoe-shaped, without albumen, cotyledons semi-terete, almost horny, the radicle very short.—Leaves ovate. Flowers in racemes, the males clustered along the rhachis, the females solitary.

Besides the Australian species, which is endemic, the genus comprises one from tropical Asia, which alone has furnished so much of the above character as relates to the female flower and fruit.

1. **P.(?) pubescens,** *Benth.* A woody climber, the young branches pubescent. Leaves petiolate, broadly ovate, shortly acuminate or rarely obtuse, 3 to 4 in. long, 5-nerved at the base, coriaceous, glabrous and shining or slightly scabrous above, pubescent underneath. Male racemes axillary, often 2 or 3 together, many-flowered but much shorter than the leaves, pubescent. Pedicels clustered, about 1 line long. Flowers glabrous, scarcely more than 1 line diameter when open. Sepals 9, in 3 series, the outer ones small and lanceolate, the next longer, the innermost still larger, narrow-ovate. Petals about half as long as the inner sepals. Stamens 6; anthers globose-didymous, almost 4-lobed. Female flowers and fruit unknown.

**Queensland.** Quail Island, *Flood* (*F. Mueller*). In the absence of the female flowers and fruit, the genus of this plant cannot be fixed with certainty. The form and venation of the leaves, the inflorescence and general structure of the male flowers, are so nearly those of the E. Indian *Pachygone ovata*, that I might have taken it for a large-leaved, more pubescent variety of that species, but for the presence of a third outer series of small sepals which are not in *P. ovata*; the inner sepals are also narrower than in that species, and not ciliate. I have only been able to examine 2 flowers; the persistent pedicels were very numerous, but almost every flower had already fallen from the only two specimens I have seen. .

## 6. PLEOGYNE, Miers.

(Microclisia, *Benth.*)

Outer sepals about 6, very small, 3 inner ones much larger, valvate in the bud, connivent at the base and recurved at the top when open. Petals 6, much shorter, the margins dilated and involute. Male fl.: Stamens 3; filaments linear-terete; anthers small, globose-didymous. Female fl. with 6 carpels (*Miers*). Drupes 3 to 6, reniform, with the scar of the style lateral, the putamen not excavated on the sides, nor with any internal process. Seed reniform, without albumen; cotyledons thick and fleshy, scarcely separable; radicle scarcely distinct.—Flowers in short axillary branching panicles.

The genus is limited to a single species. Miers had originally characterized it very shortly from female specimens only, and I failed to recognize it in the male specimens I possessed with others in fruit, which did not show the increased number of carpels mentioned by Miers. I was therefore induced to publish it as new under the name of *Microclisia*. The further more perfect fruiting specimens I have since seen have enabled me to identify it with a very imperfect fragment named by Miers in A. Cunningham's herbarium. The genus is distinguished from all, except the African *Triclisia*, by the remarkably valvate inner sepals.

1. **P. australis,** *Benth.* A climber, with a soft pubescence like that of *Pericampylus*, sometimes very copious, sometimes quite disappearing from the upper surface of the leaves. Leaves from ovate to oblong, obtuse or scarcely acute, the larger ones 3 to 4 in. long, rounded but not cordate at the base, at length rather coriaceous and shining above, reticulately penninerved. Male cymes or single flowers in little axillary solitary or clustered panicles, seldom above 1 in. long, and softly pubescent. Inner sepals about 1 line long, the outer ones very minute. Female inflorescence probably more simple. Drupes about 5 lines broad, glabrous, with a very thin endocarp.—*Microclisia*, Benth. in Benth. and Hook. Gen. Pl. part i. Addend. 435.

**Queensland.** Keppel Bay, *R. Brown;* Moreton Bay, *A. Cunningham, F. Mueller;* Fitzroy river, *F. Mueller.*

## 7. ADELIOPSIS, Benth.

Sepals 6, in 2 rows, the inner ones considerably larger, and 2 or 3 outer smaller bracts, all much imbricate in each row. Petals 3, smaller than the inner sepals, broad and slightly concave. Male fl.: Stamens 9 to 12; filaments linear-terete; anthers small, globose-didymous. Female fl.: Staminodia wanting. Carpels 3, with a large, recurved, broad and thick stigma, and 2 ovules in each carpel, inserted one above the other on the inner angle. Fruit unknown.—Flowers clustered in short axillary spikes.

The genus consists of a single species, which has the habit, imbricate sepals, and the general form of the stamens and carpels of *Pachygone*, to which I should have referred it, but for the petals reduced in number and not involute, the increased number of stamens in the males and their entire deficiency in the females, and for the 2 ovules in each carpel. The latter character appears constant, as far as I have been able to ascertain, and does not exist to my knowledge in any other Menispermaceous plants. The fruit being unknown, the tribe to which the genus must be referred cannot as yet be fixed; but it will stand either next to *Cocculus* amongst *Cocculeæ*, or more probably near *Pachygone* in *Pachygoneæ*.

1. **A. decumbens,** *Benth.* Branches rather thick, leafy, densely clothed with a soft velvety tomentum or almost hirsute, and from the name given, probably decumbent and not climbing. Leaves ovate or oval-oblong, 1½ to 2 in. long, very obtuse, rounded at the base, thickly coriaceous, softly tomentose or velvety on both sides when young, becoming nearly glabrous above when old, the thickened revolute nerve-like margin terminating at the top of the midrib on the under side in a prominent hirsute gland or tuft of hairs. Flowers small, in little clusters along the rhachis of short axillary spikes, seldom above ½ in. long, the outer bracts very small, acute, and hairy, the outer sepals also hairy, but rather larger and more obtuse, the inner sepals much larger, orbicular, and glabrous, except the ciliate edge, the petals about ⅔ as large as the inner sepals and quite glabrous.

**Queensland.** N. E. coast, near Cape Fear (Fair Cape?), *R. Brown*, described in his notes (without mention of the ovules) under the name of *Adelioides decumbens*, but, as in many other cases, the term *Adeloides* was evidently intended as a memorandum, not as a generic name, for which it is unsuited (*Hb. R. Br.*).

## Order VI. **NYMPHÆACEÆ.**

Sepals 3 to 5, petals 3 or more and stamens 6 or more, either all free and hypogynous, or the inner ones or all adnate at the base to the torus or ovary, or inserted on its summit. Anthers innate or adnate, the cells opening in longitudinal slits. Gynœcium of 3 or more carpels, either free and distinct, or immersed in the torus so as to form a several-celled ovary. Styles or stigmas free or adnate on an epigynous disk. Ovules solitary, and suspended from the apex of the cavity, or indefinite and attached to the sides of the cavity, not to its inner angle. Ripe carpels indehiscent, free or united in a fleshy or spongy fruit. Seeds immersed in a fleshy or pulpous arillus, or naked, the embryo either small, enclosed in the embryo-sac and half immersed in a cavity of a farinaceous albumen near the hilum, or without albumen, large, with thick fleshy cotyledons, and a remarkably developed plumule.—Aquatic herbs, with a submerged root or rhizome. Leaves carried by their long petioles to the surface of the water or raised above it, usually peltate or deeply cordate, or a few remaining under water and deeply cut. Flowers growing singly on long radical scapes, or axillary peduncles, either on the surface of the water or raised above it.

The Order, although not numerous in species, is found in pure, quiet, or slowly-flowing waters nearly all over the globe. The three Australian species belong to the three genera, considered as typical of as many tribes or suborders, raised by some botanists to the rank of distinct Orders. All three genera are common to the New and the Old World. They are absent, however, from the southern Australian colonies as well as from New Zealand.

Sepals and petals 3 each. Carpels 6 or more, free, on a small torus. Ovules few. Flowers small . . . . . . . . . . . . . . 1. *Brasenia.*

Sepals 4 to 6. Petals and stamens numerous, the outer ones free, the inner more and more adnate to the torus. Carpels immersed in the torus in a ring round a central conical projection . . . . . . 2. *Nymphæa.*

Sepals 4 or 5. Petals and stamens numerous, hypogynous. Carpels half immersed without order in the flat top of the torus. No albumen 3. *Nelumbium.*

### 1, **BRASENIA,** Schreb.

#### (Hydropeltis, *Mich.*)

Sepals 3, petal-like, and petals 3, hypogynous. Stamens 12 to 18, hypogynous; filaments subulate, anther-cells lateral. Carpels 6 to 18, free, on a small torus, attenuate at the top into short styles, stigmatic along the inner edge. Ovules 2 or 3, pendulous from the dorsal side of the cavity. Ripe carpels coriaceous, indehiscent. Seeds albuminous.

The genus is limited to the following species.

1. **B. peltata,** *Pursh. Fl. N. Amer.* 389. Rhizome prostrate at the bottom of the water. Stems forked, leafy, covered as well as other submerged parts, especially when young, with a thick coating of transparent jelly. Leaves floating on the surface of the water, peltately attached by their centre to long petioles, oval, entire, 3 to 4 in. long and about half as broad. Peduncles axillary, bearing solitary flowers of a dull purple on the surface of the water. Sepals and petals very much alike, about 4 or 5 lines long when they first open, but lengthening to 7 or 8 lines. Carpels shorter.—A. Gray,

Gen. Ill. t. 39 ; *Hydropeltis purpurea*, Mich. ; DC. Prod. i. 112 ; Bot. Mag. t. 1147.

**N. Australia ?**  *R. Brown.*
**Queensland.**  Lagoons near Moreton Bay, *F. Mueller.*
The species is abundant in the waters of North America and of East India.

## 2. NYMPHÆA, Linn.

Sepals 4, inserted near the base of the torus. Petals numerous, passing gradually from the sepals to the stamens, inserted on the torus or ovary, the outer petals near the base, the inner stamens almost at the top. Filaments of the outer stamens dilated and petal-like, with small lateral anther-cells, of the inner ones narrow or filiform, with longer anthers opening inwards. Carpels several, immersed in a ring in the fleshy torus, having the appearance of a several-celled ovary, with a conical or globular process in the centre. Styles thick, radiating, free or united at the base, often with an incurved appendage beyond the stigmatic portion. Ovules numerous, pendulous from the sides of the cavity. Fruit a spongy berry, breaking up irregularly when ripe. Seeds embedded in pulp, arillate, albuminous.—Rhizome perennial. Leaves floating, peltate or very deeply cordate. Flowers large, solitary, floating on the surface of the water or slightly raised above it, on long radical peduncles.

The most considerable genus of the Order, chiefly in the northern hemisphere or within the tropics, but represented also in S. Africa.

1. **N. gigantea,** *Hook. Bot. Mag. t.* 4647. Leaves orbicular or very broadly ovate, very deeply cordate, the basal lobes separated by a very acute angle, or overlapping each other, or united near the petiole, rendering the leaf partially peltate, the principal nerves radiating from the petiole, raised underneath, and in the larger specimens the whole under side covered with raised reticulations ; the margin entire or more frequently sinuate, or with short distant teeth. Flowers blue, purple, pink, or rarely white, the petals and stamens usually very numerous. Filaments nearly all filiform, or many of the outer ones flattened, but never very broad and always narrowed under the anther; connective narrow and scarcely projecting beyond the cells. Styles or stigmas thick, radiating, united at the base, either without any or with only a very short terminal appendage.—F. Muell. Fragm. ii. 141 ; *N. stellata*, F. Muell. l. c. 142.

**N. Australia.**  Lakes and marshes throughout tropical Australia, *R. Brown, F. Mueller.*
**Queensland.**  Wide Bay, *Bidwill;* Moreton Bay, *W. Hill.*
**N. S. Wales.**  Clarence river, *Beckler.*

The species is apparently confined to Australia, unless it be really a modification of the Asiatic and African *N. stellata*, Willd., as appears to have been the opinion of Brown. It varies exceedingly in size. The larger specimens have the leaves about 18 inches across, with much-raised reticulations underneath, the flowers 12 in. across, with exceedingly numerous petals, and above 200 stamens ; the smallest have leaves of 5 or 6 inches, not reticulate, the flowers 3 or 4 in. across, and the petals and stamens much fewer, but always more numerous than is usual in *N. stellata*, to which F. Mueller is disposed to refer several specimens. This Indian species may also be distinguished by the connective lengthened beyond the anther-cells into a very prominent appendage, and it appears to me that Caspary (notes in Herb. Hooker) is right in considering all the Australian specimens as forms of *N. gigantea.* In the Kew Gardens the flowers and leaves are very small in the early part of the season, and larger and larger ones are developed as the season advances. F. Mueller also distinguishes the seeds in size and shape, smaller, more ovoid, and more completely enclosed in the arillus in those he refers to *N. stellata,* than in the true *N. gigantea;* but in the true *N. stellata* the seeds are nearly globular, and usually marked with raised longitudinal costæ, not mentioned by F. Mueller. I have not myself seen the ripe seeds of Australian specimens.

The rhizome and fruits are used as an article of food by the aborigines.

### 3. NELUMBIUM, Juss.

Sepals 4 or 5, free. Petals and stamens numerous, hypogynous. Anthers opening inwards, the connective produced in a club-shaped appendage. Carpels several, half-immersed in the flat top of an obconical torus, the styles shortly projecting with somewhat dilated terminal stigmas. Ovules 1 or 2 in each carpel, suspended from the top of the cavity with a dorsal raphe. Nuts nearly globular, shortly protruding from the cells of the large flat-topped torus. Seeds with a spongy testa, without albumen; cotyledons thick and fleshy, enclosing a much-developed plumula, radicle very short.—Leaves peltate, supported above the water on erect petioles. Flowers solitary, on erect scapes above the water.

Besides the following Asiatic and Australian species, there is a second one from the West Indies.

1. **N. speciosum,** *Willd.; DC. Prod.* i. 113. Leaves orbicular, peltate, somewhat concave, 1 to 2 ft. diameter, quite entire or slightly sinuate, glabrous and often somewhat glaucous. Flowers pink, 4 to 8 in. diameter, appendage of the anthers linear-clubshaped. Fruit 2 to 4 in. diameter, the nuts from the size of a pea to that of a small cherry.—Bot. Mag. t. 3916, 3917.

**N. Australia.** Swamps in Arnhem's Land, *F. Mueller;* Lower Condamine river, *Coxon.*

**Queensland.** Mackenzie river, *F. Mueller.*

The species is widely distributed over the warmer regions of Asia, extending northwards to the Caspian Sea in the west, and to Japan in the east.

### ORDER VII. PAPAVERACEÆ.

Flowers hermaphrodite, regular, or, in *Fumarieæ,* irregular. Sepals 2 or 3, rarely 4, free, imbricate, very caducous. Petals 4, 6, or rarely 8 or 12, hypogynous, free, imbricate, and often crumpled in the bud, in 2 rarely 3 series, deciduous. Stamens hypogynous, indefinite, and free, or, in *Fumarieæ,* definite, with the filaments usually united. Anthers erect, the cells opening longitudinally. Ovary free, either 1-celled with parietal placentas often protruding into the cavity, or rarely completely several-celled by the placentas meeting in the axis, or 2-celled by a false dissepiment connecting 2 parietal placentas. Style short or none; stigmas as many as placentas, usually confluent and radiating on the disk-like or dilated top of the ovary or style. Ovules indefinite, anatropous, ascending with an inferior micropyle or horizontal. Fruit capsular, usually opening in pores or valves. Seeds globular or subreniform. Embryo minute, at the base of a fleshy albumen.—Herbs or rarely small shrubs, glabrous and often glaucous or hispid, the juice usually coloured. Leaves alternate or the floral ones almost opposite, entire, lobed or dissected without stipules. Flowers usually solitary on long peduncles, either terminal or in the upper axils.

The Order belongs almost entirely to the temperate or subtropical regions of the northern hemisphere, one only genus being represented by a single species in the southern hemisphere; but, besides the *Papaver rhœas* mentioned below, one at least of the numerous forms of the European *Fumaria officinalis* has established itself as a weed of cultivation in some parts of Victoria and S. Australia, as in S. Africa.

## 1. PAPAVER, Linn.

Sepals 2, rarely 3, Petals 4, rarely 6. Stamens indefinite. Placentas of the ovary 4 or more, covered with ovules and projecting more or less into the cavity, rarely meeting in the centre; stigmas radiating on the convex or almost conical disk-like summit of the ovary. Capsule opening in transverse pores between the placentas under the disk, with very short opercular valves. Seeds furrowed.—Herbs, with a milky juice. Leaves usually lobed or cut. Peduncles long, the buds nodding.

Except the following one, the species are all from the northern hemisphere in the Old World.

1. **P. horridum,** *DC. Syst. Veg.* i. 79. An erect annual, beset with subulate prickles or stiff bristles, but otherwise glabrous and usually glaucous. Leaves narrow-oblong or lanceolate, irregularly pinnatifid and coarsely toothed, the radical ones contracted into a petiole, the stem ones sessile or partially stem-clasping. Flowers small for the genus, of a pale brick or red colour. Sepals hispid. Petals nearly ovate, about $\frac{1}{2}$ in. long. Capsule ovoid-oblong, perfectly smooth and glabrous, the terminal disk at first pyramidal, at length nearly flat, usually with 6, 7, or 8 stigmatic rays. Placentas as many, projecting in the cavity but not meeting in the centre.—F. Muell. Pl. Vict. i. 29; Sw. Brit. Fl. Gard. ii. 173; *P. gariepinum,* DC. Syst. Veg. i. 79; Bot. Mag. t. 3623; *P. aculeatum,* Thunb. Fl. Cap. 431.

**Queensland.** Moreton Bay, *F. Mueller*; Warwick, *Beckler.*
**N. S. Wales.** Hunter's River, *R. Brown;* Hastings river, *Beckler.*
**Victoria.** Sandy localities along the Murray and Snowy rivers, *F. Mueller.*
**S. Australia.** Murray scrub, towards Mount Barker and Flinders Range, *F. Mueller.*
The species is also found in extratropical S. Africa, and is nearly allied to, but I believe really distinct from, some S. European forms of the *P. dubium,* Linn.

*P. rhœas,* Linn., the common European Corn-Poppy, distinguished by its large red flowers with broad overlapping petals, and a nearly globular or turbinate smooth capsule with about 10 stigmatic rays, has established itself in a few places in Victoria as an introduced weed.

## ORDER VIII. CRUCIFERÆ.

Flowers hermaphrodite, regular, or with the outer petals larger. Sepals 4, free, imbricate in 2 series, the outer ones often saccate at the base. Petals 4, rarely wanting, the laminæ spreading in the form of a cross; torus usually bearing 4 glands opposite the sepals. Stamens usually 6, of which 2 outer ones shorter or rarely wanting, 4 inner ones longer, in pairs alternating with the outer ones. Anthers 2-celled, attached by the base. Ovary 1-celled, with 2 parietal placentas or rarely a single one, or more frequently divided into two cells by a thin membranous septum connecting the two parietal placentas.

Style simple, often very short or none; stigmas 2, erect or divaricate, or united into a single capitate or minute stigma. Ovules 1, 2, or more in each cell, horizontal or pendulous from the parietal placenta. Fruit a pod, either long and narrow, and then called a *siliqua*, or short and broad, called a *silicule*, usually 2-celled, each cell opening by a deciduous valve, leaving persistent the thin septum surrounded by the nerve-like placentas, which form a rim called the *replum;* exceptionally the pod is 1-seeded and indehiscent, or separating into 2 indehiscent cocci or into 2 or more bead-like articles. Seeds attached in each cell in 2 rows, one proceeding from each edge of the septum, but when each seed is as broad as the cell they overlap each other, so as to appear to be, and to be described as, in a single row; testa cellular, sometimes winged, often exuding when soaked a thick coat of mucilage. Albumen usually none. Embryo usually curved, the cotyledons plano-convex with the radicle curved against their edge, when they are said to be *accumbent*, or over the back of one of them, when they are *incumbent;* in the latter case they are either flat or more or less folded over the radicle, or *conduplicate.*— Herbs or rarely undershrubs, without milky juice. Hairs simple, stellate or attached by the centre. Leaves simple, usually alternate, entire, lobed or pinnately divided, the radical ones often lyrate and the stem ones auricled. Stipules none. Flowers usually in terminal racemes, which are at first corymbose but lengthen out as the fruiting advances, and usually without bracts.

*Cruciferæ* form a very large Order, dispersed over nearly the whole globe, but most abundant in the temperate and cold regions of the northern hemisphere. They are rare within the tropics, especially in districts where there are no high mountain-ranges. The Order is one of the most easily recognized by the flowers or fruits, but, to determine the genera and species, it is absolutely necessary to have the pod and the seed in a good state.

Pods linear, at least 4 times as long as broad.
  *Pods terete or tetragonous, the valves turgid or with a very prominent midrib.*
    Seeds in a single row. Pods long.
      Cotyledons accumbent . . . . . . . . . . . 2. BARBAREA.
      Cotyledons incumbent . . . . . . . . . . . 7. SISYMBRIUM.
    Seeds in 2 rows. Pods usually short.
      Cotyledons accumbent . . . . . . . . . . . 1. NASTURTIUM.
      Cotyledons incumbent.
        Petals either obovate or, if narrow, short and erect . . . 8. BLENNODIA.
        Petals tapering into a long, subulate, often twisted point . 9. STENOPETALUM.
  *Pods flattened, usually long, the flat valves parallel with the septum. Cotyledons accumbent.*
    Stem-leaves auricled.
      Seeds smooth . . . . . . . . . . . . . . . . 3. ARABIS.
      Seeds pitted . . . . . . . . . . . . . . . . 4. CARDAMINE.
    Stem-leaves divided or rarely entire, not auricled . . . . . 4. CARDAMINE.
Pods short or oblong, rarely 4 times as long as broad.
  *Pods terete or globular, the valves very convex.*
    Cotyledons accumbent . . . . . . . . . . . . 1. NASTURTIUM.
    Cotyledons incumbent.
      Fruiting peduncles recurved, pod ripening underground . . 10. GEOCOCCUS.
      Fruiting racemes erect.
        Petals tapering into a long, subulate, often twisted point . 9. STENOPETALUM.
        Petals obovate, or if narrow, erect and short.
          Septum broader than the transverse diameter of the pod    8. BLENNODIA.

Septum narrower than the transverse diameter of the pod  12. CAPSELLA.
*Pods flattened, the flat valves parallel to the septum or to each*
*other.*
Cotyledons accumbent.  Pod with a septum.
Pod orbicular.  Seeds 2 to 4 in each cell . . . . . . .  5. ALYSSUM.
Pod elliptical.  Seeds 10 to 12 or more in each cell  . . .  6. DRABA.
Cotyledons incumbent.  No septum.  Seeds numerous, small  . 11. MENKEA.
*Pods flattened laterally, the valves boat-shaped, with their flat*
*sides at right angles to the narrow septum.*
Seeds 1 in each cell.
Pod either indehiscent or separating into 2 indehiscent cocci. 13. SENEBIERA.
Pod-valves dehiscent . . . . . . . . . . . . . 14. LEPIDIUM.
Seeds 2 to 4 or more in each cell.
Cotyledons incumbent.  Seeds, or at least ovules, 6 or more
in each cell . . . . . . . . . . . . . . . . 12. CAPSELLA.
Cotyledons accumbent.  Seeds or ovules 4 or fewer in each cell 15. THLASPI.

Besides the above genera, the following *Cruciferæ* have appeared as introduced weeds of cultivation.

*Heliophila pumila*, Linn. f., from South Africa, a slender, glabrous, erect annual, with linear or filiform leaves, small white flowers, and slender moniliform pods with flat orbicular seeds, and long, linear, twice-folded cotyledons.  Received from Swan River.

*Brassica?*, apparently *B. geniculata* (*Sinapis geniculata*, Desf.), a Mediterranean species, in Herb. Mueller, from Moreton Bay, but the specimens are too young to determine.

*Raphanus sativus*, Linn., the common cultivated Radish of Europe and Asia, has established itself as a weed in many cultivated places.

*Sinapis hastata*, Desf. Cat. Hort. Par. ed. 2, 151; DC. Prod. i. 220, described from a specimen raised in the Jardin des Plantes, supposed to have been of Australian origin, is *Diplotaxis virgata*, DC., a Spanish plant.

## 1. NASTURTIUM, R. Br.

Sepals short, equal, spreading.  Petals scarcely clawed.  Pods nearly cylindrical, short or elongated, the valves convex, slightly 1-nerved, the septum transparent; style short or long, with an entire or 2-lobed stigma.  Seeds usually distinctly ranged in 2 rows, small, turgid, with short free funicles.  Cotyledons accumbent.—Herbs, either glabrous or pubescent, with simple hairs.  Leaves entire, lobed, or pinnately divided.  Flowers small, generally yellow.

A considerable genus, dispersed over the greater part of the globe, and very difficult, both as to the discrimination of its species and as to its distinction from other genera.  The Australian species is one of the most widely diffused.

Flowers yellow . . . . . . . . . . . . . . . . 1. *N. palustre.*
Flowers white.
Half aquatic perennial.  Petals obovate . . . . . *N. officinale* (below).
Small annual.  Petals very small and narrow . . . *Cardamine eustylis* (p. 71).

*N. officinale*, R. Br. in DC. Prod. i. 137, the European Watercress, with pinnate leaves and perfectly distinct segments and white flowers, has been noticed in a few streamlets in Victoria and South Australia; but everywhere its importation from Europe could be traced (*F. Mueller*).

1. **N. palustre,** *DC. Syst. Veg.* ii. 191.  An erect or decumbent or almost trailing annual or biennial, from a few inches to 2 ft. or more in length, quite glabrous or very rarely pubescent.  Leaves toothed or pinnately lobed, or the lower ones sometimes lyrate, auriculate at the base, the lobes

ovate, oblong, or rarely lanceolate, always irregular, confluent and usually sinuate or toothed. Racemes short, loose, without bracts. Flowers small, yellow, the petals scarcely exceeding the calyx. Style short. Pod sessile, turgid, oblong, obtuse, straight, or slightly curved, generally 2 to 4 lines long and about 1½ lines broad, but occasionally rather longer and narrower. —Reichb. Ic. Fl. Germ. ii. 53; *N. terrestre*, R. Br. in Ait. Hort. Kew. ed. 2, iv. 110; Hook. f. Fl. Tasm. i. 21; F. Muell. Fl. Vict. i. 31; *N. semipin-natifidum*, Hook. Journ. Bot. i. 246.

**Queensland.** Burdekin river, *F. Mueller;* Maranoa river, *Mitchell.*
**N. S. Wales.** Port Jackson, *R. Brown;* native cabbage of the settlers, *Herb. Muel-ler;* Darling river, *F. Mueller.*
**Victoria.** Around swamps, lakes, and along the banks of rivers in many localities, *F. Mueller.*
**Tasmania.** Abundant on the wet banks of St. Patrick's river and on the Derwent river, *J. D. Hooker.*
**S. Australia.** Torrens river, near Adelaide, *F. Mueller.*

The specimen from the Darling river has narrow lobes to the almost twice pinnatifid leaves, but has the normal short pods of the species. Some specimens from the Murray river have also very narrow leaf-lobes, with a longer and more slender pod, almost like that of *N. indicum*, but not quite ripe. Mitchell's specimen has very young but slender pods, and the whole plant is hoary pubescent, and it may possibly not be correctly referred here. The species is dispersed over all temperate and subtropical regions of the globe except S. Africa. It was first published by Leysser as *Sisymbrium palustre*, and a year later by Withering as *S. terrestre.* Brown first transferred it to *Nasturtium* with Withering's specific name, and De Candolle soon afterwards with Leysser's name. Continental botanists now generally adopt *N. palustre*, DC., as the oldest absolute specific name, whilst British botanists often adopt *N. terrestre*, Br., as the oldest in the genus.

## 2. BARBAREA, R. Br.

Sepals nearly erect, equal. Petals clawed. Pod elongated, flattish-tetra-gonous; septum transparent; valves keeled or with a prominent midrib; style short; stigma capitate or 2-lobed. Seeds in a single row, oblong, not bordered; the funicles free.—Erect, branching, usually glabrous herbs, an-nual or biennial, the stem angular. Leaves entire or pinnately sinuate or lobed. Flowers yellow, sometimes bracteate. Pods usually rigid.

A genus of few species, dispersed over the temperate regions of the globe, the Australian species being the commonest over the whole range. It differs from *Nasturtium* chiefly in the robust rigid habit, the prominent midrib of the valves, and the seeds occupying the whole breadth of the pod so as to appear in a single row.

1. **B. vulgaris,** *R. Br.; DC. Prod.* i. 140. Erect, rather rigid, but often slightly branching, 1½ to 2 ft. high. Leaves lyrate-pinnatifid, the lower ones with a large terminal ovate lobe and several smaller ones more or less distinct, the upper ones often reduced to a single ovate or oblong terminal lobe, usually sinuate or toothed. Flowers bright yellow, the petals twice as long as the calyx. Pods usually numerous, in a long terminal raceme, on slightly spreading pedicels of 3 to 4 lines, in the Australian specimens usually 1 to 1½ in. long, the stigma nearly sessile or on a short style rarely exceeding ½ line.—A. Gray, Gen. Ill. t. 62; F. Muell. Pl. Vict. i. 32; *B. australis*, Hook. f. Fl. Nov. Zel. i. 14; Fl. Tasm. i. 21.

**Victoria.** Banks of the Mitta Mitta and other rivers of Gipps' Land, chiefly at an ele-vation of 1000 to 3000 feet, *F. Mueller.*

**Tasmania.**  Moist or marshy districts in the centre of the island, also near Launceston, *J. D. Hooker.*

The species is spread over Europe, North America, northern Asia, the Himalayas, and New Zealand, and as an introduced weed in South Africa.  In Australia it is evidently indigenous.  The specimens all belong to the var. *stricta* of most northern botanists (*B. præcox*, Hook. Fl. Bor. Am. i. 39, not of R. Br.), as usually defined, with nearly erect stout pods with a very short style.  European specimens are often precisely similar.

## 3. ARABIS, Linn.

(Turritis, *Linn.*)

Sepals rather short, equal or the lateral ones saccate at the base.  Petals entire, usually clawed.  Pod sessile, elongated, slender, flattened ; valves flat, keeled, or with a midrib ; septum membranous ; stigma entire or 2-lobed.  Seeds in 1 or rarely 2 rows, flattened, often bordered or winged.—Annual or perennial herbs, glabrous or tomentose with spreading, branched, or stellate hairs.  Radical leaves usually spathulate, the stem ones sessile, often auricled.  Flowers white or rarely purple, straw-coloured or pink.

The species are numerous in the temperate and colder regions of the northern hemisphere, very few inhabiting the southern one ; and none are peculiar to Australia.  *Cardamine stylosa*, which in its undivided sagittate leaves comes very near to *Arabis*, may be readily distinguished by its reticulate pitted seeds.

1. **A. glabra,** *Crantz ; Hook. f. and Thoms. in Journ. Linn. Soc.* v. 140.  Stem erect, simple, rigid, 1 to 3 ft. high, usually glabrous except at the base.  Radical leaves petiolate, narrow-oblong, entire, or sinuately toothed, 2 to 4 in. long, usually pubescent or hirsute with stellate or branching hairs ; stem-leaves erect, oblong-lanceolate, stem-clasping and usually auriculate at the base, and all except the lowest quite glabrous.  Flowers rather small, white or straw-coloured.  Fruiting racemes long, rigid, with numerous erect slender pods, mostly 2 in. long or even more, and $\frac{1}{2}$ to $\frac{3}{4}$ line broad.  Seeds small, either as broad as the septum and in 1 row, or narrower and somewhat biseriate.—*Turritis glabra*, Linn. ; DC. Syst. Veg. ii. 211 ; Reichb. Ic. Pl. Germ. ii. t. 44 ; F. Muell. Pl. Vict. i. 33 and 221.

**N. S. Wales.**  On the Severn, in New England, *C. Stuart.*

**Victoria.**  Banks of the Cobongra, Mitta Mitta, Livingstone Creek, and Snowy rivers, at an elevation of 3000 to 4000 feet, *F. Mueller.*

The range of this species extends over Europe, temperate North America and Asia, the Himalaya, and Japan.

## 4. CARDAMINE, Linn.

Sepals equal at the base.  Petals clawed.  Pod elongated, linear, compressed ; valves usually flat, without conspicuous nerves, opening elastically ; septum transparent ; style short or long ; stigma entire or 2-lobed.  Seeds flattened, not bordered, in a single row (except in *C. eustylis*).—Herbs, usually flaccid and glabrous.  Leaves entire or more frequently pinnately divided, in a few species not Australian opposite or whorled.  Flowers erect or nodding, white, purple, or lilac, not yellow.  Pods usually slender.

A large genus, widely spread over the temperate and colder regions both of the northern and southern hemisphere.  Of the 7 following species two are identical with or representatives of common northern species ; the remainder are endemic or extend only to New Zealand.

Seeds reticulate and pitted, rather large.
Leaves entire or sinuate-toothed, the stem ones sagittate.
　　Plant of 2 to 5 ft. . . . . . . . . . . . . . . 1. *C. stylosa.*
　　Lower leaves pinnate, all petiolate. Plant erect, under 2 ft. 2. *C. dictyosperma.*
Seeds smooth.
　Perennials.
　　Fruiting racemes short, leafy. Pod fully 2 lines broad . 3. *C. radicata.*
　　Fruiting racemes loose, leafless. Pod not above 1 line broad.
　　Flowers rather large, with obovate spreading petals.
　　　Style 1 to 1½ line long . . . . . . . . . 5. *C. tenuifolia.*
　　　Stigma sessile or nearly so . . . . . . . . 6. *C. hirsuta heterophylla.*
　　Flowers very small, with narrow erect petals . . . . 4. *C. laciniata.*
　Annuals.
　　Petals conspicuous, obovate, spreading . . . . . . 6. *C. hirsuta heterophylla.*
　　Petals very narrow, small, nearly erect.
　　　Seeds nearly the breadth of the septum, in a single row 6. *C. hirsuta.*
　　　Seeds numerous, small, almost biseriate. Valves of the
　　　　pod convex . . . . . . . . . ˙ . . . . . 7. *C. eustylis.*

1. **C. stylosa,** *DC. Syst. Veg.* ii. 248. A rather coarse glabrous herb, branching and decumbent or nearly erect, usually 2 to 3 ft. high and some-times attaining 5 ft. Leaves oblong-lanceolate, entire or sinuate, and mi-nutely but remotely toothed, the lower ones narrowed into a long petiole, the upper ones sessile but narrow below the middle and clasping the stem by their sagittate base, the longest 3 to 5 in. long. Flowers small, white, with obovate spreading petals. Fruiting racemes long and rather rigid, the pedi-cels very spreading, 3 to 4 lines long. Pods 1 to 1½ in. long and ¾ to 1 line broad, with a very faint nerve on the valves. Seeds oval, dark-coloured, re-ticulated with raised longitudinal nerves and transverse pits between them.— Hook. f. Fl. Tasm. i. 18 ; F. Muell. Pl. Vict. i. 34 ; *Arabis gigantea,* Hook. Ic. t. 259; *C. divaricata,* Hook. f. Fl. Nov. Zel. i. 13.

**N. S. Wales.** Mount Lindsay, *W. Hill.*
**Victoria.** Moist forest valleys, rare in open pasture land near the banks of rivers in various parts of Gipps' Land, also in the Dandenong ranges, *F. Mueller.*
**Tasmania.** Northern and eastern coasts near the sea, *J. D. Hooker ;* ascending to alpine elevations on Mount Wellington, *Oldfield ;* also in New Zealand.
　This species has as much the characters of *Arabis* as of *Cardamine,* but the habit is rather that of the latter genus.

2. **C. dictyosperma,** *Hook. Journ. Bot.* i. 246. Erect or branching and decumbent at the base, glabrous or with a few hairs at the base, under 2 ft. high. Lower leaves pinnately divided into a few distant, ovate or oblong, entire or toothed segments, the terminal one usually much the largest ; upper leaves with narrower and fewer lobes, or small, narrow, and entire, all petio-late, with the petiole scarcely dilated at the base and rarely sagittate. Flowers larger than in *C. stylosa,* the lamina narrow-obovate, usually longer than the claw. Fruiting racemes long, the pedicels very spreading, 2 to 5 lines long. Pod usually longer and more slender than in *C. stylosa,* and sometimes at-taining 2 in. but sometimes only 1 in. ; style from ¾ to 2 lines long. Seeds of *C. stylosa,* but with coarser reticulations.—Hook. f. Fl. Tasm. i. 19 ; F. Muell. Pl. Vict. i. 35 and 221 ; *C. nivea,* Hook. Comp. Bot. Mag. i. 273.

**N. S. Wales.** Moist rocky places north of Bathurst, *A. Cunningham ;* Severn river New England, *C. Stuart ;* from Clarence river, *Beckler,* to Twofold Bay, *F. Mueller.*

**Victoria.** Springy shady localities in damp valleys, from the lowlands to the alps, *F. Mueller.*

**Tasmania.** Abundant in damp ravines and by waysides throughout the island, *J. D. Hooker.*

**W. Australia,** *Drummond, n.* 94, *and* 5*th Coll. n.* 285.

In flower the smaller specimens often resemble *C. tenuifolia,* but are more erect and less branched. The seeds are very different.

3. **C. radicata,** *Hook. f. in Hook. Ic. Pl.* i. 882. Rhizomes or procumbent root-like stems elongated, cylindrical and brittle, sometimes as thick as the little finger, producing at their extremity tufts of leaves and leafy erect flowering branches 2 to 6 in. high. Leaves petiolate, obovate, coarsely toothed or almost pinnatifid, not auricled at the base, glabrous as well as the whole plant. Flowers (which I have not seen) rather large. Fruiting racemes short and dense, often leafy at the base. Pod usually ¾ in. long and fully 2 lines broad. Seeds much compressed, irregularly orbicular, not pitted. —Hook. f. Fl. Tasm. i. 18.

**Tasmania.** Summit of Mount Olympus, in crevices of basaltic columns, *Gunn ;* in crevices of rocks on a mountain westward of Mount Lapeyrouse, *Herb. F. Mueller.*

4. **C. laciniata,** *F. Muell. in Trans. Phil. Soc. Vict.* i. 34, *and Pl. Vict.* i. 35. A glabrous perennial, with a procumbent or creeping rhizome, much more slender than in *C. radicata,* the stems rather weak, ascending or erect, seldom above 1 ft. high and often leafless. Leaves chiefly radical, petiolate, linear-lanceolate or rarely obovate-oblong, pinnatifid with a few narrow lobes, or with 1 large terminal lobe and 2 or 3 smaller ones along the petiole, or rarely entire or toothed only, the stem-leaves when present few and narrow. Flowers very small, the narrow erect petals scarcely longer than the calyx. Stamens usually 4 only. Fruiting raceme very loose, with distant, slender, spreading pedicels. Pods slender, 1 to 1½ in. long. Seeds orbicular, not pitted.

**N. S. Wales.** New England, near Clifton, *C. Stuart.*

**Victoria.** In marshy places, chiefly in rich soil, not rare. Used as food by the Murray natives, *F. Mueller.*

**S. Australia.** Lake Alexandrina, Gawler river, Bugle range, the Onkaparinga and Torrens rivers, etc., rather frequent, *F. Mueller.*

5. **C. tenuifolia,** *Hook. Journ. Bot.* i. 247. Generally if not always perennial, with a slender creeping rhizome, which often dies away so as to give the tufts the appearance of an annual. Stems weak, branching and glabrous or rarely hirsute, like those of *C. hirsuta* but usually longer, sometimes attaining 1 to 1½ ft. Leaves pinnately divided, the lower ones usually with a terminal, broadly ovate, orbicular, or cordate segment, entire or coarsely toothed, the lateral segments smaller, few, distant, and all petiolate, the upper leaves or sometimes all the stem-leaves with narrow-linear segments, more numerous and more equal than in the lower ones, and usually entire and sessile ; in some specimens the leaves are all crowded at the base of the otherwise leafless scapes. Flowers rather large, white or lilac, the laminæ of the petals obovate and spreading. Fruiting racemes loose, the pedicels not very spreading. Pods usually erect, narrow, ½ to 1 in. long, tipped by a slender style often 1½ lines long. Seeds nearly orbicular, smooth.—*C. lilacina,* Hook.

Comp. Bot. Mag. i. 273 ; *C. pratensis*, Hook. f. Fl. Tasm. i. 19 ; *C. parviflora*, var., F. Muell. Pl. Vict. i. 36.

**N. S. Wales.** Interior of the colony, *A. Cunningham;* Macquarie river, *Fraser;* Hunter river, *Leichhardt;* Macleay river, *Beckler.*

**Victoria.** Swamps on Latrobe river, *F. Mueller.*

**Tasmania.** Common in marshy and wet places throughout the island, *J. D. Hooker.* This plant is united by Dr. Hooker with the European *C. pratensis,* Linn., and it certainly is a very close representative of that species, but its lax, more branching stems, give it much more the habit of *C. hirsuta.* In many respects indeed it seems almost to pass into the latter species through its variety *heterophylla,* and F. Mueller unites all these plants with *C. resedæfolia,* Linn. and others, under the Linnæan name of *C. parviflora.* But long and repeated observation of the European *C. pratensis, resedæfolia,* and *hirsuta,* in a living state in various localities, prevents my admitting their union without much more convincing proofs ; and, if they are kept distinct, it appears necessary to maintain also the Australian *C. tenuifolia.* It is, I believe, a perennial like *C. pratensis,* but that cannot always be ascertained from dried specimens.

*C. intermedia,* Hook. Ic. Pl. t. 258, can scarcely be judged of from the single specimen preserved, but the style is certainly rather long and slender, and the habit and petals are more those of *C. tenuifolia* than of *C. hirsuta.*

6. **C. hirsuta,** *Linn.; DC. Prod.* i. 152. A much-branched decumbent or tufted annual, seldom above 6 in. high, either quite glabrous or slightly hirsute with short spreading hairs. Leaves pinnately divided, the lower ones with 1 ovate or rounded terminal segment and a few smaller petiolulate lateral ones, or sometimes reduced to the terminal one, the upper leaves few with narrow lobes. Flowers very small, the petals narrow and erect or scarcely spreading. Stamens often reduced to 4 (especially in European specimens). Fruiting racemes usually short and rather dense, the pedicels not very spreading. Pods erect, slender, usually 7 to 9 lines long and scarcely more than $\frac{1}{2}$ line broad, the stigma sessile or on a style not longer than the breadth of the pod. Seeds smooth, as broad as the septum, and in a single row as in all the preceding species.—Reichb. Ic. Fl. Germ. ii. t. 26 ; Hook. f. Fl. Tasm. i. 20 ; *C. parviflora,* Linn. ; DC. Prod. i. 152 ; also F. Muell. Pl. Vict. i. 36, partly , *C. debilis,* Banks, in DC. Syst. Veg. ii. 265 ; *C. paucijuga,* Turcz. in Bull. Mosc. 1854, ii. 295.

**N. S. Wales.** Apparently common in wet places, extending northwards to Hastings river, *Beckler.*

**Victoria.** Wet meadows and along streams, dispersed over the whole colony, *F. Mueller.*

**Tasmania.** Throughout the island, abundant in many localities, *J. D. Hooker.*

**S. Australia.** As far as Flinders Range, *F. Mueller.*

**W. Australia,** *Drummond, 4th Coll. n.* 131.

The species is very abundant in the temperate regions of the northern hemisphere, in the hilly regions of the tropics, in New Zealand and the Pacific islands, and in Antarctic America. Always in the north a small-flowered annual, and sometimes glabrous. Many of the Australian specimens are precisely like the glabrous European ones, but in others there are signs of a procumbent slender rhizome, as is so frequent in the following variety or species. I have preserved the name *C. hirsuta,* in place of that of *C. parviflora* adopted by F. Mueller, because it is the one by which the plant is most universally known, both being Linnæan.

Var. (?) *heterophylla.* Rhizome apparently in some instances perennial, though very slender. Flowers rather larger, with more spreading almost obovate petals. Pod less slender, and the whole plant approaching *C. tenuifolia* in habit, but with an almost sessile stigma, as in *C. hirsuta.*—*C. heterophylla,* Hook. Ic. Pl. t. 58.—Apparently a common Tasmanian form, and would include some Victoria specimens, *Robertson,* and South Australian ones from Mount Barker creek, *F. Mueller.*

**7. C. (?) eustylis,** *F. Muell. in Trans. Vict. Inst.* i. 114 ; *Pl. Vict.* i. 37. An erect annual, much branched from the base, scarcely exceeding 6 to 8 in. in height and quite glabrous.  Leaves pinnately divided, the lower ones with ovate segments, the others with narrower ones, all usually with a few teeth or lobes.  Flowers smaller than in *C. hirsuta,* the petals narrow, erect, and scarcely exceeding the calyx.  Fruiting racemes short, leafless.  Pods rather spreading, slender, 6 to 9 lines long, tipped by a style of ½ to near 1 line, the valves convex, smooth, without nerves.  Seeds very numerous and small, much narrower than the septum, and showing 2 distinct rows.

**N. Australia.**  On the rivers flowing into the Gulf of Carpentaria, rare, *F. Mueller.*
**Victoria.**  Sandy and gravelly banks of the Murray river, *F. Mueller.*
The nearly cylindrical pod and two-rowed seeds are more those of *Nasturtium* than of *Cardamine,* but the habit and white flowers may justify the placing the species in the latter genus.  The degree of elasticity of the valves cannot be judged of in the dried specimens.

## 5. ALYSSUM, Linn.
### (Meniocus, *Desv.*)

Sepals rather short, equal at the base.  Petals rather short, entire or bifid. Stamens often bearing a tooth or small appendage on the filaments of some or all of them.  Pod short, from nearly orbicular to oblong, very flat or turgid ; the valves flat, concave, or turgid in the centre and flat on the margins, the septum membranous ; style short or long, with an entire stigma.  Seeds 2 to 10 in each cell.  Cotyledons accumbent.—Branching herbs or small shrubs, usually hoary with stellate tomentum.  Leaves undivided, usually linear.  Racemes without bracts, with white or yellow flowers.

A large genus, dispersed over the temperate regions of the Old World, but chiefly in the Mediterranean region and western Asia.  None are found in America, eastern Asia, or in the Pacific Islands.  The only Australian species is identical with one common in the eastern Mediterranean region.

**1. A. linifolium,** *Steph. in Willd. Spec. Pl.* iii. 467.  A small, but hard, wiry, and much-branched erect annual, hoary, with a minute, close, stellate tomentum.  Leaves linear, oblong-spathulate or almost obovate, mostly under ½ in., but the longest sometimes nearly 1 in. long, quite entire.  Flowers white, very small.  Pods orbicular or broadly ovate, 2 to 3 lines long, minutely hoary ; the valves flat and without nerves ; style small, subulate. Seeds 4 to 6 in each cell.—*Meniocus linifolius,* DC. Syst. Veg. ii. 325 ; Deless. Ic. Sel. ii. t. 42 ; *M. serpyllifolius,* Desv. ; DC. l. c. ; *M. australasicus,* Turcz. in Bull. Mosc. 1854, ii. 297.

**N. Australia.**  Lacrosse Island, Cambridge Gulf, N. W. coast, *A. Cunningham.*  A single specimen, with only portions of the pods remaining, but apparently belonging to this species.
**N. S. Wales.**  Darling river, *Victorian Expedition.*
**Victoria.**  Murray river, and sand-hills near Lake Hindmarsh, *F. Mueller.*
**S. Australia.**  Near Crystal Brook and about Spencer's Gulf, *F. Mueller.*
**W. Australia,** *Drummond, 4th Coll. n.* 127.
This, the only outlying representative of a genus otherwise so restricted in its range,

may possibly have been introduced from southern Europe, but it appears to be too abundant in arid desert situations to be omitted from the Flora.

# 6. DRABA, Linn.

Sepals short, equal. Petals entire. Pod elliptical or oblong, rarely almost linear, compressed, several-seeded; valves flat or nearly so, very rarely nerved; septum membranous; style short or long; stigma entire. Seeds in 2 rows, not bordered, with filiform funicles; cotyledons accumbent.—Herbs, usually small and tufted or annual, more or less hoary, with stellate tomentum. Leaves undivided and usually entire, the radical ones rosulate. Scapes leafless or flowering-stems with sessile leaves. Racemes without bracts. Flowers usually small, white or yellow, rarely pink or purple.

A large genus, chiefly distributed over the temperate and cooler regions of the northern hemisphere, very abundant in high alpine stations, and extending all along the high Andes of South America, rare in Antarctic America, entirely wanting in South Africa and New Zealand, and represented in Australia by a single species identical with a common northern one.

1. **D. muralis,** *Linn.; DC. Prod.* i. 171. A slender erect annual, 2 to 3 in. high and simple, or twice as high and branched, more or less pubescent with stellate hairs. Leaves ovate, coarsely toothed, $\frac{1}{4}$ to $\frac{1}{2}$ in. long in Australian specimens, often twice that in European ones, the radical ones petiolate, the others sessile. Flowers very small, white or pale yellow. Fruiting racemes loose, with slender spreading pedicels of 4 to 5 lines. Pod elliptical, pubescent in our specimens, about 3 lines long, containing usually above 12 seeds in each cell.—*D. nemoralis,* Ehrh.; DC. Prod. i. 171; Reichb. Ic. Fl. Germ. ii. t. 12; Hook. f. Fl. Tasm. i. 24.

**Tasmania.** Dry places near Hobarton, and on the Derwent at the Cataracts, *J. D. Hooker.*

Common in the temperate regions of the greater part of Europe and Asia, and also in some parts of North America. The usual variety in the north has glabrous pods; but the Tasmanian variety with pubescent ones, to which the name of *D. nemoralis* has been given, is also found in Europe.

# 7. SISYMBRIUM, Linn.

Sepals equal or the lateral ones slightly saccate. Petals usually elongated, with long claws. Pod linear-elongated, cylindrical or flattened, several-seeded, the valves usually convex and 3-nerved; septum membranous; style usually short, with an entire or slightly 2-lobed stigma. Seeds in a single row, not bordered, oblong, with filiform funicles. Cotyledons incumbent.—Herbs, usually annual or biennial, glabrous hirsute or tomentose. Leaves entire or pinnately lobed or divided. Flowers yellow, or rarely white or pink.

A large genus, chiefly European and Asiatic, with a few North American and a very few Antarctic species. Only one is a native of New Zealand, and none are as yet known to be truly indigenous in Australia; but the following appears now so well established as a roadside weed that it cannot be omitted from the Flora.

*1. **S. officinale,** *Scop.; DC. Prod.* i. 191. An erect annual, more or less pubescent, a foot high or rather more, with very rigid spreading branches. Leaves deeply pinnatifid, with few lanceolate slightly toothed lobes, the terminal one 1 to $1\frac{1}{2}$ in. long, the others smaller, often curved backwards towards the stem, the upper leaves sometimes undivided and hastate. Flowers very small, yellow. Pods about ¼ in. long, thick at the base, tapering to the point, more or less hairy, almost sessile, and closely pressed against the axis in long, slender, stiff racemes.—Reichb. Ic. Fl. Germ. ii. t. 72.

**S. Australia.** Abundant on roadsides and waste places about Adelaide, *F. Mueller* and others.

**W. Australia,** *Drummond.* In both colonies, introduced from Europe. The species is somewhat anomalous in the genus, the valves of the pod having a somewhat prominent midrib, and the seeds in the lower broader part showing two almost distinct rows.

## 8. BLENNODIA, R. Br.

Sepals short, open, equal at the base or slightly saccate. Petals obovate, or short and narrow. Pod linear or linear-oblong (short in a variety of *B. trisecta*), terete or 4-angled, the valves very convex, without nerves or with a prominent midrib ; septum membranous or almost spongy ; stigma capitate, sessile or on a very short style. Seeds oblong or ovoid, more or less distinctly 2-rowed, not bordered, when soaked usually emitting a copious fibrous mucus ; funicles free, filiform. Cotyledons incumbent.—Herbs or low undershrubs, glabrous or hoary-tomentose with simple or stellate hairs. Leaves entire or pinnatifid. Flowers white, yellow, or pink, the racemes without bracts.

A genus limited to extratropical or subtropical Australia, differing from *Sisymbrium*, to which some species have been referred, in the seeds never so completely overlapping each other as to form a single row, and generally in the copious mucus of the seeds, which is however not constant in all the species. From *Capsella* it differs in the longer pod, and in the dissepiment broader in proportion to the transverse diameter of the pod.

*Glabrous undershrubs. Leaves or their lobes linear-filiform. Pods slender.*
    Leaves entire . . . . . . . . . . . . . . . . 1. *B. filifolia.*
    Leaves mostly 3-cleft . . . . . . . . . . . . . 2. *B. trisecta.*
*Annuals, glabrous or with simple hairs. Leaf-lobes narrow.*
    *Pods slender, scarcely contracted at the base.*
      Glabrous . . . . . . . . . . . . . . . . . 3. *B. nasturtioides.*
      Hoary, with simple hairs . . . . . . . . . . . 4. *B. eremigera.*
*Annuals, with stellate pubescence. Leaves pinnatifid or toothed.*
    *Pods acute at the top and at the base ; valves very convex.*
      Pod rather slender, glabrous . . . . . . . . . . 5. *B. cardaminoides.*
      Pod thicker in the middle, hirsute or stellately tomentose.
        Petals scarcely exceeding the calyx.
          Flowers yellow. Pedicels about as long as the pod . . . 6. *B. curvipes.*
          Flowers white. Pedicels much shorter than the pod . . 7. *B. brevipes.*
        Petals twice as long as the calyx, white or pink.
          Calyx about 1 line long . .. . . . . . . . . . 8. *B. lasiocarpa.*
          Calyx 2½ lines long . . . . . . . . . . . . . 9. *B. canescens.*
*Perennials, with stellate pubescence. Leaves toothed or pinnatifid. Pods acute at the top and at the base ; valves very convex.*
    Hoary. Pod at least 5 times as long as broad . . . . . . 10. *B. Cunninghamii.*
    Nearly glabrous. Pod about 3 times as long as broad . . . 11. *B. alpestris.*

1. **B. filifolia,** *Benth.* Shrubby at the base and perfectly glabrous, like the *B. trisecta.* Leaves solitary or clustered, linear-filiform, entire, mostly ½ to 1 in. long. Flowers not seen. Fruiting racemes rather rigid, with spread-

ing pedicels of 4 to 5 lines. Pods shortly stipitate above the calyx-scar, slender, straight or slightly curved, seldom above $\frac{1}{2}$ in. long, the stigma raised on a very short style; valves prominently 1-nerved. Seeds obovate, rather larger than in *B. trisecta*, emitting a rather copious mucilage.—*Erysimum filifolium*, F. Muell. in Linnæa, xxv. 368; *Sisymbrium filifolium*, F. Muell. in Trans. Phil. Soc. Vict. i. 34.

**S. Australia.** Crystal Brook, *F. Mueller.*

2. **B. trisecta,** *Benth.* A perfectly glabrous often glaucous undershrub or almost a shrub, 1 to several ft. high. Leaves numerous, often clustered, linear-filiform, sometimes rather thick, divided into 3 (rarely 2 or 5) unequal linear-filiform segments, the whole leaf seldom above 1 in. long, except in very luxuriant specimens. Flowers white, scented. Sepals 1 to 1$\frac{1}{2}$ lines long. Petals obovate, spreading. Fruiting raceme 4 to 6 in. long or rarely more, with slightly spreading pedicels of $\frac{1}{4}$ to $\frac{1}{2}$ in. Pod sessile on the pedicel, usually narrow-linear, 4 to 6 lines long, but sometimes very short, straight or curved, the stigma sessile or nearly so; valves convex, with a slender longitudinal nerve. Seeds numerous, small, oblong-ovoid, those which I have soaked scarcely emitting any mucus.—*Sisymbrium trisectum*, F. Muell. in Trans. Vict. Inst. i. 114; Pl. Vict. i. 39.

**N. S. Wales.** Scrub near the Gwydir river, *Mitchell;* Darling river, *F. Mueller.*
**Victoria.** Sandy clay-soil and dry limestone plains of the Murray, *F. Mueller.*
**S. Australia.** Flinders Range, Murray river, and in the interior N.W. of Spencer's Gulf, *F. Mueller;* Cooper's Creek, *Leichhardt.*

Var. *brachycarpa.* These specimens, collected in M'Douall Stuart's Expedition, are in fruit only; the habit and foliage are precisely those of the common form gathered with them, but the pods are shortly oblong and very turgid, about 2 lines long; they may possibly be accidentally abnormal.

3. **B. nasturtioides,** *Benth.* A glabrous annual, the central scape erect and leafless, the lateral branches decumbent at the base and leafy, from 2 or 3 in. to nearly 1 ft. long. Leaves usually pinnately divided into a few linear rather thick segments, the radical ones often 2 in. long, the others much smaller. Flowers yellow, rather small. Fruiting racemes loose, 3 to 6 in. long, with slender pedicels. Pod narrow, 4 to 7 lines long, nearly straight and scarcely contracted at the base; stigma sessile or nearly so; valves slightly convex, the longitudinal nerve very slender and sometimes quite inconspicuous. Seeds small, ovate, emitting a considerable mucus when soaked.—*Erysimum nasturtium*, F. Muell. in Linnæa, xxv. 368; *Sisymbrium nasturtioides*, F. Muell. in Trans. Vict. Inst. i. 115; Pl. Vict. i. 39.

**N. S. Wales.** Inundated plains on Lachlan river, *A. Cunningham.*
**Victoria.** Plains of Murray river, towards the junction of the Darling, *F. Mueller.*
**S. Australia.** Hill, Hutt, and Rocky rivers, *F. Mueller.*

Var. *pinnatifida.* Leaves small, on long petioles, with few short lateral lobes and a larger terminal one.—Between Darling and Lachlan rivers, *Burkitt*, small specimens in fruit only, the leaves mostly withered.

4. **B. eremigera,** *Benth.* Annual and erect or branching and decumbent at the base, more or less hairy with short simple hairs, from a few in. to 1$\frac{1}{2}$ ft. high. Leaves deeply and irregularly pinnatifid, with few oblong-linear or linear, sometimes falcate lobes. Flowers small, yellow. Fruiting racemes

loose, 2 to 4 in. long, with slender spreading pedicels.    Pods like those of
*B. nasturtioides*, mostly about ½ in. long, slender, straight or curved, not contracted at the base ; stigma sessile or nearly so ; valves with a slender nerve.
Seeds small, oblong-ovate, emitting mucus when soaked.—*Sisymbrium eremi-
gerum*, F. Muell. Fragm. ii. 143.

**Queensland.**   Maranoa river, *Mitchell.*
**N. S. Wales.**   Darling river, *Victorian Expedition.*

5.  **B. cardaminoides,** *F. Muell. Herb.* (as a *Sisymbrium*).    A slender
or small annual like *B. nasturtioides*, but more or less clothed with a minute
stellate pubescence, sometimes scarcely visible without a lens.    Leaves pinnatifid, the radical ones with rather numerous small, ovate triangular or lanceolate lobes, the terminal ones confluent, the lower ones becoming distinct segments along the petiole ; stem-leaves few and small, with few short lobes.
Flowers white (or pink ?), the sepals barely 1 line long.    Petals obovate,
twice as long.    Fruiting raceme loose and slender, 2 to 4 in. long, with
slender spreading pedicels.    Pod 4 to 6 lines long, scarcely 1 line broad,
usually curved, narrowed towards the base, glabrous or with a very minute
stellate tomentum ; valves very convex and keeled.    Seeds small, ovate,
emitting mucus when soaked.

**N. S. Wales.**   Darling river, *Victorian Expedition.*
**Victoria.**   Sand-ridges and heaths on the Glenelg, *F. Mueller, Robertson.*
**S. Australia.**   Near Wellington, and other places near the mouth of the Murray, *F.
Mueller.*
Some imperfect dry specimens have a slight resemblance with the European *Sisymbrium
Thalianum*, to which F. Mueller was disposed to refer them, but the latter plant is really very
different, having the undivided leaves, the flattened pods, the single-rowed seeds, and the
whole habit of an *Arabis*, with the cotyledons less decidedly incumbent than in other *Si-
symbria.*

6.  **B. curvipes,** *F. Muell. in Trans. Phil. Soc. Vict.* i. 100, *and Pl. Vict.*
i. 42.    A small but rather coarse annual, branching from the base, seldom
above 6 to 8 in. high, hoary with a rather rough stellate or branching pubescence.    Leaves oblong-lanceolate or broadly linear, coarsely toothed or entire,
the radical ones about 1 in. long and narrowed into a petiole, the upper ones
smaller.    Flowers small, yellow, the petals scarcely longer than the calyx.
Fruiting racemes loose, 2 to 4 in. long.    Pedicels spreading or curved, 4 to
6 lines.    Pod curved, 4 to 5 lines long, turgid, 1½ line thick in the middle,
tapering into a short style at the top, contracted at the base, pubescent with
short stellate hairs ; valves very convex and keeled.    Seeds few, ovate, exuding
mucus when soaked.—*Erysimum curvipes*, F. Muell. in Linnæa, xxv. 368.

**Victoria.**   Sandy localities on the Murray, towards the junction with the Darling, *F.
Mueller.*
**S. Australia.**   Crystal Brook, to the N. W. of Lake Torrens, and about Spencer's
Gulf, *F. Mueller.*

7.  **B. brevipes,** *F. Muell. in Trans. Phil. Soc. Vict.* i. 100, *and Pl. Vict.*
i. 41.    A coarse branching annual of 1 to 2 ft., hoary with a short stellate or
branching pubescence.    Leaves lyrate-pinnatifid, 1 to 2 in. long, petiolate,
with triangular or lanceolate lobes, entire or scarcely toothed ; the upper
leaves smaller and toothed only.    Flowers very small, white, the petals
scarcely exceeding the calyx.    Fruiting racemes rigid, 3 to 4 in. long, with

erect, rigid pedicels of 1 to 2 lines. Pods mostly about ½ in. long, turgid, some-what curved, tapering into a short style at the top, contracted at the base, pubescent with stellate hairs; valves very convex, but the midrib scarcely conspicuous except at the base. Seeds few, ovate, large, but distinctly ranged in 2 rows, the mucus very copious, with radiating fibres.—*Erysimum brevipes*, F. Muell. in Linnæa, xxv. 367.

**Victoria.** Barren sandy localities on the Murray and its lower tributaries, *F. Mueller*.
**S. Australia.** Rocky River, and to the N. W. of Lake Torrens, *F. Mueller*.
**W. Australia.** South coast ?, *Drummond, n.* 128.

8. **B. lasiocarpa,** *F. Muell. in Trans. Phil. Soc. Vict.* i. 100, *and Pl. Vict.* i. 40, *t.* 2. An annual, hoary with stellate pubescence, the central scape short and erect, the lateral stems decumbent and leafy at the base, branching and attaining 1 ft. or more. Radical leaves petiolate, lyrate-pin-natifid, 1, 2, or even 3 in. long; stem-leaves smaller, pinnatifid, or the upper ones toothed only. Flowers pink or white. Calyx about 1 line, petals obovate, fully twice as long. Fruiting racemes loose, 2 to 4 in. long, with divaricate pedicels of 4 to 6 lines. Pods not above ½ in. long, turgid, curved, tapering at the top with a short slender style, contracted at the base, hispid with simple or stellate hairs; valves very convex, with the midrib scarcely conspicuous. Seeds ovate, the mucus copious.—*Erysimum blennodioides*, F. Muell. in Linnæa, xxv. 367.

**N. S. Wales.** Darling river, *Victorian Expedition*.
**Victoria.** Arid sandy plains on the Murray and its lower tributaries, *F. Mueller*.
**S. Australia.** Towards Lake Alexandrina, *Hildebrand;* Cooper's Creek, *A. C. Gregory*.

9. **B. canescens,** *R. Br. in App. Sturt. Exped.* 4. Annual, but the lateral branching stems apparently harder at the base at the close of the season, so as to be almost woody; the whole plant hoary with a short, soft, stellate pubescence. Leaves lanceolate or oblong-linear, the radical ones about 2 in. long, pinnatifid and narrowed into a petiole, the upper ones linear, toothed or entire. Flowers large, pink, resembling those of a *Mat-thiola*. Calyx 2½ lines long, hoary. Petals fully twice as long, with long claws. Fruiting racemes rather loose, 2 to 6 in. long, the pedicels short, slightly spreading. Pod linear, 1 to 1½ in. long, slightly pubescent, with con-vex valves, crowned by the large, persistent stigma. Seeds oval-oblong, smooth.

**N. S. Wales.** Darling river, *Victorian Expedition*.
**S. Australia.** Cooper's River, *A. C. Gregory;* Elizabeth river, near Lake Torrens, *Hergott*.

10. **B. Cunninghamii,** *Benth.* A tufted herbaceous perennial, more or less hoary with soft stellate hairs, occasionally mixed with simple ones; annual stems erect or decumbent at the base, from a few inches to 1 ft. high, slightly branched. Radical leaves petiolate, 1 to 2 in. long, oblong or lan-ceolate, coarsely toothed or shortly pinnatifid; stem-leaves few and small, from lanceolate to nearly obovate. Flowers small, apparently white. Fruit-ing racemes loose, 2 to 4 in. long, with spreading pedicels. Pod 4 to 5 lines long, acute at the top and at the base, tipped by a very short subulate style, pubescent with simple or stellate hairs, or nearly glabrous; valves very

convex, with a prominent midrib.   Seeds oval-oblong, smooth, the mucus rather copious.

**Queensland.**   Flats on the Maranoa, *Mitchell.*
**N. S. Wales.**   Bathurst Plains and other parts of the interior of the colony, *A. Cunningham, Fraser.*

11. **B. alpestris,** *F. Muell. in Trans. Phil. Soc. Vict.* i. 100.   A dwarf herbaceous perennial, usually tufted, sometimes at first sight glabrous, but almost always more or less pubescent with' stellate hairs visible under a lens. Flowering stems rarely 6 in. high.   Leaves chiefly radical, petiolate, obovate-oblong, with a few coarse teeth, rarely almost lyrate-pinnatifid, or sometimes nearly entire, ¾ to 2 in. long; stem-leaves few and narrow.   Flowers white or pink, often tinged with purple.   Sepals nearly 1 line, petals about twice as long.   Fruiting racemes rather dense, 1 to 2 in. long, with rigid spreading pedicels.   Pod glabrous or nearly so, slightly curved, about 3 lines long and 1 line broad in the middle, tapering at the top and the base, the valves very convex and marked with a strong midrib.   Seeds ovate, elegantly reticulate, exuding a rather thin coat of mucus when soaked.—*Capsella blennodina,* F. Muell. Pl. Vict. i. 42.

**N. S. Wales.**   Ranges near Bathurst, *W. Wools.*
**Victoria.**   Subalpine grassy meadows at the sources of the Murray and Snowy rivers, *F. Mueller.*

As observed by Dr. Mueller, this species certainly connects *Blennodia* with *Capsella,* but the habit and the broader septum in relation to the transverse diameter of the pod, appear to me to connect it much more with the former genus, where he had first placed it, than with the latter, to which he subsequently referred it.

## 9. **STENOPETALUM**, R. Br.

Sepals narrow, erect, equal at the base.   Petals shortly lanceolate above the claw, tapering to a point, often long and twisted.   Pod globular, ovoid, or shortly linear, the valves very convex, usually without any conspicuous nerve; septum membranous; stigma globular, sessile or rarely on a very short style.   Seeds several, small, in 2 rows, not bordered, with free filiform funicles; cotyledons incumbent.—Annuals, usually slender and glabrous, rarely tomentose and more rigid.   Leaves linear.   Flowers orange-yellow or white.

The genus is limited to Australia.

Pods erect, 2 to 4 times as long as broad.
  Hoary tomentose.   Pedicels as long as the pod.   Petals 3 times
    as long as the calyx . . . . . . . . . . . . . 1. *S. velutinum.*
  Glabrous or slightly tomentose.   Pedicels shorter than the pod.
    Petals about twice as long as the calyx . . . . . . 2. *S. lineare.*
  Glabrous.   Flowers almost sessile.   Petals more than twice as
    long as the calyx . . . . . . . . . . . . . 3. *S. filifolium.*
Pods spreading or pendulous, globular or ovoid.
  Sepals scarcely 1 line, petals not twice as long . . . . . . 4. *S. sphærocarpum.*
  Sepals 1½ line or more, petals more than twice as long.
    Pedicels slender, 2 or 3 times as long as the sepals.
      Slightly hoary with appressed hairs.   Leaves entire or re-
        motely toothed . . . . . . . . . . . . . 5. *S. nutans.*
      Glabrous.   Lower leaves mostly pinnatifid . . . . . 7. *S. pedicellare.*
    Pedicels shorter than the sepals . . . . . . . . . . 6. *S. robustum.*

1. **S. velutinum,** *F. Muell. Pl. Vict.* i. 49. Erect and rather rigid, 1 to $1\frac{1}{2}$ ft. high, white or hoary with a very short stellate tomentum, which disappears from the older leaves and the base of the stem. Leaves narrow-linear, rather thick, entire or with a few minute distant teeth, the lower ones $1\frac{1}{2}$ to 2 in. long, the upper ones much shorter. Flowers erect, on pedicels about as long as the calyx. Sepals about 2 lines long, tomentose. Petals yellowish, the long slender point fully 3 times as long as the calyx. Fruiting pedicels erect, 3 to 5 lines long. Pod elliptical-oblong or almost ovoid, about 3 lines long, very turgid, glabrous; valves nerveless; ovules 8 to 12 in each cell.

**N. S. Wales.** Tributaries of the Darling, *Bowman;* near Mr. Mawson's Robleck station, *Leichhardt.*

**Victoria.** Barren localities on the Murray, rare, *F. Mueller.*

**S. Australia.** Between Stokes range and Cooper's Creek, *Wheeler.*

2. **S. lineare,** *R. Br. in DC. Syst. Veg.* ii. 513. Usually erect, slender, little branched and quite glabrous, $\frac{3}{4}$ to $1\frac{1}{2}$ feet high. Leaves few, narrow-linear, 1 to $1\frac{1}{2}$ in. long, entire or occasionally pinnatifid, with 1 or 2 short linear lobes on each side. Flowers small. Sepals not $1\frac{1}{2}$ line long. Petals of a brownish-yellow, the narrow-linear exserted portion not longer than the sepals. Fruiting racemes slender but rigid, with erect pedicels not half so long as the pod. Pods erect, oblong, 2 to 3 lines long and scarcely 1 line broad, glabrous, the valves usually showing the midrib. Seeds 8 to 12 in each cell, small, ovate, smooth.—Hook. Ic. Pl. t. 618 ; Hook. f. Fl. Tasm. i. 22 ; F. Muell. Pl. Vict. i. 49.

**N. S. Wales.** Interior of the Colony, *A. Cunningham ;* between Darling and Lachlan rivers, *Burkitt.*

**Victoria.** Sandy and rocky shores of Port Phillip and Wilson's Promontory, Murray desert and sandy localities near Mount M'Ivor, *F. Mueller.*

**Tasmania.** South Esk river, thirty miles from Launceston, *Gunn.*

**S. Australia.** Near Adelaide, *F. Mueller.*

**W. Australia,** *Drummond, n.* 680.

Var. *canescens.* A low branching more robust form, the young shoots slightly hoary with a minute stellate pubescence, and the leaves rather thicker.—Port Phillip, *F. Mueller.*

3. **S. filifolium,** *Benth.* A very slender, erect, glabrous annual, 1 to $1\frac{1}{4}$ ft. high, paniculately branched in the upper part. Leaves few, in our specimens filiform and entire, the longest $1\frac{1}{2}$ in. long. Racemes slender, erect, 3 to 6 in. long. Flowers very nearly sessile, small, and apparently yellow. Sepals scarcely $1\frac{1}{2}$ lines long. Petals when opened out nearly ·5 lines, including the claw and long point. Pods oblong, $1\frac{1}{2}$ to nearly 3 lines long, $\frac{3}{4}$ to 1 line broad, the valves very convex and without any nerve, the pedicels seldom 1 line and often not $\frac{1}{2}$ line long. Ovules 6 to 8 in each cell.

**W. Australia,** *Drummond, 1st Coll.*

4. **S. sphærocarpum,** *F. Muell. in Trans. Phil. Soc. Vict.* i. 35, *and Pl. Vict.* i. 50. A slender glabrous annual, erect or branching and decumbent at the base, from a few inches to 1 ft. high. Leaves few, small, narrow-linear, entire or deeply divided into 3 to 5 narrow-linear lobes. Flowers very small, on recurved pedicels of nearly 1 line. Sepals not above 1 line long. Lamina of the petals scarcely longer. Fruiting racemes slender, one-

sided, with recurved pedicels of 2 to 3 lines. Pod nearly globular, 1½ to 2 lines long, and often rather narrower; valves very convex, without any conspicuous nerve. Ovules 6 to 8 in each cell. Seeds few, exuding abundant mucus when soaked.

**Victoria.** Sterile, chiefly humid, sandy plains on the Murray, *F. Mueller.*

**S. Australia.** Near Lake Alexandrina, Barossa Range, Crystal Brook, and around Spencer's Gulf, *F. Mueller.*

**S. Australia,** *Drummond.*

5. **S. nutans,** *F. Muell. Fragm.* iii. 27. An erect annual, about 5 in. high in the single specimen seen, slightly hoary with appressed hairs. Leaves linear, entire or remotely toothed, about 1 in. long, narrowed at each end. Racemes loose. Pedicels much longer than the calyx, slender, erect when in flower, reflexed when in fruit. Sepals about 1½ line long. Petals with a filiform point of 4 or 5 lines. Pod broadly oval-oblong, about 4 lines long, very turgid, glabrous, ripening 3 or 4 seeds in each cell.

**S. Australia.** Between Stoke's Range and Cooper's Creek, *F. Wheeler* (*a single specimen in Herb. Mueller*).

6. **S. robustum,** *Endl. in Hueg. Enum.* 4. A glabrous, erect, and branching annual, in the original form stout, 1 to 2 ft. high, with rigid, spreading branches, in the more common variety slender, ½ to 1½ ft. high, with more erect branches. Leaves few, linear, entire or the lower ones pinnatifid, with 1 to 3 narrow lobes on each side. Racemes rigid or slender, somewhat one-sided, with spreading or recurved pedicels, not longer than the calyx when in flower, often rather longer than the pod when in fruit. Calyx 1½ to near 2 lines long. Petals orange or white, the lamina more or less lanceolate at the base, tapering to a point often 3 lines long. Pods spreading or pendulous, rarely nearly erect, from nearly globular to shortly ovoid, 1½ to 2 lines long, but rarely above 1¼ lines broad. Ovules 6 to 8 in each cell. Seeds few, with not near so much mucus as those of *S. sphærocarpum.*—Hook. Ic. Pl. t. 620; *S. gracile,* Bunge, in Pl. Preiss. i. 257; *S. croceum* and *S. minus,* Bunge, l. c. 258.

**W. Australia.** Vasse river and Murchison river, *Oldfield,* the only specimens that quite agree with Endlicher's description; the more slender variety apparently much more common about Swan River, *Drummond, n.* 679, *Preiss, n.* 1936, 1938, 1939, and others.

7. **S. pedicellare,** *F. Muell. Herb.* Habit, stature, and foliage of the slender varieties of *S. robustum,* but still more slender. Racemes very loose, with filiform pedicels longer than the calyx from the first, and ½ to ¾ in. long when in fruit. Calyx rather more than 1 line long. Petals apparently yellow, with a filiform point of 5 to 6 lines. Pod nearly of *S. robustum,* globular or ovoid, but I never find more than 4 ovules in each cell.

**W. Australia.** Murchison river, *Oldfield.*

## 10. GEOCOCCUS, J. Drumm.

Sepals short, spreading, equal at the base. Petals small. Pod oblong, slightly compressed, obtuse, the valves convex, with a prominent midrib; stigma sessile, entire. Seeds few, the two series rather distinct, oblong, not bordered, with long funicles; cotyledons incumbent.—A stemless herb, with radical pinnately-divided leaves, ripening its pods underground.

The genus is limited to the following species.

1. **G. pusillus,** *J. Drumm. in Hook. Kew Journ.* vii. 52. A stemless, tufted annual. Leaves all radical, spreading, 1½ to 3 in. long, pinnately divided, with triangular or shortly lanceolate lobes, the lower ones distinct, the ultimate ones confluent. Flowers in our specimens imperfect, on short, erect, radical peduncles. Petals, according to Drummond, oblong, not clawed, shorter than the calyx. Fruiting peduncles lengthening to from ⅓ to 1 in., recurved so as to bury the pod in the ground. Our pods are irregularly ripened.

**W. Australia.** Northern districts, on the limestone part of Conolly's Station, *Drummond.*

This curious little plant, unknown from any other locality, may possibly prove to be a condition of some species having usually dimorphous flowers, in which the more perfect ones are not developed. If so, it may very likely be a *Blennodia,* of some species of which it has the radical leaves.

## 11. MENKEA, Lehm.

Sepals spreading, equal at the base. Petals short, clawed. Pod broadly oval or linear-oblong, obtuse, very flat; the valves quite flat, 1-nerved, with reticulate veins; septum none or very narrow, bordering the replum; stigma sessile. Seeds numerous, very small, in two series, suspended from free capillary funicles along the replum; cotyledons incumbent.—Small annuals. Leaves few, linear, entire. Flowers small, white.

The genus is endemic in W. Australia.

Pods ovate, about 2 lines long, in loose slender racemes . . . . . 1. *M. australis.*
Pods narrow-oblong, 4 to 5 lines long, in short dense racemes . . . 2. *M. draboides.*

1. **M. australis,** *Lehm. in Ind. Seem. Hort. Hamb.* 1843, 8. A small, slender, glabrous annual, branching at the base, very much resembling *Capsella procumbens.* Radical leaves linear-oblong or lanceolate, entire or with 1 or 2 coarse teeth, about ½ inch long including the petiole; stem-leaves small and few. Flowers white, very minute, the sepals about ¼ line long, the petals but little longer, with the lamina obovate or oblong. Fruiting racemes loose and slender, with filiform pedicels of 3 to 4 lines. Pods ovate, about 2 lines long.—Bunge, in Pl. Preiss. i. 259; *Stenopetalum procumbens,* Hook. Ic. Pl. t. 610; *Menkea procumbens,* F. Muell. Fragm. ii. 142; Pl. Vict. i. 222.

**N. S. Wales.** Darling river, *F. Mueller.*
**Victoria.** Murray desert, *F. Mueller.*
**W. Australia,** *Drummond, Coll.* 1843 *n.* 87 and 90; *Preiss, n.* 1937.

2. **M. draboides,** *Hook. f.* A smaller plant than *M. australis,* the stems seldom exceeding 2 in., but more robust and branching. Radical leaves about ½ in. long, linear-oblong or lanceolate. Flowers small, with obovate-oblong petals, apparently yellowish. Fruiting racemes short and dense, with pedicels of 1 to 2 lines. Pod narrow-oblong, acute at the base, 4 to 5 lines long and 1 to 1¼ lines broad.—*Stenopetalum draboides,* Hook. Ic. Pl. t. 617; *Menkea australis,* F. Muell. Fragm. ii. 142, not Lehm.

**W. Australia,** *Drummond, Coll.* 1843.

## 12. CAPSELLA. Mœnch.

(Microlepidium, *F. Muell.*)

Sepals spreading, equal at the base. Petals short. Pod ovoid or oblong, laterally compressed or nearly terete, the valves very turgid or boat-shaped, keeled, the septum thin; style short or stigma sessile. Seeds several, in 2 rows, not bordered, on free funicles; cotyledons incumbent or rarely accumbent.—Small or weak annuals. Radical leaves rosulate, entire or lobed. Racemes slender, with small white flowers.

A small genus dispersed over the temperate regions of both the northern and southern hemispheres. Two of the following species are exclusively Australian. The genus is nearly allied to *Blennodia*, but the pod is shorter and more compressed laterally, the septum being usually narrower than the transverse diameter.

Pod elliptical or ovoid, not much compressed, obtuse or acute at the
   top.
  Plant glabrous. Pod many-seeded, the septum about 3 times as
    long as broad . . . . . . . . . . . . . . . . . 1. *C. procumbens.*
  Plant minutely pubescent. Pod few-seeded, the septum not twice
    as long as broad . . . . . . . . . . . : . . . . 2. *C. australis.*
Pod laterally compressed, cuneate or ovate, emarginate or broadly
   truncate at the top.
  Plant much branched, of 1 to 2 in. Pod oval-oblong, emarginate,
    with rounded lobes and few seeds . . . . . . . . . . 3. *C. pilosula.*
  Plant little branched, ½ to 1½ ft. Pod cuneate-triangular, with
    numerous seeds . . . . . . . . . . . . . . . *C. Bursa-pastoris* (p. 82).

1. **C. procumbens,** *Fries, Novit. Fl. Suec. Mant.* i. 14. A small, slender, glabrous, decumbent, and much-branched annual, seldom exceeding 6 in., and often not 2 in. high. Leaves from lanceolate to nearly ovate, the lower ones petiolate, pinnatifid or toothed, rarely exceeding 1 in., the upper ones smaller, often linear and entire. Flowers white, very small, the petals scarcely exceeding the calyx. Fruiting racemes loose, with filiform spreading pedicels of 2 to 4 lines. Pod ovoid, 1½ to 2 lines long, the valves very convex and boat-shaped, the septum 3 or 4 times as long as broad, and considerably narrower than the transverse diameter of the fruit. Seeds usually 10 to 12 or sometimes more in each cell.—Reichb. Ic. Fl. Germ. ii. t. 11; *Hutchinsia procumbens*, R. Br, DC. Syst. Veg. ii. 390; Hook. f. Fl. Tasm. i. 22; *Capsella elliptica*, C. A. Mey. Verz. Pfl. Cauc. 194; F. Muell. Pl. Vict. i. 43; *Stenopetalum incisæfolium*, Hook. f. in Hook. Ic. Pl. t. 276.

**Victoria.** Boggy, slightly saline places around Port Phillip Bay, and on the Murray, *F. Mueller.*

**Tasmania.** Blackman's River, on the road to Hobarton, *Gunn.*

**S. Australia.** Near St. Vincent's Gulf and Lake Alexandrina, *F. Mueller;* Guichen Bay, *H. Edwards.*

**W. Australia.** *Drummond, 4th Coll. n.* 129.

A common plant in the northern hemisphere, especially around the Mediterranean and in Western and Central Asia, found also in N.W. America and in extratropical S. America.

2. **C. australis,** *Hook. f.* A small annual, very near *C. procumbens,* and perhaps a variety only. It is usually still smaller, and sprinkled with a minute stellate pubescence. Foliage the same. Flowers rather larger. Pod

elliptical-ovate, about 2 lines long, and less compressed than in *C. procumbens*, the septum not twice as long as broad, and as broad at least as the transverse diameter of the fruit. Ovules usually 6 to 8 in each cell, of which only 3 or 4 come to maturity.—*Hutchinsia australis*, Hook. f. Fl. Tasm. i. 23, t. 4 ; *Capsella antipoda*, F. Muell. Pl. Vict. i. 44.

**Victoria.** Mount Macedon, summit of Mount Alexander, and in the Black Forest, *F. Mueller.*

**Tasmania.** Not unfrequent in dry stony places, but easily overlooked, *J. D. Hooker.* *Draba pumilio*, R. Br. in DC. Syst. Veg. i. 353, from the minute specimens in the Banksian herbarium appears to be either *C. procumbens* or *C. australis*, in a very young dwarf state.

3. **C. pilosula,** *F. Muell. Pl. Vict.* i. 44. A small erect annual, pubescent with short simple or stellate hairs, with numerous branches, often decumbent at the base, 1 to 3 in. high. Leaves small, obovate or lanceolate, entire, toothed or with a few lobes. Flowers small, white. Fruiting racemes rather rigid, with spreading pedicels shorter than the pod. Pods oval-oblong or cuneate, emarginate with short, rounded, but not winged lobes, laterally compressed, about 2 lines long, glabrous, the stigma sessile in the notch ; septum narrow, very thin ; valves boat-shaped and keeled, but not winged. Ovules 6 to 8 in each cell. Seeds few, without mucus when soaked.— *Microlepidium pilosulum*, F. Muell. in Linnæa, xxv. 371.

**Victoria.** Sandy desert, on the Murray, rare, *F. Mueller.* I find the pod-valves hollow to the top in this species as in *C. Bursa-pastoris.*

*C. Bursa-pastoris*, Mœnch ; DC. Prod. i. 177 ; Reichb. Ic. Fl. Germ. ii. t. 11, an erect annual, often above a foot high, the radical leaves usually spreading and pinnatifid, those of the stem few, narrow, clasping with projecting auricles, the pods triangular-cuneate, much compressed in a long loose raceme ; of European or Asiatic origin, but now one of the commonest weeds nearly all over the globe without the tropics, has also established itself in cultivated places in several of the Australian colonies.

### 13. SENEBIERA, Poir.

Sepals short, spreading, equal at the base. Petals short. Pod laterally compressed, orbicular or broader than long, either indehiscent or separating into two nuts, each with a single seed. Embryo bent in a circle, or the radicle incumbent on the back of the cotyledons, but with the bend above the attenuated base of the cotyledons, not at their junction with the radicle.— Annuals or biennials, much branched and usually prostrate. Leaves entire or pinnately divided. Flowers very small, in short leaf-opposed racemes.

There are several species dispersed over the warm as well as the temperate regions both of the New and the Old World, and more especially near the sea, the following ones extending to Australia.

Pods 1 line broad, slightly wrinkled, on slender pedicels.
Leaves linear, entire . . . . . . . . . . . . . . . 1. *S. integrifolia.*
Leaves pinnate . . . . . . . . . . . . . . 2. *S. didyma.*
Pods 2 lines broad, deeply wrinkled, sessile or nearly so . . . *S. Coronopus* (p. 83).

1. **S. integrifolia,** *DC. in Mém. Soc. Hist. Nat. Par. an 7*, 144, *t.* 8, and *Syst. Veg.* ii. 522. A rigid, glabrous, somewhat glaucous annual (or biennial ?), usually decumbent, and very much branched. Leaves linear, usually acute, ½ to 1 in. long or rather more, narrowed into a petiole, quite entire or

very rarely with 1 or 2 small teeth. Flowers very small and numerous, in terminal or leaf-opposed racemes usually much longer than the leaves ; pedicels slender, rarely exceeding 1 line. Pods like those of *S. didyma*, of the same size, and reticulate when young, becoming often warted or even corky when old.—*S. linoides*, DC. ; Harv. and Sond. Fl. Cap. i. 27.

**Queensland.** Bird Island, Wreck reef, *Denham.*
The species has a wide range on the seacoasts of S. Africa and Madagascar, and we have it also from Pratas and other islands of the Chinese seas. *S. mexicana*, Hook. and Arn. Bot. Beech. 276, is the same plant, but was probably gathered in the islands of Loo Choo or Bonin, and not in Mexico.

2. **S. didyma,** *Pers. Syn.* ii. 185. A much-branched, prostrate annual, spreading on the ground from 6 in. to 1 ft. or more, glabrous, or with a few long loose hairs. Leaves pinnately divided into 7 to 11 narrow segments, which are usually again cut into 2 to 4 unequal linear or lanceolate lobes, the lower leaves often once pinnate, with oblong or obovate, entire or shortly lobed segments. Flowers very small and numerous, in leaf-opposed racemes, which seldom, even in fruit, exceed the leaves, the pedicels slender, 1 to 2 lines long. Pods about ¾ line long and 1 line broad, wrinkled, formed of 2 ovoid distinct lobes, which separate into 1-seeded nuts when ripe.—Reichb. Ic. Fl. Germ. ii. t. 9 ; *S. pinnatifida*, DC. Syst. Veg. ii. 523 ; Prod. i. 203.

A common weed in sandy soil, especially near the sea, in all warm countries, perhaps indigenous to N. Australia, and now established in the neighbourhood of towns in almost all the Colonies.

*S. Coronopus*, Poir., DC. Prod. i. 203, with rather coarser foliage, the flowers and fruits sessile or nearly so along the rhachis of the raceme, and pods about 2 lines diameter, nearly orbicular, very much wrinkled and indehiscent, a very common European weed, is mentioned by F. Mueller as introduced into Victoria, but I have not seen Australian specimens.

### 14. **LEPIDIUM,** Linn.
(Monoploca, *Bunge.*)

Sepals short, equal at the base. Petals short, equal, sometimes wanting. Pod ovate or shortly oblong, rarely orbicular, usually much compressed laterally and notched at the top, the valves boat-shaped, keeled or winged, the septum narrow ; style filiform or stigma sessile. Seeds solitary in each cell, suspended from the top of the septum with a free funicle ; cotyledons incumbent in all except one species not Australian.—Herbs, undershrubs, or even small shrubs, very variable in habit. Leaves in the Australian species narrow or entire. Flowers small, white, the racemes without bracts.

A large genus, spread over the temperate and warmer regions of the globe, but not alpine and scarcely Arctic. Of the following species, one has a very wide geographical range, the others are confined to Australia, although one has nearly allied representatives in the Pacific islands. For the opportunity of inspecting original specimens of the *Lepidia* published by Desvaux, I am indebted to the kindness of M. La Vallée, of Paris, the present possessor of his herbarium.

Leaves all quite entire. Pod usually conspicuously winged.
  Leaves broadly ovate or orbicular . . . . . . . . . 1. *L. strongylophyllum*.
  Leaves linear or lanceolate.
    Leaves linear-lanceolate. Sepals fully 2½ lines long. Pod
      with 2 acute lobes . . . . . . . . . . . 2. *L. linifolium*.
    Leaves narrow-linear. Sepals 2 lines or less. Pod-lobes
      obtuse or very small.

Petals linear. Sepals 2 lines. Stem shrubby . . . .   3. *L. leptopetalum.*
Petals oblong or ovate. Sepals 1–1½ lines. Stem her-
  baceous.
  Lobes of the pod longer than the style (about 1 line).
    Valves winged to the base . . . . . . . . .   4. *L. rotundum.*
  Lobes of the pod shorter than the style (not ½ line).
    Valves scarcely winged . . . . . . . . . .   5. *L. phlebopetalum.*
  Petals none. Stamens 4. Pod-wings almost united with
    the style . . . . . . . . . . . . . .   6. *L. monoplocoides.*
Leaves mostly toothed or lobed. Flowers very small. Pod-
  wings small or none, except in *L. papillosum.*
  Petals none. Leaves narrow-linear, the upper ones auricled.
  Stems papillose. Stamens 4. Pod about 2 lines long, with
  2 short lobes or wings . . . . . . . . . . .   7. *L. papillosum.*
  Stems glabrous. Leaves linear or cuneate, not auricled, the
  radical ones pinnatifid. Stamens 2. Pod about 1½ lines,
  scarcely lobed . . . . . . . . . . . . . .   9. *L. ruderale.*
  Petals 4, minute. Leaves oblong-cuneate. Stamens 6. Pod
  2½ to 3 lines long, with distinct lobes . . . . . .   8. *L. foliosum.*

1. **L. (?) strongylophyllum,** *F. Muell. Herb.* Apparently shrubby, quite glabrous, with the branches denuded at the base. Leaves in the upper part of the branches, broadly ovate or nearly orbicular, or the upper ones elliptical-oblong, ½ to ¾ in. long, entire, rather thick, narrowed into a short petiole. Flowers unknown. Fruiting raceme evidently dense, with spreading pedicels of about 2 lines, the thick rhachis 1 to near 2 in. long. Pods only known by the persistent replum, which is oblong-lanceolate, nearly 3 lines long, ¼ line broad in the centre, terminating in a subulate style of about 1 line, and the scars of a funicle on each side at the upper angle of the replum show that there had been a single pendulous seed in each cell as in other *Lepidia.*

  **S. Australia.** Mount Vision, on the clay-slate in the N.W. interior, *M'Douall Stuart.* A very remarkable species, of which the small remains of a pod in one of the specimens (*Hb. F. Muell.*) have been barely sufficient to give a clue to the genus.

2. **L. linifolium,** *Benth.* Glabrous and erect, 1 to 1½ ft. high or more. Leaves lanceolate or linear-lanceolate, acute, 1 to 2 inches long, entire, narrowed into a petiole. Flowers large for the genus, apparently pink or lilac. Sepals 2½ lines. Petals nearly twice as long, obovate. Fruiting racemes loose, with semi-erect or at length spreading pedicels of 4 to 5 lines. Pod without the wings nearly orbicular, rather more than 3 lines diameter, very flat, the wings at the top forming a triangular, erect, acute lobe nearly 2 lines long; the subulate style about half their length in the sinus, which is very open. Seeds compressed. Cotyledons linear.—*Lepia linifolia,* Desv. Journ. Bot. iii. 166 and 181; *Iberis linearifolia,* DC. Syst. Veg. ii. 405.

  **W. Australia.** Sharks Bay, *Herb. Mus. Par.;* Flinders Bay, *Collie;* Murchison river, *Sanford.*

3. **L. leptopetalum,** *F. Muell. Pl. Vict.* i. 48. A low, scrubby, much-branched shrub, quite glabrous. Leaves linear, thick and succulent, almost semiterete, the longer ones ½ to 1 in. long, those of the side branches much smaller. Sepals about 2 lines long. Petals scarcely longer, linear, often almost subulate. Stamens 6. Fruiting racemes short and loose, with spreading pedicels 2 to 3 lines long. Pod very flat, oval-elliptical, about 3

lines long; dorsal wings extending at least halfway down the valves, and forming at the top of the pod two short obtuse lobes, the subulate style projecting much beyond them. Seeds much compressed, exuding a viscous but clear mucilage when soaked.—*Monoploca leptopetala*, F. Muell. in Trans. Phil. Soc. Vict. i. 35.

**N. S. Wales.** Darling river, *F. Mueller.*
**Victoria.** High barren limestone rocks of the Murray, and in the surrounding district, *F. Mueller.*

4. **L. rotundum,** *DC. Syst. Veg.* ii. 537; *Prod.* i. 205. Glabrous and erect or branching and decumbent at the base, 3 to 6 in. or rarely nearly 1 ft. high. Leaves linear, obtuse or rarely acute, seldom above 1 in. long, quite entire, narrowed into a petiole. Flowers small, white. Sepals about 1 line. Petals obovate, rather longer. Fruiting racemes rigid, 2 to 4 in. long, with spreading pedicels of about two lines. Pod nearly orbicular, without the wings about two lines diameter, and not so flat as in *L. linifolium;* dorsal wings of the valves continued to their base, but much broader at the top, where they form two obtuse lobes at least 1 line long; style from ½ to ¾ their length in the sinus, which is usually narrow.—Hook. Ic. Pl. t. 609; *Lepia rotunda*, Desv. Journ. Bot. iii. 166 and 181; *Monoploca rotunda*, Bunge, in Pl. Preiss. i. 260; *Monoploca linifolia*, Bunge, l. c., without the synonyms.

**W. Australia.** Swan River, *Drummond, Preiss, n.* 1941 and 2070; Princess Royal Harbour, *Maxwell;* Murchison river, *Oldfield.*

5. **L. phlebopetalum,** *F. Muell. Pl. Vict.* i. 47. Very closely allied to *L. rotundum,* and perhaps a variety only, scarcely differing from it except in the pod, which is orbicular-ovate, 2 to 2½ lines long, with an exceedingly narrow wing extending about halfway down the back of the valves, and forming at the top two minute lobes, often not ½ line and seldom ¾ line long; with the very slender small style projecting from between them. In some specimens, however, of Burkitt's the lobes of the pod and proportions of the style are intermediate between this and *L. rotundum.*—*Monoploca phlebopetala*, F. Muell. in Linnæa, xxv. 369.

**N. S. Wales.** Darling river, *F. Mueller.* Between the Lachlan and Darling river, *Burkitt.*
**Victoria.** Barren localities on the Murray, *F. Mueller.*
**S. Australia.** Rocky Creek, *F. Mueller;* N.W. interior, *M'Douall Stuart.*

6. **L. monoplocoides,** *F. Muell. in Trans. Phil. Soc. Vict.* i. 35, *and Pl. Vict.* i. 47. An erect, branching annual, of about 6 in., glabrous or slightly rough with minute papillæ. Leaves narrow-linear, entire and not auricled, the lower ones sometimes 2 in. long, but mostly ½ to 1 in. Flowers very minute, without petals and with only 4 stamens. Fruiting racemes 2 to 3 in. long, with rigid, rather spreading, flattened pedicels, of 1½ to 2 lines Pod orbicular, scarcely 2 lines long, flat, winged all round, the wings united with the style at the top, and projecting beyond it in 2 minute, connivent, acute lobes, forming a short point to the pod. Seeds with a viscid, clear mucus, as in several of the preceding species.

**N. S. Wales.** Darling river, *F. Mueller.*
**Victoria.** Mallee scrub, on the Murray, towards its junction with the Murrumbidgee, *F. Mueller.*

7. **L. papillosum,** *F. Muell. in Linnæa,* xxv. 370, *and Pl. Vict.* i. 46. An erect, branching annual, usually under 6 in., but, according to F. Mueller, sometimes 1 ft. high or more, the stems covered with little transparent papillæ, and exhaling an unpleasant scent. Radical leaves petiolate, often 2 in. long or more, linear-oblong, coarsely toothed or irregularly pinnatifid, the upper ones lanceolate or linear-cuneate, with a few remote teeth, and clasping the stem by their auricled base, $\frac{1}{2}$ to 1 in. long, and all glabrous. Flowers very small, without petals, and with only 4 stamens. Fruiting racemes mostly 2 to 4 in. long, with rigid, flattened, rather spreading pedicels, of about 2 lines. Pod obovate, about 2 lines long, the valves winged only above the middle, forming 2 rounded terminal lobes, a little more than $\frac{1}{2}$ line long, with the stigma sessile in the rather narrow sinus. Seeds exuding a viscid, clear mucilage in great abundance.

**N. S. Wales.** Interior of the colony, *A. Cunningham.* Between the Darling and Lachlan rivers, *Burkitt.*

**Victoria.** Murray desert, in several localities, *F. Mueller.*

**S. Australia.** In great numbers on the barren hills and plains near Crystal Brook, Rocky River, and to the N.W. of Spencer's Gulf, *F. Mueller;* between Stokes' Range and Cooper's Creek, *Wheeler.*

8. **L. foliosum,** *Desv. Journ. Bot.* iii. 164 *and* 180; *DC. Prod.* i. 206. A low, straggling, glabrous herb, apparently perennial, with hard irregularly divaricate branches, sometimes attaining 2 feet, but often very much smaller. Leaves mostly oblong-cuneate, $\frac{1}{2}$ to 1 in. long, but sometimes lanceolate or almost linear and nearly 2 in. long, or short and obovate, usually with a few coarse teeth at the top, sometimes toothed from the base or pinnatifid with short entire or even toothed lobes, usually narrowed below the middle, but always with a broad half-stem-clasping base, and sometimes auricled. Flowers very small. Petals on short slender claws, with a minute white ovate lamina. Fruiting racemes 2 to 3 in. long, often becoming lateral by the elongation of leafy shoots, with spreading pedicels of about 2 lines. Pods ovate or elliptical, flat, $2\frac{1}{2}$ to 3 lines long, sometimes almost wingless, but usually the very narrow wings form 2 minute, obtuse, terminal points, between which is the very short style. Seeds exuding a not very thick mucilaginous coat.—*L. cuneifolium,* DC. Syst. Veg. ii. 545; Hook. f. Fl. Tasm. i. 25; *L. impressum,* Bunge, in Pl. Preiss. i. 260.

**N. S. Wales.** Lord Howe's Island, near the coast, and in waste places, *Milne, M'Gillivray.*

**Victoria.** On the seacoast, *Harvey.*

**Tasmania.** On the seacoast, in various places round the island, and in the islands of Bass's Straits, *J. D. Hooker.*

**S. Australia.** Kangaroo Island, *Bernier.* (*Hb. Muell.*)

**W. Australia.** Freemantle, *Collie, Preiss, n.* 1942.

This species is chiefly distinguished from *L. ruderale* by its coarser habit, usually broader leaves and more perfect flowers, and by the pods usually twice the size. It represents in Australia the *L. piscidium* of the Pacific Islands, which has a nearly similar pod and flowers, but most of its leaves are narrowed into a petiole, without the broad stem-clasping base of the Australian plant.

9. **L. ruderale,** *Linn.; DC. Prod.* i. 205. An annual, biennial, or sometimes perennial, glabrous or with a few minute scattered hairs, commencing to flower when very small, but growing out to 1 or even 2 ft., with

hard stems, and numerous divaricate, thin, wiry branches.  Radical leaves once or twice pinnatifid, with narrow-linear lobes, but soon decaying; stem-leaves linear or rarely almost oblong-cuneate, usually with a few irregular teeth, especially towards the top, sometimes almost pinnatifid, the uppermost often linear and entire.  Flowers minute, without petals, and with only 2 stamens.  Fruiting racemes usually rather loose, but rigid, 2 to 3 in. long, with slender stiff spreading pedicels of 2 or 3 lines, but sometimes the racemes remain short and dense as when in flower.  Pods ovate, 1 to near 1½ lines long, minutely 2-lobed at the top, with a short style between the lobes.  Seeds ovate, usually exuding no mucus.—Reichb. Ic. Fl. Germ. ii. t. 10 ; Hook. f. Fl. Tasm. i. 25 ; F. Muell. Pl. Vict. i. 45 ; *L. puberulum*, Bunge, Pl. Preiss. i. 261 ; *L. hyssopifolium*, Desv. Journ. Bot. iii. 164 and 179 ; *L. fruticulosum*, Desv. l. c. 165 and 180 (a tall luxuriant form).

**N. S. Wales.**  New England, *C. Stuart ;* Paramatta, *Herb. Mueller.*
**Victoria.**  Throughout the colony, except at alpine elevations, *F. Mueller.*
**Tasmania.**  Common on waysides and by the seashore in many localities, *J. D. Hooker.*
**S. Australia.**  Abundant in many localities, especially about salt-marshes and in waste places, *F. Mueller* and others.
**W. Australia.**  Apparently abundant, *Drummond, Preiss, n.* 1940, and others.
Var. *crispum.*  Usually striated and very divaricate.  Leaves short, oblong, cuneate, mostly toothed.  Pods rather long.—*S. crispum,* Desv. Journ. Bot. iii. 165 and 176 ; *L. Novæ-Hollandiæ,* Desv. l. c. 177.
Var. (?) *spinescens.*  Smaller branches becoming thorny; pods rather larger, ovate or elliptical, the notch scarcely perceptible.—Salt-marshes of S. Australia towards the mouth of the Murray, *Hildebrand, Whan,* in Herb. Mueller.  *L. ambiguum,* F. Muell. in Trans. Phil. Soc. Vict. i. 34, appears to be the same or a similar variety in a luxuriant state without the thorns.  Both are now included by F. Mueller in the *L. ruderale.*
The species has a wide range, chiefly along the seacoasts of the temperate regions of Europe, Asia, and N. Africa.

## 15. THLASPI, Linn.

Sepals erect, equal at the base.  Petals obovate, equal.  Pod short, ovate, obovate, obcuneate or oblong, much compressed laterally, notched or rarely acute at the top, the valves boat-shaped, keeled or winged, the septum narrow ; style filiform or stigma sessile.  Seeds 2 or rarely 3 or 4 in each cell, not winged ; cotyledons accumbent.—Annual or perennial herbs, the radical leaves usually spreading, entire or toothed, those of the stem often auricled at the base.  Flowers white, pink, or pale purple, rarely yellow.

A considerable genus spread over the temperate and colder regions of the northern hemisphere, with a very few S. American species, and none from S. Africa.  The Australian ones are all endemic, and differ from the generality of the northern ones in the seeds, usually 3 or 4 in each cell instead of 2 only; three of the species have not the auricled leaves of the genus, and one has yellow flowers.

Slender plant of 1 to 3 in.  Stem-leaves auricled and stem-clasping  . 1. *T. Tasmanicum.*
Stems rigid, with petiolate leaves.
  Pubescence scanty, mostly simple.
    Flowers white . . . . . . . . . . . . . . 2. *T. cochlearinum.*
    Flowers yellow . . . . . . . . . . . . . 3. *T. ochranthum.*
  Pubescence stellate . . . . . . . . . . . . . . 4. *T. Drummondi.*

1. **T (?) Tasmanicum,** *Hook. f. Fl. Tasm.* i. 23.  A small, slender,

erect, simple, or slightly-branched annual, 1 to 3 in. high, sprinkled with a few stellate hairs. Radical leaves petiolate, ovate, entire, 2 to 3 lines long; stem-leaves lanceolate or oblong, often 5 to 6 lines long, the lowest narrowed at the base, the others auricled and stem-clasping. Flowers small, white, the petals longer than the sepals. Fruiting racemes loose, with slender divaricate pedicels of 2 to 3 lines. Young pod obovate, very flat, with strongly keeled valves and 3 or 4 seeds in each cell.—*Hutchinsia Tasmanica*, Hook. Ic. Pl. t. 848.

**Tasmania.** Western mountains at Arthur's Lake, *Gunn.*

The habit of this little plant is quite that of the European species of *Thlaspi*, in which genus Dr. Hooker had at first placed it. We have since thought it might belong to the New Zealand genus *Notothlaspi*, characterized by numerous seeds and incumbent cotyledons, a point which cannot be determined till more mature seeds shall have been examined. The habit is against the association.

2. **T. cochlearinum,** *F. Muell. Pl. Vict.* i. 51. An erect, rigid, branching annual, 6 in. to 1 ft. high, slightly pubescent, with a few short, mostly simple and reflexed hairs. Leaves lanceolate or linear-oblong, entire or with 1 or 2 coarse teeth or lobes on each side, narrowed into a petiole, the lower leaves about 2 in. long, the upper ones few and smaller. Flowers white, rather large. Sepals open, 1¼ in. long. Petals much larger. Fruiting racemes loose, about 2 in. long, with half-spreading pedicels of 6 to 8 lines. Pod broadly oval, 4 to 5 lines long, obtuse at the top but not notched, pubescent with short, rigid, reflexed hairs; styles subulate, nearly 1 line long. Valves keeled, but not distinctly winged. Seeds 2 to 4 in each cell, flat, orbicular, emitting a clear, viscid mucus when soaked; cotyledons accumbent. —*Eunomia cochlearina*, F. Muell. in Linnæa xxv. 369.

**S. Australia.** Sandy hills between the Broughton and Rocky rivers, and at Crystal Brook, *F. Mueller.*

3. **T. ochranthum,** *F. Muell. mss.* From the very few specimens this appears to be a smaller plant than *T. cochlearinum*, which it approaches very nearly, with the same appressed hairs, either reflexed or attached by the centre, and a similar though smaller foliage, but the flowers are yellow, the fruiting pedicels much shorter, and the pods very broadly oval or almost orbicular, about 3 lines long.

**N. S. Wales.** On the tributaries of the Upper Darling, *Bowman.* Between the Darling and Lachlan rivers, *Burkitt*, in each case single small specimens (*Hb. F. Muell.*)

3. **T. Drummondi,** *Benth.* Stems more branching than in *T. cochlearinum*, loosely sprinkled with short stellate hairs. Upper leaves apparently linear-lanceolate, coarsely toothed and on long petioles, but the few on the specimens are in a very bad state. Fruiting racemes 2 to 4 in. long, with spreading pedicels mostly of about 2 lines. Pods obovate-oblong, 4 lines long and 2 broad, obtuse or almost notched, with a very short style, acute at the base, sprinkled with stellate hairs; the valves acutely keeled but scarcely winged. Seeds 2 to 4 in each cell, ovate, compressed, emitting a clear viscid mucus when soaked; cotyledons accumbent.

**W. Australia.** *Drummond, Coll.* 1845. The specimens are very imperfect.

## Order IX. CAPPARIDEÆ.

Flowers usually hermaphrodite. Sepals 4 to 8, either in a single series, free or united in a campanulate calyx, or 2 outer and 2 inner ones. Petals usually 4, imbricate, rarely 2 or none. Torus either small or expanded into a disk or lengthened into a straight or curved stalk to the ovary. Stamens inserted at the base or the summit of the torus or stalk of the ovary, definite or indefinite, all perfect or some reduced to staminodia. Ovary 1-celled, with 1 or usually several parietal placentas, which sometimes protrude so as to divide the ovary into imperfect cells. Stigma sessile or borne on a distinct style. Ovules usually numerous, rarely solitary, anatropous. Fruit either a capsule, with the valves separating from the persistent septum or placentas as in *Cruciferæ*, or indehiscent and succulent, or rarely dry. Seeds reniform or angular, without or with only a very thin albumen. Embryo curved, the cotyledons incumbent, folded, or convolute, very rarely flat.—Herbs or shrubs, rarely trees. Leaves alternate or very rarely opposite, simple, or consisting of 1 to 5 digitate leaflets, with or without stipules, which when present are occasionally prickly. Flowers either solitary or clustered in the axils of the leaves, or more frequently in terminal racemes.

The Order is pretty generally distributed over the warmer and tropical regions of both the New and the Old World. Of the following genera, two only, of one species each, and both anomalous in the Order, are peculiar to Australia, the other three are widely-spread tropical genera.

Herbs with a capsular fruit.
    Torus short, the stamens inserted immediately within the sepals
        and petals. Seeds several.
        Stamens 4 to 6, or rarely 8 . . . . . . . . . . . . . 1. Cleome.
        Stamens 8 to 16 . . . . . . . . . . . . . . . . 2. Polanisia.
    Torus elongated, bearing the stamens at the top under the ovary.
        Stamens all perfect, with long filaments. Leaves alternate, with
          digitate leaflets. Sepals 4. Seeds several . . . . . . . 3. Gynandropsis.
        Stamens very short, those on one side only bearing anthers.
          Leaves opposite, undivided. Calyx 5-lobed. Capsule 1-seeded. 4. Emblingia.
Shrubs or trees, with an indehiscent succulent fruit.
    Ovules and seeds many.
        Torus elongated, with a tube-like appendage at the base . . . . 5. Cadaba.
        Torus short without any basal appendage . . . . . . . 6. Capparis.
    Ovules and seeds usually solitary.
        Leaves minute or none. Flowers diœcious. Sepals imbricate.
          Torus small. Filaments long . . . . . . . . . 7. Apophyllum.
        Leaves opposite. Flowers hermaphrodite. Calyx 5-lobed. Torus
          elongated, with a lobed disk at the top, with anthers on one side 4. Emblingia.

## 1. CLEOME, Linn.

Sepals 4, sometimes united in a 4-toothed calyx. Petals 4, nearly equal. Stamens 6, rarely 4 or 8, all or some only perfect, inserted on the short torus immediately within the petals. Ovary sessile or stalked, with many ovules, the stigma sessile or on a short subulate style. Capsule usually elongated, sessile or stipitate. Seeds many, reniform, usually rough or woolly.—Herbs, either glabrous or glandular-pubescent. Leaves with 3 to 7 digitate leaflets, or in some species not Australian simple.    Flowers solitary or in terminal racemes.

A large genus chiefly abundant in the warm parts of America, and in the hot sandy districts of N.E. Africa and S.W. Asia.

Stemless, with radical leaves and 1-flowered scapes . . . . . . . . . 1. *C. oxalidea.*
Erect and leafy, with racemose flowers    . . . . . . . . . . . . 2. *C. tetrandra.*

1. **C. oxalidea,** *F. Muell. Fragm.* i. 69.    A little, glabrous, glaucous, almost stemless annual.    Leaves radical, consisting of 3 obovate or orbicular leaflets, 2 to 4 lines long, on a slender petiole longer than themselves.    Scapes filiform, 1-flowered, 1½ to 2 in. long.    Sepals about 1 line long.    Petals of a pale pink, ovate, about 2 lines long.    Stamens 6 to 8, with linear-oblong anthers attached near the base.    Capsule sessile, linear-oblong or narrow-linear, ½ to 1 in. long.

**N. Australia.**    Gravelly plains on the Upper Victoria river, and table land at the head of Sturt's Creek, *F. Mueller.*

2. **C. tetrandra,** *Banks, in DC. Prod.* i. 240.    An annual, either glabrous or sprinkled with a few short glandular hairs, the stems often several together, slender, ascending from a few inches to 1½ ft.    Leaves chiefly at the base of the stems on long petioles, with 3 or 5 linear-lanceolate or narrow-oblong leaflets sometimes above an inch long, the upper leaves few, small, with only 3 leaflets or simple.    Raceme loose and slender, with filiform pedicels.    Sepals ½ to 1 line long.    Petals narrow, 3 to 6 lines long, nearly equal.    Stamens 4 to 6.    Capsule sessile, slender, 1 to 1½ in. long, with a short subulate style, the valves thin and minutely striate.    Seeds transversely wrinkled.

**N. Australia.**    N.W. coast, *Bynoe;* Victoria river, *F. Mueller;* Port Essington, *Armstrong;* Gulf of Carpentaria, *R. Brown.*

## 2. POLANISIA, Rafin.

Sepals and petals 4 each, as in *Cleome.*    Stamens usually 8 or more, inserted on the short torus.    Ovary and capsule sessile or stalked, with many ovules and seeds, as in *Cleome.*—Herbs, with the habit of *Cleome,* from which the genus only differs in the increased number of stamens.    Flowers in terminal racemes.

The genus is distributed over the warmer and tropical regions of both the New and the Old World.    The only Australian species is a common tropical weed.

1. **P. viscosa,** *DC. Prod.* i. 242.    An erect branching annual or biennial, usually about 1 ft. high, more or less covered with short, glandular, viscid hairs.    Leaflets 3 or 5, very rarely 7, from obovate or oblong-cuneate to linear-lanceolate, the largest usually 1 to 1½ in. long, but mostly much smaller.    Flowers yellow, in terminal racemes.    Sepals about 2 lines, petals twice or thrice as long, from narrow-oblong to almost ovate.    Stamens from 8 to 16.    Capsule from oblong-linear about 1 in. long to narrow-linear and 3 in. long. strongly striate, the nerves very oblique and anastomosing in the short pods, nearly parallel in the long ones, and always glandular-pubescent.    Seeds wrinkled.—*Cleome flava,* Banks, in DC. Prod. i. 241.

**N. Australia.** Along the whole coast from westward of Victoria river to the limits of Queensland, and abundant about the Gulf of Carpentaria, *R. Brown*, and others.

**Queensland.** Moreton Bay, *F. Mueller.*

**N. S. Wales.** Clarence river, *Beckler.*

Var. *grandiflora.* Slightly pubescent. Leaflets narrow. Sepals about 4 lines, petals nearly 1 in. long. Capsule above 4 in. long. N. W. coast, *Bynoe;* Sweers Island, *Henne.*

Some specimens from the gravelly bed of the Victoria river, *F. Mueller,* have shot out from the flowering racemes, numerous branches crowded with small leaves, and very small axillary flowers almost without stamens, but producing small, slender capsules, the whole plant assuming the appearance of the *P. micrantha,* Boj., from Madagascar. Other specimens from the same locality have all the leaves entire or 3-lobed, but these have no flowers to determine the species with certainty.

The species is a common weed throughout India, extending into tropical Africa.

## 3. GYNANDROPSIS, DC.

(Rœperia, *F. Muell.*)

Sepals and petals 4 each, as in *Cleome.* Torus produced into a long slender gynophore, bearing at its summit about 6 stamens with filiform filaments. Ovary sessile or stalked within the stamens, with many ovules, the stigma sessile or on a subulate style, and the capsule sessile or stalked and many-seeded, as in *Cleome.*—Herbs, with the habit of *Cleome,* from which the genus only differs in the long stalk-like torus bearing the stamens. Flowers in terminal racemes.

*Gynandropsis,* like the last two genera, is dispersed over the tropical regions, both of the New and the Old World. The only Australian species is endemic, and remarkable for the very large size of its flowers.

1. **G. Muelleri,** *Benth.* An erect annual, covered with a glandular viscid pubescence. Leaflets 3 or 5, lanceolate or oblong-linear, those of the upper leaves ½ to 1 in. long on a long petiole. Flowers yellow, on short pedicels in the upper axils, forming a terminal leafy raceme. Sepals ½ to near 1 in. long, narrow, acuminate, unequal. Petals fully 3 in. long, oblong, narrowed into a long claw. Stamens 5 to 7, the stipes or elongated torus often 1½ in. long. Capsule linear, 2 to 2½ in. long, not striate, but rough with short glandular hairs, terminated by a slender style of nearly 1 in.—*Rœperia cleomoides,* F. Muell. in Hook. Kew Journ. ix. 15.

**N. Australia.** N.W. coast, *Bynoe.* High, rocky, sandy table-land at the sources of the river Victoria, Hooker's Creek, and Sturt's Creek, *F. Mueller.*

## 4. EMBLINGIA, F. Muell.

Calyx campanulate, 5-lobed, and split to the base on the upper side. Petals 2, united into a slipper-shaped corolla, ascending on the side opposite to the slit of the calyx. Torus produced into a linear, flat, curved stalk, ascending in the slit of the calyx, bearing a glabrous gland at the base inside. Stamens forming a spreading, disk-shaped ring at the summit of the torus, divided into 8 to 10 lobes, 4 to 6 of the outer lobes or staminodia oblong, pubescent, and without anthers, 4 or 5 on the inner side, very short, each bearing an ovoid 2-celled anther. Ovary sessile within the stamens, ovoid, shortly 2-winged at the top, with a divaricately 2-lobed stigma sessile between the wings.   Placentas 2, each bearing a single laterally-attached ovule.   After flowering, the ovary turns down into the calyx, enlarges very obliquely, the 2 wings forming 2 small points on one side near the base.   Fruit dry, inde-hiscent, with a thin pericarp.   Seed solitary, reniform, with a hard, rough, almost muricate testa.   Embryo involute, as in most *Capparideæ*.—Shrub or undershrub, with opposite leaves and axillary flowers.

This curious genus consists of only a single species peculiar to Australia.

1. **E. calceoliflora,** *F. Muell. Fragm.* ii. 3, *t.* 11.   A prostrate shrub or undershrub, harshly pubescent, resembling in habit some species of *Scævola,* and assuming a yellowish hue when dry.   Leaves mostly opposite or nearly so, lanceolate or elliptical, acute, mostly 1 to 1½ in. long, narrowed into a short petiole, wavy on the edges, and very harsh.   Stipulary spines very mi-nute, often wanting.   Flowers on very short axillary pedicels.   Calyx about 3 lines long, rather herbaceous, divided to about the middle into 5 broad lobes. Corolla about twice as long, broadly oblong, pubescent.   Torus about 4 lines long, pubescent on the thin edges, nearly glabrous along the thickened centre. Pericarp glabrous, 3 or 4 lines broad.

**W. Australia.**  Murchison river, *Oldfield.*  The specimens are too far advanced in flower for satisfactory examination.

## 5.  CADABA, Forst.

Sepals 4, free, the 2 outer ones valvate in the bud.   Petals 4, 2, or none, clawed.   Torus elongated, bearing at the base on one side a tubular, erect appendage.   Stamens 4 to 8, inserted on the summit of the torus.   Ovary on a long stalk within the stamens, 1-celled; placentas 2 or 4, with many ovules in 2 rows.   Stigma small, sessile.   Berry cylindrical.   Seeds nearly globular; cotyledons convolute.—Shrubs, unarmed or prickly.   Leaves simple, or in species not Australian 3-foliolate or wanting.   Flowers axillary, or in terminal racemes or corymbs.

The genus extends over Africa and tropical Asia; the only Australian species is also in the Indian Archipelago.

1. **C. capparoides,** *DC. Prod.* i. 244.   A tall shrub, the young branches, foliage, and inflorescence shortly pubescent.   Stipulary spines small, recurved, occasionally wanting.   Leaves simple, petiolate, from ovate to oblong-lanceolate, obtuse or the upper ones acute, 2 to 3½ in. long, membranous, penninerved, green and pubescent on both sides.   Flowers in short, loose, terminal racemes. Pedicels above 1 in. long, in the axils of small bracts.   Outer sepals herba-ceous, concave, nearly ⅓ in. long; inner ones smaller.   Petals 4, turned towards the side of the flower opposed to the stamens and pistil, 3 with slender claws longer than the calyx, and ovate laminæ of unequal size, but not exceeding 4 lines, the fourth with a shorter, broader claw, and small lamina. Stalk-like torus longer than the calyx, with a much shorter tubular process at the base.   Stamens 5 or 6, with slender filaments.   Fruit pubescent, slender, 4 or 5 in. long, on a long stalk.   Seeds numerous.—Deless. Ic. Sel. iii. 5, t. 9 (incorrect as to the sepals and petals, but accurately described in the text).

**N. Australia.**  N. coast, *Herb. Mus. Par.;* Vansittart Bay, *A. Cunningham.*  It is also found in Timor and Java.

## 6. CAPPARIS, Linn.

(Busbeckia, *Endl.*)

Sepals usually 4, rarely 5, free or the outer ones united in the bud into an entire calyx, which splits irregularly as the flower expands. Petals usually 4, imbricate. Stamens indefinite, inserted on the short torus, the filaments free, filiform. Ovary borne on a long stalk, 1 to 4-celled, with 2 to 6 placentas and several or many ovules; stigma sessile. Berry stalked, globose or elongated, very rarely dehiscent. Seeds several, immersed in pulp, with a hard or coriaceous testa and convolute embryo.—Trees or shrubs, sometimes climbing, unarmed or prickly. Leaves simple, membranous or coriaceous; stipules prickly or setaceous, often only on the young or barren shoots.

A large genus, distributed over the tropical and warm regions, both of the New and the Old World; and divisible, chiefly from remarkable differences in the calyx, into several sections, of which two only are Australian, one, *Eucapparis*, comprises the greater number of the Asiatic and African species, but is not American, the other, *Busbeckia*, is confined to Australia and Norfolk Island. The Australian species of both sections are all endemic, and many of them are remarkable for producing slender barren shoots, with very prickly stipules, and small leaves so very differently shaped from those of the flowering-branches, that where we have specimens of these barren branches only, it is impossible to identify them.

SECT. I. **Eucapparis.**—*Sepals 4, rather large, imbricate in 2 series. Berry globular or ovoid.*

Flowers on slender pedicels in terminal umbels. Outer sepals equal . 1. *C. umbellata.*
Flowers lateral or axillary, pedicels solitary or one above the other.
  One of the outer sepals larger and saccate or concave at the base.
    Stamens 12 or under. Flowers small.
      Pedicels usually 2, one over the other. Flowers very tomentose . 2. *C. lasiantha.*
      Pedicels 4 or 5, one above the other. Flowers slightly pubescent 3. *C. quiniflora.*
    Stamens numerous, or more than 15.
      Sepals very unequal, the largest ¾ in. . . . . . . . . . . 4. *C. nummularia.*
      Sepals slightly unequal, about 3 lines . . . . . . . . . 5. *C. sarmentosa.*

SECT. II. **Busbeckia.**—*Two outer sepals broad, very concave, completely united in the bud, and separating irregularly as the flower expands.*

Leaves mostly ovate or oblong.
  Leaves mostly 2 to 4 in. long. Ovary glabrous. Fruit from ½ to a little more than 1 in. diameter.
    Flowers mostly axillary, distant.
      Leaves ovate. Buds ovoid, acuminate, 1 in. long, almost woody . . . . . . . . . . . . . . . . . 6. *C. ornans.*
      Leaves ovate or oblong. Buds globular, ½ in. long, coriaceous 7. *C. nobilis.*
      Leaves ovate. Buds 4-angled . . . . . . . . . 8. *C. canescens.*
    Flowers in a terminal corymb or short raceme. Buds globular . 9. *C. lucida.*
  Leaves mostly 1 to 1½ in. long. Ovary tomentose. Fruit 2 in. diameter . . . . . . . . . . . . . . . . 10. *C. Mitchelli.*
Leaves lanceolate or long and narrow.
  Leaves obtuse at the base. Petiole very short . . . . . . . 11. *C. loranthifolia.*
  Leaves narrowed into a rather long petiole . . . . . . . 12. *C. umbonata.*

SECTION I. EUCAPPARIS, *DC. Prod.* i. 245.—Sepals 4, rather large, imbricate in two series. Berry globular or ovoid.

1. **C. umbellata,** *R. Br. in DC. Prod.* i. 247. Shrubby, with the young branches tomentose. Stipulary spines small, nearly straight or recurved. Leaves from ovate to narrow-oblong, mostly 1½ to 2 in., or when

full grown 3 in. long, at first membranous, softly pubescent or tomentóse, at length stiff and usually glabrous, on petioles of about two lines. Pedicels slender, 6 to 9 lines long, usually 6 to 8 together in terminal umbels, sessile above the last leaves, or sometimes on short, lateral, leafless branches. Buds small, globular. Outer sepals thin but stiff, equal, 2 to 2½ lines long, orbicular, concave, slightly imbricate, glabrous, inner ones scarcely longer, much imbricate. Petals about 3 lines long, pubescent. Stamens numerous. Ovary glabrous, with 8 to 10 ovules to each placenta. Berry globular, smooth, in our specimens not 1 in. diameter, on a stipes of 1 in. Seeds separated by spurious partitions.

**N. Australia.** Careening Bay, N.W. coast, *A. Cunningham;* barren plains of the Fitzmaurice and Victoria rivers, *F. Mueller;* Gulf of Carpentaria, *R. Brown;* Port Essington, *Armstrong.*

**Queensland.** Cape York, *M'Gillivray;* Port Denison, *Fitzalan.*

The species is most nearly allied to the common Indian *C. sepiaria,* differing chiefly in its sessile umbels and less numerous flowers.

2. **C. lasiantha,** *R. Br. in DC. Prod.* i. 247. A much-branched shrub, clothed with a soft tomentum, usually rust-coloured on the young branches and inflorescence, afterwards paler, and sometimes disappearing on the old leaves. Leaves from ovate to narrow-oblong or almost lanceolate, obtuse, 1 to 2 in. long, rounded at the base, with a very short petiole, thickly coriaceous when full grown, with very oblique primary nerves. Pedicels axillary, solitary or 2 together one above the other, much shorter than the leaves. Outer sepals very concave and unequal, slightly imbricate, softly tomentose, the larger one about 3 lines long and almost saccate at the base; inner sepals and petals ovate, 4 to 5 lines long, very tomentose outside. Stamens about 12. Ovary glabrous, with 10 to 12 ovules to each placenta. Young fruit ovoid, on a slender stipes of 1½ in.

**N. Australia.** N.W. coast, *A. Cunningham;* Victoria river, *F. Mueller;* Thomson river, *A. C. Gregory.*

**Queensland.** N.E. coast, *R. Brown;* Narran river, *Mitchell;* Brisbane river, *A. Cunningham* (from a specimen without flowers).

**N. S. Wales.** Tributaries of the Upper Darling river, *Bowman.*

3. **C. quiniflora,** *DC. Prod.* i. 247. Branches weak and flexuose, the young ones and very young leaves rusty-tomentose, but soon becoming glabrous. Leaves ovate, obtuse or acuminate, 3 to 4 in. long, rounded or almost cordate at the base, on petioles of 3 to 4 lines, rather coriaceous. Pedicels usually under ½ in. long, 3 to 5 together, one above the other, in lateral clusters along the leafless tops of the side-branches, or above the upper axils. Outer sepals thin, slightly pubescent, unequal, the larger one saccate at the base and about 3 lines long; inner sepals and petals longer, oval-oblong, pubescent. Stamens few. Fruit glabrous, globular, ¼ to 1 in. diameter, on a stipes of about 1 in. Some barren shoots, with very small ovate, rhomboid, or oblong leaves, assume a totally different aspect from the rest of the plant.

**N. Australia.** N. coast, *Baudin.*

**Queensland.** N.E. coast, *R. Brown, A. Cunningham;* Cape York, *M'Gillivray;* Hammond Island, Torres Straits, *Rayner.* Also in New Caledonia.

4. **C. nummularia,** *DC. Prod.* i. 246. A low glabrous shrub, prostrate

or reclining on rocks, with hard tortuous branches.  Stipular spines short, straight or recurved.  Leaves broadly ovate or orbicular, very obtuse or some-times·emarginate, with a minute point in the notch, $\frac{1}{2}$ to $\frac{3}{4}$ in. long, rather thick, on petioles of 3 to 4 lines.  Peduncles axillary, solitary, 1 in. long or more.  Outer sepals glabrous, very unequal, imbricate, the large one broadly hood-shaped, acuminate, $\frac{3}{4}$ in. long, the other much narrower and concave. Inner sepals and petals apparently longer and glabrous, but very imperfect in our specimens.  Stamens very numerous.  Berry ovoid, succulent, fully $1\frac{1}{2}$ in. long, marked with longitudinal ribs, on a stipes of at least $1\frac{1}{2}$ in.—F. Muell. Fragm. i. 143 and 244.

**N. Australia.**   Nichol Bay, *Herb. Mueller.*

**W. Australia.**   Sterile islands, *Herb. Mus. Par.*; Dirk Hartog's Island, *A. Cun-ningham, Clifton;* Abrolhos Island, *Bynoe;* Murchison river, *Oldfield, Clifton, Milne.*

5.  **C. sarmentosa,** *A. Cunn. Herb.*   A slender tree, supporting itself on the branches of others, the younger branches slightly rusty-tomentose. Stipulary spines very short and hooked.   Leaves almost sessile, broadly ovate, obovate, or orbicular, obtuse, $\frac{1}{2}$ to $\frac{3}{4}$ in. long or sometimes much smaller, thin and glabrous when full grown.   Flowers 1 or 2 together in the upper axils, on pedicels of 4 to 6 lines.   Outer sepals glabrous, slightly unequal, about 3 lines long; inner sepals and petals rather longer, slightly tomentose or pubescent.   Stamens 15 or more.   Berry ovoid, not large, on a slender stipes of about an inch.

**Queensland.**   Brisbane river, *A. Cunningham, F. Mueller;* between the Mackenzie and Archer's rivers, *Leichhardt.*

SECTION II. BUSBECKIA.—Two outer sepals broad, very concave, com-pletely united in the bud and separating irregularly as the flower expands. Two inner sepals more petal-like.  Berry globular or ovoid.

6.  **C. ornans,** *F. Muell. Herb.*   A woody climber, the branches hoary with a minute pubescence.  Leaves ovate, obtuse, 2 to 3 in. long, narrowed at the base, on petioles of $\frac{1}{2}$ to 1 in., glabrous on both sides.   Stipulary spines conical, reflexed, often wanting on the flowering branches.  Pedicels solitary in the upper axils, $1\frac{1}{2}$ to 2 in. long.  Flowers large and showy.  Outer sepals united into an ovoid acuminate bud of above 1 in. long, of a woody texture, and bursting irregularly; inner sepals orbicular, woolly inside, thick but petal-like.  Petals (4?) obovate, more than 2 in. long.  Stamens nu-merous, about 3 in. long.  Ovary glabrous.  Fruit not seen.

**Queensland.**   Port Denison, *Fitzalan.*

7.  **C. nobilis,** *F. Muell. Herb.*   A small tree, either perfectly glabrous or the young shoots and the under side of the leaves slightly covered with a close minute pubescence.  Stipulary prickles short and conical, seldom seen on the flowering-branches.   Leaves oval-oblong or oblong, acute, shortly acu-minate or obtuse, 2 to 4 in. long, coriaceous and often shining above, on pe-tioles of 3 to 6 lines.  Pedicels solitary in the upper axils or very rarely 2 together, about 1 in. long.   Buds globular, about $\frac{1}{2}$ in. diameter, often slightly emarginate at the top, showing the tips of the 2 outer sepals, which are perfectly united into a coriaceous calyx bursting or splitting irregularly;

inner sepals broadly ovate, ½ in. long, firm in the centre, thin on the edges. Petals 4, white, larger and thinner than the sepals, pubescent inside. Stamens very numerous. Fruit globular, about 1 in. diameter, with a small protuberance at the top, the stipes ½ in. to nearly 2 in. long.' Seeds numerous, embedded in a hard almost woody pulp.—*Busbeckia nobilis*, Endl. Prod. Fl. Norf. 64 ; *Busbeckia arborea*, F. Muell. Fragm. i. 163.

**Queensland.** Brisbane river, *Fraser, A. Cunningham ;* Brisbane and Fitzroy rivers, *F. Mueller.*

**N. S. Wales.** Hastings and Clarence rivers, *Beckler* and others.

Var. *pubescens*, petioles shorter, leaves more pubescent underneath, fruit scarcely umbonate. Brisbane river, *A. Cunningham.*

The same species is also found in Norfolk Island.

8. **C. canescens,** *Banks in DC. Prod.* i. 246. Habit and foliage so nearly that of *C. nobilis* that some specimens without the buds are difficult to distinguish from it, but in general they are of a paler more glaucous green, either minutely pubescent or glabrous. Stipulary prickles subulate, wanting on the flowering branches. Leaves as in *C. nobilis*, or more frequently broader and more obtuse, mostly 1½ to 2 in. long, those of the barren shoots sometimes broadly ovate-cordate with a prickly point. Pedicels solitary or 2 together in the upper axils or terminal, 1 to 2 in. long. Buds tomentose, larger than in *C. nobilis*, and prominently 4-angled. Flowers, ·of which I have only seen fragments, apparently like those of *C. nobilis*. Fruit (not yet ripe) as in *C. nobilis*, but on a longer stipes.

**Queensland.** Bay of Inlets, *Banks ;* Northumberland islands and Keppel Bay, *R. Brown ;* Burdekin and Lynd rivers, *F. Mueller.*

Var. *glauca.* Leaves 3 to 4 in. long, very thick and glaucous. Between the Flinders and Lynd rivers, *F. Mueller.*

9. **C. lucida,** *R. Br. Herb.* A shrub, very nearly allied to *C. nobilis,* but more often pubescent. Leaves ovate or oblong, obtuse, 2 to 3 or rarely 4 in. long, coriaceous and shining when old, but often thinner than in *C. nobilis* and more reticulate. Flowers white, rather smaller than in *C. nobilis*, and usually several together in a terminal cluster or short raceme, the outer ones in the axils of the uppermost leaves. Buds globular, on pedicels of about 1 in. Fruit globular, like that of *C. nobilis.*—*Thylacium lucidum*, DC. Prod. i. 254 ; *Busbeckia corymbiflora*, F. Muell. Fragm. i. 163.

**N. Australia.** N.W. coast, *A. Cunningham ;* Booby islands, Torres Straits, *Herb. Banks.*

**Queensland.** N.E. coast, *R. Brown, A. Cunningham ;* islands of Howitt's group and on the Burdekin river, *F. Mueller ;* Howitt's isles, Hope islets, and Port Molle, *M'Gillivray ;* Port Denison, *Fitzalan.*

10. **C. Mitchelli,** *Lindl. in Mitch. Three Exped.* i. 315. A much-branched shrub, more or less clothed with a minute yellowish or whitish tomentum, sometimes soft and dense, sometimes disappearing on the older leaves. Stipular prickles short, somewhat hooked, often wanting on the flowering branches. Leaves ovate or oblong, obtuse, 1 to 1½ in. long, narrowed into a petiole of 2 to 3 lines, coriaceous and rather thick, obscurely veined. Pedicels few, axillary, 1 to 1½ in. long, thickened upwards. Buds ovoid-globular, usually acuminate, nearly ½ in. long. Outer calyx thick, opening

irregularly or sometimes into 2 valvate concave sepals. Inner sepals 4 to 8 lines long, more or less pubescent, especially at the base, thin and glabrous on the edges. Petals similar, but larger. Ovary tomentose, on a long nearly glabrous stipes. Berry globular, 2 in. diameter when ripe. Seeds 4 to 5 lines long, imbedded in a hard dry pulp.—*Busbeckia Mitchelli,* F. Muell. Pl. Vict. i. 53, t. suppl. 4.

**N. Australia.** Plains of Promise, *F. Mueller.*
**Queensland.** In the interior, *Mitchell;* Burdekin river, *F. Mueller.*
**N. S. Wales.** Liverpool plains, *A. Cunningham;* plains of the Bogan, *Mitchell;* Upper Darling river, *F. Mueller.*
**Victoria.** Mallee scrub, near Eustone Cole, *F. Mueller.*
**S. Australia.** From Lake Torrens and Mount Murchison to Cooper's Creek, *F. Mueller.*

11. **C. loranthifolia,** *Lindl. in Mitch. Trop. Aust.* 220. A scrubby bush, with more or less tomentose branches. Leaves from oblong-linear to broadly lanceolate, obtuse or acute, $1\frac{1}{2}$ to $2\frac{1}{2}$ in. long, obtuse at the base, on a petiole of 1 or rarely 2 lines, coriaceous and at length glabrous. Pedicels in the upper axils about 1 in. long, thickened upwards. Buds ovoid, scarcely acuminate, the outer calyx not so thick as in the other species of the section *Busbeckia.* Inner sepals larger, thickened in the centre. Petals longer thinner, villous inside. Stamens numerous. Ovary glabrous.

**Queensland.** Scrub, near Mount Faraday, *Mitchell.*
**N. S. Wales.** Between Darling river and Cooper's Creek, *Neilson.*

12. **C. umbonata,** *Lindl. in Mitch. Trop. Austr.* 257. A shrub, with tomentose branches like the last, but the leaves usually much longer, often 7 to 8 in. long, and rarely under 3 in., always lanceolate and narrowed into a rather long petiole. Pedicels axillary, thickened upwards, 1 to $1\frac{1}{2}$ in. long. Buds ovoid, the outer calyx very thick and coriaceous. Petals as in *C. Mitchelli.* Fruit apparently small, glabrous, not always marked with the terminal protuberance which suggested the specific name ; the stipes very long.

**N. Australia.** Victoria river and dry ridges towards Fitzmaurice river, *F. Mueller :* Depuch Island, *Bynoe.*
**Queensland.** Brigalow scrub, on the Belyando, *Mitchell;* Dawson river, *Herb. F. Mueller.*

## 7. APOPHYLLUM, F. Muell.

Flowers dioecious. Sepals 3 or 4, imbricate, 2 outside the others. Petals 2 or 4, sessile, imbricate. Male fl. : Stamens 8 to 16, inserted on the short torus with filiform filaments. Ovary none. Female fl. : Stamens none, or rarely 1 to 3. Ovary stipitate with a sessile stigma ; ovules 1 or 2, attached to the sides of the cavity above the middle. Berry shortly stipitate. Seeds 1 or 2, with a smooth testa and involute cotyledons.—Leaves very few, small, alternate.

The genus is limited to the following species, and differs from *Capparis* only in its dioecious flowers and the usually solitary ovule.

1. **A. anomalum,** *F. Muell. in Hook. Kew Journ.* ix. 307. A shrub or tree, almost leafless, with cylindrical, often pendulous branches, silky-white when young, but soon becoming glabrous. Leaves on the young shoots few,

linear or linear-acute, 2 to 3 lines long and very deciduous, or rarely above $\frac{1}{2}$ in. long and more persistent. Flowers small, fragrant, either growing singly along the young shoots or in short lateral racemes or clusters. Petals 1 to $1\frac{1}{2}$ lines long. Sepals rather more than 1 line long, pubescent. Petals unequal, as long as or longer than the sepals, pubescent inside at the base. Fruit nearly globular, the size of a small pea.

**N. Australia.** Brigalow scrub, on the Burdekin, *F. Mueller ;* Cooper's river, *A. C. Gregory.*
**Queensland.** In the interior, *Mitchell.*

## Order X. VIOLARIEÆ.

Flowers usually hermaphrodite. Sepals 5, imbricate. Petals 5, imbricate, equal or unequal, with the lower one larger, or spurred or otherwise dissimilar. Stamens 5, hypogynous or nearly so, the anthers erect and connivent, or connate round the pistil, sessile or on short filaments, the connective often very broad, with the anther-cells opening inwards. Ovary free, sessile, 1-celled, with usually 3 parietal placentas, and several or rarely only 1 or 2 anatropous ovules to each placenta. Style usually simple, often thickened or curved at the top. Fruit a capsule, opening in as many valves as placentas, or rarely an indehiscent berry. Seeds with a fleshy albumen ; embryo axile, usually straight, the cotyledons usually broad and flat, the radicle next the hilum.—Herbs or shrubs. Leaves usually alternate, simple, and rarely lobed or cut, with lateral stipules. Flowers axillary, solitary, or in cymes or panicles, very rarely in racemes. Pedicels usually with 2 bracteoles. Capsules often opening elastically.

An Order generally dispersed over the globe. Of the three Australian genera, two have a very wide geographical range, the third extends from Australia to New Zealand.

Herbs or undershrubs, with very irregular flowers. Fruit capsular.
  Sepals produced into a small appendage, or at least a protuberance
    below their insertion. Lower petal spurred or saccate . . . 1. VIOLA.
  Sepals not produced at the base. Lower petal saccate or gibbous
    at the base . . . . . . . . . . . . . . . . . 2. IONIDIUM.
Shrubs with small regular flowers. Fruit a berry . . . . . . 3. HYMENANTHERA.

(The widely-spread tropical genus *Alsodeia* has not yet been detected in Australia.)

### 1. VIOLA, Linn.

Sepals produced into a small appendage or protuberance below the insertion. Petals spreading, the lowest usually larger, spurred or saccate at the base. Anthers nearly sessile, the connectives flat, produced into a membranous appendage beyond the cells, those of the 2 lower anthers usually bearing a small dorsal reflexed protuberance or spur. Style variously thickened or dilated at the top, straight with a terminal stigma, or incurved with the stigma in front. Capsule opening elastically in 3 valves. Seeds ovoid-globular with a crustaceous testa.—Herbs, with the stipules usually foliaceous and persistent. Peduncles axillary, 1-flowered. Most species, besides the perfect flowers, produce later in the season small apetalous, but very prolific flowers.

A very large genus, most of the species natives of the temperate regions of the northern hemisphere, or of the high mountains of South America, with a very few dispersed over Africa, Australia, and New Zealand. The Australian species are either quite endemic or extend only to Norfolk Island and New Zealand. They are all perennials.

Stemless, with a tufted or creeping rhizome.
  Leaves lanceolate, oblong, or scarcely ovate. No stolons. Stipules adnate. . . . . . . . . . . . . . . . . . . 1. *V. betonicæfolia.*
  Leaves nearly orbicular.
    Stolons creeping. Spur reduced to a slight protuberance. Stipules free . . . . . . . . . . . . . . . . 2. *V. hederacea.*
    No stolons. Spur prominent. Stipules adnate . . . . . 3. *V. Cunninghamii.*
Flowering-stems elongated. Leaves broad.
  Leaves scarcely cordate. Stipules adnate . . . . . . . . 3. *V. Cunninghamii.*
  Leaves deeply cordate. Stipules free . . . . . . . . 4. *V. Caleyana.*

1. **V. betonicæfolia,** *Sm.; DC. Prod.* i. 294. Glabrous or pubescent, stemless, and without stolons, and often tufted, the stock either ending underneath abruptly, with thick spreading fibres, or tapering into a horizontal or descending root. Leaves radical, from lanceolate to oblong or nearly ovate, mostly obtuse, and 1 to $1\frac{1}{2}$ in. long, entire or slightly crenate, truncate or slightly cordate, rarely narrowed at the base, with the long petiole usually dilated at the top. Stipules linear, adnate to the petiole. Scapes of the perfect flowers usually considerably longer than the leaves, with the subulate bracts below the middle. Flowers violet, rather large. Sepals lanceolate, acute, $2\frac{1}{2}$ to nearly 3 lines long, with short blunt basal appendages. Lateral petals usually copiously bearded inside, the upper ones less so, the lowest not at all; spur broad and obtuse, much shorter than the sepals. Style thickened upwards, concave at the top, not winged. Apetalous flowers on very short scapes.—Hook. f. Fl. Tasm. i. 27; F. Muell. Pl. Vict. i. 64; *V. phyteumæfolia* and *V. longiscapa*, DC. in Herb. Lamb., from the char. in G. Don, Gen. Syst. i. 322.

**Queensland.** *Mitchell;* near Brisbane, *F. Mueller.*
**N. S. Wales.** Port Jackson, *R. Brown, Sieber, n.* 180, and others; northward to Clarence and Macleay rivers, *Beckler;* southward to Twofold Bay, *F. Mueller;* and in the interior to the Lachlan river, *A. Cunningham, Fraser,* etc.
**Victoria.** Port Phillip, *R. Brown;* grassy moist ridges, sparingly scattered over the southern and eastern parts of the colony, *F. Mueller.*
**Tasmania.** Common in moist good soils throughout the island, *J. D. Hooker.*
**S. Australia.** Near Rivoli Bay and in the Bugle ranges, but rare, *F. Mueller.*
Received also from Norfolk Island, *Backhouse,* and the species is nearly allied to *V. Patrinii,* DC., which is common in India, eastern Siberia, and China, and only appears to differ from *V. betonicæfolia* in the rather longer spur and the style usually broadly winged.

2. **V. hederacea,** *Labill. Pl. Nov. Holl.* i. 66, *t.* 91. Glabrous or pubescent, densely tufted or widely creeping by its numerous stolons, very rarely emitting weak leafy stems. Leaves reniform, orbicular, or spathulate, usually under $\frac{1}{2}$ in. diameter, but when very luxuriant, 1 to $1\frac{1}{2}$ in., entire or irregularly and sometimes coarsely toothed. Stipules free, brown, lanceolate-subulate. Scapes usually longer than the leaves, the bracts about the middle. Flowers usually small, blue, rarely white, but sometimes fully $\frac{3}{4}$ in. broad. Sepals lanceolate, with only a slight protuberance below their insertion. Petals glabrous, or the lateral ones slightly pubescent inside, the spur of the lower

H 2

one reduced to a slight concavity. Lower anthers with a very slight dorsal protuberance. Style bent at the base, the upper part cylindrical, truncate at the
top, but not thickened. Seeds usually dark-coloured, but sometimes white.
—DC. Prod. i. 305; Hook. Exot. Fl. iii. t. 225; Reichb. Icon. Exot. t.
110; Hook. f. Fl. Tasm. i. 26; F. Muell. Pl. Vict. i. 65; *V. Sieberiana,*
Spreng. Syst. Cur. Post. 96; *Erpetion reniforme,* Sweet, Brit. Fl. Gard. ii.
t. 170; *E. hederaceum, E. petiolare,* and *E. spathulatum,* G. Don, Gen. Syst.
i. 335.

**Queensland.** Moreton Bay, *Fitzalan.*
**N. S. Wales.** Frequent about Port Jackson, *R. Brown, Sieber, n.* 426, and others;
northward to Clarence river, *Beckler;* and southward to Twofold Bay, *F. Mueller.*
**Victoria.** Dispersed over the whole colony, except the N.W., in sandy moist heathy
soil, along rivulets and in boggy places up to 7000 ft. elevation, *F. Mueller.*
**Tasmania.** Throughout the island, very common, *J. D. Hooker.*
**S. Australia.** Rare, near Mount Barker, on the Onkaparinga, in the Barossa ranges,
and near Rivoli Bay, *F. Mueller.*

3. **V. Cunninghamii,** *Hook. f. Fl. N. Zel.* i. 16. Glabrous, stemless,
or rarely with weak elongated stems, the stock tufted with an underground
creeping rhizome. Stipules adnate to the petiole, with a short free lanceolate-
subulate point. Leaves very broadly ovate or nearly orbicular, truncate or
slightly and broadly cordate at the base, mostly under ½ in. diameter, slightly
crenate. Peduncles of the perfect flowers longer than the leaves, the small
bracts below the middle. Flowers rather small, pale violet. Sepals oblong-
lanceolate. Lateral petals obscurely bearded; spur short and obtuse, yet
much more prominent than in *V. hederacea.* Spurs of the lower anthers
short and obtuse. Style club-shaped, emarginate at the top.—Hook. f. Fl.
Tasm. ii. 357.

**Tasmania.** In the Western Mountains, by rivulets on Cuming's Head, *Archer.* Also
in New Zealand.

4. **V. Caleyana,** *G. Don, Gen. Syst.* i. 329. Usually glabrous. Stems
weak, decumbent or half erect, from a few inches to nearly a foot long. Leaves
ovate or nearly orbicular, very deeply cordate, from ¾ to 1½ in. long, or when
very luxuriant, larger and broadly triangular, often obscurely crenate. Stipules oblong or lanceolate, leafy, free from the petiole. Peduncles of the perfect flowers usually longer than the leaves, with the bracts about the middle.
Flowers rather small, white. Sepals lanceolate. Petals glabrous or the lateral
ones slightly bearded, the spur very short and broad. Anther-spurs very
short. Style almost as in *V. biflora,* thickened upwards, concave at the top,
truncate or emarginate at the back, and open in front.—Hook. f. Fl. Tasm.
ii. 357; F. Muell. Pl. Vict. i. 64.

**N. S. Wales.** Nepean river, *R. Brown;* near Marshall's Mount, Illawarra, *Back-
house.*
**Victoria.** Banks of rivulets subject to inundation, near springs, and in wet forest
gullies, Gipps' Land, *F. Mueller.*
**Tasmania.** Deloraine, *Archer.*
Peculiar to Australia, but very nearly allied to the European and Asiatic *V. biflora,* Linn.
(*V. reniformis,* Wall.), which has more reniform leaves and yellow flowers.

## 2. IONIDIUM, Vent.

### (Pigea, *DC.*)

Sepals not produced at the base. Petals spreading, the lowest sometimes slightly larger than the others, more frequently very much larger, with a broad claw, gibbous or saccate at the base. Anthers nearly sessile, or on distinct filaments, the connectives flat, produced into a membranous appendage beyond the cells, those of the 2 lower ones bearing a dorsal reflexed protuberance, spur, or gland, the 2 rarely united into one. Style thickened and incurved at the top, with the stigma in front. Capsule opening elastically in 3 valves. Seeds ovoid-globular, with a crustaceous testa.—Herbs or small shrubs. Leaves alternate or rarely opposite, usually narrow. Stipules small and narrow. Peduncles axillary or in a terminal raceme, 1- or several-flowered.

A considerable genus, chiefly tropical, and the greater number of species American; four or five are found in tropical Asia and Africa, and one of these occurs in Australia, the others here enumerated are all endemic.

Peduncles axillary, 1-flowered, or very rarely here and there 2-flowered.
  Lower petal more than twice as long as the calyx.-
    Leaves entire, or rarely toothed. Appendages of the lower filaments
      nearly glabrous. Seeds striate . . . . . . . . . . 1. *I. suffruticosum.*
    Leaves toothed. Appendages of the lower filaments woolly-hairy.
      Seeds smooth . . . . . . . . . . . . . . . . 2. *I. aurantiacum.*
  Lower petal not half as long again as the calyx . . . . . . . 3. *I. brevilabre.*
Peduncles 1-flowered in the upper axils, the upper ones longer than the
  leaves, and forming a terminal leafy raceme . . . . . . . 5. *I. Vernonii.*
Peduncles mostly 2- to 4-flowered, not longer than the leaves. Lower
  petal small . . . . . . . . . . . . . . . . . . . . . 4. *I. floribundum.*
Peduncles slender, much longer than the leaves, with a leafless raceme
  of 2 or more flowers.
    Upper leaves often opposite. Sepals lanceolate, shorter than the
      lateral petals . . . . . . . . . . . . . . . . . 6. *I. filiforme.*
    Leaves all alternate. Sepals ovate, as long as or longer than the
      lateral petals . . . . . . . . . . . . . . . . . 7. *I. calycinum.*

1. **I. suffruticosum,** *Ging. in DC. Prod.* i. 311. Much-branched, glabrous or very slightly pubescent, and usually from 1 to 1½ ft. high, and more or less woody at the base. Leaves alternate, narrow-linear, or rarely linear-oblong or lanceolate, entire or rarely toothed, mostly 1 to 2 in. long. Peduncles axillary, filiform, 1-flowered, 2 to 4 lines long, with a pair of minute bracts under the pedicel. Sepals lanceolate, very acute, with a very prominent green midrib, 1½ to 2 lines long. Lateral petals rather longer than the calyx, with a broad ovate-falcate base, and a small, ciliate, obtuse extremity, sometimes expanded into a small lamina; upper petals smaller; lowest petal purple or rarely yellow, about ½ in. long, the claw longer than the other petals, saccate at the base, the lamina broadly ovate and longer than the claw. Filaments at least half as long as the anthers, the 2 lower ones with a thick spur, either quite glabrous or with a minute tuft of hair. Seeds elegantly marked with longitudinal striæ.—Wight, Ic. t. 308 ; *Pigea Banksiana,* DC. Prod. i. 307.

102 X. VIOLARIEÆ. [Ionidium.

**N. Australia.** Gulf of Carpentaria, *R. Brown;* Dampier's Archipelago, *A. Cunningham;* Port Essington, *Armstrong;* Arnhem's Land to lat. 32° on the E. coast, *F. Mueller.*
**Queensland.** Brisbane river, etc., Moreton Bay, *F. Mueller, Fitzalan;* Rockhampton, *Thozet;* Port Denison, *Fitzalan.*
**N. S. Wales.** Clarence and Hastings rivers, *Beckler.*

The species is widely spread over tropical Asia and Africa. The above description is taken from Australian specimens; in the majority of Indian and African ones the leaves are broader and the lower petal smaller. The flowers are almost always purple, but some specimens of Cunningham's and Brown's, said to have yellow flowers, have the seeds and foliage of *I. suffruticosum,* rather than of *I. aurantiacum.*

2. **I. aurantiacum,** *F. Muell. Herb.* Pubescent with short spreading hairs or rarely glabrous, often woody at the base, branched, 6 in. to 1 ft. high or rather more. Leaves linear or oblong-lanceolate, 1 to 1½ in. long, bordered with small, distant, acute teeth. Flowers axillary, on peduncles of 3 to 4 lines, as in *I. suffruticosum,* and nearly similar in structure, but the lower petal is smaller and always yellow, the broad lamina usually shorter than the long narrow claw, which is scarcely saccate at the base, and the appendages of the filaments of the lower stamens are covered with long woolly hairs. Seeds, in the few capsules I have seen, smooth and not striate.

**N. Australia.** N.W. coast, *A. Cunningham, Bynoe;* Victoria river, *F. Mueller.*
The distinction between this species and *I. suffruticosum* may require revision when more abundant specimens in flower and seed are obtained, and the relation of the differences of the seeds to the other characters more correctly ascertained.

3. **I. brevilabre,** *Benth.* A glabrous perennial with a woody rhizome. Stems erect, divaricately branched, 6 in. to 1 ft. high, with few small leaves, or in some specimens numerous, nearly simple, about 6 in. high, with more crowded and longer leaves, sometimes 1 in. long, always linear and entire, obtuse, or with a recurved point. Peduncles axillary, slender, 1- or rarely 2-flowered, shorter than the leaves, with a pair of small narrow bracts under the short recurved pedicels. Flowers small (blue?). Sepals narrow-ovate, acute, rather more than 1 line long. Lateral petals about the same length, very obtuse; lowest petal rather longer, the lamina broadly rhomboid, much shorter than the claw, which is broad, concave, with a short obtuse spur at the base. Stamens with the terminal appendage longer than the cells, and the 2 lower filaments distinctly spurred.

**W. Australia.** Swan River, *Drummond, 1st Coll.,* and *n.* 665 of a subsequent one.
It is possible that further specimens may prove this to be a remarkable variety of *I. floribundum.*

4. **I. floribundum,** *Walp. Rep.* ii. 767. A glabrous perennial, with the habit of some European species of *Thesium,* forming sometimes a thick woody rhizome, the stems erect, often much branched and rigid. Leaves all alternate, rather crowded, linear or lanceolate-linear, mostly with a short recurved point, ½ to 1 in. or rarely 1½ in. long, entire. Peduncles axillary, usually once or twice forked, each branch bearing 1 or 2 small violet, blue, or white flowers, on pedicels of about a line, the whole forming little cymes rarely exceeding the leaves, the lower peduncles sometimes 1-flowered, but always with several pairs of small bracts. Sepals ovate, 1 to nearly 2 lines long. Lateral petals about the same length, very obtuse; lowest petal not twice as long, the lamina broad, the short claw distinctly spurred. Two

lower stamens shortly spurred at the base.—F. Muell. Pl. Vict. i. 68, t. suppl. 8 ; *Pigea floribunda*, Lindl. in Mitch. Three Exped. ii. 165 ; *I. australasiæ*, Behr. in Linnæa, xx. 629 ; *I. multiflorum*, Turcz. in Bull. Mosc. 1854, ii. 340.

**N. S. Wales.** Eurylean scrub, *A. Cunningham.*

**Victoria.** Barren ridges and low stony and rocky ranges in the vicinity of the Murray river and its lower tributaries, *F. Mueller;* towards the Australian Pyrenees, *Mitchell.*

**S. Australia.** Not rare through the scrubby lowlands and mountain tracts from Guichen Bay to Spencer's Gulf, *F. Mueller,* and others.

**W. Australia.** South coast ?, *Drummond,* supplement to 5th Coll. n. 72, *Harvey.*

5. **I. Vernonii,** *F. Muell. Pl. Vict.* i. 223. Glabrous, with erect, slender, but stiff stems, little branched, except at the base, and usually about 1 ft. high, as in *I. filiforme,* but the branches more angular. Leaves all alternate, linear or narrow-lanceolate, rarely above 1 in. long, and the upper ones much smaller and very narrow. Peduncles 1-flowered, as in *I. suffruticosum,* but only in the upper axils, and the upper ones longer than the small floral leaves, so as to form a terminal leafy raceme. Flowers blue, very much like those of *I. filiforme,* the lower petal of the same shape and size, except that the claw is distinctly spurred at the base, and the lateral petals are more obtuse than in that species ; stamens the same, except that the subulate appendages at the top of the anther-cells are still more minute.

**N. S. Wales.** Port Jackson, *Anderson, W. Vernon, Woolls.* In the interior ?, *Leichhardt ;* Twofold Bay, *F. Mueller.*

**Victoria.** Barren plains and ridges near the Genoa river, *F. Mueller.* Specimens of this species are included by De Candolle amongst those named by him *Pigea filiformis ;* the two species are often mixed on the same sheet in the Paris and other Herbaria.

6. **I. filiforme,** *F. Muell. Pl. Vict.* i. 66. A perfectly glabrous herb, said by some collectors to be annual, but certainly in many instances forming a perennial rootstock. Stems slender, but stiff and wiry, simple or branched, usually 1 to 2 ft. high, but when eaten down, sending up numerous short erect branches. Leaves alternate or the upper ones opposite, narrow-linear, mostly 1 to 2 in. long, entire, the lowest ones shorter, broader, and petiolate. Flowers blue, in slender leafless racemes, on terminal or axillary peduncles, always much longer than the leaves, the pedicels under a line long. Sepals shorter than the lateral petals, lanceolate, acute. Lower petal usually fully $\frac{1}{2}$ in. long, ovate, narrowed into a concave claw, saccate at the base, but varying considerably in size and breadth ; lateral petals broadly falcate, acute, about 2 lines long ; upper ones smaller. Anthers with an orange ovate appendage at the top of the connective, and two minute subulate appendages on the cells themselves ; the 2 lowest have also a small glandular protuberance on the back at their base.—*Pigea filiformis,* DC. Prod. i. 307 ; *I. linarioides,* Presl, Bot. Bm. 12.

**Queensland.** Moreton Bay, *A. Cunningham, Fraser;* Glasshouse ridges, *F. Mueller.*

**N. S. Wales.** Common about Port Jackson, *R. Brown* and others, and northward to New England, ascending to 5000 ft., and Clarence and Hastings rivers, *Beckler,* and southward to the limits of the colony.

**Victoria.** Dry, grassy, or scrubby ridges near the Avon and Mitchell rivers in Gipps' Land, *F. Mueller.*

*I. monopetalum,* Rœm. and Schult. Syst. i. 400 (*Pigea monopetala,* Ging. in DC. Prod. i. 307; *Solea monopetala,* Spreug. Syst. i. 804), described from a single specimen of uncertain origin, in Rœmer's Herbarium, can only refer to the present species.

7. **I. calycinum,** *Steud.; F. Muell. Pl. Vict.* i. 224. A glabrous perennial, with the habit, narrow-linear leaves and racemose flowers on long leafless peduncles, of *I. filiforme,* but the leaves are usually all alternate, the sepals larger, ovate, with a short point, very thin and scarious on the edges, usually fully 2 and often 3 lines long. Lower petal fully as large as in *I. filiforme,* and of the same shape, except that the spur at the base is more prominent; the lateral petals scarcely exceed the calyx and are very obtuse, the upper ones rather shorter. The protuberances at the base of the lower anthers are more prominent than in *I. filiforme,* broad and very obtuse, and the subulate tips to the cells are very minute or wholly wanting.— *Pigea calycina,* DC. Prod. i. 307; *Solea calycina,* Spreng. Syst. i. 804; *Pigea glauca,* Endl. in Hueg. Enum. 5; *Ionidium glaucum,* Steud.; F. Muell. Pl. Vict. i. 67; *Vlamingia australasica,* Vriese, in Pl. Preiss. i. 399, as corrected, ii. 242.

**W. Australia.** Swan River, *Huegel, Drummond, Preiss, n.* 1449 and others; Murchison river, *Oldfield.*

## 3. HYMENANTHERA, R. Br.

Sepals nearly equal. Petals nearly equal, short. Anthers almost sessile, united in a tube round the pistil, the connectives all terminating in a membrane, and bearing on their backs an erect scale. Placentas of the ovary 2 or rarely 3, each bearing 1 ovule. Style short, with a 2- or rarely 3-lobed stigma. Berry globular, small. Seeds 1 or 2, nearly globular. Cotyledons narrow.—Rigid shrubs or small trees. Leaves alternate, often clustered, small, entire or toothed, without stipules. Flowers small, axillary, frequently polygamous.

A small genus which, besides the following species, comprises one from Norfolk Island, and another from New Zealand.

1. **H. dentata,** *R. Br. in DC. Prod.* i. 315. A glabrous, rigid, much branched shrub, often attaining many feet in height, but low and scrubby in alpine situations, the side branches often converted into strong thorns. Leaves from oblong-elliptical to linear, obtuse or acute, usually $\frac{1}{2}$ to $1\frac{1}{2}$ in. long, and marked with a few irregular distant teeth, coriaceous, sessile or narrowed into a short petiole; on some luxuriant barren shoots they become much larger, membranous, and deeply toothed or lobed. Pedicels solitary or 2 together, about 1 line long, with a pair of minute bracts. Sepals orbicular. Petals about 2 lines long, the erect portion twice as long as the sepals, the obtuse tips spreading or reflexed. Connective of the anthers with a fringed terminal membrane, involute on the edges, the dorsal scale linear, acute, as long as the cells. Female flowers in the normal form pedicellate as well as the males, but smaller, with smaller, usually imperfect anthers. Stigma occasionally 3-lobed, with 3 ovules, although usually 2 only. Berry of a purplish colour, the size of a pea.—Bot. Mag. t. 3163; *H. Banksii,* F. Muell. Pl. Vict. i. 69.

**N. S. Wales,** *R. Brown* and others; Wollondilly and Cox's rivers, *A. Cunningham;* New England, *F. Mueller.*

**Victoria.** Shady banks of rivers, creeks, and rivulets, and fissures of rocks to the highes summits of the Australian Alps, *F. Mueller.*

Var. *angustifolia.* Leaves quite entire, linear-oblong or linear-cuneate, obtuse, and not more than 1 in. long. Flowers almost sessile, the dorsal scale of the anthers broadly obovate. In all the flowers I have examined, both the anthers and the style appear to be perfect.—*H. angustifolia,* R. Br. in DC. Prod. i. 315 ; Hook. f. Fl. Tasm. i. 27.

**Tasmania.** Northern parts of the island. Port Dalrymple, *R. Brown ;* Launceston and summits of the Western Mountains to 3000–4000 ft., Arthur's Lakes, and Vale of Belvoir, *Gunn, J. D. Hooker.* From the examination of numerous specimens, wild as well as cultivated, I had retained this form as a distinct species ; but as F. Mueller assures me that in cultivation it passes into the normal form, I have followed him in uniting it with *H. dentata* as a variety only.

## Order XI. BIXINEÆ.

Flowers regular. Sepals 2 to 6, usually 4 or 5 and imbricate. Petals either none, or as many as the sepals, or indefinite, imbricate or contorted in the bud, deciduous. Stamens hypogynous or slightly perigynous, indefinite or very rarely definite. Anthers 2-celled, opening by longitudinal slits or rarely by terminal pores. Torus often bearing glands or a glandular disk. Ovary free, usually 1-celled, with 3 or more, rarely 2 or 1, parietal placentas. Styles or stigmas as many as placentas, free or united. Ovules 2 or more to each placenta, amphitropous or anatropous. Fruit succulent or dry, opening in valves, bearing the placentas in the middle, or indehiscent. Seeds usually few, with a copious and fleshy or rarely thin albumen. Embryo in the axis, straight or curved, the radicle next the hilum, the cotyledons usually broad. —Trees or shrubs, in one genus twiners. Leaves alternate, simple, and often toothed, or rarely palmately lobed or divided. Flowers axillary or terminal, solitary or in clusters, corymbs, racemes, or panicles.

A considerable Order, dispersed over the tropical or warm regions both of the old and the new world. Of the Australian genera, three are common to Asia and Africa, two of the three being also American. The species, however, are all endemic, as is also the fourth anomalous genus.

Anthers long, opening in terminal pores. Seeds curved. Trees
 or shrubs. Leaves digitate. Flowers large . . . . . 1. COCHLOSPERMUM.
Anthers small, opening longitudinally. Seeds straight. Trees or
 shrubs. Leaves simple. Flowers small.
 Sepals 4 to 6. Petals as many. Anthers with an appendage 2. SCOLOPIA.
 Sepals 4 to 6. Petals none. Anthers without any appendage 3. XYLOSMA.
Anthers long, opening longitudinally. Embryo very small.
 Stem twining. Leaves simple . . . . . . . . . . 4. STREPTOTHAMNUS.

## 1. COCHLOSPERMUM, Kunth.

Flowers hermaphrodite. Sepals 5, imbricate, deciduous. Petals 5, large. Stamens numerous. Anthers oblong or linear, opening in terminal pores or very short fissures. Placentas 3 to 5, projecting more or less into the cavity of the ovary, with numerous ovules. Style simple. Capsule 3- to 5-valved, the membranous endocarp separating from the pericarp. Seeds kidney-shaped or spirally curved, covered with wool or bordered by long hairs.— Trees, shrubs, or rarely undershrubs, usually yielding a yellow juice. Leaves palmately lobed or divided. Racemes loose, few-flowered, in the upper axils or in terminal panicles. Flowers large, yellow.

Besides the four following species, peculiar to Australia, there is 1 known from Southern India, 2 from Africa, and about 5 from South America.

Calyx and inflorescence densely tomentose . . . . . . . . . 1. *C. Fraseri.*
Calyx and inflorescence glabrous or slightly glandular-pubescent.
  Leaves tomentose, with short, rounded, obtuse lobes . . . . . 2. *C. heteroneurum.*
  Leaves glabrous, with deep ovate-lanceolate or oblong lobes . . . 3. *C. Gillivræi.*
  Leaves glabrous, divided to the base into narrow-oblong, pedate
    segments . . . . . . . . . . . . . . . . . . 4. *C. Gregorii.*

1. **C. Fraseri,** *Planch. in Hook. Lond. Journ.* vi. 307. Branches glabrous. Leaves unknown. Flowers large, the racemes short, in a loose corymbose panicle, the branches tomentose. Pedicels about ½ in. long, densely tomentose-pubescent. Sepals broadly ovate, very obtuse, tomentose within and without, unequal, the inner larger ones about ¾ in. long. Anthers about 1½ lines long.

**N. Australia.** Melville Island, *Fraser.*

In the absence of the leaves it would have been impossible to distinguish this species from the East Indian *C. gossypium*, but that the anthers are considerably shorter, which may lead one to suppose there may be other differences.

2. **C. heteroneurum,** *F. Muell. Herb.* Young branches pubescent. Leaves nearly orbicular, cordate at the base, attaining 4 or 5 in. diameter, shortly divided into 5 to 9 broad, rounded, very obtuse, and crenate lobes, tomentose-pubescent when young, nearly glabrous except the principal nerves when old, on petioles of 2 to 3 in. Panicle loose and many-flowered, glabrous, except a slight glandular pubescence on the pedicels and at the base of the calyx. Flowers not so large as in '*C. Fraseri*, on pedicels not exceeding ½ in., but lengthening to 1 in. after flowering. Sepals very unequal, quite glabrous, except at the base, with very thin edges, the inner ones about ½ in. long and very broad. Anthers as in *C. Fraseri*. Ovules exceedingly numerous, on 5 parietal placentas partially projecting into the cavity of the ovary. Young capsule slightly tomentose.

**N. Australia.** Victoria river, *F. Mueller, Wickham.*

3. **C. Gillivræi,** *Benth.* The specimens are perfectly glabrous, except a very slight pubescence on the branches of the panicle and pedicels. Leaves palmately divided to within ¼ or ½ in. of the base, into 5 or 7 ovate-lanceolate or oblong-acuminate slightly toothed lobes, of which the central largest ones are usually 2 to 3 in. long, the 2 outermost short and very acuminate. Panicles short and loose. Flowers as in *C. heteroneurum*, or the sepals rather larger. Capsule obovoid-oblong, rarely 3 in. long, truncate at the top, and very much depressed in the centre. Seeds enveloped in a very deciduous wool.

**Queensland.** Lizard Island, off the N.E. coast, *M‘Gillivray;* Burdekin river, *F. Mueller;* Port Denison, *Fitzalan.*

4. **C. Gregorii,** *F. Muell. Fragm.* i. 71. A small tree, quite glabrous, except a very slight glandular pubescence on the branches of the inflorescence and pedicels. Leaves pedately divided to the base into about 7 narrow-lanceolate entire segments, the central ones 2 to 3 in. long, the common petiole 3 to 6 in. Panicles apparently short and not much divided, or reduced to a single raceme. Pedicels about ½ in. long. Sepals and petals as in the last

2 species. Style filiform, slightly thickened towards the top. Outer stamens, as in all the other species, on longer filaments than the inner ones, but the difference is rather more decided in this species. Placentas 5. Fruit not seen.

**N. Australia.** Rocky barren hills in the S.E. part of Arnhem's Land, *F. Mueller.* The fruit described by F. Mueller from Burdekin specimens appears to belong to the *C. Gillivræi,* which has a very different foliage.

## 2. SCOLOPIA, Schreb.

### (Phoberos, *Lour.*)

Flowers hermaphrodite. Sepals 4 to 6, slightly imbricate when very young, but open long before flowering. Petals as many and nearly similar. Stamens indefinite, inserted on the thickened torus, with or without glands. Anthers short, the connective terminating in a thick process. Ovary with 3 or 4 placentas and few ovules. Style filiform, with an entire or lobed stigma. Fruit a berry. Seeds 2 to 4, with a hard testa. Cotyledons leafy.—Trees often armed with axillary spines. Leaves simple, with pinnate veins, entire or toothed. Flowers small, in axillary racemes.

The genus is dispersed over southern and eastern Africa and tropical Asia. The Australian species is endemic.

1. **S. Brownii,** *F. Muell. Fragm.* iii. 11. Perfectly glabrous in all its parts. Leaves from ovate to oblong-lanceolate, mostly acuminate, obtuse or almost acute, rarely rounded at the top, 1½ to 3 in. long, always narrowed into a petiole of 3 to 4 lines, entire or slightly undulate-toothed, rather thick and smooth, obscurely triplinerved, but all the veins less conspicuous than in most species, either without glands or with 2 or 3 marginal glands underneath. Racemes short and axillary or forming a terminal panicle of 1 to 2 in. Pedicels 2 to 3 lines. Calyx 4-cleft, smaller than in *S. crenata,* apparently persistent. Petals 4, rather longer than the calyx, deciduous. Stamens numerous, with slender filaments, surrounded by a ring of glands, either distinct and shortly club-shaped or irregularly connate. Anthers small, the process of the connective glabrous and usually as long as the cells. Placentas 3, with about 4 ovules to each. Stigma slightly 3-lobed.

**Queensland.** Cape York, *M'Gillivray.*
**N. S. Wales.** Hunter's River, *A. W. Scott;* Clarence river, *Wilcox;* Illawara, *Herb. Mueller.*
This species has much the foliage of some forms of the Indian *C. crenata,* but is readily known by the glands of the disk.

## 3. XYLOSMA, Forst.

Flowers diœcious. Sepals 4 or 5, small, imbricate. Petals none. Male fl. : Stamens indefinite, often surrounded by a glandular disk ; anthers short, without appendage. Female fl. : Ovary inserted on an annular disk, with 2 or rarely more placentas, and 2 or few ovules to each ; style entire or divided, with dilated stigmas, or rarely stigma sessile. Berry small, indehiscent. Seeds 2 to 8, with a smooth crustaceous testa. Cotyledons broad.— Trees, often thorny. Leaves toothed or rarely quite entire. Flowers small, axillary, clustered, or shortly racemose.

A genus widely dispersed over the tropical and subtropical regions of the new and the old world. The only Australian species is endemic.

1. **X. ovatum,** *Benth.* Glabrous in all its parts, the branches short and slender, rough with lenticels, and, in our specimens, without thorns. Leaves mostly ovate, obtuse, about 1½ in. long, quite entire, narrowed into a very short petiole, thinly coriaceous, with numerous fine reticulate veins; a few lower leaves short and almost orbicular and the upper ones narrow. Male fl. not seen. Female fl. very small, 5 or 6 together in very short axillary racemes. Pedicels about 1 line long, in the axils of small, ovate, ciliate bracts. Sepals 4, orbicular, ciliate, about ½ line long. Disk deeply lobed or divided. Ovary ovoid, conical, but scarcely tapering into a distinct style, with a broad, thick, slightly 2-lobed stigma. Placentas 2, very prominent, forming a complete dissepiment above the insertion of the ovules, but far from meeting below. Ovules 2 to each placenta.

**Queensland.** N.E. coast, *A. Cunningham.*

This appears to come nearest to *X. orbiculatum,* Forst., which, judging from Fiji Island specimens, has a similar almost sessile stigma, but its leaves are much larger and broader, and the ovary has 3 placentas, a 3-lobed stigma, and more than two ovules to each placenta.

### 4? STREPTOTHAMNUS, F. Muell.

Flowers hermaphrodite. Sepals 5, imbricate. Petals 5, much longer than the sepals. Stamens indefinite. Anthers oblong-linear, tipped by a small point, the cells opening longitudinally. Ovary with parietal placentas and numerous ovules; style filiform, with a peltate entire stigma. Fruit a berry. Seeds several, with a hard testa. Embryo very small, at the base of a copious albumen.—Glabrous twiners. Leaves alternate, petiolate, entire, 3-nerved. Peduncles axillary, 1-flowered.

The genus is limited to Australia. It differs from all *Bixineæ,* and approaches *Pittosporeæ* in its climbing habit and very small embryo, whilst the floral characters bring it nearer to the tribe *Oncobeæ* of *Bixineæ.* The specimens I have seen have so very few flowers that I have been unable to dissect any myself, and have taken the characters from F. Mueller.

Leaves green on both sides. Disk none . . . . . . . . . . . . 1. *S. Moorei.*
Leaves pale or whitish underneath. Disk toothed . . . . . . . . 2. *S. Beckleri.*

1. **S. Moorei,** *F. Muell. Fragm.* iii. 28. A perfectly glabrous twiner. Leaves broadly ovate or obscurely cordate, acute or shortly acuminate, 2 to 3 in. long, quite entire, 3-nerved from the base, scarcely paler underneath than above, on petioles of ½ to 1 in. Pedicels about as long as the petioles, 1-flowered. Sepals broad, about 1 line long, persistent. Petals 2 or 3 times as long, rather broad. Stamens very numerous; filaments shorter than the anthers. Berry nearly 1 in. long. Seeds ovoid-globular, about 1½ line diameter, embedded in pulp.

**N. S. Wales.** Clarence river, *C. Moore.*

2. **S. Beckleri,** *F. Muell. Fragm.* iii. 28. Closely resembles the last species, but differs in the rather more acuminate leaves, paler underneath, a deciduous calyx, the ovary surrounded by a several-toothed disk, a rather longer style, and a more ovoid berry, with smaller seeds. Flowers unknown.

**N. S. Wales.** Clarence and Hastings rivers, *Beckler.*

## Order XII. **PITTOSPOREÆ.**

Flowers hermaphrodite, regular or oblique. Sepals 5, distinct and imbricate, or rarely connate at the base. Petals 5, imbricate, the claws or narrowed base usually erect and connivent or cohering in a tube, rarely spreading from the base. Stamens 5, hypogynous, free, alternating with the petals. Torus small, rarely produced into a short gynophore, sometimes bearing 5 glands. Ovary 1-celled, with 2 or rarely 3 to 5 parietal placentas, or divided into cells by the protrusion of the placentas, which often unite in the axis, at least after flowering. Style simple, with an entire, small, capitate, or dilated stigma. Ovules several, superposed in 2 rows on each placenta, horizontal. Fruit either a capsule opening loculicidally, the valves sometimes splitting also septicidally, or succulent and indehiscent. Seeds several or rarely solitary in each cell, dry or enveloped in pulp, with a thin testa, smooth or rarely muricate, and a hard albumen. Embryo very small, in a cavity of the albumen next the hilum.—Trees, erect shrubs, or undershrubs, with flexuose, decumbent, or twining branches. Leaves alternate, entire, toothed, or rarely lobed, without stipules. Flowers white, blue, yellow, or rarely reddish, terminal or axillary, solitary and nodding, or in short racemes, or corymbose panicles.

With the exception of *Pittosporum* itself, the genera are all limited to Australia.

* *Anthers ovate or oblong. Capsule dehiscent. Petals (except in Bursaria) erect at the base.*

Trees or erect shrubs. Petals erect at the base. Capsule thick or
coriaceous. Seeds several.
  Seeds thick, not winged. Flowers usually small . . . . . . 1. Pittosporum.
  Seeds flat, horizontal, winged. Flowers large, yellow . . . . 2. Hymenosporum.
Erect shrubs, often prickly. Petals small, spreading from the base.
  Capsule thin, small, and flat. Seeds 1 or 2 in each cell, vertical,
  flat . . . . . . . . . . . . . . . . . . . . 3. Bursaria.
Undershrubs or twiners. Petals erect at the base. Capsule membranous or thinly coriaceous. Seeds thick or horizontal . . 4. Marianthus.

** *Anthers ovate or oblong. Berry indehiscent. Petals erect at the base.*

Prickly shrub, with small leaves and small sessile solitary flowers.
  Berry globular . . . . . . . . . . . . . . . . 5. Citriobatus.
Undershrubs or twiners. Flowers pedunculate. Berry ovoid or
  oblong . . . . . . . . . . . . . . . . . . . 6. Billardiera.

*** *Anthers linear, or longer than the filaments. Petals spreading from the base, or nearly so. Undershrubs or twiners.*

Fruit a berry.
  Anthers distant, recurved or revolute, opening longitudinally . . 7. Pronaya.
  Anthers connivent round the style, opening inwards . . . . . 8. Sollya.
Fruit dehiscent. Anthers turned to one side, opening in terminal
  pores . . . . . . . . . . . . . . . . . . . 9. Cheiranthera.

### 1. **PITTOSPORUM**, Banks.

Petals usually connivent or cohering in a tube at their base or above the middle. Anthers ovate-oblong. Ovary sessile or shortly stipitate, incompletely, or almost completely 2-celled, or rarely 3- to 5-celled; style short. Capsule globose, ovate or obovate, often laterally compressed ; the valves coriaceous or thick and hard, bearing the placentas along their centre.  Seeds thick or globular, not winged, often enveloped in a viscous liquor.—Shrubs or trees, glabrous, or rarely tomentose.  Leaves usually evergreen, entire or minutely toothed, the upper ones frequently collected into a false whorl.  Flowers not large, axillary or terminal, solitary or in close corymbose panicles.

A large genus, dispersed over the warmer regions of Africa, Asia, the Pacific islands, and New Zealand.  The Australian species are all endemic excepting one which is common to eastern tropical Asia and the eastern Archipelago.

Flowers numerous, small, in compound terminal corymbs, with the
   lower branches axillary.
   Leaves ovate-rhomboid, toothed.  Sepals obtuse . . . . . 1. *P. rhombifolium.*
   Leaves from obovate to oblong or lanceolate, quite entire.
     Sepals subulate or subulate-pointed.
     Young leaves and inflorescence rusty-tomentose . . . . . 5. *P. ferrugineum.*
     Plant glabrous . . . . . . . . . . . . . . . 2. *P. melanospermum.*
Peduncles all terminal, clustered, short, each bearing a short simple
   cyme or umbel.
   Glabrous, or the young shoots and inflorescence very slightly
     pubescent.  Flowers about ½ in. long . . . . . . . 3. *P. undulatum.*
   Young shoots and inflorescence rusty-tomentose or hirsute.
     Flowers about ¼ in.  Capsule ¾ in., very rough . . . . . 4. *P. revolutum.*
     Flowers 3 to 4 lines.  Capsule under ½ in.
       Leaves on long petioles, ovate to oblong-lanceolate.  To-
         mentum short and crisp . . . . . . . . . . . 5. *P. ferrugineum.*
       Leaves nearly sessile, oblong-lanceolate.  Tomentum almost
         hirsute . . . . . . . . . . . . . . . . . . 6. *P. rubiginosum.*
Pedicels axillary, solitary or clustered, 1-flowered, the uppermost
   sometimes in a terminal cluster.
   Leaves glabrous, flat.  Flowers yellow . . . . . . . 7. *P. phillyræoides.*
   Leaves revolute on the margins, glabrous above, tomentose or
     silky underneath.  Flowers purple and yellow . . . . . 8. *P. bicolor.*
Doubtful species.  Leaves very small.  Flowers terminal, 1 line
   long . . . . . . . . . . . . . . . . . 9. *P. parviflorum.*

1. **P. rhombifolium,** *A. Cunn. in Hook. Ic. Pl. t.* 621.  A tree, attaining, according to A. Cunningham, 60 to 80 ft., glabrous in all its parts. Leaves rhomboid-oval or rarely broadly oblong-lanceolate, mostly 3 to 4 in. long, coarsely and irregularly toothed from the middle upwards, narrowed into a petiole of ½ to 1 in., coriaceous and shining, but with the pinnate and netted veins prominent on both sides.  Flower white, numerous, and rather small, in a dense terminal compound corymb, the branches sometimes minutely glandular.  Sepals obtuse, rather more than 1 line.  Petals oblong, about 3 lines long, spreading from below the middle.  Ovary shortly stipitate, the thick placentas nearly meeting, each bearing about 12 to 14 ovules.  Capsule more or less obliquely pear-shaped, or almost globular, usually about 3 lines long, and ripening 2 or 3 black seeds.

**Queensland.**  Wide Bay, *Bidwill;* forests on the Brisbane river ; *A. Cunningham;* Araucaria range, between Brisbane and Dawson rivers and edge of the Killarney scrub, near Warwick, *F. Mueller.*

**N. S. Wales.**  Clarence river.  *Herb. F. Mueller.*

This has some general affinity, especially in inflorescence, with the East Indian *P. floribundum,* W. and Arn., but is quite distinct both in foliage and flowers.

2. **P. melanospermum,** *F. Muell. Fragm.* i..70.  A small tree, quite glabrous, or with a scanty minute glandular pubescence on the inflorescence. Leaves from obovate to oblong or even lanceolate, shortly acuminate, mucronate or obtuse, 2 to 4 in. long, entire and flat or slightly undulate on the margin, narrowed into a petiole of 4 to 5 lines, coriaceous, but not shining, of a pale hue and prominently veined.  Corymbs compound, terminal, many-flowered, but shorter than the last leaves.  Flowers small, the sepals subulate or lanceolate-subulate, the petals 3 or scarcely 4 lines long, spreading from about the middle.  Ovary shortly stipitate, with 10 to 12 ovules to each placenta.  Capsule obliquely globular or pear-shaped, somewhat compressed, with few or sometimes a single black seed.

**N. Australia.** York Sound, *A. Cunningham ;* low rocky hills between Victoria river and the Gulf of Carpentaria, *F. Mueller.*

**Queensland.** Keppel Bay and several points of the N.E. coast, *R. Brown.*
There is one specimen, in the Hookerian herbarium, from A. Cunningham, marked Hunter's River; but it is not in any other of the numerous collections we have from that locality, nor from any other station in N. S. Wales.

Var. (?) *lateralis.* Corymbs usually lateral.  York Sound, *A. Cunningham ;* Whitsunday Island, *Henne.*

3. **P. undulatum,** *Vent. Hort. Cels. t.* 76.  A tree, attaining in favourable situations 40 ft., or according to M'Arthur, 60 to 90 ft., although in barren exposed localities it remains a shrub, quite glabrous, except a slight appressed pubescence on the young shoots and inflorescence.  Leaves from oval-oblong to lanceolate, mostly 3 to 6 in. long and acuminate, flat or undulate on the margin, narrowed into a petiole of $\frac{1}{2}$ to $\frac{3}{4}$ in., coriaceous and shining, with the veins little conspicuous; the upper ones often almost whorled. Peduncles several, in terminal clusters, much shorter than the leaves, mostly bearing a simple cyme or umbel of 3 or 4 rather large white flowers, and one or two often 1-flowered.  Sepals lanceolate, acuminate, often connate at the base.  Petals 5 to 6 lines long, spreading from the middle.  Ovary almost sessile, hairy, the 2 placentas united at the base, each bearing numerous ovules.  Capsule nearly globular, rarely attaining $\frac{1}{2}$ in., smooth, with thick coriaceous valves and numerous seeds.—DC. Prod. i. 346 ; Andr. Bot. Rep. t. 383; Bot. Reg. t. 16 ; F. Muell. Pl. Vict. i. 71 and 224.

**N. S. Wales.** Common about Port Jackson, *R. Brown, Sieber, n.* 221 and others; northward to Hastings river, *Beckler ;* southward to Illawara, *M'Arthur,* and Twofold Bay, *F. Mueller.*
**Victoria.** Banks of rivers in humid forest districts, or rocky places about Western Port, Buchan, Tambo, Broadribb, and Snowy rivers, *F. Mueller.*

4. **P. revolutum,** *Ait. Hort. Kew. ed.* 2, ii. 27.  A tall shrub, the young shoots tomentose.  Leaves ovate-elliptical or elliptical-oblong, shortly acuminate, 2 to 4 in. long, scarcely undulate, narrowed into a petiole, usually very short, but sometimes near $\frac{1}{2}$ in., coriaceous, glabrous above when full grown, clothed underneath with a loose rusty tomentum easily rubbed off, the upper ones often almost whorled.  Peduncles terminal, few or solitary, usually decurved, bearing sometimes a single, rather large flower, but more frequently a short dense ovate or corymbose raceme.  Sepals lanceolate-subulate.  Petals nearly $\frac{1}{2}$ in. long, often united to above the middle, shortly spreading or recurved at the top.  Ovary very hirsute, with very numerous ovules to each

placenta; stigma peltate. Capsule $\frac{1}{2}$ to $\frac{3}{4}$ in. long, the hard almost woody valves rough outside. Seeds numerous, red or brown.—DC. Prod. i. 346; Bot. Reg. t. 186; F. Muell. Pl. Vict. i. 224; *P. fulvum*, Rudge in Trans. Linn. Soc. x. 298, t. 20; DC. l. c.; Sweet, Fl. Austral. t. 25; *P. tomentosum*, Bonpl. Jard. Malm. 56, t. 21; Sweet, Fl. Austral. t. 33; DC. l. c.; *P. hirsutum*, Link, according to Putterl. Syn. Pittosp. 9.

**Queensland.** Moreton Bay, *Fitzalan;* Brisbane river, *A. Cunningham.*

**N. S. Wales.** Port Jackson to the Blue Mountains, *R. Brown, A. Cunningham,* and others; northward to Hastings and Clarence river, *Beckler;* southward to Twofold Bay, *F. Mueller.*

**Victoria.** Ridges on the S.E. boundary of Gipps' Land, *F. Mueller.*

In one specimen in the Hookerian herbarium, perhaps in an abnormal condition, the flowers are in shortly pedunculate umbels, both axillary and terminal.

5. **P. ferrugineum,** *Ait. Hort. Kew. ed.* 2, ii. 27. A tree, flowering sometimes as a shrub, but attaining a height of 50 to 60 ft., the young shoots thickly clothed with a loose rusty tomentum which soon wears off. Leaves from obovate or ovate, and obtuse or scarcely acuminate, to oblong or almost lanceolate, acuminate, and 3 to 4 in. long, quite entire, narrowed into a petiole of $\frac{1}{2}$ to $\frac{3}{4}$ in., rusty tomentose on both sides when very young, but glabrous above, or on both sides when full grown. Peduncles terminal, usually clustered several together above the last leaves, each one bearing a cluster or umbel of rather small flowers, but sometimes the common peduncle grows out and the inflorescence becomes a thyrsoid or pyramidal panicle, not a corymb, as in *P. melanospermum.* Sepals lanceolate or lanceolate-subulate. Petals narrow, about 3 lines long, spreading only above the middle. Ovary villous, with 12 to 16 ovules to each placenta. Capsule sessile, nearly globular, scarcely 4 lines broad, ripening usually 3 or 4 black seeds.—DC. Prod. i. 346; Bot. Mag. t. 2075; *P. tinifolium* (*linifolium* by an error of the press), A. Cunn. in Ann. Nat. Hist. ser. 1, iv. 109; *P. ovatifolium*, F. Muell. Fragm. ii. 78.

**Queensland.** Moist rocky places, Endeavour river, and Percy Islands, *A. Cunningham;* Frankland Islands, *M'Gillivray;* dry ridges of Albany Island, *F. Mueller.*

Extends over the Malayan peninsula and adjoining islands, and the Philippines. The Australian specimens have rather larger flowers and narrower-pointed sepals than the common Malayan form; but in this respect the Malacca specimens are very variable, some of them precisely resembling some of the Australian ones; and I have never seen them so obtuse as figured in the 'Botanical Magazine,' even on old specimens preserved from the cultivated shrubs from whence the figure was taken.

6. **P. rubiginosum,** *A. Cunn. in. Ann. Nat. Hist. ser.* 1, iv. 108. Branches, petioles, and inflorescence densely clothed with a rust-coloured tomentum, consisting of much more spreading hairs than in *P. ferrugineum.* Leaves almost whorled, oblong-lanceolate, acutely acuminate, 5 to 6 in. long, entire or slightly sinuate-toothed, narrowed at the base, but almost sessile, herbaceous, glabrous above, softly pubescent underneath. Peduncles in our specimens solitary, terminal, $\frac{1}{2}$ to 1 in. long, bearing an umbel of several flowers very similar to those of *P. ferrugineum.* Fruit unknown.

**Queensland.** East side of Mount Cook, near Endeavour river, *A. Cunningham.*

7. **P. phillyræoides,** *DC. Prod.* i. 347. A small graceful tree or slender shrub, quite glabrous in all its parts. Leaves usually oblong- or

linear-lanceolate, with a small hooked point, 2 to 4 in. long, quite entire, narrowed into a petiole, thick coriaceous and indistinctly veined, but in some forms short and broadly oblong, in others long and narrow. Pedicels axillary, solitary or in sessile or shortly pedunculate clusters or umbels, or the uppermost forming a terminal cluster. Flowers yellow, usually about 4 lines long, often diœcious, the females rather larger and fewer together than the males. Sepals short and very obtuse. Petals united to the middle or still higher, spreading at the top. Ovary pubescent, almost completely 2-celled, with 6 to 8 ovules in each cell. Fruit ovate or round-cordate, much compressed, quite smooth, varying from 4 to 9 lines in length, but usually about ½ in. Seeds few, dark or orange-red.—Putterl. in Pl. Preiss. i. 192; F. Muell. Pl. Vict. i. 72 ; *P. angustifolium*, Lodd. Bot. Cab. t. 1859; *P. longifolium* and *P. Roëanum*, Putterl. Syn. Pittosp. 15, 16 ; *P. ligustrifolium*, A. Cunn. in Putterl. l. c. 16, and in Ann. Nat. Hist. ser. 1, iv. 110 ; Putterl. in Pl. Preiss. i. 190 ; *P. oleæfolium*, A. Cunn. in Putterl. Syn. Pittosp. 17 ; *P. acacioides*, A. Cunn. in Ann. Nat. Hist. ser. 1, iv. 109 ; *P. salicinum*, Lindl. in Mitch. Trop. Austr. 97 ; *P. lanceolatum*, A. Cunn. in Mitch. l. c. 272 and 291.

**N. Australia.** Upper Victoria river and Sturt's Creek, *F. Mueller.*

**Queensland.** Brigalow scrub, *Mitchell ;* and Burdekin river, Warwick, *F. Mueller.*

**N. S. Wales.** Narran river and N.W. interior, *Mitchell ;* generally dispersed over the interior, *A. Cunningham.*

**Victoria.** Sandy, barren, or stony declivities and plains dispersed through the desert, *F. Mueller.*

**S. Australia.** On the coast, *R. Brown ;* Kangaroo Island, round Spencer's Gulf and other localities, *F. Mueller.*

**W. Australia.** Swan River, *Drummond, Preiss, n.* 1297 ; Rottenest Island, *A. Cunningham, Preiss ;* Dirk Hartog Island, *A. Cunningham ;* Murchison river, *Oldfield ;* Abrolhos island, *Bynoe, Moore in Herb. Preiss. n.* 1294.

This species, apparently spread over the whole desert country of Australia, cannot be confounded with any other, notwithstanding the variability of the proportions of its leaves, flowers, and fruit. In some of the western specimens the leaves are barely 2 inches long, and fully ½ inch wide, whilst in a large number of eastern and some western ones they attain 4 or 5 inches in length with a breadth of only 2 or 3 lines.

8. **P. bicolor,** *Hook. Journ. Bot.* i. 249. A small tree, attaining in some localities a height of 40 feet, remaining a bush in others, the young branches hoary or rusty, with a close tomentum. Leaves usually crowded, oblong, lanceolate or almost linear, obtuse or with a short recurved point, mostly 1 to 2 in. long, entire, the margins much revolute, nearly sessile or on very short petioles, thick and coriaceous, glabrous above, tomentose or silky underneath. Pedicels from 2 or 3 lines to nearly 1 in. long, axillary, clustered or solitary, usually reflexed, the little bracts at their base numerous and conspicuous, the uppermost pedicels often in a terminal cluster. Sepals oblong or lanceolate. Petals purple and yellow, 4·to 5 lines long, free or nearly so, spreading from above the middle. Ovary villous, with 10 or more ovules to each placenta. Capsule rounded, somewhat compressed, 4 to 5 lines broad, tomentose, the valves not very thick. Seeds usually rather numerous. —Hook. f. Fl. Tasm. i. 38; F. Muell. Pl. Vict. i. 72 ; *P. discolor*, Regel, Gartenfl. i. 133, t. 15 ; *P. Huegelianum*, Putterl. in Endl. Nov. Stirp. Dec. 43 (from the description given).

**N. S. Wales?** E. extratropical Australia, *Huegel.* (I have not seen the specimens.)
**Victoria.** Tree-fern gullies, from Wilson's Promontory to the Delatite river, Dandenong ranges, and Mount Disappointment; also ranges towards Cape Otway and Apollo Bay, and Mount Tambo, ascending to subalpine elevations, *F. Mueller.*
**Tasmania.** *R. Brown;* throughout the island, abundant in damp ravines, ascending to 4000 ft., *J. D. Hooker.*

*Doubtful species.*

9. **P. (?) parviflorum,** *Putterl. in Pl. Preiss.* i. 189. A glabrous erect shrub of 2 ft. Leaves obovate, 4 to 5 lines long, flat or concave, entire. Peduncles terminal, solitary or 2 together, scarcely 1 line long. Flowers scarcely 1 line long. Calyx already fallen from the specimens described. Petals 5, linear-lanceolate, terminated by a dot-like gland. Stamens not seen. Ovary 3-celled, the placentas meeting in the centre, but not united; style filiform; ovules 6 to 10 in each cell. Ripe fruit not seen.

**W. Australia.** Stony sterile places, York and Wicklow districts, *Preiss, n.* 1290. I have not seen the specimen, but from the description given I much doubt its belonging to the genus or even to the Order.

## 2. HYMENOSPORUM, F. Muell.

Petals connivent or cohering in a tube to above the middle. Anthers ovate-oblong. Ovary incompletely 2-celled; style short. Capsule ovate, compressed, with thick coriaceous valves. Seeds numerous, horizontally imbricated, flat, reniform, surrounded by a membranous wing.—A shrub or tree, with the habit of *Pittosporum,* from which it only differs in its large flowers and in its seeds.

The genus is limited to a single species, endemic in Australia.

1. **H. flavum,** *F. Muell. Fragm.* ii. 77. A handsome evergreen shrub or tree, glabrous, except a loose pubescence on the inflorescence, and sometimes on the under side of the leaves. Leaves ovate-oblong or oblanceolate, acuminate, entire, from 3 to 5 or even 6 in. long, narrowed into a petiole of ½ in. or more, the upper ones often almost verticillate. Panicle terminal, loose, corymbose, often 6 to 8 in. diameter, with small linear or lanceolate bracts. Flowers large, yellow. Sepals oblong-lanceolate, 3 to 4 lines long. Petals silky-tomentose outside, the erect base or broad claws nearly 1 in., the spreading lamina nearly ½ in. long. Ovary linear, silky-tomentose, with numerous ovules. Capsule stipitate, much flattened, fully 1 in. long and nearly as broad. Seeds, including the wing, fully 4 lines broad.—*Pittosporum flavum,* Hook. Bot. Mag. t. 4799.

**Queensland.** Wide Bay district, *Bidwill;* Moreton Bay and Brisbane river, *F. Mueller;* Ipswich, *Vernet.*
**N. S. Wales.** Paterson's River and Hunter's River, *R. Brown;* Port Stephens, *A. Cunningham;* Macleay river, *Beckler;* Clarence river, *Wilcox;* Lake Macquarie, *Leichhardt.*

## 3. BURSARIA, Cav.

Petals narrow, spreading from near the base. Anthers ovoid. Ovary incompletely 2-celled; style short. Capsule shortly stipitate, flat, broadly orbicular, opening round the edge, with thinly coriaceous flat valves. Seeds 1 or 2 in each cell, flat, reniform, not winged.—Rigid, much branched shrubs or trees, often thorny. Leaves small, entire. Flowers small, in terminal panicles. Sepals very fugacious.

The genus is limited to the following one or perhaps two Australian species.

1. **B. spinosa,** *Cav. Ic.* iv. 30, *t.* 350. A shrub or small tree, occasionally attaining the height of 40 ft., in the ordinary state glabrous, and when young very bushy, the smaller branches often reduced to short subulate thorns. Leaves very variable, most frequently clustered, obovate, oblong or cuneate, obtuse, truncate or notched, $\frac{1}{2}$ to 1 in. long, narrowed at the base, and sometimes shortly petiolate, green on both sides; in luxuriant specimens they vary to oblong-lanceolate, 1 to 2 in. long; in a few others they have occasionally a few coarse teeth at the top; and in the var. *incana* they are thicker, and white underneath with a silky tomentum. Flowers white, usually very numerous, in a broad, pyramidal, terminal panicle, arranged along its branches in short racemes, on pedicels of 1 to 3 lines; occasionally the panicles are reduced to short racemes or to 1 or 2 terminal flowers. Bracts minute and very fugacious. Sepals small, also falling off long before the petals open. Petals narrow, about 2 lines long. Capsule 3 to 4 lines or, in the var. *incana*, sometimes 5 lines broad.—DC. Prod. i. 347; Bot. Mag. t. 1767; Hook. f. Fl. Tasm. i. 39; F. Muell. Pl. Vict. i. 74; *Itea spinosa*, Andr. Bot. Rep. t. 314.

**N. Australia.** About the Gulf of Carpentaria, rare, and only the var. *incana, F. Mueller;* N.E. coast, *A. Cunningham.*

**Queensland.** Brisbane river, Moreton Bay, and near Warwick, *F. Mueller.*

**N. S. Wales.** Common in all forest lands, *R. Brown, Sieber, n.* 281, and others.

**Victoria.** Common in all the lowlands as well as in the mountain districts, *F. Mueller.*

**Tasmania.** Abundant throughout the island, *J. D. Hooker.*

**S. Australia.** Extends westward at least to Streaky Bay, *F. Mueller.*

**W. Australia.** Champion Bay, *Oldfield,* only the var. incana.

Var. (?) *incana.* Young shoots, inflorescence, and under side of the leaves white or hoary, with a soft and dense, or close and thin tomentum. In the original specimens the leaves are 2 to 3 in. long, but they pass gradually, in other specimens, into small obovate or oblong ones. They are, however, usually more robust, and the flowers, and especially the fruits, rather larger than in the normal *B. spinosa.*—*B. incana,* Lindl. in Mitch. Trop. Austr. 224. This appears to be the more common variety in the tropical and subtropical regions, and the only one hitherto found in North or West Australia. It extends also southward to the desert tract on the Murray and Snowy rivers, in Victoria. I feel much hesitation in following F. Mueller in uniting the two forms in one species.

A third rather distinct variety, or perhaps a peculiar state of the common one, has very small leaves, numerous thorns, and only very few flowers, with longer and more permanent sepals. Very characteristic specimens were collected on the Glenelg river by Mr. Robertson.

## 4. MARIANTHUS, Hueg.

(Calopetalum, *Harv.*; Oncosporum, *Putterl.*; and Rhytidosporum, *F. Muell.*)

Petals connivent at the base or above the middle, spreading at the top. Anthers oblong or ovate, shorter than the filaments. Ovary sessile or shortly stipitate, usually completely 2-celled, glabrous, except very rarely in *M. laxiflorus.* Capsule ovoid or oblong, turgid or slightly compressed, membranous or slightly coriaceous, the valves sometimes splitting septicidally. Seeds ovoid, reniform or globular.—Undershrubs, with procumbent, flexuose, or more frequently twining branches. Leaves entire, toothed, or the lower ones occasionally lobed. Flowers blue, white, or reddish, in terminal compact panicles, usually corymbose or almost umbellate, rarely solitary or apparently axillary from the extreme shortness of the flowering branch.

The genus is limited to Australia. It differs from *Billardiera* solely in the capsular not baccate fruit, which is the cause of several species having been described in both genera when the fruit has not been seen. The petals are in general more spreading than in *Billardiera*, but *M. bignoniaceus* has a tubular corolla, and the cymose Billardieras have the flowers of *Marianthus*.

SERIES I. **Procumbentes.**—*Branches short, procumbent or flexuose, not twining. Leaves crowded. Pedicels 1 to 3, terminal. Sepals very pointed. Petals spreading from below the middle. Seeds ovoid-reniform, transverse, and laterally attached.*

Leaves small or heath-like, glabrous or hispid with a few setæ.
  Flowering pedicels shorter than the leaves. Seeds much
    wrinkled . . . . . . . . . . . . . . . . . . 1. *M. procumbèns.*
  Flowering pedicels much longer than the leaves. Seeds nearly
    smooth . . . . . . . . . . . . . . . . . . 2. *M. microphyllus.*
Leaves broadly obovate, ½ in. or more, very hairy. Seeds smooth 3. *M. villosus.*

SERIES II. **Oncosporeæ.**—*Twiners. Leaves distinctly petiolate, ovate-lanceolate or lanceolate, very obtuse and cordate at the base. Sepals very acute or subulate. Petals various. Seeds globular, muricate (or tuberculate ?).*

Flowers small, in loose terminal racemes or corymbs. Petals
  spreading from below the middle. Seeds muricate.
  Hairs loose, rather rusty. Ovules 3 or 4 in each cell . . . 4. *M. granulatus.*
  Hairs silky-white. Ovules numerous . . . . . . . . 5. *M. parviflorus.*
Flowers rather large, on axillary pedicels. Petals united in a tube
  above the middle. (Seeds tuberculate ?) . . . . . . 6. *M. bignoniaceus.*

SERIES III. **Normales.**—*Twiners, or rarely branches short and flexuose, or nearly straight. Leaves sessile, or narrowed into a petiole. Sepals very acute or subulate. Petals blue or white, usually connivent to the middle. Seeds (where known) smooth, nearly globular.*

Pedicels 1 to 3, sessile amongst the last leaves, or axillary.
  Leaves narrowed at the base. Ovary glabrous.
  Pedicels slender, mostly above ¼ in. Ovary distinctly stipitate 7. *M. Drummondianus.*
  Pedicels very short. Ovary scarcely contracted at the base . 8. *M. tenuis.*
Flowers in terminal corymbs or short racemes, usually numerous.
  Upper leaves sessile, obtuse at the base. Corymb or raceme
    loose and few-flowered . . . . . . . . . . . . 9. *M. laxiflorus.*
  Leaves narrowed into a petiole. Corymbs many-flowered.
    Flowers blue, often spotted. Sepals very hairy . . . 10. *M. cœruleo-punctatus.*
    Flowers white. Sepals rigid, glabrous or slightly hairy.
      Leaves lanceolate or linear. Style long and subulate . 11. *M. candidus.*
      Leaves ovate or broadly lanceolate. Style short and
        thick, with a broad stigma . . . . . . . . 12. *M. floribundus.*

SERIES IV. **Pictæ.**—*Twiners, or rarely branches short and flexuose. Leaves narrowed into a petiole. Sepals ovate or shortly lanceolate. Petals red or streaked with purple, very oblique, and connivent at the middle. Seeds (where known) smooth.*

Filaments dilated, at least at the base. Twiners with red flowers.
  Filaments dilated at the base only . . . . . . . . 13. *M. erubescens.*
  Filaments much dilated above the middle . . . . . . 14. *M. ringens.*
Filaments scarcely flattened. Branches flexuose, or slightly
  twining. Flowers streaked.
  Corymbs dense. Pedicels stout, 1 to 2 lines . . . . . 15. *M. lineatus.*
  Corymbs loose, few-flowered. Pedicels slender, 3 to 4 lines . 16. *M. pictus.*

1. **M. procumbens,** *Benth.* A low, prostrate or suberect, much branched shrub, the branches sometimes flexuose and nearly 1 ft. long, but usually much shorter, glabrous or slightly pubescent. Leaves crowded and sessile, in the northern varieties usually linear or linear-cuneate, pointed, entire or rarely toothed at the top, 4 to 6 lines long, rigid, with recurved margins ; in the southern forms usually shorter, more cuneate or even obovate or ovate, and often toothed. Flowers small, white or tinged with red, solitary or 2 or 3 together, terminal or appearing axillary from the shortness of the flowering shoots, the pedicels 1 to 2 lines long and always shorter than the leaves at the time of flowering, rather longer and recurved when in fruit. Sepals lanceolate-linear, very pointed. Petals about 3 lines long or smaller, spreading from below the middle. Filaments dilated to the middle. Ovules 6 to 8 in each cell of the ovary. Style short. Capsule truncate, 3 lines broad, and not quite so long. Seeds usually 3 or 4 in each cell, ovoid-reniform, transverse and laterally attached, deeply wrinkled.—*Pittosporum procumbens* and *P. nanum*, Hook. Comp. Bot. Mag. i. 275 ; *Bursaria procumbens*, Putterl. Syn. Pittosp. 20 ; Hook. f. Fl. Tasm. i. 39 ; *B. diosmoides*, Putterl. l. c. (from the description, I have not seen Sieber's n. 554) ; *B. Stuartiana*, Klatt, in Linnæa, xxviii. 568 ; *Rhytidosporum procumbens*, F. Muell. 1st Gen. Rep. 10 ; Pl. Vict. i. 75 ; *Campylanthera ericoides*, Lindl. in Mitch. Three Exped. ii. 277.

**N. S. Wales.** Frequent about Port Jackson and in the Blue Mountains, *A. and R. Cunningham*, and others ; extending northward to Clarence river, *Beckler*, and southward to Twofold Bay, *F. Mueller.*

**Victoria.** Barren forest ridges and heath ground, not generally common although noticed in many localities, more frequent in the eastern part of Gipps' Land, *F. Mueller.*

**Tasmania.** Common in sandy places throughout the island, *J. D. Hooker.*

2. **M. microphyllus,** *Benth.* Habit of the smaller shorter-leaved forms of *M. procumbens.* Stems apparently procumbent, branched, under 6 in. long, more or less hirsute. Leaves crowded, from obovate to oblong, obtuse, rarely 2 lines long, the margins recurved, all entire in our specimens. Pedicels solitary, terminal, about 3 lines long when in flower, and ½ in. when in fruit, and always several times longer than the last leaves. Flowers larger and apparently darker-coloured than in *M. procumbens.* Petals about 4 lines long, spreading from a little below the middle. Filaments very slightly dilated. Ovules at least 12 to each cell of the ovary. Style rather long. Capsule 3 lines long and not quite so broad. Seeds numerous, smooth or scarcely wrinkled, but not quite ripe in our specimen.—*Oncosporum microphyllum*, Turcz. in Bull. Mosc. 1854, ii. 365 ; *Marianthus rhytidosporus*, F. Muell. Fragm. ii. 145.

**W. Australia,** *Drummond, 5th Coll. n.* 242 ; also *Herb. Mueller.*

3. **M. villosus,** *Benth.* Apparently a low procumbent shrub, with short, slightly flexuose, very hispid branches. Leaves rather crowded, broadly obovate, ½ to near ¾ in. long, usually coarsely toothed, narrowed into a short petiole, softly villous on both sides, or becoming almost glabrous above when old. Pedicels terminal or on very short side-branches, solitary or 2 or 3 together, very short at first, and not 2 lines long when in fruit. Petals and stamens not seen. Ovary glabrous, with a long style. Capsule about 4 lines

long and 3 broad, with about 5 seeds in each cell, ovoid-reniform, horizontal, and laterally attached, as in *M. procumbens*, but not wrinkled.—*Oncosporum villosum*, Turcz. in Bull. Mosc. 1854, ii. 365 ?

**W. Australia,** *Drummond, Coll.* 1843, *n.* 176.

4. **M. granulatus,** *Benth.* A very slender twiner, the young shoots and leaves loosely clothed with long, soft, spreading hairs, becoming at length glabrous. Leaves distinctly petiolate, ovate-lanceolate or oval-oblong, acute or obtuse, entire, and always obtuse at the base, the larger ones above 1 in. long, those of the side-branches smaller, of a thin texture. Flowers small, 3 to 5 together, in slender racemes or cymes, on filiform pedicels of 4 to 6 lines. Sepals subulate-lanceolate, with long spreading hairs. Petals about 2 lines. Anthers very small. Ovary glabrous, with a subulate style; ovules 3 or 4 in each cell. Capsules nearly orbicular, turgid, membranous, glabrous, about 3 lines long. Seeds globular, strongly muricate.—*Oncosporum granulatum*, Turcz. in Bull. Mosc. 1854, ii. 366.

**W. Australia,** *Drummond, Coll.* 1845, *n.* 210.

5. **M. parviflorus,** *F. Muell. Fragm.* ii. 144. Very near *M. granulatus*, but not quite so slender, the young shoots silky-white, with long soft hairs. Leaves distinctly petiolate, ovate-lanceolate or almost cordate-ovate, acute or obtuse, the larger ones above 1 in. long, entire, softly hairy, with a very silky margin. Flowers several, in short terminal or leaf-opposed racemes or corymbs, not much longer than the leaves, on pedicels of 2 to 4 or rarely 6 lines. Flowers of *M. granulatus* or rather longer, the petals often 3 lines long. Ovary longer, glabrous, with a short style, and 10 to 12 ovules in each cell. Capsules very turgid, about 2 lines long. Seeds several, globular, muricate.

**W. Australia.** Plantagenet, Stirling, and Perongerup ranges, *Maxwell.*

6. **M. bignoniaceus,** *F. Muell. in Trans. Phil. Soc. Vict.* i. 6, *and Pl. Vict.* i. 77, *t.* 10. A very slender twiner, the young shoots silky-white, but soon becoming glabrous. Leaves distinctly petiolate, from ovate to oblong or lanceolate, with a rounded or cordate base, obtuse or acute, quite entire, usually ¾ to 1½ in. long, but some of the larger ones above 2 in. Pedicels terminal or from the abortion of the flowering branches, axillary, solitary or 2 or 3 together, filiform, 2 or 3 lines long. Flowers pendulous, of a yellowish or orange colour, ⅔ to nearly 1 in. long. Sepals small, lanceolate-subulate. Petals united in a tube to far above the middle and only spreading at the top, but soon separating at the base also. Anthers small. Ovary silky-villous, with a very long subulate style, and 6 to 8 ovules to each cell. Capsule oblong, turgid. Seeds globular and apparently tuberculate, but I have not seen them in a good state.

**Victoria.** Shady rivulets, springs, and cataracts, and fissures of irrigated rocks, Serra and Victoria ranges, and in the Grampians, *F. Mueller.*

**S. Australia.** Shady banks of the Onkaparinga and Mount Lofty ranges, ascending to 5000 ft., *F. Mueller.*

The inflorescence and shape of the flowers are much more those of the majority of Billardieras than of *Marianthus*, but the fruit is capsular. It is not *Billardiera latifolia*, Putterl., referred to it by Klatt, in Linnæa, xxviii. 570.

7. **M. Drummondianus,** *Benth.* A slender twiner, the young shoots and leaves clothed with long, spreading, very soft, and rather rusty hairs, or rarely glabrous. Leaves from obovate to oblong-lanceolate, mostly acute or with a small recurved point, $\frac{1}{2}$ to 1 in. long, coarsely toothed or almost entire, sessile or narrowed into a very short petiole, the lowest ones sometimes deeply cut. Pedicels terminal, 1 to 3 together, filiform, hairy, $\frac{1}{4}$ to $\frac{1}{2}$ in. long. Sepals lanceolate-subulate, hairy. Petals about $\frac{1}{2}$ in. long, spreading above the middle. Ovary stipitate, glabrous, with a slender style and 3 to 6 ovules in each cell. Capsule ovoid, very turgid, nearly $\frac{1}{2}$ in. long. Seeds small, globular, smooth.—*Oncosporum Drummondianum,* Putterl. in Pl. Preiss. i. 194.

**W. Australia.** Gravelly places, Swan River, *Preiss, n.* 1288, *Drummond,* 1*st Coll.*; Gordon river and Ironstone hills, Tone river, *Oldfield;* S.W. interior, *Maxwell* (the specimen almost completely glabrous).

8. **M. tenuis,** *Benth.* A slender twiner, the young shoots with a few soft spreading hairs, but soon glabrous. Leaves lanceolate or oblong or the lower ones almost ovate, acute, 1 to 1$\frac{1}{2}$ in. long, entire or with a few coarse distant teeth, narrowed into a distinct petiole. Flowers small, axillary, solitary or in short corymbs of 3 to 5, on pedicels of not above 1 line at the time of flowering. Sepals subulate, hairy. Petals 5 or 6 lines long, spreading from above the middle. Ovary glabrous, with a slender style. Fruit not seen.—*Billardiera parviflora,* DC. Prod. i. 346.

**W. Australia.** Geographer Bay, *Leschenault;* Flinders Bay, *Collie;* Cape Naturaliste, *Oldfield.*

9. **M. (?) laxiflorus,** *Benth.* A twiner, with the foliage nearly of *Billardiera variifolia,* the flowers and ovary more nearly those of *M. candidus* and its allies. Leaves sessile or nearly so, oblong or lanceolate, the lowest toothed, the others entire, seldom above 1 in. long, glabrous as well as the stem. Flowers apparently white, in loose pedunculate corymbose racemes, on slender pedicels, 2 or 3 times as long as the calyx, and much fewer in number and rather smaller than in *M. candidus.* Ovary glabrous or very slightly pubescent. Fruit unknown.

**W. Australia,** *Drummond;* Cape Leeuwin, *Collie;* between Perth and King George's Sound, *Harvey;* near Kalgan Bridge, Mount Barker, and Perongerup range, *Herb. Mueller.*

10. **M. cœruleo-punctatus,** *Klotzsch, in Link, Kl. and Otto, Ic. Pl.* 28, *t.* 12. Very nearly allied to *M. candidus,* and perhaps a small blue-flowered variety. Foliage the same, but usually more pubescent, at least on the under side of the leaves. Sepals smaller and more slender, and always clothed with long brown hairs. Petals as in *M. candidus,* but rather smaller, blue, the upper ones generally, but perhaps not always, spotted in the lower part with a darker colour. Style slender. Capsule oblong, with about 6 smooth globular seeds in each cell, but not seen quite ripe.—Putterl. in Pl. Preiss. i. 196.

**W. Australia.** Swan River, *Drummond, Coll.* 1843, *n.* 81, *Preiss;* also, apparently the same, but perhaps without spots, Cape Naturaliste, *Oldfield.*

11. **M. candidus,** *Hueg. Enum.* 8. A tall twiner, either glabrous or with a slight pubescence on the young shoots, under side of the leaves, and inflorescence. First leaves occasionally toothed or lobed, all the others quite

entire, the lower ones sometimes ovate-lanceolate, 3 to 4 in. long, the upper ones lanceolate or linear, 2 to 3 in. long; acuminate and narrowed into a petiole, or the uppermost almost sessile, rather firm, with recurved margins. Flowers white, usually numerous, in rather dense terminal pedunculate corymbs. Sepals lanceolate, very pointed, rather stiff, 2 to 3 lines long. Petals about 8 lines, obovate, acute, and spreading from above the middle, with narrow erect claws. Ovary glabrous, narrowed into a short stipes, with a subulate style at least as long as the ovary, and small stigma. Capsule oblong.—Putterl. in Pl. Preiss. i. 195.

**W. Australia.** Frequent about Swan River, *Huegel, Drummond, Preiss, n.* 1285, and others; Flinders Bay, *Collie.*

12. **M. floribundus,** *Putterl. in Nov. Stirp. Dec.* 61.—Allied to *M. candidus,* but a larger plant and quite glabrous. Leaves (of the flowering branches) ovate or very broadly lanceolate, acuminate, 3 to 4 in. long, 1 to 1½ in. broad, quite entire, narrowed into a petiole. Flowers usually numerous in a pedunculate corymb. Sepals lanceolate, very pointed, rigid, about 3 lines long. Petals apparently white, 9 to 10 lines long, spreading from above the middle, and acute as in *M. candidus.* Ovary sessile, narrowed at the top into a very short thick style, with a broad capitate stigma.

**W. Australia.** King George's Sound, *Huegel, Harvey;* Mair's station on the Tone river, *Clarke;* Mount Clarence, *Oldfield.*

13. **M. erubescens,** *Putterl. in Nov. Stirp. Dec.* 60, *and Pl. Preiss.* i. 197.—Twining from a woody base and quite glabrous. Leaves narrow, oblong-lanceolate or linear, obtuse or scarcely acute, 1 to 2 lines long, entire, narrowed into a petiole, almost coriaceous. Flowers red, 3 or 5, in sessile or shortly pedunculate terminal or axillary corymbs, or rarely solitary, on slender pedicels of 1 to 2 lines. Sepals broadly lanceolate, about 1½ lines long, with scarious edges. Petals about 1 in. long, the laminæ very much oblique and narrowed into long curved claws. Anthers oblong, the long slender filaments shortly and broadly membranous at the base. Ovary glabrous, with a long slender style. Young fruit as in *M. ringens.*—*M. purpureus,* Turcz. in Bull. Mosc. 1854, ii. 364.

**W. Australia.** Swan River, *Huegel, Drummond, Coll.* 1843, *n.* 78, *and Coll.* 1848, *n.* 96, *Preiss, n.* 1292; between Perth and King George's Sound, *Harvey;* Salt river, *Herb. F. Mueller.*

14. **M. ringens,** *F. Muell. Fragm.* i. 218.—Twining from a woody base, and either quite glabrous or with long silky hairs on the young leaves. Leaves from broadly lanceolate to linear-acuminate, 2 to 3 in. long, narrowed into a petiole, coriaceous and quite entire. Flowers red, in dense terminal corymbs usually shortly pedunculate. Sepals oval-oblong or broadly lanceolate, about 2 lines long. Petals very oblique, from ¾ to 1 in. long, with an obovate spreading lamina, the long erect claws rather broad and at first cohering. Filaments much dilated and petal-like, especially above the middle, and suddenly contracted into a short subulate point bearing an oblong anther. Ovary glabrous, with a long filiform style. Capsule oval-oblong. Seeds many, more or less angular.—*Calopetalum ringens,* Drumm. and Harv. in Hook. Kew Journ. vii. 53.

**W. Australia.** Chapman river, *Drummond;* Champion Bay, *Burges;* Murchison river, *Oldfield;* Greenough river, *Walcot.*

15. **M. lineatus,** *F. Muell. Fragm.* i. 217, *and* ii. 182.—Shrubby and glabrous, with rigidly flexuose or shortly twining branches. Leaves oblong-lanceolate or linear, obtuse or with a minute point, 1 to 2 in. long, narrowed into a short petiole, rather coriaceous. Flowers in dense terminal, almost sessile corymbs. Sepals ovate or ovate-lanceolate, rarely more than 1 line long. Petals 6 to 8 lines, oblique, but less so than in *M. pictus* (yellowish?) with purple streaks, obovate and spreading at the top, gradually narrowed into broad claws. Filaments subulate. Ovary sessile, with a subulate style. Capsule hard, the valves often splitting septicidally. Seeds numerous, closely packed and much flattened.

**W. Australia.** Sandy and rocky situations between White Peak and Murchison river, *Oldfield.*

16. **M. pictus,** *Lindl. Swan Riv. App.* 22.—Shrubby and glabrous, with slender twiggy, flexuose or half-climbing branches. Leaves elliptical or lanceolate, obtuse or with a small point, ½ to 1 in. long, narrowed at the base into a short petiole or almost sessile, entire or toothed, rather coriaceous. Flowers few, in short terminal racemes or corymbs, the slender pedicels usually 3 or 4 lines long. Sepals ovate, ½ to ¾ line long. Petals 6 to 8 lines, more oblique and curved than those of *M. lineatus*, streaked with purple, narrowed into a short claw. Filaments filiform. Ovary sessile, with a subulate style. Capsule ovoid-oblong, rather coriaceous, the valves splitting septicidally. Seeds nearly globular or angular.—*Oncosporum bicolor*, Putterl. Syn. Pittosp. 21, in part, as quoted in Pl. Preiss. i. 198.

**W. Australia.** Swan River, *Drummond, 1st Coll. and Coll.* 1843 *n.* 77; *Preiss, n.* 1286.

## 5. CITRIOBATUS, A. Cunn.

(1xiosporum, *F. Muell.*)

Petals connivent or connate to above the middle, in a cylindrical tube spreading at the top. Anthers oblong, shorter than the filaments. Ovary 1-celled, with 2 to 5 parietal placentas; style short. Fruit coriaceous or hard, globular, indehiscent. Seeds few or many, nearly globular, often enveloped in a viscous fluid.—Rigid, much branched shrubs, armed with short thorns or abortive branches. Leaves small, entire or toothed. Flowers small, sessile and solitary, surrounded by small sepal-like bracts.

The genus is limited to Australia.

Placentas 2, with 8 to 10 ovules each. Fruit 2 to 5 lines diameter,
with few seeds . . . . . . . . . . . . . . . . . . 1. *C. multiflorus.*
Placentas 5, with very numerous ovules. Fruit 1 in. diameter or larger,
with numerous seeds . . . . . . . . . . . . . . . 2. *C. pauciflorus.*

1. **C. multiflorus,** *A. Cunn. in Loud. Hort. Brit. (name only), and in Putterl. Syn. Pittosp.* 4.—A straggling or prostrate very much branched shrub, with slender branches, rough with a minute pubescence, and bearing numerous subulate thorns or abortive branches. Leaves sessile, ovate, orbicular, obovate, or broadly cuneate, usually 4 to 6 lines long, entire or with a few

small pointed or prickly teeth, rather thin, green and glabrous on both sides. Flowers about 2 lines long, always solitary in the axils, and not very numerous on the bush, notwithstanding the specific name. Ovary pubescent, with 2 parietal placentæ, and 8 to 12 ovules to each. Berry 2 to 5 lines diameter, containing from two to above a dozen seeds which are not viscid.

**Queensland.** Brisbane river, *A. Cunningham, F. Mueller.*
**N. S. Wales.** Damp shady woods and bushy places, Port Jackson to the Blue Mountains, *A. Cunningham* and others; northward to the Macleay, Hastings, and Clarence rivers, *Beckler ;* southward to Illawarra, *A. Cunningham* and others.

2. **C. pauciflorus,** *A. Cunn. in Loud. Hort. Brit. Suppl.* 585 (*name only*).—Habit of *C. multiflorus,* but stouter and more rigid, the branches similarly rough, with a minute pubescence, and thorny. Leaves from obovate to cuneate-oblong, rarely orbicular, mostly entire and obtuse, but occasionally mucronate or truncate and 3-toothed, rarely exceeding ½ in. in length, often petiolate and more rigid than in *C. multiflorus.* Flowers larger than in that species, the petals 4 to 5 lines long, united into a complete tube to ⅔ of their length. Ovary pubescent, with 5 parietal placentas, covered with innumerable minute ovules. Style longer than in *C. multiflorus.* Fruit attaining 1 to 1½ in. diameter, with a thick coriaceous pericarp. Seeds numerous, in a viscid pulp.—*Ixiosporus spinescens,* F. Muell. Fragm. Phyt. Austr. ii. 76.

**N. Australia.** Careening Bay, N.W. coast, *A. Cunningham.*
**Queensland.** E. coast, *R. Brown ;* in the scrub on the Fitzroy river, *Thozet ;* near the Dawson river, *F. Mueller ;* Castor creek, *Leichhardt.*
Cunningham's specimen, in leaf with the remains of a fruit, is not authentically named, but there is little reason to doubt its being the one he had in view. There are, also, in the Hookerian and in Mueller's herbaria specimens in leaf only, which may prove to be one, or perhaps two, additional species of *Citriobatus,* but they are insufficient for determination.

## 6. BILLARDIERA, Sm.

Petals connivent or cohering in a tube to above the middle, spreading at the top. Anthers oblong or ovate, shorter than the filaments. Ovary sessile or nearly so, completely or rarely imperfectly 2-celled, glabrous or pubescent. Fruit succulent or fleshy and indehiscent, ovoid or oblong. ·Seeds ovoid, reniform or globular, often enveloped in a viscid pulp.—Undershrubs, with the branches usually twining. Leaves entire or sinuate. Flowers greenish-yellow, purple or rarely blue, either solitary or clustered and pendulous, or in terminal cymes and erect.

The genus is limited to Australia. It differs from *Marianthus* only in the baccate not capsular fruit. The solitary pendulous flowers, frequent in *Billardiera,* are only in one species of *Marianthus.*

Pedicels solitary, or rarely 2 or 3 together.
  Petals elongated, slightly spreading at the top. Style long and
    filiform. Berry turgid, 1-celled . . . . . . . . . 1. *B. longiflora.*
  Petals spreading from above the middle. Style short. Berry
    oblong, 2-celled.
    Leaves ovate, linear, or rarely ovate-lanceolate, mostly wavy on
      the margin . . . . . . . . . . . . . . . . . 2. *B. scandens.*
    Leaves oval or elliptical-oblong, coriaceous, not wavy.
      Glabrous. Flowers solitary or very few . . . . . . . 3. *B. coriacea.*

Pubescent or silky-villous.   Flowers usually several  .  .  .  4. *B. cymosa,* var.
Pedicels several, clustered or corymbose (as in *Marianthus*).  [*sericophora.*
Sepals lanceolate-subulate, flowers corymbose.
 Corymbs distinctly pedunculate.   Petals about 5 lines long  .  6. *B. Lehmanniana.*
 Corymbs sessile, or very shortly pedunculate.   Petals 7 or 8
  lines.
  Sepals glabrous or silky pubescent .  .  .  .  .  .  .  .  .  4. *B. cymosa.*
  Sepals hirsute .  .  .  .  .  .  .  .  .  .  .  .  .  .  .  5. *B. variifolia.*
Sepals ovate or ovate-lanceolate.   Flowers in sessile clusters,
 usually nodding or pendulous.
 Glabrous.   Flowers solitary or very few  .  .  .  .  .  .  .  3. *B. coriacea.*
 Pubescent or silky villous.   Flowers usually several .  .  .  .  4. *B. cymosa,* var.
                 [*sericophora.*

(*B. rosmarinifolia,* DC. Prod. i. 345, described from specimens in leaf only, appears to me to be a *Mirbelia*.)

1. **B. longiflora,** *Labill. Pl. Nov. Holl.* i. 64. *t.* 89.—Stems twining, sometimes very short, but often many feet long, glabrous or silky pubescent when young.   Leaves from ovate and not above ½ in. long, to lanceolate or linear, and 1 to 1½ in. or rarely 2 in. long, obtuse or rarely acute, entire, tapering into a very short petiole or almost sessile.   Flowers greenish-yellow, often changing to purple, pendulous on solitary terminal pedicels of ½ to 1 in. Sepals lanceolate, finely pointed, 2 to 3 lines long.   Petals linear-cuneate, 1 to nearly 1½ in. long, erect and shortly spreading at the top, forming an almost tubular corolla.   Ovary glabrous or slightly pubescent, with a long subulate style.   Berry from nearly globular to narrow-ovoid, turgid, becoming unilocular from the disappearance of the half-dissepiment.   Seeds numerous, not enveloped in pulp.—DC. Prod. i. 345 ; Bot. Mag. t. 1507 ; Hook. f. Fl. Tasm. i. 37 ; F. Muell. Pl. Vict. i. 78 and 225 ; *B. ovalis,* Lindl. Bot. Reg. t. 1719 (with short badly developed flowers) ; *B. macrantha,* Hook. f. Fl. Tasm. i. 37 (with remarkably long flowers).

**N. S. Wales.**   Twofold Bay, *F. Mueller.*
**Victoria.**   Along shady rivulets and in damp mountain forests, ascending to subalpine elevations, *F. Mueller.*
**Tasmania,** *R. Brown ;* abundant throughout the island in thickets, etc., ascending to 3000 ft., *J. D. Hooker.*

2. **B. scandens,** *Sm. Bot. Nov. Holl.* i. *t.* 1.   Stems twining, often to a considerable extent, or short and flexuose, nearly glabrous or more or less silky or velvety-pubescent, or hairy.   Leaves from ovate-lanceolate to lanceolate or linear, obtuse or with a recurved point, usually 1 to 2 in. long, entire or often with undulate margins, usually narrowed into a short petiole. Flowers from greenish or pale yellow to violet or purple, pendulous on slender terminal pedicels varying from a line or two to above ½ in., solitary or very rarely 2 together.   Sepals lanceolate or lanceolate-subulate.   Petals spreading from above the middle, so as to form a narrow-campanulate corolla, 8 to 10 lines or rarely 1 in. long.   Ovary glabrous or pubescent, 2-celled, with a very short style and broad hollow stigma.   Berries cylindrical or ovoid-oblong, 2-celled, glabrous or downy.   Seeds numerous, in a close double row in each cell and embedded in pulp.—DC. Prod. i. 345 ; Bot. Mag. t. 801 ; Sweet, Fl. Austral. t. 54 ; F. Muell. Pl. Vict. i. 79 ; *B. latifolia,* Putterl. Nov. Stirp. Dec. 47, but not of Klatt, Linnæa, xxviii. 570 ; *B. grandiflora,* Putterl. l. c. 48 (all

the above referring to specimens with pubescent ovaries and fruits) ; *B. mutabilis,* Salisb. Parad. Lond. t. 48 ; Bot. Mag. t. 1313 ; DC. Prod. i. 345 ; Hook. f. Fl. Tasm. i. 37 (with glabrous ovaries and fruits) ; *B. angustifolia,* DC. Prod. i. 345 ; *B. canariensis,* Wendl. Hort. Herrenh. t. 15.

**Queensland.** Wide Bay and Moreton Bay, *F. Mueller.*

**N. S. Wales.** Port Jackson, *R. Brown, Sieber, n.* 495, etc.; northward to New England, *Stuart ;* and Hastings river, *Beckler ;* southward to Twofold Bay, *F. Mueller.*

**Victoria.** Stony and rocky declivities, chiefly amongst scrub, along rivers, and in moist forest country through the western and eastern parts of the colony ascending to the Alps, *F. Mueller.*

**Tasmania.** Stiff clayey soils in the northern parts of the island, *J. D. Hooker.*

**S. Australia.** Mount Gambier, at the S.E. extremity of the colony, *F. Mueller.*

Var. *brachyantha.* Softly hairy. Leaves narrow, undulate. Flowers about 3 together, on shorter pedicels; petals short. Ovary and fruit densely villous ; apparently connecting the species with the var. *sericophora* of *B. cymosa.*—*B. brachyantha,* F. Muell.; Klatt, in Linnæa, xxviii. 570. Buffalo range and Mount Macedon in Victoria, *F. Mueller,* whom I follow in uniting into one species the glabrous and downy-fruited forms of the common eastern *Billardiera.*

3. **B. coriacea,** *Benth.* A tall twiner, either perfectly glabrous or the young shoots slightly silky-hairy. Leaves distinctly petiolate, from broadly oval to elliptical-oblong, obtuse or shortly pointed, mostly 1½ to 2½ in. long, quite entire and coriaceous. Pedicels solitary, or 2 or 3 together, short and terminal. Flowers pendulous, apparently yellow, 8 to 9 lines long, resembling those of *B. scandens,* but more contracted in the middle, the petals slightly spreading above the middle. Sepals ovate-lanceolate, finely pointed. Ovary glabrous or slightly pubescent, 2-celled. Berry cylindrical, very obtuse.—*Pronaya latifolia,* Turcz. in Bull. Mosc. 1854, ii. 363.

**W. Australia.** S. coast towards Cape Riche, *Drummond, 5th Coll. n.* 240 ; East Mount Barren and Phillip's ranges, *Maxwell;* Point Henry, *Oldfield.*

4. **B. cymosa,** *F. Muell. in Trans. Vict. Inst.* i. 29, *and Pl. Vict.* i. 80. Shrubby with the branches more or less twining or sometimes short and flexuose, glabrous or the young parts and inflorescence silky-pubescent. Leaves usually lanceolate or oblong-linear, sessile or nearly so, obtuse or shortly pointed, 1 to 2 in. long. Corymbs, in the typical form, several-flowered, shortly pedunculate or nearly sessile. Sepals, in the same form, lanceolate-subulate, glabrous or with appressed hairs. Petals 7 to 8 lines long, spreading from above the middle, usually bluish or violet-purple. Ovary glabrous or silky-pubescent, 2-celled. Style short, with a broad hollow stigma. Berry oblong, with numerous seeds embedded in pulp.—*B. cymosa* and *B. pseudocymosa,* Klatt, in Linnæa, xxviii. 571.

**Victoria.** Desert on the Murray river and its lower tributaries, and scrubby barren ridges in Bacchus marsh, *F. Mueller.*

**S. Australia.** Barren places and scrubby arid ranges from Guichen Bay to Venus Bay and Mount Remarkable, not rare, ranging far inland, and frequent in Kangaroo Island, *F. Mueller.*

Var. (?) *sericophora.* Usually much more silky-villous, especially the young shoots. Leaves usually broader and more distinctly petiolate, sometimes almost ovate. Flowers greenish or pale yellow, few in closely sessile cymes or clusters, and often pendulous. Sepals short, ovate or ovate-lanceolate. Ovary very silky or villous. Berry usually pubescent or villous.—*B. sericophora,* F. Muell. in Linnæa, xxv. 371 ; *B. versicolor,* F. Muell.; Klatt, in Linnæa, xxviii. 571.  Victoria and chiefly South Australia, *F. Mueller.*  South coast, *R. Brown.*

I follow F. Mueller in referring this to a variety of *B. cymosa,* as he has no hesitation on the point, and it does in a few specimens appear to pass into the typical form , but the majority of specimens seem to me to be rather more nearly connected with the pubescent-fruited forms of *B. scandens,* and would have led me to adopt it as an independent intermediate species.

5. **B. variifolia,** *DC. Prod.* i. 346.  Shortly twining, with the young shoots and inflorescence more or less hirsute, with short hairs.  Leaves sessile or nearly so, oblong or lanceolate and entire, or the lower ones broader, cuneate and deeply toothed, the longest seldom above 1 in. long. Flowers blue, on very short hirsute pedicels, in terminal corymbs, usually dense and sessile, rarely looser, few-flowered, and shortly pedunculate. Sepals lanceolate-subulate, hirsute with spreading hairs.  Petals about 4 to 6 lines long, spreading from the middle.  Ovary densely villous, with a short subulate style.  Berry cylindrical, narrow, acuminate, $\frac{3}{4}$ to 1 in. long.— *Marianthus cœlestis,* Putterl. Syn. Pittosp. 23 ; *Pronaya Huegeliana,* Putterl. in Pl. Preiss. i. 204 ; *Pronaya sericea,* Turcz. in Bull. Mosc. 1854, ii. 363, and probably *P. lanceolata,* Turcz. l. c. 364, which I have not seen.

**W. Australia.**  Common about King George's Sound, *R. Brown, Labillardière, A. Cunningham* and others, to the Perongerup ranges, *Maxwell ;* also *Drummond, n.* 97.

Var. (?) *rigida.*  Branches shorter, scarcely twining.  Leaves crowded, narrow, rigid, above $\frac{1}{2}$ in. long, recurved at the top, with the margins revolute.  Perhaps a distinct species.— *Marianthus venustus,* Putterl.  Syn. Pittosp. 23, from the character given.—With the typical form, *Fraser, Drummond,* and others.

6. **B. Lehmanniana,** *F. Muell. Pl. Vict.* i. 78.  Glabrous except a very slight pubescence on the inflorescence, with numerous erect or shortly twining leafy branches.  Leaves sessile or nearly so, oblong-linear, usually obtuse, $\frac{3}{4}$ to 1$\frac{1}{2}$ in. long, rather firm and flat.  Flowers numerous, in pedunculate terminal corymbs on slender pedicels.  Sepals lanceolate-subulate.  Petals about 5 lines long, narrow-obovate, pointed, spreading from the middle. Anthers short, sometimes slightly recurved.  Ovary glabrous, 2-celled, with a short style.  Berry cylindrical.—*Marianthus angustifolius,* Putterl. in Pl. Preiss. i. 200 ; *Pronaya angustifolia,* Lehm. in Pl. Preiss. ii. 233.

**W. Australia.**  Swan River, *Drummond, Coll.* 1843, *n.* 79, and 5*th Coll. n.* 241, *Preiss, n.* 1287.

## 7. PRONAYA, Hueg.

Petals spreading nearly from the base.  Anthers narrow-oblong, about as long as the filaments, recurved or revolute as soon as the flower opens. Ovary imperfectly 2-celled, pubescent.  Fruit succulent, oblong, indehiscent. Seeds globular or angular.

The genus is limited to the following single Australian species, only differing from *Billardiera,* with which F. Mueller proposes to unite it, in the more spreading corolla and in the anthers ; the habit is that of the cymose *Billardieras* or of *Cheiranthera.*

1. **P. elegans,** *Hueg. Bot. Archiv. t.* 6.  Usually twining, with a close silky pubescence on the young shoots and inflorescence, the older leaves and branches glabrous.  Lower leaves often coarsely toothed or lobed, the others sessile or nearly so, lanceolate or linear-lanceolate, 1 to 1$\frac{1}{2}$ in. long, entire,

rather firm, the margins recurved. Flowers bluish or white, in a dense terminal corymb, sessile amongst the last leaves. Petals about ½ in. long, ovate, more spreading than in any *Billardiera* although less so at the base than in *Sollya.* Ovary tomentose, and berry oblong-cylindrical, very much like those of *Billardiera variifolia.*—Putterl. in Pl. Preiss. i. 203, Paxt. Mag. Bot. xii. 99, with a fig.; *Spiranthera Fraseri*, Hook. in Bot. Mag. under t. 3523; *Campylanthera Fraseri*, Hook. Ic. Pl. t. 82.

**W. Australia.** Common about Swan River, *Fraser, Huegel, Drummond,* and others. Var. *minor.* More slender, and smaller. Leaves mostly about ½ in. long. Flowers smaller.—*P. speciosa*, Endl. in Hueg. Enum. 9?—S. coast, *R. Brown*, whose specimens agree with the character given by Endlicher from Bauer's specimens. The other described *Pronayas* are true *Billardieras.*

## 8. SOLLYA, Lindl.

Petals spreading from the base, obovate. Anthers longer than the filaments, connivent in a cone round the pistil, and opening inwards by longitudinal slits. Ovary 2-celled, with a short style. Berry oblong. Seeds embedded in pulp.—Twiners. Leaves narrow. Flowers nodding, on slender pedicels, in terminal loose few-flowered cymes, or rarely solitary.

The genus is limited to Australia.

Peduncles several flowered. Petals 4 to 5 lines. Berry oblong-cylindrical, with closely packed seeds . . . . . . . . . . . 1. *S. heterophylla.*
Peduncles filiform, 1- to 3-flowered. Petals 3 lines. Berry slender, with few seeds . . . . . . . . . . . . . . . . . . 2. *S. parviflora.*

1. **S. heterophylla,** *Lindl. Bot. Reg. t.* 1466. Glabrous or the young shoots, under side of the leaves, and inflorescence more or less silky-hairy. Stems flexuose or twining, from a woody base. Leaves from ovate-lanceolate to ovate-oblong, and 1½ to 2 in. long or rather more, to lanceolate or oblong-linear, and 1 to 1½ in., obtuse or slightly acuminate, rather coriaceous, quite entire, usually narrowed into a very short petiole. Cymes terminal or leaf-opposed, drooping, usually 4- to 8-flowered, but sometimes with 12 or more flowers. Pedicels slender. Sepals narrow, acute, about 1 line long. Petals 4 to 5 lines. Ovary silky-pubescent. Berry cylindrical, obtuse, about ¾ in. long and fully 3 lines thick, with a thin succulent pericarp. Seeds numerous, closely packed in two rows in each cell, more or less angular or flattened by mutual pressure.—Bot. Mag. t. 3523; Putterl. in Pl. Preiss. i. 203; *Billardiera fusiformis*, Labill. Pl. Nov. Holl. i. 65. t. 90; DC. Prod. i. 345.

**W. Australia.** Common about King George's Sound, *R. Brown, Labillardière* and others, extending eastward along the coast beyond Stokes Inlet, *Maxwell;* inland to Stirling range, and perhaps to Swan River, *Drummond* and others. Var. *angustifolia.* Branches less twining. Leaves narrow-lanceolate.—*S. linearis* Lindl. Bot. Reg. 1840, t. 3. S. coast, *R. Brown, Fraser, Drummond,* etc.

2. **S. parviflora,** *Turcz. in Bull. Mosc.* 1854, ii. 361. Very much more slender and twining than *S. heterophylla*, usually sprinkled with soft loose hairs. Leaves lanceolate or oblong-linear, the larger ones often rather more than 1 in. long, but in some specimens all under ½ in., very shortly petiolate and thinner than in *S. heterophylla.* Flowers small, solitary, or 2 or 3 in a cyme, on very fine filiform pedicels. Petals about 3 lines long. Berry ½ to

¾ in. long, 1¼ to 2 lines broad, tapering at both ends.　Seeds globular, much fewer than in *S. heterophylla.*

**W. Australia,** *Drummond,* ¦*4th Coll. n.* 99, *5th Coll. n.* 238 ; Kojonerup hills, *Herb. Mueller.*

*Xerosollya Gilbertii,* Turcz. in Bull. Mosc. 1854, ii. 362, which I have not seen, may be the same plant.　The description agrees in every respect, even to the peculiar form of the fruit, except that he describes the latter as dry and 2-valved, and it appears to be succulent in *S. parviflora.*

*Sollya Drummondi,* Morren, and *S. salicifolia,* Marnock, published in gardening works, not in our botanical libraries, are unknown to me, but are most probably garden varieties of *S. heterophylla.*

## 9. CHEIRANTHERA, A. Cunn.

Petals spreading from nearly the base, obovate-oblong.　Anthers longer than the filaments, all turned towards one side, opening by two pores at the top.　Ovary 2-celled with a subulate style.　Capsule oblong, hard, opening loculicidally in 2 valves, the valves also splitting septicidally.　Seeds nearly globular.—Branches flexuose or twining.　Leaves narrow.　Flowers in terminal corymbs or cymes, or drooping from terminal solitary pedicels.

The genus is limited to Australia.

Flowers several, corymbose.
　Leaves flat or concave. Sepals lanceolate. Anthers not twice as
　　long as the filaments, and not attaining half the length of the
　　petals　.　.　.　.　.　.　.　.　.　.　.　.　.　.　.　.　.　1. *C. linearis.*
　Leaves thick or terete. Sepals narrow. Anthers fully twice as
　　long as the filaments and exceeding the half of the petals　.　.　2. *C. filifolia.*
Flowers solitary, on slender terminal pedicels.
　Leaves linear-terete or involute　.　.　.　.　.　.　.　.　.　.　.　3. *C. volubilis.*
　Leaves linear, flat, or revolute　.　.　.　.　.　.　.　.　.　.　.　4. *C. parviflora.*

1. **C. linearis,** *A. Cunn. in Bot. Reg. under t.* 1719.　A low glabrous shrub or undershrub, with erect twiggy branches of 6 in. to 1 ft., or rarely longer.　Leaves linear, acute or rather obtuse, ¾ to 1½ in., or rarely 2 in. long, entire or minutely toothed, flat, and ½ to 1 line broad, or the margins incurved, so as to be almost terete, with smaller leaves often clustered in the axils.　Flowers blue and showy.　Sepals lanceolate, 2 to 2½ lines long. Petals 8 to 10 lines.　Filaments short.　Anthers rather longer, but not reaching to the middle, and often not ⅓ of the length of the petals.　Capsule very like those of *Marianthus pictus* and *lineatus,* oblong, much flattened, hard but dehiscent when quite ripe.—Hook. Ic. Pl. t. 47 ; Fl. des Serres, viii. t. 856 ; F. Muell. Fragm. i. 97 ; Pl. Vict. i. 76 ; *C. cyanea,* Brongn. Voy. Coq. t. 77.

**N. S. Wales.**　Brushy forest country at the foot of Croker's range, frequent near Bathurst, *A. Cunningham ;* near Clifton in New England, *C. Stuart.*

**Victoria.**　Barren stony ridges and hills, Mount M'Ivor, and near the Ovens range, *F. Mueller.*

• **S. Australia.**　Mount Barker, *Whittaker ;* Flinders range, Kangaroo island, Spencer's Gulf and St. Vincent's Gulf, *F. Mueller.*

2. **C. filifolia,** *Turcz. in Bull. Mosc.* 1854, ii. 364.　Allied to *C. linearis,* but the branches are more slender and often flexuose or almost twining. Leaves very narrow, thick or almost terete, obtuse or scarcely pointed, some-

times none of them exceeding 3 or 4 lines, at others the upper ones above 1 in. long. Flowers blue, smaller than in *C. linearis.* Sepals linear or narrow-lanceolate, 1 to 2 lines long. Petals 5 to 6 lines. Anthers longer and narrower than in *C. linearis,* usually twice as long as the filaments, and exceeding the half and often reaching two-thirds of the petal.—*C. tortilis,* F. Muell. Fragm. ii. 79.

**W. Australia.** S. coast ?, *Drummond, Coll.* 1850, *n.* 94, *Oldfield;* river entering Stokes Inlet, *Maxwell.*

Var. *brevifolia.* Branches short, with crowded leaves, mostly 3 to 4 lines long.—*C. brevifolia,* F. Muell., Fragm. i. 97, and ii. 180; Phillips' range, also Plantagenet and Stirling ranges, *F. Mueller.* Drummond's specimens connect the short-leaved with the long-leaved forms.

. 3. **C. volubilis,** *Benth.* A slender glabrous twiner. Leaves narrow-linear, thick, with the margins involute or terete, with a short recurved point, mostly about ½ in. long. Peduncles slender, terminal, with a single drooping flower. Sepals lanceolate or linear-lanceolate, about 2 lines long. Petals about ½ in. Anthers scarcely so long as the filaments, very obtuse, and not reaching to the half of the petals. Ovary shorter than in *C. linearis,* with a long subulate style. Fruit not seen.

**S. Australia.** Scrub in Kangaroo Island, *Waterhouse.*

4. **C. parviflora,** *Benth.* Slender and glabrous or slightly pubescent, the branches either short and flexuose or elongated and twining. Leaves sessile or nearly so, from broadly oblong-lanceolate or almost ovate-lanceolate and 1½ in. long to linear and ½ in. or less, usually obtuse and the margins always revolute, sometimes slightly hirsute on the upper side. Flowers as in *C. volubilis,* on long terminal simple filiform peduncles, but smaller. Sepals seldom above 1 line, petals about 4 or 5 lines long. Anthers rather longer than the slender filaments and reaching to about half the length of the petals. Ovary glabrous, with a subulate style.

**W. Australia,** *Drummond, Coll.* 1843, the specimens *n.* 34, very twining, with larger and broader leaves, and *n.* 80 less twining, with smaller narrower leaves.

*C. Preissiana,* Putterl. Pl. Preiss. i. 201, if a *Cheiranthera* at all, differs from the last species in its hirsute branches and leaves, but the flowers are unknown, and the fragments I have seen are in leaf only, something like those of *Billardiera variifolia* or of *Pronaya elegans.*

ORDER XIII. **TREMANDREÆ.**

Flowers regular. Sepals 4 or 5, very rarely 3, free, valvate in the bud. Petals as many, hypogynous, spreading, induplicate-valvate in the bud. Stamens twice as many, hypogynous, free; filaments short; anthers oblong or linear, 2- or 4-celled, opening in a single terminal pore. Torus small or rarely expanded into a disk between the petals and stamens. Ovary sessile or nearly so, usually 2-celled; style filiform, deciduous, entire or minutely 2-lobed. Ovules solitary in each cell, or 2, one above the other, or rarely an additional small collateral one, pendulous, anatropous, with a ventral raphe. Capsule usually flattened, 2-celled, opening loculicidally at the edges. Seeds pendulous, the raphe usually expanded at the chalazal extremity into a twisted or strophiola-like appendage, rarely wanting; the testa crustaceous, glabrous or hairy; albumen fleshy or almost cartilaginous. Embryo small, straight, with a superior radicle.—Shrubs usually heath-like, glabrous or glandular-hairy, with small alternate opposite or verticillate leaves, rarely with a stellate tomentum and larger leaves. Flowers solitary, on axillary pedicels, usually red or purple. In many species, as in *Pittosporeæ* and *Polygaleæ*, a flower may here and there be found with a 3-merous ovary and fruit.

The Order is strictly confined to Australia, and although showing some affinity with *Cheiranthera* in *Pittosporeæ*, as well as with *Polygaleæ* proper, it is yet very different from either; the connection with *Lasiopetaleæ*, insisted upon by Steetz, appears to rest almost entirely on the valvate calyx, and on an occasional resemblance in habit, which is, however, partaken in by *Bauera* and several other genera of Australian heath-like shrubs, which have little else in common.

Anthers continuous with the filament. Leaves alternate or whorled, glabrous or glandular hairy.
Anthers 2-celled, or with 4 cells, 2 in front of the 2 others. Seeds with an appendage at the chalaza . . . . . . . . . 1. TETRATHECA.
Anthers 4-celled, the 4 cells on the same plane. Seeds without appendage . . . . . . . . . . . . . . . . 2. PLATYTHECA.
Anthers articulate on the filament. Leaves opposite, with stellate hairs. Seeds with an appendage . . . . . . . . . 3. TREMANDRA.

## 1. TETRATHECA, Sm.

Stamens apparently in a single series, the anthers continuous with the filament, 2-celled, or 4-celled with 2 of the cells in front of the 2 others, more or less contracted into a tube at the top. Disk none. Capsule opening only at the edges. Seeds with an appendage at the chalazal end usually contorted.—Glabrous or glandular-hairy. Leaves alternate, verticillate or scattered, heath-like and entire, or flat and toothed, or reduced to minute scales.

§ 1. *Stems terete, leafy (except* T. subaphylla). *Ovules 1 or 2 in each cell. Seeds hairy.* (Eastern or southern species.)

Leaves mostly verticillate. Ovules usually 2, superposed, or, if solitary, attached below the summit of the cell.
Leaves ovate, obovate, or orbicular, flat. Sepals ovate, obtuse or scarcely acute, often reflexed . . . . . . . . . . . . 1. *T. ciliata.*
Leaves ovate to lanceolate, acute, with the margins recurved. Sepals acute or acuminate, not reflexed . . . . . . . . . 2. *T. thymifolia.*
Leaves linear, the margins revolute. Sepals not reflexed . . . . 3. *T. ericifolia.*
Leaves rarely subverticillate. Ovules solitary, suspended from the summit of the cell.
Very glandular. Leaves elliptical-oblong or obovate, much narrowed at the base. Petals large, obovate . . . . . . . . . 4. *T. glandulosa.*
Glabrous or hispid, rarely glandular. Leaves linear, or, if broader, obtuse at the base. Petals oblong or scarcely obovate . . . . 5. *T. pilosa.*
Glabrous and somewhat glaucous. Leaves all, or nearly all, reduced to minute scales . . . . . . . . . . . . . . . 6. *T. subaphylla.*

§ 2. *Stems very angular or flat, almost leafless. Ovules 2 or 4 in each cell. Seeds hairy.* (Eastern and western species.)

Stems often 3-angled. Flowers 4-merous. Anther-tubes very short . 7. *T. juncea.*
Stems flat, 2-winged. Flowers 5-merous. Anther-tubes long . . . 8. *T. affinis.*

§ 3. *Stems terete, leafy, or almost leafless. Ovules solitary in each cell. Seeds glabrous and shining.* (Western species.)

Leaves minute and distant, or linear-terete and alternate.
Leaves minute and distant. Flowers 5-merous. Ovary glandular-
　　hirsute . . . . . . . . . . . . . . . . . . 9. *T. nuda.*
Leaves either minute and distant or not crowded. Flowers 4-merous.
　　Ovary glabrous . . . . . . . . . . . . . . . 10. *T. virgata.*
Leaves crowded. Flowers 5-merous. Ovary glandular-hirsute . . 11. *T. confertifolia.*
Leaves alternate, lanceolate or ovate.
　　Leaves glabrous underneath, except the setæ of the midrib . . . 12. *T. setigera.*
　　Leaves softly pubescent underneath.
　　　Leaves ovate, flat. Setæ long and numerous . . . . . . . 13. *T. hispidissima.*
　　　Leaves lanceolate, much revolute, occasionally verticillate. Setæ
　　　　rare . . . . . . . . . . . . . . . . . . 14. *T. hirsuta.*
Leaves mostly verticillate or opposite.　　　　　　　　　　.
　　Leaves villous underneath, often alternate . . . . . . . . . 14. *T. hirsuta.*
　　Leaves glabrous underneath or pubescent on the midrib, verticillate
　　　in threes or fours, very rarely alternate.
　　　Anthers purple, the tubular process as long as the cells.
　　　　Leaves glabrous or ciliate, or rarely hirsute above . . . . 15. *T. viminea.*
　　　　Leaves coriaceous, scabrous or pubescent, not ciliate . . . . 16. *T. pubescens.*
　　　Anthers yellow, contracted into a very short tube . . . . . 17. *T. pilifera.*
　　Leaves membranous, lanceolate-linear, flat, opposite or verticillate.
　　　Anthers very short and curved, with a slender tube . . . . 18. *T. filiformis.*

1. **T. ciliata,** *Lindl. in Mitch. Three Exped.* ii. 206. An undershrub with slender erect or diffuse stems, of 1 to 2 or rarely 3 ft., very shortly and roughly pubescent or glabrous. Leaves almost all verticillate in threes or fours, broadly ovate or nearly orbicular, obtuse or slightly pointed, rarely exceeding ½ in. and mostly smaller, the margins flat or scarcely recurved, ciliate or rarely glabrous. Pedicels usually longer than the leaves. Sepals broadly ovate, obtuse or scarcely acuminate, about 1 line long, more spreading than in the following species, and sometimes reflexed, bearing like the pedicels a few black glandular hairs or setæ. Petals obovate-oblong, about ½ in. long. Anther-tubes short. Ovary pubescent, with 2 superposed ovules in each cell, and occasionally a third collateral one. Capsule broad, 2 to 4 lines long. Seeds hairy.—Hook. Ic. Pl. t. 268; Hook. f. Fl. Tasm. i. 34; F. Muell. Pl. Vict. i. 181; *T. baueræfolia*, F. Muell. in Schuch. Syn. Trem. 29.

**Victoria.** Port Phillip, *R. Brown;* frequent on heathy ground and barren forest ridges in many parts of the colony, not ascending to the Alps, *F. Mueller, Mitchell,* and others.

**Tasmania.** Sandy heaths, Port Dalrymple, *R. Brown ;* mouth of the Tamar and other parts of the north of the island, *Gunn.*

2. **T. thymifolia,** *Sm. Exot. Bot.* i. 41. t. 22. Intermediate between *T. ciliata* and *T. ericifolia*, it has usually the tall habit of the former, but is much more pubescent or hirsute. Leaves almost all verticillate in threes or fours, ovate-elliptical or lanceolate, the margins more or less recurved or revolute. Flowers of *T. ciliata*, except that the sepals are usually ovate-lanceolate, more acute or acuminate than in either of the two allied species, and seldom reflexed. Ovary glabrous, or more frequently pubescent. Ovules fruit and seeds of *T. ciliata*.—DC. Prod. i. 343.

**Queensland.** Glasshouses, Moreton Bay, *F. Mueller.*
**N. S. Wales.** Port Jackson to the Blue Mountains, *Herb. Smith, A. Cunningham,*

and others ; brushy forest north of Bathurst, *A. Cunningham ;* northward to Hastings river, *Beckler,* and southward to Twofold Bay, *F. Mueller.*

**Victoria.** Heathy mountain tracts, frequent, *F. Mueller.*

F. Mueller considers this and the two following species as varieties only of *T. pilosa,* but *T. thymifolia,* especially the broad-leaved Queensland form, appears to me nearer to *T. ciliata* than to *T. ericifolia,* and I cannot find the more or less open calyx so constant a character as the foliage, indefinite as that may often be. At any rate, if the whole series be divided into two species, the one would seem rather to include *T. ciliata, thymifolia,* and *ericifolia,* with leaves mostly verticillate, pedicels usually longer than the leaves, and ovules generally two, superposed ; whilst the other, formed of *T. glandulosa* and *pilosa,* has the leaves scattered, rarely verticillate, the pedicels short, and ovules solitary in each cell, inserted at the top.

3. **T. ericifolia,** *Sm. Exot. Bot.* i. 37. *t.* 20. A heath-like undershrub, more branched and diffuse than the two preceding species, rarely attaining 1 ft., minutely and roughly pubescent or nearly glabrous, very rarely shortly hirsute. Leaves mostly verticillate, but not so regularly so as in the last two species, narrow-linear, with the margins closely revolute or rarely oblong-lanceolate and more open, mostly under ½ in. Flowers on slender pedicels, usually longer than the leaves. Sepals as in *T. ciliata,* ovate, obtuse or scarcely acute, but not reflexed. Ovary glabrous or rarely pubescent, with 2 superposed ovules in each cell, or rarely a single ovule attached below the top of the cell. Capsule obovate-cuneate. Seeds hairy.—DC. Prod. i. 343 ; Rudge, in Trans. Linn. Soc. viii. t. 11.

**N. S. Wales.** Very abundant about Port Jackson, *R. Brown, Sieber, n.* 234, and others.

Var. *rubiæoides.* Leaves broader, less revolute and more regularly verticillate, almost like those of *T. thymifolia,* but glabrous or shortly pubescent, and the sepals obtuse as in *T. ericifolia.*—*T. rubiæoides,* A. Cunn. in Field. N. S. Wales, 335.—Rocky declivities of the Blue Mountains, *A. Cunningham.*

4. **T. glandulosa,** *Labill. Pl. Nov. Holl.* i. 96. *t.* 123. Rather coarse and much branched, often exceeding 2 ft. in height, more or less densely pubescent or hirsute with glandular hairs. Leaves scattered, not verticillate, usually elliptical-oblong, acute or obtuse, 3 to 5 lines long, the margins rigidly ciliate or almost toothed and slightly revolute, always narrowed at the base. Pedicels rarely exceeding the leaves. Sepals ovate, acute, about 1 line long. Petals broad, about 4 or 5 lines. Anther-tubes often more elongated than in the allied species. Ovary glandular, with 1 ovule, suspended as in *T. pilosa* from the summit of each cell, with very rarely a second collateral abortive one. Capsule obovate. Seeds hairy.—DC. Prod. i. 343 ; Hook. f. Fl. Tasm. i. 34.

**Victoria.** Ranges near Avon river in Gipps' Land, and dry scrubby hills between Ovens and Broken River, *F. Mueller.* Some of the Avon river specimens referred here by F. Mueller, have the leaves remarkably broad, sometimes almost orbicular.

**Tasmania.** Derwent river, *R. Brown ;* heathy places abundant throughout the island, *J. D. Hooker.*

The N. S. Wales specimens, often referred to this species, belong to the following one.

5. **T. pilosa,** *Labill. Pl. Nov. Holl.* i. 95. *t.* 122. Much branched and heath-like, glabrous or hispid, but not generally glandular, and seldom much exceeding 1 or 1½ ft. in height. Leaves usually linear, with the margins much revolute, 4 to 6 lines long, but in very luxuriant shoots they are sometimes broadly lanceolate or oblong, but with an obtuse base. Flowers scarcely

so large as in *T. glandulosa*, and often much smaller with narrow petals, the pedicels usually shorter than the leaves. Sepals ovate, obtuse or acute. Ovary glabrous or pubescent, with a single ovule suspended from the summit of each cell. Capsule obovate. Seeds hairy.—DC. Prod. i. 343 ; Hook. f. Fl. Tasm. i. 35 ; *T. ericoides*, Planch. in Fl. des Serres, x. 229, t. 1065 ; *T. calva*, Schuch. Syn. Trem. 25 ; *T. ericifolia*, var., F. Muell. Pl. Vict. i. 182.

**N. S. Wales.** About Port Jackson, but apparently rare.
**Victoria.** Not frequent, *F. Mueller.*
**Tasmania.** Port Dalrymple, etc., *R. Brown;* abundant throughout the island, *J. D. Hooker.*
**S. Australia.** Lofty ranges, *Whittaker;* common towards Spencer's Gulf, *F. Mueller.*
Var. *denticulata*, with narrow revolute leaves, as in *T. pilosa*, but with a few glandular hairs on the calyx and pedicels, the leaves occasionally opposite, thus in some measure connecting *T. pilosa* with *T. ericifolia*, but the flowers and ovules are those of the former. —About Port Jackson, from several collections.—*T. denticulata*, Sieb. Pl. Exs. n. 236, and in Spreng. Syst. Cur. Post. 147 ; *T. glandulosa*, Sm. Exot. Bot. i. 39, t. 21, Rudge, in Trans. Linn. Soc. viii. 294, t. 10, but not of Labillardière.
Var. (?) *procumbens.* Glabrous, procumbent, slender, and much branched, with smaller flowers on shorter pedicels than in the common state of *T. pilosa*.—*T. procumbens*, Gunn, in Hook. f. Fl. Tasm. i. 35, t. 7, A. (with red flowers) ; *T. calva*, β, *pulchella*, Schuch. Syn. Trem. 27 ; *T. Gunnii*, Hook. f. Fl. Tasm. i. 36, t. 7, B. (with numerous white flowers).—On the Western Mountains of Tasmania, and on heathy plains near the sea, *Gunn;* Port Dalrymple, *R. Brown;* the slender white-flowered variety on the Asbestos Hills.
I have considerable doubts whether this elegant Tasmanian variety may not prove permanently distinct.

6. **T. subaphylla,** *Benth.* Stems almost leafless, erect or flexuose, rush-like, terete, branching, often 1 to 2 ft. long, glabrous and somewhat glaucous, not glandular. Leaves few, scattered chiefly on the shorter barren branches, small, lanceolate, flat, narrowed at the base; occasionally 2 or 3 attain a length of ½ in. or more ; all the rest reduced to minute distant bracts. Flowers like those of *T. pilosa*, but smaller, on very short pedicels, in the axils of minute bracts along the leafless branches.—*T. ericifolia*, var., F. Muell. Pl. Vict. i. 183.

**Victoria.** Woody mountain ranges at the sources of Genoa river, *F. Mueller.*

7. **T. juncea,** *Sm. Bot. Nov. Holl.* 5. *t.* 2. Rootstock thick and woody, with erect or ascending slender rush-like or wiry stems, 1 to 2 ft. long, with 2 or 3 acute angles or very narrow wings, the whole breadth of the stem and wings rarely exceeding 1 line. Leaves few, small and distant, linear or lanceolate, mostly minute and scale-like, rarely 3 lines long. Pedicels in the axils of the upper minute leaves, filiform, 2 to 4 lines long. Sepals 4, small, ovate, obtuse. Petals 4, about 4 lines long. Anthers tapering into very short tubes. Ovary glabrous, with 2 superposed ovules in each cell. Capsule obovate. Seeds villous.—DC. Prod. i. 343 ; Reichb. Icon. Exot. t. 78.

**N. S. Wales.** Port Jackson, *Sieber, n.* 235, *M'Arthur,* and others.

8. **T. affinis,** *Endl. in Hueg. Enum.* 7. Glabrous, with long, winged, apparently leafless branches, at first sight closely resembling *T. juncea*, but the stems have always only 2 angles or narrow wings, the leaves are still fewer and more minute, the sepals and petals are in fives, and the anthers are minutely pubescent, and suddenly contracted into a slender tubular process as

long as themselves or nearly so.　Ovary slightly glandular, with 2 ovules in each cell.　Capsule broadly ovate or obovate, shortly pointed, 3 to 5 lines long, with membranous valves.　Seeds hairy.

**W. Australia.** King George's Sound, *R. Brown, Huegel; Drummond, Coll.* 1843, *n.* 73, and others; Gordon river, *Oldfield.*

Var. *platycaula.* Branches, including the wings, often 2 lines broad. Flowers and capsules rather larger, and 4 ovules in superposed pairs in each cell of the ovary.—*Drummond, Coll.* 1843, *n.* 115 ; Blackwood and Stirling ranges, *Oldfield.*

9. **T. nuda,** *Lindl. Swan Riv. App.* 38.　Glabrous or with a few glandular hairs at the base of the stem, and sometimes on the pedicels and sepals.　Rhizome woody, with numerous erect, slender, rigid but rush-like stems, cylindrical, without prominent angles, ¼ to 1½ ft. high, often ending in an almost pungent point.　Leaves very minute and distant, or a very few linear or oblong ones 2 or 3 lines long.　Pedicels slender, 2 to 3 lines long.　Sepals and petals 5 each.　Anthers tapering into a tubular process, very short in the typical form, and of the same colour as the rest.　Ovary covered with rather long glandular hairs, with 1 ovule in each cell.　Capsule obovate, glandular-hairy.　Seeds glabrous, smooth and shining.

**W. Australia.** Darling range, *Collie, Oldfield ;* Swan River, *Drummond,* 1st *Coll., Sanford.*

Var. *spartea,* Planch. in Herb. Hook.　Tubular process of the anthers nearly as long as the cells.—*Drummond, Coll.* 1843, *n.* 101 and 104.

10. **T. virgata,** *Steetz, in Pl. Preiss.* i. 212.　Very nearly allied to *T. nuda,* and perhaps a variety, but the branches are much more slender, often filiform, glabrous or scabrous, with a few glandular hairs : the leaves are much more frequently developed, especially on the barren branches, where however they are still few and distant, linear with revolute margins, 2 to 3 lines long ; the flowers appear to be always 4-merous, and the anthers more abruptly contracted into a slender tube, usually of a paler colour, and as long as the cells.　Ovary glabrous, with uniovulate cells.　Capsule obovate, about 3 lines long, with smooth shining seeds.

**W. Australia.** Swan River, *Drummond,* 5th *Coll. n.* 243, *Preiss, n.* 1332, in part ; Mount Barker, Kalgan and Blackwood rivers, *Oldfield.*

Var. *setigera,* Steetz, l. c. 213.　Stems very scabrous, and often with reflexed bristly hairs.　Leaves more numerous.—Swan River, *Drummond, Preiss, n.* 1333.

11. **T. confertifolia,** *Steetz, in Pl. Preiss.* i. 214.　Stems numerous, erect and simple, or branched and diffuse or ascending, usually 6 to 9 in. long, roughly pubescent.　Leaves crowded but not verticillate, linear, obtuse, 2 to 3 lines long, the margins much revolute so as to be almost terete, hispid with rigid hairs.　Pedicels ¾ to nearly 1 in. long.　Flowers 5-merous.　Sepals lanceolate.　Petals rather narrow, 4 to 5 lines long.　Anthers glabrous or slightly tuberculate, tapering into a tube about as long as the cells and often of the same colour.　Ovary glandular-hispid, with 1 ovule in each cell.　Capsule glandular-pubescent, obovate-cuneate, about 3 lines long.　Seeds glabrous.

**W. Australia.** Swan River, *Drummond,* 5th *Coll. n.* 244 ; Darling ranges, *Preiss, n.* 1328, 1329.

12. **T. setigera,** *Endl. in Hueg. Enum.* 8.　Stems rather rigid, not much branched, usually about 1 ft. high, hispid with spreading bristly hairs, or,

when these are worn off, rough with their tubercular bases. Leaves sessile, not crowded, scattered, from ovate-lanceolate to linear-oblong, obtuse, mostly $\frac{1}{2}$ to $\frac{3}{4}$ in. long, the margins revolute, obtuse at the base, scabrous or setose on the upper side, glabrous and glaucous underneath, except a few setæ on the midrib. Pedicels very slender, 3 to 6, or rarely 7 or 8 lines long, more thickened and turbinate under the flower than in most other species. Flowers 5-merous. Sepals glabrous. Petals rather narrow, 4 to 6 lines long. Anthers glabrous, their tubular points rather shorter than the cells. Ovary glabrous, with 1-ovulate cells. Capsule usually ripening only 1 glabrous shining seed, with an unusually large strophiola.—*T. elongata,* Schuch. Syn. Trem. 38.

**W. Australia.** King George's Sound, *R. Brown,* and many others; Swan River, *Preiss, n.* 1322 (from a bad specimen in Herb. Sonder), *Harvey;* Blackwood and Kalgan rivers and Bald Island, *Oldfield.*

13. **T. hispidissima,** *Steetz, in Pl. Preiss.* i. 217? Branches much elongated, minutely pubescent and hispid with numerous very long spreading setæ. Leaves ovate, sessile, or very shortly petiolate, obtuse, $\frac{1}{2}$ to $\frac{3}{4}$ in. long, with flat edges, hirsute with scattered hairs above, bordered with a few long setæ, softly pubescent or villous underneath. Pedicels slender, $\frac{1}{2}$ to $\frac{3}{4}$ in. long, with the turbinate summit of *T. setigera,* glabrous or with a very few setæ. Flowers of *T. setigera.* Anther-tubes slender, fully as long as the cells. Ovary pubescent with appressed hairs.

**W. Australia.** *Drummond, Coll.* 1843, *n.* 46; King George's Sound, *Preiss, n.* 1316.

I have not seen Preiss's specimen, described by Steetz, and am therefore not quite confident of having correctly referred his name to Drummond's plant.

14. **T. hirsuta,** *Lindl. Swan Riv. App.* 38, *and Bot. Reg.* 1844, *t.* 67. Stems rather rigid and erect, $\frac{1}{2}$ to $1\frac{1}{2}$ ft. high, minutely pubescent and often hispid with a few long spreading reddish hairs. Leaves mostly alternate, but here and there a few verticillate, from ovate-lanceolate to oblong-linear, obtuse, all under $\frac{1}{2}$ in. in the smaller specimens, nearly 1 in. long when luxuriant, the margins recurved, with an obtuse base, more or less hirsute above, villous or pubescent underneath. Pedicels slender, $\frac{3}{4}$ to 1 in. long, very slightly thickened under the calyx. Flowers rather large. Sepals lanceolate. Petals oblong. Anthers smooth or slightly rough, the tube about as long as the cells. Ovary glabrous or slightly glandular, with 1 ovule in each cell. Seeds glabrous, shining.—Paxt. Mag. Bot. xiii. 53, with a fig.; *T. rubriseta,* Lindl. Swan Riv. App. 38; *T. epilobioides* and *T. aculeata,* Steetz, in Pl. Preiss. i. 218.

**W. Australia.** Swan River, *Drummond,* and many others; Harvey river, *Oldfield,* a variety with smaller flowers, apparently white, with a purple spot at the base, and shorter anthers.

15. **T. viminea,** *Lindl. Swan Riv. App.* 38. Stems rather slender, erect, little branched except at the base, sometimes only 6 in., but usually 1 to $1\frac{1}{2}$ ft. high, glabrous or with a few long spreading setæ, rarely mixed with a few short hairs. Leaves on the main stems usually ovate obovate or orbicular, 3 to 5 lines long, rather thin, nearly flat, glabrous or ciliate, or very rarely hirsute above, glabrous underneath, those of the side branches or the upper

_Tetratheca._] XIII. TREMANDREÆ. 135

floral ones often narrow-lanceolate and much revolute, all in whorls of 3 or 4, or very rarely the upper ones alternate. Pedicels slender, about ¾ in. long. Flowers 5-merous. Sepals ovate-lanceolate. Petals rather narrow. Anthers purple, short and scabrous, abruptly contracted into a tube as long as the cells. Ovary glabrous or slightly glandular, with 1 ovule in each cell. Capsule obovate. Seeds smooth and shining.—_T. gracilis_, Steetz, in Pl. Preiss. i. 215 (founded on slender side branches).

**W. Australia.** Swan River, _Drummond, 1st Coll. and_ 1843, _n._ 108, _Preiss, n._ 1327 _and_ 1335; Harvey, Preston, Blackwood, and Vasse rivers, _Oldfield._

16. **T. pubescens,** _Turcz. in Bull. Mosc._ 1852, ii. 141. Very nearly allied to _T. viminea_, and perhaps a variety only, but the slender rigid branches as well as the upper side of the leaves are often rough with a minute pubescence and the long spreading setæ very rare, the leaves, from ovate to lanceolate, are thicker and almost coriaceous, and often marked on each side with 1 or 2 coarse teeth. Pedicels shorter and not so slender. Sepals ovate, obtuse, rarely above 1 line long. Anthers more gradually attenuated into a shorter tube.—_T. tenuiramea_, Turcz. in Bull. Mosc. 1852, ii. 142.

**W. Australia.** Swan River, _Drummond_, 1845, _n._ 245 _and_ 209. The latter specimens distinguished by Turczaninow under the name of _T. tenuiramea_, only differ in their branches rather more slender.

17. **T. pilifera,** _Lindl. in Swan Riv. App._ 38. Allied to _T. viminea_, but usually smaller and more branched, and readily distinguished by the anthers. Stems 6 in. to 1 ft. high, slender, and more or less pubescent or hirsute with stiff hairs, but with few of the long setæ except at the nodes, and sometimes almost glabrous. Leaves in whorls of 3 or 4, from ovate to ovate-lanceolate, 2 to 5 lines long, often toothed, glabrous or roughly pubescent on the upper side, with a few hairs on the midrib underneath. Pedicels ½ to ¾ in. long. Flowers rather smaller than in _T. viminea_, usually 5-merous, but occasionally 4-merous. Sepals ovate or almost lanceolate. Filaments, although short, very slender. Anthers pale-coloured, nearly straight, scarcely furrowed, slightly tapering into a very short tube. Ovary slightly glandular, with 1 ovule in each cell. Seeds smooth and shining.—_T. Preissiana_, Steetz, in Pl. Preiss. i. 219; _T. micrantha_, Schuch. Syn. Trem. 43 (from the character given).

**W. Australia.** Swan River, _Drummond, 1st Coll. and_ 1843, _n._ 103, _Preiss, n._ 1323; Darling range, _Collie._ I have not seen Preiss's n. 1324 from which _T. micrantha_ was described.

18. **T. filiformis,** _Benth._ Branches in our specimens very long and slender, glabrous or bearing above the internodes a few short spreading purple hairs. Leaves opposite or occasionally in whorls of 3, very rarely 4, narrow-lanceolate or oblong-linear, ½ to ¾ in. long, thinner than in most species, flat, obtuse at the base, glabrous. Pedicels very slender, more than 1 in. long. Sepals ovate-lanceolate, about 1 line. Petals obovate-oblong, 4 to 5 lines. Anthers dark purple, short, much curved, very angular, with a straight tube as long as the cells. Ovary glabrous or slightly glandular, with 1 ovule in each cell.

**W. Australia.** Swan River, _Drummond, Coll._ 1843, _n._ 197 _and_ 181. Franklin river, _Herb. Muell._

## 2. PLATYTHECA, Steetz.

Stamens in 2 distinct series, the anthers continuous with the filament; with 4 parallel cells in a single plane, contracted into a tube at the top. Disk none. Capsule opening loculicidally at the edge, with the valves splitting septicidally. Seeds glabrous, without appendage.—A heath-like shrub, with verticillate leaves.

1. **P. galioides,** *Steetz, in Pl. Preiss.* i. 220. An erect heath-like shrub or undershrub, with slender terete branches, sometimes quite glabrous, but more frequently with a little tuft of hairs at each node, and often pubescent below the nodes. Leaves usually about 8 in a whorl, narrow-linear, sometimes very acute and pungent, sometimes almost obtuse or with slightly recurved points, about ½ in. long, with the margins often revolute so as to be almost terete or 3-angled, glabrous or rough, with a few scattered short rigid hairs. Pedicels slender, ¾ to 1 in. long. Sepals narrow-lanceolate, acute, 3 to 4 lines long. Petals nearly ½ in., blue with a dark spot at the base. Anthers short and broad, with long slender tubes. Ovary glabrous, with 2 superposed ovules in each cell. Capsule about 3 lines long.—*P. crucianella,* Steetz, l. c. 221 ; *P. crassifolia,* Steetz, l. c. 222 ; *Tetratheca verticillata,* Paxt. Mag. Bot. xiii. 171, with a fig. ; *Tremandra verticillata,* Hueg. in Walp. Ann. i. 76 (the fig. quoted from Parad. Vind. is not yet published).

**W. Australia.** Swan River, *Drummond, Coll.* 1843, *n.* 102, *Preiss, n.* 1320, 1330, 1331 (also 1321, which I have not seen) ; Preston, Kalgan, and Vasse rivers, *Oldfield.*

## 3. TREMANDRA, R. Br.

Stamens apparently in a single series, the anthers articulate on the short filiform filaments, 2-celled, not attenuated into a tube, although opening by a single terminal pore in 2 short valves. Disk crenate, almost 5-lobed, between the petals and stamens. Capsule opening at the edges. Seeds with an appendage or strophiola at the chalazal end.—Shrubs with stellate hairs or tomentum. Leaves opposite, toothed.

Densely tomentose. Leaves 1 in. or more. Pedicels shorter than the
leaves . . . . . . . . . . . . . . . . . . . . . . . 1. *T. stelligera.*
Slender, with minute scattered stellate hairs. Leaves under ¼ in. Pedicels longer, filiform . . . . . . . . . . . . . . . . . 2. *T. diffusa.*

1. **T. stelligera,** *R. Br. in DC. Prod.* i. 344. A shrub of 2 ft. or more, densely clothed with stellate hairs sometimes short and tomentose or almost floccose, sometimes long and hirsute. Leaves opposite, shortly petiolate, ovate, obtuse, 1 to 1½ in. long, coarsely and irregularly toothed or rarely entire. Pedicels shorter than the leaves. Sepals lanceolate, tomentose or villous, 2 to 3 lines long. Petals but little longer. Anthers rather longer than their filaments, dark-purple, hirsute pubescent or glabrous, truncate or oblique at the top. Ovary densely pubescent, with 2 superposed ovules in each cell. Capsule broadly ovate, pubescent. Seeds more or less silky-pubescent, with a large hooked appendage at the chalazal end.—*T. oppositifolia,* Steetz, in Pl. Preiss. i. 222.

**W. Australia.** King George's Sound, *R. Brown,* and many others.

Var. *hispida.* Branches and leaves rigidly hirsute. Anthers glabrous. Capsule narrower than in the normal form, with smaller seeds, and a shorter appendage, *Drummond, n.* 161, 194 *and* 217, *Coll.* 1843.

2. **T. diffusa,** *R. Br. in DC. Prod.* i. 344. Slender and diffuse, the branches often filiform and spreading to 1 or 1½ ft., glabrous or minutely pubescent. Leaves petiolate, broadly ovate, 3 to 5 lines long, more or less toothed, glabrous above, rough underneath, with very short scattered stellate hairs. Pedicels filiform, often longer than the leaves, although sometimes short. Sepals about 1 line. Petals 1¼ lines long. Anthers pale, almost glabrous, not longer than the filaments. Ovary villous or pubescent, with 2 superposed ovules in each cell. Capsule broader than long, didymous, pubescent. Seeds silky-pubescent, with a short straight appendage at the chalazal end.

**W. Australia.** Rocky hills, King George's Sound, *R. Brown, Drummond, n.* 216, *Oldfield.*

## ORDER XIV. POLYGALEÆ.

Flowers hermaphrodite, irregular. Sepals 5, free, much imbricate, the 2 inner ones usually larger and petal-like. Petals 3 or 5, rarely all free, most frequently 2 or 4 in pairs united at the base with the lower concave or helmet-shaped petal or keel and often with the staminal tube. Stamens 8, rarely 5 or 4, usually united to above the middle in a sheath open on the upper side. Anthers erect, 1- or 2-celled, usually opening by a single terminal or oblique pore. Torus small, or rarely expanded into a disk within the stamens. Ovary free, 2-celled or rarely 1-celled, or in a few flowers 3- to 5-celled. Style simple, usually curved at the top, with a variously shaped entire or 2-lobed stigma. Ovules usually solitary in each cell, pendulous, anatropous with a ventral raphe. Seeds pendulous, the crustaceous testa often hairy, and bearing a caruncle at the hilum or at the opposite end. Albumen fleshy or rarely deficient. Embryo straight, with flat, convex, or rarely thick and fleshy cotyledons.—Herbs, undershrubs, or small shrubs, rarely (in genera or species not Australian) tall shrubs, climbers or trees, glabrous or hairy, but without stellate hairs. Leaves usually alternate and entire, without stipules, very rarely opposite. Flowers solitary or in spikes or racemes, rarely paniculate, the pedicels usually articulate at the base, with a subtending bract, and 2 bracteoles.

A considerable Order, widely dispersed over nearly the whole globe. Of the three Australian genera, one is the largest and most extensively diffused of the whole Order, here represented by a very few species of an Asiatic or African type; another is Asiatic, of which one species extends to Australia; the third is endemic.

Sepals nearly equal. Anthers 4 or 5. Flowers minute, in terminal
spikes  . . . . . . . . . . . . . . . . . . . . . .   1. SALOMONIA.
Inner sepals larger and petal-like. Anthers 8.
  Capsule ovate or orbicular, scarcely contracted at the base. Seeds not
    comose.
    Lateral petals united with the carina (which is always crested in the
      Australian species) . . . . . . . . . . . . . . . .   2. POLYGALA.
    Lateral petals adnate to the staminal column, but distinct from the
      carina, which is not crested . . . . . . . . . . . .   3. COMESPERMA.
  Capsule cuneate, very narrow at the base. Seed hairs forming a
    long coma . . . . . . . . . . . . . . . . . . . .   3. COMESPERMA.

## 1. SALOMONIA, Lour.

Sepals nearly equal, the 2 innermost rather larger. Petals 3, united in a single corolla open on the upper side, the keel not crested. Stamens united nearly to the top into a sheath open on the upper side, and adhering to the corolla at the base ; anthers 4 or 5. Ovary 2-celled. Capsule thin, flat, obcordate or transversely oblong, usually ciliate, opening loculicidally at the edges. Seeds orbicular, with a minute or without any caruncle.—Small slender herbs, either annual or parasitical on roots. Leaves alternate, sometimes reduced to minute scales. Flowers very small, in terminal spikes.

The few species known are all natives of tropical Asia, the most common one extending into tropical Australia; but none have yet been found in Africa.

**1. S. oblongifolia,** *DC. Prod.* i. 334. A slender glabrous annual, erect and simple, or slightly branched at the base, 3 to 5, or rarely 6 in. high. Leaves sessile, the larger ones oblong, 3 to 4 lines long, and scarcely above 1 broad, the lower ones small and ovate. Flowers pink, scarcely a line long, in terminal leafless racemes or loose spikes of about an inch or rarely longer. Capsule about 1 line broad, but not so long, flattened, didymous, bordered with a fringe of hairs or slender teeth.—Deless. Ic. Sel. iii. t. 19 ; *S. obovata,* Wight, Illustr. t. 22.

**Queensland.** Endeavour river, *R. Brown (Hb. R. Br.).* Common in the warmer districts of India, from Ceylon and the Peninsula to the Archipelago and the Philippine Islands.

## 2. POLYGALA, Linn.

Sepals unequal, the 2 innermost, or wings, large and petal-like. Petals 3, united in a single corolla open on the upper side, the keel bearing a crest-like appendage on the back near the top, or rarely (in species not Australian) 3-lobed. Stamens 8, united to above the middle in a sheath open on the upper side, and adnate to the petals at the base. Ovary 2-celled. Style various. Capsule thin or rarely coriaceous, flattened, obovate, ovate, or orbicular, usually notched at the top, opening loculicidally at the edges. Seeds ovate or oblong, hairy or glabrous, but the hairs not lengthened into a coma, with or without a caruncle at the hilum.—Herbs, undershrubs, or shrubs. Leaves usually alternate or whorled. Racemes or spikes terminal or lateral, rarely axillary.

A very large genus, abundant in tropical countries, and generally also in temperate regions, except in Australia, where it is, with one exception, limited to the tropical districts, and in New Zealand, where it is entirely absent. Of the 7 Australian species, 3 are widely spread over tropical Asia, and the 4 others, although endemic, are nearly connected also with corresponding Asiatic ones.

Perennial. Style with 2 stigmatic lobes one above the other. Seeds obovate, shortly villous . . . . . . . . . . . . 1. *P. japonica.*
Annuals. Seeds oblong villous, the hairs much longer at the end furthest from the hilum.
  Racemes long, terminal. Inner sepals petaloid, obtuse. Crest fringed. Stigma simple, terminal, capitate . . . . . . 2. *P. leptalea.*
  Racemes short, very dense and hirsute, terminal or lateral. Inner sepals herbaceous, acuminate. Crest 2-horned. Style with 1 erect lobe and a lower large glandular stigma . . . . . . 3. *P. eriocephala.*

Racemes lateral. Inner sepals herbaceous, mucronate, usually falcate. Crest fringed. Style with 1 large hooked or reflexed stigmatic lobe.
  Racemes shorter than the leaves, or if longer, very dense.
    Leaves orbicular . . . . . . . . . . . . . . . 5. *P. orbicularis.*
    Leaves from obovate to linear.
      Capsules broadly winged and ciliate . . . . . . . 4. *P. rhinanthoides.*
      Capsules wingless and glabrous or nearly so . . . . . 6. *P. arvensis.*
  Racemes slender, much longer than the leaves . . . . . . 7. *P. stenoclada.*

1. **P. japonica,** *Houtt. Syst.* 8, *t.* 62, *f.* 1, *according to DC. Prod.* i. 324. Rootstock perennial, often woody with age, emitting numerous rather slender leafy stems, decumbent or erect, rarely more than 6 in. long, more or less pubescent. Leaves nearly sessile, the lower ones ovate, obtuse and small, the upper ones elliptical or lanceolate, acute, ½ to ¾ or rarely 1 in. long, of a rather firm consistence, glabrous and almost shining, distinctly veined. Racemes lateral, sometimes of 2 or 3 flowers only, and shorter than the leaves, sometimes 6- to 8-flowered and longer. Bracts small and deciduous, but less so than in most species. Outer sepals narrow-lanceolate; inner ones ovate, obtuse, 2 to 3 lines long and not oblique. Keel-petal crested. Ovary glabrous. Style thickened, incurved, with 2 unequal stigmatic lobes, the upper one arching over the lower short one. Capsule about 3 lines long and broad, including the rather broad wing. Seeds obovate, slightly pubescent, with a 3-lobed caruncle.—*P. veronicea,* F. Muell. Pl. Vict. i. 184.

**Queensland.** Dawson and Brisbane rivers, *F. Mueller.*
**N. S. Wales.** Botany Bay, *R. Brown;* Paramatta to the Blue Mountains, and shaded situations near Bathurst, *A. Cunningham;* Port Stephens, *Lady Parry;* Hastings and Macleay rivers, *Beckler;* New England, *C. Stuart.*
**Victoria.** Grassy or gravelly places on the Goulburn and Ovens rivers and their lower tributaries, *F. Mueller.*
  Also in the hilly regions of tropical Asia and northward to Japan. I can, indeed, find no difference between the Australian and the Japanese specimens, except that the flowers in the latter are rather larger: but several Khasia specimens are precisely like the Australian ones. *P. elegans,* Wall., from East India and China, differs slightly in the racemes most frequently terminal with numerous flowers.

2. **P. leptalea,** *DC. Prod.* i. 325. An erect, glabrous, slender annual, simple or slightly branched, usually 1 to 1½ ft. high. Leaves few, linear, the longer ones about 1 in., the uppermost much smaller, and the lower ones sometimes shortly oblong. Flowers small, numerous, pendulous, in a 1-sided terminal raceme, on pedicels which rarely attain 1 line. Outer sepals narrow-oblong, obtuse, the lowest rather larger and concave; inner sepals nearly twice as large, petal-like, broadly oblong, obtuse, 2 to 2½ lines long. Keel-petal crested. Style scarcely thickened, much curved, inflexed at the summit with an entire capitate stigma. Capsule broadly oblong, rather shorter than the inner sepals, with a narrow transparent wing. Seeds hirsute with reflexed hairs, the caruncle very small.—*P. oligophylla,* DC. Prod. i. 325.

**N. Australia.** Upper Victoria river, *F. Mueller;* Port Essington, *Armstrong.*
**Queensland.** Endeavour river, *R. Brown.*
  Frequent in northern and eastern India.

3. **P. eriocephala,** *F. Muell. Herb.* A more or less pubescent annual, in our specimens little branched and not exceeding 6 in. Leaves linear or

oblong-linear, some exceeding 1½ in. Racemes lateral or terminal, very dense and ovoid or oblong, ½ to 1 in. long, and very villous, the flowers nearly sessile. Outer sepals small and almost setaceous; inner sepals obliquely ovate, acuminate, about 2 lines long when in flower, nearly 4 when in fruit, herbaceous and hirsute with slender spreading hairs, completely enclosing the very fugacious corolla. Keel-petal very much shorter than the side ones, the dorsal crest consisting of 2 long simple horns. Style not thickened, 2-lobed, the upper lobe shortly filiform and incurved, the lower one expanded into a large stigmatic gland. Ovary covered with very long hairs. Capsule orbicular, emarginate, not winged, 2 to nearly 3 lines long, hirsute with long fine hairs. Seeds oblong, with reflexed hairs.

**N. Australia.** Upper Victoria river, *F. Mueller.*

4. **P. rhinanthoides,** *Soland. in Herb. R. Br.* An erect branching slightly pubescent annual, from an inch or two to above a foot high. Leaves oblong-linear, or rarely obovate-oblong, obtuse or rarely acute, ¾ to 1½ in. long, glabrous or ciliate, narrowed into a short petiole. Racemes lateral, short, rather dense, 6- to 10-flowered. Outer sepals lanceolate, with a fine point; inner sepals broadly ovate, oblique, mucronate, ciliate, 2 to 3 lines long. Keel-petal crested. Ovary broad, ciliate. Style slightly thickened, much curved, entire, with a broad almost petaloid decurved stigma, bearded underneath. Capsule 4 lines long and broad, including a broad wing, pubescent and ciliate. Seeds oblong, hirsute with reflexed hairs, the caruncle deeply 3-lobed.

**N. Australia.** Upper Victoria river. *F. Mueller.*
**Queensland.** Endeavour river, *R. Brown.*
Var. *minor.* A smaller and more glabrous plant, with narrower leaves, looser racemes, and more glabrous; capsules with narrower wings, almost connecting the species with some forms of *P. arvensis.* Upper Victoria river, *F. Mueller.*

5. **P. orbicularis,** *Benth.* An annual of 3 to 6 in., branching at the base only, glabrous or very slightly pubescent. Leaves distinctly petiolate, very broadly obovate or orbicular, or even broader than long, ¾ to 1 in. diameter, or the lower ones smaller. Racemes usually terminal, dense, ½ to 1 in. long. Outer sepals very small and lanceolate; inner sepals obliquely ovate, rounded, with a short point, glabrous, about 2½ lines long. Corolla fully as long, the lateral petals unusually large, the crest fringed. Style not thickened, with an almost petaloid uncinate-decurved stigma, glabrous, or slightly bearded underneath. Capsule orbicular, 2½ lines long, scarcely winged. Seeds hairy, the caruncle 3-lobed.

**N. Australia.** South Goulburn Island, *A. Cunningham;* Melville Island, *Fraser;* N. coast, *Armstrong.*
Allied to the var. *obovata* of *P. arvensis,* but appears to me, as far as hitherto known, too distinct in habit and foliage to be united with that species.

6. **P. arvensis,** *Willd. Spec. Pl.* iii. 876. A procumbent or rarely erect annual, branching at the base only, sometimes not exceeding a couple of inches when in full fruit, sometimes the prostrate or ascending branches extending to 6 or 8 in. or even more, and usually pubescent. Leaves from obovate to oblong or linear, ½ to ¾ in. long or rarely more. Flowers few, in short sessile racemes, usually lateral, often shorter than the leaves, and rarely

lengthening to an inch. Outer sepals very small and narrow ; inner sepals ovate-falcate, acute or mucronate, 2 to 3 lines long, herbaceous and glabrous or slightly pubescent. Corolla about as long, the lateral petals rather large, the crest of the keel fringed. Ovary glabrous. Style scarcely thickened, with an almost petaloid uncinate-decurved stigma, glabrous and glandular underneath. Capsule rather broad, glabrous or slightly pubescent, not winged. Seeds very hairy.—DC. Prod. i. 326.

**N. Australia.** Upper Victoria river, *F. Mueller ;* Goulburn Island, *A. Cunningham ;* N. coast, *R. Brown.*

**Queensland.** Endeavour river, *R. Brown.*

A very common East Indian weed, variable in foliage and stature ; the following forms appearing sometimes constant enough to be considered as distinct species :—

Var. *obovata.* Leaves all obovate, giving the plant the aspect of a young *Euphorbia helioscopia.* Cavern Island, Carpentaria, *R. Brown.*

Var. *squarrosa.* Leaves narrow. Flowers small and numerous, in oblong racemes, mostly terminal, the inner sepals narrow and falcate. *P. squarrosa,* Soland. ms. Endeavour river, *R. Brown ;* Upper Victoria river, *F. Mueller.*

Var. *stenosepala.* Leaves narrow-linear. Racemes short and few-flowered, or flowers almost solitary. Inner sepals narrow and less falcate. Capsule not above half as broad as long. Victoria river, *F. Mueller ;* and nearly the same form, but with more flowers, Arnhem Bays, *R. Brown.*

7. **P. stenoclada,** *Benth.* A slender, glabrous, erect annual, simple or little branched. Leaves distant, very narrow-linear, almost terete, obtuse or minutely pointed, ½ to 1 in. long. Peduncles lateral, slender, elongated, bearing towards the top a slender raceme of small blue flowers on very short pedicels. Outer sepals lanceolate, very acute with scarious margins ; inner sepals about 2 lines long, broadly ovate-lanceolate and falcate with a dark-coloured point. Keel-petal crested. Ovary glabrous. Style slender, much curved, with an almost petaloid deflexed blue stigma, bearded on the under side.

**N. Australia.** Upper Victoria river, *F. Mueller.*

The inflorescence is that of some specimens of the East Indian *P. Wightiana,* but besides the difference in foliage, the flowers are much smaller and narrower, and approach much more in structure the *P. arvensis,* from which *P. stenoclada* differs chiefly in inflorescence, and, in the above described specimens, in its very narrow leaves.

Var. (?) *stenosepala.* Rather taller and more branched. Leaves oblong or linear, flat, ¼ to 1 in. long. Flowers in a loose pedunculate raceme, much longer than the leaves, as in *P. stenoclada ;* but the inner sepals are narrow, pointed, and much falcate, as in the var. *stenosepala* of *P. arvensis.*—Carpentaria Point and Arnhem Bays, *R. Brown (Hb. R. Br.).*

## 3. COMESPERMA, Labill.

Sepals unequal, the 2 innermost, or wings, large and petal-like. Petals 3, the keel not crested, the two lateral ones separately attached to the staminal column, and either overlapped by the keel or outside it at the top. Stamens 8, united to above the middle in a sheath, open on the upper side and adnate to the petals at the base. Ovary 2-celled. Style incurved, obliquely stigmatic and more or less 2-lobed at the top. Capsule coriaceous or almost membranous, usually cuneate and much narrowed at the base, rarely nearly orbicular, opening loculicidally at the edges. Seeds ovate or oblong, pendulous, pubescent or hairy, the hairs lengthening into a coma whenever the capsule is narrowed at the base, without any caruncle at the hilum, but the raphe often expanded into a caruncular appendage at the opposite end.—Herbs undershrubs or shrubs, erect or twining. Leaves alternate, usually small. Racemes terminal.

A strictly Australian genus, with which was formerly united the Brazilian *Bredemeyera* (*Catocoma*, Benth.) ; but, besides the difference in habit, the latter has a more or less fleshy capsule, and the seeds have a long coma proceeding from the hilum; whilst in *Comesperma*, the coma, when present, consists of the hairs of the testa, which always extend to the base of the capsule, although the seed is often not half so long. In 2 species the capsule is that of a *Polygala*, and the seeds have no coma ; but in those the insertion of the lateral petals, very different from that of *Polygala* and approaching that of *Monnina*, is strongly marked. In *P. volubilis* (which was chiefly taken into account in verifying the characters for our ‘Genera Plantarum’), the arrangement of the petals is nearer to that of *Polygala*, but there the carpological characters are very decided. Besides that, the genus *Comesperma* is so natural a one, that it is never liable to be confounded with any of those allied to it in structure. The precise arrangement of the petals in the smaller-flowered species, very difficult to ascertain in dried specimens, requires verification from the living plant.

Capsule sessile. Seeds filling the cells, without a coma. Stems
　leafless. (Sect. **Prosthemosperma**, *F. Muell.*)
　Capsule orbicular. Flowers in a short terminal raceme . . . . 1. *C. sphærocarpum.*
　Capsule obovate or cuneate. Flowers distant.
　　Branches erect, rigid, broom-like. Seed with a broad terminal
　　　membrane . . . . . . . . . . . . . . . . 2. *C. scoparium.*
　　Branches very slender, divaricate, intricately branched. Seed
　　　with a long terminal appendage . . . . . . . . . . 3. *C. aphyllum.*
　　Branches divaricate, thorny. Seeds without any appendage . . 4. *C. spinosum.*
Capsule narrowed into a stipes, containing the long coma of the seeds,
　which only occupy the broad part of the cells.
*Outer sepals all free, much shorter than the wings.*
　Branches twining or very short and almost leafless.
　　Leaves few, mostly obtuse. Capsule not winged.
　　　Flowers blue or white. Pedicels glabrous . . . . . 5. *C. volubile.*
　　　Flowers yellow. Pedicels pubescent . . . . . . . 7. *C. integerrimum.*
　　Leaves very few and small, acute, ciliate. Bracts ciliate. Cap-
　　　sule winged. Flowers blue . . . . . . . . . . 6. *C. ciliatum.*
　Stems erect, leafy.
　　Leaves flat, ovate or oblong.
　　　Pubescent.
　　　　Leaves small, broadly ovate, mucronate, crowded. Flowers
　　　　　1 to 1¼ lines . . . . . . . . . . . . . 8. *C. secundum.*
　　　　Leaves thick, oblong, obtuse . . . . . . . . . 9. *C. Drummondii.*
　　　Glabrous.
　　　　Leaves mucronate, very glaucous . . . . . . . . 11. *C. sylvestre.*
　　　　Leaves obtuse, green . . . . . . . . . . . 10. *C. retusum.*
　　Leaves linear.
　　　Leaves pungent, strongly keeled. Keel-petal horned . . 12. *C. acerosum.*
　　　Leaves with revolute margins. Keel-petal not horned . . 13. *C. ericinum.*
　　Leaves very narrow, almost terete.
　　　Racemes elongated. Bracts comose. Flowers blue . . 14. *C. confertum.*
　　　Racemes corymbose or conical. Bracts very minute.
　　　　Flowers yellow . . . . . . . . . . . . . 15. *C. flavum.*
*Outer sepals all free, nearly as long as the wings.* (Sect. **Iso-
　calyx**, *Steetz.*)
　Stems leafy.
　　Capsule narrowed into a long stipes . . . . . . . 16. *C. calymega.*
　　Capsule elliptical or oblanceolate, shortly narrowed at the
　　　base . . . . . . . . . . . . . . . . . 17. *C. lanceolatum.*

Stems very slender, almost leafless . . . . . . . . . . 18. *C. defoliatum.*
*Two of the outer sepals connate.* (Sect. **Disepalum**, *Steetz.*)
  Flowers small, the wings not twice as long as the outer sepals.
    Leaves few, small, distant . . . . . . . . . . . 19. *C. nudiusculum.*
  Wings 3 or 4 times as long as the outer sepals. Leaves linear.
    Leaves strongly keeled. Keel-petal horned. Seeds with a
    membrane at the end furthest from the hilum . . . . 20. *C. virgatum.*
    Leaves flat, not keeled. Keel-petal not horned. Seeds with-
    out any appendage . . . . . . . . . . . . . 21. *C. polygaloides.*

1. **C. sphærocarpum,** *Steetz, in Pl. Preiss.* ii. 314. Rootstock woody but not thick, with slender, broom-like, or flexuose stems, sometimes perhaps slightly twining, ½ to 1½ ft. long, glabrous and slightly sulcate. Leaves reduced to minute distant scales, or the lower ones rarely 2 lines long, and linear. Flowers 3 to 6, in a short loose terminal raceme, on pedicels of 1 to 2 lines, the bracts very minute and deciduous. Outer sepals oblong, rather acute, almost scarious, about half the length of the inner ones, which are broadly obovate, blue and petal-like, 2 to nearly 3 lines long. Corolla and style of *C. scoparium.* Capsule nearly orbicular, about 2 lines diameter, slightly cuneate at the base or at length quite obtuse, glabrous. Seeds ovate, shortly pubescent, with a short membranous hairy appendage at the lower or chalazal end.

**N. S. Wales.** Hunter's River and Port Jackson, *R. Brown;* Mount Tomah, *R. Cunningham;* Paramatta, *Woolls;* Hastings river, *Beckler.*

2. **C. scoparium,** *Steetz, in Pl. Preiss.* ii. 309. Stems woody at the base, with numerous erect, rigid, broom-like, sulcate branches, 1 to 2 ft. high, glabrous. Leaves all reduced to minute distant scales. Flowers blue, singly scattered along the smaller branches on exceedingly short, thickened pedicels, surrounded by several minute, scale-like, obtuse, imbricate bracts. Outer sepals rather rigid, obovate-oblong, more than half the length of the inner ones, the lowest the smallest. Inner sepals petal-like, very broadly obovate, about 2 lines long; keel-petal about as long, the 2 lateral lobes broad and short; lateral petals shorter, narrow, free almost from the base, overlapping the keel. Ovary glabrous. Style not winged. Capsule sessile, cuneate-oblong, about 3 lines long, with a thickened margin. Seeds slightly pubescent, with a hairy membrane at the chalazal end, often more than half the length of the seed, and continuous with the prominent raphe.—F. Muell. Pl. Vict. i. 186.

**N. S. Wales.** Desert of the Darling, near Fitzgerald ranges, *F. Mueller.*
**Victoria.** Sandy desert, near the Murray, *Dallachy.*
**W. Australia.** Swan River, where it is known as the ' Swan-river Broom,' *Drummond;* Murchison river, *Oldfield;* Fitzgerald ranges, *Maxwell.*

3. **C. aphyllum,** *R. Br. Herb.* Tall, erect, and leafless, with very numerous slender, almost filiform, although rigid, divaricate branches, slightly sulcate, not thorny, and quite glabrous. Leaves all reduced to very minute distant scales. Flowers few and very small, singly scattered along the smaller branches. Outer sepals small and free; inner sepals scarcely above 1 line long and petals scarcely longer. Capsule sessile, obovate, about 2 lines long. Seeds without long hairs, but with a membranous appendage at the lower or chalazal end, more than half as long as the seed.

**N. Australia.**   Islands of the N. coast, *R. Brown (Herb. R. Br.).*

4. **C. spinosum,** *F. Muell. Fragm.* i. 144. A rigid, much branched, glabrous, leafless shrub, the branches scarcely sulcate, the smaller ones ending in rigid thorns. Leaves all reduced to minute subulate scales. Flowers few, scattered singly on the short branches. Outer sepals free, broad, rigid, not 1 line long; inner sepals broad, about 2 lines. Petals rather longer, lateral lobes of the keel-petal short and broad, lateral petals as long or rather longer. Capsule narrow-obovate, about 3 lines long, shortly acuminate, contracted below the middle, but scarcely stipitate. Seeds (which I have not seen) shortly and densely villous, without any appendage.

**W. Australia.**   Sandy tracts, Fitzgerald ranges, and West Mount Barren, *Maxwell.*

5. **C. volubile,** *Labill. Pl. Nov. Holl.* ii. 24, *t.* 163. A glabrous twiner, with numerous branches, sometimes extending to a considerable length, rarely short and flexuose, or almost erect. Leaves few, the lower ones oblong-linear or lanceolate, sometimes above an inch long and narrowed into a petiole, the upper ones linear or rarely obovate, small and distant. Racemes axillary or terminal, loose, 1 or rarely 2 in. long, sometimes 2 or 3 together. Flowers blue or rarely white, on pedicels of 1 to 2 lines. Outer sepals very broad, obtuse, about 1 line long; inner sepals fully 3 lines long, nearly orbicular, distinctly clawed. Keel-petal with 2 oblong lateral lobes turned inwards in æstivation and overlapped, at least at the top, by the 2 large, obovate, lateral petals. Style dilated upwards, but not winged. Capsule 4 to nearly 5 lines long, rounded, truncate and often slightly acuminate at the top, nearly 1½ lines broad, and gradually narrowed into a rather broad stipes. Seeds oblong, the long hairs forming the coma much fewer on the sides than on the edges. —DC. Prod. i. 334; Hook. f. Fl. Tasm. i. 31; F. Muell. Pl. Vict. i. 191; *C. tortuosum,* Steetz, in Pl. Preiss. ii. 303; *C. gracile,* Paxt. Mag. v. 145, with a fig.

**N. S. Wales.**   Port Jackson to the Blue Mountains, *R. Brown, Sieber, n.* 366, and others; Twofold Bay, *F. Mueller.*
**Victoria.**   Forest and scrub country, widely distributed over the colony, *F. Mueller.*
**Tasmania.**   *R. Brown;* throughout the island, abundant in light soils, climbing over bushes, etc., a most beautiful plant, well known as the 'Blue Creeper,' *J. D. Hooker.*
**S. Australia.**   *Whittaker;* Spencer's Gulf, *Warburton;* Quicken Bay, *F. Mueller.*
**W. Australia.**   King George's Sound, *R. Brown, Fraser;* Swan River, *Drummond, Coll.* 1843, *n.* 485. Some of these specimens, probably after having been eaten down, have short, flexuose, or almost erect stems.

*C. paucifolium,* Turcz. in Bull. Mosc. 1854, ii. 352, from W. Australia, Gilbert, n. 86, would appear from the character given to be very near *C. volubile* and *C. ciliatum,* but is said to have a shrubby, erect, much-branched stem. It is possible that the idea may have been suggested by stunted specimens of *C. volubile,* such as those above alluded to.

6. **C. ciliatum,** *Steetz, in Pl. Preiss.* ii. 304. Very near *C. volubile,* with similar glabrous, twining, sulcate branches. Leaves still fewer, very small, rigid and acute, usually ciliate with stiff hairs. Bracts linear-subulate, also ciliate. Flowers blue or pink, rather smaller than in *C. volubile,* much more numerous, in rather dense terminal racemes of 2 to 3 in. Outer sepals ovate-oblong, obtuse or almost acute, above 1 line long; inner sepals and petals like those of *C. volubile,* but much smaller. Capsules on long pedicels,

like those of *C. volubile*, but rather broader, owing to a membranous wing which borders them more or less, especially towards the summit.

**W. Australia.** Swan River, *Drummond;* Geographer Bay, King river, and Black-wood river, *Oldfield.*

7. **C. integerrimum,** *Endl. in Hueg. Enum.* 7. Very near *C. volubile,* with similar twining sulcate branches and few oblong-linear or lanceolate leaves, but the young shoots racemes and pedicels are usually minutely hoary-pubescent, the racemes are denser, with shorter and firmer pedicels, and the flowers yellow and rather smaller. Outer sepals broad and obtuse as in *C. volubile.* Petals similarly shaped, except that the lateral lobes of the keel are rather deeper, but I have in vain sought for the small additional petals described by Steetz. Capsule 8 to 9 lines long, $1\frac{1}{2}$ lines broad at the top, with a very prominent obtuse acumen, gradually narrowed into a stipes at the base. Seed 4 to 5 lines long, tapering almost to a point, otherwise nearly terete, the hairs of the coma proceeding from all over the surface.—Steetz, in Pl. Preiss. ii. 305; *C. scandens,* Steud. in Pl. Preiss. i. 211.

**W. Australia.** Rottenest Island, *A. Cunningham;* Swan River, *Drummond, Coll.* 1843, *n.* 486; near Mount Desmond, *Herb. F. Mueller.*

8. **C. secundum,** *Banks, in DC. Prod.* i. 334. A low, much-branched, rigid shrub, with the habit of some *Epacrideæ,* the branches softly pubescent. Leaves crowded, spreading, ovate, mucronate, 2 to 3 lines long, rigidly coriaceous, rough with minute tubercular hairs. Flowers very small and numerous, in slender one-sided racemes of 1 to 2 in., on very short pedicels. Outer sepals short, very broad and obtuse; inner sepals nearly three times as long, although scarcely exceeding 1 line, apparently pink. Keel-petal very broad, overlapping the narrow lateral ones. Style not winged. Capsule fully $\frac{1}{2}$ in. long, truncate, 3-toothed, and scarcely 1 line broad at the top, tapering into a slender stipes twice as long as the oblong part. Seed elongated, without any appendage, the long coma apparently very deciduous, but not seen quite ripe.

**N. Australia.** Islands of the north coast, *R. Brown.*
**Queensland.** Endeavour river, *R. Brown;* Cape Flinders, *A. Cunningham.*

9. **C. Drummondii,** *Steetz, in Pl. Preiss.* ii. 301. Shrubby, with short rigid branches, and all over glaucous, with a minute pubescence only visible under a lens. Leaves narrow-oblong, mostly obtuse, 3 to 4 lines long, very thick and rather concave, the midrib rarely conspicuous. Racemes many-flowered, short and almost corymbose, although the pedicels are rather long. Flowers of *C. retusum.* Capsule, according to F. Mueller, narrower, with a shorter stipes.

**W. Australia,** *Drummond;* Stirling ranges to West Mount Barren, *Maxwell.*

10. **C. retusum,** *Labill. Pl. Nov. Holl.* ii. 22, *t.* 160. Glabrous, erect, shrubby and much-branched, often several feet high, the branches mostly erect and not sulcate. Leaves oblong, obtuse, rarely above $\frac{1}{2}$ in. long, flat but rather thick, the midrib not prominent. Racemes short and dense, usually several in a terminal, leafy, flat corymb or pyramidal panicle. Outer sepals ovate, obtuse, about 1 line long; inner sepals nearly 3 lines. Petals

rather shorter, the keel not horned. Capsule usually about 5 lines long, emarginate, with rounded lobes, and about 1½ lines broad at the top, narrowed into a stipes much longer than the broad part. Seeds comose, without any membranous appendage.—DC. Prod. i. 334; Hook. f. Fl. Tasm. i. 32; F. Muell. Pl. Vict. i. 190.

**Queensland.** Moreton Island, *F. Mueller.*

**N. S. Wales.** Port Jackson, *R. Brown, Sieber, n.* 365; Blue Mountains and to the southward, *A. Cunningham;* New England, *C. Stuart.*

**Victoria.** Abundant in the sphagnum moors and along the rivulets and torrents of the Australian Alps at an elevation of 4000 to 6000 ft., *F. Mueller.*

**Tasmania,** *R. Brown;* abundant, especially in the northern parts of the island, from the sea to an elevation of 3600 ft. in the Western Mountains, *J. D. Hooker.*

11. **C. sylvestre,** *Lindl. in Mitch. Trop. Austr.* 342. A glabrous and erect shrub of several feet, resembling *C. retusum,* with which F. Mueller proposes to unite it, but much more glaucous. Leaves larger, often ¾ in. long and sometimes 3 lines broad, mucronate or pungent, often concave above. Flowers rather larger, with broader outer sepals. Capsule about ½ in. long. —F. Muell. Fragm. i. 49.

**Queensland.** Open forest, near Mounts Faraday and Pluto, *Mitchell;* sandy forest table-land on the Suttor river, *F. Mueller.*

12? **C. acerosum,** *Steetz, in Pl. Preiss.* ii. 299. Glabrous, rigid, erect, and little branched from a hard, almost woody base, 1 to 1½ ft. high. Leaves linear, erect, rigid, with a short usually pungent point, not above ½ in. long, strongly keeled. Racemes rather dense, 1 to 2 in. long, pedicels 1 to 1½ lines. Outer sepals 3, nearly equal, all free, very broad and obtuse, not 1 line long; inner petaloid sepals obovate, about 3 lines. Keel-petal with a horn-like appendage on the back as in *C. virgatum.* Capsule about 3 lines long, truncate or slightly 3-toothed at the top, narrowed into a stipes about as long as the broad part. Seeds comose, with a very short membrane at the chalazal end.

**W. Australia.** Swan River, *Drummond, n.* 431, and *Coll.* 1843, *n.* 492, mixed with *C. virgatum,* which this species closely resembles in almost all characters excepting the outer sepals, which are all free.

13. **C. ericinum,** *DC. Prod.* i. 334. Glabrous or minutely pubescent, usually erect, with rigid branches 1 to 2 or even 3 ft. high, woody at the base. Leaves linear, erect or spreading, crowded or rather distant, obtuse or acute, rarely above ½ in. long and usually shorter, the margins recurved or more frequently quite revolute. Racemes usually several and short in a leafy panicle, but longer and less dense than in *C. retusum,* rarely slender, and lengthening out to 3 or 4 in. Outer sepals all free, ovate or ovate-lanceolate, ¾ to 1 line long; inner sepals about 3 lines. Keel-petal not horned. Capsule 3 to 4 lines long, truncate, with rounded angles or entirely rounded at the top, narrowed into a stipes usually longer than the broad part. Seeds oblong, comose, with a very small membrane at the lower or chalazal end.—Hook. f. Fl. Tasm. i. 32; F. Muell. Pl. Vict. i. 190; *C. coridifolium,* A. Cunn. in Field. N. S. Wales, 337; *C. latifolium,* Steetz, in Pl. Preiss. ii. 295; *C. acutifolium,* Steetz, l. c. 296; *C. linariæfolium,* A. Cunn. in Steetz, l. c. 297.

**Queensland.** Moreton Bay, *A. Cunningham;* Glasshouses and Burnett ranges, *F. Mueller.*

**N. S. Wales.** Abundant about Port Jackson, *R. Brown, Sieber, n.* 364, 534, and *Fl. Mixt.* 550, and others ; and in the interior, *A. Cunningham ;* northward to Clarence and Hastings rivers, *Beckler ;* and southward to Twofold Bay, *F. Mueller.*

**Victoria.** Heathy tracts, as well of the lowlands as of the mountains, not rare in the southern and eastern parts of the colony, *F. Mueller.*

**Tasmania.** North coast, near the sea, and islands of Bass's Straits, in sandy soil, *J. D. Hooker.*

Var. *patentifolium.* Leaves very spreading, often pungent, very broad at the base. —Burnett ranges in the interior of N. S. Wales, *F. Mueller. C. patentifolium,* F. Muell. Fragm. i. 48. (See F. Muell. Pl. Vict. i. 190.)

Var. *oblongatum,* R. Br. Leaves oblong-linear, obtuse and mucronate, longer and with less revolute margins than usual.—East coast, *R. Brown.*

14. **C. confertum,** *Labill. Pl. Nov. Holl.* ii. 23, *t.* 161. Glabrous, erect, rigid, and usually branching above the middle, 1 to 2 ft. high. Leaves rather crowded, narrow-linear, thick, with the margins recurved so as to be almost terete, acute, often above 1 in. long. Flowers rather small, in slender but rather dense racemes of 2 to 3 in. or even more, on pedicels of 1 to 2 lines. Outer sepals free, broad and very obtuse, scarcely more than 1 line long ; inner sepals about 2 lines. Keel-petal rather shorter, not horned. Capsule 3 lines long or rather more, rounded and sometimes emarginate, but scarcely truncate at the top, narrowed into a stipes longer than the broad part. Seeds comose, the raphe projecting and membranous, but not expanded into a terminal membrane.—DC. Prod. i. 334 ; *C. longifolium,* Steud. in Pl. Preiss. i. 206 ; *C. hirtulum,* Steud. l. c. 209.

**W. Australia.** King George's Sound, *Labillardière, R. Brown, A. Cunningham, Drummond, Preiss, n.* 2359, and others ; E. Mount Barren, *Maxwell.*

15. **C. flavum,** *DC. Prod.* i. 334. Glabrous and erect, with rather crowded linear, almost terete leaves like *C. confertum,* but usually more branched and the leaves more spreading. Flowers yellow, larger than in *C. confertum,* in short, very dense, almost corymbose or shortly conical racemes, rarely above 1 in. long, the pedicels nearly 2 lines when in flower, and 3 when in fruit. Outer sepals all free, very short and obtuse ; inner sepals 2½ lines long. Keel-petal not horned. Capsule fully 4 lines long and not above 1 line broad, narrowed into a stipes much longer than the broad part. Seeds oblong, comose, without any prominent raphe.—Deless. Ic. Sel. iii. t. 20 ; *C. xanthocarpum,* Steud. Pl. Preiss. i. 209.

**W. Australia.** King George's Sound, *R. Brown, Fraser, A. Cunningham, Harvey ;* Swan River, *Drummond, Coll.* 1843, *n.* 490 ; Princess Royal Harbour, Gordon river, and Champion Bay, *Oldfield.*

16. **C. calymega,** *Labill. Pl. Nov. Holl.* ii. 23, *t.* 162. Glabrous or nearly so, with a perennial, sometimes woody rootstock, and simple or slightly branched, erect stems, from 6 in. to rather more than 1 ft. high. Leaves not numerous, the lower ones elliptical or oblong, the upper linear, rarely above ½ in. long, rather thick, flat or with slightly recurved margins, without any prominent keel. Flowers small, blue, in rather slender racemes of 1 to 3 in. Outer sepals all free, oblong or lanceolate, about 1½ lines long ; inner sepals rather longer, more deeply coloured, obovate, unguiculate. Keel-petal not horned, longer than the lateral ones. Style distinctly 2-lobed. Capsule 3 to nearly 4 lines long, truncate or 3-toothed at the top, narrowed into a stipes at

L 2

least as long as the broad part. Seeds comose, without any terminal appendage.—DC. Prod. i. 334; Hook. f. Fl. Tasm. i. 32; F. Muell. Pl. Vict. i. 188; *C. isocalyx*, Spreng. Syst. Veg. iii. 172; *C. strictum*, Endl. in Hueg. Enum. 7; *C. tenue*, Steud. in Pl. Preiss. i. 208; *C. varians* and *C. parviflorum*, Steud. l. c. 210; *C. herbaceum*, Steud. l. c. 211 (the last synonym taken from Steetz, in Pl. Preiss. ii. 307); *C. spathulatum*, Turcz. in Bull. Mosc. 1854, ii. 352 (from the character given).

**Victoria.** Bushy barren ridges and mountains, and arid heathy plains in many parts of the colony, *F. Mueller.*

**Tasmania,** *R. Brown;* common on sandy flats along the north shores of the island and in the islands of Bass's Straits, *J. D. Hooker.*

**S. Australia.** Kangaroo Island, St. Vincent's Gulf, and Lofty and other ranges in the interior, *F. Mueller, Behr,* etc.

**W. Australia.** King George's Sound to Swan River, *Drummond, Preiss, n.* 2365, 2374, etc., and others; Murchison river, *Oldfield.*

Var. *latifolium.* Lower leaves obovate, ½ to 1 in. long; upper leaves few, small, and distant. Capsule 5 lines long. Swan River, *Drummond;* King George's Sound, *R. Brown.*

17. **C. lanceolatum,** *R. Br. Herb.* Nearly allied to *C. calymega*, excepting in the fruit. Stems slender, erect, glabrous, not above 6 in. high, or branching and decumbent at the base. Leaves small, narrow-linear, rather rigid, erect and acute, mostly 2 to 3 lines long. Racemes short. Flowers blue, rather larger than in *C. calymega.* Outer sepals all free, oblong, thin, nearly 2 lines long; inner ones scarcely longer. Capsule elliptical or oblanceolate, tapering rather more at the base than at the point, nearly 3 lines long and rather more than 1 line broad. Seeds oblong, fully half as long as the capsule, comose, without any terminal appendage.

**W. Australia.** S. coast, east of King George's Sound, *R. Brown* (*Hb. R. Br.*).

18. **C. defoliatum,** *F. Muell. Pl. Vict.* i. 189. Allied in habit to *C. nudiusculum* with the flowers of *C. calymega.* Rhizome woody, with rigid and rush-like, but slender and sometimes almost filiform stems, 1 to 2 ft. high, and glabrous. Leaves very few and distant, small, narrow-linear or sometimes all reduced to small linear scales. Racemes slender, 1 to 2 in. long. Flowers rather larger than in *C. calymega.* Outer sepals all free, oblong, nearly as long as the inner ones. Capsule 3 or 4 lines long, contracted into a long narrow stipes. Seeds comose, without any terminal appendage.—*C. nudiusculum*, Steetz, in Pl. Preiss. ii. 308, not DC.

**N. S. Wales.** Port Jackson and Hunter's River, *R. Brown;* Illawarra, *Shepherd;* Clarence river, *Beckler.*

**Victoria.** Scattered over sandy heathy ridges from Port Phillip to the Broadribb river, *F. Mueller.*

**Tasmania.** South Port, *C. Stuart.*

19. **C. nudiusculum,** *DC. Prod.* i. 334. Stems elongated, slender, glabrous, with few very small distant leaves almost reduced to scales. Flowers small, blue, in a very short raceme, which after flowering lengthens to 1 in. or more. Outer sepals about 1¼ lines long, oblong, the 2 upper connate to near the top; inner sepals not twice as long, usually about 2 lines, broadly obovate, with a short claw adhering to the corolla. Keel-petal not horned, lateral ones narrow. Style much thickened above. Capsule about 3 lines long, narrowed into a stipes about twice as long as the broad part. Seeds

comose, without any terminal membrane.—*C. ramosissimum*, Steud. in Pl.
Preiss. i. 209; *C. megapteryga*, Steud. l. c. 207 (according to Steetz, in Pl.
Preiss. ii. 314).

**W. Australia.** King George's Sound, *R. Brown*, *A. Cunningham*, *Fraser*, *Preiss*, *n.*
2369, 2370, and others; Mount Barker, *Oldfield.*

20. **C. virgatum,** *Labill. Pl. Nov. Holl.* ii. 21, *t.* 159. Glabrous, with
a woody rootstock and erect, stiff, simple or somewhat branching stems, 1 to
1½ or rarely 2 ft. high. Leaves distant or rather crowded, linear or linear-
lanceolate, obtuse or scarcely pointed, rarely exceeding ½ in. in length, with
the midrib or keel prominent underneath. Flowers blue, rather numerous, in
a raceme of 1 to 3 in., often lengthening out after flowering to nearly 6 in.,
the pedicels from 1 to 2 lines. Bracts with a fine point, often comose in the
young raceme, but falling off during flowering. Outer sepals about 1 line
long, the 2 upper ones united to near the top; inner sepals nearly 3
lines long. Keel-petal very broad, with a horn-like appendage on the back
near the top, sometimes above ¼ line long, sometimes reduced to a small tu-
bercle. Style winged towards the top. Capsule about 3 lines long, truncate
or 3-toothed, and about 1 line broad at the top, narrowed into a stipes as long
as the broad part. Seeds ovate, comose with a small membranous appendage
at the chalazal end.—DC. Prod. i. 334; Steetz, in Pl. Preiss. ii. 311; *C. simplex*,
Endl. in Hueg. Enum. 7; *C. corniculatum*, Steud. in Pl. Preiss. i. 206; *C.
longebracteatum* and *C. roseum*, Steud. l. c. 207; *C. contractum* and *C. æmulum*, Steud. l. c. 208; *C. laxiusculum*, Steud. l. c. 210; *C. selaginoides*, Turcz.
in Bull. Mosc. 1854, ii. 352.

**W. Australia.** Apparently common, from the south coast to Swan River, *Labillardière*, *A. Cunningham*, *Drummond*, *n.* 215, 489, 492 (mixed in some cases with *C. acerosum*), *Preiss*, *n.* 2360, 2361, 2363, 2371, etc; Champion Bay, *Bower.*
   *C. corniculatum*, Steud., and *C. æmulum*, Steud., are both kept up by Steetz, in Pl. Preiss.
ii. 310, but the differences indicated do not appear to me to be quite borne out by the inspection of Preiss's specimens.

21. **C. polygaloides,** *F. Muell. in Trans. Phil. Soc. Vict.* i. 7, and *Pl.
Vict.* i. 187, *t.* 8. Very near *C. virgatum*, but a smaller plant, with a less
woody rootstock, and more simple stems, rarely 1 ft. high. Leaves from
linear to oblong, flat, without the prominent keel of *C. virgatum*, rarely above
½ in. long. Outer sepals slightly longer and less obtuse than in *C. virgatum*,
the two upper ones connate as in that species. Keel-petal without any dorsal
appendage. Capsule about 4 lines long, narrowed into a stipes nearly twice
as long as the broad part. Seeds very comose, without any terminal membrane.

**Victoria.** Scattered over the low ridges and barren plains of the southern and western
parts of the colony, *F. Mueller.*
   **S. Australia.** Near Adelaide, *Whittaker;* Rivoli Bay and Encounter Bay, *F. Mueller;*
Kangaroo Island, *Waterhouse;* Spencer's Gulf, *Warburton.*

## ORDER XV. **FRANKENIACEÆ.**

Flowers regular, hermaphrodite. Calyx tubular, persistent, with 4, 5, or
rarely 6 lobes, valvate in the bud, and as many prominent angles and furrows.

Petals as many, hypogynous, imbricate in the bud, free, the claws with an adnate plate or appendage on the inner face, the lamina spreading.  Stamens usually 6, sometimes 4 or 5 or indefinite, hypogynous, free or shortly united in a ring at the base; filaments filiform or flattened; anthers 2-celled, versatile.  Ovary free, sessile, 1-celled, with 3, rarely 2 or 4, parietal placentas, or very rarely a single one.  Style filiform, with as many branches as placentas, the stigmas capitate or oblique.  Ovules several, or rarely solitary, to each placenta, attached to rather long ascending funicles, amphitropous or nearly anatropous, with an inferior micropyle.  Seeds ovoid or oblong, testa crustaceous, the hilum almost terminal.  Embryo straight, in a mealy albumen, the radicle next the hilum, shorter than, or as long as, the cotyledons.—Low herbs or undershrubs, much branched and jointed at the nodes.  Leaves opposite, small, without stipules, often clustered in the axils.  Flowers usually pink or purple, sessile in the forks of the branches, forming a more or less dense, terminal, leafy cyme, sometimes contracted into a globular head.

The Order consists of a single genus, closely allied to the small group of *Diantheæ*, amongst *Caryophylleæ*, but distinguished by the parietal placentation of the ovary, and by the terminal hilum in the seed.  The species are chiefly maritime, and generally distributed over the temperate regions of the globe, more especially of the northern hemisphere, less abundaut within the tropics.

### 1. **FRANKENIA**, Linn.

Characters and distribution those of the Order.

The Australian species are all endemic, although the common one is closely allied to one of those most widely spread in the northern hemisphere.

Flowers in dense terminal heads.
  Floral leaves ovate-lanceolate, flat, several times broader than the linear-terete stem-leaves . . . . . . . . . . . . . 1. *F. bracteata.*
  Floral leaves linear-terete, like the stem-leaves . . . . . . 2. *F. glomerata.*
Flowers solitary, or in leafy terminal cymes.
  Leaves distinctly (but minutely) petiolate on the margin of the sheath.
    Petals slightly cohering by their claws.  Filaments slightly dilated and often cohering in a tube.
      Leaves much longer than their sheath.  Calyx 2 to 3 lines . 3. *F. pauciflora.*
      Leaves scarcely longer than their sheath.  Calyx about 1 line 4. *F. parvula.*
    Petals quite free.  Filaments shortly and broadly dilated at the base, free and narrow upwards . . . . . . . . . 5. *F. Drummondii.*
  Leaves sessile, the dorsal furrow continued to the base of the sheath.
    Leaves not produced below their insertion . . . . . . . 6. *F. tetrapetala.*
    Leaves produced at the base into a free, although closely appressed appendage . . . . . . . . . . . . . . 7. *F. punctata.*

(*Frankenia cymbifolia*, Hook., is *Wilsonia humilis.*)

1. **F. bracteata,** *Turcz. in Bull. Mosc.* 1854, ii. 367.  Stems, from a woody base, erect, ascending, or decumbent, 3 to 6 in. long, glabrous or slightly pubescent.  Leaves all opposite, linear-terete, 2 to 4 lines long, smooth and shining, the margins so closely revolute as to conceal the hairy under-surface, showing only a dorsal furrow, distinctly petiolate on the edge of a broad sheath, from which they early fall off, leaving a cluster of smaller similar leaves arising from within the sheath.  Cymes of flowers contracted into dense

heads, the bract-like floral leaves in whorls of 4 almost without sheaths, ovate-lanceolate or nearly ovate, flat, ciliate, and closely imbricate, so as to conceal the calyces. Calyx $2\frac{1}{2}$ to 3 lines long. Corolla and stamens of *F. pauciflora.* Style-branches and placentas 3. Ovules solitary to each placenta, attached to rather long funicles arising from near the base of the ovary.

**W. Australia,** *Drummond, Coll.* 1845, *n.* 136.

2. **F. glomerata,** *Turcz. in Bull. Mosc.* 1854, ii. 368. An apparently erect or ascending dichotomous shrub or undershrub of 6 to 8 in., glabrous or nearly so. Leaves opposite and clustered in the axils, linear-terete, 3 to 4 lines long, the margins ciliate and closely revolute so as only to show a dorsal furrow, and distinctly petiolate like those of *F. bracteata,* but the sheath shorter. Flowers in dense, terminal, leafy heads like those of *F. bracteata,* but the floral leaves are linear-terete like the stem ones. Calyx slender, about 3 lines long. Petals long and linear, slightly narrowed into long claws, with a scarcely prominent longitudinal line towards the top of the claw. Ovary in the few flowers I examined 1-ovulate, with a simple style, but perhaps not constantly so.

**W. Australia,** *Drummond,* 5*th Coll. Suppl. n.* 79.

3. **F. pauciflora,** *DC. Prod.* i. 350. Shrubby and procumbent or almost erect at the base, with ascending, erect, or divaricate dichotomous branches, nearly glabrous or hoary with a short down or scaly pubescence, often very low and spreading, sometimes above a foot high, attaining even 3 ft. according to F. Mueller. Leaves opposite or the upper ones in whorls of 4, oblong or linear, obtuse or rarely almost acute, the margins usually revolute so as only to show a dorsal furrow, when very narrow above 3 lines long, but usually much shorter, the very short sheathing petioles ciliate on the edge, with smaller leaves often clustered in the axils. Flowers closely sessile in the last forks, forming a more or less dense terminal leafy cyme and sometimes unilaterally arranged along its branches owing to the abortion of one branch of each fork. Calyx 3 to 4 lines, or rarely only $2\frac{1}{2}$ lines long. Petals with their claws cohering in an angular tube, the longitudinal appendage not very prominent, the lamina obovate, entire or crenulate. Stamens 5 or 6, with their filaments slightly dilated and usually cohering. Placentas 3 or rarely 2, with 2 to 4 ovules to each.—Bot. Mag. t. 2896 ; Hook. f. Fl. Tasm. i. 40 ; *F. scabra,* Lindl. in Mitch. Trop. Austr. 305.

**N. Australia.** Sturt's Creek, *F. Mueller ;* N. W. coast, *Bynoe.*

**Queensland.** In the interior on the Nive river, *Mitchell.*

**N. S. Wales.** Desert on the tributaries of the Darling and Murray rivers, *F. Mueller.*

**Victoria.** Saline marshes on the coast, more common in saline or sandy depressions along the Murray river and its tributaries, *F. Mueller.*

**Tasmania,** *R. Brown ;* abundant on Goose Island in Bass's Straits ; found also at Circular Head, *Gunn.*

**S. Australia.** On the coast, and particularly abundant in the saline districts in the northern part of the colony, *F. Mueller.*

**W. Australia.** Common both on the southern and western coasts, *Drummond, Coll.* 1843, *n.* 105, and 5*th Coll. n.* 77 and 78, and others ; Dirk Hartog's Island, *A. Cunningham.*

An exceedingly variable species, which F. Mueller (Pl. Vict. i. 82) unites with the common European and African *F. lœvis,* Linn. ; the latter species, however, much as it varies, has always much smaller and finer leaves, and especially very much smaller flowers, and the general aspect is so different, that it is not to be expected that the proposed union should be generally admitted. Possibly also the two following Australian varieties of *F. pauciflora* may prove sufficiently constant to be admitted as species.

Var. *serpyllifolia.* Pubescent or hirsute. Leaves, especially the lower ones short, from oblong to broadly ovate, the margins often much less recurved than in the typical *F. pauciflora.—F. serpyllifolia,* Lindl. in Mitch. Trop. Austr. 305.—Nive river, *Mitchell;* Murchison river, *Drummond.* Allied to this variety is the plant from Port Jackson, which De Candolle, Prod. i. 349, referred with doubt to the *F. pulverulenta,* Linn. The specimens in the herbarium of the Paris Museum have much the aspect of the latter species (very prostrate, with small broad flat leaves, more petiolate than is usual in *F. pauciflora*), yet I think they may prove to be only one of its numerous varieties, very near to the *serpyllifolia.*

Var. *thymoides.* More woody, erect, and much branched, with the habit of *Thymus vulgaris,* hoary all over, with a minute scaly indumentum. Leaves oblong, very obtuse, much revolute, 1 to nearly 2 lines long. Flowers rather small, the appendage of the petalclaws very prominent. Ovules 4 to 6 to each placenta.—Mount Goningbear, *Victorian expedition.—F. fruticulosa,* DC. Prod. i. 350, appears to connect this variety with the more common forms.

4. **F. parvula,** *Turcz. in Bull. Mosc.* 1854, ii. 368. Stems shortly creeping, with numerous ascending branches of 1 to 1½ in., glabrous or nearly so. Leaves crowded, oblong, obtuse, not above 1 line long, thick, but the margins closely revolute, concealing the under surface and showing only a dorsal furrow, distinctly petiolate on the margin of a broad, strongly ciliate sheath often nearly as long as the leaf, with 3 or 4 smaller leaves clustered within the sheath. Flowers terminal, solitary or in little leafy heads of 2 or 3. Calyx thickly ribbed, almost ovoid, a little more than 1 line long, strongly ciliate at the top. Petals obovate. Style 3-cleft. Ovules apparently few, but not seen in a good state.

**W. Australia,** *Drummond, 5th Coll. Suppl. n.* 81.

5. **F. Drummondii,** *Benth.* Stems prostrate and rooting at the nodes, with numerous short, erect branches, quite glabrous in our specimens. Leaves crowded, opposite or the floral ones in fours, linear-terete, about 2 lines long, distinctly petiolate, with a very short sheath, very red as well as the calyces in our specimens. Flowers small and solitary. Calyx slender, not 2 lines long. Petals all free, with a rather broad claw and a very prominent ovate-oblong scale, the lamina small and obovate. Stamens free, the filaments dilated at the base into an oval-oblong scale, filiform above. Style 3-cleft. Ovules 1 or 2 to each placenta.

**W. Australia,** *Drummond, n.* 278.

6. **F. tetrapetala,** *Labill. Pl. Nov. Holl.* i. 88, *t.* 114. Shrubby and prostrate at the base, rooting at the joints, with numerous branches, short and ascending or erect and much branched, often attaining 4 to 6 in., glabrous or minutely pubescent. Leaves crowded, but all opposite, linear-terete, acute or obtuse, 1 to 2 or rarely 3 lines long, not petiolate, but connate at the base into a short sheath, the dorsal furrow extending below their union, but without the appendage of *F. punctata.* Flowers small, like those of *F. punctata,* 5-merous in the specimens I have examined, but very likely to be occasionally 4-merous, as described by Labillardière.

**W. Australia.** King George's Sound and other points of the S. coast, *R. Brown,*

*Bauer, Bagster;* Young River and Fitzgerald range, *Maxwell;* Swan River?, *Drummond, n.* 279. Labillardière's specimens are said to have come from Tasmania; but there is very likely to have been some mistake. I have been unable to examine any flowers from them, but their habit and foliage leave no doubt as to their specific identity with those above described.

Var. (?) *brachyphylla.* Leaves, as in *F. punctata,* scarcely more than 1 line long and very obtuse, but not produced at the base. *Drummond, 5th Coll. Suppl. n.* 80.

7. **F. punctata,** *Turcz. in Bull. Mosc.* 1854, ii. 367. Shrubby and procumbent at the base, with numerous shortly ascending branches, glabrous or minutely pubescent. Leaves crowded, but all opposite, oblong or shortly linear, obtuse, 1 to 1½ lines long, not petiolate, but connate near the base, and produced below their insertion into a short obtuse appendage, closely pressed against the stem although free from it. Flowers small, on very short, leafy, lateral shoots. Calyx cylindrical, scarcely 2 lines long. Petal-claws free or scarcely cohering.

**W. Australia,** *Drummond, Coll.* 1845, *n.* 137.

## Order XVI. CARYOPHYLLEÆ.

Flowers regular, usually hermaphrodite. Sepals 4 or 5, persistent, free or united in a toothed calyx, imbricate in the bud. Petals either as many as the sepals hypogynous or slightly perigynous, entire or lobed, imbricate and frequently contorted in the bud, or rarely minute and scale-like or none. Stamens 8, 10, or fewer, inserted with the petals. Filaments filiform. Anthers 2-celled. Torus small or in a few *Silineæ* lengthened into a gynophore, or in some *Alsineæ* forming a small disk, shortly adnate to the base of the calyx, or short glands between the stamens. Ovary free, 1-celled or partially divided especially at the base into 2 to 5 cells. Styles 2 to 5, linear and stigmatic along the inside from the base or towards the top, free or more or less united into 1 branching style. Ovules 2 or more, often numerous, attached to a short or columnar placenta in the centre of the ovary, amphitropous and usually curved. Capsule membranous or crustaceous, very rarely succulent, opening at the top in as many or twice as many teeth or valves as there are styles, very rarely indehiscent. Seeds several, rarely solitary by abortion, with a membranous or crustaceous testa. Albumen mealy. Embryo curved round the albumen, or rarely straight or nearly so, and excentrical, with the radical inferior, or, when the embryo is circular, turned upwards.—Herbs, very rarely shrubby at the base, usually thickened and jointed at the nodes. Leaves opposite and entire, usually connected by a transverse line or short sheath at the base. Stipules none, or small and scarious. Inflorescence centrifugal, usually forming a terminal leafy cyme, rarely paniculate or racemose, or the pedicels all axillary.

A large Order, especially abundant in the extratropical regions of the northern hemisphere, rather less so in the high mountain-ranges of tropical America and Asia, and in the more temperate regions of the southern hemisphere, very rare in hot tropical countries. Of the Australian genera none are endemic. One, *Polycarpæa,* is chiefly tropical and almost limited to the Old World; another, *Drymaria,* is also chiefly tropical, but almost entirely American; a third, *Colobanthus,* is chiefly extratropical and limited to the southern hemisphere; a fourth, *Stellaria,* has almost as wide a range as the Order itself; the remaining genera and species, whether indigenous or introduced, are all European or East-Mediterranean.

TRIBE I. **Sileneæ.**—*Sepals united in a 4- or 5-toothed calyx. Petals and stamens hypogynous, often raised on a stalk-like torus. Styles distinct from the base. Stipules* 0.

Calyx many-nerved, with 2 or more bracts at the base. Styles 2.
   Seeds flat. Embryo straight . . . . . . . . . . . . . . DIANTHUS (p. 156).
Calyx broadly or obscurely 5-nerved.   Styles 2 . . . . . . . .   1. GYPSOPHILA.
Calyx 10-nerved. Styles 3   . . . . . . . . . . . . .   2. SILENE.
Calyx 10-nerved. Styles 5   . . . . . . . . . . . . .   LYCHNIS (p. 156).

TRIBE II. **Alsineæ.**—*Sepals free or only united by the disk at their base. Petals and stamens hypogynous or slightly perigynous, the torus not elongated. Styles distinct from the base. Stipules* 0, *or rarely small and scarious.*

Petals usually 2-cleft.
  Capsule cylindrical or conical, opening equally in twice as many
    teeth as styles. Styles 5, opposite the sepals, or rarely 4 or 3   .   3. CERASTIUM.
  Capsule globular or ovoid, opening in as many 2-cleft valves as
    styles. Styles 3, or if 5, alternate with the sepals   . . . .   4. STELLARIA.
Petals entire or none.
  Sepals 5. Styles usually 3. Capsule globular or ovoid.
    No stipules.
      Petals white, entire   . . . . . . . . . . . .   ARENARIA (p. 159).
      Petals none . . . . . . . . . . . . . . . . .   4. STELLARIA.
    Stipules small and scarious. Petals pink   . . . . . . .   7. SPERGULARIA.
  Sepals, styles, and capsular valves 4 or 5.
    No stipules. Leaves opposite.
      Stamens twice as many as sepals, or if of the same number,
        opposite to them. . . . . . . . . . . . .   5. SAGINA.
      Stamens of the same number as the sepals and alternate
        with them   . . . . . . . . . . . . . . .   6. COLOBANTHUS.
    Stipules small and scarious. Leaves clustered so as to appear
      verticillate   . . . . . . . . . . . . . . . .   SPERGULA (p. 161).

TRIBE III. **Polycarpeæ.**—*Sepals of Alsineæ. Petals usually very small or none. Stamens* 5 *or fewer, hypogynous or slightly perigynous. Style single at the base, with* 3 *or* 2 *branches or minute teeth. Stipules scarious or very minute.*

Petals lobed. Style very short. Stipules minute   . . . . . .   8. DRYMARIA.
Petals entire. Style short. Stipules scarious   . . . . . . .   9. POLYCARPON.
Petals entire or notched. Style elongated. Stipules and sepals
scarious . . . . . . . . . . . . . . . . . . . .   10. POLYCARPÆA.

TRIBE I. SILENEÆ.—Sepals united in a 4- or 5-toothed calyx. Petals and stamens hypogynous, often raised on a stalk-like torus. Styles distinct from the base. Stipules none.

## 1. GYPSOPHILA, Linn.

Calyx campanulate or turbinate-tubular, 5-toothed or 5-lobed, broadly 5-nerved, membranous between the nerves. Petals 5, with a narrow claw, and without any scale. Torus small. Stamens 10. Styles usually 2. Capsule globular or ovoid, opening to the middle or lower down in 4 valves. Seeds nearly reniform; embryo curved round the albumen.—Herbs, mostly glaucous, sometimes glandular or hirsute. Flowers usually small, numerous, and paniculate, or solitary in the forks of the stem.

A genus limited to the extratropical regions of the northern hemisphere in the Old World with the exception of the following species. It is chiefly distinguished from *Saponaria* by the calyx.

1. **G. tubulosa,** *Boiss. Diagn. Pl. Or.* i. 11. A slender erect dichotomous annual, often not above 2 or 3 in., but sometimes 8 to 10 in. high, more or less viscid-pubescent, and often slightly hirsute. Leaves linear-subulate, rarely attaining ½ in., and often much shorter. Pedicels in the forks, or sometimes appearing axillary from 1 branch only being developed, 4 to 8 lines long, erect or spreading. Calyx erect, 1½ lines long, narrower than in most *Gypsophilas,* with 5 prominent nerves, the teeth short and obtuse. Petals red, narrow-oblong, a little longer than the calyx. Capsule ovoid-oblong, rather exceeding the calyx. Seeds black, elegantly pitted under a lens.—F. Muell. Pl. Vict. i. 206 ; *Dichoglottis tubulosa,* Jaub. and Spach, Ill. Pl. Or. i. 14 *t.* 6 ; *D. australis,* Schlecht. Linnæa, xx. 631.

**N. S. Wales.** Cook's River and Nepean river, *R. Brown ;* Cox's River, *A. Cunningham.*
**Victoria.** Sandy localities, by no means rare, *F. Mueller.*
**Tasmania.** (*F. Mueller,* l. c.) I have seen no specimens from the island.
**S. Australia.** In sandy localities, near Bethanie, *Behr.*
**W. Australia,** *Drummond, n.* 93.

A native of the East Mediterranean region of Europe and Asia, possibly introduced into Australia and New Zealand, where it is also found ; yet from the localities where it was so early collected by R. Brown, and its general diffusion over extratropical Australia, it is difficult to conceive how a plant unknown in those parts of Europe whence the early colonists proceeded should have so promptly established itself. It is allied to the more common *G. muralis,* which, however, has not been detected in Australia, and is always quite distinct, especially in the form of the calyx, which is that of a true *Gypsophila,* whilst *G. tubulosa* is in this respect almost intermediate between that genus and *Saponaria.*

## 2. SILENE, Linn.

Calyx 10-nerved, rarely many-nerved, 5-toothed or 5-lobed. Petals 5, with a narrow claw, and usually with a double scale. Stamens 10. Torus usually elongated. Styles usually 3. Capsule opening in 6 or rarely 3 teeth or short valves. Seeds laterally attached ; embryo curved round the albumen.—Herbs. Flowers solitary or cymose, often forming unilateral spikes or an oblong thyrsus or panicle.

A very large genus, chiefly abundant in Europe, N. Africa, and temperate Asia, with a few N. American and S. African species, and only introduced into Australia.

*1. **S. gallica,** *Linn. ; DC. Prod.* i. 371. A hairy, slightly viscid, much branched annual, 6 in. to nearly 1 ft. high, erect or decumbent at the base. Lower leaves small and obovate, upper ones narrow and pointed. Flowers small, nearly sessile, generally all turned to one side, forming a simple or forked terminal spike, with a linear bract at the base of each flower. Calyx very hairy, with 5 slender teeth, at first tubular, afterwards ovoid and much contracted at the top. Petals very small, entire or notched, pale red or white, or in one variety with a dark spot.—*S. anglica, lusitanica, cerastoides* and *quinquevulnera,* Linn.; Reichb. Ic. Fl. Germ. vi. t. 272, 273.

A plant probably of South European origin, now common in sandy, gravelly, and waste places, especially near the sea, in most parts of the world, and established in several Australian colonies, especially about Swan River, from whence it is so frequently sent with indigenous plants, that it cannot be omitted from the Australian Flora.

*Dianthus barbatus*, Linn.; DC. Prod. i. 355, the European *Sweet-William*, and *D. Armeria*, Linn., DC. l. c., a common European species, are in F. Mueller's Herbarium as introduced plants, the latter as having been found on the stony crests of the ridges on Darebin Creek.

*Lychnis Githago*, Lam.; DC. Prod. i. 387, the *Corn Cockle*, a common cornfield weed, probably of East Mediterranean origin, has been introduced with European corn into some of the Australian colonies, as in many other countries. It is a tall, erect annual, clothed with long whitish appressed hairs. Leaves long and narrow. Flowers on long leafless peduncles, rather large and red, remarkable for the long green linear lobes of the calyx projecting much beyond the petals; the latter are broad, undivided, without scales. Stamens 10. Styles 5. Capsule opening in 5 teeth.

*Lychnis Cœli-rosa*, Dur.; DC. Prod. i. 386, is also in F. Mueller's Herbarium as an introduced plant at Shipton.

TRIBE II. ALSINEÆ.—Sepals free, or only united by the disk at their base. Petals and stamens hypogynous or slightly perigynous, the torus not elongated. Styles distinct from the base.

## 3. CERASTIUM, Linn.

Sepals 5, rarely 4. Petals as many, usually notched or 2-cleft. Stamens 10 or fewer. Styles 5 or 4, opposite the sepals, or rarely 3. Capsule cylindrical or conical, often incurved, opening at the top in twice as many teeth as styles, all equal. Seeds more or less reniform.—Herbs, usually pubescent or hirsute. Leaves rarely subulate. Cymes terminal, dichotomous, leafy, or the floral leaves reduced to small or scarious bracts. Seeds usually pitted or muricate.

A considerable genus, distributed chiefly over the temperate regions of the northern-hemisphere, more especially in the Old World, rare within the tropics except in mountain regions. The Australian species is not endemic and perhaps introduced only.

1. **C. vulgatum,** *Linn.; DC. Prod.* i. 415. A coarsely pubescent usually more or less viscid annual, branching at the base, sometimes dwarf, erect, and much branched, at others loosely ascending to 1 foot or even 2 feet, occasionally forming at the end of the season dense matted tufts, which may live through the winter, and give it the appearance of a perennial. Radical leaves small and petiolate; stem leaves sessile, from broadly ovate to narrow oblong. Sepals 2 to 2½ lines long, green and pubescent, but with more or less conspicuous scarious margins. Petals seldom exceeding the calyx, and often much shorter, sometimes very minute, or even none. Stamens often reduced to 5 or fewer. Capsule cylindrical, often curved and projecting beyond the calyx.—Reichb. Ic. Fl. Germ. v. t. 228, 229; *C. viscosum*, Linn.; DC. l. c. 416.

**Queensland.** Near Brisbane, *Henne.*

**N. S. Wales.** Port Jackson and Paramatta, but in the former case introduced, *R. Brown*, Clarence river, *Beckler;* Twofold Bay, *F. Mueller.*

**Victoria.** Common about Melbourne, also on the Murray, *F. Mueller;* Wimmera river, *Dallachy.*

**Tasmania.** Widely diffused even in almost inaccessible places, as among rocks on the North Esk river, *Gunn, J. D. Hooker.*

**S. Australia.** In good soils, *Behr.*

**W. Australia.** Common about Swan River, *Drummond, 1st Coll., 2nd Coll. n.* 698, *Coll.* 1848, *n.* 107.

Exceedingly common in the temperate regions of the northern hemisphere and now naturalized in many parts of the globe. In Australia also it is evidently introduced in many localities, but probably also indigenous. Brown, in 1802, distinguished as such his Paramatta specimens from the evidently introduced ones of Port Jackson, and Gunn found it abundant in Tasmania in localities where it was difficult to believe it to be a foreign importation. The Australian varieties are some of those most common in Europe; the var. *glomeratum,* DC. l. c., with broad orbicular leaves and compact inflorescence, most abundant in Victoria and Tasmania, and the var. *viscosum,* with oblong or narrow leaves and loose elongated cymes, in N. S. Wales and W. Australia; but very many specimens are quite intermediate. The smaller forms, with 4-merous flowers or 5 or fewer stamens, are not among the Australian specimens I have seen.

### 4. STELLARIA, Linn.

Sepals 5, rarely 4. Petals as many, usually 2-cleft, rarely wanting. Stamens 10 or fewer. Styles 3, rarely 2 or 4, or very rarely 5, and then alternate with the sepals. Capsule globular, ovoid or oblong, opening to below the middle in twice as many valves as styles. or in an equal number of 2-cleft valves.—Herbs usually diffuse, tufted or ascending, glabrous or pubescent. Leaves rarely subulate. Flowers solitary, or in loose leafless or leafy cymes. Seeds usually pitted or muricate.

A considerable genus, spread over nearly the whole globe, although within the tropics confined to mountain districts. Of the 5 Australian species 3 are endemic, one, *S. glauca,* although truly indigenous, is identical with a European species, the fifth, *S. media,* is an introduced weed.

Petals longer than or nearly as long as the sepals.
  Leaves mostly sessile, linear or lanceolate. Pedicels axillary. Perennials.
    Leaves rigid and pungent, mostly linear-lanceolate, often recurved . 1. *S. pungens.*
    Leaves linear, slender . . . . . . . . . . . . . . . 2. *S. glauca.*
  Leaves mostly petiolate, ovate or ovate-lanceolate. Pedicels axillary.
    Perennial without any pubescent line . . . . . . . . . . 3. *S. flaccida.*
  Leaves sessile or petiolate, broadly ovate. Pedicels in the forks.
    Annual, with a pubescent line down each internode . . . . . 4. *S. media.*
Petals none. Annual, with small sessile leaves . . . . . . . . 5. *S. multiflora.*

1. **S. pungens,** *Brongn. Voy. Coq. t.* 78. Perennial and very much branched, decumbent or ascending amongst bushes, often to 3 or 4 ft., with angular branches, smooth and shining, glabrous, or hirsute with loose scattered hairs. Leaves lanceolate to linear, rigid and pungent, mostly 3 to 4 lines long, and never exceeding ½ in., often spreading or recurved, all sessile or scarcely narrowed at the base, the lower ones sometimes small and crowded. Pedicels axillary, very variable in length, but usually considerably exceeding the leaves. Sepals rigid, pungent, about 3 lines long, the outer ones prominently 3-nerved. Petals about as long or rather longer, deeply cleft.—Hook. f. Fl. Tasm. i. 44.; F. Muell. Pl. Vict. i. 209; *S. squarrosa,* Hook. Journ. Bot. i. 250.

**N. S. Wales.** Blue Mountains and adjoining districts, *A. Cunningham*; New England, *C. Stuart.*
**Victoria.** Rocky, stony, or sandy places, not unfrequent throughout the greater part of the colony, ascending to the Australian Alps, but not extending into the desert, *F. Mueller.*

**Tasmania.** Port Dalrymple, *R. Brown;* common in rich and poor, moist and dry soils, *J. D. Hooker.*

2. **S. glauca,** *With.; DC. Prod.* i. 397.—Perennial, usually glabrous, smooth, and shining, with slender ascending or erect branches, often 1 to 2 ft. high, but sometimes low and intricate. Leaves linear, acute, $\frac{3}{4}$ to 1½ in. long, or the upper ones short. Pedicels axillary or terminal, slender but rigid, longer than the leaves. Sepals very acute, 3-nerved, about 3 lines long when in flower. Petals about as long, or rather longer, deeply cleft. Capsule ovate, much shorter than the calyx, which usually lengthens after flowering.—Reichb. Ic. Fl. Germ. v. t. 223 ; Hook. f. Fl. Tasm. i. 44. F. Muell. Pl. Vict. i. 210 ; *S. angustifolia,* Hook. Journ. Bot. i. 250.

**Queensland.** Plains of the Condamine river, *Leichhardt.*
**N. S. Wales.** Marshy places, Longmeadow, etc., *R. Brown ;* Lachlan river, *A. Cunningham.*
**Victoria.** Moist, rocky, grassy, or sandy localities, scattered over a considerable extent of the colony, *F. Mueller.*
**Tasmania.** Marshes in various localities, *J. D. Hooker.*
**S. Australia.** Extending to St. Vincent's Gulf, *F. Mueller.*
Var. *cæspitosa,* Hook. f. Fl. Tasm. i. 44. Stems short and very intricate, or densely tufted. Leaves lanceolate-linear. Sepals short and more obtuse.—*S. cæspitosa,* Hook. f. in Hook. Journ. Bot. ii. 411. 'Tasmania, *Gunn ;* and on the Murray in Victoria, *F. Mueller.* The specimens show a very gradual passage from this form to the elongated one, in the leaves as well as in the sepals. A similar gradation takes place in the N. American *C. longipes,* an allied species, yet, to my eyes, always distinct in inflorescence as well as in foliage.
Var. (?) *leptoclada.* Annual or, at any rate, flowering the first year, with slender, ascending, erect stems of 5 to 6 in., much branched at the base. Pedicels slender. Flowers small, as in the last variety, but the sepals more acute.—New England, *C. Stuart.*
Var. (?) *tenella.* Tufted and intricately branched, like the var. *cæspitosa,* but smaller and much more slender, with crowded, very small leaves ; one specimen, with some branches elongated, with narrow-linear leaves. Flowers few, small. Sepals rather obtuse.
**Victoria.** Near Melbourne, *Adamson ;* Glenelg river, *Robertson.*
**Tasmania.** Derwent river and Kitt's Group in Bass's Straits, *R. Brown ;* granite rocks in St. Patrick's river, *Gunn.*
The *S. glauca* is generally diffused over Europe and temperate Asia, and the Australian form, in its elongated state, cannot at all be distinguished from many European specimens grown in similar localities. The northern plant has, however, more frequently larger petals, and has sometimes a tendency to assume a paniculate inflorescence, with the floral leaves reduced to small bracts, approaching that of *S. graminea ;* the Australian plant, on the contrary, tends rather, in its extreme varieties, towards the intricate stems and habit of *S. pungens.*

3. **S. flaccida,** *Hook. Comp. Bot. Mag.* i. 275. Apparently perennial, with weak and decumbent very intricate branches, often extending to several feet, glabrous and shining, or with loose spreading scattered hairs especially about the nodes. Leaves ovate to lanceolate, very acute, thin and flaccid, often undulate on the margin, narrowed and ciliate at the base, rarely exceeding ½ in. without the petiole, which is long in the lower leaves, short or none in the upper ones. Pedicels all axillary, and usually 1 to 1½ in. long. Sepals 2 to 2¼ lines long, broadly lanceolate, acute, with a scarious border, usually 3-nerved, but the lateral nerves often very faint, often ciliate. Petals rather onger, deeply cleft. Capsule ovoid, usually exceeding the calyx.—*S. media,* var., Hook. f. Fl. Tasm. i. 43 ; F. Muell. Pl. Vict. i. 211.

**N. S. Wales.** Shoal Spit Reach, *R. Brown ;* Hastings river, *Beckler.*

**Victoria.** Shady humid places, forest lands, and gravelly banks of rivers, from the lowlands to the highest Alps, *F. Mueller.*

**Tasmania.** Dense thickets and shady places, *J. D. Hooker;* Port Dalrymple, *R. Brown.*

I cannot agree in considering this a variety of *S. media.* Besides the difference in habit, in the shape of the leaves and sepals, and in the inflorescence, the hairs, when present, are long cilia on the edges and nerves of the leaves and sepals, or on the angles of the branches, without any trace of the unilateral pubescence between two angles so constant in *S. media.*

*4. **S. media,** *Linn. DC. Prod.* i. 396. A weak, much-branched annual, glabrous with the exception of a pubescent line down one side of each internode, and a few long hairs on the petioles, and sometimes on the sepals. Leaves ovate, shortly pointed, the lowest on long petioles, short and broad, and sometimes cordate, the upper ones on shorter petioles or quite sessile, ½ to ¾ in. long, thin and flaccid. Pedicels slender, often drooping, in the forks of the branches, the upper ones usually forming a rather dense leafy cyme, very rarely one of the lowest axillary from the abortion of one fork. Sepals about 2 lines long, obtuse or rarely rather acute, thin but green, with scarcely prominent nerves, and usually pubescent. Petals about as long, deeply cleft. Capsule scarcely longer than the calyx.—Reichb. Ic. Fl. Germ. v. t. 222.

Originating, probably, in the temperate regions of the northern hemisphere in the Old World, this plant is now a common weed in cultivated places, especially gardens, as well as in waste places, almost all over the globe, and as such is found in most of the Australian colonies, especially Victoria, *F. Mueller,* and W. Australia, about Swan River, *Drummond, n.* 244.

5. **S. multiflora,** *Hook. in Comp. Bot. Mag.* i. 275. A slender, glabrous, branching annual, with decumbent or erect stems, usually under 6 in. Leaves sessile, or the lowest petiolate, mostly lanceolate, 2 to 3, or rarely 4 lines long, the upper ones very small. Pedicels axillary, sometimes all shorter than the calyx, in other specimens all filiform but rigid, 3 to 6 lines long. Sepals lanceolate, very acute, about 2 lines long, 3-nerved or strongly 1-nerved. Petals none. Stamens short, those alternating with the sepals often rudimentary or wanting. Capsule as long as or longer than the sepals. Seeds tuberculate.—Hook. f. Fl. Tasm. i. 43 ; F. Muell. Pl. Vict. i. 212.

**Victoria.** Sandy, grassy, and rocky localities, not uncommon as well in the lowlands as in the mountain regions, ascending to the Alps, *F. Mueller.*

**Tasmania.** On grassy dry pastures and rocks, etc., common, *J. D. Hooker.*

**S. Australia.** Distributed over the southern and eastern parts of the colony, *F. Mueller.* Remarkably luxuriant specimens from Rivoli Bay considerably exceed ½ ft. in length.

**W. Australia,** *Drummond, n.* 695.

*Arenaria serpyllifolia,* Linn.; DC. Prod. i. 411. A very much branched, slender, and slightly pubescent annual, seldom attaining 6 in. Leaves very small, ovate, and pointed. Pedicels from the upper axils or forks, 2 to 3 lines long, and slender. Sepals 5, acute, about 1½ lines long. Petals usually much shorter, white, obovate, entire. Stamens 10. Styles 3. Capsule short, opening in 6 narrow valves.

Common in Europe and temperate Asia, on walls and muddy, stony, or waste places, and now almost naturalized in several of the Australian colonies.

## 5. SAGINA, Linn.

Sepals 4 or 5. Petals as many, entire or scarcely notched, or none. Stamens 8, 10, or fewer. Styles as many as sepals, and alternate with them. Capsule opening to the base into as many valves as styles, alternating with the sepals.—Small matted or tufted herbs, with subulate leaves and small flowers, usually borne on long pedicels.

A small genus, dispersed over the temperate or cooler regions of the northern hemisphere, the commonest species also abundant in the southern hemisphere.

1. **S. procumbens,** *Linn. DC. Prod.* i. 389. A minute annual or rarely perennial, 1 to 2 in. or rarely 3 in. high, usually branching from the base and decumbent, forming little spreading tufts, glabrous or very minutely pubescent. Leaves small and subulate, joined by a short scarious sheath, the radical ones longer and tufted. Flowers very small, on capillary peduncles longer than the leaves. Sepals 4, about 1 line long. Petals much shorter, often wanting. Valves of the capsule as long as the sepals or rather longer. All these parts usually in fours, but occasionally met with in fives.—Reichb. Ic. Fl. Germ. t. 206 ; F. Muell. Pl. Vict. i. 208 ; *S. apetala,* Linn. ; DC. l. c. ; Reichb. l. c. t. 200.

**Victoria.** Morasses and mossy valleys between Mount Seviter and Limestone river, at an elevation of 4000 feet (the perennial form) ; the common annual form abundant about Melbourne, Port Phillip, etc., *F. Mueller.*

**S. Australia.** St. Vincent's Gulf, lofty ranges, etc., *F. Mueller.*

Very abundant, in a great variety of situations, over the whole range of the genus.

## 6. COLOBANTHUS, Bartl.

Sepals 4 or 5. Petals none. Stamens as many as sepals and alternating with them, slightly perigynous. Styles as many as sepals and opposite to them. Capsule opening in as many valves as sepals, and opposite to them.— Small tufted herbs, glabrous and often somewhat fleshy. Leaves narrow, or short and imbricate. Flowers solitary.

A small genus, spread over the mountainous or antarctic regions of South America, Australia, and New Zealand. Both the Australian species are common to New Zealand and Antarctic America. The genus has been referred by Fenzl to *Portulaceæ,* on account of the position of the stamens ; but all other characters are much more those of *Caryophylleæ.*

Leaves short and spreading. Flowers nearly sessile . . . . . . 1. *C. subulatus.*
Leaves erect or elongated. Pedicels much longer than the calyx . . . 2. *C. Billardieri.*

1. **C. subulatus,** *Hook. f. Fl. Ant.* i. 13, *t.* 93, *and* ii. *t.* 47. Stems short, with crowded leaves, forming dense moss-like tufts often covering a considerable space of ground. Leaves linear, concave and strongly keeled, with a fine almost pungent point, 2 or rarely 3 lines long, rigid and spreading. Flowers almost sessile within the tufts of leaves, and not exceeding them. Sepals 5, about $1\frac{1}{2}$ lines long, lanceolate, acute and rigid. Capsule nearly as long as the calyx.—*Spergula subulata,* Durv. Fl. Malouin. 51, not of Swartz ; *Colobanthus Benthamianus,* Fenzl, in Ann. Mus. Vind. i. 49 (the plate quoted from Endl. Atakt. never published) ; *C. pulvinatus,* F. Muell. in Trans. Phil. Soc. Vict. i. 201, and Pl. Vict. i. 213, t. 11.

**Victoria.** Bare gravelly summits of the Munyang mountains, buried the greater part of the year under snow, not occurring below 6000 ft., *F. Mueller.*

The species is also found in New Zealand and in Antarctic America. The New Zealand specimens, and some of those from Campbell's Island, are precisely like the Australian ones ; others have more elongated stems, and less rigid leaves ; and the Hermit Island specimens have always 4-merous flowers ; whilst in all others they are usually, if not always, 5-merous.

2. **C. Billardieri,** *Fenzl, in Ann. Mus. Vind.* i. 49. A small, densely tufted, almost stemless perennial. Leaves in closely crowded tufts, linear-subulate, sometimes very rigid and not ½ in. long, more frequently 1 in. long or more, somewhat flaccid, 1 line broad and sheathing at the base, and attenuated into a long point, sometimes filiform and grass-like, ½ to 1 in. long. Peduncles 1-flowered from the centre of the leaf-tufts, shorter or longer than the leaves, but always longer than the calyx, slightly thickened under the flower. Sepals 5, broadly lanceolate, very finely pointed, about 2 lines long. Capsule from globular to ovoid, shorter or longer than the calyx.—Hook. f. Fl. Tasm. i. 45 ; F. Muell. Pl. Vict. i. 212 ; *Spergula apetala,* Labill. Pl. Nov. Holl. i. 112. t. 142 ; DC. Prod. i. 395 ; *Spergula affinis,* Hook. Ic. Pl. t. 266 ; *Colobanthus affinis,* Hook. f. in Hook. Journ. Bot. ii. 410, and Fl. Tasm. i. 45.

**Victoria.** Rocky hills near Warnambool, *Hannaford.*

**Tasmania,** *Labillardière ;* Kent's Group, Bass's Straits, *R. Brown ;* northern and central parts of the island, alpine districts of the Hampshire hills, and Franklyn river, *J. D. Hooker ;* Southport, *C. Stuart.*

Two forms have been described, but they pass very much one into the other, the differences in the form of the capsules not corresponding with the variations in the leaves. The species occurs also in New Zealand and in Campbell's Island.

*Spergula arvensis,* Linn.; DC. Prod. i. 394. A slender annual, branching at the base into several erect or ascending stems, 6 in. to 1 ft. high, glabrous or slightly pubescent. Leaves almost subulate, 1 to 2 in. long, in opposite clusters and spreading so as to appear verticillate. Stipules scarious, very minute, sometimes very difficult to see. Flowers small, white, on long pedicels, in terminal forked cymes. Sepals 5. Petals 5, undivided, generally rather shorter than the calyx. Stamens 10, or occasionally 5 or fewer. Styles 5, alternate with the sepals. Capsule deeply 5-valved. Seeds slightly flattened, with or without a scarious border.

Common in Europe and temperate Asia in cultivated and waste places, and now dispersed over various parts of the world as a cornfield weed, and introduced as such into the Australian colonies, especially Swan River, *Drummond.*

## 7. SPERGULARIA, Pers.

(Lepigonum, *Fries.*)

Sepals 5. Petals 5, entire or rarely 0. Stamens 10 or fewer. Styles 3. Capsule 3-valved.—Herbs usually diffuse. Leaves linear or filiform, often clustered in the axils so as to appear verticillate. Stipules small, scarious. Flowers pedicellate, pink or white, in the forks of the stem or in terminal cymes or one-sided racemes. Seeds with or without a scarious border.

A small genus, widely dispersed over the temperate or subtropical regions of the globe, chiefly in maritime or saline localities, or heathy places, differing from *Arenaria* almost solely in the presence of stipules. The Australian species is the same as the common northern one.

1. **S. rubra,** *Pers. Syn.* i. 504 (as a subgenus of *Arenaria*). An annual, biennial or rarely perennial, glabrous or with a short viscid pubescence in the upper parts, with numerous stems branching from the base and forming spreading or prostrate tufts 3 or 4 in., or when luxuriant 6 in. long. Leaves narrow-linear, the scarious stipules at the base short but conspicuous.

Flowers very variable in size, usually pink, on short pedicels, in forked cymes, usually leafy at the base. Petals shorter, or rather longer than the sepals. Seeds more or less flattened, often surrounded by a narrow scarious border or wing.—A. Gray, Gen. Ill. t. 108 ; Hook. f. Fl. Tasm. i. 41 ; F. Muell. Pl. Vict. i. 207 ; *Arenaria rubra* and *A. media*, Linn.; DC. Prod. i. 401 ; *Lepigonum rubrum*, etc., Fries, Nov. Fl. Suec. Mant. iii. 32 ; *L. brevifolium*, Bartl. in Pl. Preiss. i. 243 ; *L. anceps* and *L. laxiflorum*, Bartl. l. c. 244 (of these last I have only seen authentic specimens of *L. anceps*) ; *Spergularia rupestris*, Fenzl, in Hueg. Enum. 9 ; Schlecht. in Linnæa, xx. 632.

**N. S. Wales.** Argyle county and Field's Plains, *A. and R. Cunningham;* New England, *C. Stuart ;* Darling river, *Victorian Expedition.*
**Victoria.** Coast meadows and subsaline tracts of the interior, on clayey and sandy soil, not unfrequent, ascending occasionally into mountainous tracts, *F. Mueller.*
**Tasmania.** Abundant on the seacoast, *J. D. Hooker.*
**S. Australia.** Near Adelaide, St. Vincent's Gulf, etc., *F. Mueller.*
**W. Australia,** *Drummond, 1st Coll.,* 5th *Coll. n.* 201 and 243, *Preiss, n.* 1944, *Oldfield,* and others.
Widely spread over Europe, temperate Asia, and North America, and some parts of South America, chiefly in maritime countries or in sandy heathy places more inland. There are two, often rather marked varieties, one chiefly occurring inland has slender leaves, small flowers, and short capsules, with the seeds less frequently bordered than in the larger variety, which has a sometimes perennial stock, thicker somewhat fleshy leaves, and larger flowers. Both forms occur in Australia and pass into each other as they do in Europe, the larger and more succulent ones are, however, the most common in Australia.

TRIBE 3. POLYCARPEÆ.—Sepals free, or only united by the disk at their base. Petals usually very small, thin and almost transparent or none, occasionally united with the stamens at the base. Stamens 5 or fewer, hypogynous or slightly perigynous. Style single, at least at the base, with 3 or 2 branches or minute teeth.

## 8. DRYMARIA, Willd.

Sepals 5, herbaceous or scarious on the edge. Petals 5, 2- to 6-cleft. Stamens 5 or fewer, slightly perigynous. Style 3-cleft. Capsule 3-valved. Seeds laterally attached ; embryo curved round the albumen.—Herbs usually diffuse, rarely erect, with dichotomous branches. Leaves flat, broad or narrow. Stipules very small, sometimes very fugacious or wanting. Flowers pedicellate, usually small, either solitary in the forks, or in little axillary or terminal cymes. Petals usually shorter than the calyx.

The genus comprises a considerable number of American species, one of which is also widely spread over the tropical regions of Asia and Africa. The Australian species is endemic, and the only one which is not American.

1. **D. filiformis,** *Benth.* A glabrous annual, very much branched at the base, with erect dichotomous very slender shining stems 6 to 8 in. high. Leaves chiefly crowded in a dense tuft at the base of the stem, narrow-linear, almost filiform, many of them above 1 in. long, the upper leaves few and small, soon passing into minute bracts. Stipules none. Pedicels in the forks, filiform, about ½ in. long. Sepals about 1 line long, narrow and acute, green, shortly connate at the base. Petals about one-third as long as the calyx, deeply divided into 2 narrow lobes, very thin and transparent, and often very difficult to find. Ovary oblong, with an exceedingly short style, divided into 3 short

oblong-linear stigmatic branches. Capsule cylindrical, from half as long again to twice as long as the calyx, opening in 3 valves, which soon split into twice that number.

**W. Australia,** *Drummond, n.* 694,

This is a very distinct plant, with something of the habit of a *Mollugo,* and the inflorescence of *Gypsophila tubulosa.* The structure is that of *Drymaria,* and in that genus it approaches nearest to *D. effusa* and *D. tenella,* A. Gr., from New Mexico, having similar narrow leaves without stipules; but the slender pedicels and cylindrical capsule distinguish it at once.

## 9. POLYCARPON, Linn.

Sepals 5, keeled, scarious on the margin. Petals 5, small, entire or notched. Stamens 3 to 5. Style short, 3-cleft. Capsule 3-valved. Seeds laterally attached near the base; embryo excentrical, curved or nearly straight, the cotyledons incumbent or oblique.—Herbs either diffuse or dichotomously branched, glabrous or pubescent. Leaves flat, usually ovate or oblong, often apparently, but not really, in whorls of 4. Stipules scarious. Flowers small, numerous, in terminal cymes, with scarious bracts.

A genus of very few species, dispersed over the temperate and tropical regions of the globe. The Australian species is identical with the commonest northern one.

1. **P. tetraphyllum,** *Linn. f.; DC. Prod.* iii. 376. A glabrous, much branched, spreading or prostrate annual, seldom more than 3 or 4 in. long. Leaves obovate or oblong, really opposite, but placed as they usually are under the forks, two pairs are so close together as to assume the appearance of a whorl of 4. Flowers very small and numerous, in loose terminal cymes. Sepals barely 1 line long. Petals much shorter and very thin. Stamens usually 3.—F. Muell. Pl. Vict. i. 205.

**N. S. Wales.** Port Jackson, *R. Brown,* and others.
**Victoria.** In light soil, widely dispersed over the colony, *F. Mueller.*
**Tasmania.** Perhaps introduced, *Gunn.*
**S. Australia.** Near Adelaide, *Herb. Mueller.*
**W. Australia,** *Drummond* and others.

Very common in sandy situations, chiefly not far from the sea, in Europe, temperate Asia, the greater part of Africa, and in many parts of North and South America; but unknown in tropical or subtropical Asia.

*P. alsinæfolium,* DC. Prod. iii. 376, a maritime variety, with thicker succulent leaves and often, but not always, 5 stamens, not uncommon in the Mediterranean region, is given as Australian on the authority of Sieber's specimens, n. 570, which I have not seen, nor have I met with the variety in any Australian collection. All the Port Jackson specimens which I have seen, although maritime, are thin-leaved and 3-androus.

## 10. POLYCARPÆA, Lour.

(Aylmeria, *Mart.*)

Sepals 5, either entirely scarious, or herbaceous in the centre and scarious on the margin, but not keeled. . Petals 5, entire or toothed. Stamens 5, hypogynous or slightly perigynous, free or united with the petals in a ring or tube. Style elongated, 3-furrowed, 3-toothed, or shortly 3-lobed at the top. Capsule 3-valved. Seeds obovoid or flattened; embryo curved or nearly straight; cotyledons usually (perhaps always) accumbent.—Annual or peren-

M 2

nial herbs, erect or diffuse. Leaves narrow-linear or rarely ovate, often clustered in the axils so as to appear verticillate. Stipules scarious. Flowers usually numerous, in terminal cymes, sometimes loose and paniculate, sometimes dense and capitate, often remarkable for the white, pink or purple scarious sepals and bracts.

The genus is dispersed over the tropical and subtropical regions of the Old World, one, the commonest species, extending also into tropical America. The 9 Australian species are all tropical; one is the above-mentioned common one, another, *P. spicata*, is also Asiatic, the 7 others are endemic.

SECT. 1. **Planchonia,** J. Gay.—*Petals and stamens united in a cup or tube, without staminodia.*

Stems hard and almost woody at the base, the radical leaves soon disappearing. Leaves all narrow. Flowers 3 to 4 lines.
    Stem tall, pubescent. Corolla-tube shorter than the free part.
      Stamens the length of the petals. Capsule short, obtuse  .  .    1. *P. longiflora.*
    Stems short, glabrous. Corolla-tube longer than the free part.
      Stamens much longer than the petals. Capsule oblong, tapering
      at the top  . . . . . . . . . . . . . . . .  2. *P. spirostyles.*
Stems herbaceous, several from a rosette of oblong or obovate radical
    leaves. Stem-leaves narrow. Flowers 1½ to 3 lines . . . .  3. *P. synandra.*

SECT. 2. **Aylmeria,** Mart.—*Petals and stamens free or nearly so, with 5 short staminodia inside the petals and opposite to them.*

Sepals purple, glabrous, nearly 3 lines long. Stamens and petals
    slightly perigynous . . . . . . . . . . . . . . .  4. *P. violacea.*
Sepals white or yellowish, hairy, about 2 lines long. Stamens and
    petals very perigynous . . . . . . . . . . . . .  5. *P. staminodina.*

SECT. 3. **Polycarpia.**—*Petals and stamens free or united in a ring at the base, without staminodia.*

Stems simple or hard and woody at the base. Radical leaves soon
    disappearing.
    Flowers 1½ lines. Petals rounded and very obtuse. Capsule much
      shorter than the sepals . . . . . . . . . . . . .  6. *P. corymbosa.*
    Flowers less than 1 line. Petals oval-oblong, acute, or toothed at
      the top. Capsule rather shorter or longer than the sepals . .  7. *P. breviflora.*
Stems herbaceous, several from a rosette of oblong or obovate radical
    leaves.
    Flower-heads pedunculate, with scarious bracts . . . . . .  8. *P. spicata.*
    Flower-heads closely sessile, surrounded by herbaceous floral leaves  9. *P. involucrata.*

SECTION 1. PLANCHONIA, *J. Gay, in Herb. Hook.*—Petals and stamens united in a cup or tube without staminodia. Sepals very scarious, often rather large.

1. **P. longiflora,** *F. Muell. in Rep. Babb. Exped.* 8. Pubescent, erect and rigid, 1 to 2 ft. high, divided at the base into several erect branches. Leaves narrow-linear, acute or ending in a hair-like point, rigid, silky-hairy, often above ½ in. long, with smaller ones clustered in their axils; the upper ones small and distant. Flowers large, brown red or purple, shortly pedicellate in dense terminal corymbose cymes or heads. Sepals fully 3 lines long, scarious, with a prominent midrib, the inner ones narrower, more acute and more deeply coloured than the outer. Petals hypogynous, united with the stamens in a campanulate tube not 1 line long, their free parts considerably

longer and shortly bifid at the point.   Filaments about as long as the petals. Ovary almost sessile.   Style long and subulate.   Capsule short ovoid, obtuse.

**N. Australia.**   Grassy flats along the Victoria river and other parts of Arnhem's Land, *F. Mueller;* N.W. coast, *Bynoe;* Nichol Bay, *Walcott.*

Var. *leucantha.*   Leaves larger, broader, and less rigid.   Sepals completely scarious and white, without any prominent midrib.—Victoria river, *F. Mueller.*

2. **P. spirostyles,** *F. Muell. in Rep. Babb. Exp.* 8.   Glabrous and often very glaucous, woody at the base, with numerous rigid opposite or dichotomous branches, our specimens not exceeding 6 in.   Leaves very narrow-linear, the margins revolute so as to be almost terete and filiform, rarely exceeding ¼ in., often clustered.   Stipules small, with subulate points.   Flowers large, on very short pedicels, either few in the upper forks, or forming at length a broad corymbose cyme.   Sepals 3 to 4 lines long, acute, white and scarious with a prominent midrib, the outer ones shorter and broader than the inner. Petals and stamens perigynous, united in a tube of fully 2 lines, with the slender filaments projecting considerably beyond the free oblong tops of the petals.   Ovary shortly stipitate, tapering into a long spirally twisted deciduous style.   Capsule stipitate, oblong, tapering at the top, nearly as long as the sepals.   Seeds numerous, very small.

**N. Australia.**   Gilbert's River, *F. Mueller.*

3. **P. synandra,** *F. Muell. in Rep. Babb. Exped.* 8.   A glabrous annual, with a rosette of petiolate spathulate or oblong radical leaves.   Stems several, erect or decumbent, not above 6 in. high, with dichotomous or clustered branches.   Leaves narrow-linear, with recurved or revolute margins, the longer ones above ½ in., but mostly shorter, and not much clustered.   Stipules small, with fine points.   Flowers rather larger than in *P. corymbosa,* in small rather loose corymbose cymes, all more or less pedicellate, the floral leaves all reduced to scarious bracts.   Sepals about 2 lines or nearly 3 lines long in the capitate variety, white and scarious with a prominent midrib often purple. Petals united with the stamens in a tube of about 1 line, their free part shorter and entire, sometimes very short, the filaments about the same length.   Ovary sessile, with a subulate style.   Capsule oblong, tapering at the top, with few seeds.

**N. Australia.**   Hooker's Creek and Sturt's Creek, *F. Mueller.*
**S. Australia.**   In the interior at Wirrawirraloo, *Babbaye's Expedition.*

Var. (?) *densiflora.*   Leaves small and few.   Flowers larger, in a dense, nearly globular head of 1 in. diameter.   Petals notched.

**Queensland.**   N.E. coast, *A. Cunningham;* Port Denison, *Fitzalan;* Rockhampton, *Thozet.*

Var. *gracilis.*   More slender.   Sepals about 1½ lines long.   Petals rather broad, notched.
**N. Australia.**   Port Essington, *A. Cunningham, Armstrong.*

SECTION 2. AYLMERIA, *Mart.*—Petals and stamens free or nearly so, with 5 short staminodia inside the petals and opposite to them.   Sepals very scarious.

4. **P. violacea,** *Benth.*   Pubescent, erect and slightly branched, 1 to 2 ft. high.   Leaves narrow-linear, flat or concave, ½ to 1 in. long, often clus-

tered in the axils, the upper ones small and distant.  Stipules scarious, lanceolate with fine points.  Flowers purple, in dense terminal leafless corymbose cymes or heads, more or less pedicellate, the floral leaves all reduced to scarious bracts.  Sepals nearly 3 lines long, with a prominent midrib, the outer ones shorter and rather less coloured.  Petals free, about ⅔ as long as the sepals, oblong-lanceolate, obtusely bifid.  Stamens about as long as the petals, the filaments filiform, united at the base in a ring, with as many minute filiform staminodia opposite the petals.  Style subulate.  Capsule short, globular, with few seeds.—*Aylmeria violacea* and *A. rosea*, Mart. in Nov. Act. Nat. Cur. xiii. 277 ; *Achyranthes violacea*, Spreng. Syst. Cur. Post. 102, and *A. rosea*, Spreng. l. c. 103.

**N. Australia.**  Croker's Island, *A. Cunningham ;* Port Essington, *Armstrong.*

5.  **P. staminodina,** *F. Muell. in Rep. Babb. Exp.* 8.  Pubescent, with erect, opposite or sometimes clustered branches, ½ to 1 ft. high.  Leaves narrow-linear or the lower ones linear-lanceolate, flat, the larger ones ½ to ¾ in., with smaller ones clustered in their axils.  Stipules with long subulate points. Flowers larger than in *P. corymbosa*, in terminal cymes or heads, forming an irregular general corymb ; the floral leaves all reduced to scarious bracts. Sepals about 2 lines long, scarious and pubescent, white or slightly yellowish, without any prominent midrib.  Petals almost free, inserted with the stamens on a thickened perigynous disk, lanceolate, entire, rather more than half the length of the sepals.  Stamens about as long, alternating with short filiform staminodia opposite the petals.  Ovary short, with a rather short style.  Capsule small, sessile or shortly stipitate, with few seeds.

**N. Australia.**  Sources of the Victoria river, Hooker's Creek and Sturt's Creek, *F. Mueller.*

SECTION 3.  POLYCARPIA.—Petals and stamens free or united in a ring at the base.  Sepals entirely or partially scarious.

6.  **P. corymbosa,** *Lam. Illustr. n.* 2798.  Minutely pubescent or rarely almost glabrous, with erect, rather slender, but stiff branches, ½ to 1 or even 1½ ft. high.  Leaves from narrow-linear to almost subulate, rarely linearlanceolate, flat or with revolute margins, the longer ones ½ to 1 in., with small ones clustered in their axils, the upper ones much smaller and often few and distant.  Stipules tapering to a fine point.  Flowers numerous, in dense terminal corymbose cymes, sometimes all forming one dense mass on the top of an otherwise simple stem, sometimes the cymes numerous and loosely paniculate.  Floral leaves all reduced to scarious bracts.  Sepals about 1½ lines long, white and scarious, without any prominent midrib, but tapering to a fine point.  Petals quite free, not ½ line long, broadly ovate, very obtuse and rather firm.  Stamens often shorter.  Style very short.  Capsule ovoid or oblong, much shorter than the sepals.—DC. Prod. iii. 374 ; Wight, Ic. Pl. Ind. Or. t. 712.

**N. Australia.**  N. coast, *R. Brown ;* Victoria river and Albany Island, *F. Mueller ;* Lizard Island, Keppel's Island, and Port Curtis, *M'Gillivray.*

The species is common in tropical Asia and Africa, and is found also in Brazil and Guiana.

7.  **P. breviflora,** *F. Muell. in Rep. Babb. Exp.* 9.  Glabrous or pubes-

cent, and very nearly allied to *P. corymbosa;* but more slender and divaricately branched, and at once known by its very much smaller flowers. Sepals scarcely 1 line long, broader and less acuminate than in *P. corymbosa,* petals much narrower, not so obtuse and usually denticulate at the top; stamens much more perigynous; capsule longer in proportion, occasionally even exceeding the sepals.

**N. Australia.** N. coast, *R. Brown;* Gulf of Carpentaria, *F. Mueller.*
**Queensland.** Islands of Moreton Bay, *F. Mueller;* Rockhampton, *Thozet.*

8. **P. spicata,** *Arn. in Ann. Nat. Hist.* iii. 91. A small glabrous annual, seldom attaining 6 in. and often not half that size. Radical leaves rosulate, obovate or oblong, on long petioles. Stems several, decumbent or erect, with few spreading dichotomous or clustered slender branches. Leaves under the branches in small false whorls, spathulate or obovate-oblong, 2 to 3 lines long, including the petiole. Stipules short, broadly scarious, with a fine point. Flowers small, white, in small dense terminal cymes or heads, the floral leaves all reduced to short obtuse scarious bracts. Sepals rather more than 1 line long, scarious, the outer one with a broad thick centre, the others with a narrow slightly thickened midrib. Petals very minute and subulate, almost free from the short stamens. Style short. Capsule small, nearly globular.—Wight, Ic. Pl. Ind. Or. t. 510; *P. staticæformis,* Steud. Nom. ed. 2, ii. 369.

**N. Australia.** N.W. coast, *Bynoe.*
The species ranges over the sandy districts of Arabia and the East India Peninsula.

9. **P. involucrata,** *F. Muell. in Rep. Babb. Exped.:* 9. Pubescent, with numerous erect or decumbent rigid dichotomous stems of 2 to 4 in. or rarely twice that length. Radical leaves rosulate, oblong or nearly obovate, narrowed into long petioles; stem-leaves more sessile, narrow-oblong or lanceolate, rather rigid, obtuse or the upper ones acute, 2 to 4 lines long, the floral ones in false whorls of 4 to 8. Flowers several together in sessile heads, in the forks or at the ends of the branches, rarely exceeding the herbaceous floral leaves. Sepals white, finely pointed, 2 to near 3 lines long; the outer ones thickened and cartilaginous at the base. Petals oblong, about ⅓ the length of the sepals, slightly united with the stamens in a ring at the base. Style very short, with a capitate slightly furrowed stigma. Capsule small, ovoid-globular.

**N. Australia.** Hooker's Creek, Sturt's Creek, and near the sources of the Victoria river, *F. Mueller.*

ORDER XVII. **PORTULACEÆ.**

Flowers regular, hermaphrodite. Sepals fewer than petals, usually 2, free or rarely adnate to the ovary at the base, usually broad, imbricate in the bud. Petals 4 or 5, rarely more, hypogynous or rarely perigynous, imbricate in the bud. Stamens inserted with the petals and often adhering to their base, of the same number or fewer and opposite to them or indefinite; anthers 2-celled. Ovary free or rarely half-inferior, 1-celled. Style more or less deeply divided into 3 or rarely 2 or more than 3 branches, stigmatic along the inner side. Ovules 2 or more, amphitropous, with an inferior micropyle, attached to funicles erect from the base of the cavity, and free or united in a central column, or in as many clusters as style-branches. Seeds several or solitary by abortion, usually more or less reniform, with a lateral hilum; testa crustaceous, sometimes with a caruncle at the hilum. Embryo more or less curved round the mealy albumen, or rarely nearly straight with very little albumen.—Herbs rarely shrubby at the base, usually glabrous and succulent or clothed with long hairs. Leaves alternate or opposite, entire. Stipules scarious or split into hairs or none. Flowers terminal and solitary, or in racemes cymes or panicles, or rarely axillary. Petals usually very fugacious or withering in a mass.

A small Order, chiefly American, with a few species dispersed over other parts of the world, especially S. Africa and Australia. The Australian genera are none of them endemic, 2 of them being chiefly American, and the other 2 generally distributed over the globe. The chief characters, derived from the ovary and seeds, are those of *Caryophylleæ*, from which *Portulaceæ* differ in habit, in the number and position of the stamens, and especially in their calyx.

Ovary half-inferior. Petals and stamens perigynous . . . . . . . 1. PORTULACA.
Ovary superior. Petals and stamens hypogynous.
  Petals free.
    Stamens 5, opposite the petals, and inserted on their base . . . 3. CLAYTONIA.
    Stamens indefinite, often numerous, rarely and irregularly reduced
    to 5 . . . . . . . . . . . . . . . . . . . 2. CALANDRINIA.
  Petals united in a simple corolla, split open on one side. Stamens
  3 to 5 . . . . . . . . . . . . . . . . . . . 4. MONTIA.

## 1. PORTULACA, Linn.

Sepals 2, united at the base in a tube adnate to the ovary, the free part deciduous. Petals 4 to 6, perigynous. Stamens indefinite, often numerous, sometimes 6 to 8, inserted with the petals. Ovary half-inferior, with several ovules. Style deeply 2- to 8-cleft. Capsule membranous, half-inferior, the free part circumsciss at maturity. Seeds reniform, shining, often granulate.— Herbs more or less succulent. Leaves alternate or opposite, often clustered in the axils, the floral ones usually forming an involucre round the flowers. Stipules scarious, or more frequently reduced to a tuft of hairs, sometimes very minute or none. Flowers terminal, sessile, or pedicellate.

The species are mostly American, with a very few tropical Australian, Asiatic, or African ones, 2 of them widely dispersed over cultivated or sandy places in various parts of the globe. One of these is included among the Australian ones, of which the remainder are all endemic.

Leaves mostly alternate.
  Stipular hairs minute or none.
    Leaves oblong-cuneate. Root slender. Capsule closely sessile . 1. *P. oleracea.*
    Leaves linear-terete. Root usually tuberous. Capsule narrowed
    into a short stipes . . . . . . . . . . . . 2. *P. napiformis.*
  Stipular hairs numerous and conspicuous.
    Leaves thick and short . . . . . . . . . . . . 3. *P. australis.*
    Leaves linear-terete, almost filiform . . . . . . . . 4. *P. filifolia.*
Leaves all opposite.
  Stipular hairs short, but conspicuous. Flowers usually 3, within
  the floral leaves, and shortly pedicellate. . Style-lobes subulate . 5. *P. digyna.*

No stipular hairs. Flowers solitary and sessile, within 4 bract-like
　floral leaves. Style-lobes flat and transparent.
　Leaves lanceolate or linear . . . . . . . . . . . . . 6. *P. oligosperma.*
　Leaves orbicular . . . . . . . . . . . . . . . . . 7. *P. bicolor.*

1. **P. oleracea,** *Linn.; DC. Prod.* iii. 353. A low, prostrate, or
spreading annual, seldom exceeding 6 in., somewhat succulent, and quite
glabrous. Leaves mostly alternate, cuneate-oblong, obtuse, very rarely
exceeding ½ in., usually narrowed into a short petiole, the stipular hairs very
minute, and sometimes quite disappearing. Flowers terminal and sessile,
between 2 or more floral leaves, rarely solitary, usually several together
in little heads which are either single or several in a dichotomous cyme.
Sepals not much more than 2 lines long. Petals 5, scarcely longer than the
calyx, slightly united at the base, yellow and very fugacious. Stamens 10 to
12 or rarely fewer. Style short, with 5 linear stigmatic lobes. Capsule sessile. Seeds minutely tuberculate, the panicles often united at the base into 5
clusters.—A. Gray, Gen. Ill. t. 99 ; F. Muell. in Rep. Babb. Exped. 10.

**N. Australia.** Victoria river, *F. Mueller.*
**Queensland.** In the interior, *Mitchell.*
**N. S. Wales.** Port Jackson, *R. Brown.*
**Victoria.** Sandy banks of Snowy River, *F. Mueller.*
**S. Australia.** Elizabeth Creek, in the interior, *Babbage's Expedition.*
Var. (?) *grandiflora.* Sepals more obtuse, 3 to 4 lines long.—Sturt's Creek, *F. Mueller.*
The species is common in maritime or sandy localities in most tropical countries, ex-
tending into the warm parts of the temperate regions, both of the northern and southern
hemispheres.

2. **P. napiformis,** *F. Muell. Herb.* Glabrous, with decumbent or erect
stems of 6 in. to near 1 ft., the tap-root thickening into an oblong tuber.
Leaves alternate, linear, succulent, apparently terete, ½ to 1 in. long. Stipular hairs exceedingly minute. Flowers smaller than in *P. oleracea,* usually
3 together, between 2 to 4 involucral leaves, but not quite sessile. Stamens
about 16. Style rather long, 4-cleft at the top. Capsule small, contracted
into a short stipes. Seeds smaller than in *P. oleracea,* black and shining,
finely granulated.

**N. Australia.** Victoria river and Beagle Valley, *F. Mueller;* N.W. coast, *Bynoe.*
The species is allied to the East Indian *P. tuberosa,* Roxb., but the flowers and fruits are
much smaller, not so closely sessile, and there are not the long stipular and involucral
hairs of that species.

3. **P. australis,** *Endl. Atakta,* 7, *t.* 6.* Apparently decumbent and
much branched, the stipular and involucral hairs copious, but otherwise
glabrous. Leaves alternate, oblong, elliptical, thick, under ½ in. long.
Flowers yellow, 1 or 2 together, sessile between 2 to 4 involucral leaves.
Stamens numerous. Style elongated, 5- or 6-cleft. Seeds shining, granulate,
the funicles united into as many clusters as styles.

**N. Australia.** Gulf of Carpentaria, *Bauer.*—I have seen no authentic specimens, and
have taken the above character from Endlicher's description and Bauer's drawing. A specimen of F. Mueller's may be the same plant, and perhaps one of R. Brown's from Broad
Sound, but neither are sufficient for determination. It is not improbable that both this
species and *P. filifolia* may prove to be forms of the tropical African *P. foliosa.*

4. **P. filifolia,** *F. Muell. Fragm.* i. 169. Annual, with erect or decum-

bent stems of $\frac{1}{2}$ to 1 ft., the stipular and involucral hairs long and copious, but otherwise glabrous. Leaves alternate, linear-terete, almost filiform, $\frac{1}{2}$ to 1 in. long.. Flowers rather large, yellow, 1 to 3 together, sessile between 2 to 4 involucral leaves. Sepals 2 to $2\frac{1}{2}$ lines, and petals twice as long. Stamens numerous. Style elongated, usually 4-cleft. Seeds shining, granulate, the funicles united in as many clusters as styles.

**N. Australia.** Sandy deserts on Sturt's Creek, *F. Mueller.*
**Queensland.** In the interior, *Mitchell.*

This may be a variety of *P. australis*, and only appears to differ from the tropical African *P. foliosa* in its more slender leaves, and from *P. tuberosa*, Roxb., in the roots not tuberous and in the large flowers.

5. **P. digyna,** *F. Muell. Fragm.* i. 170. A procumbent, glabrous annual of a few inches, with dichotomous or opposite branches. Leaves all opposite, ovate obovate or nearly orbicular, 2 to 3 lines long, very shortly petiolate. Stipular hairs very short. Flowers pink, very small, pedicellate, 1 to 3 together, between 2 or 4 involucral leaves, forming dichotomous leafy cymes. Sepals not 2 lines long. Petals 4, rather longer. Stamens about 10. Style long, with 2 long linear stigmatic branches. Ovules about 6, the funicles forming 2 clusters. Capsule elongate-conical, covered in the upper part with oblong papillæ. Seeds 1, 2, or 3, black, smooth, and shining.

**N. Australia.** Upper Victoria river, Hooker's Creek, and Sturt's Creek, *F. Mueller.*

6. **P. oligosperma,** *F. Muell. Fragm.* i. 170. A little slender annual of 2 or scarcely 3 in., with numerous opposite branches. Leaves all opposite, oblong, narrow-lanceolate or linear and semiterete, 3 to 4 lines long. Stipular hairs none or quite microscopic. Flowers very small, pink, terminal, solitary and closely sessile within 2 or 4 involucral leaves, which do not exceed the calyx-tube, so that the flower appears pedicellate, with 4 calyx-like bracts at the summit of the pedicel. Sepals scarcely 1 line long, and the petals apparently not longer. Stamens about 6, the anthers very transparent. Style divided into 2 to 4 lanceolate, transparent, and very delicate lobes. Seeds few, black, granulate.

**N. Australia.** Victoria river and Sturt's Creek, *F. Mueller.*

The Sturt's Creek specimens have smaller and rather broader leaves, and in the flower I examined the lobes of the style were broader than in those from Victoria river, but both are probably forms of one species, nearly allied to the East Indian *P. quadrifida*, but at once known by the absence of stipular hairs.

7. **P. bicolor,** *F. Muell. Fragm.* i. 171. A minute, prostrate annual, with opposite branches, rarely above $1\frac{1}{2}$ in. long. Leaves all opposite, broadly ovate or orbicular, scarcely exceeding 2 lines. Flowers as in *P. oligosperma* minute, solitary, terminal, and closely sessile between 4 bract-like floral leaves (appearing pedicellate, with 4 calyx-like bracts at the summit of the pedicel). Sepals not 1 line long. Petals minute, yellow. Stamens about 6. Style with 4 (or sometimes 2?) lanceolate, transparent, very delicate lobes. Capsule short, broad. Seeds several, small, black, granulate.

**N. Australia.** Victoria river, *F. Mueller.*
**Queensland.** Keppel Bay, *R. Brown.*

## 2. CALANDRINIA, H. B. and K.

Sepals 2, persistent or rarely deciduous. Petals 5 or more, or rarely fewer, hypogynous. Stamens indefinite, numerous or few, free or united in a ring at the base, or adhering to the petals. Ovary free, with several ovules, rarely reduced to 1 or 2. Styles 3 or rarely 4, free or united in a single style, 3- or 4-cleft, or furrowed at the top. Capsule globose, ovoid or oblong, opening in 3 or 4 valves, or almost indehiscent. Seeds reniform-globular or flattened, not strophiolate, shining or granulate. Embryo curved round the albumen. —Herbs, rarely half-shrubby at the base, glabrous or hirsute. Leaves alternate or in radical tufts, more or less fleshy. Stipules none. Flowers either solitary pedunculate and axillary, or arranged in terminal racemes or heads. Petals usually very fugacious.

A large genus, which besides numerous tropical, subtropical, or southern American species, only contains the Australian ones here described, which are all endemic. Formerly confounded with *Talinum*, it has been well distinguished from that genus chiefly by the absence of any strophiola or caruncle to the seeds, and differs from *Claytonia* in the stamens always indefinite, even when reduced to a number about the same as or fewer than that of the petals.

*Stamens numerous* (20 to 100).
 Scapes leafless, several-flowered, with numerous opposite scarious
  scales. Root tuberous . . . . . . . . . . . . . 1. *C. Lehmanni.*
 Scapes leafless, 1-flowered. Leaves radical, narrow-linear . . . 2. *C. uniflora.*
 Stems more or less leafy, several-flowered.
  Perennial. Petals very broad. Anthers linear-oblong. Styles
   united at the base . . . . . . . . . . . . 3. *C. balonensis.*
  Annuals. Petals oval-oblong. Anthers short. Styles free to
   the base.
   Styles and capsular valves 3 . . . . . . . . . . 4. *C. polyandra.*
   Styles and capsular valves 4 . . . . . . . . . . · . 5. *C. quadrivalvis.*
*Stamens few   Capsule ovoid or oblong, very readily dehiscent.*
 Stamens mostly 8 to 10. Seeds pitted (except in *C. liniflora*).
  Sepals acute or scarcely obtuse. Leaves linear-terete, the radical
   ones elongated.
   Sepals fully 2 lines. Anthers linear-oblong. Seeds smooth and
    shining . . . . . . . . . . . . . . . . 6. *C. liniflora.*
   Sepals 1 to 1½ lines. Anthers small, ovate. Seeds minutely pitted.
    Petals 5 . . . . . . . . . . . . . . . . . 7. *C. gracilis.*
    Petals about 8 . . . . . . . . . . . . . . . 8. *C. polypetala.*
  Sepals broad and very obtuse. Leaves oblong or shortly linear.
   Stems short, ascending or diffuse . . . . . . . . . 9. *C. pusilla.*
   Stems twining . . . . . . . . . . . . . . . 10. *C. volubilis.*
 Stamens mostly 3 to 5. Seeds very smooth and shining.
  Bracts leafy. Sepals 3 to 4 lines long . . . . . . . . *C. caulescens* (p. 175).
  Bracts very small. Sepals under 2 lines and often under 1 line.
   Leaves oblong or linear-oblong, thick. Racemes loose. Pedicels
    at length 3 to 5 lines, reflexed . . . . . . . . 11. *C. calyptrata.*
   Leaves small, narrow-linear. Racemes short and very numerous.
    Pedicels not 1 line, erect. Flowers very small.
    Capsule oblong, with 4 to 8 seeds. Ovules 6 to 8 . . . . 12. *C. composita.*
    Capsule narrow-cylindrical, with 1 or 2 seeds. Ovules 2 . 13. *C. corrigioloides.*
*Stamens few. Capsule globular or shortly ovoid, very smooth and shining, and scarcely dehiscent.*
 Leaves linear-terete. Stamens about 15. Anthers oblong.
  Capsular valves separating at the base . . . . . . . 14. *C. spergularina.*

Leaves linear-terete.  Stamens about 5.  Anthers globular.
Capsule indehiscent . . . . . . . . . . . . . 15. *C. granulifera.*
Leaves short and broad.  Stamens 5 to 10.  Anthers globular.
Capsule scarcely dehiscent . . . . . . . . . . 16. *C. pygmæa.*

1. **C. Lehmanni,** *Endl. in Pl. Preiss.* ii. 235.  Rootstock slender and cylindrical, bearing, when full grown, one or more tubers at the base, and at the top a few small scales, apparently the remains of leaves, and a tuft of 2 to 4 erect, slender stems, 6 to 8 in. high and quite leafless, except a number of small, opposite, sheathing scales, their fine points closely pressed against the stem.  Leaves in the very young specimens radical, small, obovate, or spathulate, soon withering away, and never more than 2 or 3.  Flowers few, in a terminal raceme, the slender pedicels of ¼ to ½ in. proceeding from the axils of the upper scales.  Sepals very broad, almost obtuse, very thin, 3-nerved, about 2 lines long.  Petals nearly 3 times as long.  Stamens short, very numerous, with short anthers.  Style simple at the base, with 3 long, linear, stigmatic branches.  Capsule ovoid, longer than the calyx, 3-valved, with numerous small granulated seeds.

**W. Australia.**  Swan River, *Preiss, n.* 1528, *Drummond, Coll.* 1844, *n.* 242; South Hutt river, *Oldfield.*

2. **C. uniflora,** *F. Muell. in Trans. Phil. Inst. Vict.* iii. 41, and *Fragm.* i. 177.  Rootstock simple, cylindrical, erect, bearing a dense tuft of narrow-linear leaves of 2 to 4 in.  Scapes numerous from amongst the leaves, 8 to 10 in. high, 1-flowered and leafless, except 1 or 2 minute scales.  Flowers rather large.  Sepals broad and thin, 3 to 4 lines long.  Petals usually 6 or 7.  Stamens very numerous, the inner ones much longer than the outer; anthers oblong.  Styles 4, erect, shortly plumose and stigmatic along their whole length.  Capsule about as long as the sepals, 4-valved.  Seeds numerous, black and shining.

**N. Australia.**  Victoria river, near the main camp, *F. Mueller.*
The species is nearly allied to two Chilian ones, *C. rupestris*, Barn., and *C. graminifolia*, Philippi.

3. **C. balonensis,** *Lindl. in Mitch. Trop. Austr.* 148.  Apparently perennial, erect, branching, 6 in. to 1 ft. high or rather more.  Leaves thick and fleshy, the lower ones oblong-spathulate or obovate, 1 in. long or less, the upper ones linear or lanceolate, often above 2 in.  Flowers large, purple, in loose terminal racemes, on pedicels of about 1 in.  Bracts scarious, acuminate, mostly opposite, but only one of each pair has a flower in its axil.  Sepals very broad and obtuse, herbaceous, obscurely veined, with a scarious margin.  Petals very broadly obovate, fully ¾ in. long.  Stamens very numerous; anthers narrow-oblong.  Style 3-lobed, the lobes thick and nearly twice as long as the entire base.

**Queensland.**  Sandy soil on the Balonne river, *Mitchell.*

4. **C. polyandra,** *Benth.*  Annual, with decumbent or ascending branches of 6 in. to 1 ft.  Leaves few, chiefly in the lower part of the stem, thick and fleshy, the lowest broadly linear or almost spathulate, the upper ones narrow-linear, occasionally almost opposite, mostly 1 to 1½ in. long.

Flowers of a red-purple, rather large, few together in a terminal raceme, the pedicels 1 in. or more. Bracts small and scarious. Sepals very broad, rather obtuse, thin and slightly coloured, with scarcely prominent veins. Petals narrow-obovate, about ½ in. long. Stamens very numerous, irregularly united at the base; anthers short. Style divided to the base into 3 linear stigmatic branches. Capsule ovoid or oblong, 3-valved. Seeds very numerous and small, black, minutely pitted.—*Talinum polyandrum*, Hook. Bot. Mag. t. 4833.

**S. Australia.** Spencer's Gulf, *Warburton;* in the interior, *Victorian Expedition.*

**W. Australia,** *Burges, Drummond, Coll.* 1848, *n.* 119; Flinders Bay, *Collie;* near Banbury, *Oldfield;* Murchison river, *Sandford;* W. coast, *Bynoe.*

Var. *leptophylla.* Slender, with very narrow leaves 2 to 3 in. long, and few, rather large flowers on long slender pedicels. W. coast, with the commoner form, *Bynoe.*

5. **C. quadrivalvis,** *F. Muell. Fragm.* i. 176. A glabrous annual, with small, oblong-spathulate radical leaves, soon disappearing, and several decumbent or ascending stems, from a few in. to 1 ft. or rather more, and sometimes much branched. Stem-leaves from linear-spathulate to oblong or lanceolate, narrowed into a petiole, the lower ones often above 1 in. long, the upper ones few and small. Flowers small, pink, in loose racemes sometimes branching into panicles; pedicels ½ to ¾ in. Bracts very small, herbaceous or slightly scarious. Sepals herbaceous, acute, about 1¼ lines long. Petals 6, fully twice as long as the calyx. Stamens numerous, with small anthers. Style divided to the base into 4 linear stigmatic branches. Capsule about as long as the calyx, 4-valved, with numerous small seeds minutely pitted.

**N. Australia.** Sandy places along the Victoria river and in the Macadam range, *F. Mueller.*

6. **C. liniflora,** *Fenzl, in Hueg. Enum.* 52. A slender annual, with a tuft of narrow-linear radical leaves of 1 to 2 in. Stems several, ascending, from a few in. to nearly 1 ft. high. Leaves few, linear, mostly small. Flowers apparently red, in a loose raceme, on pedicels of ½ to 1 inch. Bracts small and narrow, but not scarious. Sepals broadly ovate, herbaceous, acute, 2 lines or rather longer. Petals 5, obovate, fully ½ in. long. Stamens about 10, united at the base in a membranous cup; anther-cells linear, only united by a small connective in the centre. Styles or style-branches linear, very shortly united at the base. Capsule oblong, longer than the calyx, with numerous small, smooth and shining seeds.—Nees, in Pl. Preiss. i. 247.

**W. Australia.** Swan River, *Preiss, n.* 1952, *Drummond.*

Var. (?) *grandiflora.* Stems more leafy, flowers larger. Vasse river, *Mrs. Molloy.*

7. **C. gracilis,** *Benth.* A slender annual, with a tuft of narrow-linear radical leaves of 1 to 2 in., and several stems of about 1 ft., bearing few linear leaves and a loose raceme, as in *C. liniflora,* but the flowers are smaller and different in structure. Bracts minute and scarious. Sepals a little more than 1 line long, acute, thin. Petals 5, narrow, about twice as long as the sepals, apparently white. Stamens about 10, the filaments slightly dilated towards the base, but not united; anthers small. Styles divided to the base into 3 or 4 linear stigmatic branches. Capsule rather longer than the calyx, 3- or 4-valved. Seeds very minutely pitted when seen under a strong lens.

**N. Australia.** Port Essington, *Armstrong.*

8. **C. polypetala,** *Fenzl, in Hueg. Enum.* 51. A slender annual, with filiform radical leaves of 1 in. or longer. Stems ascending, simple, 3 to 6 in. high. Leaves filiform, the upper ones passing into the minute bracts. Racemes terminal, with distant, small flowers, the lower pedicels about 5 lines, the upper ones much shorter. Sepals rather obtuse, a little more than 1 line long. Petals 8 to 10, oblong, twice as long as the sepals, withering into a calyptra, as in *C. calyptrata.* Stamens 8 to 10, united in a ring at the base; anthers globular. Styles 3, filiform. Capsule half as long again as the calyx, nearly cylindrical, 3-valved, with minute, globular, black seeds, minutely granulated.—Nees, in Pl. Preiss. i. 247, excluding the var. *composita.*

**W. Australia.** Swan River, *Huegel.* I have not seen Huegel's specimens nor any others which I can refer with certainty to Fenzl's *C. polypetala.* It may possibly be the same as *C. pusilla,* but I have never seen in that species more than six petals.

9. **C. pusilla,** *Lindl. in Mitch. Trop. Austr.* 360. A small annual, the stems ascending from 1 to 3 or 4 in. or rarely higher. Leaves radical or on the lower part of the stem, about ½ to 1 in. long, much more succulent than in *C. calyptrata,* oblong or linear, mostly petiolate, but dilated and stem-clasping at the base. Racemes occupying a great part of the stems, but loose and few-flowered, with minute scarious bracts, except the lower ones, which are sometimes leafy. Flowers apparently pink, like those of *C. calyptrata,* except that the sepals are very broad and obtuse, coloured with scarious margins, attaining 1½ lines when in fruit. Petals 5 or 6, oblong. Stamens 5 to 8; anthers small. Style divided to the base into 3 short, thick, stigmatic branches. Capsule narrow, longer than the calyx, opening in 3 valves. Seeds numerous, much smaller than in *C. calyptrata* and minutely pitted.

**Queensland.** On the Maranoa, *Mitchell.*
**N. S. Wales.** Darling river, *Victorian Expedition.*
**Victoria.** On the Murray, *F. Mueller;* Wimmera river, *Dallachy.*
**S. Australia.** Mount Brown, Holdfast Bay, etc., *F. Mueller.*
**W. Australia.** Swan River, *Drummond;* Murchison river, *Oldfield.*
This and the following species are united by F. Mueller with *C. calyptrata,* but the differences in habit, calyx, and seeds appear to me to be too constant not to admit them as species.

10. **C. volubilis,** *Benth.* Allied to *C. pusilla,* and with that species considered by F. Mueller as a variety of *C. calyptrata,* but the seeds and flowers are different. Leaves crowded on a short, succulent, branching stock, linear-oblong, 1 to 1½ in. long, narrowed below the middle, but dilated at the base. Flowering branches twining, almost leafless, except minute scarious bracts. Pedicels flexuose, 2 to 6 lines long. Sepals very obtuse, broad and succulent, 1½ lines when in flower, 2 lines when in fruit. Petals about as long, withering into a calyptra on the young fruit. Stamens 8 to 10, the filaments slightly dilated at the base, but scarcely united; anthers small. Style cleft almost to the base into 3 linear stigmatic branches. Capsule acuminate, twice as long as the sepals. Seeds strongly pitted.

**N. S. Wales.** Near the Darling river, *Beckler.*
**S. Australia.** Port Lincoln, *Wilhelmi.*

11. **C. calyptrata,** *Hook. f. in Hook. Ic. Pl. t.* 296. A small annual,

with petiolate linear-oblong or linear-spathulate radical leaves. Stems branching, prostrate or ascending, from 1 or 2 to 7 or 8 in. long. Leaves few, smaller than the radical ones, varying from linear to almost obovate. Flowers very small, in a loose flexuose raceme, the pedicels 2 to 6 lines long, reflexed after flowering. Bracts very small, the upper ones often scarious. Sepals acute, about 1 line long in flower, nearly 1½ when in fruit. Petals about as long, often persistent a long time after flowering, withered into a small calyptra on the top of the young fruit. Stamens about 5, with slender, free filaments ; anthers ovate. Style very short, with 3 very short, oblong, stigmatic branches. Capsule rather longer than the calyx, 3-valved. Seeds numerous, small, very smooth and shining.—Hook. f. Fl. Tasm. i. 143 ; *Claytonia calyptrata*, F. Muell. Fragm. iii. 89.

**N. S. Wales.** Port Jackson, *R. Brown.*
**Victoria.** In the Wendu Valley, *Robertson.*
**Tasmania.** Port Dalrymple, *R. Brown ;* on basaltic rocks, near Launceston, *Gunn.*
**S. Australia.** Holdfast Bay, Mount Parker, Bugle and Barossa ranges, *F. Mueller.*
**W. Australia.** King George's Sound, *R. Brown, Baxter ;* S. coast ?, *Oldfield.*

Var. (?) *pumila,* F. Muell. A small, tufted plant, with a thick, succulent root. Leaves radical or nearly so, oblong or almost ovate, 3 to 4 lines long, but narrowed into a petiole twice that length. Flowering branches or racemes loose, 1 to 1½ in. long. Bracts small, scarious. Flowers about the size of the *C. calyptrata,* but the sepals very obtuse. Capsule ovoid-globular, the valves cohering at the summit. Seeds numerous, small, smooth, and shining.

**Queensland.** Balonne river, *Bowman.*
**N. S. Wales.** From Nangawera to Yellowinchi, *Victorian Expedition.* I am inclined to think that further specimens will prove this to be a distinct species (*Herb. F. Mueller*).

*C. caulescens,* H. B. and K. Nov. Gen. et Sp. vi. 78, t. 526, a common Peruvian weed, has established itself in waste places about Adelaide and other parts of S. Australia. Although technically the characters are nearly those of *C. calyptrata,* it is readily known by its much more leafy stems, the bracts all leaf-like, and the flowers more than twice the size, the sepals ovate, acuminate, 3 or 4 lines long. *C. compressa,* Schrad. (*C. pilosiuscula,* DC.), an equally common Chilian weed, is also very nearly allied, but is readily distinguished by the very broadly hastate sepals, as well as some differences in the foliage.

12. **C. composita,** *Nees* (*under* C. polypetala). A small diffuse annual, very densely branched, seldom exceeding 2 or 3 in. Radical leaves linear, attaining ½ in., the stem-leaves mostly 1 to 2 lines, passing into minute bracts. Flowers very small and numerous, in short racemes on pedicels rarely exceeding 1 line, and usually much shorter when in flower. Sepals ¾ line in flower, 1 line long when in fruit, obtuse and rather thick. Petals 5 or 6, scarcely exceeding the calyx, withering into a calyptra as in *C. calyptrata.* Stamens 3 to 5 ; anthers small. Style divided to the base into 3 linear stigmatic branches. Ovules about 6 to 8. Capsule ovoid-oblong, longer than the calyx, opening in 3 valves. Seeds 3 to 6, smooth and shining.—*C. polypetala,* var. *composita,* Nees, in Pl. Preiss. i. 247.

**W. Australia.** Swan River, *Drummond, Preiss, n.* 1951.

13. **C. corrigioloides,** *F. Muell. Herb.* An annual, with narrow-linear radical leaves contracted into a long petiole. Stems numerous, prostrate or slightly ascending, not much exceeding ¼ ft. Stem-leaves few, linear, petiolate. Racemes numerous, short, axillary and terminal, branching

so as to form little unilateral cymes. Bracts minute. Flowers very small, white, on pedicels which rarely exceed ½ line. Sepals not ½ line long, obtuse. Petals 5 or 6, narrow, rather longer than the sepals. Stamens usually 3; anthers small. Style divided into 3 very short stigmatic lobes. Ovules usually 2. Capsule cylindrical, slender, often above 1½ lines long, opening in 3 valves. Seed usually only 1, or rarely 2, in the base of the capsule, large in proportion, orbicular, black, and very smooth and shining.

**Victoria.** Wimmera river, *F. Mueller.*

**W. Australia.** Swan River, *Drummond;* Canning and Murchison rivers, *Oldfield.*

14. **C. spergularina,** *F. Muell. Fragm.* i. 175. A small annual, with a tuft of linear-terete leaves under 1 in. long. Stems slender, decumbent, slightly branched, 2 to 4 in. long or scarcely more. Leaves few, small, linear-terete. Flowers pink, very small, in a rather rigid often flexuose raceme on pedicels of 1 to 3 lines. Bracts very minute and scarious. Sepals acute, a little more than 1 line long in flower, 1½ lines when in fruit. Petals 6, not twice as long as the calyx. Stamens about 15; anthers oblong, the cells adhering in the centre only. Style divided to the base into 3 linear stigmatic branches. Capsule small, the valves remaining coherent at the top, separating at the base, and falling off together. Seeds small, smooth, and shining.

**N. Australia.** Sandy bed of Nicholson river, Gulf of Carpentaria, *F. Mueller.*

15. **C. granulifera,** *Benth.* A small annual, with a tuft of linear radical leaves. Stems numerous, rigid, branching, decumbent or ascending, 2 to 6 in. long. Leaves few and small. Bracts very minute. Flowers very small, in terminal one-sided racemes, on rigid pedicels of 1 or rarely 2 lines, much thickened when in fruit. Sepals little more than ½ line long and very deciduous. Petals 5, 6, or sometimes 7, apparently white, about twice as long as the calyx. Stamens scarcely as many as petals, with very short anthers. Style short, with 3 linear stigmatic branches. Capsule about 1 line long, globular-conical, black, smooth and shining, and usually indehiscent. Seeds numerous, brown, very small and obovoid.

**W. Australia.** Swan River, *Drummond.*

16. **C. pygmæa,** *F. Muell. Fragm.* i. 175. A very small annual, with numerous decumbent or erect stems, often under 1 in. and rarely exceeding 3 in. Leaves from oblong to ovate, thick and succulent, the radical ones not exceeding 5 lines and the stem ones usually 2 to 3 lines long. Racemes short and dense, with the bracts mostly leafy but small. Flowers small, on very short pedicels. Sepals succulent, obtuse, about 1½ lines long, or sometimes much larger when in fruit. Petals usually 5, 6, or 7, narrow, rather longer than the calyx. Stamens varying in number, usually 2 or 3 more than the petals, and connected in a ring at the base; anthers short. Style divided to the base into 3 long, linear, stigmatic branches. Capsule globular or ovoid, cartilaginous, very smooth and shining, and often black, the valves opening only very shortly at the top. Seeds small, minutely pitted.—*Talinum nanum,* Nees, in Pl. Preiss. i. 246.

**Victoria.** Moist rocky or sandy places in the Grampians, Mount Abrupt in the Tatiara country, Port Phillip, etc., *F. Mueller, Adamson,* and others.

**S. Australia.**  Lynedoch Valley, *F. Mueller.*
**W. Australia.**  Swan River, *Drummond, Preiss, n.* 1930 ; Vasse river, *Oldfield.*

## 3. CLAYTONIA, Linn.

Sepals 2, persistent. Petals 5, hypogynous. Stamens 5, opposite the petals and adhering to them at the base. Ovary free, with few ovules ; style 3-cleft or 3-furrowed at the top. Capsule globular or ovoid, opening in 3 valves. Seeds reniform or orbicular, flattened. Embryo curved round the albumen.—Annual or perennial herbs, usually glabrous and somewhat succulent. Radical leaves petiolate, the stem-leaves alternate or opposite, without stipules. Flowers in terminal racemes or cymes, rarely solitary.

The species are all North American or North-East Asiatic, with the exception of the following one, which is confined to Australia and N. Zealand. The genus is chiefly distinguished from *Calandrinia* by the stamens constantly of the same number as and opposite the petals, a character generally accompanied by a marked difference in aspect.

1. **C. australasica,** *Hook. f. in Hook. Ic. Pl. t.* 293, *and Fl. Tasm.* i. 144. A small tufted plant, with a creeping stem not exceeding a couple of inches in dry places, lengthening out to a foot or more in water. Leaves alternate, narrow-linear, obtuse, from 1½ in. in the small plants to 2 or 3 in. in the aquatic ones, usually narrowed below the middle, but with a widened sheathing base often scarious on the edges. Flowers white and large for the genus, terminal or leaf-opposed, solitary or 2 or 3 in a loose raceme, on long pedicels. Sepals small, orbicular. Petals several times longer, obovate-oblong. Style-lobes filiform. Capsule about as long as the calyx. Seeds usually 3, black, smooth and shining.—F. Muell. Fragm. iii. 89.

**N. S. Wales.**  Valleys of the Blue Mountains, *A. Cunningham.*
**Victoria.**  Very common in rich soils and marshy places ascending to the summits of the Australian Alps, *F. Mueller.*
**Tasmania,** *R. Brown,* common in moist places throughout the island, ascending to 4000 ft., *J. D. Hooker.*
**S. Australia.**  Rivoli Bay, *F. Mueller.*
**W. Australia,** *Drummond, n.* 220, *Oldfield.*
The species is also found in New Zealand.

## 4. MONTIA, Linn.

Sepals usually 2, persistent. Petals hypogynous, united in a 5-lobed corolla, split open on one side. Stamens 3 or rarely 5, inserted in the top of the corolla-tube. Ovary free, with 3 ovules. Capsule globular, opening in 3 valves. Seeds nearly orbicular. Embryo curved round the albumen.—A small annual. Leaves mostly opposite, without stipules. Flowers very small.

The genus consists probably of a single species, although some of its most marked varieties have been raised by some authors to the rank of species.

1. **M. fontana,** *Linn. ; DC. Prod.* iii. 362. A little glabrous, green, somewhat succulent annual, forming dense tufts from 1 to 4 or 5 in. high, the stems becoming longer and weaker in more watery situations. Leaves opposite or nearly so, obovate or spathulate, from 3 to 5 or 6 lines long. Flowers solitary or in little drooping racemes of 2 or 3, in the axils of the

upper leaves, the petals of a pure white, very little longer than the calyx. Capsules small.—Hook. f. Fl. Tasm. i. 144.

**Tasmania.** In springs on St. Patrick's River at an elevation of 1500 ft., abundantly, *Gunn.*

The species is common throughout Europe, in Northern Asia and N.W. America, and thence down the Andes to Australia, America, and in New Zealand, but not in central or tropical Asia, nor, as far as hitherto known, in any part of Africa except Algeria.

## Order XVIII. ELATINEÆ.

Flowers regular, hermaphrodite. Sepals 2 to 5, free, imbricate in the bud. Petals as many, hypogynous, imbricate in the bud, occasionally wanting. Stamens as many or twice as many, hypogynous, free; anthers 2-celled. Torus small, without any disk. Ovary free, with as many cells as there are sepals; styles as many, free from the base, with terminal capitate stigmas. Ovules several in each cell, attached to the inner angle, anatropous. Capsule opening septicidally, the valves flat or concave, with the margins inflexed, leaving more or less of the dissepiments attached to the central column. Seeds straight or curved, testa crustaceous, usually wrinkled or ribbed, albumen none or very thin. Embryo filling the seed, cotyledons short, radicle next to the hilum. —Herbs or low undershrubs, aquatic, creeping or diffuse. Leaves opposite or rarely verticillate, entire or serrate. Stipules in pairs. Flowers small, axillary, solitary or in clusters or cymes.

A small Order dispersed over nearly the whole globe, allied to *Hypericineæ* and *Caryophylleæ,* but differing from the former in habit, in the stipules, and in the perfectly isomerous flowers, from the latter chiefly in the ovary and fruit and want of albumen to the seeds; there is also considerable affinity, especially in habit, with *Lythrarieæ* and *Crassulaceæ.* The only two genera of the Order, both of them of wide geographical range, are represented in Australia.

Sepals membranous, obtuse. Capsule membranous. Glabrous, aquatic or
　creeping herbs. Flowers 2- to 4-merous . . . . . . . . . . 1. ELATINE.
Sepals herbaceous in the middle or keeled, acute. Capsule almost crusta-
　ceous. Herbs or undershrubs. Flowers usually 5-merous, rarely 3- to
　4-merous . . . . . . . . . . . . . . . . . . . . . . 2. BERGIA.

### 1. ELATINE, Linn.

Flowers 3- or 4-merous, rarely 2-merous. Sepals membranous, obtuse, not keeled. Ovary globular. Capsule membranous, the dissepiments either disappearing or remaining attached to the central column.—Small glabrous herbs, either aquatic or creeping on mud. Leaves opposite or verticillate. Flowers usually solitary in the axils, and very small.

The genus is widely dispersed over the temperate and subtropical regions of the globe. The Australian species is considered by some as endemic, by others as identical with an American one.

1. **E. americana,** *Arn. in Edinb. Journ. Nat. Sc.* i. 431, var. *australiensis.* A small, tender, glabrous annual, prostrate and creeping over mud in dense tufts, sometimes not 1 in. in diameter, sometimes extending over a considerable surface. Leaves in the ordinary form ovate, obovate, or broadly oblong, 2 to 3 lines long, thin and of a bright green; but in some luxuriant

specimens ovate-lanceolate or oblong, and exceeding $\frac{1}{2}$ in., almost always bordered by a few distant glands. Stipules very minute and deciduous, or rarely more persistent, and $\frac{1}{4}$ line long. Flowers very minute, sessile and solitary in one axil only of each pair of leaves, and in Australia almost always 3-merous. Sepals usually very minute and transparent, and the petals so very small and fugacious as to be rarely found in dried specimens, except in some western ones, where the petals are reddish and fully $\frac{1}{2}$ line long. Stamens 3. Ovary depressed-globular, with 3 cells and 3 minute, punctiform, almost sessile stigmas. Capsule often 1 line in diameter, the dissepiments sometimes complete, sometimes obliterated at maturity. Seeds cylindrical, more or less curved or nearly straight, marked with longitudinal furrows and minute, transverse wrinkles.—Hook. f. Fl. Tasm. i. 47 ; *E. minima*, Fisch. and Mey. in Linnæa, x. 73 ; F. Muell. Pl. Vict. i. 195 ; *E. gratioloides*, A. Cunn. in Ann. Nat. Hist. iii. 26.

**Queensland.** Brisbane river, *F. Mueller.*
**Victoria.** Muddy places and margins of still fresh-waters, sparingly distributed over the colony, *F. Mueller.*
**Tasmania.** Marshes in the northern and central parts of the island, *J. D. Hooker.*
**S. Australia.** Lake Torrens, *F. Mueller.*
**W. Australia,** *Drummond, n.* 604, 605, 684 ; Murchison river, *Oldfield.*
This plant, whether a distinct species or a variety of the N. American one, is found also in New Zealand and the Fiji islands, and is very variable. In the majority of specimens from various localities, I have always found 3 very thin sepals and 3 stamens, but have failed to detect the petals even in a very early stage. Amongst them Drummond's n. 605 are remarkable for the large size of the capsules ; some of Gunn's, from a lagoon at Georgetown, where they are under water, and Drummond's n. 684, probably also from under water, have elongated stems and leaves 6 to 9 lines long ; F. Mueller's, from the Brisbane river, have also long leaves and remarkably large stipules. A western specimen in Herb. Hooker, from Drummond, differs still more in the well-developed red petals, of a firm consistence and remaining long persistent. The N. American plant (*A. Gray, Gen. Ill. t.* 95) differs chiefly in the flowers almost constantly dimerous, which does not occur in any southern specimens I have examined.

## 2. BERGIA, Linn.

Flowers 5-merous, or rarely 3–4-merous. Sepals herbaceous or keeled in the centre, acute, usually membranous and transparent on the edges. Ovary ovoid or globular. Capsule somewhat crustaceous, the valves sometimes induplicate on the edges and carrying off nearly the whole of the dissepiments, sometimes nearly flat, leaving more or less of the dissepiments attached to the axis.—Herbs or undershrubs, prostrate or much branched, often pubescent. Leaves opposite, entire or more frequently serrate. Flowers axillary, solitary or clustered in cymes, small, but usually larger than in *Elatine.*

The genus is widely distributed over the warmer regions of the globe. F. Mueller proposes to unite it with *Elatine*, but slight as are the characters, they are accompanied by a very decided difference in habit, and the two genera are therefore natural. Of the three or four Australian species two are endemic, but nearly allied to corresponding S. African ones, a third *B. ammannioides*, is a common Asiatic and African weed, of which the fourth may be a mere variety.

Flowers small, clustered in the axils. Stamens of the same number
   as the petals and sepals.
   Stems pubescent . . . . . . . . . . . . . . . . . 1. *B. ammannioides.*

Stems quite glabrous . . . . . . . . . . . . .    2. *B. pusilla.*
Flowers solitary, pedicellate.  Stamens twice the number of the
 sepals and petals.
Erect annual.  Pedicels elongated.  Filaments all equal.  Styles
 short . . . . . . . . . . . . . . . . .    3. *B. pedicellaris.*
Stem woody, prostrate and tortuous.  Pedicels short.  Outer filaments much broader.  Styles filiform . . . . . . . .    4. *B. perennis.*

1. **B. ammannioides,** *Roth, Nov. Pl. Sp.* 219.  A rigid, much-branched
annual, erect or decumbent, pubescent or hirsute, with spreading hairs, usually
6 in. to 1 ft. high.   Leaves from oval-elliptical to oblong or lanceolate, the
larger ones ½ to 1 in., but mostly smaller, more or less serrate with mucronate or glandular teeth, narrowed at the base.   Stipules lanceolate, serrate.
Flowers very small, in dense axillary clusters, on very short filiform pedicels,
usually 5-merous, but sometimes 4-merous or 3-merous.   Sepals very narrow,
acute, ciliate, about ½ line long.   Petals narrow, very thin, about as long as
the sepals.   Stamens of the same number as the sepals and petals.   Capsule
rather shorter, the boat-shaped valves separating septicidally so as to leave
the axis almost wholly without any remains of the dissepiments.   Seeds very
small, ovoid, nearly straight.—*Elatine ammannioides,* Wight, in Hook. Bot.
Misc. iii. 93, t. 5 ; Wight, Ill. t. 25 *a* ; F. Muell. Fragm. ii. 147.

**N. Australia.**   Gravelly bed and banks of Victoria river, Sturt's Creek, and their
affluents, *F. Mueller.*
**Victoria.**   Junctions of the Darling and Murray rivers, *F. Mueller.*
The species is common in East India and the warmer regions of Africa.

Var. *trimera.*   Usually more procumbent and smaller.   Flowers small, 3-merous or 4-
merous.—*B. trimera,* Link, in Linnæa, x. 74 ; *B.* (or *Elatine*) *tripetala,* F. Muell. Pl. Vict.
i. 196, t. 9.   The small Victorian specimens from Dr. Mueller in Sonder's herbarium agree
precisely with some Indian ones, very properly included by Wight in the *B. ammannioides.*

2. **B. pusilla,** *Benth.*   This may be a variety only of *B. ammannioides,*
but it has a different aspect from any of the forms assumed by that species in
India and Africa.   It is perfectly glabrous, with numerous slender stems, 1
to 2 in. high, thickened at the base, with a few obovate leaves, the upper
leaves oblong-lanceolate and serrate.   Flowers small, axillary, and clustered,
as in *B. ammannioides,* but usually more sessile and 4-merous, rarely 3-
merous ; sepals more acuminate.   Capsular valves apparently less folded,
leaving a thicker central axis.—*Elatine verticillaris,* F. Muell. Fragm. ii. 148.

**N. Australia.**   Roper river in Arnhem's Land, *F. Mueller.*   The East Indian *B. verticillata,* Willd., is a very different species.

3. **B. pedicellaris,** *F. Muell. Herb.*   A more or less glandular-pubescent annual, about ½ ft. high, erect or with decumbent side-branches.   Leaves
elliptical or lanceolate, mostly acute, minutely serrate, narrowed at the base,
the larger ones above 1 in., but mostly under ½ in. long.   Stipules narrow.
Pedicels solitary, slender, longer than the leaves.   Flowers 5-merous, much
larger than in the preceding species.   Sepals keeled, 1 to 1¼ lines long.
Petals ovate-lanceolate, persistent, about as long as the sepals.   Stamens
usually 10, the filaments very thin, slightly dilated and closely pressed round
the ovary up to the middle.   Styles short.   Capsule depressed-globular, 5-
valved, leaving very little of the dissepiments attached to the axis.   Seeds

very numerous and minute, quite smooth unless seen under a very high magnifier.—*Elatine pedicellaris*, F. Muell. Fragm. ii. 145.

**N. Australia.** Careening Bay, N.W. coast, *A. Cunningham ;* gravelly beds of the Victoria and Fitzmaurice rivers, and along their affluents, *F. Mueller.* The species is closely allied to *B. polyantha*, Sond., from S. Africa, which has the same styles and stamens, but is quite glabrous, with rather larger flowers on much shorter pedicels.

4. **B. perennis,** *F. Muell. Herb.* Stems prostrate, woody, tortuous, with very short leafy branches, glabrous or with a very few short hairs. Leaves from ovate to elliptical-oblong, mostly 3 to 4 lines long, rather rigid, glabrous and glaucous, often ciliate towards the base and narrowed into a short petiole. Stigmas lanceolate, ciliate. Flowers usually 5-merous, on solitary pedicels, rarely exceeding the length of the leaves. Sepals broadly-lanceolate, keeled, with scarious margins, nearly 2 lines long. Petals longer, rather narrow. Stamens usually 10, the 5 outer filaments dilated, especially below the middle. Styles filiform. Capsule rather shorter than the calyx, the valves leaving much of the dissepiments attached to the central column. Seeds oblong, curved, slightly furrowed and transversely wrinkled like those of *Elatine.* —*Elatine perennis*, F. Muell. Fragm. ii. 146.

**N. Australia.** Banks of the rice swamps near Sturt's Creek, *F. Mueller.* The species is nearly allied to the S. African *B. anagalloides*, E. Mey., which is a perennial with the same styles and stamens, but its flowers are rather larger, on longer pedicels.

## ORDER XIX. HYPERICINEÆ.

Flowers regular, hermaphrodite. Sepals 5, rarely 4, imbricate in the bud. Petals as many, hypogynous, imbricate and usually contorted in the bud. Stamens indefinite, hypogynous, usually united or clustered into 3 or 5 bundles ; anthers 2-celled. Ovary consisting of 3 to 5 carpels more or less united, either 1-celled with the placentas on the inflexed margins of the carpels, or completely divided into cells by the union of the placentas in the axis. Styles as many as carpels, free or rarely united at the base, with terminal stigmas. Ovules usually several to each cell or placenta, anatropous. Fruit capsular, or rarely fleshy and indehiscent. Seeds straight or rarely curved, without albumen. Embryo straight or rarely curved, the radicle next the hilum.—Herbs, shrubs, or rarely trees. Leaves opposite or rarely verticillate, simple and entire or with glandular teeth. Stipules none. Flowers terminal or rarely axillary, solitary or in cymes or panicles. Leafy parts often marked with glandular, pellucid, or black dots.

The Order is dispersed over the greater portion of the globe, although represented in Australia by only one or two species, and those not endemic. It is closely allied to *Guttiferæ* and *Ternstrœmiaceæ*, none of which last Order have as yet been discovered in Australia.

### 1. HYPERICUM, Linn.

Sepals 5. Petals 5, not woolly inside. Capsule opening septicidally. Seeds not winged. Embryo oblong or cylindrical, with short cotyledons.— Herbs or shrubs. Leaves either small or thin, entire, or rarely minutely toothed. Flowers yellow or rarely white.

A large genus with nearly the same extensive geographical range as the Order.

Erect or ascending.   Leaves usually subcordate . . . . . . . .   1. *H. gramineum.*
Procumbent.   Leaves usually oblong or obovate . . . . . . .   2. *H. japonicum.*

1. **H. gramineum,** *Forst.; DC. Prod.* i. 548. A glabrous perennial, with erect or ascending angular stems, usually about 1 ft. high, but sometimes nearly twice that height, or much shorter, slender, but rather rigid, branching at the base only or in the inflorescence. Leaves closely stem-clasping, ovate to oblong-lanceolate, obtuse, rarely exceeding ½ in., entire, with numerous pellucid dots, the margins more or less revolute. Flowers 3 or more, in the forks or terminating the branches of a dichotomous cyme, with a pair of leafy bracts at the base of each fork; the pedicels erect and rigid, ¼ to ½ in. long. Sepals lanceolate, acute, appressed, 2 to 3 or rarely 4 lines long. Petals entire, longer than the sepals. Stamens very variable in number, usually rather numerous and free. Styles 3, distinct. Capsule 1-celled, 3-valved, with narrow-linear placentas and numerous small seeds.— DC. Prod. i. 548; Labill. Sert. Austr. Caled. 53, t. 53; Hook. f. Fl. Tasm. i. 53; F. Muell. Pl. Vict. i. 193; *Ascyrum involutum,* Labill. Pl. Nov. Holl. ii. 32, t. 174; *Hypericum involutum,* Chois. in DC. Prod. i. 549; *H. pedicellare,* Endl. in Hueg. Enum. 12; *Brathys Billardieri* and *B. Forsteri,* Spach, in Ann. Sc. Nat. Ser. 2, v. 367.

**N. Australia.**   Gulf of Carpentaria, *R. Brown.*
**Queensland.**   Moreton Island, *F. Mueller.*
**N. S. Wales.**   Port Jackson, *R. Brown;* Blue Mountains, *A. Cunningham;* Hastings and Clarence rivers, *Beckler.*
**Victoria.**   Common in pasture lands as well as in barren localities throughout the colony, ascending to the Australian Alps, *F. Mueller.*
**Tasmania.**   Abundant everywhere in good soil, *J. D. Hooker.*
**W. Australia.**   Swan River, *Drummond;* Murchison river, *Oldfield.* The latter specimens remarkable for their elongated inflorescence, with the flowers mostly singly axillary along its branches.
The species in the original form, above described, is common also to New Zealand and New Caledonia. The S. African *H. Lalandii,* Chois., which has been referred to it, appears to me to differ in several respects.

2. **H. japonicum,** *Thunb. Fl. Jap.* 295, t. 31. Very nearly allied to *H. gramineum,* and considered by F. Mueller as a variety only. It is much less rigid and usually very procumbent or diffuse, with ascending branches, terete or scarcely angled. Leaves smaller, flatter, and more obtuse, not so broad at the base. Flowers smaller, on shorter pedicels, the sepals less acute and the petals very seldom exceeding them.— DC. Prod. i. 548; Hook. f. Fl. Tasm. i. 53; *Ascyron humifusum,* Labill. Pl. Nov. Holl. ii. 33, t. 175; *H. pusillum,* Chois. in DC. Prod. i. 549; *Brathys humifusa,* Spach, in Ann. Sc. Nat. ser. 2, v. 367.

**N. S. Wales.**   New England, *C. Stuart;* Hastings, Macleay, and Clarence rivers, *Beckler.*
**Tasmania.**   Abundant in hilly, humid situations throughout the island, *J. D. Hooker.*
**S. Australia.**   Torrens and Onkaparinga rivers, *F. Mueller.*
The species is widely spread over tropical and eastern Asia, extending from Japan to New Zealand.

# Order XX. **GUTTIFERÆ.**

Flowers regular, usually diœcious or polygamous. Sepals 2 to 6, or rarely more, much imbricate or in decussate pairs. Petals 2 to 6, rarely more, imbricate or contorted. Male fl.: Stamens usually indefinite, free or variously united; anthers adnate, innate, or sometimes immersed in the mass of filaments. Ovary none, or rudimentary, or more or less developed. Female or hermaphrodite fl.: Staminodia or stamens usually fewer and more free than in the males. Ovary 2- or more-celled, rarely 1-celled, with 1 or more ovules in each cell, erect from the base or attached to the central angle. Stigmas as many as cells, radiating or united into one, sessile or raised on a simple or rarely branched style. Fruit usually fleshy or coriaceous, indehiscent or opening septicidally in as many valves as cells. Seeds thick, often arillate, without albumen. Embryo filling the seed, often apparently homogeneous, consisting either of a fleshy radicle, with minute or without any cotyledons, or of thick fleshy cotyledons, with a very short, usually inferior radicle.—Trees or shrubs, exuding a yellow, resinous juice. Leaves opposite or rarely verticillate, thickly coriaceous and entire. Flowers terminal or axillary, solitary, clustered or in trichotomous cymes or panicles.

A tropical Order both in the New and in the Old World, represented in Australia by a single species, apparently identical with a common Asiatic one.

## 1. **CALOPHYLLUM,** Linn.

Flowers polygamous. Sepals and petals together, 4 to 12, imbricate in 2 or 3 series. Stamens indefinite, free or nearly so; filaments shortly filiform; anthers ovate or oblong, 2-celled, opening longitudinally. Ovary 1-celled, with a single erect ovule; style elongated, with a peltate stigma. Drupe indehiscent, with a crustaceous endocarp. Seed erect, ovoid or globular, the testa thin, or thick and hard, or spongy and then often adhering to the endocarp.—Trees, with the leaves marked with numerous closely parallel, transverse veins.

The genus is tropical, chiefly Asiatic, with a few American species.

1. **C. inophyllum,** *Linn.; W. and Arn. Prod.* i. 103. A glabrous tree. Leaves petiolate, broadly oblong or obovate-oblong, rounded at the apex, about 6 in. long in well-grown specimens. Racemes in the upper axils much shorter than the leaves, loose. Flowers large for the genus, on long pedicels, the buds nearly globular. Sepals 4, the 2 inner ones more petal-like than the outer ones. Petals 4, longer than the calyx. Stamens more or less united at the base into 4 (or more?) bundles. Fruit globular, the size of a plum.—Wight, Ic. t. 77; Planch. and Tri. in Ann. Sc. Nat. Par. ser. 4, xv. 282.

**Queensland.** Percy Islands, *A. Cunningham.* From the Burdekin Expedition, *Herb. Mueller.* The latter specimens consist only of some young seedlings in leaf only, and 2 fruits. These are about 1½ in. diameter, the thick, hard, almost corky testa of the seed adhering to the endocarp. Embryo nearly globular, apparently homogeneous, slightly conical at the end furthest from the hilum. That this is the radicular end is shown by the remains of the seed still attached to one of the seedling plants. Whether the position of the radicle turned away from the hilum is accidental in that one fruit, or general in the species or variety, cannot be determined without further fruiting specimens. Cunningham's are in flower only.

## ORDER XXI. MALVACEÆ.

Flowers regular, usually hermaphrodite or rarely partially diœcious or polygamous. Sepals 5, rarely 3 or 4, more or less united in a lobed or entire calyx, the lobes valvate or very rarely slightly imbricate. Petals 5, hypogynous, usually adnate at the base to the staminal column, contorted in the bud, rarely wanting. Stamens indefinite, hypogynous, more or less united at the base, the column divided into filaments at the top or bearing the filaments outside, below or up to the top. Anthers from globose to linear, often reniform or variously waved, 1-celled or spuriously divided into two cells by a thin and incomplete longitudinal septum. Torus small or conical and protruding into the centre of the ovary, not expanded into a disk. Ovary 2- or more-celled (very rarely reduced to a single carpel), entire or lobed, the carpels verticillate round the axis or (in genera not Australian) irregularly clustered. Style simple at the base, divided at the top into as many or twice as many branches or stigmas as there are cells, or rarely entire and clavate. Ovules 1 or more in each cell, ascending or horizontal, with a ventral or superior raphe, or reversed and pendulous, with the raphe dorsal. Fruit dry or rarely baccate, the carpels separating and indehiscent or 2-valved, or united in a loculicidally dehiscent capsule. Seeds with the testa usually crustaceous, without or with very little albumen ; cotyledons usually folded and often enclosing the curved or rarely straight radicle.—Herbs, shrubs, or soft-wooded trees, the hairs usually stellate. Leaves alternate, mostly toothed, lobed or divided, with palmate nerves or divisions, rarely digitately compound. Stipules free, usually subulate or small and deciduous, rarely leafy. Peduncles usually 1-flowered and articulate above the middle, rarely bearing a bract at the joint or several-flowered, all axillary or the upper ones forming a terminal raceme or panicle. Bracteoles either none or 3 or more, free or united, forming an involucre close to or adherent to the calyx. Flowers often large, usually purple, red, or yellow.

A large Order generally dispersed over all except the coldest regions of the globe, distinguished from *Sterculiaceæ* and *Tiliaceæ* by the 1-celled anthers, and from all others by the valvate calyx and monadelphous hypogynous stamens. Of the 15 following genera, 11 are more or less tropical, 6 being common to the warmer regions of both the New and the Old World ; 3, *Malvastrum*, *Pavonia*, and *Fugosia*, chiefly American, or American and African, but not Asiatic ; and 2, *Thespesia* and *Adansonia*, African and Asiatic. *Lavatera* is a Mediterranean form, represented by one species in extratropical Australia, the remaining three are endemic or nearly so, *Plagianthus* being also represented in New Zealand and *Lagunaria* in Norfolk Island.

TRIBE I. **Malveæ.**—*Staminal column bearing filaments to the summit. Style-branches the same number as ovary-cells. Mature carpels separating more or less from the axis (imperfectly so in* Howittia *and some* Abutila).

Ovules solitary in each cell, ascending with a ventral raphe.
    Style-branches lined with decurrent stigmas.
        Bracteoles 3 to 6, united at the base . . . . . . . . . . 1. LAVATERA.
        Bracteoles 3, distinct . . . . . . . . . . . . . MALVA (p. 186).
    Stigmas terminal, capitate or truncate. Bracteoles 1 to 3 distinct,
        or none . . . . . . . . . . . . . . . . . . 2. MALVASTRUM.
Ovules solitary in each cell, pendulous or horizontal with a dorsal raphe.
    Bracteoles none.
    Styles with decurrent stigmas. Flowers more or less diœcious . . 3. PLAGIANTHUS.

Stigmas terminal, capitate, or truncate . . . . . . . . . . 4. SIDA.
Ovules 2 or more in each cell. Bracteoles none. Stigmas terminal.
Capsule 2- or 3-celled, loculicidal, the carpels scarcely separating. . 5. HOWITTIA.
Capsule 5- to 20-celled, separating or cohering at least till the seed has
   shed . . . . . . . . . . . . . . . . . . . 6. ABUTILON.

TRIBE II. **Ureneæ.**—*Staminal column truncate or 5-toothed at the summit, bearing the anthers or filaments on the outside. Style-branches twice the number of carpels. Carpels 1-seeded.*

Bracteoles 5, united at the base. Carpels muricate or glochidiate . 7. URENA.
Bracteoles 5 or more, usually free. Carpels reticulate or smooth . 8. PAVONIA.

TRIBE III. **Hibisceæ.**—*Staminal column truncate or 5-toothed at the summit, bearing the anthers or filaments on the outside, or rarely at the summit also. Style-branches or stigmas the same number as ovary-cells. Carpels united in a several-celled capsule, loculicidal or indehiscent.*

Style branched at the top or with radiating stigmas. Ovary 5-celled.
  Bracteoles 5 or more, free or united (sometimes very deciduous).
    Hairs or tomentum stellate . . . . . . . . . . . 9. HIBISCUS.
  Bracteoles 3 (sometimes very deciduous). Tomentum of scurfy scales 10. LAGUNARIA.
Style undivided, with decurrent stigmas.
  Bracteoles 3 to 5, narrow, not cordate, sometimes very small.
    Ovary 3, 4- or rarely 5-celled. Capsule coriaceous, loculicidal . 11. FUGOSIA.
    Ovary 5-celled. Capsule woody, sometimes indehiscent. . . . 12. THESPESIA.
  Bracteoles 3, broad, cordate . . . . . . . . . . . 13. GOSSYPIUM.

TRIBE IV. **Bombaceæ.**—*Staminal column, in the Australian genera divided at the top into numerous filaments, in other genera the filaments or anthers variously arranged. Style undivided, or with very short stigmatic lobes as many as ovary-cells. Carpels united in a loculicidal or indehiscent capsule.—A large tropical tribe, difficult, to distinguish from arborescent Hibisceæ by a general character, although each genus has peculiarities not found among Hibisceæ.*

Calyx entire in the bud, afterwards 3- to 5-cleft, large, woody, filled
  with mealy pulp. Leaves digitate . . . . . . . . . . 14. ADANSONIA.
Calyx truncate in the bud, afterwards 3- to 5-cleft. Capsule 5-valved,
  densely woolly inside. Leaves digitate . . . . . . . . 15. BOMBAX.

### 1. LAVATERA, Linn.

Bracteoles united into a 3- to 6-cleft involucre. Calyx 5-lobed. Staminal column divided to the top into several filaments. Ovary-cells indefinite, 1-ovulate. Style-branches of the same number as cells, filiform, stigmatic along the inner side. Fruit-carpels in a depressed circle, indehiscent, verticillate round the torus or axis, which is usually prominent beyond them, either conical or variously dilated above them. Seed ascending.—Herbs, shrubs, or trees, tomentose or hirsute. Leaves angular or lobed. Flowers pedunculate, axillary or in a terminal raceme.

The greater number of species are from Western Europe or the Mediterranean region, one extending into central Asia; there are also two from the Canary Islands, besides the subjoined Australian species, which is endemic, but nearly allied to one of the European ones.

1. **L. plebeia,** *Sims, in Bot. Mag. t.* 2269. A coarse, erect herb, becoming woody at the base and attaining the height of 5 to 10 ft., more or less scabrous or softly tomentose with minute stellate hairs. Leaves on long petioles, orbicular-cordate, 5- or 7-lobed, the lower ones sometimes attaining

6 in. diameter, the upper ones 1 to 2 in.; the lobes short, broad, very obtuse and crenate, the central·one of the upper leaves often longer than the
others.  Stipules narrow-lanceolate or triangular.  Pedicels axillary, usually
clustered, rarely solitary, sometimes very short and rarely exceeding 1 in.
Involucre deeply 3-lobed, the lobes ovate, obtuse, shorter than the 5-lobed
calyx.  Petals pale rose-colour or whitish, 1 to 1½ in. long.  Carpels of the
fruit 6 to 15, in a close ring, with flat backs and sharp angles, the receptacle
protruding from the central depression as a small conical point.—DC. Prod.
i. 439; Hook. f. Fl. Tasm. i. 47; F. Muell. Pl. Vict. i. 166; *Malva Behriana*, Schlecht. Linnæa, xx. 633; *Lavatera Behriana*, Schlecht. l. c. xxiv. 699,
and xxvii. 527; *Malva Preissiana*, Miq. in Pl. Preiss. i. 238.

**N. S. Wales.**  In the interior, W. of Peel's range, *A. Cunningham;* Darling and
Lachlan rivers, *Victorian Expedition;* common towards the Barrier Range, *W. Wills;*
Paramatta, *Herb. Mueller.*

**Victoria.**  Along watercourses and in occasionally inundated depressions, scattered over
many parts of the colony, more frequent in the N.W. portion, *F. Mueller.*

**Tasmania.**  Near the sea at Woolnorth, and in the islands of Bass's Straits, *Gunn,
J. D. Hooker.*

**S. Australia.**  St. Vincent's Gulf, Spencer's Gulf, Lake Torrens, and the country on
the eastern side of the great Australian Bight, *F. Mueller.*

**W. Australia,** *Drummond, n.* 102; King George's Sound, *R. Brown, A. Cunningham.*

The species is allied to the European *L. arborea*, Linn., which is however at once known
by its large spreading involucres.

*L. hispida*, Desf., DC. Prod. i. 438, a hirsute species with nearly sessile flowers forming
a long terminal raceme or interrupted spike, and with broad hirsute involucres, a native
of the Mediterranean region, appears to be naturalized in some islands of Bass's Straits
(*F. Mueller*).

The genus *Malva*, now restricted to the species from the temperate regions of Europe and
Asia, is only distinguished from *Lavatera* by the 3 bracteoles being quite free, and the receptacle never expanded above the carpels.  Four common European species have become
naturalized as weeds in some of the colonies, viz. 1, *M. rotundifolia*, Linn., DC. Prod. i. 432,
with decumbent or prostrate stems, small flowers, petals not twice the length of the calyx,
and carpels usually about 15, rounded on the back so as to form a disk-shaped fruit slightly
furrowed on the margin between the carpels; 2, *M. parviflora*, Linn., DC. l. c. 433, like *M.
rotundifolia* in habit and small flowers, but the carpels flat on the back with angular edges,
so that the fruit has rather projecting ribs than furrows between the carpels; 3, *M. verticillata*, Linn., DC. l. c. 433, with erect stems, small flowers in close clusters, and the carpels of *M. parviflora;* and 4, *M. sylvestris*, Linn., DC. l. c. 432, with ascending or erect
stems, large flowers, the petals 3 or 4 times as long as the calyx, the carpels angular as in
*M. parviflora.*

## 2. MALVASTRUM, A. Gray.

Bracteoles either none or 1 to 3, small and distinct.  Calyx 5-lobed.  Staminal column divided to the top into several filaments.  Ovary-cells 5 or
more, 1-ovulate.  Style-branches of the same number as the cells, filiform or
club-shaped, with terminal small or capitate stigmas.  Fruit-carpels seceding
from the short axis, indehiscent or slightly 2-valved, occasionally produced at
the top into erect connivent beaks.  Seed ascending, reniform.—Herbs or
undershrubs.  Leaves entire or divided.  Flowers red or yellow, shortly pedunculate or sessile, axillary or in terminal spikes.

A considerable genus, chiefly American, with a few South African species.  The two Australian species are both American, but now scattered over some of the warmer regions of the

Old World. The genus, formerly confounded with *Malva* and *Sida*, is readily distinguished from the former by the styles, from the latter by the ascending ovules and seeds.

Tomentum stellate. Flowers mostly in a short terminal spike  .  .  1. *M. spicatum.*
Hairs appressed, parallel. Flowers mostly axillary. Calyx broad .  2. *M. tricuspidatum.*

1. **M. spicatum,** *A. Gray, Pl. Fendl.* 22, *and Bot. Amer. Expl. Exped.* i. 147. An erect branching herb of 1 to 2 ft., becoming almost woody at the base, scabrous or softly tomentose with stellate hairs. Leaves petiolate, ovate or ovate-lanceolate, acute or obtuse, 1 to 2 in. long, irregularly serrate or crenate, very rarely obscurely 3-lobed. Flowers rather small, yellow, sessile in a dense terminal spike, rarely exceeding 1 to 1½ in. in length, and often leafy at the base. Bracts narrow, shorter than the calyx, usually 2-lobed. Bracteoles 3, filiform, closely appressed to the calyx. Calyx softly pubescent, the lobes acuminate, and often bordered by long hairs. Petals about 4 to 5 lines long. Carpels 8 to 12, not close-pressed, angular on the edges, pubescent on the top, without points.—*Malva spicata,* Linn.; Cav. Diss. t. 20, f. 4; DC. Prod. i. 430; *M. ovata,* Cav. Diss. 81, t. 20, f. 2; *M. timoriensis,* DC. Prod. i. 430; *M. brachystachya,* F. Muell. in Linnæa, xxv. 378.

**N. Australia.** Victoria river and Gulf of Carpentaria, *F. Mueller.*
**Queensland.** Broad Sound and Keppel Bay, *R. Brown;* Brisbane river, *Fraser;* subtropical interior, *Mitchell;* Moreton Bay and Gilbert river, *F. Mueller.*
**N. S. Wales.** Clarence river, *Beckler;* New England, *C. Stuart;* Darling river and other parts of the W. interior, *Victorian Expedition, Dallachy,* etc.
**S. Australia.** Flinders range, *F. Mueller.*
The species is common in tropical America, and has been found also in the Cape de Verd Islands and in Timor.

2. **M. tricuspidatum,** *A. Gray, Pl. Wright, and Bot. Amer. Expl. Exped.* i. 148. An erect branching herb, 2 to 3 ft. high, hard and almost woody at the base, although sometimes annual, the branches sprinkled or covered with closely appressed hairs. Leaves on rather long petioles, from broadly ovate to lanceolate, 1 to 2 in. long, irregularly toothed, hairy. Flowers yellow, almost sessile in the axils of the leaves, or clustered towards the ends of the branches. Calyx broadly 5-lobed, with 3 small, narrow, external bracts. Carpels 8 to 12 or even more, closely packed in a depressed ring, each one reniform, with 3 minute unequal points on the upper edge, 1 at the inner angle, 2 dorsal.—*Malva tricuspidata,* Ait.; DC. Prod. i. 430; *Sida carpinoides,* DC. Prod. i. 460.

**N. S. Wales.** Clarence river, *Beckler.* This species, probably of American origin, is much more widely scattered over the warmer regions of the Old World than the *M. spicatum.*

### 3. PLAGIANTHUS, Forst.

(Asterotrichon *and* Blepharanthemum, *Klotzsch;* Lawrencia, *Hook.;* Halothamnus, *F. Muell.*)

Bracteoles none or distant from the calyx. Calyx 5-toothed or 5-lobed. Staminal column divided at the top into several filaments. Ovary-cells 2 to 5, rarely 1 or indefinite, 1-ovulate. Style-branches as many as cells, filiform or club-shaped, stigmatic along the inner side, either the whole length or near the top. Fruit-carpels 1, 2, or more, seceding from the axis, indehiscent or irregularly breaking up.    Seed pendulous, with a dorsal raphe.—Shrubs or rarely herbs.    Leaves entire or rarely lobed.    Flowers usually small and white, more or less completely diœcious, axillary or terminal, usually clustered, rarely solitary or in short panicles.

The genus is confined to Australia and New Zealand, the several species being in each case endemic.    It was formerly referred to *Sterculiaceæ*, from a mistaken view of the anthers. It is however nearly allied to *Sida*, with which F. Mueller proposes to unite the greater number of species, but the habit is different, the flowers, although generally provided both with stamens and pistils, are nevertheless almost constantly diœcious by abortion, which has not been observed in true *Sidas*, and the character derived from the style is one of the most constant in *Malvaceæ*.

SECT. 1. **Plagianthus.**—*Calyx campanulate, the angles not prominent.  Shrubs often tall.   Leaves herbaceous, rugose, serrate or crenate, glabrous or stellate-hairy.*

Carpels 2 or 3 (1 only usually ripening) . . . . . . . . . 1. *P. sidoides.*
Carpels usually 5   . . . . . . . . . . . . . . . 2. *P. pulchellus.*

SECT. 2. **Lawrencia** (*Wrenciala*, A. Gr.).—*Calyx with 5 prominent angles.   Herbs or tortuous shrubs.   Leaves thick or small, entire or toothed at the top, nearly glabrous or scurfy.*

Flowers in dense terminal spikes.    Erect herb, glabrous or slightly
    stellate-pubescent . . . . . . . . . . . . . . . 3. *P. spicatus.*
Flowers axillary, solitary or clustered, not spicate.    Herbs either glabrous or slightly stellate-pubescent.
    Leaves cuneate-oblong.    Flowers all sessile   . . . . . . 4. *P. glomeratus.*
    Leaves small, orbicular or obovate, on long petioles.    Male flowers
      pedicellate . . . . . . . . . . . . . . . . 5. *P. diffusus.*
Tortuous shrubs, the herbaceous parts covered with scurfy scales.
    Stem-leaves petiolate, often above 1 in. long   . . . . . 6. *P. squamatus.*
    Stem-leaves sessile or nearly so, rarely exceeding ½ in. and mostly
      not ¼ in. . . . . . . . . . . . . . . . . . . 7. *P. microphyllus.*

1. **P. sidoides,** *Hook. Bot. Mag. t.* 3396.    A shrub of several feet or sometimes a small tree, the young branches, under side of the leaves, and inflorescence more or less covered with a whitish or brown stellate tomentum, sometimes very dense and floccose.    Leaves from ovate-lanceolate to lanceolate, obtusely serrate, 2 to 3 or rarely 4 in. long, rounded at the base, on petioles of 3 to 6 lines, glabrous on the upper side when full grown, with impressed veins.    Flowers small, in short axillary racemes, the males with a broad campanulate calyx about 2 lines long ; stamens about 15, the tube obscurely divided at the top into 5 clusters ; pistil small and barren, although the ovary is 2-celled, with 1 pendulous ovule in each.    In the females the calyx is almost tubular, the petals scarcely longer and persistent, the anthers small and barren, the pistil fully developed, the ovary 2-celled, the style-branches hairy at the base, much dilated from the middle upwards.    Fruit-carpels usually 1 only, apparently indehiscent, enclosed in the membranous calyx ; when both ripen they appear to separate.—Hook. f. Fl. Tasm. i. 49 ; *Sida discolor*, Hook. Journ. Bot. i. 250 ; *Asterotrichon sidoides*, Klotzsch in Link, Kl. et Otto. Ic. Pl. Rar. 19, t. 8 ; *Plagianthus Lampenii*, Lindl. Bot. Reg. 1838, Misc. 22.

**Tasmania.**    Common in ravines, etc., in the southern part of the island, *R. Brown, J. D. Hooker,* and others.

The bark, full of strong fibre, is used in Tasmania as cordage.

2. **P. pulchellus,** *A. Gray, Bot. Amer. Expl. Exped.* i. 181. A tall shrub or small tree, either quite glabrous or the young branches and under side of the leaves slightly scabrous with scattered stellate hairs. Leaves on rather long petioles, from deeply cordate-ovate to lanceolate, often acuminate, 2 to 3 in. or rarely longer, coarsely crenate, mostly membranous, glabrous above. Flowers small, clustered along the rhachis of axillary racemes, longer or shorter than the petioles. Males pedicellate, with a broadly campanulate glabrous calyx of scarcely 1½ lines. Petals twice as long. Stamens near 30. Pistil small and barren, although bearing ovules. Female flowers sessile, with a small ovoid or almost globular calyx. Petals small and persistent. Anthers small and barren. Ovary 5-celled. Style-branches much thickened and stigmatic from about the middle. Fruit much longer than the calyx, slightly tomentose, somewhat depressed, deeply divided into 5 distinct cocci, which separate from the 5-angled axis and at length open in 2 short valves. —Hook. f. Fl. Tasm. i. 49 ; *Sida pulchella,* Bonpl. Jard. Malm. t. 2 ; DC. Prod. i. 468 (character incorrect) ; F. Muell. Pl. Vict. i. 161 ; *Abutilon pulchellum,* G. Don, Gen. Syst. i. 501 ; *Blepharanthemum,* Klotzsch, in Link, Kl. and Ott. Ic. Pl. Rar. i. 20.

**N. S. Wales.** Hawkesbury river, *R. Brown.* Cox's and Macquarie rivers, *A. Cunningham ;* Illawara and Argyle county, *Backhouse.*

**Victoria.** Yarra river, *F. Mueller ;* Fitzroy river, *Robertson.*

**Tasmania.** Port Dalrymple, *R. Brown ;* abundant near Launceston and on the North Esk river, *J. D. Hooker.*

Var. *tomentosus,* Hook. f. Fl. Tasm. i. 49. More tomentose, especially the under side of the leaves and calyces. Styles elongated and slender. Cocci very tomentose.—*Sida pulchella,* Bot. Mag. t. 2753 ; *S. tasmanica,* Hook. f. in Hook. Journ. Bot. ii. 412 ; *Plagianthus tasmanicus,* A. Gray, Bot. Amer. E pl. Exped. i. 181. Tambo and Buchan rivers in Victoria, *F. Mueller ;* common in Tasmania, *Gunn.*

3. **P. spicatus,** *Benth. in Journ. Linn. Soc.* vi. 103. A tall, erect, somewhat fleshy herb, drying of a yellowish colour, and glabrous or nearly so, with a thick, hard, almost woody base, and but little branched, attaining sometimes 5 ft. in height, but sometimes only 1 or 2 ft. Leaves on long petioles, from ovate to ovate-oblong or cuneate, rarely exceeding 1 in., irregularly toothed, 3- or 5-nerved, rather thick, the upper ones smaller and more sessile, passing into leafy bracts with the stipules adnate. Flowers sessile, 1 to 3 together in the upper leaves and bracts, forming a terminal leafy spike sometimes a foot long and very dense, usually shorter, with the lower flowers distant. Calyx 5-angled, about 3 lines long. Petals scarcely longer. Stamens usually under 20. Styles long and slender. Carpels 5, glabrous, not exceeding the calyx, very angular and reticulate, terminating in short connivent points.—*Lawrencia spicata,* Hook. Ic. Pl. t. 261 ; Hook. f. Fl. Tasm. i. 48 ; *Sida Lawrencia,* F. Muell. Pl. Vict. i. 162.

**Victoria.** Salt marshes, scattered along the seacoast, and subsaline places of the N. W. desert country, *F. Mueller.*

**Tasmania.** Flinders Island, Bass's Straits, *Gunn ;* Great Swan Port, *Backhouse.*

**S. Australia.** At various points along the coast, *R. Brown, F. Mueller,* and others.

**W. Australia.** Swan River, *Drummond, Coll.* 1845, *n.* 302 ; Sussex district, *Preiss, n.* 2381 ; Hamden, *Clarke ;* Port Gregory, *Oldfield.*

Var. *pubescens.* Sprinkled with loose stellate hairs, and more branching, with the spikes more interrupted at the base, but always close at the top.—N.W. interior of Victoria, and in S. Australia.

4. **P. glomeratus,** *Benth. in Journ. Linn. Soc.* vi. 103. A glabrous or slightly hoary, decumbent and much-branched herb, with ascending branches often above 1 ft. high. Leaves cuneate-oblong, toothed at the end, resembling those of *P. spicatus,* but usually narrower and more gradually narrowed into the petiole. Flowers all axillary, usually 3 together and sessile, forming distant clusters along the leafy branches and never collected into a spike, the ends of the branches all barren. Flowers nearly those of *P. spicatus,* but smaller, and the stamens and styles much shorter.—*Lawrencia glomerata,* Hook. Ic. Pl. t. 417.

**S. Australia.** S. coast, *R. Brown.*

**W. Australia.** Swan River, *Drummond ;* Port Gregory, *Oldfield* (a hoary variety). F. Mueller unites this with *P. spicatus,* but I see no tendency to the spicate inflorescence so characteristic of that species, besides the general differences in habit and foliage.

5. **P. diffusus,** *Benth.* Herbaceous, much-branched, diffuse or prostrate, sometimes not exceeding 2 or 3 in., sometimes nearly 1 ft. long, but much more slender than *P. glomeratus,* glabrous or sprinkled with a few stellate hairs. Leaves on long petioles, orbicular or obovate, rarely exceeding ½ in. in diameter, and often much smaller, coarsely crenate. Flowers axillary, 1 to 3 together, the males on pedicels of 3 to 4 lines, the females sessile. Calyx broadly campanulate, slightly angular, not 2 lines long. Petals in the males much longer, in the females small and persistent. Stamens 10 to 15, or fewer in the females. Styles of the females long and acute. Carpels 5, glabrous, not exceeding the calyx, ending in short connivent points, and not separating very readily.

**W. Australia.** Swan River, *Drummond, n.* 104, 137, and 246 (females), and *n.* 275, *5th Coll.* (males).

6. **P. squamatus,** *Benth. in Journ. Linn. Soc.* vi. 103. A rigid tortuous shrub, the leaves and other herbaceous parts densely covered with small peltate, scurfy scales, the young branches often simple and erect, 1 ft. long or more, the short ones rarely spinescent. Leaves oblong-linear, entire, the larger ones above 1 in. long and narrowed into a long petiole with small sessile ones clustered in their axil, the floral ones rarely exceeding ⅓ in. Flowers small, closely sessile in axillary clusters, not spicate. Calyx not 2 lines long, very scurfy, with obtuse lobes. Petals narrow, scarcely exceeding the calyx, and very small in the females. Carpels 3, 4, or 5, the styles protruding considerably beyond the calyx, the stigmatic part somewhat dilated and ending in a long point. Ripe fruit not seen, but only 1 or 2 carpels appear to enlarge.—*Lawrencia squamata,* Nees, in Pl. Preiss. i. 242.

**N. Australia.** Swan River, *Drummond, 4th Coll. n.* 106, *Preiss, n.* 1231.

7. **P. microphyllus,** *F. Muell. Fragm.* i. 29. Very closely allied to *P. squamatus,* and similarly covered with scurfy scales, but a lower, more tortuous, and more branched shrub, the smaller branches slender and often spinescent. Leaves from linear to oblong-cuneate, rarely exceeding ½ in. and usually much smaller, obtuse or 3-toothed at the end, more or less tapering at the base. Flowers small, sessile or nearly so, 1 to 3 together in

*Plagianthus.*]     XXI. MALVACEÆ.     191

the axils, not spicate.  Calyx when in flower not above 1½ line long.  Carpel usually single, enclosed in the calyx and membranous as in *P. sidoides.—Halothamnus microphyllus,* F. Muell. Pl. Vict. i. 159.

**Victoria.**  Sandy, especially subsaline inland localities or in the so-called salt-bush country, thence extending through many parts of the Murray desert, *F. Mueller.*
**S. Australia.**  In the littoral tracts, *F. Mueller;* bays and islands, S. coast, *R. Brown.*
**W. Australia,** *Drummond, Coll.* 1845, *n.* 208, and 4*th Coll. n.* 252.


## 4. SIDA, Linn.

Bracteoles none, or small and distant from the calyx.  Calyx 5-toothed or 5-lobed.  Staminal column divided at the top into several filaments.  Ovary-cells 5 or more, verticillate, 1-ovulate.  Style-branches as many as cells, filiform or slightly clavate, with terminal, capitate or truncate stigmas.  Fruit-carpels either obtuse or with connivent points, seceding from the axis, indehiscent or opening shortly at the top in 2 valves.  Seed pendulous or horizontal, with a dorsal raphe.—Herbs or shrubs, usually clothed with a soft or whitish stellate tomentum.  Stipules in all the Australian species except *S. Hookeriana,* subulate and deciduous.  Flowers sessile or pedunculate, axillary or in terminal heads, spikes, or racemes, of various colours and sometimes large, but most frequently rather small, yellow, or whitish.

The genus, even as now limited to the exclusion of the *Abutilons,* is large, and widely spread over the warmer regions of the globe, but most abundant in America.  Of the Australian species three are common tropical weeds, the remainder all endemic.

§ 1. *Calyx without prominent ribs or angles.  Carpels strongly reticulate on the sides* (except *S. pleiantha*), *indehiscent, or nearly so, never aristate.  Perennials or shrubs. Leaves undivided.*
Flowers 1 or 2 together, on slender pedicels, articulate near the top.
  Calyx-lobes obtuse, not protruding beyond the broad part of the fruit.
    Carpels strongly wrinkled on the back.  Fruit 2½ to 4 lines
      diameter . . . . . . . . . . . . . . . . . 1. *S. corrugata.*
    Carpels not, or very slightly wrinkled.  Fruit not exceeding 2
      lines diameter.  Leaves and flowers very small . . . . 2. *S. intricata.*
  Calyx-lobes acute or scarcely acuminate, remaining herbaceous, and
    not much enlarged after flowering.
    Leaves ovate or ovate-lanceolate, cordate at the base . . . . 3. *S. macropoda.*
    Leaves lanceolate or oblong-lanceolate, not cordate . . . . 4. *S. virgata.*
  Calyx-lobes acuminate, with long, subulate, woolly points . . 5. *S. cryphiopetala.*
  Calyx-lobes enlarged and thinner or scarious after flowering.
    Leaves lanceolate or oblong.  Carpels 6 to 8.
      Fruiting calyx about ½ in. diameter, slightly spreading; lobes
        narrow, ovate-lanceolate . . . . . . . . . . . 6. *S. petrophila.*
      Fruiting calyx ¾ in. diameter, very spreading; lobes broadly
        ovate, scarious . . . . . . . . . . . . . . 7. *S. calyxhymenia.*
    Leaves cordate-ovate or orbicular.  Carpels above 15.  Fruiting
      calyx 2 in. diameter . . . . . . . . . . . . . 8. *S. physocalyx.*
Flowers clustered, several together.  Pedicels short, not articulate.
  Flowers nearly sessile.  Tomentum dense, or rarely scanty.  Carpels
    reticulate on the side . . . . . . . . . . . . . 9. *S. subspicata.*
  Flowers pedicellate.  Tomentum thin or floccose.  Carpels not
    reticulate . . . . . . . . . . . . . . . . . 10. *S. pleiantha.*

§ 2. *Calyx* 5-*angled, prominently* 10-*ribbed. Carpels not reticulate on the sides, and opening in* 2 *short valves at the top. Herbs or undershrubs. Leaves undivided.*

Leaves ovate or narrow, whitish with a close tomentum on both sides.
Carpels 5 . . . . . . . . . . . . . . . . . . 11. *S. spinosa.*
Leaves ovate or narrow, whitish with a close tomentum underneath.
Carpels about 10 . . . . . . . . . . . . . . . 12. *S. rhombifolia.*
Leaves broad, cordate (or rarely narrow). Tomentum soft, loose, or
velvety. Carpels about 10 . . . . . . . . . . . . . 13. *S. cordifolia.*

§ 3. *Calyx with* 15 *or* 20 *nerves prominent when in fruit. Carpels numerous. Styles free to the base. Leaves undivided.*

Calyx enlarging little after flowering, open at the top . . . . 14. *S. platycalyx.*
Fruiting calyx very large, membranous, quite closed over the fruit . 15. *S. inclusa.*

§ 4. *Calyx* 10-*ribbed at the base, each lobe having also* 2 *intramarginal veins. Annual, with deeply-lobed leaves* . . . . . . . . . . . . . 16. *S. Hookeriana.*

1. **S. corrugata,** *Lindl. in Mitch. Three Exped.* ii. 13. Rootstock and often the base of the stem woody, the branches usually diffuse or procumbent and under 1 ft. long, or in some varieties elongated, slender, and divaricate, attaining fully 2 ft., more or less hoary as well as the leaves with stellate hairs or short pubescence. Leaves orbicular, ovate or lanceolate, crenate, mostly ½ to 1 in. long, cordate or obtuse at the base, on petioles shorter than the laminæ, and sometimes very short. Pedicels axillary, 1 to 3 together, filiform or slender, rarely as long as the leaves, articulate below the top. Calyx tomentose, 2 to 2½ lines long, the lobes broad and obtuse, spreading under the fruit. Petals yellow, about twice the length of the calyx. Stamens 10 to 15. Fruit depressed-globular, varying from 2½ to near 5 lines diameter, tomentose or nearly glabrous, the obtuse often-raised centre marked with radiating furrows formed by the grooved connivent summits of the carpels, the circumference deeply wrinkled. Carpels 6 to 10, indehiscent, strongly reticulate on the sides. Seeds glabrous or slightly tomentose.—F. Muell. Pl. Vict. i. 163.

**N. Australia.** Upper Victoria river and Sturt's Creek, *F. Mueller.*
**Queensland.** On the Maranoa, *Mitchell;* in the interior, *Leichhardt.*
**N. S. Wales.** Broadland on the Hawkesbury river, *R. Brown;* desert land of the interior from Peel's range and the Bogan to the S. Australian frontier, *A. Cunningham, Fraser, Mitchell* and others.
**Victoria.** Desert tracts, basaltic downs and ridges from Bacchus Marsh to the N.W. part of the colony, *F. Mueller.*
**S. Australia.** S. coast, *R. Brown;* Flinders range, *A. Cunningham;* and N.W. interior, *Sturt.*
**W. Australia.** Between Moore and Murchison rivers, *Drummond,* 5th *Coll. n.* 106; Dirk Hartog's Island, *A. Cunningham.*

This plant assumes forms apparently so distinct that it is difficult to believe that some of them ought not to be considered as species. In attempting, however, to fix their limits, so many intermediate specimens have presented themselves, that I feel compelled to follow F. Mueller in uniting them under one name. The following appear to be the most marked:—

*a, orbicularis.* Stems short, diffuse, and tomentose. Leaves orbicular or broadly ovate, deeply and coarsely crenate, cordate at the base. Flowers and fruits rather large. *S. corrugata,* Lindl. l. c.; *S. interstans* and *S. spodochroma,* F. Muell. in Linnæa, xxv. 383. Chiefly in Victoria and N. S. Wales.

*b, ovata.* Stems usually more slender and elongated. Leaves mostly cordate-ovate, with small and regular crenatures, often softly tomentose. Petioles often short, and sometimes very short. Flowers and fruits rather small. *S. fibulifera*, Lindl. in Mitch. Three Exped. ii. 45 ; *S. filiformis*, A. Cunn. in Mitch. Trop. Austr. 361.—N. Australia (including a var. with very short pedicels), Queensland, N. S. Wales, Victoria, and S. Australia. *S. pedunculata*, A. Cunn. ms., from Peel's range, is a remarkable form, densely tomentose, with the lower leaves 2 in. long, and the lower peduncles elongated, bearing a leafless raceme of several flowers, with rigid stipulary bracts ; the inflorescence in the upper part quite normal. *S. nematopoda*, F. Muell. in Linnæa, xxv. 382, has smaller and less wrinkled fruits, although still much more so than in *S. intricata*, and the foliage is quite that of the present variety.

*c. angustifolia.* Stems slender, often nearly glabrous as well as the leaves. Leaves cordate-lanceolate, deeply toothed. Flowers and fruits small. Extends over the whole range of the species, and the only form hitherto found in W. Australia.—*S. humillima*, F. Muell. in Trans. Phil. Soc. Vict. i. 12, is a small hoary form, with larger leaves, approaching sometimes the first variety. Some specimens of A. Cunningham's from Dirk Hartog's Island have the leaves more densely white-tomentose.

*d. trichopoda.* Like the last, but the lanceolate or oblong-linear leaves are never cordate at the base, and the slender pedicels mostly exceed the leaves.—*S. trichopoda*, F. Muell. in Linnæa, xxv. 384. On nearly the whole range of the species, excepting W. Australia.

*e. goniocarpa*, F. Muell. Foliage of the last var., but the fruit larger, the angles of each carpel bordered by vertical wings, forming on the fruit as many very prominent angles as there are carpels. Nangavera in N. S. Wales, *Victorian Expedition.*

**2. S. intricata,** *F. Muell. in Trans. Phil. Soc. Vict.* i. 19, *and in Hook. Kew Journ.* viii. 9. This form also is now reduced by F. Mueller (Pl. Vict. i. 163) to the *S. corrugata.* I am inclined however to keep it distinct, as the characters appear on the dried specimens to be tolerably constant. It is a small or slender, very much branched tomentose undershrub, resembling the var. *ovata* of *S. corrugata* in general characters, but with much smaller leaves and very much smaller flowers, on short slender pedicels, the fruits not above 2 lines diameter, consisting of 5 to 8 tomentose carpels, not furrowed at their points, and smooth or only very slightly wrinkled on the back.

**N. Australia.** Stony ridges of the Upper Victoria river, *F. Mueller.*
**N. S. Wales.** From Molle's Plains, *A. Cunningham*, to the Darling and Murray rivers, *F. Mueller.*
**S. Australia.** In the interior near Mount Hope, *F. Mueller.*
**W. Australia,** *Drummond, 5th Coll. n.* 105.

**3. S. macropoda,** *F. Muell. Herb.* An erect, branching shrub, densely clothed with a stellate tomentum, thick and often yellowish on the branches, almost velvety on the leaves. Leaves ovate-cordate, obtuse, 1 to 2 in. long, crenate, thick and soft, deeply wrinkled above, prominently veined underneath. Pedicels filiform, sometimes exceeding the leaves. Calyx-lobes acuminate or acute, closed over the fruit or spreading. Petals yellow, only shortly exceeding the calyx. Fruit 3 or 4 lines diameter, with the radiating striæ in the centre and the carpels wrinkled on the back as in *S. corrugata*, from which this species differs in stature, foliage, and the acute calyx-lobes.

**N. Australia.** Summits of Sea range, head of Hooker's Creek, Arnhem's Land and Gulf of Carpentaria, *F. Mueller.* A specimen of *Leichhardt's*, from the Brigalow scrub on Bokhara Creek, appears to be the same species.
Var. (?) *cardiophylla*, F. Muell. Tomentum more dense, but closer ; leaves shorter, and nearly orbicular ; pedicels shorter.—Sturt's Creek, *F. Mueller.* This may possibly be a distinct species, but the specimens are not sufficiently advanced to determine. In other specimens in young bud only, these buds are sessile or nearly so ; the pedicel probably grows out rapidly before the flower expands, and may sometimes remain very short.

**4. S. virgata,** *Hook. in Mitch. Trop. Austr.* 361. This resembles at first sight, especially in the leaves, the *S. calyxhymenia,* and in some respects some narrow-leaved forms of *S. corrugata;* but the calyx does not enlarge as in the former, and its lobes are not obtuse as in the latter, and the stellate tomentum is dense and soft, almost woolly, and often fulvous. It appears to be an erect shrub, with long twiggy branches. Leaves shortly petiolate, lanceolate or oblong-linear, often exceeding 1 in., obtuse at the base, denticulate, less tomentose above than underneath. Pedicels slender, but rarely as long as the leaves. Calyx very tomentose, not prominently ribbed, the acute lobes about as long as the cup. Petals yellow, twice as long as the calyx, varying from 3 to 4 lines. Fruit about 3 lines diameter, depressed, with the centre slightly projecting. Carpels 6 to 8 or rarely more, their radiating summits scarcely furrowed, wrinkled on the back, strongly reticulate on the sides.

**N. Australia.** Sandstone table-land of the Upper Victoria river, *F. Mueller.*
**Queensland.** On the Maranoa, *Mitchell.*
**S. Australia.** In the interior at Depot Creek, *F. Mueller.*

Var. *phæotricha.* Stellate hairs very fulvous, almost woolly; carpels very tomentose, less wrinkled, the centre of the fruit more prominent.—*S. phæotricha,* F. Muell. in Linnæa, xxv. 382. In the interior of S. Australia.

**5. S. cryphiopetala,** *F. Muell. Fragm.* iii. 4. A shrub, nearly allied to *S. virgata,* but the tomentum longer and denser, almost woolly or floccose. Leaves ovate-lanceolate or cordate, often 2 in. long. Calyx densely woolly-hirsute, the lobes attaining 3 or 4 lines, including their long soft hirsute filiform points, exceeding the petals in the specimens seen. Carpels 5 or more, wrinkled on the back, reticulate on the sides, their summits forming a strongly projecting centre to the fruit.

**N. Australia.** Brindley's Bluff, Macdonnell ranges, *M'Douall Stuart (Herb. F. Muell.)*

**6. S. petrophila,** *F. Muell. in Linnæa,* xxv. 381. A hoary tomentose erect shrub of 2 to 4 ft., with the habit, foliage, and inflorescence of *S. calyxhymenia,* but the flowers are not nearly so broad, the unexpanded bud rather ovoid than depressed-globular, the petals longer than the calyx, and the fruiting calyx not nearly so much enlarged, the ovate-lanceolate lobes not exceeding 3 lines in length, not half so broad as in *S. calyxhymenia,* and of a much thicker consistence. Fruit depressed, tomentose, wrinkled on the circumference and furrowed between the carpels as in *S. calyxhymenia,* but the carpels are usually about 7.

**N. S. Wales.** Mount Caley, *A. Cunningham;* Peel's range, *Fraser;* Toguya hills, Darling river, *Victorian Expedition.*
**S. Australia.** Flinders range, and towards Lake Torrens, *A. Cunningham, F. Mueller;* between Stokes range and Cooper's Creek, *Wheeler;* towards Spencer's Gulf, *Warburton.*

**7. S. calyxhymenia,** *J. Gay, in DC. Prod.* i. 462. An erect shrub, hoary all over with a stellate tomentum much closer than in *S. virgata,* which this species generally resembles in habit and foliage. Leaves shortly petiolate, lanceolate or oblong-linear, or the lower ones ovate-lanceolate, mostly 1 to 1½ in. long, slightly toothed, obtuse at the base. Pedicels 1 to 3 together, mostly shorter than the leaves. Calyx tomentose, not prominently ribbed, at first campanulate as in *S. virgata*, but with the lobes more obtuse and very soon enlarging; when in fruit very spreading, fully ¾ in. diameter, the broadly ovate lobes thin and transparent. Petals yellow, rather longer than the calyx before it enlarges. Stamens 10 to 15. Fruit nearly globular, with a raised conical centre, the circumference wrinkled and grooved between the carpels. Carpels 5, reticulate on the sides.—*Fleischeria pubens*, Steud. in Pl. Preiss. i. 237; Steetz, l. c. ii. 366.

**S. Australia.** A specimen in Herb. Muell. from Margaritte river, *Babbage's Expedition*, appears to belong to this species, but the calyx is not yet sufficiently advanced to determine it absolutely.

**W. Australia.** Swan River, *Drummond;* shady rocks of Mount Mathilde, *Preiss, n.* 1662; Murchison river, *Oldfield*.

8. **S. physocalyx,** *F. Muell. Fragm.* iii. 3. A shrub, densely clothed with a soft, woolly, almost floccose tomentum. Leaves petiolate, cordate-ovate or orbicular, very obtuse, 1 to 2 in. long, crenate, thick and soft. Stipules remarkably long and filiform. Flowers not seen. Fruiting calyx pedunculate in the upper axils, very much enlarged, thin, scarious, and reticulate, broadly 5-lobed, the angles very prominent, so as to give the sides a cordate form, expanding to 2 in. diameter. Carpels numerous (above 15), glabrous, tuberculate or almost muricate, forming a depressed disk-like fruit of about 5 lines diameter.

**N. Australia.** Hammersley range, N.W. coast, *F. Gregory's Expedition*.

9. **S. subspicata,** *F. Muell. Herb.* An erect shrub, sparingly tomentose and green, or densely tomentose like *S. virgata* and *S. macropoda*, but at once known by the inflorescence. Leaves from cordate-ovate to lanceolate, 1 to 2 in. long, obtuse, crenate, cordate or rounded at the base, slightly wrinkled above, with the veins prominent underneath, scabrous, velvety or densely tomentose. Flowers small, nearly sessile, clustered or rarely solitary, the upper clusters forming often an irregular terminal spike, with few small floral leaves. Calyx not ribbed, the lobes acute, at least as long as the tube and closing over the fruit, but not covering it. Petals nearly twice as long. Stamens often under 10. Fruit nearly globular, but grooved between the carpels; carpels 5 or 6, tomentose, reticulate on the side, but not wrinkled on the back, and not acuminate.

**N. Australia.** Gulf of Carpentaria, *R. Brown;* Hooker and Sturt's Creeks, *F. Mueller.*

**Queensland.** Keppel Bay, *R. Brown;* N.E. coast, *A. Cunningham;* Brisbane river, *Fraser, F. Mueller;* Burnett and Dawson rivers, *F. Mueller;* Rockhampton, *Thozet.*

**N. S. Wales.** Kirkton, Upper Hunter river, *Backhouse;* Clarence river, *Beckler.*

10. **S. pleiantha,** *F. Muell. Herb.* A shrub or undershrub, with elongated branches, green or hoary with a loose stellate tomentum, sometimes floccose. Leaves petiolate, the smaller ones nearly orbicular, ½ in. long, the larger ones ovate or ovate-lanceolate, 1 to 2 in., toothed, rounded or scarcely cordate at the base. Flowers small, clustered several together, the pedicels 2 to 4 lines long, not articulate. Calyx broadly campanulate, when in flower about 1½ lines long, with ovate-acute tomentose lobes, somewhat enlarged when in fruit, the lobes broad, herbaceous, glabrous, and connivent over the

fruit, with projecting undulate sinuses. Stamens often not more than 10. Fruit depressed-orbicular, about 3 lines diameter, nearly glabrous, not wrinkled, but strongly grooved between the carpels. Carpels 7 to 10, not reticulate on the sides.

**Queensland.** Peak Downs, *F. Mueller.*

11. **S. spinosa,** *Linn.; DC. Prod.* i. 460. An annual or sometimes perennial, and woody at the base, with the habit and inflorescence of the narrow-leaved forms of *S. rhombifolia,* but the whole plant, including both sides of the leaves, whitish with a minute tomentum, which is soft and·more dense on the calyx. Leaves from ovate to lanceolate. Carpels almost always 5 only, more erect and less readily detached than in *S. rhombifolia,* often slightly reticulate, awnless or with short awns.—A. Gray, Gen. Ill. t. 123.

**N. Australia.** N. coast, *R. Brown;* Upper Victoria river, *F. Mueller;* Quail Island, *Flood.* The species is not uncommon in tropical Asia, more rare in America. It derives its name from the stipules in falling off often leaving a prominent tubercular base, more distinct in this than in any other species, although the character is even here not constant.

12. **S. rhombifolia,** *Linn.; DC. Prod.* i. 462. A perennial or undershrub, very variable in stature, sometimes tall and erect with the larger leaves ovate and 3 in. long, the Australian specimens more generally representing the more spreading forms, with rigid virgate minutely tomentose branches, and small narrow leaves, rarely exceeding 1 in., varying from ovate-lanceolate to narrow-lanceolate, or from nearly obovate to oblong-cuneate, always shortly petiolate, toothed, nearly glabrous above and more or less whitened underneath with a short tomentum. Pedicels mostly longer than the petiole and sometimes as long as the leaf, articulate about the middle. Flowers rather small, yellow. Calyx broad, glabrous or slightly hoary, prominently 10-ribbed at the base. Carpels about 10, with or without terminal erect-connivent awns, angled at the back, neither wrinkled nor reticulate, opening at the top in two very short valves.

**N. Australia.** Port Essington, *Armstrong.*
**Queensland.** Brisbane river, *F. Mueller.*
**N. S. Wales.** Blue Mountains, *Miss Atkinson;* Paramatta, introduced from the Mauritius, and now a troublesome weed, *C. Moore.* The species is one of the commonest tropical weeds, both in the New and the Old World, and includes *S. retusa,* Linn., *S. rhomboidea,* Roxb., *S. philippica,* and *S. compressa,* DC., and several other published forms.
Var. (?) *incana.* Leaves whitish on both sides as in *S. spinosa,* but carpels about 10, with long awns.—Nicholson river, *F. Mueller;* Comet river, *Leichhardt;* the specimens not complete.

13. **S. cordifolia,** *Linn.; DC. Prod.* i. 464. A rather coarse, branching, erect or rarely decumbent herb or undershrub, more or less clothed with a soft stellate tomentum or velvety hairs, the branches often also hirsute with spreading hairs. Leaves on rather long petioles, broadly cordate or almost orbicular or rarely ovate-lanceolate, 1 to 1½ or rarely 2 in. long, usually soft and thick. Flowers small, yellow, on short axillary pedicels or clustered into short leafy racemes. Calyx 10-ribbed at the base, softly tomentose. Carpels about 10 or sometimes fewer, smooth or slightly wrinkled, opening at the top in 2 valves, and in the usual form terminating in rather long erect-connivent awns.

**N. Australia.**   Port Essington, *Armstrong ;*  N. coast, *Bynoe.*
**Queensland.**   Peak Downs, *F. Mueller.*
The species is very abundant in almost all tropical countries, and includes *S. althæifolia,*
Lam., and several other supposed species.
    Var. (?) *mutica.*   Carpels without the awns which generally distinguish the species.
The leaves are very soft and velvety, but small and narrow, the specimens have, however,
lost those of the primary branches.—Macarthur river, Gulf of Carpentaria, *F. Mueller.*

14. **S. platycalyx,** *F. Muell. Herb.*   Shrubby and densely clothed with
a soft floccose or velvety stellate tomentum.   Leaves ovate-cordate or nearly
orbicular, obtuse, crenate, 1 in. long or more, soft and thick.   Pedicels as
long as the leaves, soft, articulate above the middle.   Calyx broadly campanulate, about 5- lines long, with a broadly obtuse base, the lobes erect or
spreading, shorter than the tube, densely tomentose outside, each sepal
marked with 3 prominent ribs, with another almost equally prominent at the
junction of the sepals.   Petals broad, shorter than the calyx.   Stamens very
numerous, the staminal tube almost truncate at the top.   Carpels about 24,
closely packed in a tomentose ring round the base of the styles, which are
free almost to the base with small capitate stigmas.   Fruit not seen.
    **N. Australia.**   Sturt's Creek, *F. Mueller.*

15. **S. inclusa,** *Benth.*   A shrub, densely velvety tomentose or almost
floccose.   Leaves ovate or orbicular, often cordate, obtuse, crenate, mostly
above 1 in. long.   Flowers not seen.   Fruiting calyx on peduncles of about
1 in., membranous and inflated, above 1 in. diameter, tomentose, marked
with numerous longitudinal veins or ribs, the short lobes connivent, so as
completely to enclose the fruit.   Carpels numerous, stellate-hirsute, echinate
with rather soft hirsute spines, forming a depressed orbicular fruit of nearly
1 in. diameter.
    **N. Australia.**   Hammersley range, N.W. coast, *F. Gregory's Expedition.*   This species
and *S. platycalyx* are distinguished in the genus by their many-ribbed calyx ; as the one is
only known in fruit, and the other in flower, or scarcely past, the distinction between the
two cannot be established with certainty, but *S. platycalyx* certainly shows no tendency to
the singular enlargement of the calyx of *S. inclusa.*

16. **S. Hookeriana,** *Miq. in Pl. Preiss.* i. 242.   An erect or decumbent annual, 1 or rarely 2 ft. high, glabrous or with a few small scattered
hairs.   Stipules narrow-lanceolate.   Leaves on long petioles, nearly orbicular
in circumscription, but deeply divided into 3 or 5 ovate or cuneate deeply
toothed lobes.   Flowers small, white, usually 2 together, one on a long pedicel articulate near the top, the other nearly sessile.   Calyx 5-ribbed, glabrous or nearly so, campanulate when in flower and about 2½ lines long ; when
in fruit broadly spreading, as in *Anoda,* about ½ in. diameter, with broadly
ovate lobes, the ribs on reaching the sinus dividing into intramarginal veins
along each lobe.   Petals about as long as the calyx.   Staminal tube slender.
Fruit depressed-orbicular, about 3 lines diameter, the centre not prominent,
glabrous and smooth.   Carpels about 10, not awned, with very thin sides,
leaving, when they fall, their dorsal filiform nerves attached to the column.—
*S. leiophloia,* Miq. in Pl. Preiss. i. 241.
    **W. Australia.**   King George's Sound, *R. Brown ;* Swan River, *Drummond ;* Rottenest
Island and Wellington district, *Preiss, n.* 1894 and 1896 ; Blackwoood and Vasse rivers,
*Oldfield.*

*S. rupestris,* Miq. l. c. 241, which I have not seen, appears from the description to be the same species with the young parts pubescent.    *S. Hookeriana* is perhaps nearer allied in appearance to *Modiola caroliniana* than to *Sida triloba,* Cav., but differs from both in the structure of the fruit; *S. triloba* is moreover a perennial, with differently-shaped leaves and a dissimilar venation of the calyx.

## 5. HOWITTIA, F. Muell.

Bracteoles none.    Calyx 5-lobed.    Staminal column divided at the top into several filaments.    Ovary-cells 3, rarely 4, with 2 collateral ovules in each.    Style elongated with as many exceedingly short branches as cells and large capitate stigmas.    Capsule depressed-globular, opening loculicidally in 3 valves bearing the dissepiments in their centre, rarely splitting also septicidally.    Seeds ascending, reniform.    Embryo involute with deeply 3-fid cotyledons.—Shrub, with the habit of a *Sida*.

The genus is limited to a single endemic species.

1. **H. trilocularis,** *F. Muell. in Hook. Kew Journ.* viii. 9, *and Pl. Vict.* i. 167. *t.* 4.    A tall, erect, sarmentose shrub, attaining sometimes 20 ft., but often much smaller, clothed with a rough stellate tomentum like that of some *Lasiopetala.*    Leaves shortly petiolate, mostly ovate-lanceolate, obtuse, 1 to 2 in. long, rounded or slightly cordate at the base, the margins recurved,. entire or slightly toothed, green, scabrous, and with impressed veins above, white or yellowish, with a denser tomentum underneath ; in luxuriant shoots they are much larger, ovate-cordate or ovate-lanceolate, and coarsely toothed.    Stipules minute and deciduous.    Pedicels axillary, shorter than the leaves.    Calyx 3 to 4 lines long, tomentose.    Petals twice as long, purple or rarely white.    Staminal column very short.    Style often apparently simple to the stigmas.    Capsule hirsute, shorter than the calyx.    Seeds glabrous.

**N. S. Wales.**    Blue Mountains, *R. Brown, A. Cunningham ;* Valley of the Grose, *Miss Atkinson ;* Wonboyn river, and near Twofold Bay, *F. Mueller.*

**Victoria.**    Coast-ridges of Gipps' Land. *F. Mueller ;* Victoria ranges, *Wilhelmi ;* Mount Arapiles, *Dallachy ;* Tattiara country, *Woods.*

## 6. ABUTILON, Gærtn.

Bracteoles none.    Calyx 5-lobed.    Staminal column divided at the top into several filaments.    Ovary-cells 5 or more, verticillate, each with 3 or more, rarely 2, ovules.    Style-branches as many as cells, filiform or club-shaped, with terminal stigmas.    Fruit-carpels united at the base or entirely seceding, rounded or angular or with diverging points (not connivent) at the top, opening in 2 valves, without internal appendages.    Seeds nearly reniform, the upper ones usually ascending, the lower ones pendulous or horizontal.—Herbs or shrubs, rarely trees, usually clothed with a soft stellate tomentum.    Leaves usually cordate, angular or lobed, rarely narrow ; petioles usually long (except in *A. crispum*).    Stipules in all the Australian species subulate and deciduous.    Flowers in the Australian species axillary, yellow or rarely white, the pedicels articulate above the middle or near the top.

A large genus, distributed over the tropical and warm regions of the globe, chiefly

American. Of the 18 Australian species, three are widely distributed over tropical Asia and Africa; one, *A. Avicennæ*, is Mediterranean and Asiatic, but scarcely tropical; one, *A. auritum*, extends only to the Indian Archipelago; one, *A. crispum*, is common to both the New and the Old World, and the remaining 12 are endemic. The genus has frequently been united with *Sida*, but the characters derived from the diverging carpels with more than 1 ovule in each, as contrasted with the converging uniovulate carpels of *Sida*, are too constant and convenient to be neglected, in groups so very numerous in species. The differential characters given to several of the following species from the tropical regions, or from the deserts of the interior, are as yet very unsatisfactory, owing to the imperfect state of many of the specimens, often mere fragments.

§ 1. *Capsule truncate or concave at the top. Carpels (usually 2- or 3-seeded) angular-pointed or awned at the upper outer edge, persistent, or rarely at length deciduous leaving the filiform placenta attached to the axis.*

Carpels (usually 10 or fewer) not exceeding the calyx-lobes, the points erect, or rarely divergent. Stems usually (perhaps always) shrubby.
Calyx-lobes shorter than the tube.
Petals adnate high up the glabrous staminal tube. Calyx tubular, 1 in. long . . . . . . . . . . . . . . 1. *A. tubulosum.*
Petals shortly adnate to the pubescent base of the staminal tube.
Calyx 1 in. long, campanulate, lobes acute, nearly as long as the tube. Petals twice as long . . . . . . . 2. *A. amplum.*
Calyx ½ to ¾ in., lobes acuminate or rather obtuse, spreading, much shorter than the tube.
Petals above 1 in. long . . . . . . . . . . 3. *A. leucopetalum.*
Petals shortly exceeding the calyx . . . . . . . . 4. *A. Mitchelli.*
Calyx about ½ in., rather inflated, truncate, sinuate, or with very short obtuse lobes.
Petals very small. Staminal column much longer than the calyx . . . . . . . . . . . . . . . . . . 5. *A. micropetalum.*
Petals very small or shortly exceeding the calyx, the staminal column not long . . . . . . . . 6. *A. cryptopetalum.*
Petals twice as long as the calyx. Leaves deeply lobed . 7. *A. geranioides.*
(The last 3 species with more slender branches and a closer hoary tomentum than *A. micropetalum.*)
Calyx-lobes longer than the tube or cup, acuminate.
Calyx-lobes very concave and prominently keeled. Carpels about 10, scarcely acuminate . . . . . . . . . . 8. *A. otocarpum.*
Calyx-ribs or angles scarcely prominent. Carpels 4 or 5, acuminate . . . . . . . . . . . . . . . . . 9. *A. subviscosum.*
Carpels usually exceeding the calyx-lobes, the points often divergent. Herbs usually tall, sometimes hard, almost woody at the base. Stems coarse and erect. Leaves broadly cordate.
Capsule truncate. Carpels numerous, the points very short. Tomentum close and dense, usually without spreading hairs.
Stipules small and subulate. Flowers mostly axillary . . 10. *A. indicum.*
Stipules broadly semisagittate. Flowers in terminal leafless racemes or panicles . . . . . . . . . . . . . 11. *A. auritum.*
Capsule truncate. Carpels about 10, with long divergent points. Pubescent or loosely tomentose . . . . . . . 12. *A. Avicennæ.*
Capsule contracted and angular at the top. Carpels numerous, without points. Tomentum dense, mixed with long spreading hairs . . . . . . . . . . . . . . . . . 13. *A. graveolens.*
Stems rather slender. Leaves ovate or cordate-lanceolate. Capsule truncate, with short divergent points . . . . . . . 14. *A. oxycarpum.*

§ 2. *Carpels (often 1-seeded by abortion) rounded or angled at the top, quite distinct, and seceding from the axis when fully ripe* (Gayoides, *Endl.*)

Carpels numerous (about 20), closely packed, very hirsute. Tall
　herbs, with large, broadly cordate leaves.

Carpels angular at the top, leaving persistent filiform placentas ． 13. *A. graveolens.*

Carpels rounded at the top, completely deciduous ． ． ． ． ． 15. *A. muticum.*

Carpels rarely more than 10, glabrous or slightly tomentose, not
　scarious. Leaves mostly cordate-orbicular.

Densely velvety-tomentose (shrubby?). Petals shortly exceeding
　the calyx ． ． ． ． ． ． ． ． ． ． ． ． ． ． ． ． ． 16. *A. Cunninghami.*

Low undershrub, shortly tomentose or pubescent, often with
　spreading hairs. Petals fully twice as long as the calyx ． ． 17. *A. Fraseri.*

Carpels 10 to 15, slightly hispid, enlarged and scarious when ripe.

Slender undershrub, with cordate, often almost sessile leaves ． ． 18. *A. crispum.*

Distinct as the two sections are in some instances, they are closely connected by *A. graveolens*, and some other intermediate species.

1. **A. tubulosum,** *Hook. ; Walp. Ann.* ii. 158. Tall and shrubby, clothed with a dense, soft, close, or velvety tomentum. Leaves deeply cordate, ovate or lanceolate, almost acuminate, crenate, attaining 3 to 4 in., very soft and velvety. Pedicels much shorter than the leaves. Buds acuminate, prominent-angled. Calyx tubular, about 1 in. long, with 10 slightly prominent ribs, softly tomentose, the lobes acuminate, much shorter than the tube. Petals (yellow?) nearly ¾ in. longer than the calyx, the claws adhering to nearly the middle of the glabrous staminal column. Capsule angular, about half the length of the calyx, softly villous ; carpels 7 to 10, strongly acuminate on their outer edge, containing each usually 3 seeds.—*Sida tubulosa,* A. Cunn. ; Hook. in Mitch. Trop. Austr. 390.

**Queensland.** Open woods on the Mooni river, *Mitchell ;* Dawson river, *F. Mueller.*
**N. S. Wales.** Rocky whinstone hills on Liverpool plains, *A. Cunningham.*

Var. (?) *breviflorum.* Petals shorter and broader, but glabrous and more adnate than in *A. leucopetalum ;* the specimen, however, scarcely sufficient for accurate determination.—Dawson river, *F. Mueller.*

2. **A. amplum,** *Benth.* Tall and shrubby, the foliage and inflorescence softly tomentose-hirsute, not so white as in the allied species, and apparently somewhat viscid. Leaves deeply cordate, ovate, acuminate, crenate, 2 to 4 in. long, soft but green. Pedicels shorter than the leaves. Buds acuminate, prominently angled. Calyx, when open, broadly tubular-campanulate, about 1 in. long, tomentose-hirsute, with 10 slightly prominent ribs, the lobes broadly lanceolate, nearly as long as the tube. Petals (yellow?) often twice as long as the calyx, much broader than in *A. tubulosum,* the claws adhering to the lower part only of the staminal column, and there very pubescent. Capsule angular, softly villous, about half the length of the calyx ; carpels about 5, scarcely acuminate.

**N. Australia.** Harding river, S.E. of Nichol Bay, *F. Gregory's Expedition.*—F. Mueller is disposed to consider this as a variety of *A. tubulosum,* but the shape of the petals and their pubescent base are more those of *A. leucopetalum,* and the calyx is different from both. Further and more complete specimens may, however, considerably modify the circumscription of *A. tubulosum, amplum, leucopetalum,* and *Mitchelli,* which are all nearly allied to each other.

3. **A. leucopetalum,** *F. Muell. Herb.* A tall shrub, clothed with a soft velvety tomentum like *A. tubulosum,* but intermixed with long spreading

hairs on the branches, and paler on the under side of the leaves. Leaves deeply cordate, from orbicular to nearly lanceolate, often shortly acuminate, irregularly crenate or almost lobed, mostly shorter than in *A. tubulosum.* Flowers large and white, on short pedicels. Calyx broadly tubular-campanulate, $\frac{1}{2}$ to $\frac{3}{4}$ in. long, 10-ribbed, scarcely acuminate in the bud, the lobes obtuse or shortly acuminate, shorter than the tube. Petals more than twice as long as the calyx, adnate only to the pubescent base of the staminal tube. Capsule as in *A. tubulosum*, but fully as long as the calyx-tube.—*Sida leucopetala*, F. Muell. Fragm. ii. 12.

**N. Australia.** Hooker's Creek and Upper Victoria river, *F. Mueller.*
**N. S. Wales.** Barrier range, *Victorian Expedition.*
**S. Australia.** Cooper's Creek, *Herb. Mueller.*

**4. A. Mitchelli,** *Benth.* Apparently shrubby, clothed with a dense, soft, velvety tomentum mixed with long spreading hairs. Leaves deeply cordate, orbicular or broadly ovate, often shortly acuminate, $1\frac{1}{2}$ to $2\frac{1}{2}$ in. long, crenate, very soft and thick. Pedicels shorter than the petioles. Calyx campanulate, 10-ribbed and somewhat 5-angled, 4 to 5 lines long, the acuminate spreading lobes shorter than the tube. Petals (yellow?) shortly exceeding the calyx, pubescent at the base. Ovary-cells and style-branches about 10. Fruit not seen.

**Queensland.** Gullies in the ranges on the Maranoa, *Mitchell.* The plant has at first sight the aspect of *A. muticum*, but the calyx and ovary are quite different.
Var. (?) *mollissima.* Tomentum very dense and soft, but without the long hairs of the other specimens. Stony Ridge, *Mitchell.*
*Abutilæa cryptantha*, F. Muell. in Linnæa, xxv. 379, from a specimen without flower from Cudnaka, S. Australia, *F. Mueller* in Herb. Sonder, and from the description given, appears to be a form of *A. Mitchelli*, with semiabortive petals.

**5. A. micropetalum,** *Benth.* Shrubby, very densely and softly tomentose or velvety. Leaves deeply cordate, acuminate, 2 to 4 in. long, crenate. Pedicels short, in the upper axils. Calyx loosely campanulate, almost inflated, very shortly sinuate-toothed or almost truncate, 4 to 5 lines long, tomentose, slightly 5-angled and 10-ribbed. Petals, in some flowers at least, very small. Stamens very numerous, the slender column much longer than the calyx. Capsule as long as the calyx, truncate at the top; carpels about 10 to 12, persistent, angular, or scarcely pointed at the upper outer edge.— *Sida micropetala*, R. Br. Herb.

**Queensland.** Hills about Shoalwater Bay, *R. Brown.*
**N. S. Wales.** Bowen river, *Herb. Mueller (Herb. R. Br. and F. Muell.).*

**6. A. cryptopetalum,** *F. Muell. Herb.* Shrubby, but much more slender than the preceding species, clothed with a whitish tomentum, often intermixed on the young branches with a loose pubescence, the older branches nearly glabrous. Leaves cordate, from orbicular to ovate-lanceolate, obtuse, crenate, often under 1 in., the larger ones above 2 in. long, sometimes obscurely lobed, soft with a rather dense velvety tomentum. Pedicels rarely exceeding the leaves and sometimes very short. Calyx about 4 to 6 lines long, somewhat inflated, softly canescent with 10 prominent veins or ribs, the lobes much shorter than the tube. Petals often very small, but sometimes shortly exceeding the calyx. Capsule pubescent, about the length of the calyx-tube. Carpels about 10, angular or shortly acuminate on the outer edge. Seeds 3 or fewer.—*Sida cryptopetala*, F. Muell. Fragm. ii. 11.

**N. S. Wales.** Mount Murchison, *Herb. Mueller.*
**W. Australia.** Swan River, *Drummond*; near White Peak, Champion Bay, *Oldfield.*

7. **A. geranioides,** *Benth.* A shrub, with slender branches like *A. cryptopetalum*, hoary with a close rather soft tomentum, without spreading hairs. Leaves deeply cordate, ovate to ovate-lanceolate, obtuse, 1 to 2 in. long, deeply 5-lobed with the middle lobe much longer, all deeply crenate or lobed, and often crisped. Pedicels axillary, ½ to 1 in. long. Calyx ovoid, inflated, above ½ in. long, softly hoary, with 10 prominent veins or ribs, almost truncate with very short obtuse lobes. Petals nearly twice as long as the calyx. Fruit not seen.—*Sida geranioides*, DC. Prod. i. 474.

**W. Australia.** Sterile islands, *Baudin's Expedition.*

8. **A. otocarpum,** *F. Muell. in Trans. Phil. Soc. Vict.* 1855, 13, *and in Hook. Kew Journ.* viii. 10. A tall shrub, densely clothed with a soft velvety tomentum, the branches and petioles almost villous. Leaves deeply cordate, orbicular or broadly ovate, mostly 1½ to 2½ in. long, rarely acuminate, crenate, very soft and thick. Pedicels much shorter than the leaves, often crowded at the ends of the branches. Calyx 4 to 6 lines long, very prominently 5-angled, deeply divided into very concave, almost boat-shaped, strongly keeled, acuminate lobes, making the calyx intruded at the base. Petals slightly exceeding the calyx. Capsule villous, shorter than the calyx-lobes, narrowed at the top, depressed in the centre; carpels about 10, rather obtuse or scarcely pointed on the upper outer edge. Seeds 3 or fewer.

**N. Australia.** In the desert on Sturt's Creek, and on Gilbert river, *F. Mueller;* Nichol Bay, *F. Gregory.*
**Queensland.** Stokes range, *Wheeler.*
**N. S. Wales.** Mount Murchison, *Dallachy and Godwin;* Barrier range, Mount Goningbear, etc., *Victorian Expedition.* In these specimens the tomentum is closer, the flowers rather smaller, and the capsule closely tomentose, with the carpels more acute than in the Western ones, but they have the same remarkable calyx.

9. **A. subviscosum,** *Benth.* Apparently shrubby, with much of the aspect of *A. indicum*, but the branches, petioles, and pedicels greener and clothed with a viscid stellate pubescence intermixed with longer hairs. Leaves broad, deeply cordate, abruptly acuminate, 3 to 4 in. long, irregularly toothed, softly but sparingly pubescent above, tomentose and whitish underneath. Pedicels short. Calyx with slightly prominent angles, pubescent, deeply divided into acuminate lobes about ½ in. long. Petals exceeding the calyx, but imperfect in our specimens. Capsule shorter than the calyx-lobes, consisting of about 5 erect carpels, acuminate with rather long points.

**Queensland?** Subtropical regions of the interior, *Mitchell.*
There are in Herb. Muell. two shrubby *Abutila*, allied to *A. indicum*, which it is difficult to refer to any of the above species, but of which the specimens are insufficient to characterize as distinct. With the foliage of *A. indicum*, they are said to be shrubby; in one, the tomentum is close and white without spreading hairs as in *A. indicum*, the other, with the same tomentum, has also long spreading hairs as in *A. graveolens* and *A. subviscosum.* The flowers and fruit in both are very near those of *A. indicum*, but smaller, and the carpels fewer (about 10) and less hirsute. They are both from Victoria river.

10. **A. indicum,** *G. Don, Gen. Syst.* i. 504. A tall biennial or peren-

nial, clothed with a whitish tomentum, usually very close and short. Leaves cordate-orbicular, irregularly crenate, toothed or almost lobed, usually acuminate, attaining sometimes 5 to 6 in., the upper ones much smaller. Pedicels shorter than the leaves. Calyx campanulate, 5 to 6 lines long, angular in the bud, the ribs scarcely prominent when in flower, deeply divided into acuminate lobes. Petals yellow, longer than the calyx. Capsule hairy, exceeding the calyx, truncate, and attaining sometimes 7 or 8 lines diameter at the top ; carpels about 20, acute-angled or minutely acuminate at their upper outer edge, like all the preceding species not readily separating at maturity. Seeds 3 or fewer in each carpel.—*Sida indica,* Linn. ; DC. Prod. i. 471 ; Wight, lc. Pl. t. 12 ; *Sida asiatica,* Linn. ; DC. Prod. i. 470 ; *Abutilon asiaticum,* G. Don, Gen. Syst. i. 503.

**N. Australia.** Point Cunningham and Cyguet Bay, *A. Cunningham ;* Gulf of Carpentaria, *Landsborough.*

**Queensland.** Keppel Bay and Shoalwater Bay, *R. Brown ;* Percy Island, *A. Cunningham ;* Port Denison, *Fitzalan.*

The species is widely spread over tropical Asia and Africa.

11. **A. auritum,** *G. Don, Gen. Syst.* i. 500. A tall herb or perhaps undershrub, softly clothed with a soft tomentum. Stipules broad, semi-sagittate, often 4 to 6 lines long, and persistent. Leaves deeply cordate, acuminate, denticulate, 2 to 4 in. long, softly pubescent-tomentose above, white underneath. Flowers rather small, of a brown-reddish yellow, on very short pedicels, in almost leafless, terminal, branching racemes or panicles, with a broad, whitish, deciduous, stipular bract under each pedicel. Calyx obtusely 5-angled, softly tomentose, deeply divided into broad acuminate lobes. Petals not twice as long. Stamens not very numerous. Capsule longer than the calyx, hirsute, truncate ; carpels numerous, with short divaricate points. —*Sida aurita,* Wall. ; DC. Prod. i. 468 ; Bot. Mag. t. 2495.

**N. Australia.** Keppel Bay, *R. Brown ;* Percy Island, *A. Cunningham.*

The species is also found in Java and in the Philippine Islands.

12. **A. Avicennæ,** *Gærtn. Carp.* ii. 251, *t.* 135. A coarse, erect, branching annual, from 1 to 2 ft. high, softly and more or less densely tomentose-pubescent, without spreading hairs. Leaves broadly orbicular-cordate, acuminate, often 3 to 4 in. long, nearly entire or toothed, or obscurely lobed. Flowers yellow, rather small, on pedicels usually short. Calyx about 3 lines long, somewhat longer when in fruit, rather prominently 5-ribbed, deeply lobed. Petals exceeding the calyx. Capsule exceeding the calyx, pubescent or hirsute, truncate, and often $\frac{3}{4}$ in. diameter at the top ; carpels usually 10 to 15, with subulate diverging points, persistent till after the seeds are fallen, and then leaving at least the filiform placentas attached to the axis.—*Sida Abutilon,* Linn. ; DC. Prod. i. 470 ; F. Muell. Pl. Vict. i. 164 ; *Abutilon Behrianum,* F. Muell. in Trans. Phil. Soc. Vict. 1855, 13, and in Hook. Kew Journ. viii. 10.

**N. S. Wales.** On the Darling and many of its tributaries, *F. Mueller.*

**Victoria.** Dry beds of lagoons adjoining the Murray, *F. Mueller.*

**S. Australia.** Cooper's Creek, *Wright.*

A native of the Mediterranean region and of the neighbouring districts of Asia, also perhaps of northern China and Amur-land, where it is said to be cultivated for textile purposes. It has also naturalized itself as a weed over many parts of Asia, Africa and N. America, and includes *A. californicum*, Benth., and *Sida tiliæfolia*, Fisch. The Australian plant is be-lieved to be indigenous.

13. **A. graveolens,** *W. and Arn. Prod. Fl. Pen. Ind. Or.* i. 56. A coarse annual or perhaps perennial, from 1 to 5 ft. high, clothed with a viscid strong-scented tomentum, intermixed, especially on the branches and petioles, with long spreading hairs. Leaves broadly orbicular-cordate, resembling those of *A. Avicennæ*, but softer. Flowers yellow, rather large, on pedicels about as long as the petioles. Calyx about 5 lines long, deeply divided into acuminate lobes, each with a prominent midrib. Petals twice as long. Cap-sule exceeding the calyx, 8 to 10 lines diameter, hirsute, contracted at the top so as to approach in form that of *A. muticum*, and the carpels are nu-merous and closely packed as in that species, but angular or very shortly pointed at the top and less deciduous, generally leaving the filiform placentas attached to the axis, the species thus connecting the true *Abutila* with the section *Gayoides*.—Hook. Comp. Bot. Mag. i. t. 2 ; *Sida graveolens*, Roxb.; DC. Prod. i. 473.

**Queensland.** Piper's Island, off the N.E. coast, *M'Gillivray.*
The species is widely spread over East India and tropical Africa. The petals have there usually a dark spot at the base which does not appear in our Australian specimens.

14. **A. oxycarpum,** *F. Muell. Herb.* Herbaceous, diffuse or erect, at-taining 2 or 3 ft., clothed with a close tomentum or soft velvety pubescence, sometimes almost hirsute, the branches usually slender and divaricate. Leaves from cordate-ovate to ovate-lanceolate, crenate, obtuse or acuminate, 1 to 3 in. long. Pedicels slender, often 2 together, 1 to 2 in. long. Flowers small, yellow. Calyx deeply cleft, about 2 lines long. Petals not twice as long. Capsule closely tomentose or pubescent, about 4 lines long, truncate and somewhat dilated at the top ; carpels rarely above 10 and often much fewer, with short divaricate points at the outer angle, not separating till the seeds shed, and then leaving the filiform placentas attached to the axis. Seeds 2 or rarely 3.—*Sida oxycarpa*, F. Muell. Fragm. ii. 12.

**N. Australia.** Fitzroy and Mackenzie rivers, *F. Mueller.*
**Queensland.** Keppel Bay, *R. Brown;* Brisbane river, *Fraser, F. Mueller;* Rock-hampton, *Thozet.*
**N. S. Wales.** Portland Head and Richmond district, *R. Brown;* from Hastings river, *Beckler;* Clarence river, *Wilcox;* to Illawara, *Backhouse;* and in the interior to the Blue Mountains, *Miss Atkinson;* Liverpool plains, *A. Cunningham;* Macquarie river, *Mitchell;* Darling river, *F. Mueller.*
**W. Australia.** Swan River, *Drummond.*
There are two principal forms in our herbaria : 1, *acutatum*, softly tomentose, pubescent or almost hirsute; leaves ovate-lanceolate, or lanceolate, acuminate; the most common Brisbane and N. S. Wales form; and 2, *incanum*, tomentum close and white; leaves broadly cordate-ovate, obtuse or acuminate; chiefly within the tropics and in the west. Both are readily recognized by the small calyx, usually not half so long as the capsule.
Var. (?) *malvæfolium.* Less tomentose, but hirsute with long spreading hairs. Leaves cordate-ovate, very obtuse, crenate, and more or less distinctly 3-lobed. Sepals almost as long as the carpels.—Mount Murchison in N. S. Wales, *Dallachy.* This may prove to be a distinct species.

15. **A. muticum,** *G. Don, Gen. Syst.* i. 502. Tall and erect, with the habit of *A. graveolens*, with which it is often confounded, but differs in the fruit. Tomentum dense and soft, but not usually mixed with spreading

hairs. Leaves cordate-orbicular, often acuminate and irregularly toothed, 2 to 3 in. diameter, thick and soft. Pedicels rarely exceeding the petioles. Calyx ½ in. long, the lobes equal to or longer than the tube, the ribs not very prominent. Petals not twice as long, often with a dark base as in *A. graveolens.* Capsule longer than the calyx, depressed-globular with a concave centre, 7 to 8 lines diameter, densely villous; carpels about 20, closely packed, rounded or very obtuse at the top, and separating completely without leaving the persistent placentæ of *A. graveolens.*—*Sida mutica,* Delil. ; DC. Prod. i. 470.

**Queensland.** Keppel Bay, *R. Brown ;* Percy Island, *A. Cunningham;* Sources of the Burdekin and on the Dawson, *F. Mueller;* Rockhampton, *Thozet.*

The specimens are not complete, but agree well with those from tropical Africa, where the species is common, and generally referred to *A. asiaticum,* but is not *Sida asiatica* of Linnæus. *S. tomentosa,* Roxb., appears to be an E. Indian form of the same species, with the tomentum mixed with spreading hairs as in *A. graveolens,* from which it cannot always be distinguished without good fruit. It is this form which is represented as *Sida graveolens,* Bot. Mag. t. 4134.

16. **A. Cunninghamii,** *Benth.* Allied to *A. Fraseri,* but apparently shrubby, much branched, and densely clothed with soft, short, but velvety tomentum, without spreading hairs. Leaves cordate-orbicular, very obtuse, crenate, 1 to 2 in. diameter, thick and soft. Flowers on rather long peduncles in the upper axils. Calyx 4 to 5 lines long, densely tomentose, deeply divided into broad acuminate lobes. Petals about ½ in. long. Carpels 10 or fewer, distinct and seceding completely from the axis, rounded at the top, densely but closely tomentose, and not scarious.

**N. Australia.** Enderby Island, N.W. coast, *A. Cunningham ;* Albert river, *Henne.*
**Queensland.** Estuary of the Burdekin, *Herb. Mueller.*

17. **A. Fraseri,** *Hook.; Walp. Ann.* ii. 158. A low branching undershrub, rarely exceeding 1 ft., shortly tomentose or pubescent, with longer hairs occasionally intermixed. Leaves cordate, from orbicular to ovate, crenate, often all under 1 in. diameter, but sometimes 1½ in. Pedicels rarely exceeding the petioles. Flowers rather large. Calyx 3 to 4 lines long, tomentose-pubescent and sometimes hirsute, divided to about the middle. Petals more than twice as long. Fruit usually exceeding the calyx, slightly tomentose or pubescent, 3 to 4 lines diameter, depressed in the centre; carpels 6 to 10, very distinct, and seceding completely from the axis, obtuse or almost pointed at the top, not scarious. Seeds 1 or 2 in each carpel, glabrous or minutely pubescent.—*Sida Fraseri,* Hook. in Mitch. Trop. Austr. 368.

**N. Australia.** *M'Douall Stuart's Expedition.*
**Queensland.** On the Maranoa, *Mitchell ;* Sutton river and Broad Sound, *F. Mueller;* Comet river, *Leichhardt.*
**N. S. Wales.** Peel's range, *A. Cunningham ;* Darling river, *Dallachy and Goodwin ;* Goginya mountains, *Victorian Expedition.*
**S. Australia.** Subsaline barren plains and hills from Flinders range to Spencer's Gulf, *F. Mueller.*
**W. Australia.** Murchison river ?, from a single specimen in leaf only, and therefore doubtful, in *Herb. Mueller.*

Var. *parviflora.* Leaves very obtuse. Flowers much smaller.—*A. diplotrichum,* F. Muell. in Linnæa, xxv. 380.—S. Australia.

Var. *halophilum.* Leaves usually orbicular, very obtuse, often truncate or retuse, the carpels 5 or 6 lines long, and very broad and obtuse.—*A. halophilum*, F. Muell. in Linnæa, xxv. 381.—N. S. Wales, S. Australia, and W. Australia?

18. **A. crispum,** *G. Don, Gen. Syst.* i. 502. A herb or undershrub, with slender spreading branches, closely tomentose, often viscid, with long spreading hairs intermixed. Leaves cordate, acuminate, crenate, softly tomentose, the upper ones on short petioles or quite sessile. Pedicels slender, often exceeding the upper leaves. Flowers small, yellow. Calyx 2 or rarely 3 lines long, deeply divided into lanceolate or triangular acuminate lobes, reflexed under the fruit. Petals not much longer. Fruit nearly globular, hispid with scattered hairs, 4, 5, or sometimes above 6 lines diameter; carpels about 10 to 15, distinctly separating from the axis, very thin, shining inside and almost scarious when ripe, and almost always 1-seeded, although the ovary has 2 or 3 ovules.—A. Gray, Gen. Ill. t. 126; Wight, Ic. Pl. t. 68; *Sida crispa*, Linn.; DC. Prod. i. 469; *Bastardia crispa*, St. Hil. Fl. Bras. Mer. i. 194.

**N. Australia.** Sources of Hooker's Creek, and Macarthur river, *F. Mueller;* Maitland river, *F. Gregory's Expedition.*

The species is widely spread over tropical America, and is also found in East India and tropical Africa.

## 7. URENA, Linn.

Bracteoles 5, united in a 5-cleft involucre, adnate to the calyx at the base. Calyx 5-toothed or 5-lobed. Staminal column bearing several filaments or almost sessile anthers outside, below the truncate or 5-toothed summit. Ovary-cells 5, 1-ovulate; style branches 10, with terminal capitate stigmas. Fruit-carpels seceding from the axis, indehiscent, muricate or covered with hooked bristles. Seeds ascending.—Rigid tall herbs or shrubs, more or less scabrous-tomentose. Leaves usually angled or lobed, at least the lower ones. Flowers sessile or on very short peduncles, often clustered, axillary or in terminal leafy racemes.

Besides the one or two species common in all tropical regions, the genus comprises two or three tropical Asiatic ones which appear distinct. As a genus, *Urena* scarcely differs from *Pavonia.*

1. **U. lobata,** *Linn.; DC. Prod.* i. 441, var. *grandiflora.* A hard, erect herb or shrub of 2 to 4 ft., covered on the stems and under side of the leaves with a whitish close often scabrous tomentum. Leaves petiolate, the lower ones nearly orbicular, the upper ones ovate or lanceolate, palmately 3- to 7-veined, irregularly toothed, angular, or broadly and shortly lobed, glabrous above or slightly scabrous-tomentose. Flowers sessile or nearly so. Involucre deeply cleft into narrow-lanceolate lobes, in the single Australian specimen nearly ½ in. long, and fully twice as long as the calyx, but often not longer than the calyx or shorter. Petals pink, about 1 in. long in this specimen, but often much smaller. Carpels in our specimen shortly muricate.—Bot. Mag. t. 3043 (with short involucres).

**Queensland.** Sutton and Burdekin rivers, *Leichhardt.*

The species is widely spread over tropical America, Africa, and Asia, and is very variable in the shape of the leaf and proportions of the involucre, calyx, and petals, as well as in the carpels, more or less glochidiate or muricate; and most probably the *U. sinuata*, Linn., almost equally common, is only a variety with deeply-cut leaves.

## 8. PAVONIA, Cav.

(Greevesia, *F. Muell.*)

Bracteoles 5 or more, free or united at the base. Calyx 5-toothed or 5-lobed. Staminal column bearing several filaments on the outside, below the truncate or 5-toothed summit. Ovary-cells 5, 1-ovulate; style-branches 10, with terminal capitate stigmas. Fruit-carpels seceding from the axis, indehiscent or 2-valved at the top, with or without 1 or 3 awns or points, but not covered by the hooked bristles of *Urena*. Seeds ascending.—Herbs or shrubs, tomentose, hirsute, or glabrous. Leaves often angled or lobed. Flowers on axillary pedicels or in terminal heads or clusters.

A large genus, chiefly South American, with a few species scattered over the warmer regions of the Old World. The Australian species is the same as one of the South American ones.

1. **P. hastata,** *Cav. Diss.* 138, *t.* 47, *f.* 2. A low spreading shrub, more or less hoary, with a minute close stellate tomentum. Stipules subulate. Leaves petiolate, from ovate-cordate to oblong-hastate, obtuse, 1 to 2 in. long, coarsely crenate, scabrous above, hoary-tomentose underneath; when hastate, the lateral lobes short and obtuse. Pedicels usually shorter than the leaves. Bracteoles 5, ovate, herbaceous, nearly as long as the calyx. Calyx tomentose, 2 to 3 lines long, divided to the middle into 5 ovate lobes. Petals in the perfect flowers twice as long as the calyx, of a reddish-purple with a dark centre, but in other flowers, equally fertile, they are very small and closed over the stamens, which are then reduced to 5, whilst they are much more numerous in the perfect flowers. Carpels obovoid, indehiscent, usually pubescent, strongly reticulate and with a slightly raised dorsal rib.—DC. Prod. i. 443; Reichb. Icon. Exot. t. 227; *Greevesia cleisocalyx,* F. Muell. in Kew Journ. viii. 8 (founded on clandestine-flowered specimens).

**Queensland.** Moreton Bay, *F. Mueller;* Brisbane river, *Hill;* Expedition Range, *Leichhardt.*

**N. S. Wales.** Nepean, Hawkesbury and Patterson rivers, *R. Brown;* Hunter's river, *U. S. Exploring Expedition;* Liverpool Plains, *A. Cunningham;* Clarence river, *Beckler.*

Also a native of Montevideo in South America, where, as well as in Australia, it produces both kinds of flowers, although the clandestine ones appear never to have been observed until pointed out by F. Mueller.

## 9. HIBISCUS, Linn.

(Abelmoschus, *Medik.;* Paritium, *A. St. Hil.*)

Bracteoles several, rarely reduced to 5 or fewer, usually narrow, free or more or less united, sometimes very small. Calyx 5-lobed or 5-toothed. Staminal column bearing usually numerous filaments on the outside below the truncate or 5-toothed summit. Ovary 5-celled, with 3 or more ovules in each cell; style-branches 5, spreading, or rarely erect and subconnate or exceedingly short, with terminal dilated or capitate stigmas. Capsule membranous or coriaceous, loculicidally 5-valved, the endocarp not usually separating, and rarely produced into spurious dissepiments apparently doubling the number of cells. Seeds reniform or nearly globular, glabrous pubescent or woolly.—Herbs, shrubs, or trees, hispid tomentose or glabrous, the hairs almost always stellate. Leaves various, often deeply divided. Stipules in the Australian species subulate or small and deciduous, except in *H. tiliaceus.*

Flowers usually large, the petals almost always marked with a deeper colour at the base. Filaments usually short and numerous, crowded along the greater part of the elongated staminal column, rarely elongated, fewer and placed close round the top of the short column. Bracteoles usually persistent, but in a few species so deciduous as only to be seen on the very young buds.

A very large genus, widely dispersed over the tropical regions of the globe, a few extending into more temperate climates both in the northern and southern hemispheres. Of the Australian species four are generally distributed over E. India and Africa; of three others belonging to the section *Abelmoschus*, one is found in the Indian Peninsula, another is cultivated, if not wild, in the Indian Archipelago, the third is nearly allied to a corresponding E. Indian species, but in some respects distinct, an eighth species, of the section *Paritium*, is a common maritime tropical tree; the remaining 18 are all endemic.

§ 1. *Bracteoles free (sometimes very deciduous). Calyx 5-toothed, splitting open on one side and deciduous. Tall annuals.* (Abelmoschus, *Medik.*)

Glabrous or the inflorescence tomentose. Bracteoles small, falling
off from the young bud. Flowers white . . . . . . . 1. *H. ficulneus.*
Hispid. Bracteoles 8- to 12, linear, persistent. Flowers red . 2. *H. rhodopetalus.*
Glabrous or slightly setose. Bracteoles 5, broad-lanceolate, persistent . . . . . . . . . . . . . . . . 3. *H. Manihot.*

§ 2. *Bracteoles free. Calyx shortly 5-lobed, inflated. Herb
with deeply lobed leaves.* (Trionum, *Medik.*) . . . . . 4. *H. trionum.*

§ 3. *Bracteoles free. Calyx deeply 5-lobed, the lobes 1- or 3-nerved, without thickened
margins. Seeds bordered or covered by long woolly hairs. Low or slender shrubs or undershrubs.* (Bombicella, *DC.*)

Staminal tube short with long filaments round the summit . . 5. *H. brachysiphonius.*
Staminal tube slender, the short filaments extending to the middle
or lower.
    Plant loosely scabrous-hispid. Leaves deeply divided . . . 6. *H. Drummondii.*
    Plant densely and rigidly velvety-tomentose. Leaves ovate or
        lanceolate, mostly undivided. Bracteoles small . . . . 7. *H. microchlænus.*
    Plant closely and densely tomentose. Leaves orbicular, mostly
        broadly 3-lobed . . . . . . . . . . . . . 8. *H. Pinonianus.*

§ 4. *Bracteoles free. Calyx deeply 5-lobed, the lobes with a central nerve and thickened
nerve-like margins. Seeds glabrous. Tall herbs or shrubs, often more or less armed with
short prickles (except the last two species).*

Herb, glabrous or with scattered hairs. Calyx ribs ciliate.
    Flowers white or pink . . . . . . . . . . . . . 9. *H. radiatus.*
Tall shrubs, glabrous or with scattered hairs.
    Flowers axillary, without bracts under the pedicels.
        Flowers yellow. Calyx ciliate or setose . . . . . . . 10. *H. divaricatus.*
        Flowers white. Calyx densely tomentose . . . . . . 11. *H. heterophyllus.*
    Flowers in a terminal raceme, with a trifid bract under each
        pedicel. Calyx densely hirsute . . . . . . . . 12. *H. diversifolius.*
Tall shrub, densely velvety-tomentose or villous. Flowers large,
    pink. Calyx densely hirsute . . . . . . . . . . 13. *H. splendens.*
Tomentose or densely villous shrubs, without prickles. Calyx
    tomentose or villous.
        Flowers 1½ to 2 in. long . . . . . . . . . . . 14. *H. zonatus.*
        Flowers about ¾ in. long . . . . . . . . . . . 15. *H. Coatesii.*

§ 5. *Bracteoles free. Calyx deeply 5-lobed, the lobes 1- or 3-nerved, without thickened
margins. Seeds glabrous or shortly pubescent.*

Low or slender shrubs or undershrubs, glabrous, scabrous-pubes-
    cent or bristly hispid.
Leaves undivided.
    Scabrous-pubescent. Leaves ovate-lanceolate or oblong . . 16. *H. leptocladus.*

Glandular viscid and rigidly setose.  Leaves broad-cordate
   or orbicular . . . . . . . . . . . . . . . 17. *H. setulosus.*
Leaves deeply divided.
   Glabrous or nearly so.  Calyx ¾ in. long.  Capsule hispid . 18. *H. pentaphyllus.*
   Hirsute and densely setose.  Calyx not ½ in.  Capsule gla-
    brous . . . . . . . . . . . . . . . . 19. *H. geranioides.*
Small velvety-tomentose shrubs or undershrubs.  Leaves shortly
   lobed.
   Bracteoles several, subulate . . . . . . . . . . 23. *H. Krichauffianus.*
    (See also 8, *H. Pinonianus,* and 7, *H. microchlænus.*)
   Bracteoles 5, broadly ovate . . . . . . . . . . 22. *H. Normani.*
Tall shrub, scabrous, tomentose or hirsute.  Leaves deeply divided 25. *H. Huegelii.*
Tall coarse herbs or shrubs, densely tomentose and often setose.
   Bracteoles small, subulate.  Capsule very prominently angled . 20. *H. vitifolius.*
   Bracteoles dilated above the middle.  Capsule not angled . . 21. *H. panduriformis.*

     § 6. *Bracteoles united at least at the base.  Calyx 5-lobed.*

Tomentose shrubs or undershrubs.  Leaves crenate or broadly and
   shortly lobed.
   Involucral teeth or lobes short or broad.  Filaments long and
    few.  Calyx lobes obscurely nerved . . . . . . . 24. *H. Sturtii.*
   Involucral bracts united at the base only.  Filaments short and
    numerous.  Calyx lobes 1-nerved, with thickened margins . 14. *H. zonatus.*
Tall shrub, glabrous, scabrous or tomentose-hirsute.  Leaves
   deeply divided . . . . . . . . . . . . . . 25. *H. Huegelii.*
Glabrous tree.  Leaves broad-cordate, entire . . . . . . 26. *H. tiliaceus.*

   1. **H. ficulneus,** *Linn.; DC. Prod.* i. 448.  An erect annual of several
feet, glabrous except a few scattered hairs on the leaves, and a velvety pubescence on the racemes and calyces.  Leaves orbicular, 2 to 3 in. diameter, the
lower ones with 5 or 7 short broad lobes, the upper ones more deeply divided,
with obovate or oblong lobes, all usually crenate.  Flowers white, turning at
length reddish, on short pedicels, in a terminal leafless raceme.  Bracteoles
few, small and so deciduous as only to be seen on the very young buds.
Calyx about ½ in. long, shortly 5-toothed, splitting laterally and deciduous.
Petals 1 in. or rather more, glabrous.  Capsule ovoid-oblong, acute, 5-angled,
pubescent.  Seeds hairy.—*Abelmoschus ficulneus,* W. et Arn. Prod. i. 53;
Wight, Ic. t. 154; *A. alborubens,* F. Muell. Fragm. i. 67.

**N. Australia.**  In basaltic tropical and subtropical plains, *F. Mueller.*
**Queensland.**  Fitzroy plains, *F. Mueller ;* Rockhampton, *Thozet.*
  The species is common in some parts of the E. Indian peninsula, and includes *H. strictus,*
Roxb. Fl. Ind. iii. 206, and probably also *H. prostratus,* Roxb. l. c. 208.  The plant figured
by Reichenbach, Icon. Exot. t. 161, with persistent broad bracts, is a different species.

   2. **H. rhodopetalus,** *F. Muell. Herb.*  An erect or decumbent coarse
annual, of 1½ to 3 ft., more or less hirsute with long bristly hairs.  Leaves
(except the lowest) more or less deeply 5-lobed, the lobes of the lower ones
short and broad, of the upper ones oblong or lanceolate, often 2 to 3 in. long,
more or less toothed, the lowest leaves often entire and cordate, and the
uppermost lanceolate-hastate.  Flowers large, red, on axillary pedicels longer
than the petioles.  Bracteoles 8 to 12, linear, distinct, persistent, usually
shorter than the calyx.  Calyx pubescent, 6 to 7 lines long, minutely 5-
toothed, splitting laterally and deciduous.  Petals 1½ to above 2 in. long.

Capsule oblong-ovoid, acute, 5-angled, longer than the bracteoles, very hispid. —*Abelmoschus rhodopetalus,* F. Muell. Fragm. ii. 112.

**N. Australia.** Arnhem's Land, *R. Brown;* Port Molle, *M'Gillivray* (with very narrow leaf-lobes).

**Queensland.** Woody streams, Point Pearce and Brisbane river, *F. Mueller.*

This species is very nearly allied to the common East Indian *H. Abelmoschus,* Linn., differing chiefly, as observed by F. Mueller, in the colour of the flowers, red not yellow, and in smaller, more divided leaves.

3. **H. Manihot,** *Linn.; DC. Prod.* i. 448. A tall herb, sprinkled with a few pungent bristly hairs, more copious on the peduncles, otherwise, glabrous. Leaves deeply palmate; lobes 5 to 9, lanceolate, the larger ones narrow, 4 to 5 in. long, more or less toothed. Flowers large, yellow with a purple eye, on rather long pedicels in the axils of the upper reduced leaves. Bracteoles 5, herbaceous, broadly lanceolate, fully 1 in. long, roughly pubescent, persistent long after the flower has fallen. Calyx shorter than the bracteoles, shortly 5-toothed, tomentose, deciduous. Petals fully $2\frac{1}{2}$ in. long. Capsule oblong, $1\frac{1}{2}$ to 2 in. long, 5-angled, hispid especially on the angles with stiff bristly hairs.—Bot. Mag. t. 3152; *Abelmoschus Manihot,* Walp. Rep. i. 311; *Hibiscus pentaphyllus,* Roxb. Fl. Ind. iii. 212.

**Queensland.** Shoalwater Bay, *R. Brown.* The species is frequently cultivated in eastern tropical Asia, and in the islands of the Archipelago and the Pacific, but we have no certain record of it in a wild state.

4. **H. trionum,** *Linn.; DC. Prod.* i. 453. An erect annual or perennial of short duration, usually 1 to 2 ft. high, scabrous-pubescent or shortly hirsute. Leaves 2 to 3 in. long, deeply 3- or 5-lobed with oblong or lanceolate irregularly-toothed lobes. Flowers rather large, pale-yellow with a dark purple centre, on axillary pedicels. Bracteoles 7 to 12, linear-setaceous. Calyx about $\frac{1}{2}$ in. long when in flower, twice that size in fruit, inflated, membranous with about 20 raised veins, glabrous or slightly hirsute, very shortly 5-lobed. Capsule ovoid-globose, hirsute, enclosed in the calyx. Seeds glabrous.—Reichb. Fl. Germ. v. 181; F. Muell. Fragm. ii. 115; *H. Richardsoni,* Sweet; Lindl. Bot. Reg. t. 875; *H. trionioides,* G. Don, Gen. Syst. i. 483; *H. tridactylites,* Lindl. in Mitch. Three Exped. i. 85.

**N. Australia.** Victoria river and Sturt's Creek, *F. Mueller.*

**Queensland.** Between the Burnett and Dawson rivers, *F. Mueller.*

**N. S. Wales.** Hunter's and Nepean rivers, *R. Brown;* Clarence and Hastings rivers, *Beckler;* Darling river, *Dallachy and Goodwin.*

**S. Australia.** Cooper's Creek, *Herb. F. Mueller.*

Common throughout Africa and southern Asia, extending northwards to China and the Amur. Found also in New Zealand.

5. **H. brachysiphonius,** *F. Muell. Fragm.* i. 67 and 243. A low perennial or undershrub, with erect or decumbent stems, rarely above 1 ft. long, slightly hirsute with short stiff stellate hairs. Lower leaves small, orbicular, undivided, crenate; upper ones divided into 3 obovate or oblong-cuneate coarsely crenate or lobed segments or deep lobes, mostly 1 to $1\frac{1}{2}$ in. long. Flowers rather small, pink, on axillary or terminal pedicels, sometimes very long. Bracteoles about 10, rather rigid, linear, shorter than the calyx. Calyx ciliate with a few stiff hairs, deeply divided into lanceolate 1-nerved lobes, not thickened at the margin. Petals about $\frac{1}{2}$ in. long. Sta-

minal column short, bearing round the summit about 20 filaments much longer than in most species. Style-branches long, with large capitate stigmas. Capsule nearly globular, glabrous, 4 to 6 lines diameter. Seeds 4 to 6 in each cell, tomentose-villous.

**Queensland.** Mooni river, *Mitchell;* Peak Downs, *F. Mueller;* Comet river, *Leich-hardt.*

**N. S. Wales.** Macquarie river, *Mitchell;* on the Murray, *F. Mueller;* Darling river, *Dallachy and Goodwin;* Goyinga mountains, *Victorian Expedition.*

6. **H. Drummondii,** *Turcz. in Bull. Mosc.* 1858, i. 195. A slender branching shrub or undershrub, scabrous or hispid with short rigid stellate hairs. Leaves mostly divided into 3, rarely 5, cuneate, oblong-linear or rarely obovate segments, coarsely toothed or lobed, and usually hispid under-neath, rarely much exceeding 1 in., the lower leaves smaller, broader, and more entire. Flowers few in the upper axils, rather large, purple with a dark centre. Bracteoles 8 to 10, linear, hispid, often as long as the calyx. Calyx ¾ to 1 in. long, very hirsute, deeply divided into lanceolate, acuminate, 3-nerved lobes, the lateral nerves not marginal. Capsule ovoid, acute, hispid. Seeds numerous, ciliate or covered with long woolly hairs when quite ripe.— *H. Elliottii*, F. Muell. Fragm. i. 220.

**W. Australia,** *Drummond, n.* 90; between Moore and Murchison rivers, *Drummond, 5th Coll. n.* 101; Murchison and Greenough rivers, *Walcott and Oldfield.*

7. **H. microchlænus,** *F. Muell. Fragm.* ii. 116 (under *H. solanifo-lius*). Apparently shrubby, densely clothed with a scabrous, rigid-velvety, or softer and almost floccose stellate tomentum. Leaves on rather short petioles, from ovate to oblong-lanceolate, 1 to 1½ in. long, obtuse, slightly toothed, thickly and rigidly tomentose. Flowers apparently pink or purple, on pedi-cels rather longer than the petioles. Bracteoles 7 to 9, sometimes very mi-nute, sometimes half as long as the calyx. Calyx ½ in. or rather more, densely scabrous-tomentose, deeply divided into lanceolate 1-nerved lobes. Petals 1 to 1½ in. long, more or less stellate-tomentose outside where ex-posed in the bud. Capsule globular, glabrous or slightly hairy. Seeds more or less bordered or covered with long woolly hairs.—*H. brachychlænus,* F. Muell. Fragm. iii. 5.

**N. Australia.** Upper Victoria river, *F. Mueller;* Maitland river, Nichol Bay, *Wal-cott;* Fortescue river, *M. Brown.*

8. **H. Pinonianus,** *Gaudich. in Freyc. Voy. Bot.* 476, *t.* 100. Shrubby, clothed with a close, short, soft, or scarcely scabrous tomentum. Leaves on rather long petioles, mostly nearly orbicular, ½ to above 1 in. long and broad, shortly and broadly 3-lobed, crenately toothed, undulate and often crisped on the margin, strongly reticulate underneath, the lower ones almost entire. Flowers rather large, on short pedicels in the upper axils. Bracteoles 5 to 10, linear, short. Calyx 6 to 8 lines long, tomentose, deeply divided into lanceo-late 3- or 5-nerved lobes. Petals 1½ to near 2 in. long, softly tomentose out-side where exposed in the bud. Style-branches filiform, with large, often pe-nicillate stigmas, connivent at first, then spreading, and often closing again when withering, so as to give the style a simple clavate appearance. Capsule tomentose outside, glabrous inside. Seeds covered with long woolly hairs.— *H. solanifolius,* F. Muell. Fragm. ii. 116.

**N. Australia.** Mount Denison, *M'Douall Stuart.*

**W. Australia.** Sharks Bay, *Gaudichaud;* between Moore and Murchison rivers, *Drummond, 5th Coll. n.* 104. The flowers in Gaudichaud's specimens are larger than in the others.

9. **H. radiatus,** *Cav. Diss.* 150, *t.* 54, *f.* 2. An erect annual (or rarely perhaps perennial) of 2 to 3 ft., glabrous or hispid in the lower part with a few rigid hairs, and often bearing also small conical prickles. Lower leaves broad and shortly lobed, upper ones deeply 3- to 5-lobed or the uppermost undivided, the lobes narrow, toothed and unequal, the central one often 2 to 3 in. long. Flowers white or pink with a dark centre, on axillary pedicels usually very short, rarely attaining 1 in. Bracteoles about 10, narrow-linear, often spreading or reflexed, and ciliate with a few rigid hairs. Calyx about ¾ in. long, deeply divided into lanceolate acuminate lobes, of a thin texture, but marked with a prominent midrib and thickened marginal nerves, more or less rigidly ciliate. Petals 1 to 1½ in. long. Capsule globose, glabrous in the Australian specimens. Seeds few, glabrous.—DC. Prod. i. 449; Bot. Mag. t. 1911; F. Muell. Fragm. ii. 117.

**N. Australia.** Arnhem's Land, islands of Carpentaria Bay, etc., *R. Brown;* Victoria and Fitzmaurice rivers, Macadam range, etc., *F. Mueller.*

**Queensland.** Percy Islands and other points of the N.E. coast, *A. Cunningham ;* Palm Islands and Curtis Island, *Henne.*

The species extends over F. India and tropical Africa, but the extra-Australian specimens I have seen have always hirsute and less obtuse capsules. *H. Lindleyi,* Wall. Pl. As. Rar. i. 4, t. 4, is probably a purple-flowered variety. *H. cannabinus,* Linn., cultivated in Asia and Africa for its fibre, differs from *H. radiatus* only in the glands on the calyx.

10. **H. divaricatus,** *Grah. in Edinb. Phil. Journ. Jul.–Oct.* 1830. A tall, erect, glabrous shrub, with the foliage of some varieties of *H. heterophyllus* and the flowers of *H. radiatus,* the branches often beset with small conical prickles. Leaves on short petioles, entire or deeply 3-lobed, from round-cordate to ovate-lanceolate or oblong, often fully 4 in. long, more or less toothed. Flowers large, yellow with a crimson eye, on short pedicels in the axils of the upper reduced leaves. Bracteoles 10 to 12, linear, rigid, ciliate. Calyx deeply divided into lanceolate lobes, with prominent midribs and margins as in *H. radiatus,* rigidly ciliate or rarely minutely tomentose. Petals 2 to 2½ in. long. Capsule ovoid-globose, densely silky-hairy.—*Abelmoschus divaricatus,* Walp. Rep. i. 309 ; *Hibiscus magnificus,* F. Muell. Fragm. ii. 118.

**Queensland.** Shoalwater Bay, *R. Brown;* N.E. coast, *A. Cunningham ;* Newcastle range, Mackenzie and Dawson rivers, *F. Mueller.*

One of F. Mueller's specimens, with the calyx not ciliate but minutely tomentose, seems to connect this species with some forms of *H. heterophyllus.*

11. **H. heterophyllus,** *Vent. Hort. Malm. t.* 103. A tall shrub, glabrous, except a stellate tomentum on the inflorescence and very young shoots, the branches often bearing small conical prickles. Leaves entire or deeply 3-lobed, linear, lanceolate or elliptical-oblong, often 5 to 6 in. long, usually serrulate or crenulate, in some specimens white underneath. Flowers large, white with a purple centre, on short pedicels in the upper axils. Bracteoles about 10, linear, rigid, not ciliate. Calyx often above 1 in. long, deeply divided into lanceolate lobes, densely covered with a stellate tomentum often

concealing the venation, which, as in *H. radiatus,* consists of a midrib and the thickened margins of each lobe. Petals nearly 3 in. long. Capsule ovoid-globular, acute, densely setose or silky-hairy. Seeds glabrous.—Bot. Reg. t. 29; DC. Prod. i. 450; *H. grandiflorus,* Salisb. Par. Lond. t. 22.

**Queensland.** Broad Sound, Shoalwater Bay, *R. Brown;* Percy Isle and Port Curtis, *M'Gillivray;* Brisbane river, *Fraser, A. Cunningham, F. Mueller,* etc.; Rockhampton, *Thozet.*

**N. S. Wales.** Macleay and Hastings river, *Beckler;* Hawkesbury river, *Paterson;* Kiama, *Harvey;* Port Stephens, *Lady Parry;* Port Macquarie, *Thozet.*

The northern specimens belong mostly to a broader-leaved form, distinguished by A. Cunningham under the name of *H. Margeriæ.*

12. **H. diversifolius,** *Jacq.; DC. Prod.* i. 449. A tall, rigid herb or undershrub, sprinkled with a rigid pubescence, the branches and petioles more or less beset with small conical prickles. Leaves broadly cordate or nearly orbicular, irregularly toothed, angular or more or less 5-lobed. Flowers in a terminal raceme, on very short pedicels in the axils of small lanceolate or 3-fid floral leaves, often reduced, especially the upper ones, to small linear bracts. Bracteoles linear, and calyx with marginate lobes, as in *H. radiatus,* but the lobes are narrower, and usually densely hispid with rigid bristly hairs. Capsule acuminate, very hispid. Seeds glabrous.—Bot. Reg. t. 381; *H. Beckleri,* F. Muell. Fragm. ii. 117.

**Queensland.** Rockhampton, *Thozet?*

**N. S. Wales.** Hunter's river, *R. Brown;* Clarence river, in woods, *Beckler;* along the river, not common, *Wilcox.*

The species is chiefly found in S. Africa, Mauritius, and Madagascar, but is also common in waste places in the Fiji and other S. Pacific islands. In E. India it appears to be in gardens only. Thozet's specimen is somewhat doubtful, it is much more hispid, but insufficient for determination.

13. **H. splendens,** *Fraser; Grah. in Edinb. Phil. Journ., Apr.–June,* 1830. A tall shrub, of great beauty, attaining 12 to 20 ft., densely clothed with a soft velvety tomentum, the branches and petioles armed with small scattered prickles or bristles. Leaves on long petioles, broadly ovate-cordate or palmately 3 or 5-lobed, often 6 or 7 in. long, the lobes oblong-acuminate or lanceolate, often narrowed at the base. Stipules often 2 on each side. Flowers very large, rose-coloured, on pedicels about as long as the petioles. Bracteoles 10 to 15 or sometimes many more, linear-subulate, as long as the calyx, densely hispid or softly villous. Calyx at least 1 in. long, densely tomentose or hispid, deeply divided into lanceolate lobes, with a dorsal and marginal nerve, as in *H. radiatus.* Petals 3 in. long or more, glabrous. Capsule silky-hairy. Seeds glabrous.—Bot. Mag. t. 3025; Bot. Reg. t. 1629; *Abelmoschus splendens,* Walp. Rep. i. 309.

**Queensland.** Percy Island, N.E. coast, *A. Cunningham;* Rockhampton, *Thozet;* Moreton Bay, *F. Mueller.*

**N. S. Wales.** Clarence and Hastings rivers, *Fraser, Beckler.*

14. **H. zonatus,** *F. Muell. Fragm.* i. 221. A shrub with a scabrous tomentum, sometimes short and close, sometimes dense and velvety, the rather slender branches occasionally hirsute or bristly. Leaves from orbicular-cordate to ovate, the larger ones attaining 3 or 4 in., and shortly and broadly 3-, 5-, or 7-lobed, the upper ones entire or toothed and often narrow.

Flowers rather large, pink, on very short pedicels in the upper axils. Bracteoles narrow and rigid, rarely exceeding half the length of the calyx, free or slightly united at the base. Calyx nearly ¾ in. long, densely tomentose, deeply divided into lanceolate lobes, prominently 1-nerved and with thickened margins, as in the preceding species. Petals 1½ to 2 in. long, nearly glabrous. Style-branches short, spreading. Capsule very hispid, nearly globular, shorter than the calyx. Seeds glabrous.

**N. Australia.** Islands of the Gulf of Carpentaria, *R. Brown ;* W. coast of the Gulf, *Leichhardt ;* rocky banks of the Seven Emu, Macarthur and Nicholson rivers, *F. Mueller.*

15. **H. Coatesii,** *F. Muell. Fragm.* iii. 5. A shrub, evidently very nearly allied to *H. zonatus,* with the same shaped leaves and flowers, but much more densely tomentose, hirsute with rather long rigid or woolly hairs, and the flowers much smaller. Calyx about ½ in. long, very hirsute, the lobes much narrower than in *H. zonatus,* the corolla apparently about ¾ in. long.

**N. Australia.** Hammersly range, near Nichol Bay, *F. Gregory's Expedition.* The specimen is very incomplete. It may possibly prove to be a variety of *H. zonatus* (*Herb. F. Muell.*)

16. **H. leptocladus,** *Benth.* Apparently a low herb or ʲundershrub, with slender branches, rough with short rigid stellate hairs. Leaves on rather long petioles, ovate-lanceolate, lanceolate or oblong, 1 to 2 in. long, irregularly toothed, narrowed or rounded at the base, roughly pubescent on both sides with rigid stellate hairs. Flowers apparently pink, on rather long pedicels in the upper axils. Bracteoles about 7 to 9, linear-subulate, rarely exceeding half the length of the calyx. Calyx about ½ in. long, pubescent or hispid with stiff stellate hairs, deeply divided into lanceolate-acuminate, 1- or 3-nerved lobes, without thickened margins. Petals 1 to 1½ in. long, glabrous. Capsule nearly globular. Seeds 2 or 3 in each cell, glabrous.

**N. Australia.** Islands of Carpentaria Bay, *R. Brown ;* Victoria river, *Bynoe, F.Mueller.* This species resembles in some respects *H. microchlænus,* but is much more slender and less tomentose, and both petals and seeds appear to be quite glabrous.

17. **H. setulosus,** *F. Muell. Fragm.* i. 221. , A much-branched, viscid, strong-scented shrub of several feet, covered with resinous glands, the branches very hispid with long spreading bristles. Leaves broadly cordate or orbicular, mostly 1 to 1½ in. long, toothed, more or less hirsute or pubescent with scattered rigid stellate hairs. Flowers rather large, pink with a dark centre, on axillary pedicels about as long as the petioles. Bracteoles linear, rigid, about as long as the calyx. Calyx about ¾ in. long, pubescent and glandular like the leaves, deeply divided into lanceolate 3-nerved lobes. Petals about 1½ in. long. Staminal column conspicuously produced above the filaments and 5-toothed. Capsule globular, hispid, shorter than the calyx. Seeds glabrous or minutely scabrous.

**N. Australia.** Rocks on the Macarthur and Seven Emu rivers, Gulf of Carpentaria, *F. Mueller.*

18. **H. pentaphyllus,** *F. Muell. Fragm.* ii. 13. An erect or diffuse annual of a few feet, glabrous except a few rigid hairs on the upper leaves and inflorescence. Leaves divided into 5 or rarely 7 oblong or lanceolate toothed segments, mostly 1 to 2 in. long. Flowers rather large, yellow with

a brown centre, the pedicels in the upper axils longer than the petioles. Bracteoles linear, rigid, fully as long as the calyx. Calyx ¾ to 1 in. long, deeply divided into broadly lanceolate acuminate lobes, glabrous or slightly ciliate, 1- or 3-nerved. Petals 1 to 1½ in. long. Capsule globular, scarcely acuminate, hirsute. Seeds glabrous.

**N. Australia.** Victoria river and Arnhem's Land, *F. Mueller ;* gathered also in *Leich-hardt's* and *M'Douall Stuart's Expeditions.*

19. **H. geranioides,** *A. Cunn. Herb.* A low branching annual of 1 to 2 feet, densely hispid with long rigid stellate hairs or bristles. Leaves deeply divided into 3 or 5 oblong-linear or cuneate segments, mostly about 1 in. long, lobed or coarsely toothed, the lobes or teeth obtuse, hispid on both sides. Flowers small for the genus, on hispid pedicels often as long as the leaves. Bracteoles 8 to 10, linear-subulate, hispid. Calyx 4 to 5 lines long, hirsute, deeply divided into lanceolate-acuminate, 3-nerved lobes. Petals about ¾ to 1 in. long, dark at the base. Filaments short, along the upper part of the column. Stigmas capitate. Capsule small, globular, glabrous. Seeds glabrous.

**N. Australia.** Islands of the Gulf of Carpentaria, *R. Brown ;* Vansittart's Bay, N.W. coast, *A. Cunningham.*

20. **H. vitifolius,** *Linn. ; DC. Prod.* i. 450. A coarse, erect, divari-cately-branched herb of several feet, in India usually shortly tomentose, more hispid in Africa, and in the Australian specimens still more beset with rigid hairs. Leaves broadly cordate, 2 to 3 in. long and broad, usually broadly 3- or 5-lobed and toothed, very densely and softly villous-tomentose. Flowers rather large, pale yellow with a purple centre, on short pedicels, the upper ones forming a short dense leafy raceme. Bracteoles 7 to 10, linear-subulate, shorter than the calyx. Calyx deeply divided into broadly lanceolate lobes, often enlarging after flowering. Capsule depressed globular, beaked in the centre, 5 to 8 lines diameter, hirsute with scattered hairs, the 5 acute angles raised into wings and transversely veined. Seeds glabrous.—F. Muell. Fragm. ii. 114.

**Queensland.** Keppel Bay, *R. Brown ;* Percy Island, *A. Cunningham ;* Dawson river *F. Mueller ;* Palm Islands, *Henne ;* outskirts of the northern brush, *Leichhardt.* A very common species in E. India, extending into the warmer regions of Africa, and introduced into the W. Indies, readily known by its winged capsules.

21. **H. panduriformis,** *Burm. Fl. Ind.* 151, *t.* 47, *f.* 2. A tall, coarse herb or shrub, densely covered with a tomentum, usually thick and velvety on the upper side of the leaves, closer and whiter on the under side and on the petioles and branches, where it is often intermixed with long spreading bristly stellate hairs. Leaves broad-cordate, 3 or 4 in. long and broad, or rarely nar-row, usually 5-angled or broadly lobed and irregularly crenate. Flowers yellow, on very short pedicels in the axils of the upper reduced leaves, the side-branches often assuming the appearance of several-flowered peduncles. Bracteoles 6 to 8, linear or linear-spathulate, often as long as the calyx, more herbaceous than in most species and always dilated above the middle. Calyx 7 to 9 lines long, densely tomentose-hirsute, the lobes lanceolate, 1-nerved. Petals 1 to 2 in. long, densely hirsute where exposed in the bud　Capsule

ovoid-globular, very hispid.   Seeds shortly pubescent or rarely glabrous.—
DC. Prod. i. 455 ; F. Muell. Fragm. ii. 115 ; *H. tubulosus,* Cav. Diss. 161, t.
68, f. 2 ; DC. Prod. i. 447.

**N. Australia.**   Victoria river, *F. Mueller ;* Maitland river, *F. Gregory's Expedition ;*
Albert river, *Henne.*   The species is widely spread over tropical Asia and Africa.   Bur-
mann's figure represents a narrow-leaved form, not as yet found in Australia, and rare in
India.

22. **H. Normani,** *F. Muell. Fragm.* iii. 4.   An undershrub, with apparently simple erect stems of about 1 ft., densely velvety-tomentose.   Leaves
petiolate, from ovate to lanceolate, acute or obtuse, 2 to 3 in. long, obscurely
sinuate-toothed, tomentose on both sides, especially underneath.   Peduncles
1½ to 2 in. long.   Involucre of 5 broadly-ovate or rhomboidal leafy bracteoles, nearly as long as the calyx, distinct or scarcely united at the base.
Calyx tomentose, about ½ in. long, deeply divided into ovate-lanceolate 3-
nerved lobes.   Petals about twice as long or rather more, glabrous.

**Queensland.**   Palm Island, *Henne ;* Fitzroy Island, *M'Gillivray.*

23. **H. Krichauffianus,** *F. Muell. Rep. Babb. Exped.* 7.   An undershrub, with the habit and foliage of some varieties of *H. Sturtii,* but the
tomentum closer and whiter.   Leaves ovate or ovate-lanceolate, obtuse, 1 to
1½ in. long, irregularly and usually rather deeply crenate-toothed.   Flowers
rather larger than in most forms of *H. Sturtii.*   Bracteoles linear-subulate,
almost free, shorter than the calyx and sometimes very short.   Calyx very
tomentose.   Petals 1 to 1½ in. long.   Seeds slightly pubescent.

**N. S. Wales.**   Darling river, *Victorian Expedition.*
**S. Australia.**   Lake Gregory, *Babbage's Expedition ;* Cooper's Creek, *Victorian Expedition ;* towards Spencer's Gulf, *Warburton.*

24. **H. Sturtii,** *Hook. in Mitch. Trop. Austr.* 363.   A rather rigid,
simple or branched undershrub, rarely exceeding 1 ft., clothed with a whitish
tomentum, either short and rather close, or dense and velvety or sometimes
almost floccose.   Leaves broadly cordate or ovate, rarely ovate-lanceolate,
mostly 1 to 1½ in. long, obtuse, irregularly crenate-toothed, usually rather
thick and soft.   Flowers few in the upper, axils rather small, white or pink.
Involucre obconical or campanulate, with 7 or 8 teeth or short lobes, very
variable in shape, but usually nearly as long as the calyx.   Calyx very tomentose, the lobes shorter or rarely longer than the cup, thick and soft, obscurely
3-nerved.   Petals varying from ¾ to fully 1½ in. long.   Staminal column
slender, with scattered filaments as in most species, but the filaments not so
numerous and longer than usual, showing an approach to those of *H. brachysiphonius.*   Capsule globular, silky.   Seeds glabrous or rarely woolly.—F.
Muell. Fragm. ii. 13.

**N. Australia.**   N.W. coast, *A. Cunningham ;* Victoria river, *F. Mueller ;* N. of
M'Donnell range, *M'Douall Stuart.*
**Queensland.**   Mackenzie, Burdekin, Suttor, and Dawson rivers, Peak Downs, etc.,
*F. Mueller ;* Fitzroy Island, *M'Gillivray ;* Maranoa and Belyando rivers, *Mitchell.*
**N. S. Wales.**   In marshes and meadows of the interior, *Sturt, Fraser, Mitchell,* etc.;
Clarence river, *Beckler ;* New England, *C. Stuart.*

This very variable species, remarkable for its cup-shaped short-lobed involucre, presents
in our specimens the following principal forms :—

*a. grandiflora.*   Involucre shorter than the calyx, with triangular or lanceolate, somewhat acute, erect teeth.　Petals above 1 in., and often 1½ in. long.—Mount Goningbear in N. S. Wales.

*b. Muelleri.* Involucre of the preceding variety with the small flowers of the following one.—Gathered by most collectors, as well as the following variety.

*c. Sturtii.* Involucre as long as the calyx, dilated, and spreading at the top, with short broad rounded lobes.　Calyx 3 to 4 lines long, with rather short lobes.　Petals rarely exceeding 1 in., and often much smaller.—The most common N. S. Wales form.

*d. campylochlamys*, F. Muell.　Both involucre and calyx more or less deeply divided into lanceolate acuminate lobes.　Calyx otherwise rather longer than in the preceding varieties.—Victoria river and Sturt's Creek, *F. Mueller;* Dampier's Archipelago, *A. Cunningham.* In the latter specimens the seeds are woolly, but in the Victoria river plant they appear to be glabrous, as in the other varieties.

*e. platychlamys.* Very densely clothed with a somewhat rigid, velvety tomentum.　Involucre very spreading, often above 1 in. diameter, with broad lobes.　Calyx exceeding ½ in., with large ovate or ovate-lanceolate lobes.—Victoria river, *F. Mueller.*

25. **H. Huegelii,** *Endl. in Hueg. Enum.* 10.　A tall shrub, more or less scabrous or tomentose with scattered stellate hairs, or rarely glabrous, and never hoary.　Leaves deeply 3- or 5-lobed, 1 to 2 or even 3 in. long, the lobes obovate, oblong, cuneate or rarely lanceolate, more or less pinnatifid, 3-lobed or coarsely toothed, often undulate, and the lobes or teeth obtuse or rarely rather acute.　Flowers large, violet purple (or rarely yellow ?), the pedicels rather long, bearing sometimes a small bract, and still more rarely a second flower at the joint.　Involucral bracteoles more or less united at the base into a short broad cup, with 7 to 10 linear or subulate teeth or lobes very variable in length, rarely nearly free to the base.　Calyx ¾ to nearly 1 in. long, tomentose or softly villous, deeply divided into lanceolate-acuminate 3- or 5-nerved lobes.　Petals 2 to 3 in. long, softly tomentose or villous outside where exposed in the bud.　Styles united almost to the stigmas, which are large and spreading.　Capsule ovoid-globose, tomentose or villous, the cells hairy inside.　Seeds glabrous.

**S. Australia.** Goose Island Bay and Memory Cove, *R. Brown;* Mount Arden, Mount Remarkable, and Port Lincoln, *F. Mueller;* Streaky Bay, Venus Bay, etc., *Warburton.*

**W. Australia.** From Cape Riche, *Preiss, n.* 1340, to Swan River, *Fraser, Drummond, Preiss, n.* 1336, 1339, 1341, and others; and Murchison river, *Drummond, Oldfield,* etc.

A variable species, of which the following are the most conspicuous forms in our herbaria:—

*a. angulatus.* Glabrous, except a close tomentum on the flowers; branches strongly angular, by prominent lines decurrent from the stipules.　Flowers large.—Murchison river.

*b. glabrescens.* Stem and leaves glabrous or slightly tomentose, the branches terete or with slightly raised angles.　Flowers large, tomentose, drying of a pale colour.—Swan River. To this form should be referred the original specimen of *H. Huegelii.* My suspicion that the statement that it had a yellow flower, purple in the centre, originated in a mistake, has been fully confirmed by Dr. Fenzl, who has sent me full notes on the varieties exemplified in the Vienna Herbarium.

*c. Wrayæ.* More or less abundantly sprinkled or clothed with a scabrous tomentum or stellate hairs.　Flowers large, of a bluish-purple.　Bracts united.　Calyx densely tomentose-villous,—*H. Wrayæ*, Lindl. Bot. Reg. 1840, t. 69; *Paritium Wrayæ*, Walp. Rep. i. 311; *H. Huegelii*, Miq. in Pl. Preiss. i. t. 239; *H. Pinonianus*, Miq. l. c. 240, but not of Gaudichaud.—S. and W. Australia.

*d. leptochlamys.* Like the last, but more villous, and the bracteoles longer, free to the base.—Murchison river.　The stigmas appear to be erect and closed, almost as in *Fugosia,* but I am not sure that they are perfect in the very few flowers we have.

*e. grossulariæfolius.* Like *Wrayæ*, but the flowers rather smaller.　Leaves often, but not always smaller, with broader and shorter lobes.—*H. grossulariæfolius*, Miq. in Pl. Preiss. i. 240; Bot. Mag. t. 4329; *H. Meisneri*, Miq. l. c., *H. geraniifolius*, Turcz. in Bull. Mosc. 1858, i. 195.—Swan River and S. coast.

26. **H. tiliaceus,** *Linn.; DC. Prod.* i. 454. A small tree. Leaves on long petioles, orbicular-cordate, shortly acuminate, entire or crenulate, white or hoary underneath with a close short tomentum, nearly glabrous above, 3 to 5 in. diameter. Stipules large, broadly oblong, very deciduous. Flowers large, yellow with a dark crimson centre, on short peduncles in the upper axils or at the ends of the branches. Involucre campanulate, divided to about the middle into 10 to 12 lobes, about half the length of the calyx. Calyx nearly 1 in. long, with lanceolate 1-nerved lobes. Petals 2 to 3 in. long, slightly tomentose outside. Capsule nearly 1 in. diameter, the valves bearing the dissepiments in their centre, and their thin margins turned inwards so as to make the capsule appear 10-celled.—*Paritium tiliaceum,* St. Hil. Fl. Bras. Mer. i. 256; Wight, Ic. Pl. t. 7.

**N. Australia.** Islands of the Bay of Carpentaria, *R. Brown, Henne;* Port Molle, *M'Gillivray.*

**Queensland.** *Burdekin Expedition;* Rockhampton, *Thozet.*

A common seacoast tree in most tropical countries, particularly abundant in the islands of the Pacific.

## 10. LAGUNARIA, G. Don.

Bracteoles 3 or 4, broad and united at the base, often very deciduous. Calyx very shortly 5-lobed. Staminal column bearing numerous filaments on the outside below the 5-crenate summit. Ovary 5-celled, with several ovules in each cell. Style clavate at the top, with 5 distinct ovate radiating stigmas. Capsule loculicidally 5-valved, the endocarp villous inside and separating from the pericarp. Seeds reniform, thick, glabrous.—A tree. Leaves entire, sprinkled or curved, with scurfy scales. Flowers large, axillary, on short thick pedicels.

The genus, scarcely perhaps sufficiently distinct from *Hibiscus,* is limited to a single species, represented, however, by two distinct varieties, one Australian, the other peculiar to Norfolk Island.

1. **L. Patersoni,** *Don, Gen. Syst.* i. 485, var. *bracteata.* A tree, the young parts and inflorescence more or less covered with minute scurfy scales, but otherwise glabrous. Leaves petiolate, oblong or broadly lanceolate, rarely ovate-oblong, 3 to 4 in. long, entire, somewhat coriaceous, white underneath when young, glabrous and pale-green on both sides when full grown, the scales of the under surface almost disappearing. Pedicels very short and angular. Bracteoles 3 to 5, very obtuse, united in a broad, shortly-lobed cup, usually persistent at the time of flowering in the Australian variety, but sometimes even these falling off early. Calyx 4 to 5 lines long. Petals narrow, above 1½ in. long, slightly tomentose outside.

**Queensland.** Port Denison, *Fitzalan;* Port Cowper, *T. Sutherland;* Cumberland Islands, *Herb. Mueller.*

The Norfolk Island form (*Hibiscus Patersonius*, Andr. Bot. Rep. t. 286; *H. Patersoni,* DC. Prod. i. 454; *Lagunæa Patersonia,* Bot. Mag. t. 769; *L. squamea,* Vent. Jard. Malm. t. 42) is much more scaly-tomentose, the leaves are broader and very white underneath, and the bracteoles fall off at so very early a stage that they have always been said to be entirely wanting.  I had, on that account, at first considered the Australian plant as distinct, but I have since seen the bracts on very young buds of the Norfolk Island one, and observe them to be here and there very deciduous on Australian specimens, and the other characters, although as far as hitherto known constant, may not be sufficient to distinguish the two as more than varieties or races.

## 11. FUGOSIA, Juss.

Bracteoles 3, distinct and narrow, or several united in a 3- to 6-toothed involucre.  Calyx 5-lobed.  Staminal column bearing numerous filaments on the outside, below the truncate or 5-toothed summit, or rarely quite to the top.  Ovary 3- to 5-celled, with 3 or more ovules in each.  Style thickened towards the top, grooved or divided into short, erect lobes, with decurrent stigmas.  Capsule loculicidally 3- to 5-valved.  Seeds obovoid-globular or slightly reniform, usually pubescent or woolly.  Cotyledons much folded over the radicle.—Shrubs or undershrubs, with the habit of *Hibiscus*, but usually more glabrous.  Leaves entire or lobed, rarely divided.  Stipules small or subulate and deciduous.  Flowers usually large, yellow or purple.  Calyx often marked with black dots, but not the cotyledons.

The genus comprises several species from tropical and subtropical regions of America and one from Africa, but none from Asia.  The Australian ones are all endemic.  It is very nearly allied on the one hand to *Hibiscus*, on the other to *Gossypium*, differing from the former chiefly in the style, from the latter in the bracteoles.

Involucre minutely toothed, placed a little below the calyx.  Glabrous
 or nearly so.  Ovary-cells 5.
  Leaves entire, cuneate-oblong or broadly linear . . . . . . .  1. *F. cuneiformis.*
  Leaves narrow-linear or almost terete, mostly deeply divided  . .  2. *F. hakeæfolia.*
Bracteoles 3, distinct, on the base of the calyx.  Ovary-cells usually 3
 or 4.
 Whole plant softly tomentose.
  Calyx-lobes linear or lanceolate.  Bracteoles linear . . . . . .  3. *F. australis.*
  Calyx truncate, minutely 5-toothed.  Bracteoles setaceous, minute  4. *F. thespesioides.*
 Plant glabrous or very slightly hoary-tomentose.
  Calyx deeply divided into lanceolate lobes.
   Leaves ovate or lanceolate, narrowed at the base, on very short
    petioles . . . . . . . . . . . . . . . .  5. *F. punctata.*
   Leaves orbicular, 5-nerved, on petioles of 1 in. ,. . . . .  6. *F. latifolia.*
  Calyx truncate, with small linear lobes.  Leaves on long petioles,
   cordate, acuminate . . . . . . . . . . . . . .  7. *F. populifolia.*

(Some varieties of *Hibiscus Huegelii* appear to have sometimes the stigmatic lobes erect, but the bracteoles and other characters are more those of *Hibiscus*.)

1. **F. cuneiformis,** *Benth.*  Shrubby and glabrous.  Leaves cuneate-oblong or broadly linear, obtuse, 1 to 2 in. long, entire, thick and somewhat fleshy.  Peduncles short and thick.  Involucre very small, minutely 5 or 6-toothed, placed a little below the calyx.  Calyx ¾ to 1 in. long, glabrous or minutely tomentose, and occasionally glandular-dotted, deeply divided into lanceolate 1-nerved lobes.  Petals about 1½ in. long, slightly tomentose.  Capsule 5-celled, ovoid-oblong, acuminate, woolly.  Seeds numerous, covered with long woolly hairs.—*Hibiscus cuneiformis*, DC. Prod. i. 454; *Lagunaria cuneiformis*, G. Don, Gen. Syst. i. 485.

**W. Australia.**  Seacoast, Dirk Hartog's Island, *A. Cunningham, Milne;*  Sharks Bay, *Sanford.*

2. **F. hakeæfolia,** *Hook. Bot. Mag. t.* 4261. An erect shrub, flower-ing young, but attaining 8 to 10 ft., entirely glabrous, or tomentose on the flower only. Leaves from deeply bipinnatifid to trifid only, or the upper ones entire, often several inches long, the whole leaf or lobes narrow-linear, some-what fleshy, grooved above or almost terete. Flowers large, of a purple lilac, on axillary peduncles, articulate, and often bearing a small bract about the middle. Involucre placed a little below the calyx, very small, divided into 3 to 6 short, rigid, unequal teeth. Calyx ¾ to 1 in. long, deeply divided into lanceolate-acuminate 3-nerved lobes. Petals 1½ to 2 in. long. Capsule tomentose, ovoid, with a short point, 5-celled. Seeds woolly.—*Hibiscus hakeæfolius*, Giord.; Endl. in Hueg. Enum. 10 ; *H. multifidus*, Paxt. Fl. Gard. vii. 103, with a fig.

**S. Australia.** Goose Island Bay, S. coast, *R. Brown ;* in the interior, *M'Douall Stuart.*

**W. Australia.** From King George's Sound, *Fraser*, to Swan River, *Drummond, Preiss, n.* 1342, and Murchison river, *Drummond, Oldfield.*

Var. *coronopifolia.* Leaf-segments often somewhat dilated and deeply toothed. *Hibiscus lilacinus*, Lindl. Bot. Reg. t. 2009 ; *H. coronopifolius*, Miq. in Pl. Preiss. i. 239 (from the description) ; *Lagunaria lilacina*, Walp. Rep. i. 311. W. Australia.

3. **F. australis,** *Benth.* An undershrub of several feet, hoary with a dense but very short tomentum. Leaves broadly or narrow-ovate, obtuse, 1½ to 2½ in. long, entire or more or less sinuate or 3-lobed. Flowers rather large, pink, on very short pedicels, which are often clustered 2 or 3 together at the top of axillary peduncles, with a bract or small leaf under each. Bracteoles 3, linear, distinct. Calyx from ½ to ¾ in. long, tomentose and marked with black glandular dots, the lobes lanceolate or almost linear, vary-ing very much in length. Petals 1½ in. long, slightly tomentose outside. Capsule obovoid-oblong, shortly acuminate, tomentose, 3- or 4-valved. Seeds numerous, woolly.—*Gossypium australe*, F. Muell. Fragm. i. 46, and iii. 6.

**N. Australia.** Barren plains, not rare, *F. Mueller ;* N.W. coast, *Bynoe ;* Maitland river, *F. Gregory's Expedition ;* Gulf of Carpentaria, *Landsborough.*

In habit and foliage this much resembles the Brazilian *F. phlomidifolia*, St. Hil., which has, however, more numerous bracteoles and yellow flowers.

4. **F. thespesioides,** *Benth.* Habit nearly that of *F. australis*, but larger and more tomentose, especially the inflorescence and under side of the leaves, which are somewhat rust-coloured. Leaves orbicular or broadly ovate, 2 to 4 in. long, softly tomentose. Flowers large, on short pedicels, or the lower ones on longer peduncles, articulate and bracteate below the summit. Bracteoles 3 or rarely 5, usually minute and setaceous. Calyx broadly cup-shaped, truncate, with 5 minute distant teeth, about ½ in. diameter, tomen-tose. Petals above 2 in. long, tomentose outside. Capsule nearly globose, twice as long as the calyx, hard and almost woody, 3-celled and 3-valved. Seeds apparently pubescent, but not seen ripe.—*Hibiscus thespesioides*, R. Br. Herb.

**N. Australia.** N. coast, without any precise locality indicated, *R. Brown* (*Hb. R. Br.*).

5. **F. punctata,** *Benth.* Apparently shrubby, with tall erect branches, the whole plant glabrous or very minutely hoary. Leaves on very short pe-tioles, from ovate to lanceolate, mostly acute, 2 to 3 in. long, penninerved or

obscurely 3-nerved at the base. Flowers large, on rather long pedicels in the axils of the uppermost reduced leaves. Bracteoles 3, lanceolate, persistent. Calyx about 1 in. long, deeply divided into lanceolate, obscurely 1 or 3-neɪved lobes, marked with a few black dots. Petals fully 2 in. long. Capsule small, nearly globose but rather acute, 3-valved. Seeds apparently globose, but not seen ripe —*Hibiscus punctatus*, A. Cunn. Herb.

**N. Australia.** Port Essington, *A. Cunningham.*

6. **F. latifolia,** *Benth.* Habit and general characters of *F. punctata*, but the leaves are orbicular or broadly ovate, 5-nerved, on petioles of 1 in. or more, and the calyx-lobes are marked each with 3 strongly raised nerves, which unite into 10 prominent ribs on the tube. The whole plant is also somewhat hoary with a minute pubescence, especially the inflorescence and younger leaves. Petals and stamens not seen. Capsule of *F. punctata*.

**N. Australia.** Careening Bay, N.W. coast, *A. Cunningham.*

7. **F. populifolia,** *Benth.* Apparently shrubby, with slender, perhaps procumbent branches, quite glabrous or with a minute pubescence on the under side of the leaves. Leaves on long petioles, cordate, long-acuminate, entire, rarely above 2 in. long, green on both sides. Flowers rather large, on pedicels longer than the petioles. Bracteoles 3, linear-lanceolate, reflexed. Calyx not ½ in. diameter, marked with black dots, almost truncate, with linear-acuminate lobes about as long as the tube. Petals nearly 1½ in. long, minutely tomentose outside. Capsule globular, glabrous, 3-valved, but not seen fully ripe.

**N. Australia.** Greville Island, Montague Sound, Isles of King George IV.'s Sound, N.W. coast, *A. Cunningham ;* N.W. coast, *Bynoe.*

## 12. THESPESIA, Corr..

Bracteoles 1 to 5, small or deciduous. Calyx truncate, minutely 5-toothed or rarely 5-lobed. Staminal column bearing numerous filaments on the outside, below or up to the summit. Ovary 5-celled, with few ovules in each cell. Style club-shaped at the top, 5-furrowed or obscurely divided into erect stigmatic lobes. Capsule hard, almost woody, indehiscent or loculicidally 5-valved. Seeds obovoid, glabrous or woolly. Cotyledons very much folded, enclosing the radicle, often black-dotted.—Trees or tall herbs. Leaves large, entire or angularly lobed. Flowers large, usually yellow.

A small genus, limited to tropical Asia, the Pacific isles, and eastern Africa, the Australian species being one which extends over the whole range. Closely allied to *Hibiscus, Fugosia,* and *Gossypium,* it differs from the former chiefly in the style, from the two latter generally either in the calyx or bracts, and from all in the more woody capsule.

1. **T. populnea,** *Corr. ; DC. Prod.* i. 456. A tree, with the young parts and under side of the leaves sprinkled with minute rust-coloured scales, otherwise glabrous. Leaves broad-cordate, acuminate, entire, 4 or 5 in. long. Flowers reddish-yellow, rather large, on axillary pedicels usually shorter than the petioles. Bracteoles 1 to 3, lanceolate and deciduous, or sometimes wanting. Calyx very open, 6 to 8 lines diameter, truncate, with minute teeth. Petals broad, 1½ to 2 in. long. Capsule fully 1½ in. diameter, hard

and woody, indehiscent or opening longitudinally when very dry.—Wight, Ic. t. 8.

**N. Australia.** Islands of the Gulf of Carpentaria, *R. Brown, Henne.*
**Queensland.** N.E. coast, *A. Cunningham, M'Gillivray.*

The species is widely spread over the seacoasts of tropical Asia, extending from eastern Africa to the Pacific Islands. It is also introduced into the West Indies.

## 13. GOSSYPIUM, Linn.

### (Sturtia, *R. Br.*)

Bracteoles 3, large and cordate. Calyx much shorter, truncate or shortly 5-lobed. Staminal column bearing numerous filaments outside, below or up to the top. Ovary 5-, rarely 4-celled, with several ovules in each cell. Style club-shaped at the top, furrowed, with decurrent stigmas. Capsule loculicidally 5-, rarely 4-valved. Seeds angular or nearly globular, very woolly or nearly glabrous; cotyledons very much folded, enclosing the radicle. —Tall herbs, shrubs, or almost trees. Leaves 3- to 9-lobed, or rarely entire. Flowers large, yellow or purple. Bracteoles entire, toothed or cut, usually, as well as the calyx and cotyledons, marked with black dots.

The genus, besides the Australian species, which is endemic, comprises the cultivated *Cotton,* whose various forms, described as species, races, or varieties, are distributed either as indigenous or introduced plants over the warmer regions both of the New and the Old World, but not hitherto found in a wild state in Australia.

1. **G. Sturtii,** *F. Muell. Fragm.* iii. 6. A shrub of several feet, glabrous and more or less marked with black dots. Leaves on rather long petioles, broadly ovate, entire, 1 to 2 in. long, rather coriaceous and glaucous. Flowers large, purple with a dark centre, on short pedicels in the upper axils. Bracteoles cordate, entire, ¾ to 1 in. long, many-nerved and black-dotted. Calyx not half so long, broad, truncate with minute or narrow-linear teeth, copiously black-dotted. Petals fully 2 in. long. Capsule ovoid, shortly acuminate, much longer than the calyx, usually 4-celled, glabrous but copiously black-dotted. Seeds very sparingly and shortly woolly.— *Sturtia gossypioides,* R. Br. App. Sturt. Exped. 5.

**S. Australia.** In the interior; Barren Range, *Sturt;* Elder's Range, *F. Mueller;* Mutanié Ranges, *Beckler;* Flinders Range, *Victorian Expedition;* towards Spencer's Gulf, *Warburton.*

## 14. ADANSONIA, Linn.

Calyx ovoid or oblong, deeply splitting into 3 to 5 lobes. Staminal column divided at the top into numerous filaments. Ovary 5- to 10-celled, with many ovules in each cell. Style shortly divided at the summit into as many radiating stigmas as there are cells. Fruit oblong, woody, indehiscent, the cells filled with a mealy pulp. Seeds reniform-globular, embedded in the pulp; cotyledons very much folded, enclosing the radicle.— Trees with a comparatively short trunk, acquiring an immense girth, the wood soft and spongy. Leaves digitate, with entire leaflets. Peduncles axillary, 1-flowered, bracteate. Flowers large, white, pendulous. Fruits large, pendulous.

Besides the Australian species, which is endemic, the genus only contains one other, the celebrated *Baobab* of tropical Africa, which extends into the western districts of East India.

1. **A. Gregorii,** *F. Muell. in Hook. Kew Journ.* ix. 14. A large tree, not lofty in proportion to its size, with an enormous gouty stem, attaining from 30 to 80 ft. in circumference, and usually contracted under the main branches of the head. Leaflets 5, 7, or rarely 9, oblong-lanceolate, acuminate, the larger ones 4 to 5 in. long, narrowed at the base but rarely petiolulate, minutely pubescent above, white-tomentose underneath. Flowers of a yellowish-white, on pedicels of 1 to 1½ inch. Calyx oblong and entire in the bud, and little more than ½ in. diameter, attaining 3 in. in length, and splitting into 3 to 5 lobes as the flower opens, tomentose outside, silky-villous inside. Petals 5 or rarely 4, cuneate-oblong, fully 4 in. long, silky-villous outside in their upper portion. Staminal column pubescent outside, rather shorter than the filaments. Fruit resembling a small gourd, in our specimens about 6 in. long and 3 to 4 in. diameter, but probably often larger, of a brownish-red colour, densely tomentose, exuding a dark red gum.

**N. Australia.** Sandy plains and low stony ridges, from the Glenelg to the western shores of Arnhem's Land, and rarely above 100 miles inland, *F. Mueller, G. Bennett,* and others. The interior substance of the fruit has an agreeable acidity, and, boiled with sugar, is of material service in scorbutic complaints. (See G. Bennett, ' Gatherings of a Naturalist,' 292, t. 5.)

The African *A. digitata,* which is closely allied, and, according to G. Bennett, has precisely the same fruit (above a foot long in our specimens), differs chiefly in broader leaflets, a broader calyx more regularly 5-cleft, broader petals, and still more numerous and shorter filaments.

## 15. BOMBAX, Linn.

(Salmalia, *Schott.*)

Calyx cup-shaped, truncate, or splitting into 3 to 5 lobes. Staminal column divided into numerous filaments, of which the inner ones, or nearly all, are more or less connected in pairs and united at the base into 5 or more bundles. Ovary 5-celled, with several ovules in each cell; style club-shaped, or shortly 5-lobed at the top. Capsule woody or coriaceous, opening loculicidally in 5 valves, the cells densely woolly inside. Seeds obovoid or globular, enveloped in the wool of the pericarp; albumen thin; cotyledons much folded round the radicle.—Trees. Leaves digitate, with leaflets usually entire. Peduncles 1-flowered, axillary or terminal. Flowers white or red.

The species are chiefly South American, with one from tropical Africa, and another from tropical Asia extending also into Australia.

1. **B. malabaricum,** *DC. Prod.* i. 479. A large tree, the trunk covered with short conical prickles. Leaves on long petioles, deciduous; leaflets 5 to 7, petiolulate, elliptical-oblong, acuminate, 4 to 6 in. long, coriaceous, entire, glabrous. Flowers large, red, on short pedicels, clustered towards the ends of the branches which are then destitute of leaves. Calyx above 1 in. long, thick and coriaceous, glabrous outside, silky-hairy inside, dividing into short broad obtuse lobes. Petals fully 3 in. long, oblong, tomentose outside, nearly glabrous within. Staminal column short, filaments much longer, but shorter than the petals, five innermost forked at the top, each branch bearing an anther, about 10 intermediate ones simple, and the numerous outer ones shortly united in 5 clusters. Capsule large, oblong, and woody. —*Salmalia malabarica,* Schott, Meletem. 35 ; *Bombax heptaphylla,* Cav. ; Roxb. Pl. Corom. iii. 43, t. 247 ; Wight, Ill. t. 29.

**N. Australia.** Careening Bay, N.W. coast, *A. Cunningham.* The specimen consists of a single flower; the foliage and fruit are therefore described from East Indian specimens, where the species has a considerable range.

## Order XXII. STERCULIACEÆ.

Flowers regular, hermaphrodite or unisexual. Calyx usually persistent, more or less deeply divided into 5 or rarely 4 or 3 valvate lobes or segments, or rarely splitting irregularly, or the sepals entirely free. Petals either 5, hypogynous, free, or adhering to the staminal column, contorted-imbricate in the bud, or small and scale-like, or none. Stamens usually united into a ring, a cup, or tube, with 5 terminal teeth or lobes (staminodia) alternating with the petals, and one or more anthers sessile or stipitate (on distinct filaments) in each interval, the anthers 2-celled and opening outwards, in longitudinal slits, or exceptionally the anthers are numerous or the staminodia wanting, or the stamens 5, free and alternate with the sepals or the anther-cells confluent or opening in terminal pores. Ovary free, 2- to 5-celled, with the carpels more or less united, rarely 10- or 12-celled, or reduced to a single carpel. Style entire, or divided into as many branches as there are cells, or rarely styles as many, nearly or quite free. Fruit various. Seeds sometimes hairy but not woolly, sometimes enveloped in pulp or strophiolate, the testa coriaceous, occasionally enclosed in an outer membranous integument; albumen fleshy or none; cotyledons usually foliaceous, flat or folded, the radicle shorter, next the hilum or rarely distant from it.—Herbs, shrubs, or trees, the tomentum or hairs stellate, rarely mixed with simple hairs. Leaves alternate or irregularly opposite, simple and pinnately or palmately nerved, entire toothed or lobed, or digitately compound. Stipules rarely wanting.

A large Order, chiefly tropical, dispersed over the New and the Old World, with some extratropical genera in S. Africa or Australia, and very few species without the tropics in the Northern hemisphere. Of the 19 Australian genera 10 are common to the tropical regions of the Old World or both of the Old and the New World, the remaining 9 are endemic, with the exception of single species of *Rulingia* and *Keraudrenia*, found in Madagascar.

Anthers 5 to 15, sessile or stipitate, surrounding the ovary at the
    top of a column or gynophore
*Flowers unisexual or polygamous. No petals.* Anthers sessile.
    No staminodia. Fruit-carpels separate, sessile or stipitate.
    Trees. Leaves simple or digitate. (Tribe **Sterculieæ.**)
    Anthers irregularly clustered. Seeds albuminous.
        Ovules 2 or more in each cell. Carpels follicular or open-
          ing along the inner edge . . . . . . . . . . .  1. STERCULIA.
        Ovules single in each cell. Carpels winged, indehiscent . .  2. TARRIETIA.
    Anthers 5, in a ring. Ovules solitary. Carpels large, indehis-
        cent. Albumen none . . . . . . . . . . . .  3. HERITIERA.
*Flowers hermaphrodite. Petals 5, clawed.* Anthers on short
    filaments, surrounding or alternating with 5 teeth of the co-
    lumn or staminodia. Leaves simple. (Tribe **Helictereæ.**)
    Anther-cells divaricate or confluent into one. Fruit-carpels dis-
        tinct, or spirally twisted . . . . . . . . . . .  4. HELICTERES.
    Anther-cells parallel. Fruit woody, 5-valved. Seeds winged .  5. PTEROSPERMUM.
Stamens 5 (or in *Abroma* more), united at the base in a short cup or
    ring, or rarely free, with or without intervening staminodia, and
    surrounding the sessile ovary.

*Petals flat, longer than the calyx.*
  Stamens 5, united in a cup, with 5 intervening elongated flat
    staminodia . . . . . . . . . . . . . . . . . . 6. MELHANIA.
  Stamens 5, united at the base without intervening staminodia.
    (Tribe **Hermannieæ**.)
      Ovary 5-celled . . . . . . . . . . . . . . . 7. MELOCHIA.
      Ovary 2-celled . . . . . . . . . . . . . . . 8. DICARPIDIUM.
      Ovary of one 1-celled carpel . . . . . . . . . 9. WALTHERIA.
*Petals with a short, broad, very concave base, and a sessile or sti-
  pitate lamina.* (Tribe **Buettnerieæ**.)
  Lamina of the petals stipitate, longer than the calyx.  Stamino-
    dia 5, obcordate, with 2 to 4 stamens between each . . . 10. ABROMA.
  Lamina of the petals short, sessile, stamens 5.
    Staminodia single between each 2 stamens, lanceolate . . . 11. RULINGIA.
    Staminodia 3 between each 2 stamens, all linear-spathulate, or
      the central one lanceolate, and the lateral ones subulate . 12. COMMERSONIA.
*Petals small and scale-like or none.* (Tribe **Lasiopetaleæ**.)
  Anthers (linear-oblong) opening outwards in parallel slits.
    Calyx herbaceous, scarcely enlarged, and not coloured after
      flowering.  Staminodia large.  Carpels membranous, winged 13. SERINGIA.
    Calyx enlarged after flowering, thin and coloured.  Staminodia
      single or none.  Capsule or carpels membranous, rounded
      or rarely winged . . . . . . . . . . . . . . 14. KERAUDRENIA.
    Calyx strongly ribbed after flowering.  Staminodia 3 between
      each 2 stamens.  Capsule hard or woody . . . . . . 15. HANNAFORDIA.
  Anthers (often obtusely sagittate or acuminate) opening in ter-
    minal or inwardly oblique pores, or in slits, extending more
    or less down the sides.
    Calyx divided to above or a little below the middle, enlarged,
      and coloured after flowering, each sepal with the midrib
      either very prominent inside or deeply coloured.  Stipules
      leafy or rarely none . . . . . . . . . . . . . 16. THOMASIA.
    Calyx divided to the middle or lower, each sepal with 3 or 5
      ribs, very prominent after flowering.  Stipules leafy . . 17. GUICHENOTIA.
    Calyx divided almost to the base, scarcely enlarging, obscurely
      several-veined at the base.  Stipules none . . . . . 18. LASIOPETALUM.
    Sepals entirely free, narrow and petal-like.  Stipules very
      small or none . . . . . . . . . . . . . . . 19. LYSIOSEPALUM.

## 1. STERCULIA, Linn.

(Brachychiton, Trichosiphon, *and* Pœcilodermis, *Schott;* Delabechea, *Lindl.*)

Flowers unisexual or polygamous.  Calyx more or less deeply 5-cleft, rarely 4-cleft, usually coloured.  Petals none.  Staminal column adnate to the gynophore, bearing at the summit 15 or rarely 10 stamens, irregularly clustered in a head.  Carpels of the ovary 5, distinct or nearly so, with 2 or more ovules in each.  Styles united under the peltate or lobate stigma.  Fruit-carpels distinct, spreading, either firm or woody, and scarcely opening along the inner edge, or thinner, and opening as follicles, even long before they are ripe.  Seeds 1 or more in each carpel, rarely winged; albumen adhering to the cotyledons, often splitting in two, assuming the aspect of fleshy cotyledons; real cotyledons flat or nearly so, and thin, the radicle next the hilum or at the opposite end, or intermediate.—Trees.  Leaves undivided or lobed, or digitately compound.  Flowers in panicles or rarely racemes, mostly axillary, sometimes very short; terminal flowers usually female, in these the staminal column is shorter and the anthers less perfect than in the males, surrounding the base of the ovary; in the males the ovary is often entirely abortive.

A large genus, almost entirely tropical, and more abundant in Asia than in Africa or America, where however several species are found. The Australian ones are all endemic, except *S. fœtida*, which is a widely-spread Asiatic one.

The species of this genus were distributed by Schott into a number of genera, founded chiefly on the flowers and habit, afterwards reduced and rearranged by R. Brown, chiefly on carpological characters, without reference to habit or calyx. The majority of the Australian ones belong to the group distinguished by R. Brown chiefly by the seeds having a loose outer coating covered with hairs, which in some species are so adhesive that the seeds fall out in their inner coating only, leaving the outer coating adhering to the equally hairy endocarp, with the appearance of the cells of a beehive; and by the radicle next to the hilum. The seeds do not appear to cohere in all the species, in some they are hitherto unknown, and in flowers and habit, *S. ramiflora* and *S. rupestris*, *S. fœtida* and *S. quadrifida* are more different from each other than from species belonging respectively to other groups. Among species not Australian, the position of the radicle unites two very heteromorphous ones under *Firmiana*, and would (as observed to me by M. Poinsot, of the Paris Herbarium) lead to separate *S. mexicana* from other digitate-leaved American species. I have therefore, with Endlicher and others, considered Schott and Brown's genera as sections only.

SECT. 1. **Sterculia.**—*Radicle at the end remote from the hilum. Seeds and inside of the carpels glabrous.*

Leaves digitate. Calyx-lobes 5, spreading. Staminal column long and
    incurved . . . . . . . . . . . . . . . . . . . . . . . 1. *S. fœtida.*
       (See 12. *S. rupestris,* which has the leaves sometimes digitate.)
Leaves large, entire. Calyx-lobes 4, cohering at the tips . . . . 2. *S. quadrifida.*

SECT. 2. **Brachychiton.**—*Radicle next the hilum. Seeds and inside of the carpels usually villous, often cohering. Leaves entire or lobed (digitate only on some branches of* S. rupestris). *Calyx-lobes spreading.*

Calyx-lobes (where known) with induplicate margins. Seeds (where known) scarcely coher-
    ing. Leaves tomentose or pubescent, at least underneath. Flowers large, sessile.
    (*Brachychiton,* Schott.)
    Leaves green and softly tomentose or pubescent on both sides.
      Leaves broad, entire or obscurely 5- or 7-lobed. Calyx broadly
        campanulate . . . . . . . . . . . . . . . . . 3. *S. ramiflora.*
      Leaves 3-lobed. Calyx tubular-campanulate . . . . . . . 4. *S. Bidwilli.*
      Leaves palmately 5- or 7-lobed . . . . . . . . . . . 7. *S. lurida.*
    Leaves white underneath.
      Leaves angular or obscurely 5- or 7-lobed . . . . . . . 5. *S. discolor.*
      Leaves palmately 5- or 7-lobed, with acuminate lobes . . . . 6. *S. incana.*
Calyx-lobes strictly valvate. Outer coating of the seeds usually re-
    maining adherent to the endocarp. Leaves glabrous. Flowers in
    short panicles.
    Calyx narrow, lobes lanceolate, shorter than the tube. Leaves pal-
      mately 5- or 7-lobed (*Trichosiphon,* Schott) . . . . . . 8. *S. trichosiphon.*
    Calyx broadly campanulate, deeply lobed (*Pœcilodermis,* Schott).
      Leaves large, palmately 5- or 7-lobed. Flowers quite glabrous . 9. *S. acerifolia.*
      Leaves entire, ovate or cordate, or 3-lobed, acuminate. Flowers
        tomentose outside when young, glabrous inside. Follicles sti-
        pitate . . . . . . . . . . . . . . . . . . 10. *S. diversifolia.*
      Leaves cordate-acuminate, entire. Flowers tomentose outside,
        hirsute inside at the base. Follicles nearly sessile . . . . 11. *S. caudata.*
      Leaves entire and lanceolate, or digitate. Flowers tomentose outside.
        Follicles long-stipitate . . . . . . . . . . . . 12. *S. rupestris.*

1. **S. fœtida,** *Linn.; DC. Prod.* i. 483. A tall stout tree, glabrous, except

the very young leaves. Leaves crowded at the ends of the thick branchlets, deciduous, digitately compound on long petioles; leaflets 5 to 11, elliptical oblong or almost lanceolate, 4 to 8 in. long, mostly acuminate, entire, coriaceous, contracted into short petiolules. Flowers rather large, of a dull red, coming out with the young leaves in loose, simple or branched racemes, not exceeding the petioles. Calyx deeply divided into 5 lanceolate spreading segments, about ½ in. long, glabrous outside, tomentose inside. Staminal column or gynophore slender and curved, both in the males and females. Ovary very villous, 5-celled, with many ovules in each cell. Follicles large, woody, glabrous outside, fibrous within. Seeds 10 to 15, oblong, the radicle remote from the hilum.—R. Br. in Benn. Pl. Jav. Rar. 227; Wight, Ic. t. 181 and 364.

**N. Australia.** N. coast (*R. Brown*).
**N. S. Wales.** Hastings and Mackay rivers, *Beckler.*
I have not seen R. Brown's specimens, and Beckler's are leaves only. I insert the species therefore on Brown's authority, describing it from Indian specimens. It ranges over the East Indian and Malayan peninsulas and the Archipelago.

2. **S. quadrifida,** *R. Br. in Benn. Pl. Jav. Rar.* 233. Glabrous, except the inflorescence. Leaves petiolate, ovate or cordate, obtuse or acuminate, mostly 3 to 5 in. long. Racemes several, crowded within the uppermost leaves, 1 to 2 in. long, clothed with a stellate tomentum. Bracts broad, acuminate, very deciduous. Pedicels 2 to 4 lines. Calyx about 4 lines long, tomentose, cleft to the middle, the lobes usually 4, lanceolate, connivent and cohering at the tips. Staminal column short. Follicles sessile, ovoid, 2 to 3 in. long, hard and almost woody, minutely tomentose or glabrous. Seeds 2 to 4, ovoid, black, the radicle remote from the hilum.

**N. Australia.** Sims Island, *A. Cunningham;* Arnhem's Land, *F. Mueller;* Port Essington, *Armstrong;* Cape Upstart, *M'Gillivray.*
**Queensland.** Delta of the Burdekin and Port Denison, *Fitzalan;* Wide Bay, *Bidwill;* Moreton Bay, *F. Mueller.*
The northern specimens have longer and more acute leaves, and rather smaller flowers on longer pedicels than the eastern ones.

3. **S. ramiflora,** *Benth.* A shrub or small tree, clothed with a soft stellate tomentum or pubescence, which rarely disappears on the upper surface of the older leaves. Leaves on long petioles, broadly ovate-cordate or nearly orbicular, mostly acuminate, entire, angular or obscurely 3- or 5-lobed, often attaining 5 or 6 in. Flowers few, large, red, nearly sessile, and clustered in the axils of the upper leaves. Calyx broadly campanulate, 1 to 1½ in. long, the lobes shorter than the tube, spreading, obtuse, 3-nerved in the centre, with broad induplicate margins; inside the tube at the base are 5 small, inflexed, and very villous double scales. Staminal column slender, hirsute at the base. Ovary pubescent; stigmas recurved. Follicles shortly stipitate, 3 to 4 in. long, glabrous outside, villous inside, stipitate (according to R. Brown), with very numerous seeds; I have not seen them perfect.—*Brachychiton paradoxum,* Schott, Meletem. 34; *Brachychiton ramiflorum,* R. Br. in Benn. Pl. Jav. Rar. 234.

**N. Australia.** Brunswick and Vansittart's Bays, N.W. coast, *A. Cunningham;* Victoria river and Point Perron, *F. Mueller.*

**4. S. Bidwilli,** *Hook. Herb.* A shrub or tree, softly pubescent or tomentose in all its parts, closely allied to *S. ramiflora,* but differing in the leaves almost always deeply 3-lobed with acuminate lobes, green, and softly villous on both sides, and especially in the calyx, which is narrow, tubular-campanulate, 1 to 1½ in. long; the red colour and induplicate lobes are the same as in *S. ramiflora.*—*Brachychiton Bidwilli,* Hook. Bot. Mag. t. 5133.

**Queensland.** Wide Bay, *Bidwill;* Burdekin Expedition, *Herb. Mueller;* also in *Leichhardt's* collection.

F. Mueller's herbarium contains a leaf gathered by C. Moore in the mountains near Ipswich, precisely like some of those of *S. Bidwilli,* but with a memorandum by C. Moore that the flower is only ⅓ in. long. If that be the case, it probably forms a distinct species, named by F. Mueller *S. pubescens.*

**5. S. discolor,** *F. Muell.* A tall tree, the young shoots tomentose. Leaves very broadly cordate, nearly orbicular, shortly acuminate, angular or very shortly and irregularly 5- or 7-lobed, glabrous above, white underneath with a very close tomentum, mostly 4 to 6 in. diameter. Flowers (if correctly matched) like those of *S. ramiflora,* and similarly clustered. Calyx 1½ to 2 in. long, broadly campanulate, tomentose inside and out, divided to the middle into broad lobes with induplicate margins. Follicles very shortly stipitate, 4 to 6 in. long, acuminate; densely rusty-tomentose outside.—*Brachychiton discolor,* F. Muell. Fragm. i. 1.

**N. Australia.** Buckland's Table Land, *A. C. Gregory.*
**Queensland.** Pine river, *Hill.*
**N. S. Wales.** Clarence and Richmond rivers, *C. Moore, Beckler.*
The specimens I have seen are in leaf only, with loose flowers and fruits.

**6. S. incana,** *Benth.* A tree, densely clothed with a close, soft tomentum, very white on the under side of the leaves. Leaves deeply divided into 5 or 7 palmate broadly lanceolate lobes, the larger leaves fully 8 in. diameter. Flowers not known. Follicles sessile, ovoid, shortly acuminate, thick and woody, softly tomentose outside, densely tomentose-hirsute inside as well as the seeds, which however do not appear to cohere as in some species.—*Brachychiton incanum,* R. Br. in Benn. Pl. Jav. Rar. 234 ; *Sterculia acerifolia,* A. Cunn. in Loud. Hort. Brit. 392 (in part).

**W. Australia.** Cambridge Gulf, N.W. coast, *A. Cunningham.* The specimens are in leaf and fruit.

**7. S. lurida,** *F. Muell.* A tree. Leaves on long petioles, deeply 5- or 7-lobed, the lobes sinuate or even lobed as in *S. acerifolia,* and of the same size, but softly pubescent, especially underneath. Flowers like those of *S. discolor,* of a livid variegated colour. Calyx campanulate, 1½ to 2 in. long, divided to the middle into broadly ovate lobes, with the margins thin and induplicate. Follicles (according to F. Mueller) shortly stipitate, large, tomentose, many-seeded.—*Brachychiton luridum,* F. Muell. Fragm. i. 1, and ii. 177.

**N. S. Wales.** Clarence river, *C. Moore.* The specimens I have seen are in leaf, with loose flowers. The real distinctions between *S. ramiflora, S. Bidwilli, S. discolor, S. incana,* and *S. lurida,* which alone enter into the section *Brachychiton* as originally defined by Schott, cannot be well ascertained until we have more complete specimens, with the leaves, flowers, and fruits properly matched. These can only be procured by residents in the country itself, as these organs are generally developed at different seasons.

8. **S. trichosiphon,** *Benth.* A tree, quite glabrous, leafless when in flower. Leaves 4 to 8 in. long and broad, more or less deeply cut into 5 or rarely 7 palmate lobes, sometimes broad and shortly acuminate, sometimes lanceolate with long points, and glabrous on both sides. Racemes short, mostly simple. Calyx narrow, tubular-campanulate, about ¾ in. long, the lobes lanceolate, spreading, much shorter than the tube. Staminal column swollen and hairy in the middle. Stigma peltate. Follicles shortly stipitate, glabrous, oblong-triangular, 2 to 3 in. long.—*Trichosiphon australe*, Schott, Melet. 34 ; *Brachychiton platanoides*, R. Br. in Benn. Pl. Jav. Rar. 234.

**N. Australia.** Abel Tasman river, *F. Mueller ;* Nicol Bay, *F. Gregory.*
**Queensland.** Northumberland Island (*R. Brown*), Burdekin and Suttor and Dawson rivers, *F. Mueller ;* Wide Bay, *Bidwill.* The few flowers I have seen were much damaged by insects. I have not seen R. Brown's specimens.

9. **S. acerifolia,** *A. Cunn. in Loud. Hort. Brit* 392 (*partly*). A large timber-tree, quite glabrous. Leaves on long petioles, deeply 5- or 7-lobed ; lobes oblong-lanceolate or almost rhomboid, occasionally deeply sinuate, the whole leaf often 8 or 10 in. diameter, thin but shining, and glabrous on both sides. Flowers of a rich red, in loose axillary racemes or small panicles of 2 to 3 in. Calyx broadly campanulate, ¾ in. long, quite glabrous, with short broad lobes, valvate in the bud. Ovary raised on a short column, quite glabrous, the carpels quite distinct, and the styles scarcely cohering at the broad radiating stigmas. Follicles large, on long stalks, quite glabrous.—*Brachychiton acerifolium*, F. Muell. Fragm. i. 1, and ii. 177.

**N. S. Wales.** Illawarra, *A. Cunningham, M'Arthur,* where it is known by the name of " Flame-tree ;" Macleay and Clarence rivers, *Beckler.*

10. **S. diversifolia,** *G. Don, Gen. Syst.* i. 516. A tree of from 20 to 60 ft., quite glabrous except the flowers. Leaves on long petioles, glabrous and shining, either entire and from ovate to ovate-lanceolate, or more or less deeply 3- or rarely 5-lobed, the 2 lateral lobes sometimes very short, sometimes all lanceolate, 2 or 3 in. long, the simple leaves or their lobes always ending in long points. Flowers in axillary panicles, rarely exceeding the leaves. Calyx very broadly campanulate, slightly tomentose when young, attaining when fully out 7 to 9 lines diameter, acutely lobed to the middle, of a yellowish-white and glabrous except the ciliate margins outside, reddish and glabrous within. Staminal column also glabrous. Ovary slightly tomentose. Follicles nearly ovoid, 1½ to 2 or even 3 in. long, thick and glabrous, on stalks of 1 to 2 in., the endocarp and outer coating of the seeds very shortly hirsute and cohering.—*Pœcilodermis populnea*, Schott, Melet. 33 ; *Brachychiton populneum*, R. Br. in Benn. Pl. Jav. Rar. 234 ; F. Muell. Pl. Vict. i. 156, and Suppl. 5.

**Queensland.** Dawson river, *F. Mueller ;* Rockhampton, *Thozet ;* in the interior, *Mitchell,* according to whom the natives eat the pods.
**N. S. Wales.** From New England, *C. Stuart,* and Macleay river, *Beckler,* to Twofold Bay, *F. Mueller ;* in the interior, *Fraser ;* Lachlan river, *A. Cunningham.*
**Victoria.** Granite ranges on Snowy River and its tributaries, and Hume river, *F. Mueller.*

Var. (?) *occidentalis.* Leaves mostly deeply 3-lobed with narrow lobes, with the addition sometimes of short lateral lobes. Calyx rather smaller and more tomentose than in the eastern form, but not fully out in our specimens, and quite glabrous inside.—*Brachychiton Gregorii,* F. Muell. in Hook. Kew Journ. ix. 199.

**W. Australia.** Murchison river, *Gregory, Drummond, 5th Coll. n.* 93.

11. **S. caudata,** *Heward, in Herb. Cunn.* A tree, quite glabrous except the flowers. Leaves ovate-cordate, entire, long-acuminate, mostly 3 or 4 in. long, the veins more transverse than in any other species, some occasionally narrow-oblong or linear. Flowers rather small, in short axillary panicles, the rachis and pedicels quite glabrous. Calyx broadly campanulate, deeply lobed, 6 to 7 lines diameter when fully out, very tomentose outside, pubescent inside especially at the bottom, but without appendages. Staminal column slender in the males, short in the females, pubescent at the base. Ovary very tomentose. Follicles glabrous, ovoid, rather large and thick, almost sessile.—*Brachychiton diversifolium,* R. Br. in Benn. Pl. Jav. Rar. 234.

**N. Australia.** Careening Bay, N.W. coast, *A. Cunningham ;* Victoria river and Point Pearce, *F. Mueller.* I have been unable to retain R. Brown's specific name, which had been previously applied by G. Don to the last species.

12. **S. rupestris,** *Benth.* A considerable tree, the trunk often swelling out to a large size, contracted at the top and bottom. Leaves quite glabrous, either quite entire, oblong-linear or lanceolate, 3 to 6 in. long, or digitate, consisting of 5 to 9 linear-lanceolate sessile leaflets, often above 6 in. long. Panicle tomentose, usually longer than the petioles. Calyx about 4 lines long, campanulate, deeply lobed, tomentose both inside and out. Staminal column short, hirsute at the base. Follicles ovoid, acuminate, about 1 in. long, on stalks longer than themselves. Seeds, when deprived of the outer coating which remains adherent to the endocarp, smooth and shining, marked with a large scar at the chalazal end, but the radicle in those I have opened always next to the true hilum.—*Delabechea rupestris,* Lindl. in Mitch. Trop. Austr. 155 ; *Brachychiton Delabechii,* F. Muell. Pl. Vict. i. 157.

**Queensland.** Isolated summits of the Grafton range, *Mitchell ;* Wide Bay, *Bidwill ;* Dawson, Mackenzie, and Burnett rivers, Rockhampton and Peak Downs, *F. Mueller.* The colonists give it the name of " Bottle-tree," on account of the singular shape the trunk often assumes. The digitate leaves appear to grow on luxuriant barren branches, for I have never seen them on flowering specimens.

## 2. TARRIETIA, Blume.

(Argyrodendron, *F. Muell.*)

Flowers unisexual. Calyx 5-cleft. Petals none. Staminal column short, adnate to the gynophore, bearing at the summit 10 to 15 anthers irregularly clustered in a head. Carpels of the ovary 3 to 5, nearly distinct, 1-ovulate rarely 2-ovulate. Styles as many, shortly filiform, stigmatic on the inner edge. Fruit-carpels or samaras distinct, spreading, indehiscent, produced at the back into a wing. Seed oblong, albumen splitting in two, cotyledons flat.—Tall trees. Leaves digitately compound, glabrous or scurfy. Flowers small and numerous, in axillary or lateral panicles.

Besides the Australian species, which is endemic, there is another from the Indian Archipelago.

1. **T. argyrodendron,** *Benth.* A tall tree, glabrous except minute scurfy scales on the young shoots and inflorescence, and often on the under

side of the leaves.  Leaflets 3, or on the younger trees often 5, petiolulate, oblong or lanceolate, obtuse or acuminate, 3 to 4 in. long, coriaceous.  Panicles
dichotomous, the upper ones sometimes exceeding the leaves.  Flowers very
numerous.  Calyx broadly campanulate, about 3 lines diameter.  Carpels
with a semiorbicular wing about 1 in. long.—*Argyrodendron trifoliolatum,* F.
Muell. Fragm. i. 2, ii. 177.

**Queensland.**  Common in shady woods on the Brisbane, *A. Cunningham ;* Pine river,
*W. Hill.*

**N. S. Wales.**  Richmond and Clarence rivers, *C. Moore.*

The timber of this tree is said to be hard, and valuable for building.  The flowers in the
Japanese species are much smaller and more numerous, but the structure is the same, its
carpels having a wing of 2 to 3 in.

Var. *grandiflora.*  Calyx 4 lines diameter.  Stigmas short and broad.  Port Denison,
*Fitzalan.*

### 3. HERITIERA, Ait.

Flowers unisexual.  Calyx 5-toothed or 5-cleft.  Petals none.  Staminal
column slender, bearing on the outside below the summit a ring of 5 anthers
with parallel cells.  Carpels of the ovary 5, nearly distinct, 1-ovulate ; style
short, with 5 rather thick stigmas.  Fruit-carpels woody, indehiscent, keeled
or almost winged on the back.  Seeds without albumen, cotyledons very
thick, the radicle next the hilum.—Trees.  Leaves undivided, coriaceous,
scurfy underneath, penninerved.  Flowers small, in axillary panicles.

The genus consists of two tropical Asiatic seacoast trees, of which the one extending to
Australia has the widest range.

1. **H. littoralis,** *Ait. ; DC. Prod.* i. 484.  A tree, attaining a considerable size.  Leaves very shortly petiolate, oval or oblong, the larger ones
fully 8 in. by 4, but often much smaller, quite entire, coriaceous, glabrous
above, silvery underneath with a close scaly tomentum.  Flowers small, numerous, in loose tomentose panicles in the upper axils much shorter than the
leaves.  Calyx about 2 lines long.  Staminal column in the males, pistil in
the females, much shorter than the calyx.  Fruit carpels sessile, ovoid, 2 to 3
in. long, thick and almost woody, with a slightly projecting inner edge, and
a strong, projecting, almost winged keel along the outer edge.

**Queensland.**  N.E. coast, *A. Cunningham.*  Widely dispersed over the seacoasts of
tropical Asia.

### 4. HELICTERES, Linn.

(Methorium, *Schott.*)

Calyx tubular, 5-cleft at the top, often oblique.  Petals 5, equal or the 2 upper
ones broader, the claws elongated, and all or two of them often with a lateral
appendage.  Staminal column adnate to the gynophore, truncate at the
top, or more frequently bearing 5 teeth or small lobes (staminodia), with 1 or
2 stipitate anthers between each, anther-cells divaricate, often confluent into
one.  Ovary nearly sessile on the top of the staminal column, 5-lobed, 5-celled,
with several ovules in each cell.  Styles 5, subulate, more or less connate,
slightly thickened and stigmatic at the top.  Fruit-carpels distinct or separating, opening along their inner edge, straight or spirally twisted.  Seeds with
little albumen, cotyledons leafy, folded round the radicle.—Trees or shrubs,

with stellate or branched tomentum. Leaves entire, serrate or obscurely lobed. Flowers axillary, solitary or clustered. Bracteoles none or distant from the calyx. Capsules usually tomentose, the clusters of tomentum often forming long woolly processes. The appendages on the claws of the petals appear to vary in different flowers of the same species.

A considerable genus, dispersed over the tropical regions both of the New and the Old World, but chiefly American. Of the Australian species one is a common Asiatic one, the two others endemic. The frequently unilocular anthers closely connect the genus with *Malvaceæ*. The other characters are however more of *Sterculiaceæ*, and in some species the anthers are distinctly bilocular.

Calyx ½ in. long. Carpels spirally twisted . . . . . . . . . . . 1. *H. Isora.*
Calyx not above 2 lines long. Carpels straight.
    Leaves obtuse, entire . . . . . . . . . . . . . . . . 2. *H. cana.*
    Leaves toothed, mostly acute . . . . . . . . . . . . . 3. *H. dentata.*

1. **H. Isora,** *Linn.; DC. Prod.* i. 475. A shrub or small tree, with a rather rough stellate pubescence. Leaves on short petioles, broadly obovate or orbicular, often oblique, irregularly toothed or the lower ones obscurely 3-lobed, mostly about 4 in. long, scabrous above, more or less tomentose underneath or sprinkled with short stellate hairs. Pedicels short, usually 2 or 3 together. Calyx ½ in. long or rather longer, obliquely and unequally 5-toothed. Petals red, twice as long as the calyx, 2 of them much broader than the 3 others. Anthers 10, on short filaments, alternating in pairs with the linear staminodia round the ovary. Fruit about 1 in. long, on a stalk of 1½ to 2 in., slightly tomentose, the carpels spirally twisted.—Wight, Ic. t. 180; Bot. Mag. t. 2061.

**N. Australia.** Roper river, *F. Mueller.* Common in East India and the Archipelago.

2. **H. cana,** *Benth.* A shrub, densely clothed with a short, soft or velvety whitish tomentum. Leaves on short petioles, oval or oval-oblong, obtuse, 1½ to 2½ in. long, entire or very obscurely toothed towards the top. Flowers small, in very short axillary sessile cymes or clusters. Calyx about 2 lines long, with short acute teeth. Petals not twice as long, nearly equal or the upper ones rather broader. Anthers 10, small, the filaments rather long, alternating in pairs with the shorter ovate, very thin and transparent staminodia. Fruit ovoid, under ½ in. long, on a stalk of about 2 lines, loosely woolly, the carpels straight.—*Methorium canum,* Schott, Meletem. 29, t. 5; *M. integrifolium,* F. Muell. Trans. Phil. Soc. Vict. iii. 40.

**N. Australia.** Brunswick Bay and York Sound, *A. Cunningham;* Upper Victoria river, *F. Mueller.*

3. **H. dentata,** *F. Muell. Herb.* Apparently a small shrub or undershrub, the slender branches, inflorescence, and under side of the leaves whitish with a close stellate tomentum. Leaves shortly petiolate, from orbicular to ovate or oblong-elliptical, rather acute, rarely exceeding 1 in., more or less toothed, greener and less tomentose above than underneath. Flowers pink or purple, rather smaller, more numerous, and in looser cymes than in *H. cana.* Calyx rarely attaining 2 lines. Petals and stamens as in *H. cana,* but the staminodia much shorter and broader, and exceedingly delicate. Fruit small, with straight carpels.

**N. Australia.** Upper Victoria river, *F. Mueller.*

Var *procumbens.* Branches procumbent, ½ to 2 ft. long; tomentum looser; leaves smaller and rounder velvety-villous on the upper side; staminodia longer. Macadam range, *F. Mueller.*

Var. (?) *flagellaris.* Branches prostrate, 1 to 2 ft. long; leaves nearly sessile, cordate or orbicular, 1 to 1½ in. long; cymes on long slender peduncles. Port Essington, *Armstrong.*

## 5. **PTEROSPERMUM,** Schreb.

Bracteoles 3, entire or laciniate, sometimes very deciduous, or perhaps none. Calyx tubular, 5-cleft, deciduous. Petals 5, often very long, deciduous. Staminal column adnate to the gynophore, divided at the top into 5 linear-clavate staminodia, with 3 stipitate anthers between each; anther-cells linear, parallel. Ovary sessile in the top of the column, 5-celled with several ovules in each cell. Style undivided, club-shaped, and 5-furrowed at the top. Capsule woody or coriaceous, ovoid or oblong, terete or angular, opening loculicidally in 5 valves. Seeds ascending, produced into a wing at the top; albumen little or none; cotyledons wrinkled or folded; radicle inferior, rather long.—Trees or shrubs, clothed with a stellate tomentum or scurfy scales. Leaves coriaceous, often oblique, entire, cuneate-toothed or angled at the upper end, penninerved or several-nerved at the base. Peduncles short, axillary, 1-flowered. Flowers often several inches long.

The genus is limited to East India and the Archipelago, the Australian species being probably the same as one of the Asiatic ones.

1. **P. acerifolium,** *Willd.; W. and Arn. Prod.* 69? I have seen a fragment only in very young bud, which agrees with this species in the very angular rusty-tomentose young calyx, and in the bracteoles divided into narrow-linear lobes, and falling off at a very early stage. There are 3 leaves only, the largest is, as in *P. acerifolium,* coriaceous, broad at the end, cordate at the base, nearly glabrous above, tomentose underneath, with about 11 prominent nerves radiating from the petiole; but it is much narrower than usual in that species, measuring 9 in. by 4. The 2 others are as yet not half developed, but are broader in proportion, and although the specimen is insufficient for identification, it shows no character to separate it from *P. acerifolium.*—Wight, Ic. t. 631.

**N. S. Wales.** Illawarra? *Vernon (Herb. F. Mueller).*

## 6. **MELHANIA,** Forsk.

Bracteoles 3, persistent. Calyx divided almost to the base into 5 segments. Petals 5, persistent. Staminal cup very short, bearing 5 ligulate staminodia, and 5 stipitate anthers alternating with them, the anther-cells parallel. Ovary sessile, 5-celled with 1 or more ovules in each cell. Style usually short, with 5 subulate branches, stigmatic along the inner side. Capsule opening loculicidally in 5 valves. Seeds with albumen; cotyledons folded, 2-cleft; radicle inferior.—Herbs, undershrubs, or small shrubs, softly tomentose. Leaves ovate or cordate, serrate-crenate. Peduncles axillary, 1- or few-flowered. Bracteoles often exceeding the calyx. Flowers yellow.

The genus extends over the tropical and subtropical regions of the Old World, but is most abundant in Africa. The Australian species is the same as an Indian one. The habit is that of some *Malvaceæ.*

1. **M. incana,** *Heyne; W. and Arn. Prod.* 68.   A rather slender shrub of 1 or several ft., hoary or white except the upper side of the leaves with a close or velvety tomentum.   Leaves shortly petiolate, oblong or ovate-lanceolate, obtuse, scarcely toothed, 1 to 2 or even 3 in. long, tomentose on both sides, or nearly glabrous above.   Peduncles bearing 1, 2 or rarely 3 or 4 flowers, the pedicels very short.   Bracteoles narrow-linear or subulate, rather shorter than the calyx.   Sepals lanceolate-subulate, tomentose, about 4 to 6 lines long.   Petals rather longer, broad, yellow.   Staminodia linear, often 3 lines long; anthers shorter, linear, on short filaments.   Style elongated. Capsule tomentose, shorter than the calyx, with 2 or 3 seeds in each cell.— *M. oblongifolia*, F. Muell. Fragm. i. 69.

**N. Australia.**   York Sound, Cygnet Bay, and Dampier's Archipelago, *A. Cunningham;* Upper Victoria river and Sturt's Creek, *F. Mueller;* islands of the Gulf of Carpentaria, *R. Brown;* Albert river, *Henne.*

**Queensland.**   Broad Sound, *R. Brown;* Rockhampton and Burdekin rivers, *F. Mueller;* Port Curtis, *M'Gillivray;* Port Denison, *Fitzalan.*

The species is also found in the East Indian peninsula, and a slight variety or closely allied species in tropical Africa.

## 7. MELOCHIA, Linn.

### (Riedleia, *Vent.*)

Calyx 5-lobed or 5-toothed, campanulate or inflated.   Petals 5, spathulate or oblong.   Stamens 5, united at the base, without any or with very minute tooth-like intervening staminodia; anther-cells parallel.   Ovary sessile or shortly stipitate, 5-celled with 2 ovules in each cell, styles 5, free, or united at the base, often thickened at the stigmatic top.   Capsule opening loculicidally in 5 or fewer valves, some of the cells occasionally abortive.   Seeds usually solitary in each cell, ascending, with more or less of albumen; embryo straight, with flat cotyledons.—Herbs, shrubs, or rarely trees, the stellate tomentum occasionally mixed with spreading hairs.   Leaves serrate. Flowers small, axillary or terminal, clustered or in cymes or panicles.

A large genus, dispersed over the warmer regions of the globe, the herbaceous and suffruticose species chiefly American.   The two Australian species are both herbaceous; one belongs to the American series, the other is Asiatic.

Capsule very angular, pyramidal, much longer than the calyx   .   .   .   1. *M. pyramidata.*
Capsule small, globular   .   .   .   .   .   .   .   .   .   .   .   2. *M. corchorifolia.*

1. **M. pyramidata,** *Linn.; DC. Prod.* i. 490.   Herbaceous, with a hard almost woody base, although sometimes annual only.   Branches slender, divaricate, often 2 or 3 ft. long, slightly pubescent in a decurrent line or all over.   Leaves petiolate, lanceolate, or the lower ones ovate, the larger ones 1 to 2 in. long, serrate, usually glabrous.   Flowers small, purplish, 2 to 4 together in little almost sessile axillary umbels.   Calyx 10-ribbed.   Petals about 2 lines long.   Capsule 3 to 4 lines long, acuminate, the very prominent angles produced into short horizontal points, giving each valve a rhomboidal, and the whole capsule a pyramidal shape.—A. Gray, Gen. Ill. t. 134.

**N. Australia.**   Victoria river, *F. Mueller.*
**Queensland:**   Rockhampton, *Wallace.*

The species is very generally distributed over tropical America, and occurs also in E. Africa, the Mauritius, and the Pacific islands.

2. **M. corchorifolia,** *Linn. Spec.* 944.　Herbaceous, with the habit of *M. pyramidata*, but usually more erect, glabrous or with slightly pubescent decurrent lines.　Leaves petiolate, from broadly ovate to lanceolate, mostly 1 to 2 in. long, serrate or crenate, glabrous.　Flowers small, purplish, nearly sessile in clusters, usually several together in a broad, terminal, sessile cyme, rarely a few smaller clusters in the upper axils.　Calyx 5-angled.　Petals about 2 lines long.　Capsule small, depressed-globular, with scarcely prominent angles, sprinkled with a few hairs, the valves very rarely splitting septicidally. —*Riedleia corchorifolia*, DC. Prod. i. 491 ; W. and Arn. Prod. i. 66.

**N. Australia.**　Port Essington, *Armstrong ;* Sturt's Creek and Macadam range, *F. Mueller.*

The species is common in E. India, and includes *M. concatenata*, Linn., and *M. supina*, Linn., with all the synonyms referred to these plants respectively by Wight and Arnott (l. c., under *Riedleia*).　Some of the Australian specimens are much starved, with small, occasionally axillary, heads of flowers, apparently approaching *M. nodiflora*, Sw., another widespread tropical species, which however not only has all the flowers in axillary clusters, but the capsule is much more deeply furrowed, and usually septicidal as well as loculicidal, the carpels often entirely separating.

## 8. DICARPIDIUM, F. Muell.

Calyx 5-lobed.　Petals oblong-spathulate, persistent.　Stamens 5, very shortly united at the base, without intervening staminodia, anther-cells parallel.　Ovary sessile, 2-celled with 2 ovules in each cell ; styles 2, distinct, thickened upwards.　Fruit-carpels separating, 2-valved, with 1 or 2 seeds in each.　Seeds ascending ; albumen fleshy ; embryo straight, with flat cotyledons.—An undershrub, with the habit of *Waltheria*, from which the genus only differs in the carpels, two instead of one.　The flowers are also more or less unisexual, but that is perhaps sometimes the case in *Waltheria*.

The genus is limited to the single Australian species.

1. **D. monoicum,** *F. Muell. in Hook. Kew Journ.* ix. 302.　An undershrub of 1 to 2 ft., hirsute all over with rigid stellate hairs, the branches rather slender, diffuse or erect.　Leaves nearly sessile, oblong, mostly about 1 in. long, toothed, plicate, and densely hirsute.　Flowers small, almost sessile, solitary, or 2 or 3 together in the upper axils, each within a bract and 2 bracteoles, the males with small carpels and short styles, the ovules, although apparently perfect, not setting ; the female flowers rather smaller, with smaller anthers, but perfecting their fruit.　Carpels small, tomentose.

**N. Australia.**　Macarthur river and Seven Emu creek, *F. Mueller.*

## 9. WALTHERIA, Linn.

Calyx 5-lobed.　Petals 5, spathulate, persistent.　Stamens 5, united at the base, without intervening staminodia ; anther-cells parallel.　Ovary sessile, consisting of a single 1-locular, 2-ovulate carpel, style excentrical, thickened or fringed upwards.　Capsule 2-valved, 1-seeded.　Seed ascending, albumen fleshy ; embryo straight, cotyledons flat.—Herbs, undershrubs, or rarely trees, the stellate tomentum usually mixed with spreading hairs.

Leaves serrate    Stipules narrow.    Flowers usually small, axillary or terminal in clusters, heads, cymes, or panicles.

The species are mostly American, two are African, and two from the Pacific islands. The Australian species is one which is very generally dispersed over the tropical regions of both the Old World and the New.

1. **W. americana,** *Linn. ; DC. Prod.* i. 492. A perennial or under-shrub, 1 to 2 ft. or more high, densely tomentose or softly villous in every part. Leaves shortly petiolate, from ovate to oblong, 1 to 1½ in. long, obtuse, toothed and plicately veined. Flowers small, yellow, in dense heads, almost sessile in the axils of the leaves, or the upper ones clustered in a short spike, or irregularly collected into dense cymes or leafy corymbs. Bracts narrow. Calyx 1½ to 2 lines long. Petals nearly twice as long, narrow.—*W. indica,* Linn.; DC. Prod. i. 493.

**N. Australia.** Cambridge Gulf, *A. Cunningham ;* Victoria river and Arnhem's Land, *F. Mueller ;* Port Essington, *Armstrong ;* Gulf of Carpentaria, *R. Brown, Landsborough.*
**Queensland.** Cape Flinders, *A. Cunningham ;* Port Denison, *Fitzalan.*
The species is common within or near the tropics all round the globe.

## 10. ABROMA, Jacq.

Calyx 5-cleft. Petals 5, the claw dilated and concave at the base, the lamina stipitate, ovate, plane. Staminal cup with 5 obcordate lobes (stami-nodia) alternating with the petals, anthers 2 to 4 in each sinus, nearly sessile, with divaricate cells. Ovary sessile, 5-celled with several ovules in each cell ; styles 5, short, connivent. Capsule membranous, truncate, 5-angled, the angles winged and produced at the top into as many horn-like points, open-ing at the top loculicidally and septicidally. Seeds several, albuminous ; embryo straight, with flat cotyledons.—Tall shrubs or small trees, with stel-late pubescence. Leaves entire or palmately lobed. Peduncles leaf-opposed or terminal, few-flowered. Dissepiments of the capsule fringed at the inner edge with long hairs.

A genus of two or three species from tropical Asia, one of them the same as the Austra-lian one.

1. **A. fastuosa,** *R. Br. ; DC. Prod.* i. 485. A tall shrub, the branches softly pubescent, and bearing a few minute conical prickles. Leaves shortly petiolate, obliquely cordate-ovate, acuminate, 4 to 6 in. long, undivided, slightly sinuate-toothed, nearly glabrous above, softly pubescent underneath. Peduncles very much shorter than the leaves, bearing a cluster of 3 to 5 shortly pedicellate flowers, one only usually fertile. Bracts linear, deciduous. Sepals narrow-lanceolate, about ½ in. long. Petals rather exceeding them, the broadly ovate lamina supported above the concave base by a filiform stipes. Capsule hirsute with a few rigid hairs, or at length glabrous, 1¼ in. long, the wings of the angles nearly ½ in. broad, besides the long incurved points of their upper angle. Seeds 10 to 12 in each cell.—Gærtn. Fr. i. t. 64 ; Salisb. Parad. Lond. t. 102.

**Queensland.** Endeavour river, *R. Brown (Hb. R. Br.).*
The species is widely distributed over the Eastern Archipelago.

## 11. RULINGIA, R. Br.

(Achilleopsis, *Turcz.*)

Calyx 5-lobed. Petals 5, broad and concave or convolute at the base, with a small, broad, or linear ligula at the top. Stamens shortly or scarcely connate at the base, 5 without anthers (staminodia), linear-lanceolate and petal-like, alternate with the petals and connivent or spreading; 5 short, opposite the petals, and perfect, the anther-cells parallel. Ovary sessile, 5-celled with 2 or rarely 3 ovules in each cell, styles connate, at least at the top, or rarely quite free. Capsule tomentose or beset with prickles or soft setæ, opening loculicidally in valves, or the carpels separating. Seeds 1 or 2 in each cell or carpel, ascending, usually strophiolate. Albumen fleshy; cotyledons flat.—Shrubs or undershrubs, with stellate tomentum or hairs. Leaves entire, toothed, or lobed. Stipules narrow, deciduous, the upper ones often laciniate. Flowers mostly white, small, in leaf-opposed or terminal, rarely axillary cymes. Petals shorter than the calyx. Strophiola of the seeds small, variable in shape in the same species.

The genus is confined to Australia, with the exception of one Madagascar species.

A. *Leaves of the flowering branches or their lobes lanceolate or ovate-lanceolate, mostly above 1 and often 2 or 3 in. long, entire or serrate, not undulate, crenate or crisped. Capsule loculicidal.*

Leaves or their lobes quite entire, softly hoary-tomentose   .   .   .   1. *R. salvifolia.*
Leaves or their lobes serrate, velvety or hirsute, at least underneath.
  Capsule scarcely dehiscent, nearly glabrous, with rigid prickly
    setæ   .   .   .   .   .   .   .   .   .   .   .   .   .   .   2. *R. pannosa.*
  Capsule dehiscent, tomentose with soft pubescent setæ   .   .   .   3. *R. rugosa.*

B. *Leaves ovate or oblong, irregularly crenate or lobed, often undulate or crisped, mostly above 1 in. and often 2 or 3 in. long. Calyx very prominently angled in the bud (except R. loxophylla). Capsule loculicidal and often septicidal also.*

Buds obtuse.
  Petals gibbous at the base, abruptly ligulate. Leaves glabrous
    or pubescent above.
    Calyx-lobes erect or connivent. Leaves large, little lobed . .   4. *R. corylifolia.*
    Calyx-lobes rounded, very spreading. Leaves smaller, much-
      lobed   .   .   .   .   .   .   .   .   .   .   .   .   .   7. *R. platycalyx.*
  Petals not gibbous, tapering into a short linear ligula. Leaves
    little-lobed, hoary-tomentose .   .   .   .   .   .   .   .   .   .   5. *R. grandiflora.*
  Petals not gibbous. Ligula short, oblong-spathulate. Leaves
    oblique, densely velvety   .   .   .   .   .   .   .   .   .   10. *R. loxophylla.*
Buds acute. Petals tapering into a slender ligula about as long
  as the calyx. Leaves much-lobed, often crisped, nearly glabrous
  or pubescent above .   .   .   .   .   .   .   .   .   .   .   .   .   6. *R. malvæfolia.*

C. *Leaves (except R. loxophylla) crenate, more or less undulate, and crisped or bullate, but little lobed, and rarely exceeding 1 in. Buds small, scarcely angular. Capsule loculicidal, sometimes also septicidal, or the carpels separating.*

Cymes pedunculate. Leaves glabrous or scabrous above.
  Buds acute. Ligules long and slender   .   .   .   .   .   .   .   .   6. *R. malvæfolia.*
  Buds obtuse.
    Leaves narrow-oblong and crenate, or, when luxuriant, ovate-
      lanceolate and slightly lobed   .   .   .   .   .   .   .   .   9. *R. hermanniæfolia.*
    Leaves mostly ovate and lobed.
      Calyx about 3 lines diameter. Petals not gibbous at the
        base .   .   .   .   .   .   .   .   .   .   .   .   .   .   .   8. *R. parviflora.*

Calyx 5 or 6 lines diameter, lobes very broad. Petals gib-
　　bous at the base . . . . . . . . . . . . . . 7. *R. platycalyx.*
Cymes sessile or nearly so. Leaves hoary-tomentose or velvety on
　　both sides.
Leaves very oblique, densely velvety, ¾ to 2 in. Ligules of the
　　petals shortly oblong . . . . . . . . . . . . . 10. *R. loxophylla.*
Leaves small, hoary-tomentose.
　　Ligules linear, rather broad. Leaves ½ to 1 in. . . . . . 11. *R. cuneata.*
　　Ligules obovate or spathulate. Leaves under ½ in. . . . . 12. *R. rotundifolia.*

　　D. *Leaves pinnatifid. Flowers in dense terminal corymbose
cymes. Carpels separating, crested on the back.* (**Achilleopsis,**
*Turcz.*) . . . . . . . . . . . . . . . . . . . . . 13. *R. densiflora.*

1. **R. salvifolia,** *Benth.* An apparently erect shrub, clothed with a soft
but dense and close whitish tomentum. Leaves on very short petioles, lanceolate or lanceolate-linear, 2 to 4 in. long, entire or deeply divided into 3
lanceolate lobes, the middle one the longest, all quite entire and softly tomentose on both sides, especially underneath. Cymes pedunculate, but
shorter than the leaves. Calyx spreading, about 3 lines diameter. Ligula
of the petals linear, usually pubescent. Stamens very shortly united. Fruit
not seen.—*Thomasia (?) salvifolia,* A. Cunn. Herb. ; Steetz, in Pl. Preiss. ii.
333.

　　**Queensland.** Brisbane river, *A. Cunningham ;* Minto's Craig, *Fraser.*

2. **R. pannosa,** *R. Br. in Bot. Mag. t.* 2191. A shrub of several feet,
but flowering young so as to appear an undershrub, softly hirsute with velvety
stellate hairs. Leaves on the full-grown plant shortly petiolate, ovate-lanceolate or lanceolate, mostly 2 to 3 in. or sometimes longer, toothed, rounded
or cordate at the base, scabrous-pubescent above, with impressed veins,
densely velvety or hirsute underneath ; on the younger plants they are
broader and often 3 or 5-lobed. Cymes shortly pedunculate. Calyx tomentose, spreading to 3 or 4 lines diameter. Ligula of the petals linear, rather
short. Staminodia pubescent, united with the perfect stamens higher up
than in most species. Ovary glabrous, granulate. Capsule nearly glabrous,
globular, hard and almost indehiscent, beset with rigid subulate bristles, glabrous except a stellate tuft at the tip.—Steetz, in Pl. Preiss. ii. 351 ; F.
Muell. Pl. Vict. i. 150 ; *Commersonia dasyphylla,* Andr. Bot. Rep. t. 603 ;
*Buettneria dasyphylla,* J. Gay, in DC. Prod. i. 486, and in Mem. Mus. Par. x.
200, t. 12 ; *B. pannosa,* DC. Prod. i. 486.

　　**Queensland.** Glasshouses, Moreton Bay, *F. Mueller.*
　　**N. S. Wales.** Port Jackson, *R. Brown, Sieber, n.* 217, and *Fl. Mixt. n.* 546, and
others ; northward to Clarence and Hastings rivers, *Beckler ;* and New England, *C. Stuart ;*
southward to Twofold Bay, *F. Mueller.*
　　**Victoria.** Amongst granite boulders in the Buffalo range, and near Mount Imlay, *F.
Mueller.*

3. **R. rugosa,** *Steetz, in Pl. Preiss.* ii. 352. A shrub, so closely resembling *R. pannosa* in indumentum, foliage, and apparently in flowers, that it is
difficult to distinguish it without the fruit. Leaves usually narrower, more
rugose, and almost bullate. Flowers (which I have only seen very young)
fewer in the cymes. Ovary tomentose. Capsule about 4 lines diameter
without the setæ, not so hard as in *R. pannosa* and readily dehiscent, beset

with soft pubescent setæ, which are long in Cunningham's specimens, shorter in Stuart's.

**N. S. Wales.** Wellington Valley and to the westward, *A. Cunningham ;* New England, *C. Stuart.*

4. **R. corylifolia,** *Grah. in Bot. Mag. t.* 3182. An erect shrub, roughly tomentose-villous with stellate hairs. Leaves broadly ovate, 2 to 3 in. long, irregularly toothed or broadly lobed, wrinkled, green and roughly pubescent above, more densely tomentose-villous or pubescent underneath. Cymes dense and sessile, forming dense terminal leafy corymbs. Bracts and stipules lanceolate. Calyx prominently 5-angled, villous, deeply lobed, the segments about 4 lines long, erect or connivent. Petals gibbous at the base, the margins of the erect broad part involute, but not united above their attachment as represented by mistake in the plate, the ligula linear, rather short. Stamens shortly united. Ovary prominently 5-angled, styles quite distinct. Capsule depressed-globular, 5-furrowed, covered with rigid stellate hairs, deeply loculicidal and sometimes septicidal also.—Steetz, in Pl. Preiss. ii. 358 ; *Commersonia Preissii,* Steud. in Pl. Preiss. i. 237.

**W. Australia.** King George's Sound, *R. Brown, A. Cunningham, Drummond, Preiss, n.* 1652, and others ; *Leschenault, Oldfield.*

5. **R. grandiflora,** *Endl. in Hueg. Enum.* 12. A shrub or undershrub of 2 or 3 ft., clothed with a whitish, close or velvety tomentum. Leaves broadly or narrow-ovate, obtuse, mostly 1½ to 2 in. long, irregularly toothed or slightly lobed, tomentose on both sides but whiter underneath. Cymes dense and nearly sessile, but not so much so as in *R. corylifolia.* Calyx prominently angled, scarcely spreading, tomentose. Petals concave at the base, but not gibbous, more gradually narrowed into the ligula than in most species. Staminodia and stamens very short. Capsule globose, longer than the calyx, 4 to 5 lines diameter, densely hirsute with stellate hairs borne on very short setæ, the cells or carpels usually 2-seeded.—Steetz, in Pl. Preiss. ii. 355 ; *R. altheæfolia,* Turcz. in Bull. Mosc. 1852, ii. 151 ; *Commersonia cinerea,* Steud. in Pl. Preiss. i. 238.

**W. Australia.** King George's Sound, *Menzies, Huegel, Drummond, n.* 268, etc. Perongerup ranges and road to Cape Riche, *Maxwell, Preiss, n.* 1664.

6. **R. malvæfolia,** *Steetz, in Pl. Preiss.* ii. 356. A low diffuse or ascending shrub or undershrub, resembling *R. platycalyx* and the larger specimens of *R. parviflora,* but readily known by the calyx and petals. Leaves ovate or rarely oblong, obtuse, ¾ to 1½ in. or even 2 in. long, mostly 3- or 5-lobed, the lateral lobes short, all coarsely crenate or obtusely lobed and often undulate or crisped, glabrous or pubescent above, more or less hirsute underneath as well as the branches. Cymes shortly pedunculate. Buds angular and rather acute. Calyx spreading to at least 4 lines diameter, the lobes very acute, hairy outside especially at the base. Petals with a very short broad base, tapering into a very narrow ligula nearly or quite as long as the calyx. Capsule rather large, beset with long glandular-hairy setæ.—*Commersonia cygnorum,* Steud. in Pl. Preiss. i. 237.

**W. Australia.** Swan River, and to the northward, *Drummond, Preiss, n.* 1642, also King George's Sound, *Menzies, Oldfield* (a narrow-leaved variety).

7. **R. platycalyx,** *Benth.* Shrubby and apparently diffuse, the branches hirsute-tomentose with rigid stellate hairs. Leaves broadly ovate, mostly under 1 in. long and deeply 3-lobed, the lobes crenate or almost pinnatifid, undulate and often crisped, glabrous or scabrous-pubescent above, tomentose and hirsute underneath. Cymes pedunculate. Buds obtuse, slightly angular. Calyx spreading to 5 or 6 lines diameter, the lobes broad and very obtuse. Petals gibbous at the base, almost as in *R. corylifolia,* the ligula linear, rather short. Capsule densely beset with short hirsute setæ, but not seen fully ripe.

**W. Australia,** *Drummond, 5th Coll. n.* 269.

8. **R. parviflora,** *Endl. in Hueg. Enum.* 12. A low shrub or under-shrub, with prostrate or ascending branches of ½ to 1½ ft., the young ones hirsute with stellate hairs. Leaves very shortly petiolate, ovate or ovate-lanceolate, obtuse, rarely 1 in. long, deeply crenate and mostly lobed, with undulate often crisped margins, glabrous or nearly so above, tomentose or hirsute underneath. Cymes shortly pedunculate. Buds small, obtuse, scarcely angular. Calyx spreading to about 3 lines, hirsute outside especially at the base, the lobes obtuse. Petals broad and very open at the base, with a rather long ligula, yet much shorter than in *R. malvæfolia.* Capsule about 2 lines diameter, slightly hirsute, with stellate hairs on very short setæ.— Steetz, in Pl. Preiss. ii. 356 ; *R. corylifolia,* Steud. in Pl. Preiss. i. 237, not Grah. ; *R. nana,* Turcz. in Bull. Mosc. 1852, ii. 150.

**W. Australia.** King George's Sound, *Menzies, Huegel, Drummond, n.* 270, *Preiss, n.* 1650, and others. Readily distinguished from the last two, of which it has nearly the foliage, by the calyx and petals ; it is much more nearly allied in character to the eastern *R. hermanniæfolia,* from which the chief differences consist in habit and foliage difficult to describe in words.

9. **R. hermanniæfolia,** *Steetz, in Pl. Preiss.* ii. 353. A shrub, often of several ft., with slender but rigid divaricate branches, hirsute when young but soon nearly glabrous. Leaves in most specimens narrow-oblong and not above ½ in. long, in more luxuriant ones often ovate-lanceolate, or with short broad basal lobes, always obtuse, crenate, much wrinkled with revolute margins of a firm consistence, at length glabrous above, white-tomentose under-neath. In young plants the leaves are often broader and more lobed. Cymes shortly pedunculate. Buds small, obtuse, scarcely angular. Calyx tomentose, opening to nearly 3 lines diameter. Petals broad and open at the base, the ligula linear, rather short. Capsule 2 or rarely 3 lines diameter, pubescent and densely beset with very short hirsute setæ.—*Buettneria hermanniæfolia,* J. Gay, in DC. Prod. i. 486, and Mem. Mus. Par. x. 204, t. 13 ; *Rulingia cristifolia,* A. Cunn. Herb., (usually miswritten *cistifolia*) ; Steetz, in Pl. Preiss. ii. 354 ; *R. oblongifolia,* Steetz, l. c. 353 ; *Lasiopetalum dumosum,* Lodd. Bot. Cab. t. 1564.

**N. S. Wales.** Port Jackson, *R. Brown, Backhouse,* and others ; Hunter's River, *Paterson, A. Cunningham.*

10. **R. loxophylla,** *F. Muell. Fragm.* i. 68. An erect shrub of 1½ ft., densely velvety tomentose, almost hirsute. Leaves obliquely ovate or cordate, obtuse, ¾ to 2 in. long, crenate, soft and thick, the tomentum rather harsh on the upper side, very dense and whitish underneath. Cymes small, sessile or

nearly so.  Calyx tomentose inside and out, spreading to about 2 lines diameter, the lobes acute.  Petals broad, concave, with an oblong rather short ligula.  Staminodia glabrous.  Fruit not seen.

**N. Australia.**  Table land between Victoria river and Hooker's and Sturt's Creeks, *F. Mueller.*

11. **R. cuneata,** *Turcz. in. Bull. Mosc.* 1852, ii. 151.  A low shrub or undershrub, with prostrate or ascending branches of ½ to 2 ft. or rather more, whitish with a close tomentum without spreading hairs.  Leaves petiolate, from obovate to nearly orbicular, ½ to 1 in. long, very obtuse, irregularly and coarsely crenate, and often undulate or crisped on the margin, whitish with a close tomentum on both sides.  Cymes small, nearly sessile.  Calyx tomentose, spreading to nearly 3 lines diameter, the lobes obtuse.  Petals broad and expanding into involute lobes at the base, the ligula cuneate-oblong or almost obovate, rather shorter than the calyx.  Fruit not seen.

**W. Australia.**  S. coast (?) *Drummond, n.* 61, 271, and 273 ; Fitzgerald river, *Herb. Mueller.*  Some larger-leaved specimens were described by Turczaninow (Bull. Mosc. 1852, ii. 151), under the name of *R. hexamera,* given to them probably from having examined an abnormally hexamerous flower.

12. **R. rotundifolia,** *Turcz. in Bull. Mosc.* 1852, ii. 152.  Shrubby, with elongated slender branches, tomentose when young, but soon becoming glabrous.  Leaves on short petioles, nearly orbicular, rarely ½ in. long, very obtuse, crenate or rarely lobed, undulate or crisped on the margin, whitish-tomentose on both sides, especially underneath.  Cymes nearly sessile.  Flowers the smallest of the genus, when expanded scarcely measuring above 2 lines diameter.  Calyx tomentose, with obtuse lobes.  Petals very concave, but not gibbous or saccate at the base, the ligula linear, rather broad and nearly as long as the calyx.  Stamens almost free.  Carpels of the ovary almost free.  Fruit not seen.

**W. Australia,** *Drummond, n.* 270 ; Fitzgerald river, *Herb. Mueller.*

13. **R. densiflora,** *Benth.*  An erect shrub of several feet, densely hirsute with stellate hairs.  Leaves from ovate to linear, mostly lanceolate, 1 to 2 in. long, pinnatifid, the lobes short, obtuse, and coarsely crenate, or longer and again lobed, very rugose and convex, almost bullate, scabrous or hispid above, white-tomentose or hirsute underneath.  Flowers numerous, white, crowded in a terminal compound corymbose cyme, often many inches in diameter.  Calyx spreading to about 6 lines diameter, with petal-like, rather acute, softly pubescent lobes.  Petals with a broad concave base, often produced into shortly involute lateral lobes, the ligula linear but very short.  Stamens nearly or quite free ; staminodia pubescent.  Fruit-carpels quite separating, keeled and crested on the back, 1-seeded.—*Achilleopsis densiflora,* Turcz. in Bull. Mosc. 1849, ii. 10.

**W. Australia.**  Murchison river and Wangan hills, *Drummond, n.* 100, 38, *Oldfield* and others.

## 12. COMMERSONIA, Forst.

Calyx 5-lobed.  Petals 5, broad and concave at the base, with a small broad or linear ligula at the top.  Stamens united in a short cup at the base, 5 perfect with short filaments opposite the petals, alternating with staminodia in threes, the central one of each three lanceolate or spathulate, the lateral ones linear or spathulate, attached at the base either to the central one or to the adjoining anther-bearing filament. Ovary sessile, 5-celled, with 2 to 6 ovules in each cell; styles distinct or united at least at the top. Capsule beset with soft pubescent setæ, opening loculicidally in 5 valves. Seeds usually 2 or 3, ascending, with a small strophiola; albumen fleshy; cotyle-dons flat.—Trees or shrubs, with stellate tomentum or hairs. Leaves toothed or lobed, often oblique. Flowers small, in terminal, leaf-opposed, or axillary cymes.

The species are all Australian, one is also widely dispersed over Eastern India, the Archi-pelago and Pacific islands, the others are endemic.

Tall shrubs or trees. Leaves mostly above 3 in. long, acuminate.
　Ligula of the petals linear or oblong.
　Staminodia all linear-spathulate, elongated, the lateral ones attached
　　to the central . . . . . . . . . . . . . . . . . . 1. *C. Fraseri*.
　Central staminodia lanceolate, lateral ones filiform.
　　Lateral staminodia attached to the central one. Ligula of the
　　　petals oblong, rather short . . . . . . . . . . . . 2. *C. Leichhardtii*.
　　Lateral staminodia attached to the anther-bearing filaments.
　　　Ligula of the petals long and linear . . . . . . . . 3. *C. echinata*.
Small shrubs. Leaves obtuse, undulate and crenate, usually small.
　Ligula of the petals short and broad.
　Lateral staminodia attached to the anther-bearing filaments. Leaves
　　very unequally cordate . . . . . . . . . . . . . 4. *C. Gaudichaudi*.
　Lateral staminodia (very small) attached to the central lanceolate
　　one. Leaves equal at the base.
　　Calyx-lobes rather acute. Leaves much crisped . . . . . 5. *C. crispa*.
　　Calyx-lobes very obtuse, broad and white.
　　　Leaves scabrous or tomentose, mostly ½ to 1 in. long . . . 6. *C. pulchella*.
　　　Leaves glabrous above, rarely above 3 lines . . . . . . 7. *C. microphylla*.

1. **C. Fraseri,** *J. Gay, in Mem. Mus. Par.* x. 215, *t.* 15. A tall shrub, with tomentose or hirsute branches. Leaves cordate-ovate, acuminate, 3 to 6 in. long, irregularly toothed, often oblique at the base, glabrous or slightly pu-bescent above, white-tomentose or softly hirsute underneath, the lower ones in the young plants broad and 3- or 5-lobed. Cymes loosely dichotomous, many-flowered, but shorter than the leaves. Calyx tomentose, fully 3 lines diameter, the lobes acute. Petals with a very short broad concave base, the ligula oblong-spathulate, nearly as long as the calyx. Staminodia linear-spathulate, as long as the petals, the central one of each three rather broader and lanceolate at the base, the lateral ones filiform at the base and shortly adnate to the central one; anther-bearing filaments very short. Capsule large, densely beset with soft villous setæ.—Steetz, in Pl. Preiss. ii. 359; F. Muell. Pl. Vict. i. 148.

**N. S. Wales.** Banks of the Hawkesbury, *R. Brown;* Port Jackson, *Sieber, n.* 270, and others; northward to Hunter's River, *Fraser, Beckler;* southward to Illawara, *A. Cunningham;* Twofold Bay, *F. Mueller.*

**Victoria.** Genoa river and valleys under Mount Imlay, *F. Mueller.* The southern form is very tomentose-hirsute, with rather larger flowers.

2. **C. Leichhardtii,** *Benth.* Probably a tall shrub, with the habit of *C. Fraseri;* branches densely velvety-tomentose or hispid. Leaves ovate-lan-ceolate or cordate, 2 to 3 in. long in the specimens seen, unequally toothed,

rather harshly velvety-tomentose on both sides. Cymes nearly sessile, few-flowered. Calyx very tomentose, spreading to about 5 lines diameter, the lobes broad and acute. Petals with an oblong ligula much shorter than the calyx. Central staminodium of each three lanceolate and fine-pointed, lateral ones filiform, attached to it near the base. Anther-bearing filaments very short. Ovary glabrous.

**Queensland.** Head of Boyd river, *Leichhardt, in Herb. F. Muell.*

3. **C. echinata,** *Forst.; DC. Prod.* i. 486. A tall shrub or small tree, the young branches and inflorescence whitish-tomentose. Leaves ovate or cordate, acuminate, 3 to 6 in. long or even more, irregularly toothed or nearly entire, often oblique at the base, glabrous or slightly tomentose above, more densely whitish-tomentose underneath. Cymes pedunculate, many-flowered, but shorter than the leaves. Calyx tomentose, nearly 3 lines diameter, the lobes acute. Petals with a very short concave broad base, the ligula narrow-linear, nearly as long as the calyx. Central staminodium of each three lanceolate, pubescent, much shorter than the petals, lateral ones small, filiform, recurved, attached to the very short anther-bearing filaments. Anther-cells less divaricate than in the other species. Capsule often $\frac{1}{2}$ in. diameter, without the long, soft, villous setæ which cover it.

**Queensland.** Cape York, *M'Gillivray*; Endeavour river, *Banks*; Pine river, *Hill*; Upper Brisbane river, *F. Mueller.*
**N. S. Wales.** Clarence river, where the natives use the stony fibre of the bark for kangaroo and fishing nets, *Beckler.*
The species is widely spread over the Indian Archipelago and the Pacific islands. The Australian whitish-tomentose form is like the original one described by Forster from the Pacific; the more common one in the Archipelago, often distinguished as a species under the name of *C. platyphylla,* Andr. Bot. Rep. t. 519 (as corrected under n. 603), Bot. Mag. t. 1813, is very villous-tomentose, and has often larger and broader leaves.

4. **C. Gaudichaudi,** *J. Gay, in DC. Prod.* i. 486, *and Mem. Mus. Par.* x. 213, *t.* 14. A low shrub, the young branches tomentose-hirsute. Leaves on very short petioles, obliquely ovate or orbicular, very obtuse, $\frac{1}{2}$ to 1 in. long or rather more, very unequally cordate at the base, the lower broad lobe sometimes quite overlapping the short upper one, scabrous-pubescent or rarely glabrous above, densely tomentose-hirsute and white underneath. Cymes pedunculate, few-flowered. Calyx densely hirsute, spreading to about 3 lines diameter. Petals broad with involute lobes at the base, the ligula very broad and nearly as long as the calyx. Central staminodium of each three lanceolate, the lateral ones filiform, uncinate, attached to the anther-bearing filaments. Capsule densely covered with soft, hispid, almost golden setæ.—Steetz, in Pl. Preiss. ii. 358.

**W. Australia.** Shark's Bay, *Gaudichaud*; Dirk Hartog's Island, *A. Cunningham*; Murchison river, *Drummond, Oldfield.*

5. **C. crispa,** *Turcz. in Bull. Mosc.* 1846, ii. 501. A low shrub, with elongated, perhaps procumbent branches, hispid with stellate hairs. Leaves shortly petiolate, ovate, obovate or oblong, crenate or irregularly lobed, very much undulate or crisped on the margin, glabrous or nearly so above, white-tomentose and often hirsute underneath. Cymes nearly sessile, few-flowered. Calyx tomentose-hirsute, spreading to 4 or 5 lines diameter, the lobes rather

acute. Petals broad with involute lobes at the base, the ligula obovate or spathulate. Central staminodium of each three lanceolate, lateral ones attached to it, linear-filiform and recurved. Capsule densely covered with short, soft, hirsute setæ.—*Rulingia crispa*, Turcz. in Bull. Mosc. 1849, ii. 10.

**W. Australia,** *Drummond, n.* 110.

6. **C. pulchella,** *Turcz. in Bull. Mosc.* 1846, ii. 502. A low shrub or undershrub, the upper branches scabrous-tomentose or hispid with rust-coloured stellate hairs. Leaves shortly petiolate, ovate or oblong, ½ to 1 in. long, coarsely and obtusely sinuate-toothed or lobed, undulate or often crisped on the margin, glabrous or scabrous above, white-tomentose underneath. Cymes pedunculate, few-flowered. Calyx rusty-tomentose at the base, spreading to 4 or 5 lines diameter, the lobes petal-like, white (or pink?), broad, and very obtuse. Petals with a cuneate concave base, and a short broad ligula. Central staminodium of each three lanceolate, the lateral ones attached to it, filiform and recurved. Fruit not seen.—*Rulingia pulchella*, Turcz. in Bull. Mosc. 1849, ii. 10.

**W. Australia,** *Drummond, Coll.* 1845, *n.* 111, and Murchison river, *n.* 97.

7. **C. microphylla,** *Benth.* Apparently a low shrub, with divaricate branches, tomentose when young. Leaves often clustered, very shortly petiolate, ovate or oblong, obtuse, 2 to 4 lines long, entire or sinuately lobed, very convex, glabrous above, white-tomentose underneath. Cymes pedunculate, few-flowered. Calyx tomentose at the base, spreading to 3 or 4 lines diameter, the lobes petal-like, white, broad, and very obtuse. Petals with a cuneate concave base, and a very short broad ligula. Central staminodium of each three lanceolate, lateral ones attached to it, filiform and recurved as in *C. pulchella*, but much smaller. Capsule about 4 lines diameter, villous with short soft not crowded setæ.

**W. Australia.** Murchison river, *Drummond, n.* 98. This species has most of the characters of *C. pulchella*, but the foliage is too widely different to unite it without having seen intermediate forms.

## 13. SERINGIA, J. Gay.

Calyx deeply 5-lobed, scarcely enlarged after flowering, and neither scarious nor coloured. Petals none. Stamens 5, alternate with the calyx-lobes, alternating with 5 subulate staminodia, and slightly united with them at the base; anther-cells parallel, opening by dorsal slits. Ovary 5-celled, with 2 or 3 ovules in each cell; styles cohering at the summit or nearly from the base. Fruit-carpels distinct, winged on the back, opening in 2 valves. Seeds strophiolate, albuminous, embryo straight, with flat cotyledons.—Shrub, with the habit nearly of a *Commersonia*. Flowers in dense, terminal or leaf-opposed cymes. Bracteoles none.

The genus is now limited to a single Australian species.

1. **S. platyphylla,** *J. Gay, in Mem. Mus. Par.* vii. 443, *t.* 16, 17. A tall shrub, with the habit nearly of *Commersonia Fraseri*, the young branches loosely whitish- or rusty-tomentose. Leaves ovate to ovate-lanceolate, acuminate, coarsely toothed, 3 to 4 or even 5 in. long, often oblique at the base, glabrous or sprinkled with minute stellate hairs above, densely tomentose underneath.

Cymes rather dense and many-flowered, but much shorter than the leaves. Calyx angular in the bud, attaining, when fully out, about 2 lines in length. Filaments and staminodia nearly similar, rather thick. Anthers oblong. Carpels about as long as the calyx, densely pubescent, the short broad vertical wing truncate at the top.—DC. Prod. i. 488; Steetz, in Pl. Preiss. ii. 349; *Lasiopetalum arborescens*, Ait. Hort. Kew. ed. 2, ii. 36.

**N. S. Wales.** Port Jackson, *R. Brown;* Blue Mountains, *Miss Atkinson;* Hastings river, *Beckler*

## 14. KERAUDRENIA, J. Gay.

Calyx 5-lobed, enlarged and scarious or thin and coloured after flowering, the midrib of each sepal usually thickened without lateral ribs. Petals none, or minute and scale-like. Stamens 5, alternate with the sepals, free or shortly united at the base, with or without intervening staminodia, anther-cells parallel, opening by dorsal slits. Ovary 3- to 5-celled, with 3 or more ovules in each cell; styles cohering at the summit. Capsule membranous, villous or shortly setose, opening loculicidally, and usually separating into distinct carpels. Seeds strophiolate, albuminous; embryo straight or curved, with flat cotyledons.—Shrubs more or less stellate-tomentose. Leaves entire or sinuate-lobed. Stipules narrow, or small and deciduous. Cymes terminal or opposite the upper leaves, few-flowered. Bracteoles none.

Besides the Australian species, there is one other from Madagascar, which on a further examination proves more nearly allied to *K. lanceolata* than had appeared to us when preparing the 'Genera Plantarum.' The genus has the anthers of *Seringia* and *Hannafordia*, with the calyx nearly of *Thomasia*, and must include species, in which as in the Madagascar one, the carpels do not appear to separate, as well as those in which they are quite distinct.

Bracts narrow. Carpels several-seeded, not always separating, the
  seeds nearly straight. Leaves mostly lanceolate, 1 to 3 in.
  Leaves quite glabrous and smooth above. Capsule scarcely septicidal.
    Leaves broad-lanceolate. Carpels angular, villous and setose . 1. *K. lanceolata*.
    Leaves narrow-lanceolate or linear. Carpels rounded on the
      back, very villous, but not setose . . . . . . . . . 2. *K. Hillii*.
    Leaves very rugose and pubescent above . . . . . . . . 3. *K. Hookeriana*.
Lower bracts broad scarious and coloured, very deciduous. Carpels
  1–2-seeded, the seeds reniform. Leaves ovate or oblong.
  Leaves thick and soft, very rugose, tomentose above, mostly 1 to
    2 in. long . . . . . . . . . . . . . . . . . 4. *K. nephrosperma*.
  Leaves smooth or slightly rugose, mostly under 1 in.
    Leaves undulate, crenate or crisped . . . . . . . . . 5. *K. hermanniæfolia*.
    Leaves quite entire . . . . . . . . . . . . . . . 6. *K. integrifolia*.

**1. K. lanceolata,** *Benth.* A tall shrub, the young branches rusty-tomentose. Leaves shortly petiolate, oblong-lanceolate, 3 to 4 in. long, rather thick, entire, glabrous above and smooth, or with the veins slightly impressed, white-tomentose underneath. Cymes short, few-flowered, very tomentose. Bracts narrow, deciduous. Calyx tomentose, spreading to 4 or 5 lines diameter, divided to about the middle, the midribs prominent and pubescent inside, the lobes of the fruiting calyx attaining 3 or 4 lines or more. Petals none. Filaments rather long, with slender staminodia intervening. Anthers linear. Ovary 5-celled, hirsute. Capsule truncate at the top, fully $\frac{1}{2}$ in. diameter, scarcely septicidal, but distinctly furrowed between the carpels, each

carpel very angular on the edges, so as to make the capsule appear almost
10-winged, but it is so hispid and beset with short, soft, hirsute setæ as
almost to disguise its form.   Seeds, several in each cell, obovoid ; embryo
straight.—*Seringia lanceolata*, Steetz, in Pl. Preiss. ii. 349.

**Queensland.**   Port Bowen, *R. Brown, A. Cunningham*, also in Leichhardt's collection.
It is this species which is closely allied to one from Madagascar, which I had formerly referred to *Thomasia*, on account of its capsule not separating into distinct carpels.

2. **K. Hillii,** *F. Muell. Herb.*   Very near to *K. lanceolata*, with the
same inflorescence and flowers.   Leaves much narrower, linear-lanceolate or
linear, 1½ to 3 in. long, coriaceous, glabrous without impressed veins above,
white-tomentose, and often sprinkled with rusty stellate hairs underneath.
Anther-bearing filaments scarcely dilated.   Ovary of *K. lanceolata*.   Capsule
not so large, very hirsute, but without prominent setæ, furrowed between the
carpels, which are rounded on the back, and not angular.   Seeds of *K. lanceolata.*

**Queensland.**   Glasshouses, Moreton Bay, *F. Mueller* and *W. Hill.*
**N. S. Wales.**   Port Macquarie and Port Stephens, *Fraser.*

3. **K. Hookeriana,** *Walp. Ann.* ii. 164.   Branches rusty-tomentose
or hirsute.   Leaves mostly oblong-lanceolate, 1½ to 3 in. long, entire, green,
very rugose and velvety-pubescent above, densely white-tomentose underneath ; the lower leaves or those of some branches often broader and shorter,
almost ovate.   Cymes or racemes 2- to 4-flowered, terminal or opposite the
upper leaves, on very short peduncles.   Bracts narrow, deciduous.   Calyx
divided nearly to the base into acute lobes, 3 or 4 lines long when in flower,
5 or 6 when in fruit.   Petals small and scale-like or none.   Filaments short,
alternating with subulate staminodia.   Anthers linear, much incurved.   Ovary
5-celled, tomentose.   Capsule very hirsute, 4 to 5 lines diameter, the carpels
distinct and separating, each opening in 2 valves.   Seeds several in each cell,
obovoid ; embryo straight.—*Seringia corollata*, Steetz, in Pl. Preiss. ii. 330 ;
*Keraudrenia integrifolia*, Hook. in Mitch. Trop. Austr. 341, not Steud. ;
*K. Hookeri*, F. Muell. Fragm. i. 28, 242.

**N. Australia.**   Arnhem's South Bay, *R. Brown ;* Nicholson river, *F. Mueller.*
**Queensland.**   Keppel Bay, *R. Brown* ; Suttor, Burnett, Upper Pine, and Brisbane
rivers, *F. Mueller.*   On the Maranoa, and southward to Lindley's range, *Mitchell ;* Robin-
son's range, *Leichhardt.*
The petals are certainly present in those Carpentaria specimens which I have examined,
and as certainly wanting in the flowers I opened of the more southern specimens, and the
two are distinguished under different names in R. Brown's herbarium and notes, but I can
discover no other character whatever.

4. **K. nephrosperma,** *Benth.*   A shrub, with the branches very
densely clothed with a soft, velvety, sometimes almost floccose tomentum.
Leaves ovate or oblong, very obtuse, 1 to 2 in. long, entire, sinuate or almost
lobed at the base, often slightly cordate, green, and minutely tomentose above,
densely white or rusty-tomentose underneath.   Cymes very short, several-
flowered.   Bracts ovate, membranous, very deciduous.   Calyx tomentose,
the lobes very broad and obtuse, attaining about 3 lines, very thin and
coloured.   Filaments as long as the ovary, with subulate staminodia intervening ; anthers oblong.   Ovary 5-celled.   Fruit carpels separating, nearly

globular, very tomentose.  Seeds 1 or 2 in each, globose, reniform.—*Seringia nephrosperma*, F. Muell. in Hook. Kew Journ. ix. 15.

**N. Australia.**  Desert at the sources of Victoria river, Sturt's and Hooker's Creeks, *F. Mueller ; Forster's Range, M'Douall Stuart.*

5. **K. hermanniæfolia,** *J. Gay, in Mem. Mus. Par.* vii. 462, *t.* 23. A small rigid shrub, the branches tomentose or hirsute with white or rust-coloured stellate hairs.  Leaves petiolate, ovate or oblong, very obtuse, rarely above 1 in. long, and often much smaller, mostly sinuate-crenate or undulate and crisped on the margin, glabrous or sprinkled with short, rigid, stellate hairs above, white-tomentose underneath.  Cymes loosely several-flowered, almost sessile.  Bracts ovate and very thin, but very deciduous.  Calyx tomentose, the lobes broad, rather acute, attaining from 3 to near 6 lines, thin and coloured.  Filaments dilated at the base, almost free, with 1 or 2, or without any intervening staminodia.  Anthers linear-oblong.  Ovary 3- to 5-celled, with 3 or 4 ovules in each cell.  Capsule often reduced to 1 or 2 carpels, with 1 or 2 reniform-globose seeds.—DC. Prod. i. 490 ; Steetz, in Pl. Preiss. ii. 346 ; *K. microphylla*, Steetz, l. c. 347 ; *Seringia microphylla*, F. Muell. Fragm. ii. 5.

**W. Australia.**  Sharks Bay, *Gaudichaud ;* Swan River, and northward to Murchison river and Champion Bay, *Drummond, Collie, Oldfield*, etc.

6. **K. integrifolia,** *Steud. in Pl. Preiss.* i. 236, *and Steetz, l. c.* ii. 347. A small much-branched shrub, the young shoots white or rusty with a close tomentum.  Leaves petiolate, oblong, very obtuse, 4 to 8 lines long, entire, glabrous or nearly so above, white-tomentose underneath.  Cymes rather loose, several-flowered.  Bracts ovate, thin and very deciduous.  Calyx tomentose ; lobes broad, rather acute, attaining 3 or 4 lines under the fruit, or sometimes more.  Filaments dilated and shortly connate at the base, recurved at the top, without any or rarely with 1 or 2 intervening staminodia.  Ovary 5-lobed, with about 4 ovules in each.  Capsule globular, softly villous ; carpels 1- or 2-seeded, not very readily separating.—*Seringia integrifolia*, F. Muell. Fragm. ii. 5.

**W. Australia.**  Swan River, *Drummond, Preiss, n.* 1651 ; S.W. coast, *Maxwell.*
Var. *velutina.*  Leaves rather larger, minutely velvety-tomentose above, densely tomentose underneath.  Flowers larger, filaments longer.—*K. velutina*, Steetz, in Pl. Preiss. ii. 348 ; *Seringia velutina*, F. Muell. Fragm. ii. 5 ; *S. grandiflora*, F. Muell. Fragm. i. 142. To this belong Drummond's specimens, n. 109, and Maxwell's, from East Mount Barren. The specimen described by Steetz, which I have not seen, was gathered by Roe, between Swan River and King George's Sound.
*Actinostigma lanceolatum*, Turcz. in Bull. Mosc. 1859, i. 259, from 'New Holland, Brogden,' is described as closely resembling *K. lanceolata* in habit, foliage, and most of the characters, but with axillary, not leaf-opposed inflorescence, 10 stamens all perfect and free, 5 biovulate carpels, the styles connate, with 5 radiating stigmas.  I am quite unable to identify any Lasiopetalous plant with this description.  It may belong to some very different Natural Order, possibly *Rutaceæ.*

## 15. HANNAFORDIA, F. Muell.

Calyx 5-lobed, somewhat enlarged after flowering, with prominent raised ribs, 3 to each sepal, besides those connecting the sepals.  Petals 5, lanceolate, slightly concave, shorter than the calyx.  Stamens 5, opposite the petals ; staminodia 3 or fewer between each 2 stamens, linear-subulate, all slightly connected in a ring at the base; anther-cells parallel, opening by dorsal slits. Ovary 3- or 4-celled, with 3 or 4 ovules in each cell. Style simple. Capsule hard, almost woody, opening loculicidally in 3 or 4 valves. Seeds strophiolate, albuminous; embryo straight, with flat cotyledons.—Shrub, with the habit of a *Thomasia*, but without stipules. Bracteoles 3, persistent.

The genus is limited to a single species. It has the anthers of *Keraudrenia* and *Seringia*, with the calyx nearly of *Guichenotia*.

1. **H. quadrivalvis,** *F. Muell. Fragm.* ii. 9. A much-branched shrub of 3 or 4 ft., densely clothed with a soft velvety tomentum, often rusty on the young shoots. Leaves on rather long petioles, obliquely ovate-cordate, obtuse, 1 to $2\frac{1}{2}$ in. long, coarsely sinuate-toothed or broadly lobed, thick and soft. Cymes leaf-opposed, short, and few-flowered. Bracteoles linear, much shorter than the calyx. Calyx about $\frac{1}{2}$ in. long, divided to below the middle into narrow acuminate lobes. Petals about as long as the calyx-tube, but variable. Staminodia in Mueller's specimens 3 between each 2 stamens, but in one of Douglas's I formerly examined I found them singly alternating with the stamens. Capsule shorter than the calyx, most frequently 4-celled, but often also 3-celled.

**W. Australia.** Murchison river, *Oldfield, Drummond, n.* 100.


## 16. THOMASIA, J. Gay.

(Leucothamnus, *Lindl.*; Rhynchostemon, *Steetz*; Asterochiton, *Turcz.*)

Calyx 5-lobed, much enlarged and scarious or coloured after flowering, the sepals 1-nerved and reticulately veined, with the midrib usually thickened, spreading or erect-connivent, closing over the fruit. Petals none or minute and scale-like. Stamens 5, alternate with the sepals, free or shortly connate at the base; staminodia none, or 5 alternating with the stamens. Anthers opening at the top towards the inside in short slits, which at length extend more or less down the sides. Ovary 3- or rarely 4- or 5-celled, with 2 or more ovules in each cell; style simple. Capsule enclosed in the calyx, usually crustaceous, opening loculicidally in 3 to 5 valves. Seeds usually strophiolate, albuminous; embryo straight, with flat cotyledons.—Shrubs more or less tomentose or hirsute with stellate hairs, rarely quite glabrous. Leaves entire or lobed. Stipules leafy, usually semihastate or reniform, in one species similar to the leaves, in others small, and in a few entirely wanting. Racemes leaf-opposed, simple or rarely cymosely branched. Bracts narrow, deciduous. Bracteoles under the calyx 3, slightly connate at the base or free. Calyx usually purple bluish or white.

The genus is confined to Australia. It differs from *Lasiopetalum* more constantly in the calyx than in the anthers, the opening of the latter in some *Thomasias* being little more than oblong pores, and in a few *Lasiopetala* extending at length down the sides to the base. The two genera are natural, and the majority of species distinguished by a variety of characters, although there is no one to which there is not some exception. The presence or absence and size of the scale-like petals, the presence or absence and number of staminodia, are liable in all these genera to great variation in individual species.

A. *Stipules leafy. Stamens and staminodia in a distinctly perigynous ring.* (**Leucothamnus.**)

Leaves angular or shortly lobed, scarcely wrinkled, whitish pubescent
   above, tomentose underneath . . . . . . . . . . . . . 1. *T. macrocarpa.*

Leaves lobed, very much wrinkled, roughly stellate-hairy above,
　densely tomentose underneath . . . . . . . . . . . 2. *T. rugosa.*

　　B. *Stipules leafy. Stamens and staminodia united in a hypogynous cup as long as the ovary.*
Leaves ovate or broadly oblong, almost entire . . . . . . . 3. *T. montana.*

　　C. *Stipules leafy (in* T. foliosa *sometimes wanting). Stamens hypogynous, free or slightly connected at the base, with or without staminodia.*
Leaves mostly ovate-cordate, often sinuate-lobed.
　Leaves closely hoary-tomentose on both sides, without rigid hairs,
　　and scarcely lobed . . . . . . . . . . . . . . 4. *T. tenuivestita.*
　Leaves glabrous or hirsute above, tomentose underneath, usually
　　lobed.
　　Tall shrubs. Leaves 1½ to 3 in.
　　　Filaments very short. Leaves scabrous or hirsute above. Ra-
　　　　cemes rarely branched.
　　　　Bracteoles small, linear. Calyx divided to the middle or
　　　　　lower, lobes acute . . . . . . . . . . . . 5. *T. solanacea.*
　　　　Bracteoles broadly lanceolate. Calyx not divided to the
　　　　　middle, lobes rather obtuse . . . . . . . . 6. *T. brachystachys.*
　　　Filaments about as long as the anthers. Leaves nearly gla-
　　　　brous above. Racemes branched. Calyx-lobes acute. . 7. *T. discolor.*
　　Low shrubs. Leaves mostly under 1 in. Flowers small.
　　　Calyx-lobes short, broad and obtuse. Stipules reniform . . 8. *T. quercifolia.*
　　　Calyx deeply divided, lobes acute. Stipules very small . . 9. *T. foliosa.*
Leaves glabrous on both sides, or sprinkled or hispid with rigid stel-
　late hairs, usually lobed.
　Stipules very small. Calyx-lobes deep and acute . . . . . . 9. *T. foliosa.*
　Stipules rather large, reniform or lobed. Calyx-lobes short and
　　rather obtuse.
　　Leaves small, glabrous, nearly equally 3-lobed. Flowers rather
　　　small . . . . . . . . . . . . . . . . . . 10. *T. triloba.*
　　Leaves 1 to 2 in., more or less stellate-hispid. Flowers large . 11. *T. triphylla.*
Leaves (except the lowest) oblong, lanceolate or linear, entire or
　hastate with very short lateral lobes, the margins often crisped
　or revolute.
　Stipules reniform or semihastate. Leaves flat or crisped. Ovary
　　3- or rarely 4-celled.
　　Ovary and style glabrous.
　　　Flowers rather small. Filaments very short . . . . . 12. *T. purpurea.*
　　　Flowers large. Filaments about as long as the anthers . . 13. *T. macrocalyx.*
　　Ovary and often the base of the style tomentose.
　　　Calyx thin, except the prominent midribs.
　　　　Flowers rather small. Filaments short . . . . . . 14. *T. pauciflora.*
　　　　Flowers large. Bracteoles broad. Filaments nearly as
　　　　　long as the anthers . . . . . . . . . . . . 15. *T. rhynchocarpa.*
　　　Calyx large, the lobes broadly thick in the centre, with broad,
　　　　thin, undulate margins . . . . . . . . . . . 16. *T. grandiflora.*
　Stipules semihastate. Leaves crisped or revolute on the margins.
　　Ovary 5-celled.
　　　Leaves petiolate, crisped. Ovary villous . . . . . . 17. *T. cognata.*
　　　Leaves sessile, the margins revolute. Ovary glabrous . . . 18. *T. rulingioides.*
　Stipules semihastate. Leaves wrinkled, with revolute margins.
　　Ovary 3- or rarely 4-celled.
　　　Bracteoles linear-lanceolate . . . . . . . . . . . 19. *T. angustifolia.*
　　　Bracteoles broadly lanceolate or ovate . . . . . . . 20. *T. petalocalyx.*
Stipules like the leaves, narrow, heath-like, with revolute margins 21. *T. sarotes.*

D. *Stipules none.   Stamens hypogynous, free or slightly connected at the base, without staminodia.* (**Rhynchostemon.**)

Tomentum close or dense, not scaly.   Leaves 1 to nearly 3 in.   Ra-
  cemes or cymes several-flowered.   Bracteoles subulate, distant
  from the calyx.
Racemes mostly simple.   No petals.   Anthers long-acuminate   .  22.  *T. glutinosa.*
Racemes mostly branched.   Petals present.   Anthers shortly acu-
  minate . . . . . . . . . . . . . . . . . . . . 23.  *T. laxiflora.*
Tomentum scaly.   Leaves under 1 in.   Racemes 1- to 3-flowered.
  Bracteoles small under the calyx.
Leaves oblong-lanceolate or linear, ½ to 1 in. . . . . . . . . 24.  *T. stelligera.*
Leaves cordate-orbicular, under ½ in. . . . . . . . . . . 25.  *T. pygmæa.*

1. **T. macrocarpa,** *Hueg. in Endl. Nov. Stirp. Dec.* 32.   A tall shrub,
the branches whitish with a loose tomentum.   Leaves broadly ovate-cordate,
obtuse, 1½ to 2 in. long, irregularly angular-toothed or shortly lobed, pubescent above when young, at length glabrous, tomentose underneath.   Stipules
small, oblique or rarely ½ in. long and reniform.   Racemes tomentose-hirsute,
with few large flowers.   Bracteoles broadly ovate-lanceolate, woolly.   Calyx
opening to about 1 in. diameter, loosely woolly-hirsute outside.   Stamens
and staminodia united at the base in a very perigynous ring, hirsute outside, glabrous within.   Filaments and staminodia longer than the anthers.
Ovary tomentose, 3-celled with 2 erect ovules in each cell.   Style glabrous.—
Steud. Pl. Preiss. i. 235 ;   *T. stipulacea,* Bot. Mag. t. 4111, not Lindl. ;
*Leucothamnus montanus,* Lindl. Swan Riv. Ap. 19 ; Steetz, in Pl. Preiss.
ii. 336.

**W. Australia.**   Swan River, *Drummond, 1st Coll. ; Preiss, n.* 1654.

2. **T. rugosa,** *Turcz. in Bull. Mosc.* 1846, ii. 501.   Branches densely
tomentose-villous.   Leaves cordate-ovate, obtuse, 1 to 3 in. long, sinuate-
lobed, very much wrinkled and scabrous with stellate hairs above, very
densely tomentose underneath.   Stipules reniform.   Racemes simple, with
rather large nearly sessile flowers.   Bracteoles ovate-lanceolate, obtuse, thick
and very villous-tomentose.   Calyx above ½ in. diameter, softly pubescent,
divided to nearly the base into obtuse connivent lobes.   Filaments nearly as
long as the anthers, inserted with the staminodia in a slightly perigynous
ring.   Anthers scarcely acuminate.   Ovary tomentose, 3-celled with 6 to 8
ovules in each cell ; style glabrous.—*Leucothamnus polyspermus,* Turcz. in
Bull. Mosc. 1849, ii. 11.

**W. Australia.**   Swan River, *Drummond, n.* 101 *and* 105.

3. **T. montana,** *Steud. in Pl. Preiss.* i. 230 ; *Steetz, l. c.* ii. 331.
Branches tomentose-hirsute.   Leaves petiolate, ovate-cordate or broadly
oblong, obtuse, mostly under 1 in. long, entire or slightly undulate-crenate,
green on both sides, glabrous or sprinkled with short stellate hairs.   Stipules
broadly oblique or reniform.   Racemes on long peduncles, rather closely
tomentose-pubescent.   Bracteoles linear-oblong or slightly spathulate.   Calyx
about ½ in. diameter, tomentose, divided to about the middle, the sepals
broadly thickened as in *T. grandiflora,* but with a very narrow thin undulate
margin.   Petals minute.   Stamens and staminodia united in a cup as long as
the ovary ; anthers attached by the middle and nearly sessile on the margin

of the cup, between the short tooth-like staminodia. Ovary tomentose, 3-celled ; style glabrous.

**W. Australia.** Rocky summits of Mount Bakewell, Swan River, *Preiss, n.* 1661.

4. **T. tenuivestita,** *F. Muell. Fragm.* ii. 7. Hoary all over with a close minute but soft tomentum, without rigid hairs. Leaves on slender petioles, ovate-cordate, obtuse, ¾ to 1½ in. long, hoary-tomentose on both sides. Stipules broad, oblique or reniform. Racemes slender, with rather small flowers. Bracteoles oblong-linear, hoary-tomentose. Calyx opening to 5 or 6 lines diameter, slightly tomentose ; the lobes not reaching to the middle, broad with a prominent midrib. Petals usually present. Anthers shortly acuminate. Staminodia none. Ovary tomentose, 3-celled ; style glabrous.

**W. Australia.** Murchison river, *Walcot and Oldfield.*

5. **T. solanacea,** *J. Gay, in Mem. Mus. Par.* vii. 456, *t.* 21. A tall shrub or small tree, the branches densely tomentose or shortly hirsute. Leaves deeply cordate-ovate, obtuse, mostly 1½ to 3 in. long, rather deeply sinuate-lobed, scabrous or hirsute above with stellate hairs, more softly and densely tomentose or hirsute underneath. Stipules rather large, reniform, often petiolate. Racemes pedunculate, several-flowered, occasionally branched. Bracteoles small, linear. Calyx more or less tomentose, spreading to about ½ in. diameter, divided to rather below the middle into acute lobes. Petals usually none. Filaments very short ; anthers shortly acuminate. Staminodia usually 4, sometimes bearing small anthers. Ovary tomentose, 3-celled ; style glabrous.—DC. Prod. i. 489 ; Steetz, in Pl. Preiss. ii. 327 ; *Lasiopetalum solanaceum,* Sims, Bot. Mag. t. 1486.

**W. Australia.** King George's Sound, *R. Brown, Fraser,* and others ; Bald Island and Princess Royal Harbour, *Oldfield, Maxwell.*
Some monstrous specimens from King George's Sound are very villous, with more or less developed petals, and the stamens and carpels mostly deformed.

6. **T. brachystachys,** *Turcz. in Bull. Mosc.* 1852, ii. 143. Very nearly allied to the more hirsute specimens of *T. solanacea,* and perhaps a variety of that species. Leaves rather less obtuse and less deeply cordate. Racemes apparently all simple, very hirsute-tomentose. Pedicels very short. Bracteoles broadly lanceolate, thick and rusty-hirsute. Calyx more tomentose than in *T. solanacea,* less deeply divided into more obtuse lobes. Petals usually present.

**W. Australia,** *Drummond, 5th Coll. n.* 262.

7. **T. discolor,** *Steud. in Pl. Preiss.* i. 233 ; *Steetz, l. c.* ii. 326. A tufted shrub of 2 to 4 ft., the branches densely tomentose. Leaves cordate-ovate, obtuse, 1 to 2 in. long, sinuately lobed, coriaceous, glabrous or scabrous above, white or rusty-tomentose underneath. Stipules reniform, occasionally petiolate. Cymes pedunculate, several-flowered. Bracteoles linear, rather thick, tomentose. Calyx spreading to about ¾ in. diameter, sprinkled with stellate hairs outside, glabrous within, deeply divided into acute lobes, less coloured than in most *Thomasias,* but with prominent midribs. Petals none. Filaments as long as the anthers, without intervening staminodia. Ovary very villous, 3-celled, with 2 ovules in each cell ; style glabrous.

**W. Australia.** King George's Sound, *Drummond ;* rocks at Williamstone, *Preiss, n.* 1658; Mount Elphinstone, *Oldfield.*

8. **T. quercifolia,** *J. Gay, in Mem. Mus. Par.* vii. 459, *t.* 21. A low shrub, with numerous branches, rigidly hirsute-tomentose. Leaves ovate, usually deeply 3-lobed, the lateral lobes short, divaricate and often obtusely 3-lobed, the middle one longer, often 3-lobed, the whole leaf rarely exceeding 1 in., coriaceous, sprinkled above with rigid stellate hairs, tomentose and often rigidly hirsute underneath. Stipules reniform. Racemes simple. Flowers rather small. Bracteoles linear. Calyx-lobes not reaching the middle, broad and obtuse. Petals none. Filaments about as long as the ovary, the anthers rather short, obtuse, opening to the base ; staminodia usually present. Ovary tomentose, 3-celled ; style glabrous.—DC. Prod. i. 489 ; Steud. in Pl. Preiss. ii. 329 ; *Lasiopetalum quercifolium,* Andr. Bot. Rep. t. 459 ; Bot. Mag. t. 1485 ; *T. hypoleuca,* Steud. in Pl. Preiss. i. 234.

**W. Australia,** *Drummond ;* King George's Sound, *R. Brown, Preiss, n.* 1646, and others; Franklin river, *Maxwell.*

9. **T. foliosa,** *J. Gay, in Mem. Mus. Par.* vii. 454, *t.* 22. A shrub, with numerous rather slender branches, tomentose and hirsute when young. Leaves petiolate, ovate-cordate, rather deeply sinuate-lobed, rarely exceeding 1 in., sprinkled with stellate hairs above, more densely hirsute underneath. Stipules very small, rarely attaining 2 lines and sometimes almost wanting. Racemes numerous, often branched, slender, hirsute. Flowers small, on slender pedicels. Bracteoles small, linear. Calyx hirsute, about 3 lines diameter, deeply divided into acute usually connivent lobes. Petals none. Filaments as long as the ovary, without intervening staminodia ; anthers short and obtuse, almost didymous, the cells opening laterally almost their whole length. Ovary tomentose, 3-celled. Style glabrous.—DC. Prod. i. 489 ; Steetz, in Pl. Preiss. ii. 325 ; *T. viridis,* Steud. in Pl. Preiss. i. 234 ; also most probably *T. diffusa,* G. Don, Gen. Syst. i. 527, which I have not seen.

**W. Australia.** Geographer Bay, *Leschenault, Baudin ;* Swan River, *Fraser, Drummond, Preiss, n.* 1630, 1649, 1653 ; Gordon, Salt, Kalgan, and Phillips rivers, *Oldfield.*

10. **T. triloba,** *Turcz. in Bull. Mosc.* 1846, ii. 500. A low shrub, with slender branches, quite glabrous or slightly tomentose towards the top. Leaves on long petioles, broadly cordate, mostly $\frac{1}{2}$ to $\frac{3}{4}$ in. long, nearly equally 3-lobed, lobes broad, obtuse, often sinuate-crenate and undulate, glabrous or rarely sprinkled with a very few stellate hairs. Stipules reniform or 3-lobed. Racemes long and slender, usually glabrous. Bracteoles linear, slightly ciliate. Calyx spreading to about $\frac{1}{2}$ in. diameter, divided to about the middle into broad rather obtuse lobes, glabrous or nearly so, the midrib not much thickened. Filaments short. Ovary densely tomentose, 3-celled; style glabrous.

**W. Australia,** *Drummond, n.* 106.

11. **T. triphylla,** *J. Gay, in Mem. Mus. Par.* vii. 458. Branches scabrous-tomentose and sometimes hispid. Leaves petiolate, ovate-cordate, $1\frac{1}{2}$ to 2 in. long, sinuate-pinnatifid, with short broad very obtuse lobes, more or less sprinkled with very rigid stellate rusty hairs, but otherwise glabrous. Stipules petiolate, broad, obliquely 2 or 3-lobed, or reniform. Flowers large, in short

hispid racemes. Bracteoles linear-lanceolate, hispid. Calyx opening to nearly 1 in. diameter, hispid at the base only, divided to about the middle into broad lobes with thick midribs. Petals none. Filaments rather long; anthers shortly and obtusely acuminate, staminodia often present. Ovary tomentose, 3-celled.—DC. Prod. i. 489 ; Steetz, in Pl. Preiss. ii. 328 ; *Lasiopetalum triphyllum*, Labill. Pl. Nov. Holl. i. 63, t. 88 ; *Thomasia stipulacea*, Lindl. Swan Riv. App. 18 ; *T. glabrata*, Steud. in Pl. Preiss. i. 234.

**W. Australia.** Cape Leeuwin, *Labillardière ;* Swan River, *Drummond, 1st Coll.*, *Preiss, n.* 1635, 1636, *Oldfield*, and others.

*T. Gilbertiana,* Turcz. in Bull. Mosc. 1849, ii. 10, which I have not seen, would appear from his description to be the same as *T. triphylla.*

12. **T. purpurea,** *J. Gay, in Mem. Mus. Par.* vii. 452, *t.* 21. A small shrub or undershrub, the slender branches more or less tomentose or hirsute. Leaves oblong or nearly linear, obtuse, ½ to 1 in. long, entire, sprinkled with stellate hairs above, more hirsute underneath, or rarely nearly glabrous. Stipules broad and oblique, or almost reniform. Racemes longer than the leaves. Flowers rather small, on very short pedicels. Bracteoles linear. Calyx slightly tomentose, expanding to about ½ in. diameter, divided to about the middle into ovate lobes. Petals small, occasionally wanting. Filaments very short, anthers slightly acuminate. Ovary glabrous, 3- or 4-celled with 2 ovules in each cell ; style glabrous.—DC. Prod. i. 489 ; Steetz, in Pl. Preiss. ii. 318 ; *Lasiopetalum purpureum,* Ait. Hort. Kew. ed. 2, ii. 36 ; Bot. Mag. t. 1755 ; *Thomasia rupestris,* Steud. in Pl. Preiss. i. 231.

**W. Australia.** King George's Sound, *R. Brown ; Fraser* and others ; Mount Elphinstone, *Preiss, n.* 1648.

Var. *undulata.* Larger in all its parts and slightly hoary-tomentose. Leaves mostly 1 to 1½ in. long. Flowers larger, the racemes more pedunculate. Petals usually none.—*T. undulata,* Steetz, in Pl. Preiss. ii. 320. Swan River, *Drummond, 1st Coll. and 2nd Coll. n.* 58.

Steetz describes the capsule of this and the following species as stipitate, but the stipes, if any, is so short as to be scarcely perceptible.

13. **T. macrocalyx,** *Steud. in Pl. Preiss.* i. 230 ; *Steetz, l. c.* ii. 319. A shrub of 1½ to 2 ft., nearly allied to *T. purpurea,* but differing chiefly in the large, inflated, fruiting calyx. Branches tomentose and hirsute with stiff stellate hairs. Leaves petiolate, oblong-lanceolate, obtuse, 1 to 1½ in. long, scabrous-pubescent above, tomentose or hirsute underneath. Stipules oblique or semicordate. Racemes long, several-flowered. Bracteoles linear-lanceolate. Calyx expanding to nearly ½ in. diameter, with broad short lobes, the midribs much thickened, when in fruit much inflated, depressed-globose, somewhat 5-angled, fully ½ in. diameter, although the lobes are closely connivent. Filaments as long as the anthers, which are more obtuse than in *T. purpurea.* Ovary and style glabrous as in *T. purpurea.*

**W. Australia.** Preston river, Wellington district, *Preiss, n.* 1657 ; S. W. coast, *Maxwell.*

14. **T. pauciflora,** *Lindl. Swan Riv. App.* 18. Scabrous-tomentose or hirsute. Leaves lanceolate, often cordate, and sometimes hastately 3-lobed at the base, 1 to 2 in. long, green and sprinkled with short, rigid, stellate hairs on both sides. Stipules broad, semihastate or reniform. Racemes

several-flowered. Bracteoles linear or scarcely lanceolate, rather thick, rusty-tomentose or hirsute. Calyx expanding to $\frac{1}{2}$ in. or rather more, divided to below the middle, the midribs prominent. Petals usually but not always present. Filaments short. Ovary tomentose, usually 3-celled; style tomentose at the base, glabrous upwards, the tomentose base often persisting on the ripe capsule.—Steetz, in Pl. Preiss. ii. 329; *T. subhastata,* Steud. in Pl. Preiss. i. 232; Steetz, l. c. ii. 330.

**W. Australia.** Swan River, *Drummond, 1st Coll., Preiss, n.* 1633, 1647; King George's Sound, *Harvey.*

*T. paniculata,* Lindl., Swan Riv. App. 18; Steetz, in Pl. Preiss. ii. 323, from Swan River, appears to be only a luxuriant form of *T. pauciflora,* with rather larger flowers and the glabrous part of the style rather longer. A still more luxuriant variety, with leaves 3 in. long, and the calyx 7 lines diameter, was gathered by Maxwell in the moist valleys of Franklin river.

15. **T. rhynchocarpa,** *Turcz. in Bull. Mosc.* 1852, ii. 142. Very near *T. pauciflora,* with a similar foliage, but the indumentum more ferruginous and denser, the bracteoles and flowers rather differently shaped. Racemes 2- or 3-flowered. Bracteoles oblong or broadly lanceolate, obtuse, thick, and densely rusty-tomentose. Calyx opening to nearly 1 in. diameter, scarcely divided to the middle, with broad obtuse lobes, much replicate on the margins over the fruit, the midribs very prominent inside. Petals minute. Filaments rather long. Ovary tomentose; style also tomentose, excepting quite the extremity, and usually persistent. Fruiting calyx closing over to about $\frac{1}{2}$ in. diameter.—F. Muell. Fragm. ii. 8.

**W. Australia,** *Drummond, 5th Coll. n.* 261; Kojonerup valley and Salt river, *Maxwell.*

16. **T. grandiflora,** *Lindl. Swan Riv. App.* 18. A shrub or undershrub of 1 or 2 ft., with the habit and foliage of *T. pauciflora,* but at once known by the flowers. Leaves mostly ovate-lanceolate, or oblong, or the lowest ovate, obtuse, $\frac{1}{2}$ to 1 in. long, entire, cordate or obscurely 3-lobed at the base, glabrous or sprinkled with a few stellate hairs. Stipules oblique or semihastate. Flowers large, in terminal racemes. Bracteoles broadly lanceolate, thick, and tomentose-hirsute. Calyx spreading to about 1 in. diameter, not divided to the middle, the broad thick centre of each sepal hirsute-tomentose outside and short-tomentose inside, the broad margins thin, glabrous, and undulate. Petals none. Filaments very short; anthers acuminate. Ovary tomentose, 3-celled, with 8 to 20 or even more ovules in each cell.—Steetz, in Pl. Preiss. ii. 324; *T. cycnopotamica* and *T. lucida,* Steud. in Pl. Preiss. i. 231.

**W. Australia.** Swan River, *Drummond, 1st Coll., Preiss, n.* 1645 and 1667; Murchison river and Champion Bay, *Oldfield.*

17. **T. cognata,** *Steud. in Pl. Preiss.* i. 232; *Steetz, l. c.* ii. 320. A low shrub, very hispid with rigid stellate hairs. Leaves petiolate, oblong or lanceolate, obtuse, rarely exceeding 1 in., wrinkled, and very much crisped on the margin, green and hispid on both sides. Stipules broadly semihastate. Racemes slender, with small, nearly sessile flowers. Bracteoles linear-lanceolate. Calyx hispid, opening to about $\frac{1}{2}$ in. diameter, the angles very prominent, divided to about the middle into broad lobes not undulate on the

margin. Petals usually present, very concave and hirsute. Filaments rather long; anthers not acuminate. Ovary very villous, 5-celled, deeply furrowed; style glabrous.

**W. Australia.** Swan River, *Drummond, n.* 68; Rottenest Island, *Preiss, n.* 1660 and 1666; Freemantle and King George's Sound, *Oldfield.*

18. **T. rulingioides,** *Steud. in Pl. Preiss.* i. 232; *Steetz, l. c.* ii. 322. A very hispid shrub, at first sight closely resembling *T. cognata*, or the hispid forms of *T. purpurea*, and with the 5-celled ovary of the former, but the leaves are narrower, almost or quite sessile, the crisped margins much revolute, and narrowed at the base. Stipules broadly semihastate or sometimes hastate, 3-lobed. Flowers nearly sessile in the raceme and hispid as in *T. cognata*, but rather smaller. Calyx similar. Petals usually smaller and less hirsute. Filaments rather shorter. Ovary glabrous, granulate.

**W. Australia.** Swan River, *Preiss, n.* 1663.

19. **T. angustifolia,** *Steud. in Pl. Preiss.* i. 232; *Steetz, l. c.* 322. The whole plant clothed with a hoary tomentum, somewhat scabrous on the upper side of the leaves, denser and often rusty underneath, without spreading hairs. Leaves narrow-oblong or rarely lanceolate, obtuse, mostly about 1 in. long, wrinkled with deeply impressed veins, the margins revolute, rounded at the base. Stipules broadly semihastate or semicordate. Racemes slender, with about 4 to 8 small flowers. Bracteoles linear-lanceolate, tomentose. Calyx opening to about 5 lines diameter, divided much below the middle, the margins flat, the principal branching veins of each sepal sometimes prominent as well as the midrib, but not starting from the base, as in *Guichenotia*. Petals generally present, and often a few staminodia. Anthers shortly and obtusely acuminate. Ovary densely tomentose, usually 3-celled; style glabrous.

**W. Australia.** Southern districts, *Drummond, n.* 107, *Preiss, n.* 1634; near Cape Riche, *Harvey;* King George's Sound, *Oldfield;* Kojonerup and Fitzgerald ranges, *Maxwell.*

In foliage and habit this has much resemblance to *Lysiosepalum rugosum,* but the flowers are very different.

20. **T. petalocalyx,** *F. Muell. in Trans. Phil. Soc.* i. 35, *and Pl. Vict.* i. 147. Very near *T. angustifolia*, and perhaps a variety. Tomentum more copious, looser, and mixed with long stellate hairs. Leaves often larger, attaining 1½ in., the margins less revolute. Flowers larger. Bracteoles usually broadly lanceolate or almost ovate. Calyx-lobes broader and very obtuse; in other respects the characters are those of *T. angustifolia.—T. macrocalyx,* Schlecht. Linnæa, xx. 633, not Steud.

**Victoria.** Stony coast ridges, Wilson's Promontory, *F. Mueller.*
**S. Australia.** Light and Gawler rivers, *Behr;* Barossa and Bugle ranges, *F. Mueller;* Kangaroo Island, *Waterhouse.*
**W. Australia.** Between King George's Sound and the Great Australian Bight, *Maxwell.*

21. **T. sarotes,** *Turcz. in Bull. Mosc.* 1852, ii. 145. Branches slender, minutely tomentose. Leaves almost sessile, linear, obtuse, rarely exceeding ½ in., quite entire, the margins closely revolute, minutely tomentose or glabrous above, more rusty-tomentose underneath. Stipules similar to the leaves and

often nearly as long, giving the plant a heath-like aspect.   Racemes long and
slender.   Bracteoles narrow-linear or slightly spathulate.   Calyx 5 or 6 lines
diameter, deeply lobed, the lobes almost acute.   Petals small and broad;
staminodia also occasionally present.   Ovary tomentose; style glabrous.

**W. Australia,** *Drummond, 5th Coll. n. 256.*

22. **T. glutinosa,** *Lindl. Swan Riv. App.* 18.   Branches tomentose or
slightly hispid, viscid towards the top.   Leaves petiolate, the lower ones or
sometimes nearly all ovate-cordate, the upper ones or nearly all lanceolate or
hastately 3-lobed, the middle lobe often 1 to 2 in. long, the lateral ones very
short, all obtuse, glabrous or sprinkled with stellate hairs above, loosely tomentose underneath.   Stipules none.   Racemes on long peduncles, hirsute
and very glutinous.   Bracteoles filiform, inserted on the pedicel at some distance from the calyx.   Calyx spreading to 6 to 8 lines diameter, slightly
pubescent or sometimes hirsute at the base, divided to about the middle into
broad acute lobes, petal-like, as in most species, and the central vein of each
sepal deeply coloured, but scarcely thickened.   Petals none.   Filaments very
short; anthers produced into a rather long light-coloured point.   Ovary villous, 3-celled, with 2 ovules in each cell.   Style glabrous or slightly tomentose at the base.—*Rhynchostemon glutinosum,* Steetz, in Pl. Preiss. ii. 334.

**W. Australia.**   Swan River, *Drummond, 1st Coll., Preiss, n.* 1632 and 1668, and
others.

Var. *latifolia.*   Leaves mostly ovate-cordate, entire or obscurely 3-lobed.   Indumentum
of the branches and under side of the leaves tomentose only and scarcely hispid.   Flowers
sometimes, but not always, smaller.—*T. canescens,* Lindl. Swan Riv. App. 18; *T. æmula*
and *T. lasiopetaloides,* Steud. in Pl. Preiss. i. 233; *Rhynchostemon canescens,* Steetz, in Pl.
Preiss. ii. 335.   Swan River, *Drummond, 1st Coll., Preiss, n.* 1636 and 1641.

23. **T. laxiflora,** *Benth.*   Young branches densely clothed with a close
tomentum.   Leaves, like those of the broad-leaved variety of *T. glutinosa,*
from ovate-cordate to broadly lanceolate, acuminate, entire or obscurely
3-lobed, 1½ to 2½ in. long, almost coriaceous and glabrous above when full-
grown, densely and softly tomentose underneath.   Stipules none.   Racemes
elongated, pedunculate, apparently viscid and subulate; bracteoles distant
from the calyx, as in *T. glutinosa.*   Calyx very angular, divided to below
the middle into ovate, cordate, acuminate segments, glabrous inside at
the base with prominent midribs, the broad thin margins tomentose inside.
Petals small, broad.   Anthers acuminate, but much less so than in *T. glutinosa.*   Ovary villous, 3-celled with 2 ovules in each cell.   Style glabrous.

**W. Australia.**   Swan River, *Drummond, Coll.* 1843, *n.* 25.

24. **T. stelligera,** *Benth.*   A low shrub, with slender wiry branches,
covered, as well as the under side of the leaves, with a whitish, almost silvery,
scaly tomentum.   Leaves shortly petiolate, the upper ones sometimes opposite, from oblong to lanceolate or almost linear, very obtuse, ½ to 1 in. long,
glabrous and smooth on the upper side.   Stipules none.   Flowers rather
large, pink, 2 or 3 in the raceme.   Bracteoles small, close to the calyx.
Calyx sprinkled outside with a few scale-like stellate hairs, slightly tomentose
inside, divided to about the middle, angular and almost 5-saccate at the base,
the lobes broad and acute, the midribs richly coloured, but scarcely prominent.   Petals small.   Anthers shortly acuminate.   Ovary densely covered

with scaly stellate hairs, 3-celled, with 2 ovules in each cell; style glabrous.—
*Lasiopetalum stelligerum*, Turcz. in Bull. Mosc. 1852, ii. 147.

**W. Australia,** *Drummond, 5th Coll. n.* 257.

25. **T. pygmæa,** *Benth.* Not much branched and only 3 or 4 in. high,
but woody, the young shoots and under side of the leaves covered with a
minute scaly tomentum. Leaves shortly petiolate, orbicular-cordate, 2 to 4
lines diameter, entire, coriaceous and glabrous above when full-grown. Stipules none. Flowers large, solitary or 2 together, on peduncles longer than
the leaves. Bracteoles very small, close to the calyx. Calyx very angular,
sprinkled with stellate hairs more or less united into scales, deeply divided
into broadly ovate-cordate, acute segments, attaining fully 5 lines, thin and
petal-like, with the midribs prominent inside. Petals none. Filaments rather
long; anthers very obtuse. Ovary covered with scale-like papillæ, 5-celled,
with 2 ovules in each cell. Style glabrous, prominently 5-angled, almost
5-winged to near the summit; stigmas at length separating.—*Asterochiton
pygmæus*, Turcz. in Bull. Mosc. 1852, ii. 139.

**W. Australia,** *Drummond, 5th Coll. n.* 258.

Notwithstanding the curious style, this plant is too closely allied to *T. stelligera* to be
separated from it generically. I had formerly referred them both to *Lasiopetalum*, but they
have the calyx of *Thomasia,* a character which, after a detailed review of all the species, appears
to be the best for distinguishing naturally the two genera.

## 17. GUICHENOTIA, J. Gay.

(Sarotes, *Lindl.*)

Calyx 5-lobed, enlarged and membranous after flowering, with raised ribs,
3 or 5 to each sepal. Petals 5, small and scale-like. Stamens 5, opposite
the petals, slightly connected at the base or free; staminodia none or rarely
1 to 5, very small, alternating with the stamens. Anthers opening at the top
towards the inside in short slits, which at length extend more or less down
the side. Ovary 5-celled, with 2 to 5 ovules in each cell. Style simple.
Capsule shorter than the calyx, opening loculicidally in 5 valves. Seeds
usually strophiolate, albuminous; embryo straight, with flat cotyledons.—
Shrubs, more or less tomentose with stellate hairs. Leaves narrow, entire,
with revolute margins. Stipules leafy, either similar to the leaves or short
and oblique. Racemes simple, leaf-opposed. Bracts small and deciduous.
Bracteoles small, and not so close to the calyx as in most *Lasiopetaleæ.*

The genus is confined to Australia. It differs from *Thomasia* chiefly in the calyx. The
leaves and stipules of those species where they are similar are sometimes described as verticillate leaves.

Style glabrous or tomentose at the base only.
  Stipules like the leaves and scarcely smaller.
    Flowers several in the raceme. Calyx not above 4 lines . . . 1. *G. ledifolia.*
    Flowers 2 or 3. Calyx ¾ to 1 in. . . . . . . . . . . 2. *G. macrantha.*
  Stipules semihastate, much smaller than the leaves . . . . . 3. *G. semihastata.*
Style glabrous at the base, thickly stellate-hairy in the upper half.
  Stipules like the leaves, but smaller. Calyx above 5 lines. Anthers
    acuminate. No staminodia . . . . . . . . . . . 4. *G. Sarotes.*
  Stipules small, semicordate. Calyx about 3 lines. Anthers truncate.
    Staminodia 5 or fewer . . . . . . . . . . . . 5. *G. micrantha.*

1. **G. ledifolia,** *J. Gay, in Mém. Mus. Par.* vii. 449, t. 20. A shrub clothed with a soft whitish tomentum, either close, or dense and velvety, or almost floccose. Leaves on very short petioles, oblong-linear, obtuse, mostly 1 to 1½ in. long, the margins much revolute, wrinkled, thick, and soft. Stipules similar, but usually rather shorter and more sessile. Racemes several-flowered. Calyx 2½ to 4 lines long, scarcely membranous, tomentose, the 3 prominent ribs on each sepal giving it a rigid striate appearance. Filaments rather short; anthers acuminate. Ovary densely tomentose, usually 5-celled, with 3 to 5 ovules in each cell; style glabrous.—DC. Prod. i. 489; Steetz, in Pl. Preiss. ii. 318.

**W. Australia.** Swan River, *Drummond, n.* 67; *Preiss, n.* 1670; and northward to Rottenest Island, *A. Cunningham;* Sharks Bay, *Leschenault;* Murchison river, *Oldfield.*

2. **G. macrantha,** *Turcz. in Bull. Mosc.* 1846, ii. 500. A shrub with the foliage and indumentum of *G. ledifolia,* but very much larger pendulous flowers, on thickened pedicels, in racemes of 2 or 3. Bracteoles sometimes, but not always, larger and closer to the calyx. Calyx at length ¾ to 1 in. long, sprinkled with stellate tomentum, more membranous than in *G. ledifolia,* divided to about the middle into broad, acute lobes, with 3 prominent ribs to each sepal. Filaments rather long. Ovary 5-celled, densely glandular-tomentose, with 4 or 5 ovules in each cell; style tomentose at the base, glabrous upwards.—Bot. Mag. t. 4651.

**W. Australia.** Swan River, *Drummond, n.* 133; in the interior, *Roe.*

3. **G. semihastata,** *Benth.* A low shrub, with the aspect nearly of *G. Sarotes,* but the tomentum usually closer and thinner, sometimes disappearing from the upper surface of the leaves. Leaves on short petioles, oblong-linear, obtuse, ½ to 1 in. long, the margins much revolute. Stipules semihastate, sometimes very small, sometimes half as long as the leaves. Flowers pendulous, solitary, or 2 or 3 in a short raceme, large, like those of *G. Sarotes.* Bracteoles small, cordate-acuminate or lanceolate. Calyx ½ to ¾ in. long, sprinkled with a slight tomentum, divided to below the middle into broad, almost cordate lobes, with 3 or 5 raised ribs to each sepal. Filaments shorter than the anthers. Ovary tomentose and glandular, 5-celled, with 2 ovules in each cell; style glabrous.—*Sarotes semihastata,* F. Muell. Fragm. ii. 4; *Ditomostrophe angustifolia,* Turcz. in Bull. Mosc. 1846, ii. 499.

**W. Australia,** *Drummond, n.* 102; White Peak and Champion Bay, *Oldfield.*

4. **G. Sarotes,** *Benth.* A low, much-branched, softly tomentose shrub. Leaves almost sessile, oblong-linear, obtuse, the margins much revolute, mostly about ½ in., but sometimes 1 in. long. Stipules similar, but smaller, sometimes scarcely above half as long. Racemes pedunculate, with few rather large pendulous flowers. Bracteoles small. Calyx thin, 5 or 6 lines long, or nearly 9 when in fruit, divided to below the middle into broad, almost cordate lobes, with 3 raised ribs to each sepal. Filaments very short; anthers acuminate. Ovary densely glandular-tomentose, 5-celled with 2 or 3 ovules to each cell. Style glabrous at the base, densely covered with stellate hairs in the upper half.—*Sarotes ledifolia,* Lindl. Swan Riv. App. 19; Hook. Journ. Bot. ii. 381, t. 16; *Thomasia pumila,* Steud. in Pl. Preiss. i. 233, according to Steetz, l. c. ii. 345.

**W. Australia.** Swan River, *Drummond,* 1*st Coll.,* 2*nd Coll. n.* 59, *Preiss, n.* 1643.

5. **G. micrantha,** *Benth.* Smaller and more branched than the other species, but equally tomentose. Leaves on very short petioles, oblong-linear or almost lanceolate, obtuse, mostly under ½ in. and rarely 1 in. long, the margins much revolute. Stipules obliquely ovate-cordate, usually very small, and sometimes wanting. Racemes 3- to 6-flowered. Bracteoles linear-filiform. Calyx about 3 lines long, very broad and angular, the sepals united much above the middle, with 3 or 5 prominent ribs to each. Filaments very short, alternating with small staminodia; anthers very truncate and usually tipped with a tuft of short hairs. Ovary tomentose, 5-celled; style densely covered with stellate hairs from below the middle to the top.—*Sarotes micrantha,* Steetz, in Pl. Preiss. ii. 346; *Thomasia pogonanthera,* F. Muell. Fragm. ii. 7.

**W. Australia,** *Drummond, n.* 63; *Burges;* White Peak, Champion Bay, *Oldfield.*

## 18. LASIOPETALUM, Sm.

(Corethrostylis, *Endl.*)

Calyx 5-lobed nearly to the base, not much enlarged after flowering, without prominent ribs, the sepals obscurely several-veined at the base. Petals small and scale-like, or rarely none. Stamens 5, opposite the petals, slightly connected at the base or free, without intervening staminodia; anthers opening in terminal or inwardly oblique pores or short slits, which rarely extend down the sides. Ovary 3-celled, or rarely 4- or 5-celled, with 2 or very rarely more ovules in each cell; style simple. Capsule shorter than the calyx, opening loculicidally. Seeds usually solitary in each cell, erect, strophiolate, albuminous; embryo straight, with flat cotyledons.—Shrubs, more or less tomentose or pubescent with stellate hairs. Leaves entire or rarely lobed, often coriaceous and glabrous on the upper side, in one species nearly all opposite, and in some others occasionally so. Stipules none. Flowers in small drooping cymes contracted into heads, or in looser-branched cymes, or rarely in simple racemes. Bracteoles 3 or fewer, in some species very small.

The genus is entirely Australian. It differs from *Thomasia* chiefly in the calyx and generally in habit. The want of stipules is constant in *Lasiopetalum,* but occurs also in the section *Rhynchostemon* of *Thomasia,* to which some *Lasiopetala* of the section *Corethrostylis* bear much affinity. They are, however, readily known by the peculiar hairs of the style in that group, which never occur in *Thomasia.*

A. *Style glabrous.*

Bracteoles longer than the calyx, forming an involucre round the soft woolly flower-heads. Leaves cordate-ovate, white-tomentose underneath . . . . . . . . . . . . . . . 1. *L. discolor.*
Bracteoles not exceeding the calyx, or subulate and loose.
　Calyx-segments glabrous inside (except the edges).
　　Calyx-segments mostly 3 lines long or more.
　　　Cymes dense. Calyx very angular.
　　　　Leaves cordate-ovate or lanceolate . . . . . . 2. *L. dasyphyllum.*
　　　　Leaves linear. Calyx-segments acuminate . . . . 5. *L. parviflorum.*
　　　Cymes few-flowered, not dense. Leaves oblong or linear, thickly coriaceous. Calyx scarcely angular.
　　　　Sepals narrow-lanceolate, loosely woolly tomentose . . 3. *L. indutum.*

s 2

Sepals broad, thick, closely tomentose . . . . . .   4. *L. Behrii.*
Calyx-segments rarely exceeding 2 lines and mostly smaller.
   Leaves opposite, linear-lanceolate . . . . . . . .   6. *L. oppositifolium.*
   Leaves linear, coriaceous, smooth above. Calyx very angular   5. *L. parviflorum.*
   Leaves linear, rugose. Anther-cells opening laterally to the
     base . . . . . . . . . . . . . . . . . .   7. *L. micranthum.*
   Leaves cordate-lanceolate, rugose. Cymes loose, many-
     flowered. Calyx very spreading . . . . . . . .   3. *L. macrophyllum.*
Calyx-segments tomentose or pubescent inside.
   Leaves linear. Cymes or racemes almost simple.
     Racemes several-flowered, reflexed. Calyx-segments not
      above 3 lines, acute . . . . . . . . . . .   9. *L. Baueri.*
     Flowers on slender pedicels, scarcely racemose. Calyx-seg-
      ments 4 to 5 lines, broad and scarcely acute . . . . 10. *L. rufum.*
   Leaves cordate or lanceolate.
     Cymes dense, nearly sessile. Calyx very angular. Anthers
      not acuminate . . . . . . . . . . . . . . 11. *L. ferrugineum.*
     Cymes pedunculate, loose. Calyx-segments thick. An-
      thers acuminate . . . . . . . . . . . . . 12. *L. acutiflorum.*

B. *Style densely covered from the top of the anthers to the summit or near the summit with stellate horizontal or reflexed hairs, forming a cylindrical or conical mass.* (**Corethrostylis,** *Endl.*)

Leaves oblong-lanceolate or linear. Bracteoles usually 3.
   Cymes densely capitate. Bracteoles and calyx-segments linear
     and softly plumose-villous . . . . . . . . . . . 13. *L. Drummondii.*
   Cymes few-flowered. Bracteoles short. Calyx hoary-tomentose 14. *L. rosmarinifolium.*
Leaves cordate-ovate. Bracteoles 1 or 2.
   Bracteoles linear or subulate.
     Pedicels short. Bracteoles close to the calyx.
      Leaves coriaceous, glabrous above, tomentose underneath.
       Calyx 2 to 3 lines . . . . . . . . . . . . 15. *L. cordifolium.*
      Leaves densely tomentose, villous underneath. Calyx 3 to
       4 lines . . . . . . . . . . . . . . . . 16. *L. Schulzenii.*
     Pedicels longer than the calyx. Bracteoles distant. Calyx-
      segments narrow-lanceolate, 2 lines . . . . . . . . 17. *L. floribundum.*
   Bracteoles ovate, membranous and coloured.
     Bracteoles close to the calyx. Pedicels short. Tomentum
      short and soft . . . . . . . . . . . . . . . 18. *L. molle.*
     Bracteoles below the middle of the pedicel.
      Leaves sprinkled with rigid stellate hairs, otherwise glabrous. 19. *L. membranaceum.*
      Leaves tomentose underneath . . . . . . . . . . 20. *L. bracteatum.*

A. LASIOPETALUM proper.—Style glabrous.

**1. L. discolor,** *Hook. Comp. Bot. Mag.* i. 276, *and Journ. Bot.* ii. 414. A shrub of several feet, the branches tomentose. Leaves petiolate, ovate-cordate, obtuse, 1 to 2 or rarely 3 in. long, coriaceous, loosely tomentose above when young, but soon glabrous, white-tomentose underneath. Cymes contracted into dense softly tomentose heads, on short recurved peduncles. Bracteoles longer than the calyx, oblong-linear, petal-like, with a broad thick central nerve, tomentose, arranged in a kind of radiating involucre round the head. Calyx-segments thin and petal-like, 3 to 4 lines long, softly tomentose outside, glabrous inside. Petals often shortly clawed. Anthers rather long. Ovary very villous; style glabrous.—*L. confertiflorum,* F. Muell. in Linnæa, xxv. 377 ; *L. capitellatum,* Turcz. in Bull. Mosc. 1852, ii. 148.

**Tasmania.** Hummock or Prime Seal Island, Bass's Straits, *Backhouse, Gunn,* and others.

**S. Australia.** Port Lincoln, *R. Brown, Wilhelmi;* Memory Cove, *R. Brown;* Venus Bay and Kangaroo Island, *Waterhouse.*

**W. Australia.** S. coast, *Drummond,* 5*th Coll. n.* 263; sandy hills, S.W. Bay and Doubtful Island Bay, *Oldfield.*

2. **L. dasyphyllum,** *Sieb.; Hook. Journ. Bot,* ii. 414. A tall shrub, the young branches rusty-tomentose. Leaves from ovate-cordate to cordate- or oblong-lanceolate, in luxuriant specimens 3 or 4 in. long and acute, in others much smaller and obtuse, entire, coriaceous, glabrous or slightly scabrous above, white or rusty-tomentose underneath. Cymes reflexed, nearly sessile, very dense, almost capitate, and densely rusty-tomentose. Bracteoles lanceolate, shorter than the calyx. Calyx-segments ovate-lanceolate, 3 to 4 lines long, tomentose outside, glabrous inside. Anthers truncate, about as long as the filaments. Ovary tomentose, usually 4-celled (almost constantly 3-celled in all other species of the genus). Style glabrous or tomentose at the base only.—Steetz, in Pl. Preiss. ii. 341; F. Muell. Pl. Vict. i. 144; *L. Gunnii,* Steetz, l. c. ii. 342; Hook. f. Fl. Tasm. i. 51; *L. Wilhelmi,* F. Muell. in Trans. Phil. Soc. Vict. ii. 65.

**N. S. Wales.** Grose river, *R. Brown;* Port Jackson, *Sieber, n.* 240, and others; Blue Mountains, *A. Cunningham* and others; southward to Twofold Bay, *F. Mueller.*

**Victoria.** Highest declivities of the Grampians, *Wilhelmi;* granitic ridges, more frequent towards the eastern boundary, *F. Mueller.*

**Tasmania.**· Rocky Cape and islands of Bass's Straits, *Gunn.*

3. **L. indutum,** *Steud. in Pl. Preiss.* i. 235; *Steetz, l. c.* ii. 340. A much-branched shrub of 2 to 4 ft., the young shoots rusty-tomentose. Leaves from broadly oblong to linear, obtuse, 1 to 3 in. long, not cordate at the base, the margins slightly recurved, coriaceous, glabrous above when full grown, densely tomentose underneath. Cymes shortly pedunculate, compact, reflexed, few-flowered. Bracteoles linear, tomentose-villous, shorter than or sometimes as long as the calyx. Calyx-segments lanceolate, acute, 2 to 3 lines long, densely tomentose or softly villous outside, glabrous or nearly so inside. Filaments short; anthers contracted at the top, with oblique pores. Ovary tomentose-villous; style glabrous.

**W. Australia.** Konkoberup hills, *Preiss, n.* 1655; towards Cape Riche, *Drummond;* S.W. interior, *Maxwell* (with less woolly flowers): sand-hills by the south coast; *R. Brown* (with small flowers).

4. **L. Behrii,** *F. Muell. in Trans. Phil. Soc. Vict.* i. 36, *and Pl. Vict.* i. 143, *t.* 3. An erect or diffuse shrub, of several feet, nearly allied to *L. indutum,* but with larger more rigid flowers. Young branches rusty-tomentose. Leaves shortly petiolate, from oblong to linear, obtuse, mostly 1½ to 2½ in. long, not cordate at the base, the margins recurved, coriaceous, glabrous above when full grown, hoary or rusty underneath with a close tomentum. Cymes shortly pedunculate, rather loose. Bracteoles linear, much shorter than the calyx. Calyx-segments ovate, acute, 3 to 4 lines long, rather thick, obscurely 3-nerved and white-tomentose outside, glabrous inside. Filaments very short; anther-pores small, terminal. Ovary tomentose; style glabrous.

**N. S. Wales.** Darling and Murrumbidgee rivers, *F. Mueller.*

**Victoria.** In the N.W. district, *F. Mueller.*
**S. Australia.** S. coast, *R. Brown ;* from the Murray river to Kangaroo Island and the E. extremity of the Great Australian Bight, and northward to Lake Torrens, *F. Mueller.*

5. **L. parviflorum,** *Rudge, in Trans. Linn. Soc.* x. 297, *t.* 19. A tall shrub, the young branches hoary or rusty-tomentose. Leaves on short petioles, linear, obtuse, mostly 1½ to 3 in. long, coriaceous, glabrous above, white or rusty-tomentose underneath. Cymes shortly pedunculate, corymbose and several-flowered, but much shorter than the leaves. Bracteoles small, the 2 lateral ones sometimes minute or even wanting. Calyx-segments 1¼ to 2 lines long, minutely white-tomentose outside, glabrous inside. Filaments very short, anthers ovate, truncate. Ovary tomentose; style glabrous.— J. Gay, in Mem. Mus. Par. vii. 447, t. 19 ; DC. Prod. i. 489 ; Steetz, in Pl. Preiss. ii. 339 ; F. Muell. Pl. Vict. i. 142.

**N. S. Wales.** Port Jackson, *R. Brown, A. Cunningham ;* Twofold Bay. *F. Mueller.*
**Victoria.** Granite banks of watercourses towards the eastern frontier, *F. Mueller.*
Var. *major.* Calyx-segments 2 to 3 lines long. Cymes denser. To this belong most of the southern specimens.
Var. (?) *occidentale.* Leaves smaller, rarely above 1 in. long. Flowers small. Bracteoles very small. Scarcely, however, to be distinguished from some of the smaller-flowered Port Jackson specimens.
**W. Australia,** *Drummond, 5th Coll. n.* 267.

6. **L. oppositifolium,** *F. Muell. Fragm.* ii. 5. A diffuse shrub, with slender, rigid, divaricate branches, whitish with a close tomentum. Leaves mostly opposite, shortly petiolate, lanceolate or oblong-linear, 2 to 4 in. long, slightly cordate at the base, glabrous above, minutely tomentose underneath. Cymes shortly pedunculate, reflexed, glandular-hispid. Flowers small. Bracteoles linear, longer than the calyx. Calyx-segments lanceolate, 2 to 3 lines long, hirsute outside, glabrous within. Anthers opening in terminal pores, but at length splitting also laterally. Ovary hirsute ; style glabrous.

**W. Australia.** Murchison river, *Oldfield.* The specimens have only very imperfect withered flowers.

7. **L. micranthum,** *Hook. f. Fl. Tasm.* i. 51. A small shrub, branches tomentose. Leaves petiolate, oblong-linear, obtuse, 1 to 2 in. long or rarely more, the margins revolute, glabrous, or slightly scabrous and wrinkled with impressed veins above, densely tomentose underneath. Cymes shortly pedunculate, corymbose, reflexed. Bracts broadly ovate, concave, the lateral ones small. Calyx-segments lanceolate, acuminate, about 2 lines long, slightly tomentose outside, glabrous within. Filaments short ; anthers truncate, the cells opening laterally to the base more readily than in any other species. Ovary tomentose; style glabrous.

**Tasmania.** Eastern Tier, near Oyster Bay, and S.E. of Launceston, *Gunn ;* near Swan Port, *C. Stuart.*

8. **L. macrophyllum,** *Grah. in Bot. Mag. t.* 3908. A tall shrub, the branches densely rusty-tomentose. Leaves petiolate, ovate-lanceolate or lanceolate, mostly acute, 2 to 4 in. long, glabrous, or slightly scabrous, and much wrinkled with impressed veins above, densely tomentose underneath. Cymes shortly pedunculate, corymbose, reflexed, rather loose, with numerous small

flowers.   Bracteoles linear, nearly as long as the calyx.   Calyx-segments very spreading or almost reflexed, under 2 lines long in the wild specimens, lanceolate, acuminate, tomentose outside, glabrous within.   Filaments rather long; anthers oblong, their terminal pores light-coloured and very conspicuous.   Ovary tomentose, style glabrous.

**N. S. Wales.**   Paramatta and Sydney, *R. Brown*; Southward of the colony, *A. Cunningham.*   I have not seen Graham's specimens, but the figure quoted well represents this plant, except that the flowers are larger than in the specimens I have seen.

9. **L. Baueri,** *Steetz, in Pl. Preiss.* ii. 339.   A shrub of several feet, the branches hoary or rusty with a close tomentum.   Leaves on short petioles, linear or oblong-linear, obtuse, mostly 1 to 2 in. long, the margins revolute, coriaceous, glabrous or minutely tomentose above, white or rusty-tomentose underneath.   Flowers few, in short pedunculate reflexed racemes, rarely branching into cymes.   Bracteoles small, oblong or linear.   Calyx-segments $2\frac{1}{2}$ to 3 lines long, acute, tomentose outside and slightly so inside.   Filaments very short; anthers contracted at the top.   Ovary tomentose. Style glabrous or occasionally bearing a few stellate hairs.—F. Muell. Pl. Vict. i. 142.

**N. S. Wales.**   Blue Mountains, *Miss Atkinson;* Darling and Murrumbidgee rivers, *F. Mueller.*

**Victoria.**   Murray scrub and Sandy Desert near Brighton, but rare, *F. Mueller.*

**S. Australia.**   Memory Cove, *R. Brown;* sand ridges from the Murray river to St. Vincent's Gulf, Kangaroo Island, and Spencer's Gulf, *F. Mueller.*

This is a very variable plant, difficult to define from dried specimens.   Some of the numerous forms, especially in Mr. Brown's collection, seem to connect it on the one hand with the large-flowered varieties of *L. parviflorum,* and on the other hand, in some measure, with some forms of *L. ferrugineum.*

10. **L. rufum,** *R. Br. Herb.*   A slender much-branched shrub of $1\frac{1}{2}$ to 2 ft., the young branches minutely tomentose.   Leaves, as in *L. parviflorum,* linear, obtuse, 1 to $1\frac{1}{2}$ in. long, coriaceous, the margins revolute, glabrous above, white-tomentose underneath.   Flowers solitary or 2 or 3 together in very loose simple racemes, the pedicels 2 to 4 lines long.   Bracts linear-subulate, not close to the calyx.   Calyx broad, slightly tomentose both within and without, the segments broader and less acute than in most *Lasiopetala,* but faintly several-veined, not 1-nerved as in *Thomasia.*   Petals scale-like, filaments short and anthers contracted at the top as in *L. Baueri,* to which the species is in many respects nearly allied.

**N. S. Wales.**   St. George's River, *R. Brown (Herb. R. Br.).*

11. **L. ferrugineum,** *Sm. in Andr. Bot. Rep. t.* 208.   A tall shrub, the young branches hoary or rusty with a short tomentum.   Leaves on very short petioles, the longer ones narrow-lanceolate or oblong-linear, 3 or 4 in. long, the margins slightly recurved, entire sinuate or hastate with short basal lobes, coriaceous, glabrous above, tomentose underneath, the lower ones often shorter and broader and sometimes cordate-ovate.   Cymes dense, nearly sessile and reflexed.   Calyx very angular, the segments ovate, acute, 3 or rarely 4 lines long, rather thick and tomentose inside as well as out.   Anthers about as long as the filaments.   Ovary tomentose; style glabrous, except at the base.—DC. Prod. i. 489; Vent. Jard. Malm. t. 59; Bot.

Mag. t. 1766 ; J. Gay, in Mem. Mus. Par. vii. 446, t. 18 ; Steetz, in Pl. Preiss. ii. 337 ; F. Muell. Pl. Vict. i. 141.

**N. S. Wales.**   Port Jackson, *R. Brown, Sieber, n.* 572, and others ; Blue Mountains, *A. Cunningham.*

**Victoria.**   Granite ridges of the E. extremity near Mount Imlay, *F. Mueller.*

Var. *cordatum.*   Leaves shorter, from cordate-ovate to cordate-lanceolate. Cymes looser. —*L. Sieberi,* Steetz, in Pl. Preiss. ii. 338 ; *L. rubiginosum,* A. Cunn. in Field. N. S. Wales, 354 ; Steetz, l. c.   To this variety belong the Blue Mountain and Victorian specimens ; the small-flowered ones described by Steetz do not otherwise differ from the larger-flowered ones gathered by Cunningham in the same locality.

12. **L. acutiflorum,** *Turcz. in Bull. Mosc.* 1852, ii. 145.   Branches densely rusty-tomentose.   Leaves petiolate, cordate-lanceolate, obtuse, 1½ to 2½ in. long, tomentose above when young, at length glabrous, densely tomentose underneath, coriaceous, with impressed veins, the margins recurved. Cymes pedunculate, little branched or reduced to simple racemes.   Bracteoles linear-filiform, softly villous.   Calyx-segments about 3 lines long, but slightly united at the base, lanceolate, thick, softly tomentose-villous outside, tomentose inside.   Petals thicker than in most species, truncate and almost gland-like.   Filaments very short ; anthers shortly acuminate and opening in short oblique slits as in most *Thomasias.*   Ovary villous, 5-celled according to Turczaninow, 3-celled in our specimens ; style glabrous.

**W. Australia,** *Drummond, 5th Coll. n.* 254.

Var. *Oldfieldi.*   Leaves shorter and broader, sometimes ovate-cordate.  Petals villous, whilst in Drummond's specimens they are only slightly so or glabrous.—*L. Oldfieldi,* F. Muell. Fragm. ii. 6.—Murchison river, *Oldfield.*

Var. *quinquenervium.*   Leaves ovate-cordate, 1 to 2 in. long.  Cymes looser.  Flowers larger, the calyx segments fully 4 lines long.  Petals more or less villous.  Filaments as long as the petals.—*L. quinquenervium,* Turcz. in Bull. Mosc. 1852, ii. 146.—South coast ? *Drummond, 5th Coll. n.* 260 ; Point Henry and Doubtful Island Bay, *Oldfield* ; W. Mount Barren, *Maxwell.*

B.  CORETHROSTYLIS, Endl.—Style, so-called *scopiform,* that is, covered from below the middle to the summit or near the summit with a dense mass of prominent horizontal or reflexed stellate hairs, the lower ones often longer and covering the tips of the closely appressed anthers.

This group, proposed as a genus by Endlicher, appears to me quite artificial.  Some species have also a looser inflorescence and single bracteoles, but in the first two, the habit, inflorescence, and 3 bracteoles, are quite those of the true *Lasiopetala.*

13. **L. Drummondii,** *Benth.*  Branches densely rusty-tomentose.  Leaves petiolate, oblong-lanceolate, obtuse, 1 to 2 in. long, coriaceous, with recurved margins, glabrous above when full-grown, densely and softly tomentose underneath.  Cymes contracted into dense heads, on short recurved peduncles, softly plumose-villous and white as in *L. discolor.*  Bracteoles 3, linear-filiform, softly villous, as long as the calyx.  Calyx-segments lanceolate-linear, about 4 lines long, softly villous outside, glabrous inside.  Filaments very short.  Ovary villous ; style scopiform, the tip often glabrous.

**W. Australia,** *Drummond,* a single specimen.

14. **L. rosmarinifolium,** *Benth.*  A much-branched shrub, the young shoots hoary or rusty, with a close tomentum.  Leaves shortly petiolate, linear, obtuse, 1 to 2 in. long, coriaceous, with revolute margins, glabrous

above when old, tomentose underneath.   Cymes shortly pedunculate, reflexed, few-flowered.   Bracteoles 3, linear, short.   Calyx-segments lanceolate, 3 to 4 lines long, tomentose outside, glabrous within.   Anthers almost acuminate, with oblique pores.   Ovary densely tomentose, occasionally 4-celled; style scopiform, the tip glabrous.—*Sarotes rosmarinifolia,* Turcz. in Bull. Mosc. 1852, ii. 149.

**W. Australia.**   Swan River, *Drummond, 5th Coll. n.* 266, *Roe.*
Var. *latifolia.*   Leaves shorter and broader, mostly linear-oblong, 1 to 1½ in. long. Flowers rather smaller.—*Sarotes latifolia,* Turcz. in Bull. Mosc. 1852, ii. 150; *Drummond, 5th Coll. n.* 265.

15. **L. cordifolium,** *Endl. in Hueg. Enum.* 10.   A low, erect shrub, the young branches hoary with a minute tomentum.   Leaves petiolate, broadly cordate, obtuse or shortly acuminate, rarely above 1½ in., and mostly under 1 in. long, coriaceous, glabrous above and not wrinkled, tomentose underneath.   Cymes shortly pedunculate, but scarcely exceeding the leaves, rather dense and few-flowered.   Bracteoles linear, solitary, or rarely with 1 or 2 lateral small ones.   Calyx very angular, the segments about 2 lines long, ovate-lanceolate or cordate-acuminate, rather thick, tomentose outside.   Petals none.   Anthers nearly sessile.   Style scopiform.—*Corethrostylis cordifolia,* Steetz, in Pl. Preiss. ii. 344; *C. microphylla,* Turcz. in Bull. Mosc. 1852, ii. 148.

**W. Australia.**   King George's Sound, *A. Cunningham* and others; Mount Melville, *Preiss, n.* 1659; south coast? *Drummond, 5th Coll. n.* 259, and *Suppl. n.* 39.

16. **L. Schulzenii,** *Benth.*   A shrub of several feet, the branches densely tomentose-villous.   Leaves petiolate, deeply cordate, broadly ovate or almost orbicular, obtuse or scarcely acute, mostly 1½ to 2 in. long, pubescent or tomentose above, or at length glabrous, densely but loosely tomentose underneath, sometimes almost floccose..   Cymes loose, many-flowered, but rarely exceeding the leaves.   Bracteoles 1 or 2, linear-filiform, small.   Calyx softly tomentose, the segments 3 or 4 lines long, rather thin, glabrous inside, except on the margins.   Petals usually present.   Filaments short.   Ovary tomentose.   Style scopiform,—*Corethrostylis Schulzeni,* F. in Muell. Trans. Phil. Soc. Vict. i. 36, and Pl. Vict. i. 145.

**Victoria.**   Cape Nelson, *Allitt;* entrance of the Glenelg river, *J. E. Woods.*
**S. Australia.**   Memory Cove, *R. Brown;* Mount Benson, near Cape Bernouille and Guichen Bay, *Schulzen;* Kangaroo Island, *Waterhouse.*

17. **L. floribundum,** *Benth.*   Branches slender, more or less tomentose or hirsute, or rarely nearly glabrous.   Leaves petiolate, broadly ovate-cordate, obtuse, mostly ¾ to 2 in. long, entire or irregularly sinuate or broadly lobed, thin, but rigid, glabrous, scabrous, or sprinkled with stellate hairs above, more or less stellate-hairy or sometimes tomentose underneath, rarely quite glabrous.   Cymes slender, often twice forked, longer than the leaves, hirsute or tomentose.   Bracteoles solitary or rarely 2, minute and filiform, inserted on the slender pedicel above or below the middle.   Calyx tomentose or hirsute at the base, the segments 2 to 3 lines long, narrow-lanceolate, acuminate, glabrous inside.   Petals none.   Anthers nearly sessile.   Style scopiform.—*Corethrostylis parviflora,* Turcz. in Bull. Mosc. 1847, i. 174 (from the character given); *C. oppositifolia,* F. Muell. Fragm. ii. 6.

**W. Australia,** *Drummond, n.* 28 and 156, *Oldfield, Maxwell:* Darling Range, *Collie;* between Perth and King George's Sound, *Harvey.* The upper leaves are often here and there opposite, as in a few other *Lasiopetala.*

18. **L. molle,** *Benth.* Branches hoary or rusty-tomentose. Leaves petiolate, cordate, from orbicular and very obtuse to ovate-acuminate or almost lanceolate, but never acute, 1 to 2 in. long, or in some specimens under 1 in., entire or sinuate, thick, soft and much-wrinkled, scabrous-pubescent above, densely tomentose underneath. Cymes little-branched, loose, and several-flowered, but scarcely exceeding the leaves. Bracteoles solitary, broadly ovate, membranous and coloured, 3 to 4 lines long, close to the calyx. Calyx loosely tomentose-villous, the segments fully 3 lines long, including their long points, glabrous inside. Petals none. Anthers nearly sessile. Ovary tomentose; style scopiform.

**W. Australia,** *Drummond, n.* 26 and 108.

19. **L. membranaceum,** *Benth.* A low shrub, the young branches hispid with stipitate stellate hairs, and slightly tomentose. Leaves petiolate, deeply cordate, ovate or orbicular, obtuse, 1 to 2 in. long, rigidly membranous, much wrinkled, green on both sides, and more or less sprinkled with rigid stellate hairs. Cymes forked, pedunculate, usually longer than the leaves, hirsute and apparently viscid. Bracteoles solitary, ovate, acuminate, membranous and coloured, inserted below the middle or near the base of the pedicel. Calyx tomentose-villous outside, the segments ovate-lanceolate, less acuminate than in *L. bracteatum,* to which this species is closely allied, differing chiefly in the indumentum.—*Corethrostylis membranacea,* Steud. in Pl. Preiss. i. 236; Steetz, l. c. ii. 343.

**W. Australia,** *Drummond, n.* 155, *Oldfield;* sandy woods, Port Leschenault, *Preiss, n.* 1656.

20. **L. bracteatum,** *Benth.* A shrub of 2 ft. or more, the branches tomentose and hirsute with long brown stellate hairs. Leaves broadly ovate-cordate, usually rather acute, 1 to 2 in. long, entire or the margins slightly crisped, thinly coriaceous, scabrous or glabrous, and not wrinkled above, tomentose and sometimes hirsute underneath. Cymes forked, many-flowered, longer than the leaves, hirsute. Bracteoles solitary, ovate, membranous and coloured, inserted below the middle or near the base of the slender pedicels. Calyx hirsute outside at the base, the segments about 4 lines long, with long fine points, glabrous within, dark coloured and somewhat thickened at the base, less deeply separated in this and the last species than in most others of the genus. Petals usually none. Anthers nearly sessile. Ovary tomentose; style scopiform.—*Corethrostylis bracteata,* Endl. Nov. Stirp. Dec. 1; Steetz, in Pl. Preiss. ii. 343; Bot. Reg. 1844, t. 47; *C. coriacea,* Steud. in Pl. Preiss. i. 236.

**W. Australia.** Swan River, *Huegel, Drummond, 1st Coll.* and *2nd Coll. n.* 65; *Preiss, n.* 1637.

## 19. LYSIOSEPALUM, F. Muell.

Sepals 5, petal-like, quite free, valvate in the bud, and then enclosed in the thick valvate bracteoles. Petals 5, minute and scale-like. Stamens 5, free, opposite the petals, without intervening staminodia. Anthers opening at the top in pores or short slits, sometimes extending at length down the sides. Ovary 3-celled, with several ovules in each cell; style simple, glabrous. Capsule shorter than the calyx, opening loculicidally in 3 valves, tomentose. Seeds (not seen ripe) like those of *Thomasia.*—Shrubs, with nearly the habit of *Thomasia angustifolia* and its allies. Stipules very small and cordate or none.

The genus is limited to Australia, and remarkable for the calyx-like bracts and petal-like sepals.

Leaves almost or quite sessile. Bracteoles ovate . . . . . . . 1. *L. Barryanum.*
Leaves distinctly petiolate. Bracteoles oblong or lanceolate . . . . 2. *L. rugosum.*

1. **L. Barryanum,** *F. Muell. Fragm.* i. 143. A small shrub, densely clothed with a short soft velvety tomentum. Leaves sessile or nearly so, oblong-linear, obtuse, mostly ½ to ¾ in. long, the margins much revolute, wrinkled and tomentose. Stipules very small and cordate or none. Racemes loose, few-flowered, much longer than the leaves. Bracteoles ovate, 2 to 3 lines long, thick and densely tomentose, completely enclosing the bud, like a valvate 3-sepaled outer calyx. Sepals oblong-lanceolate, 3 or 4 lines long, coloured and petal-like. Anthers acuminate, nearly sessile. Ovary covered with closely-packed oblong scales, with 8 to 10 ovules in each cell.—*Thomasia involucrata,* Turcz. in Bull. Mosc. 1852, ii. 143.

**W. Australia.** Swan River, *Drummond, 5th Coll. n.* 255; in the interior, *Roe:* Flats of Phillips River, *Maxwell.* In these specimens the flowers are smaller than in Drummond's.

2. **L. rugosum,** *Benth.* A small shrub, closely allied to *L. Barryanum,* and much resembling in habit, foliage, and indumentum, *Thomasia angustifolia.* Branches hoary-tomentose. Leaves shortly, but distinctly petiolate, narrow-lanceolate, obtuse, ½ to 1 in. long, much wrinkled, the margins revolute, slightly hoary above, more densely tomentose underneath. Stipules very small. Racemes slender, several-flowered. Involucre at first ovoid, but lengthening much before the flower expands, the bracteoles at length lanceolate, 4 lines long, thick and tomentose as in *L. Barryanum.* Sepals petal-like, about as long as the bracteoles, broadly oblong. Filaments very short; anthers shortly acuminate. Ovary covered with a close scaly tomentum, with several ovules in each cell.

**W. Australia.** Swan River, *Drummond.*

ORDER XXIII. **TILIACEÆ.**

Flowers regular, hermaphrodite or rarely unisexual. Sepals 5, rarely 3 or 4, free or more or less cohering, usually valvate. Petals as many or fewer or none, alternate with the sepals, inserted round the base of the torus. Stamens indefinite, rarely reduced to very few, inserted on the torus, which is often raised or disk-like. Filaments free or slightly united at the base. Anthers 2-celled, with parallel or rarely divaricate cells, opening in longitudinal slits or in terminal pores. Ovary free, sessile, 2- or more celled. Style simple and entire, or divided at the top into as many stigmatic teeth or lobes as there are cells. Ovules 1, 2, or more in each cell, erect, pendulous, or horizontal. Fruit capsular or indehiscent, with single- or several-seeded cells, where several-seeded the cells often subdivided by spurious vertical or transverse partitions. Seeds without any arillus, the testa usually coriaceous or crustaceous. Albumen fleshy, rarely deficient. Embryo straight or rarely curved or slightly folded. Cotyledons leafy or rarely fleshy, the radicle next to the hilum, usually shorter than the cotyledons.—Trees, shrubs, or rarely herbs. Leaves alternate or very rarely opposite, simple, with pinnate or palmate nerves, entire, toothed, or rarely lobed. Stipules usually free, and small or deciduous. Flowers axillary, terminal or leaf-opposed, usually in little cymes, often almost umbellate, either solitary and sessile or pedunculate, or arranged in panicles.

A large Order, chiefly tropical or subtropical, spread over both the New and the Old World, with one extratropical genus (*Tilia*) in the northern and another (*Aristotelia*) in the southern hemisphere. The Australian genera are none of them endemic, the extratropical *Aristotelia* is common to Chili and New Zealand. The others are all tropical and Asiatic, *Grewia* extending into Africa and *Corchorus* also partially into America, whilst *Triumfetta* belongs equally to the New and the Old World.

Anthers short, with confluent cells. Calyx irregularly 3- to 5-lobed.
 Petals entire. Capsule loculicidal, each valve 2-winged . . . . 1. BERRYA.
Anthers short, with 2 parallel distinct cells opening longitudinally.
 Sepals distinct. Petals entire.
Drupe indehiscent, not echinate, entire or 2-lobed. Petals narrow,
 short, with a foveolate base. Trees or shrubs . . . . . . 2. GREWIA.
Fruit globular, echinate, indehiscent, or separating into 1-seeded
 cocci. Petals narrow, with a foveolate or pubescent base.
 Shrubs or herbs . . . . . . . . . . . . . . . 3. TRIUMFETTA.
Capsule 2-to 5-celled, with several seeds in each, opening in valves,
 usually long and smooth, rarely short and echinate. Petals
 usually obovate or broad, without a foveola. Shrubs or herbs . 4. CORCHORUS.
Anthers elongated, opening in terminal valves or pores. Sepals distinct. Petals (except in one species) lobed or fringed.
Sepals 4, imbricate in 2 series. Capsule echinate, 4-valved . . 5. ECHINOCARPUS.
Sepals 4 or 5, valvate. Fruit a berry . . . . . . . . . 6. ARISTOTELIA.
Sepals 4 or 5, valvate. Fruit a drupe . . . . . . . . . 7. ELÆOCARPUS.

## 1. BERRYA, Roxb.

Calyx campanulate, irregularly 3- to 5-lobed. Petals 5, without any foveola at the base. Stamens numerous, free, without staminodia ; anthers subglobose, the cells at length confluent into one. Torus not raised. Ovary (2- ? or) 3-celled, with 4 ovules in each cell ; style subulate (2- ? or) 3-lobed (or the styles distinct ?). Capsule nearly globular, opening loculicidally in 2 or 3 valves, each valve bearing 2 vertical, diverging, coriaceous wings. Seeds 1 or 2 in each cell, densely covered with rigid hairs; albumen fleshy; cotyledons leafy, flat.—Trees. Leaves entire, 5- or 7-nerved. Flowers small, white, the umbel-like cymes arranged in a terminal panicle.

The genus consists of a single species, common to tropical Australia and Asia.

1. **B. Ammonilla,** *Roxb. Pl. Corom.* iii. 60, *t.* 264, var. *rotundifolia.* A small tree, the young branches slightly tomentose. Leaves cordate-orbicular, very obtuse, 3 or 4 in. diameter, rigidly membranous, glabrous when full-grown. Flowers of the Australian variety unknown, except from some

fragments remaining about the fruits seen by R: Brown, in which he ascertained that the calyx was lobed and the stamens numerous. Capsule (always?) 2-celled, the wings broadly obovate, about ½ in. long, sinuate-crenate on the margin. Seeds 1 or 2 in each cell.

**Queensland.** Cumberland Islands, *R. Brown (Hb. R. Br.).* The shape of the fruit and its wings and the seeds are the same as in the Asiatic *B. Ammonilla,* Roxb., DC. Prod. i. 517, Wight, Ill. t. 34; but as that species has acuminate leaves and a 3-celled capsule, I had at first thought that this one might be distinct. I find, however, some Ceylon specimens with the same rounded leaves, and the Australian specimens are not sufficient to show whether the reduced number of carpels is more than accidental.

## 2. GREWIA, Linn.

Sepals 5, distinct. Petals 5, with a foveola or thickened cavity at the base, usually shorter than the calyx, inserted round the base of the torus. Stamens indefinite, inserted on the raised torus. Ovary 2- to 4-celled, with 2 or more ovules in each cell; style subulate, minutely toothed or lobed. Drupe containing 1 or 4 pyrenes or nuts, entire or 2- or 4-lobed, the nuts either 1-seeded or 2- or more seeded, and then divided by transverse partitions between the seeds. Seeds ascending or horizontal, the albumen usually copious, the cotyledons flat.—Trees or shrubs, the hairs or tomentum stellate. Leaves entire or serrate, 3- to 7-nerved. Stipules narrow, deciduous. Flowers usually yellow, the umbel-like cymes axillary or terminal. In the Australian species (except *G. breviflora*) the ovary is 2-celled, but each cell is subdivided by a vertical, nearly complete partition, so as to appear 4-celled, with two or rarely more superposed ovules in each half-cell, each half-cell forming in the fruit a separate nut, with 1 or rarely more superposed seeds in each.

The genus is a large one, widely spread over the tropical and subtropical regions of the Old World. Of the Australian species, 3 extend over tropical Asia, the remaining 5 are endemic.

Leaves glabrous or nearly so, 3-nerved at the base. Flowers hermaphrodite.
  Sepals 7 to 9 lines. Petals small, the foveola very large. Torus elongated. Fruit depressed-globose, not lobed, ⅓ in. diameter or more . . . . . . . . . . . . . . . . . . . . . . . . 1. *G. orientalis.*
  Sepals about 4 lines. Petals very small, the foveola large. Torus short. Fruit small, 2-lobed (unless reduced to one carpel) . . 2. *G. multiflora.*
  Sepals about 2 lines. Petals more than half as long, the foveola very small. Torus short. Fruit small, entire . . . . . 3. *G. breviflora.*
Leaves softly velvety-tomentose underneath, 3 or 5-nerved. Flowers hermaphrodite. Petals small, foveola large . . . . . . . 4. *G. latifolia.*
Leaves white-tomentose underneath or scabrous, 3- or 5-nerved. Flowers polygamo-diœcious.
  Leaves obovate-oblong to lanceolate. Foveolate base of the petals broader than the lamina . . . . . . . . . . . . . . . 5. *G. polygama.*
  Leaves ovate or orbicular. Foveolate base of the petals small.
    Leaves ovate-cordate, acuminate, often 3 to 4 in. Staminodia in the female flowers numerous, clavate, without anthers . . 6. *G. xanthopetala.*
    Leaves small, ovate-obtuse. Stamens in the female flowers 1 or 2 apparently perfect, without staminodia. Buds not striate . 7. *G. scabrella.*
    Leaves small, orbicular, very scabrous. Buds striate. (Female flowers unknown.) . . . . . . . . . . . . . . . 8. *G. orbifolia.*

270 XXIII. TILIACEÆ. [*Grewia.*

1. **G. orientalis,** *Linn.; W. and Arn. Prod.* 76. A tall, rather weak shrub, glabrous, except a minute tomentum on the young shoots, or sparingly sprinkled on the under side of the leaves and more abundant on the inflorescence. Leaves shortly petiolate, from oval-oblong to oblong-lanceolate, acuminate, 3 to 4 in. long, minutely crenulate, 3-nerved at the base. Peduncles 1- or 2-flowered, axillary or the upper ones forming a short terminal panicle. Sepals rusty-tomentose, 7 to 9 lines long. Petals not half so long, the foveolate base broader than and almost as long as the lamina, pubescent round the edge. Torus elongated. Stamens very numerous. Drupe depressed-globular, $\frac{1}{2}$ to $\frac{3}{4}$ in. diameter, flat-topped, slightly furrowed but not lobed, minutely tomentose with a few short straight hairs intermixed, containing usually 4 nuts, each with 2 or 3 horizontal, superposed seeds, separated by transverse partitions.

**N. Australia.** Van Diemen's Gulf, *A. Cunningham;* islands of the Gulf of Carpentaria, *R. Brown.*

**Queensland.** N.E. coast, *Banks and Solander;* Northumberland Islands, *R. Brown.* The species is not uncommon in Ceylon and a part of the Indian peninsula.

Var. *latifolia.* Leaves ovate-cordate, crenate, fruit more densely pubescent. Port Denison, *Fitzalan.*

2. **G. multiflora,** *Juss. in Ann. Mus. Par.* iv. 89, *t.* 47, *f.* 1. A shrub or tree, with rather slender branches, glabrous or sprinkled with a few appressed simple or stellate hairs. Leaves from ovate-acuminate to elliptical-oblong or almost lanceolate, 3 or 4 in. long or sometimes more, serrate, 3-nerved at the base. Peduncles axillary, usually 2 or three together, 2- to 5-flowered. Sepals lanceolate, about 4 lines long, minutely tomentose. Petals very short, the broad foveolate base villous round the edge, not longer than the short torus, the lamina still smaller. Stamens numerous. Ovary hirsute, with 2 superposed ovules in each half-cell. Drupe small, sprinkled with a few rigid hairs, deeply 2-lobed or entire by the abortion of one carpel, with 2 nuts in each carpel, each containing a single seed.—DC. Prod. i. 508.

**N. Australia.** Port Essington, *Armstrong.*
**Queensland.** Percy Islands, *A. Cunningham.*
The species was originally described from Philippine Island specimens; our Australian ones agree well with Jussieu's figure, as well as with Cuming's specimens, n. 461, 701, and 1515. The common East Indian *G. sepiaria,* Roxb., as well as *G. prunifolia,* A. Gray, Bot. Amer. Expl. Exp. i. 77, said to be a common shrub on the leeward coast of the Fiji Islands, appear from our specimens to be the same species, which we have also from Java and Singapore, although not included in Miquel's Flora. It is, however, frequently confounded with *G. lævigata,* Vahl, which differs in longer flowers, a more raised torus, and several other points.

3. **G. breviflora,** *Benth.* A large spreading shrub or small tree, the young shoots slightly tomentose, otherwise nearly glabrous. Leaves petiolate, obliquely ovate, acuminate, 2 to 4 in. long, glabrous or slightly scabrous. Peduncles usually 3 or 4 together, 3- to 5-flowered, unequal in length, but rarely exceeding the petioles. Sepals elliptical-oblong, more obtuse than in any other species, not exceeding 2 lines, rather thick, tomentose outside. Petals more than half as long, with a very small foveolate base. Stamens numerous. Ovary hirsute, 2-celled, with 2 superposed ovules in each half-cell. Drupe depressed-globular, quite entire, about 3 lines diameter, glabrous or slightly hairy, broad and flat-topped, the hard almost woody endocarp scarcely

separating into 2 nuts, each one containing when perfect 2 superposed pairs of seeds placed singly in separate compartments, but often fewer by abortion.

**N. Australia.** Cygnet Bay, *A. Cunningham;* N.W. coast, *Bynoe;* islands of the Bay of Carpentaria, *R. Brown.*

4. **G. latifolia,** *F. Muell. Herb.* A shrub or tree, the branches stellate-tomentose. Leaves petiolate, broadly cordate, ovate, 3 or 4 in. long, irregularly serrate, scabrous-pubescent above and wrinkled, softly tomentose or hirsute underneath. Peduncles 2 or 3 together, 2- to 5-flowered, of unequal length, but scarcely exceeding the petioles. Sepals softly villous, 4 to 5 lines long, acute. Petals about one-third as long, the broad foveolate base as long as the small lamina. Torus considerably elevated. Stamens numerous. Ovary hirsute, 2-celled, with 2 superposed ovules in each half-cell. Fruit depressed-globular, 5 or 6 lines diameter, hirsute when young, at length shining and nearly glabrous, 2-lobed, each lobe containing 2 1-seeded nuts and slightly furrowed between them.—*G. Richardiana,* Hook. in Mitch. Trop. Austr. 383 ; not Walp.

**Queensland.** Islands off the N. coast, *R. Brown;* Bustard Bay, *Banks;* Brisbane river, *Fraser, F. Mueller;* Moreton Island and Peak Downs, *F. Mueller;* St. George's Bridge on the Balonne, *Mitchell.*

The foliage is nearly that of *G. asiatica,* Linn., with the fruit of *G. polygama,* Roxb., and the flowers different from both. In some flowers, I have seen the style divided some way below the dilated fringed stigmas.

5. **G. polygama,** *Roxb. Fl. Ind.* ii. 588. An erect shrub, the branches tomentose or softly hirsute. Leaves almost sessile, from obovate-oblong to oblong-elliptical or almost lanceolate, 2 to 3 in. long, serrate, wrinkled and softly pubescent or scarcely scabrous above, velvety-tomentose underneath. Flowers diœcious, 3 or 4 together on very short peduncles. Sepals about 4 lines long, silky-tomentose outside. Petals about one-third as long, the oblong lamina twice as long as the broad foveolate base. Male fl. : Stamens about 20, on the very hirsute torus, with a very rudimentary pistil or none at all. Female fl. : Stamens very short, with small anthers. Ovary very hirsute, with 2 superposed ovules in each half-cell. Style short, with broad, spreading, fringed stigmatic lobes. Drupe depressed-globular, 5 or 6 lines diameter, hirsute when young, at length smooth and shining, 2-lobed, each lobe containing 2 1-seeded nuts and slightly furrowed between them.

**N. Australia.** Victoria and Fitzmaurice rivers and Arnhem's Land, *F. Mueller;* Goulburn Island, *A. Cunningham;* islands of the Gulf of Carpentaria, *R. Brown;* Swears Island, *Henne.*

**Queensland.** Cape York and Port Molle, *M'Gillivray;* Bay of Inlets, *Banks;* Keppel Bay, *R. Brown;* Percy Islands, *A. Cunningham;* Rockhampton, *F. Mueller;* Port Denison, *Fitzalan.*

The species spreads over a great part of East India.

6. **G. xanthopetala,** *F. Muell. Herb.* Young branches tomentose-hirsute. Leaves shortly petiolate, broadly ovate-cordate, acuminate, the larger ones 3 or 4 in. long, serrate, minutely pubescent above, tomentose underneath or at length nearly glabrous, 5-nerved at the base or rarely 3-nerved. Stipules rather broader than in the other species. Flowers diœcious, in sessile, several-flowered clusters, the males not seen. Female fl. : Sepals 2 to 3 lines long, obtuse, softly villous outside. Petals about half as long, some-

times notched at the top, the foveola at the base small. Torus short. Staminodia very numerous, scarcely exceeding the ovary, clavate, without anthers. Ovary villous, with 2 superposed ovules in each half-cell. Style short, with broad, spreading, fringed, stigmatic lobes. Fruit (only seen young) small, depressed-globular, with 2 1-seeded nuts.

**N. Australia.** Sandstone rocks, Upper Victoria river, *F. Mueller.*

7. **G. scabrella,** *Benth.* A shrub with the habit of *G. orbifolia,* but the tomentum rather more sparing. Leaves broadly ovate, but not so rounded as in that species nor quite so rigid, 1 to 1½ in. long. Flowers in small sessile clusters, apparently dioecious, the males not seen. Female fl.: Sepals softly tomentose, 2 to 2½ lines long, the buds not striate, as in *G. orbifolia.* Petals nearly as long as the sepals, glabrous, with a small foveola at the base, less distinct than in most species. Stamens 1 or sometimes 2 or 3, apparently perfect, without staminodia. Ovary oblong, villous, with 2 superposed ovules in each half-cell. Style very short, with broad, fringed, spreading, stigmatic lobes.

**Queensland.** Mackenzie and Dawson rivers, *F. Mueller.*

8. **G. orbifolia,** *F. Muell. Herb.* A much-branched shrub, the young shoots tomentose. Leaves petiolate, nearly orbicular or broadly obovate, very obtuse, irregularly serrulate, ¾ to 1½ in. diameter, 3- or 5-nerved at the base, scabrous-tomentose on both sides. Flowers dioecious, in small sessile clusters. Male fl.: Sepals 2 to 2½ lines long, tomentose, the buds striate. Petals fully ¾ as long, pubescent outside, the foveola of the base not broader than the lamina. Stamens rather short, inserted on the hirsute torus round the small very rudimentary pistil. Female flower and fruit not seen.

**N. Australia.** Sandstone rocks of the Upper Victoria river, *F. Mueller.*


## 3. TRIUMFETTA, Linn.

Sepals 5, distinct, usually concave, or with a dorsal point or appendage at the top. Petals 5, thickened and globular, or foveolate at the base, inserted round the base of the torus, rarely wanting. Stamens indefinite, or rarely reduced to 5 or 10, free, inserted on the raised torus; anther-cells opening longitudinally. Ovary 2- to 5-celled, with 2 collateral ovules in each cell; style filiform, stigma minutely 2- to 5-toothed. Fruit globular or nearly so, echinate or bristly, indehiscent or (in species not Australian) separating into cocci. Seeds in each coccus or cell solitary, or, if 2, separated by vertical dissepiments, pendulous, albuminous; embryo straight; cotyledons flat, leafy.—Herbs, undershrubs, or shrubs, with the hairs or tomentum stellate. Leaves serrate, entire, or 3- or 5-lobed. Flowers yellow, in little pedunculate or almost sessile cymes or clusters, either leaf-opposed or lateral, rarely strictly axillary. Petals usually narrow and not exceeding the calyx, especially in the Old World species.

A considerable genus, widely spread over the tropical regions of both the New and the Old World. Of the Australian species, one, a maritime plant, extends to several of the South Pacific islands, the others are all endemic.

Ovary 3- to 5-celled. Fruit 3- to 8-celled, with 1 seed in each cell,
Leaves round-cordate, entire or lobed. Fruit rather large, with
two cells and seeds to each carpel.
Stems prostrate. Leaves mostly lobed. Sepals 4 to 5 lines
with minute pointed appendages . . . . . . . . . 1. *T. procumbens.*
Shrub densely woolly-tomentose. Leaves undivided. Sepals
above ½ in. with leafy toothed appendages . . . . . . 2. *T. appendiculata.*
Leaves ovate to lanceolate. Fruit small, with as many cells and
seeds as carpels. Erect tomentose shrubs.
Broader leaves obovate-rhomboid. Calyx appendages ovate-
peltate. (Fruit unknown.) . . . . . . . . . . . 3. *T. glaucescens.*
Broader leaves cordate-lanceolate. Calyx appendages small,
obtuse. Fruit depressed-globular, 5-celled, nearly gla-
brous . . . . . . . . . . . . . . . . . 4. *T. denticulata.*
Broader leaves obovate-cordate. Calyx appendages small,
acute. Fruit ovoid-globular, 4-celled, very tomentose . . 5. *T. micracantha.*
Ovary 2-celled. Fruit 1-seeded.
Fruit covered with long soft plumose setæ . . . . . . 6. *T. plumigera.*
Fruit very small, scarcely pubescent, shortly echinate . . . 7. *T. parviflora.*

*A. Cunningham's* herbarium contains also specimens from the N.W. coast of two other
species apparently either of *Triumfetta* or *Grewia*, but too imperfect to determine.

1. **T. procumbens,** *Forst.; DC. Prod.* i. 508. Stems procumbent or
prostrate and rooting at the joints, often attaining several feet, the branches
shortly ascending, tomentose. Leaves petiolate, broadly ovate-cordate or orbicular, obtuse, 1 to 2 in. long, entire, crenate, or more or less deeply divided
into 3 or 5 lobes, nearly glabrous above, more tomentose underneath. Peduncles short, few-flowered. Sepals 4 or 5 lines long, with small pointed
appendages. Ovary hirsute and papillose, 3- or 4-celled, each cell again
divided into 2. Fruit globular, about ½ in. diameter, glabrous or villous,
covered with hard conical prickles; endocarp hard, divided into 6 or 8 one-
seeded cells.—Guillem. in Ann. Sc. Nat. Par. ser. 2, vii. 365 ; Hook and Arn.
Bot. Beech. 60.

**Queensland.** Maritime sands, Northumberland Islands, *R. Brown ;* Fitzroy Island, *A.
Cunningham ;* Frankland Islands, *M'Gillivray ;* Howick Islands, *F. Mueller.*
The species is found in several islands of the Eastern Archipelago, and the Pacific, where
the leaves are usually entire or not very deeply 3-lobed; Cunningham's specimens agree
very well with these, in all the others (generally far advanced) the leaves are deeply 3- or
5-lobed, with glabrous fruits.

2. **T. appendiculata,** *F. Muell. Fragm.* iii. 7. Shrubby, the whole
plant densely tomentose-villous, or almost woolly. Leaves petiolate, broadly
ovate-cordate, obtuse, 2 to 4 in. long, crenate, and sometimes sinuate-lobed,
very soft and thick. Peduncles mostly 3-flowered. Sepals above ½ in. long,
woolly-tomentose, the dorsal appendages leafy and toothed, spreading, and
forming on the bud a kind of cup. Ovary hirsute, 3-celled. Capsule globular, hard, indehiscent, very villous, about as large as in *T. procumbens,* but
the prickles not so rigid, and sometimes wearing off ; endocarp hard, divided
into 6 one-seeded cells.

**N. Australia.** Nichol Bay, N.W. coast, *F. Gregory ;* islands of the Gulf of Carpentaria, *R. Brown.*

3. **T. glaucescens,** *R. Br. Herb.* Shrubby, with tomentose branches.
Leaves petiolate, from obovate-rhomboid to lanceolate, acute, or somewhat

VOL. I. T

obtuse, $1\frac{1}{2}$ to 2 in. long, serrate-crenate, rather thick, roughly tomentose-pubescent above, hoary-tomentose underneath. Peduncles very short, few-flowered. Sepals about 3 lines long, tomentose, the dorsal appendages broadly ovate-peltate, thick and gland-like, forming a radiating disk on the thick truncate buds. Ovary 3-celled. Fruit not seen.

**N. Australia.** Islands of the Gulf of Carpentaria, *R. Brown.* (*Hb. R. Br.*)

4. **T. denticulata,** *R. Br. Herb.* An erect, much-branched shrub, of 2 or 3 ft., the branches tomentose-villous. Leaves petiolate, from ovate-cordate to lanceolate, acute, $1\frac{1}{2}$ to 3 in. long, slightly serrate-crenate, scabrous-pubescent above, tomentose underneath. Peduncles very short, several-flowered. Sepals about 3 lines long, pubescent, the small glabrous dorsal appendages obovate or cuneate, and quite distinct round the top of the truncate buds. Fruit depressed-globular, about 2 lines diameter, umbilicate, glabrous or nearly so, tuberculate or muricate with minute fine prickles, 5-celled, with 1 ovule in each cell.

**N. Australia.** Cavern Island and Groote Eyland in the Gulf of Carpentaria, *R. Brown.* (*Hb. R. Br.*)

5. **T. micracantha,** *F. Muell. Fragm.* iii. 7. Shrubby, erect, the branches closely tomentose or almost villous. Leaves petiolate, from broadly obovate-cordate to narrow rhomboid-oblong, or the upper ones lanceolate, the larger ones 2 to 3 in. long, serrate, with the lower teeth glandular, soft and thick with a close whitish tomentum or villous-tomentose on both sides. Peduncles about as long as the petioles, with about 4 pedicellate flowers. Sepals about 3 lines long, the dorsal appendage small and pointed. Petals none (in the buds I have opened). Fruit ovoid-globose, scarcely above 3 lines long, densely tomentose, the small prickles often scarcely exceeding the wool; endocarp hard, 4-celled, with one seed in each cell.

**N. Australia.** Victoria river, Hooker's and Sturt's creeks and Abel Tasman river, *F. Mueller.* There are two forms, differing in the tomentum either close and whitish or dense and almost woolly. The species much resembles in aspect the common African *T. glandulosa,* Forsk. (*T. Vahlii,* Poir., *T. glandulosa,* Lam., and *T. velutina,* Vahl), which however has separable cocci.

6. **T. plumigera,** *F. Muell. Fragm.* i. 69. Shrubby, with erect nearly simple branches, closely or loosely tomentose. Leaves from cuneate-oblong to lanceolate, 1 to 2 in. long, rather thick, and either closely whitish tomentose or roughly tomentose-villous on both sides. Cymes or clusters several-flowered, nearly sessile, or the lower ones pedunculate. Sepals scarcely above 2 lines long, the dorsal appendages small and pointed. Stamens rather numerous. Ovary 2-celled. Fruit 1-seeded (or sometimes 2-seeded?), small, but covered with long soft plumose-villous setæ, which often attain 4 lines when the fruit is ripe.

**N. Australia.** Montague Sound, *A. Cunningham;* N.W. coast, *Bynoe;* Depot Creek, sources of the Victoria river, *F. Mueller.* The species is nearly allied to the S. African *T. trichocarpa,* Sond., which has however larger flowers, and the setæ of the fruit more than $\frac{1}{2}$ in. long.

7. **T. parviflora,** *Benth.* An erect, rigid herb, the branches stellate-pubescent. Leaves shortly petiolate, narrow-oblong, obtuse, 1 to $1\frac{1}{2}$ in. long,

crenate, wrinkled and roughly pubescent above, tomentose underneath. Flowers minute, in nearly sessile clusters. Sepals scarcely more than 1 line long, with a small dorsal point. Petals rather shorter. Stamens very few (5 according to Brown's notes). Ovary hispid, stigma 2-lobed. Fruit globular, 1-seeded, about 1 line diameter, shortly echinate and slightly pubescent.

**N. Australia.** Islands of the Gulf of Carpentaria, *R. Brown.* (*Hb. R. Br.*)

## 4. CORCHORUS, Linn.

Sepals 5, rarely 4. Petals as many, without any cavity at the base. Stamens indefinite, rarely few, inserted on a torus scarcely raised, but occasionally expanded in a disk round their base; anther-cells opening longitudinally. Ovary 2- to 5-celled, with several ovules in each cell; style short, simple. Capsule either long without prickles, or short or globular and more or less warted, muricate or echinate, opening loculicidally in 2 to 5 valves, with several seeds in each cell, rarely separated by transverse partitions. Seeds pendulous or horizontal, albuminous; embryo usually curved, with leafy cotyledons.—Herbs, undershrubs, or shrubs, with simple or stellate hairs. Leaves serrate. Peduncles very short, lateral or leaf-opposed, bearing 1 or several flowers. Bracts small. Flowers usually yellow.

A considerable genus, of which a few species appear to be limited to tropical America or to Australia, the remainder generally dispersed over various tropical regions in the Old as well as the New World. Of the 13 Australian species 9 are endemic, the remaining 4 are common weeds in tropical Asia and Africa. The fruit in this genus is often indispensable for determining the species.

*Annuals (or biennials), glabrous or loosely pubescent.*
　Capsule globular or ovoid, very obtuse.
　　Capsule echinate, 3- or 4-celled . . . . . . . . . . 1. *C. echinatus.*
　　Capsule slightly warted, 2- or 3-celled . . . . . . . 2. *C. hygrophilus.*
　Capsule (½ to ¾ in. long) rather thick, angular or winged.
　　Capsule acute or acuminate, angular but not winged. Stamens
　　　numerous . . . . . . . . . . . . . . . . . 3. *C. Cunninghamii.*
　　Capsule 3-winged, truncate at the top, with 3 diverging
　　　points. Stamens under 20. Flowers very small . . . 6. *C. acutangulus.*
　Capsule linear, not winged.
　　Capsule ¾ to 2 in. Lower pair of serratures of the leaves
　　　ending in fine setæ. Stamens numerous.
　　　Capsule 5-celled, the transverse partitions conspicuous . . 4. *C. olitorius.*
　　　Capsule 3- or 4-celled, without transverse partitions . . 5. *C. tridens.*
　　Capsule under ½ in., 2- or 3-celled. Leaves without setæ.
　　　Flowers very small. Stamens few. Pubescent plants.
　　　Capsule 3-celled, erect or spreading, 3-toothed at the top . 7. *C. fascicularis.*
　　　Capsule 2-celled, reflexed, very hirsute, rather acute . . . 8. *C. pumilio.*
*Undershrubs or shrubs more or less tomentose or hirsute.*
　Fruiting pedicels recurved. Capsule linear, curved or twisted,
　　more or less torulose, 2- or 3-celled.
　Low diffuse shrubs or undershrubs. Capsule few-seeded.
　　Sepals under 2 lines. Stamens about 10. Capsule 3 or
　　　4 lines long, very hispid, slightly curved . . . . . 8. *C. pumilio.*
　　Sepals under 2 lines. Stamens about 20. Capsule elongated, much twisted, torulose, glabrous . . . . . . 9. *C. vermicularis.*
　　Sepals 3 to 4 lines. Stamens numerous. Capsule tomentose, slender but not twisted . . . . . . . . 10. *C. tomentellus.*

Erect shrubs.  Capsule tomentose, elongated, many-seeded.
  Tomentum scabrous or almost villous.  Sepals 2 or 3 liues.
    Petals narrow . . . . . . . . . . . . .    11. *C. sidoides.*
  Tomentum soft, close and hoary.  Sepals 4 lines or more.
    Petals broad  . . . . . . .. . . . . . . . .    12. *C. leptocarpus.*
Capsule erect, straight, not twice as long as the calyx, 5-celled.    13. *C. Walcottii.*

1. **C. echinatus,** *Benth.*  An erect annual, glabrous or nearly so.
Leaves petiolate, from ovate-lanceolate to oblong, rather obtuse, serrate, the
lowest pair of serratures rarely bearing short setæ.  Pedicels 1 to 3, often
as long as the petioles, on a very short common peduncle.  Buds globular.
Sepals 2¼ to 3 lines long.  Petals broad, rather longer.  Ovary obtuse,
tuberculate.  Capsule ovoid-globular, 3 to 5 lines long, very obtuse, glabrous, densely echinate with short recurved soft prickles, 3- or 4-celled.
Seeds usually 6 to 8 in each cell, in two rows, without transverse partitions.
—*Triumfetta macropetala,* F. Muell. Fragm. iii. 8.

**N. Australia.**  Hooker's and Sturt's creeks, sources of the Victoria river, *F. Mueller.*

2. **C. hygrophilus,** *A. Cunn. Herb.*  A tall, erect, glabrous herb, apparently annual.  Leaves petiolate, ovate or ovate-lanceolate, acuminate, 3 to
5 in. long, acutely and irregularly toothed.  Cymes several-flowered, reflexed,
shortly pedunculate, but rarely equalling the petioles.  Flowers small,
the buds obovoid, contracted at the base.  Petals the length of the calyx.
Stamens numerous, on a raised torus.  Capsule globular or ovoid-oblong,
very obtuse, 2 to 4 lines long, more or less tuberculate, 2- or 3-celled.  Seeds
8 or more in 2 rows in each cell, without transverse partitions.

**Queensland.**  Cleveland Bay, *A. Cunningham.*

3. **C. Cunninghamii,** *F. Muell. Fragm.* iii. 8.  A tall erect glabrous
herb, annual, or sometimes perhaps perennial.  Leaves petiolate, from cordate-ovate to lanceolate, acuminate, 2 to 4 in. long, coarsely serrate, without
setæ.  Peduncles short, bearing a cyme of 3 to 7 or 8 flowers, on rather
long pedicels.  Buds obovoid, narrowed at the base.  Stamens numerous,
on a raised torus.  Ovary narrowed at the top.  Capsule narrow-oblong,
acute, ½ to ¾ in. long, slightly 3- or 4-angled, 3- or 4-celled, with numerous
seeds in each cell.

**Queensland.**  Dawson and Burnett rivers and Moreton Bay, *F. Mueller;* Brisbane
river, *Fraser.*

4. **C. olitorius,** *Linn.; W. and Arn. Prod.* 73.  An erect annual, of
2 ft. or more, glabrous or nearly so.  Leaves from ovate-acuminate to lanceolate, 2 to 3 in. long or more, serrate, the lowest pair of serratures ending in spreading or recurved setæ.  Flowers single or 2 together, on very
short pedicels, the buds obovoid-globular.  Stamens numerous, on a small
torus.  Capsule linear, often above 2 in. long, slightly 5-angled, 5-celled,
with numerous seeds, separated by almost complete transverse partitions.

**N. Australia.**  Van Diemen's Gulf, *A. Cunningham.*  The species is common in
tropical Asia and Africa.

5. **C. tridens,** *Linn.; W. and Arn. Prod.* 73.  A glabrous, hard annual,
with decumbent ascending or erect branches.  Leaves mostly lanceolate,
rather obtuse, 2 to 3 in. long, crenate-serrate, the lowest pair of serratures

terminating in setæ as in *C. olitorius.* Flowers small, nearly sessile, usually 2 or 3 together. Stamens numerous, the torus scarcely raised. Capsule linear, rigid, ¾ to 1½ in. long, straight or curved, glabrous, 3- or 4-celled, and often terminating in as many very short spreading points or teeth. Seeds numerous, without any, or with very imperfect, transverse partitions.

**N. Australia.** Islands of the N. coast, *R. Brown;* Upper Victoria river, Sturt's and Hooker's creeks, *F. Mueller.* The species is widely spread over tropical Asia and Africa.

6. **C. acutangulus,** *Lam. ; W. and Arn. Prod.* 73. An annual, sometimes very small, but attaining 2 ft., decumbent or erect, slightly pubescent and often sprinkled with a few rigid hairs. Leaves petiolate, ovate, serrulate, without setæ. Flowers 1 to 3, nearly sessile, and very small. Sepals little more than 1 line long. Stamens 15 to 20. Capsule straight, ½ to ¾ in. long, rather thick, prominently 3-angled, or with 3 longitudinal wings, truncate at the top, with 3 spreading points or teeth, 3-celled. Seeds numerous. Very rarely the capsule has 4 cells, and as many wings and teeth.—Wight, Ic. t. 739.

**N. Australia.** Upper Victoria river, *F. Mueller.* The species is common in tropical Asia and Africa, and occurs also, perhaps introduced, in some parts of S. America.

7. **C. fascicularis,** *Lam.; W. and Arn. Prod.* 72. A small annual, with procumbent or ascending branches, rarely attaining 1 ft., glabrous or loosely pubescent. Leaves petiolate, oblong or lanceolate, ¾ to 1½ in. long, or the lower ones small and broad, slightly serrate without setæ. Flowers very minute, in clusters of 3 to 6, on very short peduncles. Sepals about 1 line long. Stamens 5 to 10. Capsule nearly cylindrical, rarely ½ in. long, usually slightly hairy, terminating in 3 teeth, 3-celled. Seeds several, without transverse partitions.

**N. Australia.** Victoria river, Sturt's and Hooker's creeks, *F. Mueller;* Arnhem's Bay, *R. Brown* (the latter somewhat doubtful, the specimen very imperfect). The species extends over tropical Africa and Asia, from Senegal to Bengal, and includes *C. brachycarpus,* Guill. and Perr.

8. **C. pumilio,** *R. Br. Herb.* A small rigid, much-branched herb or undershrub, not much more than ½ ft. high, hirsute with spreading stellate hairs, the slender branches appearing almost woody at the base, although the plant flowers the first year. Leaves petiolate, ovate or oblong, obtuse, rarely above ½ in. long, crenate, rugose and plicate, sprinkled with rigid stellate hairs. Flowers very small, in sessile clusters. Buds narrow-oblong. Sepals very narrow, acute, hirsute, 1 to 1½ lines long. Petals narrow. Stamens about 10. Ovary very hirsute. Capsules reflexed, linear, 3 to 4 lines long, slightly curved, rather acute, very hirsute, 2-celled, with few oblong seeds.

**N. Australia.** Islands of the Gulf of Carpentaria, *R. Brown;* Upper Victoria river, *F. Mueller.*

9. **C. vermicularis,** *F. Muell. Fragm.* iii. 11. A low shrub or undershrub, with numerous slender branches, like *C. pumilio* in habit, but more diffuse, and rather stellate-tomentose than hirsute. Leaves petiolate, oblong, obtuse, ½ to ¾ in. long, serrate-crenate, rugose and plicate, rather roughly stellate-tomentose. Flowers very small, 2 or 3 together, and shortly pedicellate. Buds obovoid. Sepals linear-cuneate, 1½ lines long. Petals broadly obovoid. Stamens

about 20. Ovary glabrous. Capsule reflexed, linear, slender and very much twisted, contracted between the seeds, 2-valved. Seeds ovoid-oblong, few and distant, the cells usually closed between them.

**N. Australia.** Upper Victoria river, *F. Mueller.*

10. **C. tomentellus,** *F. Muell. Fragm.* iii. 10. A low, diffuse, stellate-tomentose shrub or undershrub. Leaves petiolate, from ovate to ovate-ob-long, obtuse, ½ to 1 in. long, crenate, slightly plicate and rugose, rather loosely stellate-tomentose, especially underneath. Flowers pedicellate, in nearly sessile clusters, much larger than in *C. vermicularis.* Buds obovoid. Sepals 3 to 4 lines long. Stamens numerous, the torus expanded into a pro-minent disk round their base. Capsule very slender, tomentose, ½ to ¾ in. long, 3-valved, with few distant seeds, but scarcely contracted between them.

**Queensland.** Mackenzie river, *F. Mueller.* It is possible that this may prove a form of the very variable *C. sidoides,* but besides the difference in habit and foliage, the flowers appear to be larger and the disk much more developed.

11. **C. sidoides,** *F. Muell. Fragm.* iii. 9. An erect shrub of several feet, the branches densely but rather loosely tomentose. Leaves shortly petio-late, from oval-oblong to oblong-lanceolate, obtuse, 1 to 2 in. long, rather thick, crenate, plicate and rugose or on luxuriant specimens longer and thinner, scabrous-tomentose above, more densely tomentose underneath. Flowers in nearly sessile clusters. Calyx tomentose-villous, 2 to 3 lines long, the buds often tipped by the tooth-like points of the sepals. Petals narrow, in some flowers very small. Stamens numerous, on a small torus. Capsule slender, ¾ to near 2 in. long, tomentose or villous, more or less torulose, 2- or 3-celled. Seeds oblong, often distant in each cell, although rather nu-merous on the whole.

**N. Australia.** N.W. coast, *Bynoe ;* Cygnet Bay, *A. Cunningham ;* frequent in sterile places on the Victoria river, *F. Mueller ;* islands of the Gulf of Carpentaria, *R. Brown.*

12. **C. leptocarpus,** *A. Cunn. Herb.* An erect shrub of several feet, hoary all over with a minute close tomentum. Leaves petiolate, from ovate to lanceolate, obtuse, 2 to 4 in. long, irregularly crenate, cordate at the base, minutely but softly tomentose on both sides. Flowers in nearly sessile clus-ters. Buds tomentose, angular, tipped with the long points of the sepals. Sepals 4 lines long or rather more. Petals broad, almost foveolate at the base. Stamens numerous, on a prominent torus. Capsule slender, incurved at the base, densely tomentose, 1 to 1½ in. long, slightly torulose, 3-celled. Seeds numerous, oblong, with incomplete transverse partitions between them.

**N. Australia.** Water Island, N.W. coast, *A. Cunningham.*

13. **C. Walcottii,** *F. Muell. Fragm.* iii. 9. A shrub or undershrub of 1 to 3 ft., densely and softly tomentose or woolly. Leaves petiolate, from broadly ovate to ovate-oblong, 1½ to 2½ in. long, coarsely toothed or crenate, not cordate, thick, soft and densely tomentose. Peduncles nearly as long as the petioles, with 3 to 6 rather large pedicellate flowers. Sepals woolly-tomentose, 5 or 6 lines long, lanceolate with long soft subulate points. Petals broad. Stamens numerous. Capsule erect and straight, from about

the length of the calyx to twice as long, very tomentose, 5-celled. Seeds few, without partitions between them.

**N. Australia.** Enderby Island, N.W. coast, *A. Cunningham ;* Hearson Island and Nichol Bay, *F. Gregory.*

Var. (?) *parviflora.* Leaves smaller, often narrow, tomentum closer, and flowers much smaller. N.W. coast, *Bynoe ;* Nichol Bay, *F. Gregory.*

Var. (?) *densiflora.* Foliage various, pedicels very short, flowers intermediate in size. Gulf of Carpentaria, *F. Mueller.* The specimens are insufficient for correctly estimating the constancy of these forms.

## 5. ECHINOCARPUS, Blume.

Sepals 4, imbricate in two rows.   Petals 4, broad, short, imbricate. Stamens numerous, free, covering the broad, thick, pitted disk from the petals to the ovary; anthers linear, the cells placed back to back and opening from the top in a slit extending more or less down the sides.   Ovary 3- or 4-celled, with several ovules in each cell; style subulate.   Capsule thickly coriaceous or woody, densely echinate or covered with setæ, 3- or 4-celled or 1-celled by abortion, opening in 3 or 4 valves.   Seeds several or solitary and pendulous, ovoid; testa hard; albumen fleshy; cotyledons broad, flat.—Trees.   Leaves entire or sinuate-toothed, with pinnate veins.   Peduncles axillary, 1-flowered, solitary or clustered, rarely forming terminal racemes.

A tropical Asiatic genus, represented in Australia by a single endemic species.

1. **E. australis,** *Benth. in Journ. Linn. Soc.* v. *Suppl.* 73.   A tree, at-taining 80 ft., glabrous in all its parts.   Leaves obovate-oblong, ½ to 1 ft. long, shortly acuminate, more or less sinuate-toothed, much narrowed to-wards the base, but obtuse or slightly cordate at the petiole, coriaceous. Flowers pendulous, on erect pedicels of 1 to 2 in., the upper ones forming terminal racemes shorter than the last leaves.   Sepals ovate-oblong, about 4 lines long.   Petals not seen.   Anthers scarcely pointed.   Capsule opening in 4 hard almost woody valves, about ½ in. long, external setæ short and ex-ceedingly densely crowded.   Fully expanded flowers and seeds not seen.

**Queensland.** Scrub near Dunuduni, Moreton Bay, *W. Hill.*

**N. S. Wales.** Hastings river, *Beckler* (capsules only seen) ; Kiama, *Harvey.*

## 6. ARISTOTELIA, L'Hér.

### (Friesia, *DC.*)

Sepals 4 or 5, valvate.   Petals as many, imbricate, 3-lobed, toothed or entire, inserted round the base of the thickened torus.   Stamens indefinite, inserted on the torus, within a glandular ring ; anthers linear, the cells placed back to back and opening from the top in short confluent slits.   Ovary 2- to 4-celled, with 2 ovules in each cell; style subulate.   Fruit a berry.   Seeds few, as-cending or pendulous ; testa hard, often pulpy outside ; albumen fleshy ; embryo straight, with flat or undulate cotyledons.—Shrubs.   Leaves mostly opposite or nearly so, entire or toothed.   Flowers axillary or lateral, in ra-cemes, or in the Australian species solitary or 2 or 3 together, often polyga-mous.

Besides the 2 Australian species, which are endemic, the genus has 2 from New Zealand and 1 from Chili.

Leaves oblong-lanceolate or rarely ovate-lanceolate, glabrous . . .   1. *A. peduncularis.*
Leaves ovate, acuminate, pubescent underneath . . . . . . . . .   2. *A. australasica.*

1. **A. peduncularis,** *Hook. f. Fl. Tasm.* i. 52. A weak straggling shrub of 2 to 4 or 5 ft., glabrous in all its parts. Leaves irregularly opposite or here and there alternate or in whorls of 3, shortly petiolate, from ovate-lanceolate to oblong or lanceolate, acuminate, 1½ to 3 in. long, serrate. Peduncles slender, 1-flowered, solitary or 2 to 3 together, with a few small leafy bracts at the base. Flowers white, pendulous. Sepals 4 or rarely 5, 3 or 4 lines long. Petals rather longer, broadly cuneate, 3-lobed. Torus tomentose. Stamens 10 to 12, the filaments ciliate at the base; anthers shortly pointed. Berry varying in size from that of a pea to a small cherry. —*Elæocarpus peduncularis,* Labill. Pl. Nov. Holl. ii. 15, t. 155; *Friesia peduncularis,* DC. Prod. i. 520; Bot. Mag. t. 4246.

**Tasmania.** Derwent river, *R. Brown;* southern and mountainous parts of the island, common in shady places ascending to 4000 ft., *J. D. Hooker.*

2. **A. (?) australasica,** *F. Muell. Fragm.* ii. 79. A slender shrub of several feet, with a few soft hairs on the young branches, petioles, and principal veins on the under side of the leaves, otherwise glabrous. Leaves opposite, on slender petioles, ovate, acuminate, 2 to 3 in. long, serrate, 3-nerved at the base. Pedicels slender, really axillary, although sometimes apparently terminal before the intermediate bud has grown out. Flowers unknown. Berry globular, about 4 lines diameter, nearly dry.

**N. S. Wales.** Mountain woods at the mouth of the Clarence river, *Beckler.* Until the flowers have been seen, the generic identity cannot be considered as certain.

## 7. ELÆOCARPUS, Linn.

### (Monocera, *Jack.*)

Sepals 4 or 5, usually valvate. Petals as many, fringed, lobed or rarely entire, inserted round the base of the torus, induplicate-valvate, and embracing some of the outer stamens in the bud. Stamens indefinite, inserted on the torus, within a glandular ring; anthers oblong or linear, opening at the top in 2 valves (that is, the cells placed back to back and opening in short, terminal, confluent slits). Ovary 2- to 5-celled, with 2 or more ovules in each cell; style subulate. Fruit a drupe, with a hard often bony putamen, 2- to 5-celled or 1-celled by abortion. Seeds solitary in each cell, pendulous (or rarely erect?); testa hard; albumen fleshy; cotyledons broad, flat or undulate.—Trees. Leaves alternate or rarely opposite, entire or serrate. Flowers in axillary racemes, sometimes polygamous.

A large tropical Asiatic genus, extending to the Pacific islands, New Caledonia and New Zealand. The Australian species are all endemic.

Leaves tomentose underneath. Petals entire or slightly crenate . .   1. *E. holopetalus.*
Leaves glabrous. Petals lobed or fringed.
 Flowers about 2 lines. Petal-lobes about 7, obtuse. Anthers short,
  obtuse . . . . . . . . . . . . . . . . . . . . .   2. *E. obovatus.*
 Flowers about 4 lines. Petal-lobes 10 to 12, acute. Anthers linear,
  pointed. Leaves strongly reticulate . . . . . . . . . .   3. *E. cyaneus.*
 Flowers 6 or 7 lines. Petals silky on the edges, with about 5 deeply
  fringed lobes. Anthers linear, with short setæ on their points .   4. *E. grandis.*

1. **E. holopetalus,** *F. Muell. Fragm.* ii. 143 ; *and Pl. Vict.* i. 153. A tree attaining 80 ft., the young shoots rusty-tomentose or villous. Leaves on very short petioles, oblong-lanceolate or slightly obovate, acute or acuminate, 2 to 4 in. long, sinuate-serrate, coriaceous, reticulate and glabrous above, loosely tomentose underneath or almost glabrous with age. Racemes in the upper axils, tomentose-villous. Pedicels rather long. Sepals 2 to 2¼ lines long. Petals rather longer, entire or slightly crenate. Stamens 15 to 20, within a prominent almost cup-shaped disk ; anthers pubescent, short, obtuse. Ovary 2-celled.

**Victoria.** Eastern Gipps' Land, at an elevation of 2000 to 4000 ft., *F. Mueller.*

2. **E. obovatus,** *G. Don, Gen. Syst.* i. 559. A tree attaining 60 ft., glabrous in all its parts. Leaves from oval-elliptical to obovate-oblong or almost lanceolate, obtuse or obtusely acuminate, 2 to 4 in. long, irregularly sinuate-crenate, narrowed at the base, thinly coriaceous, the smaller veins much less numerous and less conspicuous than in *E. cyaneus.* Racemes solitary or clustered, many-flowered, but shorter than the leaves. Flowers small, white. Sepals acute, 1½ lines long. Petals rather longer, divided to about the middle into about 7 linear obtuse lobes. Anthers short, obtuse or scarcely pointed. Ovary glabrous, 2-celled, with 4 ovules in each cell. Drupe globular or ovoid, often blue, the putamen rugose or tuberculate.—F. Muell. Fragm. ii. 80 ; *E. parviflorus,* A. Rich. Sert. Astrol. 67, t. 24 ; *E. pauciflorus,* Walp. Rep. i. 364 (a mistake in the name and a wrong station).

**Queensland.** Brisbane river, Moreton Bay, *W. Hill ;* Ipswich, *Nernst.*
**N. S. Wales.** Port Jackson and Hunter's river, *R. Brown* and others ; Hastings and Clarence rivers. *Beckler ;* Port Macquarie, *Fraser.*
Var. (?) *foveolatus.* Drupe larger, ovoid, very prominently tuberculate. Flowers not seen.
**N. Australia.** Islands of the N. coast, *R. Brown ;* Liverpool river, *A. Cunningham ;* Fitzmaurice river and Macadam range, *F. Mueller.*

3. **E. cyaneus,** *Ait. Epit. Hort. Kew.* A tree, usually small, but attaining sometimes 60 ft. or more, glabrous in all its parts. Leaves elliptical-oblong or oblong-lanceolate, acuminate, 3 or 4 in. long or more when luxuriant, more or less serrate, acute at the base, coriaceous and very conspicuously reticulate. . Racemes loose, shorter than the leaves. Sepals acute, 3 to 4 lines long, glabrous. Petals as long or rather longer, divided into 10 to 12 acute lobes, here and there united in pairs. Stamens numerous, within the undulate glandular disk. Anthers linear, the upper valve with a short point. Ovary glabrous, 2-celled, with 8 to 10 ovules in each cell. Drupe usually 1-seeded, globular or ovoid, blue outside, the putamen 4 to 6 lines long, hard and rugose.—DC. Prod. i. 519 ; Bot. Mag. t. 1737 ; F. Muell. Pl. Vict. i. 152 ; *E. reticulatus,* Sm. in Rees' Cycl. xii. ; Bot. Reg. t. 657.

**Queensland.** Moreton Bay, *F. Mueller ;* Pine river, *Fitzalan.*
**N. S. Wales.** Port Jackson , *R. Brown* and others ; northward to Mount Lindsay, Macleay and Clarence rivers, *Beckler ;* southward to Twofold Bay, *F. Mueller.*
**Victoria.** Forest gullies of Wilson's Promontory and wooded ranges from the Tambo river to the E. boundary, *F. Mueller.*
Some specimens from the Illawarra, *M'Arthur,* in fruit only, are remarkable for their thick branches, with leaves of 6 to 8 in.

4. **E. grandis,** *F. Muell. Fragm.* ii. 81. A tree of moderate size, gla-

brous, except the young shoots, slightly silky-hairy. Leaves on short petioles, oblong or lanceolate, obtuse or scarcely acuminate, 4 to 6 in. long,
crenulate, narrowed at the base, scarcely coriaceous, the smaller veins not
prominent. Flowers large, in short dense racemes. Sepals fully ½ in., including their long subulate points. Petals longer, divided into about 5
deeply fringed lobes, silky-pubescent on the margin towards the base. Stamens very numerous ; anthers linear, the upper valve pointed and ending in
1 or 2 short, fine setæ. Ovary silky-tomentose (5-celled?), with about 4
ovules in each cell. Drupe (which I have not seen) globular, 1 in. diameter.

**Queensland.** Pine river, *Herb. F. Mueller.* The large flowers, pubescent petals, and
pointed anthers, refer this species to the section *Monocera*, usually considered as a distinct
genus, but the group is neither natural nor accurately defined.

## ORDER XXIV. LINEÆ.

Flowers regular, hermaphrodite. Sepals 5, rarely 4, free or united at the
base, imbricate or rarely almost valvate. Petals as many, hypogynous or
rarely slightly perigynous, imbricate, usually contorted. Stamens as many as
petals or twice or rarely thrice as many, united into a ring or short tube at
the base ; anthers 2-celled, with parallel cells opening longitudinally. Glands
5, adnate to or embedded in the outside of the staminal tube or rarely wanting.
Disk none (besides the staminal tube). Ovary free, entire, 3- to 5-celled.
Ovules 2 or rarely 1 in each cell, pendulous, anatropous, with a ventral raphe.
Styles 3 to 5, distinct or more or less united, with terminal usually capitate
stigmas. Fruit either a capsule, separating into cocci, usually dehiscent, or a
drupe, with as many pyrenes as carpels, or more frequently reduced by abor-
tion to 1. Seeds 1 or 2 in each coccus or pyrene ;. testa membranous or al-
most coriaceous ; albumen fleshy, abundant or thin or entirely wanting.
Embryo usually straight, with flat, ovate cotyledons ; radicle superior.—
Herbs, shrubs, or rarely trees, glabrous or rarely hirsute or tomentose.
Leaves alternate or very rarely opposite, simple and entire or slightly serrate.
Stipules lateral or within the petiole, sometimes minute or wanting.

An Order, formerly almost limited to the genus *Linum*, but lately extended to include
several small Orders or genera, chiefly tropical, from both the New and the Old World.
The two Australian genera are the only two large ones, both of them widely dispersed, one
chiefly in temperate regions, the other within the tropics.

Herbs. Petals without appendages. Capsule apparently 10-celled,
    with 1 seed in each cell. . . . . . . . . . . . . 1. LINUM.
Shrubs or trees. Petals with an appendage at the base of the lamina.
    Drupe 1-seeded . . . . . . . . . . . . . . . . 2. ERYTHROXYLON.

## 1. LINUM, Linn.

Sepals 5. Petals 5, contorted, without appendages. Stamens 5, perfect ;
staminodia as many, alternating with the stamens, minute, tooth-like or hair-
like, or sometimes scarcely conspicuous. Glands 5, small, scarcely prominent
on the staminal tube, opposite the petals. Ovary 5-celled, with 2 collateral
ovules in each cell. Capsule dividing into 5 cocci, with 2 seeds in each

separated by an imperfect partition, or into 10 1-seeded cocci when the partition is more complete. Albumen thin.—Herbs. Leaves narrow, entire. Stipules none or minute and gland-like.

A large genus, widely distributed over the temperate or warmer extratropical regions of the globe, with a few tropical American species. The Australian species are endemic, but very closely allied to some of the commonest blue-flowered species of the northern hemisphere.

Sepals acute or acuminate . . . . . . . . . . . . . . 1. *L. marginale.*
Sepals very obtuse . . . . . . . . . . . . . . . . 2. *L. suædæfolium.*

Besides these, *L. gallicum,* Linn.; Planch. in Hook. Lond. Journ. vii. 168, a slender erect annual, with very small yellow flowers in a terminal corymb, a common plant in the Mediterranean region, has established itself as an introduced weed in the neighbourhood of Paramatta.

1. **L. marginale,** *A. Cunn.; Planch. in Hook. Lond. Journ.* vii. 169. A glabrous herb, forming a thick perennial rootstock, but also sometimes apparently annual, with erect or ascending slender stems of 1 to 2 ft., corymbosely branched above the middle. Leaves linear or linear-lanceolate, acute or the lowest almost obtuse, often all under ½ in., but the upper ones sometimes 1 in. long. Stipular glands wanting. Flowers blue, on erect pedicels, forming a loose, irregular, terminal corymb. Sepals ovate or ovate-lanceolate, acute or cuspidate, 2 to 3 lines long, with a strong midrib, the margins thin and often with a narrow scarious border. Petals from a little longer to twice as long. Styles united to above the middle. Capsule dividing into 10 1-seeded cocci.—Hook. f. Fl. Tasm. i. 46; F. Muell. Pl. Vict. i. 178; *L. angustifolium,* DC. Prod. i. 426 (as to the New Holland locality); Bartl. in Pl. Preiss. i. 161.

**N. S. Wales.** Port Jackson, *R. Brown* and others; northwards to Hastings river, *Beckler;* and in the interior, *A. Cunningham.*

**Victoria.** Port Phillip, *R. Brown;* throughout the colony, ascending to the Alps, *F. Mueller.*

**Tasmania.** Abundant throughout the island, *J. D. Hooker.*

**S. Australia.** From the Murray to Lofty Range and Spencer's Gulf, *F. Mueller* and others.

**W. Australia.** King George's Sound, *Bagster;* Swan River, *Drummond, 1st Coll.;* Doubtful Island, *Oldfield.* The species very much resembles the northern *L. angustifolium,* Huds., with which many authors have confounded it; but it appears to be constantly distinct in the union of the styles.

2? **L. suædæfolium,** *Planch. in Hook. Lond. Journ.* vii. 168. Apparently an annual, with numerous short erect stems. Leaves crowded, linear, obtuse, 3 or 4 lines long, without stipular glands. Flowers and fruit of the small varieties of *L. marginale,* except that the sepals are very obtuse, those of the lower flowers almost dilated at the top.

**Queensland.** Balonne river, *Mitchell (Herb. Lindl.).* The specimen is very imperfect. It is probably a variety of *L. marginale,* with which some specimens in F. Mueller's Herbarium with less pointed sepals than usual would seem to connect it.

## 2. ERYTHROXYLON, Linn.

Sepals 5, rarely 6, united into a lobed calyx, or free. Petals as many, with a 2-lobed appendage inside below the lamina. Stamens 10, rarely 12, the basal tube short, without glands, or more or less thickened into 10 glands, the filaments attached inside just below the crenulate top. Ovary 3- rarely 4-celled, with 1 or rarely 2 ovules in each cell. Drupe usually 1-seeded. Albumen copious, or thin, or none.—Trees or shrubs. Leaves entire. Stipules united into one within the petiole, deciduous, or persistent especially on the leafless base of the young shoots. Flowers small, whitish, solitary or clustered in the axil of leaves or of leafless stipules.

A large tropical genus, abundant in S. America, less so in Africa and Asia. The two Australian species are perhaps endemic, but there is so much general similarity in the species of this genus, and their characters so vague and variable, that it is exceedingly difficult to determine their limits.

Leaves oblong or narrow-elliptical, 1 in. long or less, or the smaller ones cuneate-obovate, the veins few . . . . . . . . . . . . . 1. *E. australe*.
Leaves obovate or ovate-elliptical, 1½ to 2½ in. long, or the smaller ones rarely 1 in., the veins numerous and finely reticulated . . . . . 2. *E. ellipticum*.

1. **E. australe,** *F. Muell. in Trans. Vict. Inst.* iii. 22. A glabrous shrub, with slender divaricate branches. Leaves elliptical-oblong, or the smaller ones cuneate or almost obovate, in some specimens all under ½ in. long, in more luxuriant ones about 1 in., the pinnate veins fewer and less reticulate than in many other species. Stipules small and deciduous. Pedicels solitary or rarely clustered, short or rarely attaining 3 lines, with minute bracteoles at their base. Flowers very small. Calyx not 1 line long, divided to below the middle, the lobes almost or quite valvate. Inner appendage of the petals with 2 very short crested lobes. Styles free or shortly cohering at the base. Drupe oblong, 3 to 3½ lines long, 3-celled, but with only 1 seed. Albumen thin; radicle slender, shorter than the ovate cotyledons.

**Queensland.** Brigalow scrub on the Burdekin, Suttor, and Dawson rivers, *F. Mueller;* Comet river, *Leichhardt;* Rockhampton and Fitzroy river, *Thozet.*

2. **E. ellipticum,** *R. Br. Herb.* A glabrous shrub of above 5 ft., the young branches flattened. Leaves obovate or ovate-elliptical, very obtuse, 1½ to 2½ in. long or the smaller ones rarely only 1 in., on petioles of about 1 line, rather thin, with very numerous and finely reticulated veins. Stipules usually about 2 lines long, and always longer than the petioles, deciduous. Flowers nearly of *E. australe,* very small, in clusters of 3 to 6, the pedicels 2 or 3 lines long, with minute bracts at their base. Calyx about 1 line long, divided nearly to the base into lanceolate acute lobes, very slightly imbricate or almost valvate. Petals slightly exceeding the calyx. Styles quite free. Drupe oblong, 3 to 4 lines long, 1-seeded.

**N. Australia.** Gulf of Carpentaria, on the mainland opposite Groote Eyland, *R. Brown.* The foliage is nearly that of the largest and broadest-leaved specimens of *E. indicum,* but the styles are quite free.

## ORDER XXV. **MALPIGHIACEÆ.**

Flowers usually hermaphrodite. Calyx 5-cleft, the segments imbricate or rarely valvate, all, or more frequently 4 only (or rarely 3 or none of them), bearing 2 glands outside. Petals 5, usually equal, concave, toothed or notched, on slender claws. Disk scarcely prominent. Stamens usually 10, all perfect, or some of them deformed or without anthers, or sometimes wanting, the filaments usually united at the base; anthers 2-celled. Ovary usually 3-celled, or the 3 carpels distinct, with 1 ovule in each, ascending from a pendulous ventral funicle. Styles distinct, or united, or one only developed, with small terminal stigmas. Fruit-carpels 3 or fewer, either united in a berry, drupe, or hard capsule, or more frequently forming separate indehiscent nuts, or winged samaræ. Seeds without albumen, the testa usually membranous and double. Embryo straight or curved; cotyledons thin or fleshy, often unequal; radicle short, superior.—Trees, shrubs, or rarely undershrubs, frequently climbing. Hairs usually closely appressed and fixed by the centre. Leaves mostly opposite, with glands at the top of the petiole, and often on the margin underneath. Stipules usually small, deciduous, or none. Flowers usually yellow, red, or white, in racemes either simple and terminal, or collected in corymbs or umbels, the pedicels articulate on the common peduncle.

A large tropical and subtropical Order, abundant in S. America, much less so in Africa and Asia. The only two Australian species belong to small genera spread over the Eastern Archipelago and S. Pacific islands. Both genera are exceptional as being deprived of the calycine glands so general in the Order.

Carpels with 1 vertical, large, oblong or incurved wing. Flowers in
    irregular corymbs. Styles 3 . . . . . . . . . . . . 1. Ryssopterys.
Carpels with several (7 or more) small linear, stellately spreading
    wings. Flowers in simple racemes. Styles 1 or 2, unequal . . 2. Tristellateia.

## 1. RYSSOPTERYS, Blume.

Calyx without glands. Petals scarcely clawed. Stamens all perfect, the filaments thickened at the base; anthers without appendages. Ovary 3-lobed, 3-celled, villous; styles 3, slender, with capitate stigmas. Samaras 1 to 3, expanded at the summit into a wing, of which the upper margin is thickened, tuberculate on the sides below the wing. Seed oblong, with a slightly curved embryo.—Woody climbers. Leaves opposite. Inflorescence terminal or apparently axillary from the reduction of the flowering branches, compound, irregularly corymbose. Peduncles bracteate at the base, with 2 bracteoles at the articulation of the pedicels.

A small genus, dispersed over the Eastern Archipelago, one of the species extending into Australia.

1. **R. timorensis,** *Blume; A. Juss. Malpigh.* 133. A tall climber, the young shoots hoary-pubescent. Leaves on rather long petioles, broadly cordate-ovate or orbicular, obtuse or rather acute, 3 to 5 in. long, somewhat coriaceous, glabrous above when full grown, hoary-pubescent underneath, with 1 or 2 prominent glands at the top of the petiole, those on the margin of the leaf very small. Flowers on pedicels of 2 or 3 lines, in short racemes arranged in irregular corymbs. Bracts and bracteoles very small. Fruit-carpels or samaras pubescent, the lateral tubercles very prominent, the wing broadly semicircular, about ¾ in. long and 5 or 6 lines broad.—Deless. Ic. Sel. iii. t. 35.

**Queensland.** Cape Cleveland, *A. Cunningham;* Fitzroy river, *Thozet.* The specimens are in fruit only, but agree perfectly with those we have in the same state from Timor. Some other species from the Archipelago are closely allied, but differ chiefly in the longer and narrower wing of the samaras.

## 2. TRISTELLATEIA, Thouars.

Calyx without any or with very minute glands.  Petals distinctly clawed. Stamens all perfect, filaments rigid, truncate, and articulate at the top ; anthers acute.  Ovary 3-lobed ; style single or 2, or very rarely 3 unequal ones, the others reduced to small papillæ.  Fruit-carpels 3, each one bearing about 7 small linear stellately spreading wings.  Seeds obovoid ; testa membranous, cotyledons fleshy, hooked.—Woody climbers.  Leaves opposite or whorled, the petiole bearing 1 or 2 glands at the top, and minute stipules at the base. Flowers yellow, in terminal or lateral racemes.

A small genus ranging over Madagascar and the Indian Archipelago, one species from the latter region extending into Australia.

1. **T. australasica,** *A. Rich. Sert. Astrol.* 38, *t.* 15.  A tall, glabrous climber.  Leaves opposite, on rather short petioles, ovate, acute, 2 to 4 in. long, membranous, the glands of the petiole usually single and sometimes wanting.  Racemes terminal, loose, 4 to 6 in. long.  Pedicels opposite, $\frac{1}{4}$ to 1 in. long, articulate, with 2 minute bracteoles below the middle.  Petals 3 or 4 lines long, spreading, the lamina ovate-cordate, the claw slender.  Filaments much thickened below the middle, and very shortly united.  Fruit (only seen in Archipelago specimens) quite glabrous, the wings of the carpels unequal, the longest often 3 lines long.

**Queensland.**  Brown's River, *M'Gillivray.*

The species is found in various islands of the Indian Archipelago.  The specimens described under the name of *Platynema laurifolium* by Wight and Arnott, in Jameson's Journal, and inserted in their ' Prodromus,' p. 107, as of doubtful Ceylonese origin, proved afterwards to have been from Singapore.

## Order XXVI. ZYGOPHYLLEÆ.

Flowers usually hermaphrodite and regular.  Sepals 5 or 4, very rarely 6, free or connate at the base, imbricate or rarely valvate in the bud.  Petals as many, free, imbricate or contorted, rarely valvate or wanting.  Disk convex or depressed, rarely annular or undeveloped.  Stamens usually the same or twice the number of the petals, the filaments most frequently with a scale or wings at or below the middle ; anthers 2-celled, opening longitudinally. Ovary sessile or shortly stalked, often angular, with as many cells as petals or sepals, rarely more or fewer ; style simple, with a simple or rarely lobed stigma. Ovules 2 or more in each cell, rarely solitary, pendulous or ascending, with a ventral raphe.  Fruit sometimes drupaceous, never baccate, more usually separating into indehiscent or 2-valved cocci, the endocarp occasionally separating. Seeds solitary or rarely several, pendulous ; testa membranous, crustaceous, or thick and mucilaginous when wetted ; albumen usually thin.  Embryo as long as the seed, green, straight, or rarely curved ; cotyledons oblong or linear, radicle short, superior.—Shrubs, undershrubs, or herbs, the branches usually divaricate and articulate at the nodes.  Leaves opposite, or rarely alternate by the abortion of one of each pair, 2-foliolate or pinnate, rarely simple, the leaflets usually entire.  Stipules in pairs.  Peduncles axillary, 1-flowered, or rarely branching into cymes.  Flowers mostly white, yellow, or red.

A small Order, nearly allied on the one hand to *Malpighiaceæ*, on the other to *Geraniaceæ* and *Rutaceæ*, dispersed chiefly over the subtropical regions of both the Old and New World, and most abundant in dry desert or saline regions. The three Australian genera are all common to Africa and Asia, and one of them extends also to Europe and America.

Leaves pinnate. Petals 5, flat. Fruit of 5 hard, indehiscent, usually
    prickly or tuberculate cocci . . . . . . . . . . . . . 1. TRIBULUS.
Leaves simple. Petals 5, concave. Fruit a drupe with a hard
    1-seeded nut . . . . . . . . . . . . . . . . . . 2. NITRARIA.
Leaves with 2 leaflets or lobes. Petals 4 or 5, flat. Fruit a 4- or
    5-angled or winged capsule . . . . . . . . . . . . . 3. ZYGOPHYLLUM.

## 1. TRIBULUS, Linn.

*(Tribulopis, R. Br.)*

Sepals 5, rarely 6. Petals as many, flat. Disk annular, 10-lobed or sinuate, with a gland at the base of each of the inner stamens, alternating with the petals. Stamens twice as many as petals, the filaments filiform, without appendages. Ovary of 5 or sometimes more cells, with 1 or 2 to 5 superposed ovules in each cell. Fruit separating into as many cocci as carpels, hard, indehiscent, and each usually bearing 2 or more prickles or tubercles. —Herbs, usually prostrate or divaricate and hairy. Leaves abruptly pinnate, opposite, with one of each pair smaller than the other, or sometimes abortive or all alternate. Stipules small, lanceolate, or falcate. Pedicels solitary in the axil of the smaller leaf of each pair, or opposed to the leaf when alternate. Flowers white or yellow.

The genus is dispersed over the greater part of the tropical and warm regions of the globe, extending into Europe and N. America. Of the Australian species, one is abundant in Asia, Africa, and S. Europe, another is most common in tropical America, less so in Asia and Africa, and the other 9 are all endemic.

Leaves, at least the upper ones, opposite. Glands of the disk not
    very prominent. Ovules 2 or more in each cell. (*Tribulus*
    proper.)
Cocci rounded at the back, without angular or winged edges.
    Cocci with 2 or 4 prickles, rarely minute or deficient.
        Leaves almost all opposite. Ovules 3 or 4 in each cell.
            Annual. Flowers small. Petals about ¼ in. . . . . . 1. *T. terrestris.*
            Perennial. Flowers large. Petals about ¾ in. . . . 2. *T. cistoides.*
        Lower leaves alternate. Ovules 2 in each cell. Flowers
            large . . . . . . . . . . . . . . . . . 3. *T. ranunculiflorus.*
    Cocci covered with numerous nearly equal prickles . . . . 4. *T. hystrix.*
    Cocci with prominent almost winged angles, and 2 prickles on
        the back between them . . . . . . . . . . . . 5. *T. macrocarpus.*
    Cocci broadly winged at the angles, without prickles.
        Plant glabrous except the inside of the sepals. Cocci smooth 6. *T. platyptcrus.*
        Plant hirsute. Cocci strongly reticulate on the back and
        sides . . . . . . . . . . . . . . . . . 7. *T. hirsutus.*
Leaves (except *T. minutus*) all alternate. Glands of the disk
    prominent. Ovules solitary. Fruit pyramidal, the cocci
    with 2 or 4 tubercles or small prickles below the middle.
    (*Tribulopis, R. Br.*)
Leaflets 2 pairs, the lowest much smaller. Perfect stamens
    usually 5 . . . . . . . . . . . . . . . . . . 8. *T. pentandrus.*

288 XXVI. ZYGOPHYLLEÆ. [*Tribulus.*

Leaflets about 3 pairs, ovate, the lowest not far from the stem.
Anthers 5 short, 5 oblong or linear . . . . . . . . 9. *T. bicolor.*
Leaflets about 3 pairs, ovate or lanceolate, the lowest distant
from the stem. Anthers 10, nearly similar. Flowers
small . . . . . . . . . . . . . . . . . . . . 10. *T. Solandri.*
Leaflets 4 to 6 pairs, linear. Anthers 10, similar. Flowers
large . . . . . . . . . . . . . . . . . . . . . 11. *T. angustifolius.*
Leaflets 3 to 6 pairs, small ovate or lanceolate. Leaves mostly
opposite. Anthers 10, similar. Flowers very small . . 12. *T. minutus.*

1. **T. terrestris,** *Linn.; DC. Prod.* i. 703. A prostrate annual or biennial, more or less hirsute or silky-hairy, especially the young shoots, the
stems extending often to 1 or 2 ft. Leaves opposite, unequal; leaflets of
the larger one usually 5 to 7 pairs, obliquely oblong, 3 to 5 lines long. Pedicels shorter than the opposite larger leaf. Flowers small, the sepals rarely
attaining 2 lines and often much less, the petals rather longer, but very rarely
nearly twice as long. Anthers 10, all small and perfect. Ovules 3 or 4 in
each cell. Cocci 5, hard, 2 to 3 lines long, glabrous or hairy, rounded on
the back, with 2 marginal, divaricate, horizontal, subulate or conical prickles
about halfway up, and often 2 smaller reflexed ones lower down, the rest of
the surface usually tuberculate or shortly muricate. Seeds 2 to 4 in each
coccus, horizontal and separated by transverse partitions.—Reichb. Ic. Fl.
Germ. v. t. 161; F. Muell. Pl. Vict. i. 99 ; *T. lanuginosus,* Linn.; DC.
Prod. i. 704 ; Wight, Ic. t. 98 ; *T. acanthococcus,* F. Muell. in Trans. Phil.
Soc. Vict. i. 9.

**N. Australia.** Sturt's Creek, *F. Mueller.*
**Queensland.** Gilbert river, *F. Mueller.*
**N. S. Wales.** Darling river, *Dallachy.*
**Victoria.** At the junction of the Murray and Murrumbidgee, *F. Mueller.*
**S. Australia.** N. of Lake Torrens, *M'Douall Stuart's Expedition.*
The species is a common weed in S. Europe, temperate Africa, and S. Asia.

2. **T. cistoides,** *Linn.; DC. Prod.* i. 703. A perennial, forming at
length a thick rootstock. Branches procumbent or ascending, attaining 1 to
2 ft. Indumentum more silky than in *T. terrestris.* Larger leaf of each
pair with frequently 7 or 8 pairs of leaflets. Flowers large, on longer peduncles than in *T. terrestris ;* the sepals 3 or 4 lines long, very acute, silky-
hairy ; the petals obovate, at least ¾ in. long. Anthers usually (perhaps not
always) oblong or linear. Fruit like that of *T. terrestris* or rather larger,
with 2 or very rarely 4 prickles to each coccus.—A. Gray, Ill. Gen. N. Am.
t. 145.

**N. Australia.** Gulf of Carpentaria, *R. Brown ;* Port Essington, *Armstrong ;* Albert
river and Swears Island, *Henne.*
**Queensland.** Northumberland Island, *R. Brown ;* Port Curtis and Port Molle,
*M'Gillivray ;* Lord Howick's group, *F. Mueller ;* Port Denison, *Fitzalan.*
**N. S. Wales.** Liverpool plains, *Leichhardt.*
The species is frequent in the West Indies and many parts of tropical America, and in the
Pacific islands, rare in tropical Asia and Africa.

3. **T. ranunculiflorus,** *F. Muell. Fragm.* i. 48. An annual, with
procumbent or ascending stems, hirsute with spreading hairs. Lower leaves
alternate, upper ones opposite, the larger one of each pair with about 8 or 10
pairs of obliquely lanceolate leaflets, more or less silky-hairy. Flowers

large, on rather long pedicels.  Sepals very acute, 3 to 4 lines long.  Petals more than twice as long.  Anthers short, ovate.  Ovary very hirsute, with only 2 ovules in each cell.  Fruit about 3 lines long, the cocci slightly muricate and often with 2 short prickles, containing each 1 or 2 seeds separated by a transverse partition.

**N. Australia.**  Dry sandy pastures on the Upper Victoria river, *F. Mueller.*

4. **T. hystrix,** *R. Br. in App. Sturt. Exped.* 6.  A diffuse or prostrate perennial or undershrub, the branches densely tomentose-hirsute or woolly. Lower leaves (at least in some specimens) alternate, upper ones opposite, the larger one of each pair with 6 to 8 or even more pairs of leaflets, rather broad and softly silky-hairy.  Flowers smaller than in *T. cistoides*, but much larger than in *T. terrestris*, the petals generally about ½ in. long.  Ovary very hirsute, with 3 or 4 ovules in each cell.  Cocci very villous, covered all over with hairy prickles, either subulate from the base or more or less thickened and conical.—*T. occidentalis*, R. Br. l. c. (from the short diagnosis given).

**N. Australia.**  N.W. coast, *A. Cunningham;* on sandy soil, in the interior from Nichol Bay, *F. Gregory.*
**S. Australia.**  Towards Spencer's Gulf, *Warburton.*

The specimens I have seen are most of them very incomplete, and those described by R. Brown unfortunately mislaid.  The few fruits on Gregory's specimens show, however, that the character relied on by R. Brown for the distinction of *T. hystrix* and *T. occidentalis*, the subulate or conical prickles of the fruit, does not hold good.  In M'Douall Stuart's collection is a fragmentary specimen from Fink river, with a much larger flower, which may possibly be a variety of the same species, but is indeterminable without the fruit.

5. **T. macrocarpus,** *F. Muell. Herb.*  Foliage and flowers unknown. Cocci quite glabrous, nearly ¾ in. long, the edges bordered by narrow, vertical, hard, slightly denticulate wings, with 2 straight, horizontal, conical prickles on the back about halfway up and a vertical prominent rib between them, the sides smooth.

**N. Australia.**  In the interior, from Nichol Bay, *F. Gregory.*

6. **T. platypterus,** *Benth.*  A shrub of 2 to 3 feet, glabrous, except the inside of the sepals, the older branches in one specimen corky.  Leaves opposite, the larger one of each pair with about 5 or 6 pairs of obtuse leaflets. Pedicels rather short.  Sepals very acute, at least 4 lines long, bright-green and glabrous outside like the rest of the plant, woolly-hairy inside.  Petals narrow, about ½ in. long.  Fruit about ½ in. long and ¾ in. broad, including the wings, truncate rather than cordate at the base; the cocci glabrous or very slightly hairy, bordered by broad, vertical, semicircular, membranous wings, and smooth between the wings and on the sides.

**N. Australia.**  Hammersley range, *F. Gregory.*

7. **T. hirsutus,** *Benth.*  A shrub allied to *T. platypterus* and considered by F. Mueller as a variety, but the branches, leaves, and inflorescence are hirsute with long fine spreading hairs; the flowers are rather smaller; the sepals hirsute outside, not woolly inside, but hirsute along the middle with straight hairs; the wings of the fruit form rounded auricles at the base, giving the outline a cordate form; and the cocci are prominently reticulate, almost muricate, both on the back between the wings and on the sides next to the adjoining cocci.

**N. Australia.**  Nichol Bay, *F. Gregory.*

8. **T. pentandrus,** *Benth.*  A slender, prostrate, branching annual, often attaining 1 ft. in length, more or less hairy.  Leaves all alternate, with 2 pairs of oblong-lanceolate leaflets, the terminal ones 4 to 8 lines long, the lower pair much smaller, usually not half the size.  Flowers small, on slender pedicels.  Petals oblong.  Stamens usually 5 with globular or ovoid perfect anthers, and 5 small with imperfect capitate anthers, or entirely wanting.  Ovules solitary in each cell of the ovary.  Fruit pyramidal, 1 to 1½ lines long, with 2 small tubercles at the base of each coccus.—*Tribulopis pentandra,* R. Br. in App. Sturt, Exped. 7 ; F. Muell. Fragm. i. 48.

**N. Australia.**  Victoria river and Sea Range, *F. Mueller ;* islands of the Gulf of Carpentaria, *R. Brown.*

9. **T. bicolor,** *F. Muell. Pl. Vict.* i. 99.  A prostrate annual, often attaining 1 ft. or more, pubescent or hairy.  Leaves all alternate, with 2 or 3 pairs of rather broad leaflets, 2 to 4 lines long, nearly equal, the lowest pair usually close to the stem, glabrous except the ciliate margins.  Flowers small, on rather short pedicels.  Petals oblong, red at the base according to F. Mueller, but the difference of colour does not show in the dried state.  Stamens usually 5 with small short anthers, and 5 with linear anthers, some of them occasionally imperfect.  Ovules solitary in each cell of the ovary.  Fruit pyramidal, about 2 lines long, pubescent, with 2 pairs of tubercles below the middle of each coccus.—*Tribulopis bicolor,* F. Muell. Fragm. i. 47.

**N. Australia.**  Sandy shores of the Victoria river, *F. Mueller.*  This species appears to me to be much more nearly allied to *T. Solandri* than the *T. angustifolius.*

10. **T. Solandri,** *F. Muell. Pl. Vict.* i. 99 (partly).  An annual, with prostrate or ascending stems, pubescent or nearly glabrous.  Leaves alternate ; leaflets usually 3 pairs, rarely 2 pairs, obliquely ovate or oblong-falcate, 3 to 6 lines long, the lowest pair distant from the stem and nearly of the size of the others, all glabrous except the ciliate margins or slightly hairy, those of the upper leaves sometimes narrower and lanceolate.  Flowers small.  Stamens usually all perfect, with small anthers.  Fruit pyramidal, about 3 lines long, glabrous or slightly tomentose, with 2 pairs of prominent reflexed tubercles below the middle of each coccus.—*Tribulopis Solandri,* R. Br. in App. Sturt, Exp. 7.

**N. Australia.**  Victoria river and Gilbert river, *F. Mueller.*
**Queensland.**  Endeavour river, *Banks ;* Lizard island, *M'Gillivray.*

11. **T. angustifolius,** *Benth.*  An annual or according to some specimens forming a perennial rootstock, with procumbent, ascending, or erect stems, glabrous or silky-pubescent.  Leaves all alternate ; leaflets 4 or 5 pairs or sometimes more, linear, attaining 1 in. in length, more or less silky-pubescent.  Flowers much larger than in the other species, the petals usually exceeding ½ in.  Stamens all perfect, with small anthers.  Fruit 3 lines long, besides the rigid persistent style which is about as long, with 2 minute tubercles at the base of each coccus.—*Tribulopis angustifolia,* R. Br. in App. Sturt, Exped. 7 ; *Tribulopis Solandri,* F. Muell. Fragm. i. 47 (partly) ; *Tribulus Solandri,* var. *angustifolia,* F. Muell. Pl. Vict. i. 99.

**N. Australia.** N.W. coast, *Bynoe;* Victoria river, *F. Mueller;* islands of the Gulf of Carpentaria, *R. Brown, Henne.*

12. **T. minutus,** *Leichh. in Herb. F. Muell.* Pubescent, apparently prostrate, and more slender than any other species. Leaves mostly opposite, those of each pair unequal or one occasionally abortive, the larger one of 3 to 5 pairs of obovate or oblong leaflets, about 2 or rarely 3 lines long. Flowers very small. Stamens 10, with the anthers all similar. Glands prominent. Ovules solitary (or sometimes 2 ?) in each cell. Fruit nearly of *T. Solandri,* but smaller; each carpel bearing a pair of small, reflexed, conical spines about the middle and a pair of minute tubercles lower down.

**Queensland (?),** *Leichhardt's Expedition.* This species connects the two groups, having the opposite leaves of *Tribulus* proper, with the fruit of *Tribulopis.*

## 2. NITRARIA, Linn.

Calyx small, 5-lobed. Petals 5, concave with inflexed points, induplicate-valvate in the bud. Disk not prominent. Stamens 15, rarely 10 to 14, the filaments free, without appendages. Ovary sessile, 2- to 6-celled, terminating in a short thick style, with 2 to 6 adnate stigmas; ovules solitary in each cell, ascending from pendulous funiculi, which are more or less adnate to their inner face. Fruit a drupe, with a berry-like sarcocarp; putamen ovoid-acute, hard, marked outside with irregular depressions, and opening at the top in 6 short, pointed valves, of which 3 inner ones smaller. Seeds solitary, pendulous, without albumen.—Rigid shrubs, often thorny. Leaves alternate or clustered, undivided, succulent. Stipules small. Flowers small, white, in once- or twice-forked scorpioid cymes.

The genus, besides the widely-spread Australian species, comprises one other from Northern Africa. The raphe of the seed is described as dorsal by Spach, but we have always found it ventral in the ovary, although the seed sometimes hangs obliquely.

1. **N. Schoberi,** *Linn.; DC. Prod.* iii. 456. A rigid spreading shrub, attaining 3 to 6 ft., glabrous or hoary with a very minute down, the smaller branches occasionally spinescent. Leaves from cuneate-oblong to lanceolate or linear, the lower ones obtuse and often 1 in. long, those of the smaller branches smaller and more acute, all entire, thick and fleshy. Cymes usually shortly pedunculate, the flowers sessile or shortly pedicellate along the scorpioid branches. Petals about 1½ lines long. Ovary 3-celled. Drupe varying from ovoid-globular to ovoid-oblong, the putamen from ¼ to more than ½ in. long, the depressions in the lower part round or oblong, the upper part marked with 6 furrows, along which the valves ultimately open. Only 1 seed or very rarely 2 come to maturity.—Andr. Bot. Rep. t. 529; *N. Billardieri,* DC. Prod. iii. 456; F. Muell. Pl. Vict. i. 92, t. Suppl. 7; *N. Olivieri,* Jaub. and Spach, Ill. Pl. Or. iii. 143, t. 295; *Zygophyllum australasicum,* Miq. in Pl. Preiss. i. 164.

**N. S. Wales.** Darling and Murrumbidgee rivers, *Dallachy* and *Goodwin.*

**Victoria.** Saline tracts on the Murray river, and in the N.W. part of the colony, *F. Mueller.*

**S. Australia.** Goose Island bay, *R. Brown;* along the coast, and northward to Lake Torrens, *F. Mueller* and others.

**W. Australia,** *Drummond, n.* 227; Cornac Island, *Preiss, n.* 2397; Murchison river, *Oldfield,* and in several other collections.

The species is spread over the hot, more or less saline, tracts of western Asia and northern Africa. A careful examination leaves no doubt of the identity so often suggested of the Australian and northern plants.

# 3. ZYGOPHYLLUM, Linn.

Sepals 4 or 5. Petals as many, flat, contracted into a short claw. Disk concave, angular or cup-shaped. Stamens twice as many as petals, inserted at the base of the disk; filaments filiform, with an adnate scale or wing-like appendage at the base, which however is wanting in some of the Australian species. Ovary sessile, 4- or 5-angled, narrowed at the top into an angular style, 4- or 5-celled, with 2 or more superposed ovules in each cell. Fruit capsular; with 4 or 5 angles or vertical wings, indehiscent or separating into cocci or opening loculicidally, the endocarp sometimes separating. Seeds 1 or more in each cell, pendulous; albumen scanty.—Shrubs or undershrubs, often prostrate. Leaves opposite, with 2 distinct leaflets or rarely 2-lobed, frequently fleshy. Stipules small. Peduncles 1-flowered, axillary, solitary or rarely 2 together. Flowers white or yellow.

A considerable and widely-spread genus, though confined, with one exception, to the Old World, and chiefly numerous in the desert or saline regions of central and western Asia, North and South Africa. The Australian species are all endemic.

Filaments winged at the base. Capsule angular, loculicidal.
Capsule broad and truncate at the top, the angles usually produced into short appendages. Flowers mostly 5-merous . 1. *Z. apiculatum.*
Capsule equally rounded at the top and the base.
Capsule 4 to 8 lines long, the cells 2- to 4-seeded. Wings of the filaments toothed. Flowers usually 4-merous . . . 2. *Z. glaucescens.*
Capsule 2 to 3 lines long, the cells 1-seeded. Wings of the filaments small and entire. Flowers usually 5-merous . . 3. *Z. iodocarpum.*
Capsule oblong, the angles produced at the top into erect appendages . . . . . . . . . . . . . . . . . 4. *Z. prismatothecum.*
Filaments subulate, not winged.
Capsule angular, loculicidal, broad and truncate at the top, narrow at the base . . . . . . . . . . . . . . . . . . . 5. *Z. Billardieri.*
Capsule indehiscent, the angles produced into broad membranous wings . . . . . . . . . . . . . . . . . . . . . . 6. *Z. fruticulosum.*

Varieties with leaves 2-lobed instead of 2-foliolate occur in *Z. iodocarpum, Z. prismatothecum, Z. Billardieri,* and *Z. fruticulosum;* with lobed or crenate leaflets in *Z. glaucescens* and *Z. iodocarpum;* and forms or states with minute flowers in several of the species.

**1. Z. apiculatum,** *F. Muell. in Linnæa,* xxv. 373, *and Pl. Vict.* i. 101. A diffuse, glabrous undershrub. Leaflets 2, obliquely obovate or rarely oblong, ¼ to 1 in. long, on a short common petiole. Flowers usually 5-merous. Filaments with rather broad wings, adnate to above the middle and toothed at the top. Capsule about 4 lines long, opening loculicidally, broader and truncate at the top, the angles very obtuse, and produced at the upper outer corner into a short obtuse appendage. Seeds usually solitary in each cell.— *Ræpera latifolia,* Hook. f. Fl. Tasm. i. 60; *Zygophyllum terminale,* Turcz. in Bull. Mosc. 1858, i. 437.

**Queensland.** Mackenzie and Dawson rivers, *F. Mueller* (a very small-flowered variety).

**N. S. Wales.** Molle's Plains, *A. Cunningham.*
**Victoria.** Along the Murray river, from the Murrumbidgee downwards, *F. Mueller.*
**Tasmania.** Islands of Bass's Straits, *Gunn.*
**S. Australia.** Broughton river, Flinders Range, Spencer's Gulf, *F. Mueller;* Stevenson river, *M'Douall Stuart.*
**W. Australia,** *Drummond, 5th Coll., n.* 90; towards Sharks Bay, *Oldfield.*

2. **Z. glaucescens,** *F. Muell. Pl. Vict.* i. 228. Herbaceous, diffuse or erect and glabrous. Leaves of 2 broad leaflets as in *Z. apiculatum,* the petiole occasionally winged at the base. Flowers usually 4-merous. Filaments with toothed wings as in *Z. apiculatum.* Capsule usually above ½ in. long, opening loculicidally, the angles equally rounded at the top and the base. Seeds 2 or 3 or sometimes 4 or 5 in each cell.—*Z. glaucum,* F. Muell. in Trans. Vict. Inst. i. 29, and Pl. Vict. i. 102 ; not of Sonder.

**N. S. Wales.** Erskine river, *A. Cunningham;* Lachlan river, *Fraser;* Darling river, *Herb. F. Mueller.*
**Victoria.** Subsaline deserts on the Murray, Wimmera, and Avoca rivers, *F. Mueller.*
**S. Australia.** Barossa Range, St. Vincent's and Spencer's Gulf, Venus Bay, *F. Mueller.*
Var. *lobulatum.* Leaflets irregularly 2- or 3-lobed or deeply crenate. Flowers and fruit precisely as in the ordinary form.—*Z. crenatum,* F. Muell. in Linnæa, xxv. 374, and Pl. Vict. i. 103, t. 6. On the Lachlan and Murray rivers, and in the interior of S. Australia, *F. Mueller.*

3. **Z. iodocarpum,** *F. Muell. in Linnæa,* xxv. 372, *and Pl. Vict.* i. 105. A small, much-branched, diffuse annual. Leaflets oblong-cuneate or almost linear, very obtuse, rarely ½ in. long, the petiole often 2-winged, especially towards the top. Flowers very small, usually 5-merous, the petals not 2 lines long. Filaments dilated at the base into short, narrow, entire wings, entirely adnate or very shortly free. Capsule 2 or rarely 3 lines long, loculicidal, the angles equally rounded at the top and the base. Seeds solitary in each cell.

**N. S. Wales.** Between the Darling and the Lachlan rivers, *Burkitt.*
**Victoria.** On the Murray river below the Murrumbidgee, *F. Mueller.*
**S. Australia.** Flinders Range and near Lake Torrens, *F. Mueller.*
Var. *lobulatum.* Leaflets irregularly 2- or 3-lobed or toothed.
**W. Australia.** Champion Bay, *Oldfield.*
Var. *bilobum.* Leaflets continuous with the petiole, as in *Z. prismatothecum.* Capsule rather longer than in the other varieties. W. Australia, *Drummond.*

4. **Z. prismatothecum,** *F. Muell. in Linnæa,* xxv. 375. A much-branched, small annual. Leaves rather thick, the leaflets, in the few specimens seen, short and confluent with the more or less dilated petiole, so as to form a single 2-lobed leaf. Flowers, which I have not seen, small and 4-merous, according to F. Mueller, the filaments dilated at the base and toothed or entire. Capsules nearly sessile, oblong, 4-angular, about 4 lines long, of equal breadth at the base and the top where the angles terminate in small erect leafy appendages. Seeds solitary in each cell.

**S. Australia.** Dry hills near Arkaba, *F. Mueller.* The very few specimens seen have all the foliage of the 2-lobed varieties of *Z. iodocarpum, Z. Billardieri* and *Z. fruticulosum,* but as in those species there is probably also a variety with normally 2-foliolate leaves.

5. **Z. Billardieri,** *DC. Prod.* i. 705. Herbaceous, prostrate or diffuse and much-branched. Leaflets oblong, cuneate or linear, rarely obovate, ½ to

1 in. long, the petioles not usually winged. Flowers usually 4-merous, the size of those of *Z. apiculatum*. Sepals narrow, very acute. Petals about 3 lines long. Filaments subulate or slightly flattened, but not winged. Capsule 3 to 5 lines long, loculicidal, broad and truncate at the top, narrowed to the base, the angles acute or shortly pointed or scarcely rounded at the upper outer corner. Seeds 1 or rarely 2 in each cell.—Hook. f. Fl. Tasm. i. 60; F. Muell. Pl. Vict. i. 104; *Rœpera Billardieri*, A. Juss. in Mem. Mus. Par. xii. 454 (by inference); *Z. ammophilum*, F. Muell. in Linnæa, xxv. 376, in adnot.

**Victoria.** Coast rocks and drift sands from Port Phillip to the Glenelg, and on the lower Murray river, *F. Mueller.*

**Tasmania.** Islands of Bass's Straits, *R. Brown, Gunn*, and others.

**S. Australia.** Spencer's Gulf, Goose Island Bay and Kangaroo Island, *R. Brown;* southern shores and towards Lake Torrens, *F. Mueller;* Cooper's Creek, *A. Gregory.*

**W. Australia.** From King George's Sound to Swan River and Champion Bay, *Drummond, Oldfield,* and others.

Var. *bilobum.* Leaflets narrow, continuous with the petiole, as in *Z. prismatothecum.* On the Murray river and Holdfast Bay, *F. Mueller.*

A minute-flowered form occurs also on the Murray and in West Australia.

6. **Z. fruticulosum,** *DC. Prod.* i. 705. A low diffuse or divaricately-branched shrub. Leaflets obliquely oblong or lanceolate, rarely ovate. Flowers 4-merous, the size of those of *Z. apiculatum*. Filaments subulate, without wings. Capsule $\frac{1}{2}$ in. long, indehiscent, or at length separating septicidally into cocci opening inside, the angles expanded into broad membranous wings, rounded at both ends and not splitting. Seeds solitary in each cell. —F. Muell. Pl. Vict. i. 105; *Rœpera fabagifolia*, A. Juss. in Mem. Mus. Par. xii. 525, t. 15; Deless. Ic. Sel. iii. t. 42; Miq. in Pl. Preiss. i. 164.

**W. Australia.** Arthur's Head, Swan River, *Preiss, n.* 1953; Port Gregory and Murchison river, *Oldfield.*

Var. *bilobum.* Leaflets narrow, continuous with the petiole, as in *Z. prismatothecum.—Rœpera aurantiaca*, Lindl. in Mitch. Three Exped. ii. 70; *Z. aurantiacum*, F. Muell. in Linnæa, xxv. 376 (note).

**N. S. Wales,** Mitchell; Darling river, *Goodwin* and *Dallachy.*

**Victoria.** Murray Desert, from the Murrumbidgee downwards, *F. Mueller.*

**S. Australia.** From Spencer's Gulf, *Warburton,* to Flinders Range and Lake Torrens, *F. Mueller.*

**W. Australia.** Dirk Hartog's Island, *A. Cunningham.*

Var. (?) *platypterum.* Leaflets obovate, as in the broad-leaved specimens of *Z. glaucum.* Fruits winged, as in *Z. fruticulosum,* but very much larger, attaining more than 1 in. diameter. Port Jackson, *Leichhardt (Herb. F. Mueller).* The specimen insufficient for accurate diagnosis.

ORDER XXVII. **GERANIACEÆ.**

Flowers usually hermaphrodite, regular or irregular. Sepals 5, or rarely fewer, free, or rarely connate at the base, imbricate or (in genera not Australian) valvate in the bud. Petals as many or rarely wanting, hypogynous or slightly perigynous, variously imbricate in the bud. Torus more or less expanded into a disk, often bearing 5 glands alternate with the petals, and usually protruding into a short axis in the centre of the ovary. Stamens usually twice the number of the petals, 5 of them occasionally without anthers, or rudimentary, or in irregular flowers, 3 or more without anthers or wanting; filaments either free and filiform, or dilated or connate at the base; anthers with 2 parallel cells. Ovary usually 3- to 5-lobed, with as many cells, the carpels adnate to the axis up to the insertion of the ovules, and often produced above that into a beak bearing the style or stigmas; stigmas as many as cells, either raised on the style or sessile on the carpels, radiating from a connate base or rarely entirely connate. Ovules either 1 in each cell or 2 inserted nearly at the same point, 1 ascending, the other pendulous, or several in 1 or 2 rows. Fruit either a lobed capsule, the lobes 1-seeded, separating from the axis with the seed, and elastically rolled upwards along the beak, leaving the placentiferous portion attached to the axis, or the lobes several-seeded, remaining attached to the axis, but opening loculicidally, or, in genera not Australian, the fruit is a berry or separates into indehiscent cocci. Seeds pendulous or ascending; testa thin or rarely crustaceous; albumen usually scanty or none. Embryo straight or curved, radicle short and straight or long and curved or forked over the cotyledons.—Herbs or shrubs, or rarely (in genera not Australian) trees. Leaves opposite or alternate, toothed, lobed, or divided, very rarely quite entire. Stipules usually 2. Peduncles axillary, 1- or 2-flowered, or bearing an umbel of several flowers, very rarely a cyme or raceme.

The Order is chiefly dispersed over the temperate regions of the northern hemisphere, very abundant in Southern Africa, with a few extratropical South American and tropical species. Of the four Australian genera, two are common in the northern hemisphere, a third, although chiefly American, is represented in Australia by species of an extratropical European as well as American type, and the fourth is almost entirely South African. The Order is very closely allied to *Zygophylleæ*.

Capsule beaked, the lobes 1-seeded, and elastically rolled upwards along the beak. Leaves toothed, lobed, or divided.
  Flowers regular.
    Anthers usually 10. Tails of the carpels glabrous inside . . 1. GERANIUM.
    Anthers 5. Tails of the carpels bearded inside . . . . . 2. ERODIUM.
  Flowers irregular, with a linear tube or spur adnate to the pedicel.
    Anthers 5, 6, or 7 . . . . . . . . . . . . . 3. PELARGONIUM.
Capsule opening loculicidally, the valves adhering to the axis. Leaves
  with 3 leaflets . . . . . . . . . . . . . . . 4. OXALIS.

## 1. GERANIUM, Linn.

Flowers regular. Sepals 5. Petals 5. Glands 5, alternating with the petals. Stamens 10, all usually bearing anthers. Ovary 5-lobed, beaked, the beak terminating in the style, with 5 short stigmatic lobes. Ovules 2 in each cell. Capsule-lobes 1-seeded, separating from the placenta-bearing axis, enclosing the seed, and curled upwards on a long awn detached from the beak, and glabrous inside. Radicle of the embryo turned back on the folded or convolute cotyledons.—Herbs, rarely undershrubs. Leaves opposite or alternate, toothed, lobed, or divided, the lobes or segments palmate, or rarely (in species not Australian) pinnate. Peduncles axillary or in the forks, 1- or 2-flowered.

A large genus, widely distributed over nearly the whole globe, but more abundant in the northern hemisphere, and rare within the tropics. Both the Australian species are also in New Zealand and S. America, and one of them extends up the whole length of that continent to the N.W., and in a slight variety also over most temperate parts of the northern hemisphere. Neither of them occurs in S. Africa.

Flowering-stems elongated and leafy. Seeds reticulate    . . .   1. *G. dissectum.*
Flowering-stems undeveloped or short.   Seeds quite smooth  . .  2. *G. sessiliflorum.*

Besides these, *G. molle,* Linn.; DC. Prod. 1643, a European annual weed, with orbicular leaves divided to the middle only, small flowers with deeply notched petals, wrinkled capsule-lobes, and smooth seeds, has established itself in some parts of Tasmania.

1. **G. dissectum,** *Linn.; DC. Prod.* i. 643, var. *australe.* Usually perennial, forming at length a thick rootstock, descending into a taproot. Stems diffuse, procumbent or shortly erect, more or less hairy with spreading or reflexed hairs, or hoary with a short pubescence. Leaves on long petioles, nearly orbicular in their circumscription, deeply divided into 5 or 7 segments, each one again more or less cut into 3 or more lobes, varying from broadly cuneate-oblong to linear, and usually pubescent or hairy, especially underneath. Peduncles 2-flowered, or rarely 1- or 3-flowered. Sepals 3-nerved, obtuse, acute, or very shortly mucronate; usually 2 or 3 lines long. Petals cuneate-obovate, entire or slightly notched, from rather longer than the sepals to twice as long. Anthers all perfect. Lobes of the capsule sprinkled with hairs, not wrinkled. Seeds covered with minute reticulations or rarely smooth. —Hook. f. Fl. N. Zeal. i. 39, and Fl. Tasm. i. 56; F. Muell. Pl. Vict. i. 173; *G. pilosum,* Forst.; DC. Prod. i. 642; Nees, in Pl. Preiss. i. 162; *G. parviflorum,* Willd.; DC. Prod. i. 642; *G. philonothum,* DC. Prod. i. 639 (from the character given); *G. potentilloides,* L'Hér., DC. Prod. i. 639; Hook. f. Fl. N. Zeal. i. 40; Fl. Tasm. i. 57; *G. australe,* Nees, in Pl. Preiss. i. 162.

**N. S. Wales.** Port Jackson, *R. Brown;* common in the colony extending northwards to New England, *C. Stuart;* and Clarence river, *Beckler.*

**Victoria.** Port Phillip, *R. Brown;* frequent throughout the colony, ascending to alpine situations, *F. Mueller.*

**Tasmania.** Abundant throughout the colony, *J. D. Hooker.*

**S. Australia.** Common in the colony, *F. Mueller.*

**W. Australia,** *Drummond, Preiss,* n. 1900, 1907; *Oldfield* and others.

The original form of *G. dissectum,* as generally diffused over the temperate regions of the northern hemisphere, in the Old World, is an annual, with the petals very rarely exceeding the sepals, and the seeds very prominently reticulate. In the eastern United States of N. America, under the name of *G. carolinianum,* Linn., it is also annual or biennial, but has the petals often rather larger and the reticulations of the seeds are finer and less prominent. West of the Rocky Mountains the stock often appears to be perennial, and then it is undistinguishable from some Australian forms. The commonest Australian form is frequently seen from extratropical S. America, and extends all along the mountainous regions of that continent to Mexico and the Rocky Mountains, often apparently together with and passing into the northern annual variety. The Australian plant again, both in that country and in New Zealand, is very variable, and may be generally subdivided into two principal races, although I have, after repeated trials, found it impossible to distribute our numerous specimens quite satisfactorily into the two groups, viz.: —

*a. pilosum.* Root thick. Stems erect, ascending or procumbent, usually hirsute. Seeds strongly reticulate.

*b. potentilloides.* Root and stock less thickened. Stems more slender and prostrate, less hairy, and usually only slightly hoary with more appressed pubescence. Seeds more finely reticulate, or rarely almost smooth. To this variety belongs generally the *G. potentilloides* of authors, and *G. australe,* Nees. It appears to be rather the more common form in the East, whilst the var. *pilosum* is more frequent in the West. But both are found throughout extratropical Australia.

**2. G. sessiliflorum,** *Cav. Diss.* 198, *t.* 77, *f.* 2.　Perennial, with the rootstock thick, descending into a taproot.　Hairs of the peduncles and sepals long and silky, spreading or reflexed as in the var. *pilosum* of *G. dissectum.*　Leaves mostly radical, on long petioles, divided, as in *G. dissectum,* into 5 or 7 lobed segments.　Flowering-stems undeveloped or very short, rarely as long as the leaves, very hirsute.　Peduncles short.　Sepals much more prominently mucronate or awned than in *G. dissectum.*　Petals small.　Anthers all perfect.　Capsule-lobes sprinkled with hairs, not wrinkled.　Seeds perfectly smooth or minutely punctulate under a strong lens.—*G. brevicaule,* Hook. Journ. Bot. i. 253 ; Hook. f. Fl. Tasm. i. 57.

**Victoria.**　In alpine situations, *F. Mueller.*

**Tasmania.**　Common in alpine districts, at an elevation of 3000 to 4000 ft., *J. D. Hooker.*　Also in New Zealand, in Fuegia and Chili.　Considered by F. Mueller as an alpine form of *G. dissectum,* but, besides the habit, the smoothness of the seeds seems to be constant.

## 2. ERODIUM, L'Hér.

Flowers regular or nearly so.　Sepals 5.　Petals 5.　Glands 5, alternating with the petals.　Stamens 5 bearing anthers, opposite the sepals, and 5 staminodia, usually scale-like, alternating with them.　Ovary 5-lobed, beaked, the beak terminating in the style, with 5 short stigmatic lobes.　Ovules 2 in each cell.　Capsule-lobes 1-seeded, separating from the placenta-bearing axis, enclosing the seed and curled upwards on a long elastic awn, which separates from the beak, and is usually twisted and bearded inside with long hairs.　Radicle of the embryo turned back on the folded or convolute cotyledons.—Herbs or rarely undershrubs.　Leaves unequally opposite or alternate, pinnately or rarely ternately lobed or divided.　Peduncles axillary, bearing an umbel of several flowers, or rarely 1-flowered.

The species are numerous in Europe, North Africa, and temperate Asia, 2 or 3 are natives of S. Africa, and 2 or 3 more are now widely dispersed as weeds over many parts of the globe.　Two of these are in Australia, one of them perhaps indigenous, but the common Australian species is endemic.

Leaves of 3 lobed or divided segments, the middle one the largest　.　1. *E. cygnorum.*
Leaves pinnate with deeply-lobed narrow segments　.　.　.　.　.　.　2. *E. cicutarium.*

Besides these, *E. moschatum,*Willd., a coarser plant than *E. cicutarium,* usually smelling of musk, the leaves pinnate as in that species, but with ovate, toothed, or scarcely lobed segments, has established itself as a weed in some parts of Victoria, S. Australia, and W. Australia.

**1. E. cygnorum,** *Nees, in Pl. Preiss.* i. 162.　An annual or biennial, with the habit of the coarser forms of *E. cicutarium,* sometimes slightly pubescent, sometimes very hispid, with the hairs of the stem spreading or reflexed.　Leaves deeply 3-lobed or divided to the base into 3 lobes or segments, usually obovate or cuneate, and more or less deeply toothed or again 3-lobed, the central lobe larger, broader, and more lobed than the lateral ones.　Flowers blue, usually 2 to 5 in the umbel.　Sepals pointed.　Petals obovate, scarcely exceeding the calyx or shorter.　Filaments broad at the base, with subulate points ; staminodia scale-like, often toothed.　Capsule-lobes glabrous, hairy or hispid ; beak usually above 2 in. long.—F. Muell. Pl. Vict. i. 172.

**Queensland.**　Peak Downs, *F. Mueller ;* Maranoa river, *Mitchell.*

**N. S. Wales.**  Nepean river, *R. Brown*; Lachlan and Darling rivers, *A. Cunningham, Dallachy*; on the Murrumbidgee, *M'Arthur*.

**Victoria.**  Not rare in many parts of the colony, *F. Mueller*.

**S. Australia.**  From Kangaroo Island and Spencer's Gulf to Lake Torrens, and further north, *F. Mueller* and others.

**W. Australia.**  Swan River, *Drummond, Preiss, n.* 1902; and thence to Murchison river, *Oldfield*.

2. **E. cicutarium,** *L'Hér.; DC. Prod.* i. 646.  Usually an annual, but often forming a dense tuft, with a thick taproot, which may last over a second year, always more or less covered with spreading hairs, which are sometimes viscid.  Stems sometimes exceedingly short, but lengthening out to near 1 ft. Leaves mostly radical, pinnate, the segments distinct and deeply pinnatifid, with narrow, more or less cut lobes.  Peduncles erect, bearing an umbel of from 2 or 3 to 10 or 12 small purple or pink flowers.  Sepals pointed, about the length of the obovate entire petals.  Filaments and staminodia lanceolate-subulate.  Lobes of the capsule slightly hairy, the beak $\frac{1}{2}$ to $1\frac{1}{2}$ in. long.— Nees, in Pl. Preiss. i. 161; Reichb. Ic. Fl. Germ. v. t. 183.

**N. S. Wales.**  Between the Lachlan and Darling rivers, *Dallachy*; Twofold Bay, *F. Mueller*.

**Victoria.**  On the Murray, and now rather frequent in many parts of the colony, *F. Mueller*.

**Tasmania.**  Along roadsides, evidently introduced, *J. D. Hooker*.

**S. Australia.**  Towards Spencer's Gulf, *F. Mueller*.

**W. Australia,** *Drummond, Preiss, n.* 1899.

A very common weed in Europe and temperate Asia, and found in many other parts of the world, in many cases introduced, as in several or perhaps all of the Australian localities, but too widely spread now to be omitted from the Flora, even if it be not really indigenous.

### 3. PELARGONIUM, L'Hér.

Flowers irregular.  Sepals 5, shortly united at the base and produced into a tube or spur, adnate to the pedicel.  Petals 5 or fewer, the 2 upper ones different from the others (usually larger), and inserted on the sides of or behind the spur.  Disk without glands.  Stamens usually 10, hypogynous, shortly united, 5 to 7 or rarely only 2 or 3 bearing anthers, the remainder without anthers or rudimentary.  Ovary and fruit of *Erodium*.  Cotyledons flat or folded.—Herbs, undershrubs, or shrubs.  Leaves opposite or rarely alternate, entire, toothed, lobed, or variously divided.  Peduncles usually axillary, bearing an umbel of several flowers.

A very large genus, but which, with the exception of 3 N. African or Levant species and the 2 Australian ones, is confined to S. Africa.  One of the Australian species appears identical with a S. African one, and extends to New Zealand; the other, whether species or variety, is endemic.

Leafy stems usually elongated.  Peduncles rarely twice as long as the
  leaves.  Petals from a little longer to half as long again as the
  calyx . . . . . . . . . . . . . . . . . . . . . . . 1. *P. australe*.
Stems short and erect.  Peduncles much longer, erect.  Petals fully
  twice as long as the calyx . . . . . . . . . . . . . 2. *P. Rodneyanum*.

1. **P. australe,** *Willd.; DC. Prod.* i. 654.  Herbaceous, often flowering the first year, but forming a perennial rootstock, either horizontal and

almost creeping, or short and thick.    Leafy stems decumbent or erect, sometimes short, but usually attaining 1 ft. or more, generally pubescent or hirsute with spreading hairs.    Leaves reniform-cordate, or very rarely broadly ovate-cordate, crenate, or very shortly lobed, very obtuse, rarely 2 in. diameter, and usually much smaller, softly pubescent or hirsute.    Stipules broad.    Peduncles usually longer than the leaves, but not so long as in *P. Rodneyanum*, and sometimes very short.    Flowers small, in an umbel, sometimes very dense, almost reduced to a head, sometimes loose with pedicels of ½ in. or more. Sepals acute, 2 to 3 lines long, usually very hairy, the decurrent tube rarely so long, and sometimes very short.    Petals from a little longer than the sepals to about half as long again.    Capsule-lobes pubescent, the beak from ½ to ¾ in. long, the awns of the lobes bearded inside as in *Erodium*.    Seeds smooth. —Sweet, Geran. t. 68 ; Hook. f. Fl. Tasm. i. 57 ; F. Muell. Pl. Vict. i. 170 ; *P. glomeratum*, Jacq.; DC. Prod. i. 659 ; *P. inodorum*, Willd. ; DC. l. c. ; Sweet, Geran. t. 56 ; *P. littorale*, Hueg. Bot. Arch. t. 5 ; *P. crinitum*, Nees, in Pl. Preiss. i. 163 ; *P. stenanthum*, Turcz. in Bull. Mosc. 1858, i. 149 ; *P. Drummondi*, Turcz. l. c. 421 (a robust form with large flowers).

**N. S. Wales.**   Port Jackson, *R. Brown, Sieber, n.* 252; northward to Clarence river, *Beckler*, and New England, and inland to the Blue Mountains and Lachlan river, *A. Cunningham* and others.

**Victoria.**   Frequent on sandy shores, desert land, river banks, mountains, etc., *F. Mueller.*

**Tasmania,** *R. Brown.*   Abundant in many parts of the colony, especially near the sea, *J. D. Hooker.*

**S. Australia.**   Chiefly near the sea, *F. Mueller* and others.

**W. Australia.**   King George's Sound, *R. Brown ;* and thence to Swan River, *Drummond, 1st Coll., Coll.* 5, *n.* 191, 192, 193, etc. ; *Preiss, n.* 1905, 1906 ; *Oldfield*, and others.

Var. *erodioides ;* small and slender, pubescent, the leaves not above ½ in. diameter, and flowers small, the sepals varying from 1 to 2 lines.    *P. erodioides*, Hook. Journ. Bot. i. 252 ; *P. acugnaticum*, Thou. ; Hook. f. Fl. Tasm. i. 58.—Tasmania, and occasionally in Victoria, and especially in N. S. Wales, where is also a more robust form, but with flowers at least as small.    This is the *P. clandestinum*, L'Hér.; Hook. f. Fl. N. Zeal. i. 41, and is the most common form in New Zealand.    *P. acugnaticum*, Thou., from Tristan d'Acunha, is also a form of the same species scarcely to be distinguished from the var. *erodioides*, and all these collectively cannot be separated from the S. African var. *anceps* of *P. grossularioides*, Ait., or *P. anceps*, Ait.    But although the Australian *P. australe* and S. African *P. grossularioides* thus coincide in this particular form, the more common varieties are in each case endemic, the ordinary *P. australe* described above never occurring in S. Africa, where the most common form is one with deeply-cut leaves, which is never to be met with in Australia. See Harv. and Sond. Fl. Cap. i. 289.

2. **P. Rodneyanum,** *Lindl. in Mitch. Three Exped.* ii. 144.    A perennial, forming a thick rootstock and a very short erect stem, hirsute with spreading hairs.    Leaves chiefly radical, petiolate, from broadly orbicular-cordate to ovate, obtuse, 1 to 2 in. long, crenulate and sometimes shortly lobed, glabrous or minutely hoary-tomentose.    Peduncles erect, 4 to 8 in. long, bearing an umbel of 4 to 8 showy reddish-purple flowers.    Sepals about 3 lines long, rather obtuse; the adnate calyx-tube usually longer, but sometimes rather shorter than the sepals. Petals fully twice as long as the calyx, the two upper ones larger than the others.    Fruit not seen.—F. Muell. Pl. Vict. i. 171, t. suppl. 11.

**Victoria.** Near the Murray river, *Mitchell*; Forest Creek and towards Mount Alexander, *F. Mueller*; in the Grampians, *Wilhelmi*.

**S. Australia.** Lynedoch Valley, *Behr*; near Skipton, *Whan*.

**W. Australia.** In the interior from Swan River, *Drummond, Roe.* In these specimens the leaves are more decidedly cordate, almost reniform.

The species much resembles in habit and flowers, and in the shape of the foliage *P. reniforme*, Curt., from South Africa, but wants the dense whitish velvety tomentum of the under side of the leaves of that species. F. Mueller thinks it a variety only of *P. australe*, but of that we have not as yet sufficient evidence to justify the union.

## 4. OXALIS, Linn.

Flowers regular. Sepals 5. Petals 5. Disk without glands. Stamens 10, free or united at the base, all bearing anthers. Ovary 5-lobed, 5-celled, without any beak or with a very short one; styles 5, with terminal stigmas, capitate or lobed; ovules 1, 2, or several in each cell. Capsule opening loculicidally, the valves persistent on the axis. Seeds with an outer fleshy coating, opening elastically, with the appearance of an arillus; testa crustaceous; albumen fleshy; embryo straight.—Herbs. Leaves alternate or radical, compound; leaflets 3, digitate, or, in species not Australian, 3 or more and pinnate. Stipules scale-like or none. Peduncles axillary or radical, 1-flowered or bearing an umbel of several flowers.

A large genus, especially abundant in South America and extratropical South Africa, with a very few species widely dispersed over the temperate or tropical regions of the globe. Of the two Australian species, one is common to New Zealand and Antarctic America, and perhaps not different from a common northern one, the other is a widely-spread weed in various parts of the world.

Flowers white. Peduncles radical, 1-flowered . . . . . . . . 1. *O. magellanica.*
Flowers small, yellow. Stem elongated. Peduncles axillary, 1- or more-flowered . . . . . . . . . . . . . . . . 2. *O. corniculata.*

1. **O. magellanica,** *Forst.; DC. Prod.* i. 700. Rootstock shortly creeping, slender, but often knotted with thickened scale-like persistent stipules. Leaves radical, sprinkled with a few hairs; leaflets 3, obcordate, of a delicate green, on a long common petiole. Peduncles radical, long and slender, bearing a single rather large white flower, with a pair of narrow bracts above the middle. Sepals small, ovate, obtuse, thin. Petals obovate, 4 to 6 lines long. Capsule ovoid, with 1 or 2 shining black seeds in each cell. —Hook. f. Fl. Tasm. i. 59; Fl. N. Zeal. i. 42, t. 13; F. Muell. Pl. Vict. i. 176; *O. lactea*, Hook. Comp. Bot. Mag. i. 276, and Journ. Bot. ii. 416; *O. cataractæ*, A. Cunn.; Hook. Ic. Pl. t. 418.

**Victoria.** Humid subalpine forests and alpine streams in the western parts of Gipps' Land, at an elevation of 2500 to 5000 ft., *F. Mueller.*

**Tasmania.** Mountain woods and streams in various parts of the colony at an elevation of 1500 to 3000 ft., *J. D. Hooker.*

The species is also in New Zealand and in Fuegia and S. Chili. Some of the Victorian specimens can also scarcely be distinguished from the *O. Acetosella*, Linn., a widely-spread species in the temperate or mountainous regions of the northern hemisphere. The stipules are rather larger, the bracts longer and narrower, and the leaflets more deeply notched, the minute glandular appendage in the notch being often more or less visible in the northern plant. The Tasmanian form, like the New Zealand and S. American ones, is smaller and more stunted.

2. **O. corniculata,** *Linn. ; DC. Prod.* i. 692. A decumbent, prostrate or ascending, much-branched, delicate perennial or sometimes annual, more or less pubescent, of a pale green, from a few inches to a foot long. Stipular scales small, adnate to the petiole. Leaves alternate; leaflets 3, broadly obcordate, usually 3 or 4 lines long, but sometimes half that size. Peduncles axillary, about the length of the petioles, bearing an umbel of several small yellow flowers, rarely reduced to 1 or 2, on reflexed pedicels. Capsule column-like, often above ¼ in. long, with several seeds in each cell, rarely short and few-seeded.—Reichb. Ic. Fl. Germ. v. t. 199; Wight, Ic. t. 18; Hook f. Fl. Tasm. i. 59; F. Muell. Pl. Vict. i. 177; *O. microphylla,* Poir.; DC. Prod. i. 692; *O. perennans,* Haw.; DC. l. c. 691 (from the character given); *O. Preissiana* and *O. cognata,* Steud. in Pl. Preiss. i. 160.

**Queensland.** Islands of the coast as well as on the mainland, Keppel Bay, *R. Brown ;* Percy Island, *A. Cunningham* and others; and in the interior as far north as the Burdekin, *F. Mueller, Mitchell,* etc.

**N. S. Wales.** Port Jackson, and northward to Clarence and Hastings rivers, *Beckler ;* southward to Twofold Bay, *F. Mueller ;* and in the interior.

**Victoria.** Common throughout the colony, except the alpine tracts, *F. Mueller.*

**Tasmania.** Common in pastures, waste places, etc., throughout the island, *J. D. Hooker.*

**S. Australia.** Extending over the colony inland to Lofty Range, *F. Mueller* and others.

**W. Australia.** From the S. coast to Swan River, *Drummond, Preiss, n.* 1915, 1916, and others; and to Murchison river, *Oldfield.*

## ORDER XXVIII. RUTACEÆ.

Flowers regular and hermaphrodite, or very rarely unisexual. Calyx usually small, 4- or 5-lobed, or divided into as many distinct imbricate sepals, rarely large, or with fewer or more numerous or valvate lobes. Petals of the same number as sepals, free or rarely cohering, hypogynous or slightly perigynous, imbricate or valvate in the bud. Stamens usually free, either equal in number to the petals and alternate with them, or double the number, or rarely more numerous, when twice as many as petals the sepaline ones (those opposite the sepals) usually longer than the others. Anthers usually versatile, with 2 parallel cells opening longitudinally, the connective occasionally tipped by a gland or projecting appendage. Torus usually more or less thickened into an entire crenate or lobed disk, within the stamens, under or round the ovary. Gynœcium of 4 or 5, rarely more or fewer carpels, more or less united into a single lobed or entire ovary, or rarely quite distinct, with one cell to each carpel. Styles as many as carpels, either free at the base but united upwards, or united from the base; stigma terminal, entire or lobed. Ovules usually 2 in each cell, superposed or rarely collateral or solitary, or more than 2; the micropyle superior. Fruit separating into 2-valved or rarely indehiscent cocci, or the carpels united in an indehiscent berry or drupe, or rarely in a loculicidally dehiscent capsule, the endocarp frequently separating from the pericarp. Seeds usually solitary in each cell; testa crustaceous and often shining, or rarely coriaceous or membranaceous; albumen fleshy or none. Embryo straight or curved, large in proportion to the seed; cotyledons flat or rarely folded; radicle superior.—Trees or shrubs, very rarely herbs, marked with glandular pellucid dots on the leaves and other thin herbaceous parts. Indumentum usually stellate, if any. Leaves opposite or alternate, simple or compound, entire or rarely toothed or lobed. Stipules none. Flowers axillary or terminal, solitary, clustered, cymose, or paniculate, very rarely racemose and seldom if ever spicate.

A large Order, ranging over the hotter and temperate regions of the whole world, but chiefly abundant within the tropics, in South Africa and in Australia. Among the Australian genera, the large tribe of *Boronieæ* is entirely endemic, with the exception of one New Zealand and one New Caledonian species. The monotypic genera, *Bosistoa*, *Medicosma*, and *Pentaceras*, and the small genus *Geijera*, are also endemic. *Melicope* extends to the Pacific islands, and the remaining genera range over tropical Asia, three of them extending into Africa. *Zanthoxylum* alone, a wide-spread tropical genus, is common to America and Australia, and even here the Australian species belong to the exclusively Australasian section *Blackburnia*.

Difficult as it is to distinguish *Rutaceæ* by well-marked floral or carpological characters from *Geraniaceæ*, *Zygophylleæ*, or *Simarubeæ*, they are so readily known by their dotted exstipulate leaves, that the ambiguous genera are remarkably few. They have usually been distributed into 3 or 4 Orders, *Rutaceæ* (including or not *Diosmeæ*), *Zanthoxyleæ*, and *Aurantieæ*, upon characters which break down upon a close scrutiny; the *Toddalieæ* being much nearer to the *Aurantieæ* than to the *Zanthoxyleæ* proper, which again have only vague differences to distinguish them from *Boroniæ*. We therefore, in our ' Genera Plantarum,' proposed the union of the whole into 1 Order, divided into 2 series, according as the ovary is lobed or entire, and subdivided into 7 tribes, of which 4 only are Australian.

TRIBE I. **Boroniæ.**—*Shrubs, very rarely arborescent. Leaves simple, 3-foliolate or rarely pinnate, with opposite small leaflets. Ovary lobed. Fruit separating into distinct, 2-valved cocci. Endocarp separating elastically. Seeds albuminous. Embryo usually terete.*

Leaves opposite (except in one *Zieria*) simple or compound.
  Petals 4, united or connivent in a cylindrical or campanulate
    corolla. Leaves petiolate, simple . . . . . . . . 12. CORREA.
  Petals 4, free, spreading.
    Stamens 4, inserted on 4 prominent glands or lobes of the disk  1. ZIERIA.
    Stamens 8. Disk without prominent glands (excepting *B. tetrandra*) . . . . . . . . . . . . . . . . 2. BORONIA.
  Petals 5, rarely more, free, spreading . . . . . . . . 3. ACRADENIA.
Leaves alternate, simple.
  Flowers in dense pedunculate reflexed heads. Stamens much exserted.
    Bracts subulate. Sepals 5. Petals narrow. Leaves lobed  . 14. CHORILÆNA.
    Bracts ovate or lanceolate, numerous and imbricate. Sepals 0.
      Petals very narrow. Leaves entire . . . . . . . 15. DIPLOLÆNA.
      (See also *Phebalium Ralstoni*.)
  Flowers distinct or in sessile, erect heads.
    Petals united or conniveut in a tubular corolla . . . . . . 13. NEMATOLEPIS.
    Petals free. Stamens twice as many, monadelphous.
      Stamens all perfect . . . . . . . . . . . . 9. PHILOTHECA.
      Stamens 5 perfect, 5 without anthers . . . . . . . 10. DRUMMONDITA.
    Petals free. Stamens twice as many, free.
      Calyx inconspicuous or none. Petals induplicate-valvate,
        tomentose outside . . . . . . . . . . . . 11. ASTEROLASIA.
      Calyx distinct but shorter than the petals.
        Petals broad, much imbricate, not scurfy, without inflexed
          tips. Filaments hairy.
          Anthers minutely or not at all apiculate . . . . . 5. ERIOSTEMON.
          Anthers tipped with long, horn-like, hairy appendages . 4. CROWEA.

Petals valvate or slightly imbricate, with inflexed valvate
tips, glabrous or scaly.
Ovary of 5, rarely fewer carpels, the styles attached below
the middle. . . . . . . . . . . . . . . 6. PHEBALIUM.
Ovary of 2 carpels, the style attached above the middle.
Flowers small, in sessile, terminal heads . . . . 7. MICROCYBE.
Calyx of coloured petal-like sepals longer than the petals . 8. GELEZNOWIA.
Petals free. Stamens of the same number, free . . . . 21. GEIJERA.

TRIBE II. **Zanthoxyleæ.**—*Trees or shrubs. Leaves pinnate or 3-foliolate with opposite leaflets, or 1-foliolate (truly simple in Geijera), the leaflets usually large. Ovary
lobed. Fruit separating into distinct 2-valved cocci. Endocarp persistent, or separating elastically. Seeds with or without albumen. Cotyledons usually flattened and
broader than the radicle.*

Stamens twice as many as petals.
Leaves all or mostly opposite. Cocci dehiscent.
Leaves pinnate. Petals valvate or slightly imbricate. Seeds
without albumen . . . . . . . . . . . . . 16. BOSISTOA.
Leaves 3-foliolate. Petals valvate or slightly imbricate, with
inflexed lips . . . . . . . . . . . . . . 17. MELICOPE.
Leaves 1-foliolate. Petals large, broadly imbricate, not inflexed 19. MEDICOSMA.
Leaves alternate, pinnate. Petals valvate. Cocci winged, inde-
hiscent . . . . . . . . . . . . . . . . 22. PENTACEROS.
Stamens the same number as petals. Cocci dehiscent.
Leaves all or mostly opposite, usually 3-foliolate . . . . . . 18. EVODIA.
Leaves alternate, simple . . . . . . . . . . . . 21. GEIJERA.
Leaves alternate, pinnate . . . . . . . . . . . . 20. ZANTHOXYLUM.
(See also *Flindersia* among *Meliaceæ*.)

TRIBE III. **Toddalieæ.**—*Trees or shrubs, with the habit of Zanthoxyleæ. Ovary not
lobed. Fruit several-celled, indehiscent, or rarely loculicidally dehiscent. Seeds albuminous* (in the Australian genus).
Leaves 1-foliolate. Stamens twice as many as petals . . . . 23. ACRONYCHIA.

TRIBE IV. **Aurantieæ.**—*Trees or shrubs. Leaves pinnate, with usually alternate
leaflets, or 1-foliolate or simple. Stamens twice as many as petals or more. Ovary
not lobed. Fruit indehiscent. Seeds without albumen.*

Leaves all or mostly pinnate. No thorns.
Flowers in terminal, flat, corymbose panicles. Filaments subulate.
Petals valvate or nearly so. Cotyledons much folded. Flowers
small . . . . . . . . . . . . . . . . 25. MICROMELUM.
Petals imbricate, erect. Cotyledons flat. Flowers large . . 26. MURRAYA.
Flowers in oblong, pyramidal, or loose axillary or terminal pani-
cles. Filaments dilated at the base or middle.
Ovules solitary. Leaflets few . . . . . . . . . 24. GLYCOSMIS.
Ovules 2 in each cell. Leaflets numerous . . . . . . . 27. CLAUSENA.
Leaves all simple or 1-foliolate, coriaceous. Thorns axillary.
Ovary 5- or fewer celled, with 1 or 2 ovules in each cell . . . 28. ATALANTIA.
Ovary 6- or more celled, with 4 or more ovules in each cell . . 29. CITRUS.

TRIBE I. BORONIEÆ.—Shrubs, very rarely arborescent. Leaves simple,
3-foliolate or rarely pinnate, with opposite small leaflets. Ovary lobed. Fruit
separating into distinct 2-valved cocci. Endocarp separating elastically.
Seeds albuminous. Embryo usually terete.—The tribe differs from the S.
African *Diosmeæ* chiefly in the presence of albumen.

## 1. ZIERIA, Sm.

Calyx 4-cleft. Petals 4, imbricate or almost valvate in the bud, spreading.

Disk with 4 distinct gland-like lobes, alternating with the petals. Stamens 4, inserted on the outside of the glands of the disk. Carpels 4, distinct or nearly so; styles nearly terminal, short and united at least at the top; stigma capitate, 4-furrowed or shortly 4-lobed. Ovules 2 in each carpel, superposed. Cocci 4, 2-valved, the endocarp cartilaginous and separating elastically. Seeds solitary, or rarely 2 in each coccus, oblong; testa crustaceous.—Shrubs or rarely small trees, glabrous hirsute or tomentose. Leaves usually opposite, with 3 leaflets, rarely alternate or simple. · Flowers white, usually small, axillary, in small trichotomous cymes or rarely solitary.

The species are all endemic in Australia, and F. Mueller considers them as forming a section only of *Boronia ;* but the characters and habit appear to me sufficiently distinct to justify the maintenance of so old-established and generally adopted a genus.

Anthers distinctly apiculate. Plant glabrous or slightly pubescent.
   Leaflets with revolute margins. Cymes pedunculate.
     Branchlets angular, glabrous. Leaflets ½ to 1 in. on a distinct
      common petiole . . . . . . . . . . . . . . . . . 1. *Z. lævigata.*
     Branchlets terete, pubescent. Leaflets under ½ in., sessile, appearing verticillate . . . . . . . . . . . . . . . 2. *Z. aspalathoides.*
Anthers minutely apiculate, Plant pubescent or hirsute, rarely tomentose. Flowers 1 to 3, small. Calyx-segments very narrow, nearly as long as the petals . . . . . . . . . . 3. *Z. pilosa.*
Anthers not apiculate. Calyx-lobes short.
   Flowers 1 to 3, on short axillary pedicels. Leaves densely pubescent or tomentose.
     Leaflets 3, small, obovate or obcordate. Flowers very small . 4. *Z. obcordata.*
     Leaves simple, ovate or oblong . . . . . . . . . . 5. *Z. veronicea.*
   Flowers in pedunculate cymes or heads, with leafy bracts. Leaves densely tomentose or villous.
     Upper leaves simple. Cymes contracted into dense heads, with imbricate bracts . . . . . . . . . . . . . . 6. *Z. involucrata.*
     Leaves all 3-foliolate. Cymes not capitate . . . . . . 7. *Z. cytisoides.*
   Flowers in loose pedunculate cymes, with small bracts.
     Densely tomentose or velvety. Leaflets flat, lanceolate. Petals almost valvate . . . . . . . . . . . . . . . 8. *Z. furfuracea.*
     Glabrous or slightly pubescent.
      Leaflets flat, lanceolate. Petals distinctly imbricate . . . 9. *Z. Smithii.*
      Leaflets narrow-linear. Flowers small, the petals almost valvate . . . . . . . . . . . . . . . . . 10. *Z. granulata.*

1. **Z. lævigata,** *Sm.; DC. Prod.* i. 723. A glabrous, erect shrub, the branchlets angular. Leaflets 3, on a common petiole of 1 to 3 lines, linear, pointed, ½ to 1 in. long, the margins closely revolute. Cymes few-flowered, mostly about as long as the leaves. Calyx-lobes short and broad. Petals fully 3 times as long as the calyx, broad, imbricate, slightly tomentose outside. Connective of the anthers distinct, produced beyond the cells into a short point or appendage. Style very short. Cocci and seeds of *Z. Smithii.* —Deless. Ic. Sel. iii. t. 49 ; Paxt. Mag. Bot. ix. 77, with a fig. ; *Boronia lævigata,* F. Muell. Fragm. i. 101 ; *Z. revoluta,* A. Cunn. in Field, N. S. Wales, 330.

**Queensland.** Sandstone rocks near Mount Pluto, *Mitchell.*

**N. S. Wales.** Port Jackson, *R. Brown* and others ; Blue Mountains, *A. Cunningham ;* Mount Lindsay, *Fraser.*

Var. *laxiflora.* Leaflets longer (1 to 1½ in.), on a longer common petiole. Flowers much smaller, in a looser cyme. Petals not twice as long as the calyx.—Stradbrooke Island, *Fraser;* Moreton Island. *F. Mueller.*

2. **Z. aspalathoides,** *A. Cunn. Herb.* A heath-like shrub, the branches terete and pubescent, but usually with a decurrent glabrous line. Leaflets 3, sessile or with the common petiole so exceedingly short that they appear verticillate, lanceolate or linear, rarely above 3 lines long, or when very luxuriant 4 or 5 lines, the margins revolute, glabrous or slightly pubescent. Cymes usually 3-flowered, rather longer than the leaves. Calyx-lobes broad, obtuse or acute. Petals about 2 or 3 times as long. Anthers tipped with a small obtuse appendage.—*Boronia lævigata,* F. Muell. Pl. Vict. i. 111 (in part).

**N. S. Wales.** Wellington Valley, Blue Mountains, and W. branches of Hunter's River, *A. Cunningham;* Peele's ranges, *Fraser.*

**Victoria.** Grampians, *A. Cunningham;* barren ridges near Goulburn river, *F. Mueller.*

3. **Z. pilosa,** *Rudge, in Trans. Linn. Soc.* x. 293, *t.* 17. A shrub or undershrub, the branches terete and densely pubescent or hirsute. Leaflets 3, with a short common petiole, linear, oblong or lanceolate, obtuse, ½ to ¾ in. or rarely 1 in. long, the margins recurved or revolute, slightly pubescent or glabrous above, more or less hirsute or tomentose underneath. Flowers small, solitary and nearly sessile or 2 or 3 together on short pedicels. Calyx hirsute, with linear-subulate or narrow-lanceolate lobes, nearly as long as the petals and always much narrower than in any other species. Anthers minutely apiculate. Cocci hirsute, broader than in most species.—DC. Prod. i. 723; *Z. pauciflora,* Sm. in Rees, Cycl. xxxix.; DC. l. c.; *Z. hirsuta,* DC. l. c.; Deless. Ic. Sel. iii. t. 50; *Boronia hirsuta,* F. Muell. Fragm. i. 101.

**N. S. Wales.** Port Jackson and Botany Bay, *Banks, R. Brown, Sieber, n.* 283 (partly mixed with *Boronia polygalifolia,* var. *triphylla*), and many others.

Var. *parviflora.* Less pubescent; leaves smaller; flowers and fruit much smaller. Both in *Banks'* and in *R. Brown's* collections.

Var. (?) *canescens.* More tomentose-hirsute; leaves narrow, very tomentose underneath, the margins scarcely recurved; inflorescence looser, the peduncles rather lengthened and 3-flowered, but with the calyx of *Z. pilosa.*—*Z. canescens,* R. Br. Herb.—Hills in the interior, *Caley.*

*Z. microphylla,* Bonpl. Jard. Malm. 64, DC. Prod. i. 723, only known by an exceedingly short diagnosis, is probably this species. I did not find it in the Paris herbarium. *Z. trifoliata,* Bonpl., mentioned in gardening works, is probably this or one other of the common species met with in gardens.

4. **Z. obcordata,** *A. Cunn. in Field, N. S. Wales,* 330. A shrub of low growth, with elongated diffuse branches, terete and softly hirsute. Leaflets 3, with a very short common petiole, obovate or obcordate, 2 to 4 lines or rarely ½ in. long, softly pubescent or tomentose above, more hirsute or velvety and whitish underneath, the margins recurved or revolute. Flowers 1 to 3 in the axils, very small, on short slender pedicels, the petals not above 1 line and the calyx about half as long with broad and obtuse segments. Anthers not apiculate. Cocci small, glabrous.—*Boronia minutiflora,* F. Muell. Fragm. i. 100.

**Queensland.** Glasshouse Mountains, *F. Mueller.*
**N. S. Wales.** Macquarie river, *A. Cunningham.*

5. **Z. veronicea,** *F. Muell. Trans. Phil. Soc. Vict.* i. 11. A low shrub, clothed all over with a soft close or velvety tomentum. Leaves all simple,

opposite or alternate, sessile or nearly so, ovate or oblong, obtuse, mostly 3 to 4 lines and rarely ½ in. long, the margins revolute. Flowers solitary or 2 or 3 together, on short pedicels. Bracts small but leafy. Calyx tomentose. Petals about twice as long, tomentose outside, much imbricate. Filaments hairy. Anthers obtuse or obscurely apiculate. Ovary and style stellate-pubescent. Cocci tomentose.—*Boronia veronicea*, F. Muell. Pl. Vict. i. 228.

**Victoria.** Sandy Mallee scrub along the lower Wimmera, *Dallachy.*
**S. Australia.** Encounter Bay and Kangaroo Island, *F. Mueller* and others.

6. **Z. involucrata,** *R. Br. Herb.* The whole plant densely and softly tomentose-hirsute. Lower leaves simple, oblong, obtuse, 1 to 1½ in. long, flat; upper ones 3-foliolate with a short common petiole; leaflets similar to the simple leaves or smaller. Flowers several together, sessile, in dense heads on axillary peduncles. Bracts ovate, leafy, softly villous, nearly as long as the flowers and imbricate with them. Sepals ovate-lanceolate, acute, more than half as long as the petals. Anthers not apiculate.

**N. S. Wales.** Valleys of the Blue Mountains, *Backhouse.*

7. **Z. cytisoides,** *Sm.; DC. Prod.* i. 723. A much-branched shrub, hoary all over with a soft close or more or less velvety tomentum. Leaflets 3, with a common petiole of 1 to 3 lines, obovate-oblong, about ½ or rarely ¾ in. long, obtuse or minutely pointed, the margins revolute, narrowed at the base. Cymes dense but few-flowered, rarely much exceeding the leaves. Bracts leafy, as long as the pedicels or often nearly as long as the flowers. Calyx rather short, with broad acute segments. Petals rarely twice as long, much imbricate in the bud. Anthers not apiculate.

**N. S. Wales.** In the mountains, *Caley;* high granitic ranges near Bathurst, *Fraser, A. Cunningham;* Twofold Bay, *Huegel, F. Mueller;* Castle Creek, *Leichhardt.*

8. **Z. furfuracea,** *R. Br. Herb.* A tall shrub, so nearly resembling some forms of *Z. Smithii* in the shape and size of the leaves and in inflorescence that F. Mueller suggests it may be only a remarkable variety. Whole plant densely clothed with a soft velvety stellate tomentum, the tubercular glands also tomentose and often projecting on the branches and under side of the leaves, and the dots quite opaque or rarely pellucid. Leaflets lanceolate, flat. Flowers numerous in the cymes, much smaller than in *Z. Smithii,* and the petals less imbricate or almost valvate. Cocci hairy.

**N. S. Wales.** N.W. interior, *Fraser;* Hastings river, *Beckler.*

9. **Z. Smithii,** *Andr. Bot. Rep. t.* 606 (1810). A tall shrub or small tree, glabrous or slightly pubescent with a very minute usually stellate down, the branches terete or compressed, occasionally covered with glandular tubercles. Leaflets 3, with a distinct common petiole, lanceolate or the larger ones oblong, elliptical, acute or rarely obtuse, 1 to 2 in. long in the original form, flat or the margins slightly recurved. Flowers usually about 3 lines diameter, in axillary 2–3-chotomous cymes, shorter than the leaves. Calyx-lobes broad and short. Petals fully 3 times as long as the calyx, tomentose outside. Anthers obtuse, not apiculate. Cocci about 2 lines long, glabrous, usually glandular-tuberculate. Seeds shining, finely reticulate-striate.—Bot. Mag. t. 1395; Bonpl. Jard. Malm. 62, t. 24; *Z. lanceolata,* R. Br.; DC.

Prod. i. 723 ; Hook. f. Fl. Tasm. i. 65 ; *Boronia arborescens,* F. Muell. Fragm. i. 100, and Pl. Vict. i. 111.

**Queensland.** Brisbane river, *A. Cunningham ;* Stradbrooke Island, *Fraser.*

**N. S. Wales.** Port Jackson, *R. Brown, Sieber, n.* 280, and others; Blue Mountains, *Fraser* and others; northward to Hastings river, *A. Cunningham* and others; and Mount Lindsay, *W. Hill ;* southward to Twofold Bay, *F. Mueller.*

**Victoria.** From the Grampians and Cape Otway ranges eastward, along humid forest valleys, ascending to high mountain ravines, *F. Mueller.*

**Tasmania.** Port Dalrymple and King's Island, *R. Brown ;* common in rich soil throughout the island, *J. D. Hooker.*

Var. *parvifolia.* Leaflets rarely exceeding 1 in.; cymes often as long.—Sandy Bay and Cape Hervey, *R. Brown ;* New England, *Stuart.*

Var. *macrophylla.* More arborescent ; leaflets often 3 in. long; flowers larger than in the ordinary form ; seeds broader and less reticulate.—*Z. arborescens,* Sims ; Hook. Journ. Bot. i. 256 ; *Z. macrophylla,* Bonpl. ; Deless. Ic. Sel. iii. t. 48 ; Bot. Mag. t. 4451. To this variety belong the Tasmanian and many of the Victorian specimens.

The stamens in this and other *Zierias* are figured in Delessert's ' Icones,' by some mistake, as attached inside instead of outside the glands or lobes of the disk. The name of *Z. lanceolata* was adopted by Smith (in Rees' Cycl. xxxix.), on the consideration that the synonym quoted in the Bot. Mag. was a sufficient publication ; Andrews' name had, however, been published a year previous to the plate in Bot. Mag.

10. **Z. granulata,** *C. Moore, in Herb. Hook.* A tall shrub or small tree, glabrous or very minutely pubescent, and densely covered with glandular tubercles as in some varieties of *Z. Smithii,* with which F. Mueller proposes to unite it. It differs chiefly in the narrow-linear leaflets, 1 to 2 in. long, the margins revolute and whitish underneath, and in the very small flowers, with the petals almost strictly valvate. Cocci glabrous.—*Boronia granulata,* F. Muell. Fragm. i. 101.

**N. S. Wales.** Near Goulburn, *C. Moore,* woods of Paris Exhibition, n. 204 ; Kiama, *Harvey.*

## 2. BORONIA, Sm.

Calyx 4-cleft. Petals 4, either much imbricate or valvate in the bud, spreading. Disk thick, entire or (in one species only) with 4 gland-like lobes. Stamens 8, inserted outside the disk ; anthers either all similar and perfect or 4 different from the others and imperfect. Carpels of the ovary 4, distinct or nearly so ; styles terminal, united ; stigma entire or 4-lobed. Ovules 2 in each carpel, superposed or rarely collateral. Cocci usually 4, 2-valved, the endocarp cartilaginous and separating elastically. Seeds solitary or rarely 2 in each coccus, oblong ; testa crustaceous.—Shrubs, undershrubs, or rarely annuals, glabrous pubescent or hirsute, rarely tomentose. Leaves opposite, simple, pinnate with a terminal leaflet, or once or twice ternately compound, the rhachis usually articulate at each pair of leaflets and often dilated between them. Peduncles axillary or terminal, either 1-flowered and jointed with a pair of minute bracts at the joint, or bearing an umbel or dichotomous cyme of several flowers with small bracts at the base of the pedicels. Flowers red, white, purple, or blue. Calyx-segments or sepals usually valvate when the petals are valvate and sometimes also when they are imbricate, but in the latter case the sepals are usually also imbricate at the base. In some species the anthers and stigma are different in different individuals of the same variety. In most of the species the filaments of the sepaline

x 2

stamens (those alternating with the petals) are longer and more distinctly clavate or capitate and glandular at the top than the petaline ones. Anthers usually very shortly stipitate, rather below the obtuse summit of the filament. The species are all limited to Australia.

SERIES I. **Valvatæ.**—*Petals strictly valvate. Sepals usually valvate.*

Sepals as long as or longer than the petals, enclosing them in the bud. (Plants tomentose or pubescent.)

Sepals longer than the petals.

Leaves all simple. Sepals 5 to 6 lines . . . . . . . . 1. *B. grandisepala.*

Leaves mostly or all pinnate. Sepals 3 to 4 lines . . . . 2. *B. artemisiæfolia.*

Sepals (about 2 lines) of the size of the petals. Leaves pinnate.

Leaflets small, ovate, numerous. Pedicels slender . . . . 4. *B. filicifolia.*

Leaflets linear. Pedicels very short . . . . . . . . 3. *B. affinis.*

Sepals much smaller than the petals.

Inflorescence entirely or mostly terminal.

Cymes terminal, leafy. Leaves pinnate. Flowers large . . 5. *B. alata.*

Flowers small, 1 to 3 together in the forks of spreading dichotomous stems. Common petiole very short.

Leaflets usually 5, obovate, about 2 lines, thick, glabrous and green on both sides. Flowers almost sessile . . . 6. *B. algida.*

Leaflets 3, obovate-oblong, about 3 lines, pale underneath. Pedicels slender . . . . . . . . . . . . . 7. *B. Edwardsii.*

Inflorescence entirely axillary.

Peduncles 1-flowered.

Leaflets 3, sessile.

Leaflets small, obovate, coriaceous, flat.

Leaves glabrous. Peduncles as long as the leaves . . 8. *B. calophylla.*

Leaves tomentose. Flowers almost sessile . . . . 9. *B. ternata.*

Leaflets linear, revolute at the margin. Flowers almost sessile . . . . . . . . . . . . . . . . 10. *B. ericifolia.*

Leaflets 3 or more, with a distinct common petiole.

Leaflets (about 5) linear, thick, but flat. Flowers glabrous, minute . . . . . . . . . . . . . . 11. *B. inconspicua.*

Leaflets (usually 5 or 7) obovate or cuneate, glabrous, complicate. Flowers tomentose, rather large . . . 12. *B. eriantha.*

Leaflets 7 to 13 or more, small, linear or oblong, the margins revolute. Sepals lanceolate, subulate-acuminate . . . . . . . . . . . . . . . . . . 13. *B. alulata.*

Leaflets 3, rarely 5, the margins recurved or revolute, tomentose or hoary underneath . . . . . . . 14. *B. ledifolia.*

Leaves simple.

Leaves linear or linear-lanceolate. Flowers about 4 lines 14. *B. ledifolia.*

Leaves oblong-lanceolate. Flowers about 2 lines . . . 15. *B. lanceolata.*

Peduncles bearing an umbel of several flowers.

Leaves simple, lanceolate, tomentose underneath. Flowers small . . . . . . . . . . . . . . . . . . 15. *B. lanceolata.*

Leaves mostly pinnate, with few distant leaflets. Flowers 3 to 6 lines.

Glabrous or slightly hoary . . . . . . . . . 16. *B. Fraseri.*

Softly hirsute or tomentose . . . . . . . . 17. *B. mollis.*

SERIES II. **Heterandræ.**—*Petals imbricate. Sepaline anthers different from the others, and often imperfect. Stigma usually thick and fleshy. Leaves mostly pinnate. Leaflets linear. Peduncles axillary, 1-flowered.*

Sepaline anthers large, black, or purple.

Glabrous. Leaflets 1 to 3, nearly sessile, heath-like. Petals dark purple outside, yellowish inside . . . . . . . . . 18. *B. megastigma.*

Glabrous. Leaflets single, long and linear, or 3 with a long petiole.
Flowers pink . . . . . . . . . . . . . . . . . . . 19. *B. heterophylla.*
Branches hirsute. Leaflets several, in distant pairs. Flowers
pink . . . . . . . . . . . . . . . . . . . 20. *B. elatior.*
Sepaline anthers very small.
Branches hirsute. Leaflets several, in rather distant pairs. Se-
paline filaments long and inflected . . . . . . . . . . 21. *B. tetrandra.*
Glabrous or slightly pubescent. Leaflets crowded on a short
petiole. Sepaline filaments very short . . . . . . . 22. *B. crassifolia.*

SERIES III. **Pinnatæ.**—*Petals imbricate. Anthers nearly uniform. Leaves pinnate. Peduncles mostly axillary.*

Peduncles all 1-flowered. (Western species.)
Low or diffuse undershrubs or shrubs. Leaflets linear-cuneate,
obtuse, crowded on a short petiole.
Branches hirsute. Flowers nearly sessile. Stigma conical . 23. *B. albiflora.*
Branches pubescent. Flowers shortly pedicellate. Stigma
depressed, 4-lobed, radiating . . . . . . . . . 22. *B. crassifolia.*
Stems erect, virgate, hirsute. Leaflets linear-terete. Flowers
nearly sessile. Sepals usually lanceolate-subulate . . . 24. *B. lanuginosa.*
Erect shrubs. Leaflets in distant pairs. Flowers pedunculate.
Sepals broad.
Pedicels shorter than the leaves, thickened upwards . . . 25. *B. pulchella.*
Pedicels long and slender . . . . . . . . . . 26. *B. gracilipes.*
Peduncles mostly 3- or several-flowered. (Eastern species.)
Glabrous. Leaflets small, thick, obovate . . . . . . . . 27. *B. microphylla.*
Glabrous. Leaflets linear or oblong in distant pairs . . . . 28. *B. pinnata.*
More or less pubescent. Leaflets crowded, the lowest pair close
to the stem . . . . . . . . . . . . . . . . . 29. *B. pilosa.*

SERIES IV. **Cyaneæ.**—*Petals imbricate. Anthers nearly uniform. Leaves simple or 3-foliolate, or the terminal leaflet or all three again 3-foliolate. Flowers axillary, blue or bluish. Filaments usually much flattened. (Cyanothamnus, Lindl.)*

Leaves or leaflets short, oblong or cuneate, thick. Appendage of
the anthers small.
Lower branchlets divaricate, spinescent. Sepals leafy . . . 30. *B. spinescens.*
No thorns. Sepals usually small . . . . . . . . . . . 31. *B. cærulescens.*
Leaves or leaflets narrow-linear or subulate. Appendage of the
anthers long and broad.
Flowers pedicellate.
Annual. Leaves all simple . . . . . . . . . . . . 32. *B. tenuis.*
Undershrub or shrub. Leaves mostly compound . . . . 33. *B. ramosa.*
Flowers sessile, or nearly so. Leaves simple, linear-terete . . 39. *B. subsessilis.*

SERIES V. **Variabiles.**—*Petals imbricate. Anthers nearly uniform. Leaves simple or 3-o-liolate, or the terminal leaflet or all three again 3-foliolate. Flowers axillary, red or pi k.*

Terminal leaflets or all three dentate, or again 3- or 5-foliolate.
Erect or spreading shrub. Peduncles usually 3- to 5-flowered 35. *B. anemonifoli .*
Leaves mostly 3-foliolate.
Common petiole distinct.
Leaflets flat, linear oblong or obovate. Anthers apiculate.
Pedicels 1-flowered . . . . . . . . . . . . . 34. *B. polygalifolia.*
Leaflets linear-terete, mucronate. Anthers not apiculate.
Pedicels 1- to 3-flowered . . . . . . . . . . 36. *B. falcifolia.*
Leaflets sessile. Flowers minute. Appendage of the anthers
broad, ciliate . . . . . . . . . . . . . . . 37. *B. penicillata.*
Leaves all simple.
Leaves flat.
Leaves obovate or broadly cuneate, often denticulate . . . 42. *B. crenulata.*

Leaves linear or lanceolate, acute, or the lower ones rarely
  cuneate.
  Low undershrub. Flowers all axillary. Sepals short  .  34. *B. polygalifolia.*
  Virgate shrub. Flowers all axillary. Sepals lanceolate-
    subulate, elongated . . . . . . . . . . . .  38. *B. crassipes.*
  Small branching shrub. Flowers many of them terminal .  46. *B. viminea.*
Leaves linear-terete.
  Flowers all axillary. Appendage of the anthers large  . . .  39. *B. subsessilis.*
  Flowers many of them terminal. Anthers minutely or not at
    all apiculate . . . . . . . . . . . . . . . .  41. *B. nematophylla.*

SERIES VI. **Terminales.**—*Petals imbricate. Anthers nearly uniform. Leaves all simple (except in* B. filifolia, inornata, *and* oxyantha). *Flowers mostly or all terminal, sessile or on short 1-flowered peduncles.*

Terminal flowers sessile, capitate.
  Leaves linear-terete.
    Branches hirsute. Leaves very obtuse . . . . . . .  40. *B. capitata.*
    Glabrous. Leaves mucronate or acute . . . . . . .  41. *B. nematophylla.*
  Leaves obovate or spathulate, often crenulate . . . . .  42. *B. crenulata.*
  Leaves rhomboidal, serrulate . . . . . . . . . . .  43. *B. serrulata.*
Terminal flowers solitary, or rarely 2 or 3, sessile or shortly pedi-
  cellate.
  Leaves obovate-orbicular, coriaceous . . . . . . . .  44. *B. rhomboidea.*
  Leaves linear or lanceolate, rarely oblong-cuneate, flat.
    Small undershrub. Filaments nearly glabrous. Anthers not
      apiculate . . . . . . . . . . . . . . .  45. *B. parviflora.*
    Slender shrub. Filaments woolly. Anthers apiculate  . .  46. *B. viminea.*
  Leaves or leaflets linear-terete.
    Leaves simple, or leaflets 3 on a distinct petiole. Pedicels
      slender . . . . . . . . . . . . . . . .  47. *B. filifolia.*
    Leaflets mostly 3 or 5, small, clustered on a very short
      common petiole.
      Sepals broad, short. Petals slightly pointed . . . . .  48. *B. inornata.*
      Sepals lanceolate-subulate. Petals mucronate . . . .  49. *B. oxyantha.*

SERIES VII. **Pedunculatæ.**—*Petals imbricate. Anthers nearly uniform. Leaves all simple. Peduncles terminal, several-flowered, or very rarely 1-flowered.*

Leaves (usually numerous) small, sessile or nearly so, with revolute
  margins.
  Leaves linear or oblong.
    Roughly pubescent or hirsute. Peduncles slightly exceeding
      the last leaves. Sepals subulate-acuminate . . . .  50. *B. scabra.*
    Glabrous or slightly pubescent. Peduncles much longer than
      the last leaves. Sepals broad, short . . . . . . .  51. *B. thymifolia.*
  Leaves ovate-cordate. Peduncles long . . . . . . .  52. *B. ovata.*
Leaves flat, usually thick, glabrous, contracted at the base.
  Leaves small, obovate or oblong, mostly denticulate. Cymes
    umbel-like. Peduncles short. Pedicels long. Sepals large  53. *B. fastigiata.*
  Leaves elongated, mostly denticulate. Cymes shortly peduncu-
    late, loose. Sepals small . . . . . . . . . .  54. *B. denticulata.*
  Leaves entire, thick ; lower ones spathulate ; upper ones narrow
    or linear, distant. Flowers large, few, or in a loose dichoto-
    mous panicle . . . . . . . . . . . . . . .  55. *B. spathulata.*
Leaves linear-terete.
  Leaves few, thick, and small. Sepals lanceolate-subulate, nearly
    as long as the small petals . . . . . . . . . .  56. *B. juncea.*
  Leaves numerous. Cymes many-flowered, on long peduncles.  57. *B. cymosa.*
  Branches slender, divaricate. Leaves slender. Peduncles short,
    mostly 1-flowered . . . . . . . . . . . . . .  47. *B. filifolia.*

SERIES I. VALVATÆ.—Petals valvate.

1. **B. grandisepala,** *F. Muell. Fragm.* i. 66.   A shrub with tomentose branches.   Leaves simple, nearly sessile, oblong-lanceolate, obtuse, 1 to 1½ in. long, softly hoary-tomentose on both sides, the edges flat, the midrib very prominent underneath.   Pedicels axillary, solitary, short, 1-flowered. Sepals ovate or ovate-lanceolate, tomentose, valvate, attaining 5 or 6 lines. Petals valvate and tomentose like the sepals, but smaller, and enclosed in them in the bud.   Filaments slightly hirsute, clavate and glandular at the top.   Anthers scarcely apiculate.   Ovary pubescent.

**N. Australia.** M'Adam range, *F. Mueller.*

2. **B. artemisiæfolia,** *F. Muell. Fragm.* i. 66.   A shrub, clothed all over with a soft hoary close or velvety tomentum.   Leaves all or nearly all pinnate.   Leaflets 7 to 11 or more, crowded on a short common petiole, linear, obtuse, rarely exceeding ½ in. and often much shorter, the margins closely revolute.   Peduncles axillary, solitary, short, 1-flowered.   Sepals lanceolate, tomentose, valvate, attaining 3 to 4 lines.   Petals lanceolate, valvate and tomentose like the sepals, but smaller and enclosed in them in the bud. Filaments slightly hirsute, clavate and glandular at the top.   Anthers scarcely apiculate.   Ovary pubescent.   Seeds smooth but scarcely shining.

**N. Australia.** Islands of the Gulf of Carpentaria, *R. Brown ;*  M'Adam, Fitzroy, and Sea ranges, *F. Mueller.*

Var. *Wilsoni,* F. Muell.   Branches more villous.   Leaflets short, oblong, and less crowded. — N.W. coast, *Bynoe ;*  Vansittart's Bay, *A. Cunningham ;*   Victoria river, *Wilson.*

F. Mueller, Fragm. ii. 179, refers this species as a variety to *B. grandisepala,* and some of R. Brown's specimens have some of the leaves undivided ; yet I have seen no approach to the large flowers of *B. grandisepala,* and I retain the two as distinct until really intermediate specimens shall have been observed.

3. **B. affinis,** *R. Br. Herb.*   A shrub, with numerous slender divaricate branches, pubescent when young, at length glabrous.   Leaves pinnate ; leaflets 7 to 15, linear, obtuse, mostly 3 to 4 lines long, the margins revolute, pubescent when young, glabrous at least above when full-grown, the pairs distant. Pedicels very short, axillary, 1-flowered.   Sepals broadly lanceolate, subulate-acuminate, 2 to nearly 3 lines long, slightly pubescent, very thin but apparently valvate.   Petals similar to the sepals, and about the same length, but narrower, valvate.   Filaments clavate, and glandular at the top.   Anthers scarcely apiculate.   Seeds smooth, but scarcely shining.

**N. Australia.** Islands of the Gulf of Carpentaria, and mainland opposite Groote Eyland, *R. Brown.* (*Hb. R. Br.*)

4. **B. filicifolia,** *A. Cunn. Herb.*   Branches rather slender, tomentose-pubescent or villous.   Leaves pinnate ; leaflets 12 to 20 pairs, with a terminal odd one, ovate or oblong, 1 to 2 lines long, pubescent, the margins slightly recurved.   Peduncles axillary, slender, often ½ in. long, bearing a single small flower.   Sepals lanceolate-valvate, tomentose, attaining about 2 lines.   Petals lanceolate, valvate and tomentose, like the sepals, and of the same size.   Filaments clavate and glandular upwards.   Anthers shortly apiculate.   Style pubescent.

**N. Australia.** York and Montague sounds, N.W. coast, *A. Cunningham.*

5. **B. alata,** *Sm. in Trans. Linn. Soc.* viii. 283. A shrub, usually quite glabrous and somewhat glaucous, but occasionally sprinkled with a slight pubescence, especially on the under side of the leaves. Branches angular. Leaves pinnate; leaflets usually 7, 9, or 11, obovate or broadly oblong, often ½ in. long, very obtuse, entire or crenate. Flowers large, in terminal cymes not exceeding the last leaves. Sepals small, lanceolate. Petals attaining 5 lines, acute, valvate in the bud, glabrous outside without prominent midribs, minutely tomentose with a ciliate midrib inside, the young buds very angular. Filaments ciliate, obtuse and glandular at the top. Anthers minutely apiculate. Ovary pubescent. Seeds opaque but smooth.—Sweet, Fl. Austral. t. 48; Bartl. in Pl. Preiss. i. 169; *Zanthoxylum oppositifolium,* DC. Prod. i. 728.

**W. Australia.** King George's Sound, *R. Brown, Fraser,* and others; Champion Bay, *Bowen;* Bald Island and Harvey river, *Oldfield;* Mount Manypeak, *Maxwell;* Rocky Bay and Rottenest Island, *Preiss, n.* 2012 *(Bartling).* I have not myself seen Preiss's specimens.

6. **B. algida,** *F. Muell. in Trans. Phil. Soc. Vict.* i. 100. A glabrous stunted shrub, with numerous dichotomous or divaricate branches. Leaves pinnate, with a very short common petiole; leaflets usually 5, the lowest pair close to the stem, obovate, rarely 2 lines long, thick and rigid. Flowers solitary at the ends of the branches or in the forks, on very short pedicels. Sepals small, acute. Petals ovate-lanceolate, valvate, attaining nearly 3 lines. Filaments glabrous or nearly so, thickened and glandular upwards; anthers minutely apiculated. Stigma globular.

**N. S. Wales.** Upper Clarence river, also Mounts Latrobe, Hotham, and Kosciusko, *F. Mueller.*

7. **B. Edwardsii,** *Benth.* A dichotomous shrub, nearly allied to *B. algida,* and possibly a variety. Branches pubescent. Leaflets 3, almost sessile, obovate or oblong, obtuse, attaining sometimes 3 lines, glabrous or slightly pubescent, pale underneath. Flowers solitary or 2 or 3 together, terminal or in the forks of the branches, on distinct slender pedicels. Petals valvate. Filaments glabrous. Anthers tipped with recurved points or appendages. Stigma globular.

**S. Australia.** Mount Barker, *Edwards.* I have seen only a single small specimen. (*Hb. F. Muell.*)

8. **B. calophylla,** *Turcz. in Bull. Mosc.* 1852, ii. 160. A glabrous, rigid, much-branched shrub. Leaves 3-foliolate or rarely simple, the common petiole exceedingly short; leaflets sessile, obovate, very obtuse, 2 or rarely nearly 3 lines long, glabrous, thick and rigid. Flowers rather large, hoary-tomentose, on 1-flowered peduncles, longer than the leaves, hoary-tomentose as well as the branchlets. Sepals small, ovate. Petals attaining 3 lines or rather more, valvate in the bud, with the midrib prominent outside. Filaments slightly ciliate, obtuse and glandular at the top; anthers minutely apiculate. Ovules almost collateral. Cocci glabrous. Seeds smooth but opaque.

**W. Australia,** *Drummond,* 5th *Coll. n.* 205.

9 ? **B. ternata,** *Endl. Nov. Stirp. Dec.* 6. Branches rigid, with a minute ashy pubescence. Leaflets 3, sessile, obovate, very obtuse, not above 2 lines

long, densely hoary-tomentose on both sides. Pedicels axillary, solitary, scarcely ¼ line long. Sepals tomentose, ovate, acute, about 1 line long. Petals twice as long as the calyx, pale pink, tomentose-pubescent. Filaments dilated upwards; anthers apiculate. Cocci stellate-tomentose.

**W. Australia.** In the interior, *Roe.* I have not seen this plant; the æstivation of the petals is not described; if it be valvate, the species must be closely allied to *C. calophylla*, differing chiefly in the tomentose leaves and almost sessile flowers.

10. **B. ericifolia,** *Benth.* An erect, branching, heath-like shrub, the young branches hoary-tomentose. Leaves 3-foliate or simple; leaflets sessile, linear with the margins closely revolute so as to be almost terete, obtuse, 3 or 4 lines long, glabrous. Flowers axillary, nearly sessile, hoary-tomentose. Sepals lanceolate, valvate. Petals about twice as long, attaining 3 lines or rather more, valvate, with the midrib prominent outside. Filaments glabrous, glandular and obtuse at the top; anthers with a minute recurved appendage. Style glabrous, with a more or less capitate stigma. Cocci rather large. Seeds opaque, but not seen quite ripe.

**W. Australia,** *Drummond, Coll.* 1843, *n.* 46.

11. **B. inconspicua,** *Benth.* A glabrous, rigid shrub. Leaves pinnate; leaflets 3, 5, or 7, linear, very obtuse, rarely ½. in long, thick and rigid, the pairs distant, the rhachis thick and somewhat dilated between the leaflets. Peduncles axillary, short, bearing single, minute, glabrous flowers. Sepals rather thin, ovate, obtuse. Petals 2 or 3 times as long, in our specimens not exceeding 1 line, but perhaps not fully developed, valvate, somewhat concave, slightly inflexed at the tip. Filaments flattened, ciliate, not thickened at the top; anthers all very small, not apiculate. Ovary glabrous. Style very small, with a rather large globular stigma. Cocci about 2 lines long, glabrous. Seeds opaque, glandular-tuberculate.

**W. Australia,** *Drummond, n.* 212. The immediate affinities of this species are not very clear. It is in some respects nearer to some of the *Pinnatæ* than to the *Valvatæ* generally, but as far as our specimens go I cannot trace any immediate connection with any species of either group.

12. **B. eriantha,** *Lindl. in Mitch. Trop. Austr.* 298. A glabrous shrub, the branches angular. Leaves pinnate; leaflets 3 to 9, obovate or oblong-cuneate, obtuse or with a recurved point, rarely above 3 lines long, rather thick, and often folded upwards lengthwise, the margins never recurved. Peduncles axillary, short, 1- or rarely 2-flowered. Sepals ovate, acute, glabrous outside, minutely tomentose inside. Petals more than twice as long, attaining 3 or 4 lines, rather narrow, valvate, hoary-tomentose outside, with a prominent midrib. Filaments usually ciliate; anthers apiculate.

**Queensland,** *Bidwill;* near Mount Pluto, *Mitchell.* With the aspect of *B. microphylla* this has the floral characters of *B. ledifolia*, with which F. Mueller proposes to unite it, but besides a totally different habit, the leaflets are thick, equally green on both sides, with the margins flat or folded upwards, not recurved with a pale or hoary-tomentose under-surface as in *B. ledifolia*.

13. **B. alulata,** *Soland. in Herb. Banks.* Apparently a divaricate or diffuse shrub, the young branches glandular-tomentose. Leaves pinnate; leaflets 7 to 13 or even more, oblong or linear, rarely almost ovate, obtuse, 2 to 3 lines long, the margins revolute, glabrous above when full-grown, hoary-tomentose

underneath. Peduncles very short, axillary, 1-flowered. Sepals lanceolate, subulate-acuminate, from ⅓ to nearly as long as the petals. Petals about 3 lines long, mucronate, valvate in the bud but rather broad, glabrous outside with a prominent midrib, slightly tomentose inside. Filaments clavate and glandular upwards.

**Queensland.** Endeavour river, *Banks* and *Solander, R. Brown.* (*Hb. Brit. Mus. and R. Br.*)

14. **B. ledifolia,** *J. Gay; DC. Prod.* i. 722. An erect shrub, the young branches glandular-tomentose. Leaves simple, 3-foliolate, or rarely pinnately 5- or even 7-foliolate; leaflets linear, oblong-linear, lanceolate or rarely broadly oblong, when single often above 1 in. long, when several rarely above ½ in., the margins recurved or revolute, glabrous above when full-grown, hoary or rusty underneath with a minute tomentum. Peduncles axillary, 1-flowered, shorter than the leaves. Sepals broad, obtuse but valvate. Petals twice as long or more, attaining 4 or 5 lines, valvate in the bud, minutely tomentose outside, with a prominent midrib. Filaments short, as in several allied species, slightly ciliate or glabrous, clavate and glandular upwards; anthers more or less apiculate. Ovules usually, as in some allied species, almost or quite collateral. Style clavate, with a slightly furrowed stigma. Seeds smooth but not shining.—Reichb. Icon. Exot. t. 74; *Lasiopetalum ledifolium,* Vent. Jard. Malm. under n. 59; *Eriostemon paradoxum,* Sm. in Rees, Cycl. xiii.; *Boronia (?) paradoxa,* DC. Prod. i. 722.

**Queensland.** Burnett river, *F. Mueller;* Moreton Bay and islands, *A. Cunningham, Fraser, etc.*

**N. S. Wales.** Port Jackson and Blue Mountains, *R. Brown, Sieber. n.* 297 and 303, and *Fl. Mixt. n.* 531 and 534, and others.

Var. *rosmarinifolia.* Leaves rigid, usually narrow, small, and all simple. Peduncles very short.—*B. rosmarinifolia,* A. Cunn. in Hueg. Enum. 16. To this form belong especially most of the Moreton Bay specimens.

Var. (?) *triphylla.* Leaves mostly or all 3-foliolate, or the lower ones pinnate.—*B. triphylla,* Sieb. in Spreng. Syst. Cur. Post. 148; Reichb. Icon. Exot. t. 73; apparently as common about Port Jackson as the large simple-leaved form. A subvariety, with broader leaflets, is figured Bot. Reg. 1841, t. 47, and Paxt. Mag. viii. 123.

Var. (?) *rubiginosa.* Leaflets 3 or 5, still broader, almost obovate. Peduncles, according to Endlicher, 3-flowered, but 1-flowered in our specimens.—*B. rubiginosa,* A. Cunn. Herb., Endl. in Hueg. Enum. 16; Hunter's River, *A. Cunningham.*

*B. ledifolia* is enumerated also (Pl. Preiss. ii. 226) amongst W. Australian plants, a very unlikely station. I have not seen Preiss's specimen n. 2644, nor any western species agreeing with the character given, and therefore have no clue to the plant referred to. F. Mueller, presuming like myself that it cannot be Gay's plant, proposes (Fragm. i. 67) to give it the name of *B. ledophylla;* but without seeing specimens it is impossible to characterize it.

15. **B. lanceolata,** *F. Muell. Fragm.* i. 66. A tall shrub with tomentose branches. Leaves simple, petiolate, oblong-lanceolate, obtuse or mucronulate, 1 to 2 in. long, flat or the margins recurved, glabrous above, tomentose underneath. Peduncles very short, bearing an umbel of 3 to 5 small flowers, rarely reduced to a single flower. Sepals small, ovate, with a subulate point, sometimes very short, sometimes nearly as long as the petals. Petals broad, attaining about 2 lines in length, valvate in the bud, tomentose outside with a prominent midrib. Filaments glabrous, thickened and glan-

dular at the top; anthers scarcely apiculate.   Cocci glabrous.   Seeds smooth but not shining.

**N. Australia.**  Islands of the Gulf of Carpentaria, *R. Brown;* Port Essington, *Armstrong, Leichhardt.*  Stony places in Arnhem's Land and Carpentaria, *F. Mueller.*

16. **B. Fraseri,** *Hook. Bot. Mag. t.* 4052.   A shrub of 3 or 4 ft., the branches glabrous, angular or compressed.   Leaves pinnate; leaflets 3 or 5, in distant pairs, oblong-lanceolate, obtuse, the terminal one usually 1 to 1½ in. long, the others smaller, all glabrous but pale underneath.   Peduncles axillary, short, bearing an umbel of 3 to 6 flowers.   Sepals very small. Petals attaining fully 3 lines, valvate, hoary outside, with a prominent midrib. Filaments glabrous, much thickened and glandular at the top; anthers minutely apiculate.   Disk very thick.   Stigma capitate but small.—*B. anemonifolia*, Paxt. Mag. Bot. ix. 123, with a fig., not A. Cunn.

**N. S. Wales.**  Ravines on the Nepean river, *Fraser.*

17. **B. mollis,** *A. Cunn.; Lindl. Bot. Reg.* 1841, *under t.* 47.   A shrub, with the habit of *B. Fraseri*, but the branches and petioles densely and softly hirsute.   Leaflets usually 3 or 5, in distant pairs, the terminal one oblong or lanceolate, obtuse, 1 to 1½ in. long, the others much shorter and broader in proportion, all glabrous or nearly so above, tomentose-pubescent or villous underneath.   Peduncles axillary, very short, bearing an umbel of several flowers larger than those of *B. Fraseri.*   Sepals linear.   Petals ovate-acuminate, attaining 5 or 6 lines, valvate.   Stamens and style of *B. Fraseri.*

**N. S. Wales.**  Nepean river, *A. Cunningham;* near Sydney, *Lyall.*

SERIES II. HETERANDRÆ.—Sepaline anthers usually different from the petaline ones, and often imperfect.

18. **B. megastigma,** *Nees, in Pl. Preiss.* ii. 227.   A shrub, with erect virgate branches, glabrous or nearly so.   Leaflets 3 or rarely 5, sessile or with a very short common petiole, linear, obtuse, rarely ½ in. long, rather thick and rigid, glabrous.   Peduncles axillary, 1-flowered, the pedicel much thickened under the flower.   Sepals short, broad, obtuse.   Petals attaining about 3 lines, broad and much imbricate, of a dark purple outside, drying almost black, yellowish inside.   Filaments glabrous, rather attenuate and incurved at the top, the 4 longer ones opposite the sepals with large purple anthers, the 4 smaller opposite the petals with small yellow anthers close under the stigma.   Stigma purple, very broad and thick, truncate at the top, expanded laterally into 4 thick prominent lobes.—F. Muell. Fragm. ii. 97 ; *B. tristis*, Turcz. in Bull. Mosc. 1852, ii. 162.

**W. Australia.**  King George's Sound and neighbouring districts, *Milne, Preiss, n.* 1232; *Drummond, 5th Coll. n.* 201, and others.   In this and the two following species the large purple or black anthers are said to be barren, and the pollen perfect only in the very small yellow petaline anthers, a point I am unable to ascertain positively from dried specimens.

19. **B. heterophylla,** *F. Muell. Fragm.* ii. 98.   A tall glabrous shrub, with numerous slender branches.   Leaves either simple and linear, 1 to 2 in. long, or pinnate, with 3 or 5 linear leaflets on an elongated common petiole.   Peduncles axillary, 1-flowered, slender below the bracts, thickened

under the flower. Sepals very short and orbicular. Petals attaining about 4 lines, broad and imbricate, glabrous outside, pubescent inside, apparently pink. Filaments glabrous, the larger ones opposite the sepals, thickened and much incurved at the top, with large ovoid black anthers, the smaller ones with small yellow anthers. Ovary ciliate-hirsute, with a thick conical deciduous style. Cocci pubescent. Seeds smooth and shining.

**W. Australia,** *Drummond, n.* 117. In places sometimes inundated, on the Kalgee river, *Maxwell.*

20. **B. elatior,** *Bartl. in Pl. Preiss,* i. 170. Apparently a tall shrub, the branches hirsute with long spreading hairs. Leaves pinnate; leaflets 5 to 13 or more, linear, flat, rather rigid, often $\frac{1}{2}$ in. long or even more, glabrous or hirsute, the pairs rather distant with the rhachis often dilated between them. Peduncles axillary, often as long as the common petiole. Sepals broad, mucronate, usually ciliate and often coloured. Petals attaining 3 to 4 lines, usually mucronate, glabrous or slightly pubescent, much imbricate. Filaments slightly ciliate, 4 opposite the sepals, thick, attenuate at the top, with large black anthers, 4 opposite the petals incurved at the top, with minute yellowish anthers close under the stigma. Ovary hirsute with a very large glabrous, thick, obtusely pyramidal stigma, 4-lobed at the base. Seeds smooth and shining.—*B. semifertilis, F. Muell. Fragm.* ii. 98.

**W. Australia,** *Drummond, n.* 36, 43, and 118. King George's Sound, *R. Brown;* Darling Range, *Preiss, n.* 2013; Wilson's Inlet, *Oldfield;* Franklin river, *Maxwell.*

*B. psoraleoïdes,* DC. Prod. i. 721, from the S. coast, is unknown to me; it is described as having pinnate leaves, with 3 or 5 linear obtuse leaflets, glabrous and glandular-dotted as well as the branches; peduncles short, 1-flowered; flowers small, pale, tetrandrous. It would therefore rank among the *Heterandræ*, which however have usually rather large flowers.

21. **B. tetrandra,** *Labill. Pl. Nov. Holl.* i. 98, *t.* 125. An erect much-branched shrub, the branches more or less hirsute with spreading hairs. Leaves pinnate; leaflets usually 7 to 13, linear, obtuse, the largest rarely above 4 lines long, the upper ones of each leaf usually gradually shorter, all flat or the margins slightly recurved, glabrous or slightly hirsute, the pairs rather distant. Peduncles axillary, 1-flowered, very short. Sepals broadly ovate. Petals attaining 3 or 4 lines, glabrous, imbricate. Filaments slightly ciliate, 4 opposite the petals short, thick, with perfect anthers, shortly apiculate, 4 opposite the sepals longer and more slender, inflected at the top, with minute apparently imperfect anthers. Disk with 4 lobes inside the sepaline stamens, almost as in *Zieria.* Ovary small, glabrous; stigma very large and thick, truncate at the top with 4 very prominent almost winged lateral lobes.—*B. bicolor,* Turcz. in Bull. Mosc. 1852, ii. 163.

**W. Australia,** *Drummond, 5th Coll. n.* 200. Cape Leeuwin, *Labillardière.* Labillardière's specimen has very few small flowers, which I have been unable to examine, but Drummond's agrees with it in every other respect, as well as with his figure and description, except that the flowers are rather larger.

22. **B. crassifolia,** *Bartl. in Pl. Preiss.* i. 169. A dwarf much-branched shrub or undershrub, not exceeding 8 in. in any of our specimens, glabrous or minutely pubescent. Leaves pinnate; leaflets 3, 5, or 7, on a short common petiole, linear-cuneate or oblong-linear, very obtuse, rarely $\frac{1}{2}$ in. long and often much smaller, rather thick and coriaceous. Peduncles axil-

lary, 1- or rarely 2-flowered, short, and often recurved. Sepals ovate, minutely pubescent. Petals attaining 3 lines, imbricate and nearly glabrous. Filaments 4 opposite the sepals, thick, attenuate at the top, with perfect shortly apiculate anthers, 4. opposite the petals shorter, clavate, glandular, with anthers usually minute and less perfect. Ovary minutely pubescent. Stigma very large and thick, broadly conical and peltate.—*B. humilis*, Turcz. in Bull. Mosc. 1852, ii. 160.

**W. Australia,** *Drummond, 5th Coll. n.* 199, and *Coll.* 1843, *n.* 59 ; *Bynoe, Preiss, n.* 2033.

*B. multicaulis*, Turcz. in Bull. Mosc. 1852, ii. 160, appears to refer to some unnumbered specimens in Drummond's 5th Coll., agreeing in every respect with *B. crassifolia*, except that the anthers of the sepaline stamens are more perfect, and the stigma is reduced to 4 glabrous radiating lobes, closely adnate on a pubescent surface, not distinguishable from the apparently imperfect ovary. I have seen but few flowers of this form, but believe the differences from *B. crassifolia* to be rather sexual than specific.

SERIES III. PINNATÆ.—Anthers uniform. Leaves pinnate. Peduncles axillary.

23. **B. albiflora,** *R. Br. Herb.* A dwarf, much-branched, erect undershrub or shrub, hirsute with short spreading hairs. Leaves pinnate ; leaflets 7 to 11, crowded on a short common petiole, oblong-linear, slightly cuneate, very obtuse, rather coriaceous, the margins often recurved, the lowest of each leaf often 4 or 5 lines long, the others gradually smaller. Flowers small, axillary, nearly sessile. Sepals ovate or lanceolate, ciliate. Petals attaining about 2½ to 3 lines, imbricate, glabrous. Filaments glabrous, clavate and glandular at the top. Anthers all perfect, distinctly apiculate. Ovary pubescent. Style conical with a small stigma. Cocci pubescent or glabrous. Seeds smooth.

**W. Australia.** South coast, *R. Brown;* King George's Sound, *Baxter;* Garden Range, hills N. of Stirling range, and Cheynye Beach, *Maxwell.*

Some specimens of this plant, with fewer and less crowded leaflets, have the aspect of *B. crassifolia*, but the larger filaments are not attenuate at the top, the anthers more distinctly apiculate, and the style quite different.

24. **B. lanuginosa,** *Endl. in Hueg. Enum.* 16. Stems erect, simple or with erect virgate branches, 1 to 2 ft. high, hirsute with spreading hairs, hard and woody at the base. Leaves pinnate ; leaflets 5 to 9 or rarely more, linear-terete or slightly flattened and cuneate, mostly acute, rarely ½ in. long, glabrous or hirsute, somewhat crowded on a rather short common petiole. Peduncles axillary, short, or the flowers almost sessile. Sepals usually lanceolate-subulate, more than half as long as the petals. Petals attaining nearly 4 lines, mucronate, imbricate, slightly pubescent, deeply coloured in the centre. Filaments glabrous or ciliate, the longer ones especially thickened and glandular at the top ; anthers all perfect, shortly apiculate. Stigma small.—*B. stricta*, Bartl. in Pl. Preiss. i. 169.

**W. Australia,** *Drummond, Coll.* 1845, *n.* 9 ; King George's Sound and neighbouring districts, *R. Brown; Preiss, n.* 2034; *Maxwell.* I have not seen specimens named by Endlicher, but this is the only species of R. Brown's (with whom F. Bauer collected) which answers to the short diagnosis given.

Var. (?) *brevicalyx.* Sepals very small, without the long point of the common form.— Phillips River, *Herb. Mueller.*

*B. pubescens,* Bartl. in Pl. Preiss. ii. 227 ; from W. Australia, *Preiss, n.* 2643, is un-known to me, but from the description given it would appear to be a small-flowered variety of *B. lanuginosa.*

25. **B. pulchella,** *Turcz. in Bull. Mosc.* 1852, ii. 162. An erect branching shrub, perfectly glabrous, or the young branches minutely pubes-cent or shortly hirsute. Leaves pinnate ; leaflets usually 7 to 11, linear, rather obtuse, rarely above 4 lines long, rather rigid, flat or the margins slightly recurved, the pairs not crowded, the rhachis often dilated. Flowers large, of a rich pink, on axillary peduncles usually shorter than the leaves and rather thickened under the flowers. Sepals short, broad and acute. Petals attaining 3 to 4 lines, imbricate, glabrous. Filaments glabrous, capi-tate and glandular at the top ; anthers scarcely apiculate. Ovary slightly hirsute. Stigma capitate, rather large. Cocci glabrous. Seeds smooth, opaque, but not seen quite ripe.—*B. Drummondii,* Planch. in Fl. des Serres, ix. 65, t. 881 ; *B. tetrandra,* Lindl. and Paxt. Fl. Gard. i. 35, t. 8, not Labill. ; and perhaps also Paxt. Mag. xvi. 227.

**W. Australia,** Drummond, *n.* 13 ; *5th Coll. n.* 202 ; S.W. interior, *Maxwell.*

26. **B. gracilipes,** *F. Muell. Fragm.* ii. 99. An erect shrub, the branches pubescent or hirsute with spreading hairs. Leaves pinnate ; leaf-lets usually 5 or 7, rarely 9, oblong-linear or lanceolate, rarely exceeding 4 lines, the margins entire, or when broad often denticulate, flat or slightly re-curved. Peduncles slender, axillary, 1-flowered, often as long as the leaves and scarcely thickened under the flower. Sepals broad, short and acute. Petals attaining about 3 lines, imbricate, glabrous. Filaments ciliate, capi-tate and glandular at the top ; anthers minutely apiculate. Ovary pubescent. Stigma ovoid-capitate, rather large, almost sessile.

**W. Australia.** Franklin and Mount Manypeak rivers, Plantagenet and Stirling ranges, *Herb. Mueller.* This may prove to be a variety of *B. pulchella,* but, as far as our specimens go, the hirsute branches, broader leaflets, and slender pedicels appear to be constant.

27. **B. microphylla,** *Sieb. in Spreng. Syst. Cur. Post.* 148. A low stunted shrub, glabrous but often very glandular. Leaves pinnate ; leaflets 5 to 11, obovate or oblong-cuneate, obtuse or acute, rarely above 3 lines long, and usually about 2 lines, thick and rigid. Peduncles in the upper axils 1- to 3-flowered. Flowers of *B. pinnata,* or rather smaller, the anthers often conspicuously apiculate. Stigma slightly enlarged. Seeds in our specimens shining and reticulate.—Reichb. Icon. Exot. t. 72.

**N. S. Wales.** Blue Mountains, *Sieber, n.* 302 ; *A.* and *R. Cunningham ;* Para-matta and Upper Clarence river, *Herb. Mueller.* The latter station rather doubtful, the specimen being very incomplete. F. Mueller unites this species with *B. pinnata ;* but, as far as I have seen, the difference in foliage appears constant.

28. **B. pinnata,** *Sm. Tracts,* 290, *t.* 4. A glabrous shrub, attaining several feet, but sometimes dwarf or diffuse, the small branches more or less angular. Leaves pinnate ; leaflets 5 to 9 or rarely more, linear or oblong-lan-ceolate, acute, rigid, the pairs rather distant and the common petiole often dilated between them. Flowers rather large, usually 3 or more together, in loose axillary or subterminal corymbose cymes. Sepals small, acute. Petals attaining 3 to 5 lines, imbricate, glabrous or minutely tomentose inside,

usually mucronate. Filaments woolly-hairy, especially towards the thickened summit; anthers very minutely or not at all apiculate. Style short. Seeds smooth and shining.—DC. Prod. i. 721; Andr. Bot. Rep. t. 58; Vent. Jard. Malm. t. 38; Bot. Mag. t. 1763; F. Muell. Pl. Vict. i. 115; *B. floribunda*, Sieb. in Spreng. Syst. Cur. Post. 148; Reichb. Icon. Exot. t. 71.

**N. S. Wales.** Port Jackson and Blue Mountains, *R. Brown, Sieber, n.* 300, 301, and *Fl. Mixt. n.* 533, and others. These specimens appear to be sexually dimorphous. In some I find the stamens densely woolly, the anthers small, 4 of them perhaps imperfect, and the very short style bearing a thick globular stigma as large as or larger than the ovary. In other specimens the filaments are shorter and not quite so woolly, the anthers larger and more perfect, the style cylindrical, with the stigma scarcely thickened.

Var. *Muelleri.* Leaflets in distant pairs. Flowers nearly as large as in the Port Jackson specimens, but the filaments much less hairy, the anthers not at all apiculate, and I am unable to detect any dimorphism; the stigma minute or slightly capitate.

**Victoria.** Sources of the Bunyip river, in the Grampians, near Portland Bay, and towards the mouth of the Glenelg, *F. Mueller.*

Var. *Gunnii.* Leaflets more crowded, but the lowest pair always distant from the stem. Flowers smaller than in the Port Jackson plant, with the filaments much less hairy, and the anthers and style (as far as I have been able to ascertain) homomorphous, as in the var. *Muelleri.*—*B. tetrandra*, var. *grandiflora*, Hook. Journ. Bot. ii. 419; not Labill.; *B. Gunnii*, Hook. f. Fl. Tasm. i. 68, t. 10.

**Tasmania.** Near Port Dalrymple, *R. Brown;* S. Esk river, near Launceston, *Gunn.* *B. citriodora*, Gunn, in Hook. f. Fl. Tasm. i. 68, common in alpine situations in Tasmania (*J. D. Hooker, Gunn*); is generally of smaller stature, with the leaflets often reduced to 3; but it is often not distinguishable from the var. *Gunnii* in the dried state, when the peculiar lemon-scent, which it is said to be so easily known by, has entirely disappeared.

29. **B. pilosa,** *Labill. Pl. Nov. Holl.* i. 97, t. 124. A shrub, very nearly allied to *B. pinnata*, with which F. Mueller proposes to unite it, and perhaps with reason, but the aspect is different. Branches almost always more or less pubescent. Leaflets crowded on a short common petiole, the lowest pair close to the stem, usually narrower and more obtuse than in *B. pinnata.* Cymes compact, 3- or rarely 5-flowered and often reduced to single flowers, which are generally smaller than in *B. pinnata.* Filaments ciliate rather than woolly; anthers not at all apiculate. Stigma slightly enlarged, never large and globular, nor yet very minute.—DC. Prod. i. 721; Hook. f. Fl. Tasm. i. 67.

**Victoria.** In the Grampians, *Wilhelmi;* Portland Bay and mouth of the Glenelg' *Robertson.*

**Tasmania,** *R . Brown;* abundant throughout the colony, *J. D. Hooker.* In a very few Tasmanian specimens the leaflets are not quite so crowded, but their narrow form and the pubescent branches are those of *B. pilosa.*

SERIES IV. CYANEÆ.—Flowers usually blue or bluish. Foliage of the *Variabiles.*

30. **B. spinescens,** *Benth.* A glabrous undershrub with erect or ascending rigid stems of 1 to 1½ ft., the lower branchlets often converted into divaricate leafless thorns of 1 to 2 in. Leaves nearly sessile, simple, entire or 3-lobed, either ovate or lanceolate and scarcely 2 lines or rarely almost linear and 3 or 4 lines long. Peduncles axillary, 1-flowered, 2 or 3 lines long. Sepals leafy, obtuse, often fully 2 lines long. Petals not twice as long, apparently bluish. Filaments slightly dilated at the base, ciliate, terete and

glandular upwards, attenuate at the top. Appendage of the anthers much shorter than the cells and not so broad as in the other blue species.

**W. Australia,** *Drummond, n.* 78.

31. **B. cærulescens,** *F. Muell. in Trans. Phil. Soc. Vict.* i. 11, and *Pl. Vict.* i. 117. An undershrub of a pale green, glabrous or minutely pubescent. Leaves simple, sessile, linear or linear-cuneate, obtuse, rarely attaining ½ in., and often only 2 or 3 lines long, rather thick, often tuberculate underneath. Pedicels 1-flowered, mostly axillary, 1 to 2 or 3 lines long. Sepals ovate. Petals twice or thrice as long as the sepals, attaining 3 to 4 lines, imbricate, glabrous, or pubescent outside along the centre. Filaments ciliate, not clavate ; anthers with a short broad obtuse recurved appendage. Stigma capitate. Seeds reticulate.

**Victoria.** Desert of the Murray and its lower tributaries, and sterile plains at the foot of the Grampians, *F. Mueller.*

**S. Australia.** Sandy coast of Guichen Bay and Cape Jaffa, St. Vincent's and Spencer's Gulf, *F. Mueller* and others.

**W. Australia.** Salt river, S. Hutt river, and Chapman river to E. Mount Barren, *Herb. Mueller.*

32. **B. tenuis,** *Benth.* Apparently annual, quite glabrous, with slender ascending or erect branches ½ to 1½ ft. high. Leaves simple, slender, linear-terete, acute or obtuse, mostly ½ to 1 in. long. Flowers blue, on axillary pedicels of 1 to 4 or 5 lines. Sepals ovate-lanceolate, with white membranous margins. Petals about twice as long as the sepals, attaining 3 lines or rarely more, imbricate, glabrous. Filaments flat, ciliate, narrowed at the top ; anthers with broad recurved appendages, nearly as long as the cells. Cocci glabrous. Seeds reticulate, striate.—*Cyanothamnus tenuis,* Lindl. Swan Riv. App. 18.

**W. Australia.** Swan River, *Drummond, 1st Coll. ;* Ballgarup ranges W. of Kojonerup, *Herb. Mueller.*

33. **B. ramosa,** *Benth.* An erect or diffuse heath-like glabrous shrub. Leaves once or twice ternately compound ; leaflets linear-terete, usually not thicker than the common petiole, ¼ to ½ in. long. Peduncles axillary, 2 to 4 or 5 lines long, bearing a single blue flower. Sepals broad and short. Petals varying from about 2 to above 3 lines long, imbricate, glabrous. Filaments broad, flat and ciliate at the base, terete, obtuse, and glandular at the top. Appendage of the anthers very broad and obtuse, usually longer than the cells. Stigma in some specimens capitate, in others small and 4-lobed. —*Cyanothamnus ramosus,* Lindl. Swan Riv. App. 18.

**W. Australia.** Swan River, *Drummond, 1st Coll.,* also *n.* 84 and 180 ; Murchison river, *Oldfield.*

Var. *anethifolia.* Leaves mostly 3-foliolate. Flowers smaller than in the original form, not so blue, at least in the dried state, with much smaller appendages to the anthers. *Cyanothamnus anethifolius,* Bartl. in Pl. Preiss. i. 179.—*Boronia subcœrulea,* F. Muell. Fragm. ii. 100. Swan River, *Drummond ;* Canning river, *Preiss, n.* 2035 ; Murchison river, *Oldfield ;* Champion Bay, *Walcott.*

SERIES V. VARIABILES.—Anthers uniform. Leaves simple or ternately compound. Flowers axillary, not blue.

34. **B. polygalifolia,** *Sm. Tracts,* 297, *t.* 7. Usually a low glabrous

undershrub with a thick rhizome as in *B. parviflora,* or a small shrub, rarely stouter and 1 to 2 ft. high. Leaves either simple with lanceolate or linear-lanceolate acute leaflets, mostly under ½ in., but sometimes nearly 1 in. long, or 3-foliolate with small acute leaflets, on a short common petiole. Pedicels axillary, solitary, and 1-flowered. Sepals short. Petals 2 or 3 times as long, imbricate, pink, and glabrous. Filaments hairy and glandular towards the top. Anthers conspicuously apiculate, the appendage erect or recurved. Seeds opaque and usually minutely tuberculate.—DC. Prod. i. 722 ; F. Muell. Pl. Vict. i. 114 ; *B. hyssopifolia,* Sieb. in Spreng. Syst. Cur. Post. 148 ; Hook. f. Fl. Tasm. i. 66 ; *B. tetrathecoides,* DC. Prod. i. 722 ; Hook. Comp. Bot. Mag. i. 277.

**Queensland.** Stradbrooke Island, *Fraser.*
**N. S. Wales.** Port Jackson, *R. Brown, Sieber, n.* 296, and others; northward to Hastings and Clarence rivers, *Beckler ;* New England, *C. Stuart.*
**Victoria.** Not rare, as well in swamps and alpine localities as in dry forest-land or on stony ridges, *F. Mueller.*
**Tasmania.** Abundant throughout the colony, *J. D. Hooker.*
**S. Australia.** Stringybark Forest, between Mount Lofty and the Onkaparinga, *F. Mueller.*
Var. *trifoliolata.* Stems short, glabrous. Leaves 3-foliolate, with linear leaflets.—*B. nana,* Hook. Ic. Pl. t. 270.—In Victoria and Tasmania. In some of the Victorian specimens, simple and trifoliolate leaves occur on different branches of the same plant.
Var. *robusta.* Leaves 3-foliolate as in the last var., but stems stout and more shrubby, attaining 2 ft. or more.—Port Jackson, *Sieber, n.* 283 ; Blue Mountains, *A. Cunningham ;* Moreton Island, *F. Mueller.*
Var. (?) *pubescens.* More or less pubescent. Leaves 3-foliolate. Leaflets very small, ovate or obovate. Flowers small, the pedicels usually longer than the leaves.—In the Grampians, *Wilhelmi, Robertson.*

35. **B. anemonifolia,** *A. Cunn. in Field, N. S. Wales,* 330. A shrub of 2 or 3 ft., glabrous or pubescent, and often glaucous. Leaves either simply 3-foliolate with the leaflets 3-toothed, or all 3 leaflets or the terminal one only again 3-foliolate or pinnately 5-foliolate, or sometimes some of them a third time divided, and all usually thick, linear-cuneate or, if entire, acutely linear. Flowers in axillary cymes of 3, 5, or even more, very rarely reduced to single flowers. Stamens and fruit of *B. polygalifolia.*

**Queensland.** Newcastle and Burnett rivers, *F. Mueller ;* near Lindley's Range, *Mitchell.*
**N. S. Wales.** E. coast, *R. Brown ;* Hunter's River and Blue Mountains, *A. Cunningham* and others.
**Victoria.** Mountains of Gipps' Land, *F. Mueller.*
**Tasmania.** Derwent river, King's Island, *R. Brown ;* northern parts of the island near the coast, *J. D. Hooker.*
**W. Australia.** Canning river, *Preiss, n.* 2628.
This species, which F. Mueller thinks ought to be united with *B. polygalifolia* as a variety, has by others been subdivided into 3, which may be considered as tolerably distinct races, viz. :—
*a. dentigera.* Pubescent or rarely glabrous. Leaflets usually 3, linear-cuneate, thick, 3-toothed at the top. Flowers 1 to 3 on each peduncle.—*B. dentigera,* F. Muell. in Trans. Vict. Inst. 1855, 82 ; *Cyanothamnus tridactylites,* Bartl. in Pl. Preiss. ii. 227.—N. S. Wales, Victoria ; Tasmania, E. coast, *C. Stuart ;* W. Australia, *Preiss, n.* 2628.
*b. variabilis.* Usually glabrous. Leaves irregularly compound, more or less twice ternate, but scarcely bipinnate. Leaflets oblong, obtuse, or linear-cuneate. Flowers rather small, 3 or more in the cyme.—*B. variabilis,* Hook. Comp. Bot. Mag. i. 277 (partly) ; Hook. f. Fl. Tasm. i. 67.—The common Tasmanian form.

*c. anethifolia.* Leaves still more compound, often bipinnate, and leaflets narrower and more acute than in the last var. Flowers 3 or more in the cyme.—*B. anethifolia,* A. Cunn.; Endl. in Hueg. Enum. 16; Lindl. Bot. Reg. 1841, under n. 47; *B. bipinnata,* Lindl. in Mitch. Trop. Austr. 225.—The common form in the interior of Queensland and N. S. Wales.

36. **B. falcifolia,** *A. Cunn.; Lindl. in Bot. Reg.* 1841, *under n.* 47. A glabrous, erect, heath-like shrub, with virgate branches. Leaves rather crowded, 3-foliolate; leaflets linear-terete, mucronate, mostly $\frac{1}{4}$ to $\frac{1}{2}$ in. long, on a common petiole rather shorter than themselves. Pedicels 1- to 3-flowered, in the upper axils. Bracts linear-subulate. Sepals lanceolate, subulate-pointed. Petals rather longer than the sepals, attaining 3 to 4 lines, acute, imbricate, glabrous. Filaments clavate and glandular upwards; anthers not apiculate. Stigma in some specimens capitate, in others not thicker than the style.— *B. paleifolia,* Endl. in Hueg. Enum. 16 (through a misreading of Cunningham's label).

**Queensland.** Moreton Bay and islands, *A. Cunningham, F. Mueller,* and others; Wide Bay, *Bidwill.*

**N. S. Wales.** Port Macquarie and Port Stephens, *Backhouse.*

37. **B. penicillata,** *Benth.* An erect, rather rigid shrub or undershrub, more or less pubescent. Leaves simple or 3-foliolate; leaflets sessile, linear or linear-cuneate, flat, rather thick, rarely above $\frac{1}{2}$ in. long, Flowers axillary, very small, on short pedicels. Sepals broad, acute, glabrous or ciliate, very glandular. Petals about twice their length, but not exceeding $1\frac{1}{2}$ lines, rather thick and glandular, with thin transparent imbricate edges. Filaments slightly flattened, ciliate, rather thickened at the top; anthers tipped with a short broad appendage, ciliate with a few rather long stiff hairs. Stigma slightly thickened. Cocci glabrous, rather longer than the petals. Seeds not seen.

**W. Australia.** Between Swan River and King George's Sound, *Drummond.* The species resembles *B. inconspicua* in the minuteness of its flowers, but is quite different in their structure as well as in foliage.

38. **B. crassipes,** *Bartl. in Pl. Preiss.* i. 168. Shrubby and glabrous, with elongated, rather slender, virgate branches. Leaves simple, linear, rather acute, $\frac{1}{2}$ to 1 in. long, entire or serrulate. Pedicels axillary, 1-flowered, shorter than the leaves, thickened under the flower. Sepals lanceolate-subulate with long points. Petals about twice as long as the sepals, attaining fully 4 lines, acute or mucronate. Filaments slender, slightly ciliate, obtuse and glandular at the top; anthers minutely apiculate. Seeds opaque and scabrous, but not seen quite ripe.—Dietr. Fl. Univ. N. Ser. ii. t. 2.

**W. Australia,** *Drummond, Coll.* 1845. n. 10; Near Mount Wuljenup, *Preiss, n.* 2040; King George's Sound and Mount Barker, *Oldfield.*

39. **B. subsessilis,** *Benth.* Glabrous, with rigid twiggy branches. Leaves simple, sessile, linear-terete, rather obtuse, mostly $\frac{1}{2}$ to $\frac{3}{4}$ in. long. Flowers sessile or on very short thickened pedicels, glabrous, apparently red. Sepals short, broadly ovate. Petals attaining 3 lines, imbricate, obtuse, of a firm consistence. Filaments flattened, ciliate, slightly thickened and obtuse and glandular at the top; anthers tipped with a large, broad, recurved appendage.

**W. Australia,** *Drummond.* The stamens are nearly those of some of the blue-flowered species, but the flowers are much more sessile than in any blue species and apparently red.

SERIES VI. TERMINALES.—Anthers uniform.  Leaves simple or rarely 3–5-foliolate.  Flowers terminal, usually solitary.

40. **B. capitata,** *Benth.*  Apparently a rigid divaricate shrub.  Branches pubescent or hirsute.  Leaves simple, linear, obtuse, under ½ in. long in our specimen, pubescent, thick, terete or almost flat above and convex underneath. Flowers several, almost sessile, in terminal heads.  Sepals rather broad, pubescent.  Petals twice as long as the sepals, rather exceeding 3 lines, imbricate, glabrous.  Filaments densely ciliate at the edges, slightly thickened and glandular at the top; anthers tipped with a small recurved appendage.  Style pubescent, with a small stigma.

**W. Australia.**  In the eastern regions of the colony, *Drummond.*  It is possible that this may prove to be an extreme variety of *B. nematophylla,* differing chiefly in inflorescence and indumentum.

41. **B. nematophylla,** *F. Muell. Fragm.* ii. 100.  An erect, virgate, or diffuse, glabrous shrub.  Leaves all simple, linear-terete, obtuse acute or mucronulate, mostly ½ to 1 in. long.  Flowers axillary or terminal, nearly sessile or shortly pedunculate, solitary or the terminal ones in clusters of 3 to 5.  Sepals short and broad, ciliate.  Petals 3 or 4 times as long as the sepals, attaining 3 or 4 lines, imbricate, glabrous.  Filaments more or less woolly on the edges, clavate and glandular at the top; anthers minutely apiculate or sometimes quite obtuse.  Style slender, occasionally exceeding the stamens, with a small capitate stigma.

**W. Australia.**  King George's Sound, *Oldfield;* N. side of Stirling range and Gordon plains, *Maxwell.*

42. **B. crenulata,** *Sm. in Trans. Linn. Soc.* viii. 284.  A glabrous bushy shrub.  Leaves obovate or cuneate, rounded and usually (but not always) crenulate at the upper end, rarely exceeding ½ in., narrowed into a very short petiole, coriaceous and nerveless.  Flowers terminal and solitary or few together, on very short pedicels or almost sessile, and also frequently solitary in the upper axils.  Sepals ovate, scarious at the edges and minutely ciliate.  Petals about twice as long, attaining 3 lines, broad, imbricate, and glabrous.  Filaments densely woolly at the sides, obtuse at the top; anthers apiculate.  Style short, often slightly pubescent.  Seeds smooth and shining. —DC. Prod. i. 721; Bot. Mag. t. 3915; Bot. Reg. 1838, t. 12; Bartl. in Pl. Preiss. i. 169.

**W. Australia.**  King George's Sound, *R. Brown, Menzies, Drummond,* and others; Stirling range, *Preiss, n.* 2010; Kalgan and Gordon rivers, *Oldfield;* and eastward to Stokes Inlet, *Maxwell.*

Var. *pubescens.*  Branches pubescent.  Leaves more sessile and less narrowed at the base, ciliate on the edge.  Sepals narrower.—W. Australia, *Drummond;* Vasse river, *Oldfield.*

43. **B. serrulata,** *Sm. Tracts,* 292, *t.* 5.  A glabrous shrub.  Leaves crowded, simple, almost sessile, broadly obovate or rhomboidal, acute, rarely exceeding ½ in., serrulate, narrowed at the base, coriaceous and nerveless. Flowers rather large, terminal, nearly sessile or very shortly pedicellate, several together in a leafy compact cyme or head or rarely solitary.  Sepals

Y 2

acute. Petals 2 or 3 times as long as the sepals, attaining 4 lines, broad, imbricate, mucronate, glabrous. Filaments more or less hairy, clavate-globular and hispid at the top; anthers minutely apiculate. Ovary glabrous. Style short, with a large globular 4-lobed stigma. Seeds black and shining.— Sw. Fl. Austral. t. 19; Bot. Reg. t. 842; Paxt. Mag. Bot. i. 173, with a figure.

**N. S. Wales.** Port Jackson, *R. Brown; Sieber, n.* 298, and others. Said to be known as "Native Rose" by the colonists.

44. **B. rhomboidea,** *Hook. Ic. Pl. t.* 722. A small, glabrous, much-branched, rigid shrub. Leaves simple, sessile, broadly rhomboid, obovate orbicular or almost reniform, obtuse, not exceeding ½ in., quite entire, coriaceous and nerveless. Flowers rather smaller than in *B. serrulata*, almost sessile, terminal and solitary or few together, or occasionally 1 or 2 in the axils of the next pair of leaves, surrounded by 1 or 2 pairs of floral leaves or bracts, usually spathulate and petiolate. Sepals ovate. Petals not twice as long as the sepals, attaining about 3 lines. Filaments glabrous, glandular-tuberculate, thickened upwards; anthers not apiculate. Ovary glabrous. Style rather long. Seeds apparently black and shining, but not seen quite ripe.— Hook. f. Fl. Tasm. i. 66.

**Tasmania.** North-west River near Hobarton and Western Mountains, *Gunn ;* ascending to 3000 or 4000 ft., *C. Stuart.*

45. **B. parviflora,** *Sm. Tracts,* 295, *t.* 6. A small, glabrous undershrub, forming a thick woody rhizome with numerous prostrate, ascending, or erect branching stems, usually under 6 in., but sometimes nearly 1 ft. long. Leaves all simple, from oblong to linear-lanceolate, rather acute, rarely ½ in. long. Flowers small, terminal, solitary or few in a leafy terminal cyme, on short thickened pedicels, one or two rarely axillary by the abortion of the flowering branch. Sepals acute, 1½ to 2 lines long. Petals not much exceeding them, imbricate, glabrous. Filaments glabrous or slightly hairy and glandular towards the top; anthers very minutely or not at all apiculate. Ovary glabrous; style short and thick. Cocci small. Seeds smooth and shining.—DC. Prod. i. 721; F. Muell. Pl. Vict. i. 113; *B. pilonema*, Labill. Pl. Nov. Holl. i. 98, t. 126; DC. Prod. i. 722; Hook. f. Fl. Tasm. i. 66.

**N. S. Wales.** Port Jackson, *R. Brown, Sieber, n.* 299, and others; northward to Hastings river, *Beckler.*

**Victoria.** Heathy and sandy moors at Port Albert, towards Wilson's Promontory, and near Cape Liptrap, *F. Mueller.*

**Tasmania.** Port Dalrymple, *R. Brown ;* common in heaths and sandy places throughout the island, *J. D. Hooker.*

Some specimens much resemble at first sight some of the smaller forms of *B. polygalifolia*, but a careful examination of the inflorescence will always suffice to distinguish them, independently of the seeds.

46. **B. viminea,** *Lindl. Swan Riv. App.* 17. A small or slender glabrous shrub. Leaves all simple, usually linear-lanceolate or linear-cuneate, flat, in some specimens ½ to 1 in. long, in others all under ½ in. Pedicels mostly axillary but also terminal, 1-flowered, short, thickened under the flower. Sepals ovate or lanceolate, short. Petals attaining 2 to 3 lines, glabrous, imbricate. Filaments densely woolly, glabrous glandular and obtuse

at the top; anthers tipped with a prominent erect or recurved appendage. Style hirsute. Seeds smooth and shining.

**W. Australia.** Swan River, *Drummond, 1st Coll.* The smaller specimens often much resemble elongated ones of *B. parviflora,* but have a more axillary inflorescence and apiculate anthers.

Var. *latifolia.* Leaves rather shorter, the lower ones broader and cuneate. Flowers rather larger with mucronate petals.—*B. tenuifolia,* Bartl. in Pl. Preiss. i. 168.—Canning river, *Preiss, n.* 2022 ; S. coast, *Gilbert, n.* 108 ; Fitzgerald range and E. Mount Barren, *Herb. F. Mueller.*

Var. *gracilis.* Leaves small. Flowers small, mostly axillary.—*Drummond, Coll.* 1848, *n.* 92.

*B. colorata,* Lehm. in Pl. Preiss. ii. 226 ; Herb. Preiss. n. 2627, which I have not seen, appears from the description to be referable to *B. viminea.*

47. **B. filifolia,** *F. Muell. Fragm.* i. 3, *and* ii. 177. A low glabrous shrub, with short slender but rigid branches. Leaves either simple and sessile, linear-terete, about ½ in. long, or 3-foliolate with 3 smaller linear-terete leaflets on a distinct common petiole. Peduncles slender, terminal, bearing a single flower, or short with a cyme of 3 flowers on pedicels of 2 or 3 lines thickened under the flower. Sepals small, lanceolate. Petals attaining 2 or 3 lines, glabrous, imbricate. Filaments slightly ciliate, clavate and glandular at the top ; anthers not apiculate. Seeds smooth and shining.

**S. Australia.** Sandy plains near Encounter Bay, *F. Mueller;* Kangaroo Island, *Waterhouse;* Tatiara country, *Woods;* near Adelaide, *Herb. Hooker.*—F. Mueller (Pl. Vict. i. 229) thinks that this may prove to be a variety of *B. pinnata,* a species with which however it appears to me to have very little connection.

48. **B. inornata,** *Turcz. in Bull. Mosc.* 1852, ii. 164. A low, much-branched, rather slender but rigid shrub, usually glabrous and often tuberculate with prominent glands. Leaves usually compound, with a very short common petiole ; leaflets 3, 5, or rarely 1, linear-terete, very obtuse, rarely above 3 lines long, and often much shorter. Flowers terminal, solitary or 2 or 3 together on short pedicels. Sepals broad and short, usually ciliate. Petals attaining 2 to 3 lines, nearly glabrous, much imbricate, rather acute but not prominently mucronate. Filaments glabrous, clavate-glandular ; anthers apiculate. Stigma capitate in some specimens, minute in others where the anthers are longer. Seeds smooth but not shining.—*B. leptophylla,* Turcz. l. c. ; *B. clavellifolia,* F. Muell. in Trans. Phil. Soc. Vict. i. 12 ; Fragm. i. 99 ; Pl. Vict. i. 117.

**Victoria.** Sandy desert towards Lake Albert, *F. Mueller.*
**S. Australia.** Mallee scrub near the Murray, *F. Mueller.*
**W. Australia,** *Drummond, 5th Coll. n.* 196 *and* 197; Phillips ranges, Gardiner's River, and Middle Mount Barren, *Herb. F. Mueller.*

49. **B. oxyantha,** *Turcz. in Bull. Mosc.* 1852, ii. 165. Nearly allied to *B. inornata* and with the same habit and foliage, but with the branches minutely hoary-pubescent and not glandular. Leaflets 3 or 5, linear-terete and frequently 3 lines long. Flowers rather larger than in *B. inornata.* Sepals lanceolate, with long subulate points, or almost subulate from the base. Petals distinctly pointed. Filaments densely ciliate ; anthers apiculate. Style short, with a small stigma.—*B. brachyphylla,* F. Muell. Fragm. ·i. 99 ; ii. 180.

**W. Australia,** *Drummond, 5th Coll. n.* 198 ; Fitzgerald ranges, *Herb. F. Mueller.*

SERIES VII. PEDUNCULATÆ.—Anthers uniform. Leaves simple. Peduncles terminal, elongated, usually several-flowered.

50. **B. scabra,** *Lindl. Swan Riv. App.* 17. A much-branched erect shrub of 1 to 1½ ft., roughly pubescent or hirsute with short spreading hairs. Leaves all simple, nearly sessile, linear or oblong, very obtuse, rarely exceeding 4 or 5 lines, the margins much revolute and usually pale underneath. Flowers terminal, solitary or few in shortly pedunculate cymes, or in cyme-like leafy clusters. Sepals with a very short broad base and filiform hispid points. Petals rather narrow, but imbricate, 2 to 3 lines long, finely mucronate. Filaments ciliate, almost capitate and glandular at the top; anthers tipped with a rather large recurved appendage. Style rather thick, glabrous or pubescent. Cocci usually pubescent. Seeds smooth and apparently opaque, but not seen quite ripe.

**W. Australia.** Swan River, *Drummond, 1st Coll.; Fraser.* The young leaves are often clustered in the axils, but, as far as I have seen, always simple.

51. **B. thymifolia,** *Turcz. in Bull. Mosc.* 1852, ii. 165. A much-branched, rather slender shrub, glabrous or slightly pubescent with short spreading hairs. Leaves all simple, nearly sessile, linear, obtuse, rarely attaining 4 lines, the margins much revolute. Flowers 1 to 3, on rather long terminal peduncles, or sometimes more numerous, forming a showy corymbose cyme. Sepals broad, shortly acuminate, glabrous or hirsute. Petals attaining about 3 lines, imbricate, glabrous. Filaments slightly ciliate, clavate and glandular at the top ; anthers tipped with a prominent recurved white appendage. Style short, pubescent; stigma oblong.—*B. fasciculifolia,* F. Muell. Fragm. i. 99 ; ii. 99.

**W. Australia,** *Drummond, 5th Coll. n.* 195, *A. Gregory ;* Salt river, Fitzgerald river, etc., *Maxwell.*—The species differs from *B. scabra,* chiefly in the long peduncles, short sepals, and in the want of the long points to the petals.

52. **B. ovata,** *Lindl. in Bot. Reg.* 1841, *under n.* 47. A glabrous undershrub or shrub, forming a thick stock and erect dichotomous stems, usually under 1 ft. Leaves almost sessile, cordate-ovate or the upper ones lanceolate, obtuse or acute, under ½ in. long, the margins entire and recurved. Flowers few, in loose terminal pedunculate dichotomous cymes, the branches and pedicels slender. Sepals short, acuminate. Petals attaining about 4 lines, imbricate, glabrous. Filaments glabrous, capitate and glandular at the top; anthers tipped with an obtuse recurved appendage. Style rather thick, glabrous or hairy.

**W. Australia.** Swan River, *Drummond, 1st Coll. ;* Darling range, *Collie.*

53. **B. fastigiata,** *Bartl. in Pl. Preiss.* i. 167. A glabrous glaucous shrub or undershrub, with erect and rigid or weak and decumbent branches. Leaves obovate, spathulate or oblong, rarely attaining ½ in., very obtuse, entire or denticulate, narrowed at the base. Flowers in loose umbel-like simple cymes, terminal or in the upper axils, the common peduncle short, with usually 4 to 6 rather long pedicels, thickened upwards. Sepals ovate or ovate-lanceolate, acute, herbaceous and almost valvate. Petals rarely twice as long, attaining about 3 lines. Filaments ciliate, narrowed upwards, slightly glandular ; anthers oblong, almost terminal, not apiculate. Cocci truncate. Seeds smooth and shining.

**W. Australia,** *Drummond, n.* 119; Plantagenet district, *Preiss, n.* 2028; Gordon river, *Oldfield;* S.W. interior, *Maxwell.*

Var. (?) *tenuior.* Leaves thin, almost lanceolate, serrate.—W. Australia, *Gilbert, n.* 3 *and* 18.—Weak drawn-up specimens of this and of *B. viminea* have much general resemblance, although the species generally are widely distinct.

54. **B. denticulata,** *Sm. in Trans. Linn. Soc.* viii. 284. Shrubby, erect, glabrous and somewhat glaucous. Leaves nearly sessile, simple, linear or lanceolate, rarely oblong-cuneate, flat but rather thick, $\frac{1}{2}$ to $1\frac{3}{4}$ in. long, often bordered by a few small glandular teeth or more distinctly denticulate when broad. Flowers rather large, in loose terminal shortly pedunculate cymes or corymbs, the pedicels thickened upwards. Sepals very acute, usually short but variable. Petals attaining about 3 lines or rather more, imbricate, glabrous. Filaments ciliate and flattened towards the base, terete and glandular upwards, obtuse at the top; anthers short, not apiculate.—DC. Prod. i. 721; Bot. Reg. t. 1000; *B. chironiifolia,* Bartl. in Pl. Preiss. i. 167.

**W. Australia.** King George's Sound, *R. Brown* and others; and other parts of the southern districts, *Drummond, n.* 22, *Preiss. n.* 2027, *Oldfield,* and others; eastward to Phillips river and E. Mount Barren, *Maxwell.*

55. **B. spathulata,** *Lindl. Swan Riv. App.* 17. A glabrous glaucous undershrub, forming a thick stock, with erect simple or branched stems, $\frac{1}{2}$ to $1\frac{1}{2}$ ft. high, or when very luxuriant attaining 3 ft. Leaves not numerous, from obovate or oblong-spathulate to linear-cuneate or lanceolate, obtuse or rarely acute, $\frac{1}{2}$ to 1 in. or rarely longer, thick, nerveless, quite entire. Flowers few, rather large, in irregular terminal pedunculate cymes. Pedicels glabrous or glandular. Sepals usually very acute. Petals attaining 4 lines or more, imbricate, glabrous. Filaments ciliate; anthers often minutely apiculate.— Bartl. in Pl. Preiss. 167; *B. flexuosa,* Bartl. l. c. i. 166; *B. macra,* Bartl. l. c. 167.

**W. Australia.** Apparently common from King George's Sound, *R. Brown* and others, to Swan River, *Drummond* and others; Canning river, *Preiss, n.* 2024, 2025; Darling range, *Preiss, n.* 2026; Preston river, *Oldfield.*

Var. *ramosa.* More branched, flowers more numerous, in long pedunculate cymes.—Swan River, *Drummond;* King George's Sound, *Baxter, Collie;* eastward to E. Mount Barren, *Maxwell.*

Var. *elatior.* Tall, with elongated branches, the upper leaves linear and distant, occasionally slightly dilated at the base. Flowers in very loose dichotomous cymes.—*B. dichotoma,* Lindl. Bot. Reg. 1841, under n. 47.—Vasse river, *Mrs. Molloy;* Swan River, *Drummond, Coll.* 1843, *n.* 38.

56. **B. juncea,** *Bartl. in Pl. Preiss.* i. 166. An undershrub, with erect virgate or rush-like stems, glabrous and little branched or dichotomous upwards. Leaves few, linear-terete, rather thick, the lower ones sometimes 1 in. long, the upper ones few, small and distant, and some specimens almost leafless. Flowers small, terminal, solitary or few together, on short glabrous or woolly pedicels. Sepals lanceolate-subulate, nearly as long as the petals. Petals about 2 or rarely 3 lines long, mucronate, imbricate, often slightly pubescent outside along the centre. Filaments glabrous, slightly ciliate, glandular and obtuse; anthers not apiculate. Cocci small, truncate. Seeds smooth and shining.—*B. laniflora,* Bartl. in Pl. Preiss. i. 165 (specimens with woolly calyces).

**W. Australia.** King George's Sound, *R. Brown* and others; southern districts, *Preiss, n.* 2030, 2036, and 2037. Some specimens from near Tone Bridge, in Herb. F. Mueller are remarkable for their large flowers. In all others they rarely much exceed 2 lines.

57. **B. cymosa,** *Endl. in Hueg. Enum.* 16. A glabrous, often glaucous undershrub or shrub, forming a thick stock with erect virgate branches. Leaves sessile, linear-terete, often crowded towards the upper part of the branches or clustered in the axils, ½ to 1 in. or rather longer, sometimes fewer and more distant, the larger ones rarely flattened with revolute margins but always narrow-linear and quite entire. Flowers rather small, usually numerous and cymose, on long terminal peduncles. Pedicels short. Sepals short and broad. Petals attaining about 3 lines. Filaments ciliate, slightly dilated at the base, terete and glandular upwards; anthers minutely apiculate.— *B. teretifolia*, Lindl. Swan Riv. App. 17; Bartl. in Pl. Preiss. i. 166; F. Muell. Fragm. ii. 101.

**W. Australia.** Swan River, *Drummond, 1st Coll.* and *(2nd Coll.) n.* 88, *Preiss, n.* 2023, 2029; Vasse river and Darling Range, *Oldfield.*

### 3. ACRADENIA, Kipp.

Calyx 5-cleft, rarely 6- or 7-cleft. Petals 5, rarely 6 or 7, imbricate. Disk thick, entire. Stamens 10, rarely 12 or 14, inserted outside the disk; anthers all similar and perfect. Carpels usually 5, united almost to the top, each terminating in a glabrous gland. Styles terminal, united in one filiform style, with a small stigma. Ovules 2 in each carpel, collateral or almost superposed. Cocci 5 or fewer, 2-valved; endocarp and seeds unknown.— Leaves opposite, 3-foliolate. Flowers white, in a terminal trichotomous cyme.

The genus is limited to a single species, endemic in Tasmania. It is evidently nearly allied to *Boronia* and especially to *Zieria*, from which it differs in the flowers, usually 5-merous, with all the stamens perfect and no glands to the disk. The endocarp has been described, on the authority of Kippist, as not separating, but the only fruits known are open and have already shed their seed; and, on comparing them carefully with those of other *Diosmeæ* in a similar state, I cannot but conclude that, as is usual in the tribe, the endocarp has been cast with the seed.

1. **A. Frankliniæ,** *Kipp. in Trans. Linn. Soc.* xxi. 207, *t.* 22. A shrub of 8 to 12 ft., glabrous or the young shoots minutely pubescent. Leaves mostly opposite, 3-foliolate, with a short common petiole; leaflets oblong-lanceolate, obtuse, 1 to 2 in. long, more or less crenately toothed, coriaceous, green on both sides, usually scabrous, with prominent glands. Cymes nearly sessile at the ends of the branches, loosely trichotomous. Sepals distinct, short. Petals 2½ to 3 lines long, pubescent outside. Filaments filiform, glabrous, scarcely shorter than the petals; anthers not apiculate. Ovary very villous, except the small glands terminating each carpel. Cocci hard, truncate, scarcely beaked, transversely wrinkled.—Hook. f. Fl. Tasm. i. 69.

**Tasmania.** Macquarie Harbour and river, *Milligan.*

### 4. CROWEA, Sm.

Calyx 5-cleft. Petals 5, imbricate in the bud. Disk annular. Stamens 10, shorter than the petals; filaments flattened, ciliate or woolly; anthers linear, hirsute, tipped with long hirsute appendages. Ovary 5-lobed; styles inserted above the middle of the carpels, immediately united into one filiform style with a small or globular stigma.   Ovules 2, superposed or almost collateral.   Cocci 2-valved, rounded or truncate at the top, the endocarp cartilaginous and separating elastically.—Glabrous shrubs or undershrubs.   Leaves alternate, simple.   Flowers rather large, red purple or green, glabrous, solitary, axillary or terminal.

The genus is confined to Australia.   It is united by F. Mueller with *Eriostemon*, from which it differs chiefly in the long hairy appendages of the anthers.

Peduncles terminal or, if axillary, leafy at the base.   Branches scarcely
  angular . . . . . . . . . . . . . . . . . . . . . . . 1. *C. exalata.*
Peduncles all axillary, without leafy bracts.   Branches very angular or
  almost winged.
  Leaves quite entire.   Style short.   Stigma globular . , . . . 2. *C. saligna.*
  Leaves mostly or all denticulate.   Style long.   Stigma short.
    Branches erect, almost herbaceous.   Leaves linear or narrowed
    at both ends . . . . . . . . . . . . . . . 3. *C. angustifolia.*
    Rigid shrub.   Leaves from broadly cuneate to oblong, truncate
    or very obtuse . . . . . . . . . . . . . . . 4. *C. dentata.*

1. **C. exalata,** *F. Muell. in Trans. Phil. Soc. Vict.* i. 11.   Shrubby, with the branches more slender than in *C. saligna,* and scarcely angular.   Leaves numerous, narrow-linear, mostly obtuse, often all under 1 in. and rarely attaining 1½ in., all entire.   Flowers smaller than in *C. saligna,* on short peduncles, almost terminal, or if axillary by the abortion of the flowering branch, the peduncle usually bears 1 or more small leaves at its base.   Petals rarely ½ in. long, red or rarely green.   Stamens as in *C. saligna,* the petaline filaments shorter than the others.   Ovary very short; style very short, with a large globular stigma.   Cocci small, free from the base.—*Eriostemon Crowei* (partly), F. Muell. Pl. Vict. i. 119.

**N. S. Wales.**   Paramatta, *Wilson;* Yowaka river, Mount Tambo, etc., near Twofold Bay, *F. Mueller.*
**Victoria.**   Mount Macfarlane, near Omeo, Mitta-Mitta, Livingston and Genoa rivers, and Boggy Creek, towards Lake King, *F. Mueller.*
This plant is now considered by F. Mueller as specifically identical with *C. saligna,* and it may possibly prove to be a variety of that species; but, besides the general habit, foliage, and less angular stems, the inflorescence appears to me to be different in all the specimens I have seen.

2. **C. saligna,** *Andr. Bot. Rep. t.* 79.   Shrubby and erect, the branches prominently angular.   Leaves mostly lanceolate, narrowed at each end, acute or obtuse, 1 to 2 in. long, of a much thinner consistence than those of *Eriostemon salicifolius,* which this species sometimes resembles, in some specimens passing into a broadly oblong or elliptical-ovate shape, in others almost linear, like those of *C. exalata.*   Flowers red, on axillary pedicels shorter than the leaves, thickened upwards, with 2 very minute bracts at their base.   Sepals short and broad.   Petals 7 to 9 lines long.   Appendage of the anthers longer than the cells themselves.   Style very short, with a large globular stigma.   Cocci short, united to near the top.   Seeds reticulate, somewhat shining.—Vent. Jard. Malm. t. 7 ; Bot. Mag. t. 989 ; DC. Prod. i. 720 ; *C. latifolia,* Lodd. in G. Don, Gen. Syst. i. 792 ; *Eriostemon Crowei* (partly), F. Muell. Pl. Vict. i. 119.

**N. S. Wales.** Port Jackson, *R. Brown, Sieber, n.* 295 (the names or numbers of this and n. 294, *Eriostemon salicifolius*, interchanged in some collections), and others.

*C. latifolia,* Paxt. Mag. Bot. xiv. 222, with a fig., is one of the commonest forms of this species. In some specimens from Manly Reach, *Woolls (Herb. Muell.)*, the leaves are nearly twice as broad. In others from between Richmond river and Raymond Terrace, *A. Ralston (Herb. Muell.)*, they are linear, elongated, mostly rounded or truncate at the top. Again, in numerous specimens collected by *R. Brown* on the Hawkesbury river, they are linear, but smaller and more crowded, approaching those of *C. exalata ;* but in all, the pedicels are axillary and leafless.

3. **C. angustifolia,** *Turcz. in Bull. Mosc.* 1849, ii. 13. Apparently an undershrub with virgate erect branches of 1 to 2 ft., less woody than in other species, acutely angled and almost winged. Leaves sessile, linear, mostly acute, 1 to 2 in. long, entire or minutely serrulate. Flowers red or white, rather smaller than in *C. saligna,* all axillary, solitary or rarely 2 together, on very short pedicels, thickened upwards, with minute bracts at the base. Sepals very short. Petals not exceeding $\frac{1}{2}$ in. Filaments glabrous or slightly ciliate ; anthers with longer cells and a shorter, less hairy, and flatter appendage than in *C. saligna.* Style elongated, with a small stigma. Cocci broad, transversely wrinkled.—*Eriostemon Turczaninowii,* F. Muell. Pl. Vict. i. 120.

**W. Australia.** King George's Sound, *R. Brown ;* southern districts, *Drummond* and others.

Var. (?) *platyphylla.* Leaves ovate-elliptical, narrowed at each end, minutely and regularly crenate-serrate.—Franklin river, *Maxwell.*

4. **C. dentata,** *R. Br. Herb.* A rigid erect branching shrub, the young branches very angular. Leaves sessile with a broad base, from broadly cuneate and truncate to narrow-oblong, $\frac{1}{2}$ to 1 in. long, strongly and acutely serrate, coriaceous and rigid. Peduncles 1-flowered, axillary, short and thick, slightly hoary as well as the petals. Sepals very short. Petals 4 to 5 lines long. Filaments glabrous or slightly ciliate ; anthers with an appendage as long as the cells, very hairy, as in *C. saligna.* Cocci obtuse or obscurely beaked.

**W. Australia.** King George's Sound, *Baxter (Hb. R. Brown).*

## 5. ERIOSTEMON, Sm.

Calyx 5-cleft or rarely 4-cleft. Petals 5, rarely 4, imbricate. Disk usually more or less thickened. Stamens 10, rarely 8, shorter than the petals ; filaments hairy, attenuate or rarely obtuse at the top ; anthers usually tipped with a very small point or appendage. Carpels 5, rarely 4 or fewer, distinct from the base (or in one species united to the middle), usually produced into a short appendage above the cells ; styles inserted below the middle and immediately united into one ; stigma small. Ovules 2 in each cell, superposed. Cocci 2-valved, usually more or less beaked at the top or at the outer angle ; the endocarp cartilaginous and separating elastically. Seeds solitary.—Shrubs, either glabrous or slightly pubescent, without scurfy scales. Leaves alternate, simple, entire, the glands often large and prominent. Inflorescence axillary or terminal ; peduncles bearing a single flower, or an umbel of few, white pink or rarely blue flowers. Calyx small, with short broad lobes or sepals, except in *E. nodiflorus.*

Besides the Australian species, which are all endemic, the genus comprises one from New Caledonia. F. Mueller proposes to extend its limits so as to include *Phebalium, Microcybe, Geleznovia, Crowea, Philotheca, Drummondita,* and *Asterolasia,* which are all no doubt nearly enough related to it to be equally well regarded as sections or as substantive genera; but as the majority of them have been long established and universally adopted, and are distinguished by characters easily recognized, their union into one vast genus seems to me to be scarcely justified.

Inflorescence axillary.
  Filaments clavate and glandular at the top.
    Leaves linear or lanceolate, thick, obscurely 1-nerved. Bracts
      on the pedicel several, imbricate . . . . . . . . . 1. *E. salicifolius.*
    Leaves oblong, finely 3-nerved. Bracts on the pedicel 1 to 3,
      distant . . . . . . . . . . . . . . . . 2. *E. Banksii.*
  Filaments subulate at the top, usually flattened below.
    Flowers 4-merous . . . . . . . . . . . . . . 3. *E. virgatus.*
    Flowers 5-merous.
      Leaves oblong or lanceolate, 1 to 3 or 4 in. long, flat, 1-
      nerved.
        Pedicels slender, 1-flowered. Carpels of the ovary united
          to above the middle, and not rostrate when ripe . . . 4. *E. trachyphyllus.*
        Pedicels rigid, usually several-flowered. Carpels free from
          the base, rostrate when ripe . . . . . . . . . 5. *E. myoporoides.*
      Leaves linear or linear-spathulate, mucronate, with recurved
        margins and a prominent midrib . . . . .· . . . . 6. *E. hispidulus.*
      Leaves short, cordate-ovate or obovate, the margins thickened
        or recurved, the midrib prominent . . . . . . . . 7. *E. buxifolius.*
      Leaves obovate or spathulate, thick, flat or concave, the
        midrib faint or none . . . . . . . . . . . 8. *E. obovalis.*
      Leaves narrow-linear, convex underneath or terete.
        Filaments flat . . . . . . . . . . . . . . 9. *E. scaber.*
        Filaments subulate . . . . . . . . . . . . . 10. *E. linearis.*
Inflorescence terminal, appearing sometimes lateral by the elongation
  of the side shoots.
  Flowers solitary or rarely 2 or 3 together.
    Leaves small, flat or with recurved margins.
      Leaves not above 2 lines long, thick, warted or crenate with
        large prominent glands . . . . . . . . . . 11. *E. difformis.*
      Leaves flat, oblong or linear, 3 to 4 lines, crenate, with a
        prominent midrib . . . . . . . : . . . . . 11. *E. difformis,* var.
                                         [*Smithianus.*
      Leaves flat, linear-cuneate, 2 to 4 lines, slightly crenate,
        nerveless . . . . . . . . . . . . . . . 12. *E. parvifolius.*
    Leaves linear-terete.
      Leaves warted with large glands. Flowers not above 3 lines 11. *E. difformis,* var.
                                         [*brevifolius.*
      Leaves smooth. Flowers nearly 5 lines . . . . . . . 13. *E. ericifolius.*
  Flowers (usually blue) densely clustered or capitate . . . . . 14. *E. nodiflorus.*
  Flowers (usually pale blue) in loose racemes . . . . . . . 15. *E. spicatus.*
(*Eriostemon dentatus,* Colla, is *Elæocarpus dentatus,* Vahl, a New Zealand plant.)

1. **E. salicifolius,** *Sm.; DC. Prod.* i. 720. An erect shrub, the branches rigid and often angular, glabrous or minutely hoary. Leaves linear or linear-lanceolate, mostly ₁ 1 to 2 in. long, rather thick and rigid, glabrous when full-grown, obscurely 1-nerved. Peduncles axillary, short and 1-flowered, with a few broad scale-like imbricate bracts at the base, hoary with a minute tomentum as well as the calyx and petals. Sepals short, orbicular, rigid. Petals

pink, attaining about ½ in. Filaments flattened, densely fringed with woolly hairs, clavate and glandular at the top, bearing the anthers on a short stipes as in *Boronia ;* anthers tipped with a very short broad recurved appendage. Ovary glabrous ; style slightly pubescent below the middle. Cocci truncate at the top, but not beaked, transversely wrinkled. Seeds smooth and shining. —Rudge, in Trans. Linn. Soc. xi. t. 26 ; Deless. Ic. Sel. iii. t. 46 ; Bot. Mag. t. 2854 ; *E. lanceolatus,* Gærtn. f. Fr. iii. 154, t. 210 ; *Crowea scabra,* Grah. in Edinb. Phil. Journ. 1827, 174.

**N. S. Wales.** Port Jackson, *. Brown, Sieber, n.* 294 (the names or numbers of this and *Crowea saligna,* 295, interchanged in many herbaria), and *Fl. Mixt. n.* 536, and others.

The synonym often quoted of *E. australasia,* Sm., is an error. Smith mentions no species in Trans. Linn. Soc. iv. 221, but in describing the genus gives the station Australasia, which has been mistaken for a specific name.

2. **E. Banksii,** *A. Cunn.; Endl. in Hueg. Enum.* 15. A large shrub, the young branches angular and loosely hairy. Leaves from obovate-oblong to oblong-lanceolate, often oblique, obtuse, 1 to 1½ in. long, contracted into a very short petiole, thinly coriaceous, finely veined and obscurely 3-nerved, glabrous or slightly hairy. Peduncles very short, axillary, 1- or rarely 2-flowered, usually with 2 or 3 scale-like distant bracts. Sepals small, ciliate. Petals attaining about 3 lines, hoary outside, with a prominent midrib. Filaments slightly flattened, woolly outside, clavate and glandular at the top as in *E. salicifolius ;* anthers not apiculate. Ovary glabrous, style pubescent._ Carpels of the fruit 4 or 5 lines long, truncate, very shortly beaked.

**Queensland.** Sandy shores of the Endeavour river, *Banks and Solander, R. Brown, A. Cunningham.* The leaves have very much the aspect of the phyllodia of some *Acacias.*

3. **E. virgatus,** *A. Cunn.; Hook. f. in Hook. Journ. Bot.* ii. 417. An erect, glabrous shrub, with virgate branches. Leaves rather crowded, cuneate-oblong, obtuse, mucronate, mostly about ½ in. long, flat, almost shining above, pale underneath, with a prominent midrib, the tubercular glands small. Pedicels axillary, 1-flowered, shorter than the leaves, but rather slender. Flowers 4-merous. Sepals small. Petals glabrous, 2½ to 3 lines long. Filaments flattened, ciliate, attenuate at the top ; anthers minutely apiculate. Cocci glabrous, rostrate.—Hook. f. Fl. Tasm. i. 64.

**Tasmania.** Rocky shores of Macquarie Harbour, *A. Cunningham ;* Rocky Cape, *Gunn ;* hills on Huon river, *Oldfield.* This is the only species with 4-merous flowers, and appears to be constantly so. *Phebalium Oldfieldi,* F. Muell., referred to it in Hook. f. Fl. Tasm. ii. 358, from specimens in leaf only, is very different in inflorescence and flowers, and even the leaves differ in being never mucronate.

4. **E. trachyphyllus,** *F. Muell. in Trans. Phil. Soc. Vict.* i. 99, *and Pl. Vict.* i. 121. A tall glabrous shrub, with prominent tubercular glands. Leaves from cuneate-oblong to narrow-lanceolate, shortly mucronate, 1 to 2 in. long, much narrowed at the base, flat or the margins slightly recurved, the midrib prominent underneath. Pedicels axillary, 1-flowered, slender, but shorter than the leaves. Petals white, glabrous, about 3 lines long. Filaments somewhat flattened, ciliate, attenuate at the top ; anthers minutely apiculate . Ovary glabrous, the carpels united to ¾ of their height but

deeply depressed in the centre, the style attached below the middle. Capsule obtuse, 5-angled, the carpels not rostrate, separating at length to below the middle. Seeds smooth and shining.

**N. S. Wales.** Forest gullies near Twofold Bay, and about the sources of the Yowaka river, *F. Mueller.*

**Victoria.** Rocky declivities on Snowy River, near Pinch river, *F. Mueller.*

This species differs from all others in the united carpels of the ovary; but the habit, æstivation of the petals, stamens, and other characters, are those of *Eriostemon;* and even the ovary is different in shape from that of *Asterolasia* and other genera where the carpels are more or less united.

Var. (?) *Leichhardtii.* Foliage of the typical form. Flowers much larger; filaments much dilated and shortly ciliate to the middle, fringed with long hairs in the upper part; anthers larger; lobes of the ovary produced into long appendages, and carpels therefore probably beaked.—" From Brroa " (N. S. Wales?), *Leichhardt.*

5. **E. myoporoides,** *DC. Prod.* i. 720. A stout, usually tall, glabrous shrub, with the habit of a *Myoporum,* the glandular tubercles sometimes very prominent, sometimes almost inconspicuous. Leaves sessile, from obovate-oblong to lanceolate or linear-lanceolate, obtuse or rarely acute, always mucronate, 1 to 3 or rarely above 4 in. long, rather firm and sometimes coriaceous, flat with the midrib prominent underneath. Peduncles shorter than the leaves, usually bearing an umbel of 3 to 9 flowers, very rarely reduced to 1 or 2, especially on the smaller-leaved branches. Flowers white ' or pink, rather large, the petals attaining about 4 lines. Filaments flat, more or less ciliate, attenuate at the top. Ovary glabrous. Cocci beaked.—Bot. Mag. t. 3180; Deless. Ic. Sel. iii. t. 47; F. Muell. Pl. Vict. i. 122; *E. cuspidatus,* A. Cunn. in Field, N. S. Wales, 331; *E. neriifolius,* Sieb. in Spreng. Syst. Cur. Post. 164; *E. lancifolius,* F. Muell. in Trans. Vict. Inst. i. 32.

**Queensland.** Glasshouse Mountains, *F. Mueller.*

**N. S. Wales.** Port Jackson to the Blue Mountains, *R. Brown, Sieber, n.* 306, *A. Cunningham,* and others; northward to New England, *Herb. Mueller;* in the interior to Lachlan river, *A. Cunningham.*

**Victoria.** Upper valleys of the Mitta-Mitta river, mounts Hotham, Latrobe, Tambo, and Macfarlane, *F. Mueller.*

Var. *minor.* Leaves rarely much above 1 inch long, peduncles mostly 1- or 2-flowered.—*E. intermedius,* Hook. Bot. Mag. t. 4439.—To this form belong the Queensland and Lachlan river specimens. I cannot, however, see in them any near approach to *E. buxifolius.*

6. **E. hispidulus,** *Sieb. in Spreng. Syst. Cur. Post.* 164. Shrubby, with elongated branches, more or less pubescent. Leaves sessile, linear or linear-spathulate, mucronate with a straight or recurved point, ½ to 1 in. long, the margins revolute, usually pubescent especially underneath, rarely glabrous, often tuberculate with prominent glands. Peduncles axillary, shorter than the leaves, 1- or rarely 2-flowered, the pedicel thickened under the flower. Petals attaining 3 or 4 lines. Stamens, style, and fruit of *E. buxifolius.*

**N. S. Wales.** Port Jackson to the Blue Mountains, *R. Brown, Sieber, n.* 305, *A. Cunningham,* and others. F. Mueller considers this as a variety of *E. buxifolius.* The foliage appears to me, however, to be constantly distinct.

7. **E. buxifolius,** *Sm.; DC. Prod.* i. 720. Shrubby, with rigid pubescent branches. Leaves sessile, small, cordate-ovate or obovate, usually mucronate, under ½ in. long, thick and usually tuberculate with prominent glands, the margins thickened or recurved, the midrib prominent underneath.

Peduncles short, axillary, 1- or very rarely 2-flowered with very minute bracts below the middle or at the base, thickened upwards. Petals broadly oblong, attaining 4 or 5 lines. Filaments flattened, slightly ciliate, the longer ones or all attenuate and glabrous at the top; anthers minutely apiculate. Carpels much elongated above the cells; style glabrous. Cocci ovate, beaked on the upper outer edge.—Deless. Ic. Sel. iii. t. 45; Bot. Mag. t. 4101.

**N. S. Wales.** Port Jackson, *R. Brown, Sieber, n.* 304, and others. This species seems occasionally almost to run into *E. obovalis* in the shape of its leaves, but is then always known by the recurved margins and prominent midrib.

8. **E. obovalis,** *A. Cunn. in Field, N. S. Wales,* 331. A glabrous shrub of 2 to 3 ft. Leaves obcordate, obovate or oblong-spathulate, very obtuse or truncate, rarely attaining ½ in., much narrowed at the base and often petiolate, thick but flat or concave above, the midrib little conspicuous, usually strongly tuberculate with prominent glands. Pedicels axillary, 1-flowered, short and thickened upwards. Flowers rather smaller than in *E. buxifolius*. Petals glabrous, attaining 3 or 4 lines. Filaments flattened, ciliate. Cocci beaked, at least when young (not seen ripe).—*E. verrucosus,* A. Rich. Sert. Astrol. 74, t. 26; Hook. f. Fl. Tasm. i. 64; F. Muell. Pl. Vict. i. 123; *E. obcordatus,* A. Cunn. in Hook. Journ. Bot. i. 254; Hook. Ic. Pl. t. 60.

**N. S. Wales.** Verge of Regent's Glen, Blue Mountains, *A. Cunningham;* Bluff's Head, *Caley.*
**Victoria.** Barren ranges and forest land, not common, *F. Mueller.*
**Tasmania.** Derwent river, *R. Brown;* common in gravelly and sandy soil throughout the island, *J. D. Hooker.*
A. Richard gives Moreton Bay as the station of his plant, but that is probably owing to some mistake of Lesson's in labelling the plants received from Fraser. In A. Cunningham's diagnosis the flowers are said to be terminal, but I find them always axillary in his specimens, although sometimes proceeding from the upper axils so as to appear terminal without close examination.

9. **E. scaber,** *Paxt. Mag. Bot.* xiii. 127, *with a figure.* A shrub, with the general aspect of *E. hispidulus,* but with glabrous or very minutely pubescent branches. Leaves sessile, narrow-linear, acute and mucronulate, under 1 in. long, thick and very convex underneath, flat or channelled above and often almost terete, the margins never revolute, more or less tuberculate with prominent glands. Inflorescence and flowers of *E. obovalis.* Carpels much compressed, prominently rostrate.

**Queensland.** Glasshouse Mountains, *F. Mueller.*
**N. S. Wales.** St. George's river, *R. Brown;* Paramatta, *Woolls;* Port Jackson, *Caley;* near Liverpool, *Leichhardt.*
This is considered by F. Mueller as a variety of *E. buxifolius.* It appears to me to be nearer to *E. obovalis,* and differs from both chiefly in foliage.

10. **E. linearis,** *A. Cunn.; Endl. in Hueg. Enum.* 16. A rigid heath-like shrub, quite glabrous or the branches minutely pubescent. Leaves sessile, linear-terete, obtuse or scarcely mucronate, sometimes all under ½ in., but attaining ¾ in. when very luxuriant, more or less tuberculate with prominent glands. Pedicels short, axillary, 1-flowered. Flowers white or pink. Petals glabrous, attaining 2½ or scarcely 3 lines. Filaments filiform, very hairy; anthers minutely apiculate. Ovary glabrous; stigma slightly dilated and lobed. Cocci glabrous, beaked.—*E. halmaturorum,* F. Muell. in Linnæa, xxv. 376.

**N. S. Wales.** Mount Boyne, *Fraser;* Goulburn and Peel ranges, *A. Cunningham;* Mount Murchison and Ebers ranges, *F. Mueller.* United by F. Mueller with *E. difformis;* it differs in the inflorescence, which is that of the last 3 species, from which it is distinguished by the filaments quite filiform or scarcely perceptibly flattened. The leaves are more slender than in either species.

11. **E. difformis,** *A. Cunn.; Endl. in Hueg. Enum.* 15. A much-branched compact shrub, glabrous or the younger branches minutely pubescent. Leaves in the normal form small, numerous, obovate, oblong, or almost rhomboidal, very obtuse, rarely above 2 lines long, usually tuberculate or as it were crenate, with 2 or 3 very large prominent glands, thick and convex, the margins often recurved, glabrous on both sides. Flowers small, terminal, solitary or 2 or 3 together, on very short pedicels. Calyx very small. Petals 2 to nearly 3 lines long, usually pubescent outside. Filaments flattened, densely ciliate; anthers shortly apiculate. Ovary villous; style short. Cocci very shortly beaked.—F. Muell. Pl. Vict. i. 123; *E. rhombeus,* Lindl. in Mitch. Trop. Austr. 293.

**Queensland.** Mantua downs, *Mitchell;* between Mackenzie and Dawson rivers, *F. Mueller;* near Warwick, *Beckler;* near Broad Sound, *Herb. Mueller.*
**N. S. Wales.** Lachlan river, *A. Cunningham.*
**Victoria.** Murray river and Grampian Mountains, *F. Mueller.*
**W. Australia.** *Drummond, n.* 55.

Var. (?) *Smithianus.* Quite glabrous. Leaves flat, thin, oblong or linear, glandular crenate, 3 to 4 lines long, with a conspicuous midrib. Petals usually glabrous.—*E. Smithianus,* Hill, in Herb. Muell.
**Queensland.** Wide Bay, *W. Hill;* near Brisbane, *Henne.*
**N. S. Wales.** Macleay river, *Beckler.*

Var. (?) *teretifolius.* Glabrous or pubescent. Leaves linear-terete, more or less crenate or tuberculate, with large prominent glands, usually short and crowded, but sometimes 3 or nearly 4 lines long. Petals glabrous. Ovary glabrous or pubescent.—*E. brevifolius,* A. Cunn.; Endl. in Hueg. Enum. 16.
**N. S. Wales.** Peel's range, *A. Cunningham.*
**S. Australia.** Lynedoch valley, *Behr.;* Lofty Range and near Gawler river, *F. Mueller.*
**W. Australia,** *Drummond, 5th Coll. n.* 204 (with rather larger flowers). Phillips and Fitzgerald rivers, *Maxwell.*

Endlicher describes the leaves of Cunningham's plant as revolute and pubescent underneath, which I do not find in any of his specimens. This and the last variety appear in our herbaria so distinct in foliage from the ordinary form of *E. difformis,* that I should have admitted them as substantive species, had it not been for the authority of F. Mueller, who observes that they pass much one into the other.

12. **E. parvifolius,** *R. Br. Herb.* A low, erect, compact, much-branched, glabrous shrub. Leaves crowded, linear-cuneate, obtuse, 3 to 4 lines long, slightly glandular-crenate, flat, coriaceous, without any conspicuous midrib. Flowers small, terminal, solitary, shortly pedicellate, glabrous. Sepals small. Petals 2 to 2½ lines long. Filaments flattened, ciliate; anthers minutely apiculate. Cocci short, truncate, obscurely beaked. Seeds minutely tuberculate.
**Queensland.** Shoalwater Bay, *R. Brown (Herb. R. Br.).*

13. **E. ericifolius,** *A. Cunn. Herb.* An erect, heath-like, glabrous shrub. Leaves crowded, linear-terete, obtuse or nearly so, much longer than in *E. difformis,* although rarely exceeding ½ in., slightly glandular but not tuberculate. Flowers terminal, solitary or 2 or 3 together on short pedicels, sometimes apparently lateral by the elongation of the side shoot. Sepals

broad-lanceolate. Petals attaining 5 lines, glabrous or ciliate, with a prominent midrib. Filaments flattened, woolly-ciliate on the edges, attenuate at
the top, the longer ones bearing a long tuft of rigid hairs behind the anthers;
anthers shortly apiculate. Ovary very hairy. Carpels of the fruit beaked.

**N. S. Wales.** Skirts of Liverpool plains, *A. Cunningham.* This species has the foliage nearly of *E. nodiflorus,* but larger usually solitary flowers, and is remarkable for the
long hairs covering the anthers.

14. **E. nodiflorus,** *Lindl. Swan Riv. App.* 17. A heath-like shrub,
with virgate branches, glabrous or slightly pubescent. Leaves narrow-linear
or almost terete, acute or rather obtuse, under $\frac{1}{2}$ in. long, glabrous, the glands
not tubercular. Flowers usually blue, several together in dense terminal
heads, which become lateral by the elongation of one or more side shoots.
Pedicels short. Sepals linear-lanceolate, nearly glabrous or hirsute, often
more than half as long as the petals. Petals attaining $2\frac{1}{2}$ to $3\frac{1}{2}$ lines. Filaments slightly flattened, ciliate, attenuate at the top; anthers scarcely apiculate. Ovary glabrous. Cocci acutely beaked.—Bartl. in Pl. Preiss. i. 171.

**W. Australia.** King George's Sound to Swan River, *Drummond, 1st Coll., 4th Coll.*
n. 95, *5th Coll. n.* 203, *Preiss, n.* 2049; Mount Barker and Kalgan river, *Oldfield;* W.
Mount Barren, *Maxwell.*

There are two principal forms which at first sight look very distinct, one with small flowers
and very villous calyces, the other with larger almost glabrous flowers, but they are connected by so many intermediates that they cannot be well defined even as varieties. *E. ca-
lycinus,* Turcz. in Bull. Mosc. 1849, ii. 14, founded on Drummond's specimens, n. 93 of
the 4th Coll., appears to be the same species, although the petals in the dried state show
nothing of the blue tinge. I can find no other difference.

15. **E. spicatus,** *A. Rich. Sert. Astrol.* 76, *t.* 27. A heath-like shrub
or undershrub of 1 or 2 ft., with virgate erect branches, glabrous or slightly
pubescent. Leaves erect or spreading, very narrow-linear or almost terete,
rarely much exceeding $\frac{1}{2}$ in. Flowers blue according to most collectors, pink
according to Oldfield (in Herb. Muell.), generally drying pale-blue or almost
white, in loose terminal usually pubescent racemes of 1 to 3 in., with a leafy
bract of $\frac{1}{2}$ to $1\frac{1}{2}$ lines at the base of each pedicel at a very early stage, but
these bracts fall off usually long before the raceme is fully developed, and are
only very rarely persistent till after the first flowers open. Sepals small.
Petals rather broad, about 3 lines long. Filaments flattened, densely ciliate,
attenuate at the top; anthers shortly apiculate. Ovary glabrous. Carpels
of the fruit slightly beaked.—Bartl. in Pl. Preiss. i. 171.—*E. racemosus* and
*E. ebracteatus,* Endl. in Hueg. Enum. 15; *E. effusus,* Turcz. in Bull. Mosc.
1849, ii. 14.

**W. Australia.** From King George's Sound to Swan River, *Drummond, Preiss, n.*
2021, *Harvey,* and others. I have not seen Gilbert's specimens n. 95, described by Turczaninow, but refer them to this species from the character given.

## 6. PHEBALIUM, A. Juss.

Calyx small, 5-cleft or 5-toothed. Petals 5, valvate or laterally imbricate,
but always with valvate inflexed tips. Disk narrow or angular. Stamens
10, shorter or longer than the petals; filaments glabrous or rarely slightly
ciliate, filiform or rarely flat, subulate at the top; anthers tipped with a small
gland or not at all apiculate. Carpels 5, rarely 4 or fewer, distinct from the

base or nearly so, usually produced into a short or long appendage above the cells ; styles inserted below the middle and immediately united into one ; stigma small ; ovules 2 in each cell, superposed.  Cocci 2-valved, usually more or less beaked at the top or the outer angle ; the endocarp cartilaginous and separating elastically.  Seeds usually solitary.—Shrubs either glabrous or slightly stellate-pubescent or clothed with scurfy scales, very rarely hirsute.  Leaves alternate, simple, entire or slightly toothed, the glands often large and prominent.  Inflorescence axillary or terminal, peduncles rarely 1-flowered, usually forming an umbel-like short raceme, rarely reduced to a compact head.  Flowers small, white or yellow, very rarely and exceptionally 4-merous or 6-merous.

Besides the Australian species, which are all endemic, the genus comprises one from New Zealand, nearly allied to, but apparently distinct from one of the Australian ones.  F. Mueller unites the genus with *Eriostemon,* but the æstivation of the corolla, besides the habit and a number of smaller characters, appear to me sufficient to warrant the maintaining it as distinct.  Practically, the section *Leionema* may be at once distinguished from *Eriostemon* by the strictly valvate corolla, and *Phebalium* proper by the scurfy scales always present at least on the flower and ovary.

SECT. 1. **Leionema,** F. Muell.—*Glabrous or pubescent plants without scurfy scales. Petals strictly valvate, glabrous.*

Flowers axillary.
  Peduncles short, 1-flowered.  Stamens not exserted.
    Leaves flat, linear or linear-lanceolate, rigid, pungent . . . 1. *P. pungens.*
    Leaves linear-terete, obtuse, channelled above . . . . . 2. *P. montanum.*
    Leaves linear, obtuse, the margin revolute . . . . . . 3. *P. lachnoides.*
  Peduncles several-flowered.  Stamens slightly exserted.
    Leaves linear, with revolute margins, crowded, not exceeding ½ in.  Peduncles short, few-flowered.  Ovary tomentose . 4. *P. phylicifolium.*
    Leaves linear, 1 to 3 in.  Peduncles several-flowered.  Ovary glabrous . . . . . . . . . . . . . . 5. *P. dentatum.*
Flowers terminal.  Stamens usually exserted.
  Leaves flat or nearly so.  Flowers umbellate.
    Leaves truncate, notched or 2-lobed at the top.
      Umbels pedunculate and reflexed.  Petals erect . . . . 6. *P. Ralstoni.*
      Umbels erect, nearly sessile . . . . . . . . . 7. *P. bilobum.*
    Leaves acute or obtuse.
      Leaves oblong or lanceolate.
        Leaves acute, under ½ in. long . . . . . . . 8. *P. lamprophyllum.*
        Leaves obtuse, ½ to ¾ in., thinly coriaceous . . . . 9. *P. elatius.*
        Leaves crowded, under ½ in., coriaceous, very obtuse, the margins recurved . . . . . . . . . . . . 10. *P. Oldfieldii.*
      Leaves small, obovate or orbicular.
        Leaves rigid but not thick, flat or concave . . . . . 11. *P. rotundifolium.*
        Leaves very small, thick, convex . . . . . . . . 12. *P. brachyphyllum.*
    Leaves linear, with closely revolute margins.  Flowers capitate 13. *P. diosmeum.*

SECT. 2. **Euphebalium.**—*The whole plant or at least the inflorescence and calyx, and often the petals and ovary, more or less covered with scurfy peltate scales, often fringed at the edge, those of the ovary often closely imbricate in one mass.  Petals laterally imbricate or rarely almost valvate in the bud, with inflexed valvate tips.*

Umbels terminal.  Leaves small or rarely exceeding 1 in.
  Calyx truncate or very shortly toothed.  (Eastern species.)
    Leaves obovate with recurved margins, coriaceous, shining above, scaly underneath . . . . . . . . . . . 14. *P. ozothamnoides.*

Leaves very small, obcordate or broadly cordate, silvery-scaly 15. *P. obcordatum.*
Leaves linear-cuneate, truncate or emarginate . . . . . 16. *P. glandulosum.*
Leaves oblong or linear, rounded or obtuse at the top, ½ to
   1½ in. long . . . . . . . . . . . . . . 17. *P. squamulosum.*
Calyx-teeth as long as the tube. (Western species.)
Leaves narrow-linear, channelled above, keeled underneath,
   very glandular . . . . . . . . . . . . . . 18. *P. tuberculosum.*
Leaves small, oblong, with revolute margins, coriaceous,
   shining above . . . . . . . . . . . . . . 19. *P. microphyllum.*
Leaves small, oblong, flat, silvery underneath . . . . . 20. *P. Drummondii.*
Leaves linear-filiform, hoary-scaly . . . . . . . . . 21. *P. filifolium.*
Umbels terminal and lateral, loose. Leaves oblong or lanceolate
   or linear, 1 to 2 in. long or more.
Leaves silvery-white underneath. Petals distinctly imbricate,
   not scaly . . . . . . . . . . . . . . . . 22. *P. Billardieri.*
Leaves green on both sides when full-grown. Petals valvate or
   nearly so, densely scaly . . . . . . . . . . 23. *P. argenteum.*
Peduncles all axillary, short, 1- to 3-flowered. Leaves small.
Leaves ovate, white underneath . . . . . . . . . 24. *P. ovatifolium.*
Leaves obcordate or broadly cuneate, truncate or 2-lobed . . 25. *P. rude.*
Leaves linear-cuneate, thick, notched or 2-lobed.
Flowers distinctly pedicellate, about 2 lines long. Leaves
   slightly 2-lobed . . . . . . . . . . . . . . 26. *P. amblycarpum.*
Flowers almost sessile, 4 or 5 lines long. Leaves divaricately
   2-lobed . . . . . . . . . . . . . . . . 27. *P. Baxteri.*

SECT. 1. LEIONEMA.—Glabrous or pubescent plants without scurfy scales. Petals strictly valvate, glabrous.

1. **P. pungens,** *Benth.* A small rigid, erect or diffuse shrub, with the aspect of some Epacrideous plants, glabrous or the branches slightly hairy. Leaves linear or linear-lanceolate, rigid, with a strong pungent point, usually ½ in. long or shorter, rarely nearly ¾ in., flat, with the midrib prominent underneath. Peduncles short, axillary, 1-flowered. Flowers white, glabrous. Calyx small. Petals rather more than 2 lines long, valvate. Stamens shorter than the petals; filaments glabrous or slightly ciliate, somewhat flattened, obtusely contracted at the top into a short stipes; anthers not apiculate. Ovary glabrous. Cocci beaked.—*Eriostemon pungens,* Lindl. in Mitch. Three Exped. ii. 156; F. Muell. Pl. Vict. i. 125.

**Victoria.** Near Mount Hope, *Mitchell;* Murray river and its lower tributaries and Grampian Mountains, *F. Mueller.*

**S. Australia.** Towards Mount Lofty and Glen Osmond, *F. Mueller.*

2. **P. montanum,** *Hook. Journ. Bot.* i. 255, *and Ic. Pl. t.* 59. A dwarf, rigid, diffuse or prostrate shrub, glabrous or the branches very minutely stellate-pubescent. Leaves crowded, linear, obtuse, rarely above ½ in. long, thick and nearly terete or very convex underneath and channelled above. Flowers in the upper axils on very short thick pedicels. Sepals very short. Petals about 2 lines long, valvate, glabrous. Stamens not exserted. Filaments glabrous, filiform or slightly flattened. Ovary glabrous, with short, oblong, terminal appendages to the carpels; style glabrous. Cocci very minutely beaked.—Hook. f. Fl. Tasm. i. 63.

**Tasmania.** Highest part of the Western Mountains, Arthur's Lake, etc., at an elevation of 3500 to 4500 ft., *Gunn.*

3. **P. lachnoides,** *A. Cunn. in Field, N.S. Wales,* 332. A tall heath-like shrub, glabrous or the branches minutely stellate-pubescent. Leaves crowded, narrow-linear, obtuse or scarcely mucronate, rarely exceeding ½ in., the margins revolute, glabrous above, hoary underneath. Flowers on short axillary pedicels, usually crowded near the ends of the branches. Calyx very short. Petals 2 to 2½ lines long, glabrous, valvate. Stamens not exserted; filaments filiform, glabrous; anthers not apiculate. Ovary glabrous, with long terminal appendages to the carpels. Style glabrous.

**N. S. Wales.** Barren rocky situations in the Blue Mountains, *A. Cunningham.*

4. **P. phylicifolium,** *F. Muell. in Trans. Vict. Inst.* i. 32. A dwarf, robust, diffuse shrub, glabrous or the branches and under side of the leaves minutely stellate-pubescent. Leaves crowded, linear, obtuse, under ½ in. long, the margins revolute. Flowers pale-yellow, usually 2 or 3 together in shortly pedunculate umbels, all axillary but crowded towards the summit of the branches. Calyx very short. Petals about 2 lines long, valvate. Stamens exserted; filaments filiform, glabrous; anthers not apiculate. Ovary pubescent, the terminal appendages of the carpels short and obtuse; style glabrous. Cocci glabrous, ovate, minutely beaked.—*Eriostemon phylicifolius,* F. Muell. Fragm. i. 105.

**Victoria.** Summits of the Munyang, Cobberas, Mitta-Mitta, and other mountains, at an elevation of 4000 to 6000 ft., *F. Mueller.* In Pl. Vict. i. 128, F. Mueller unites this with *P. dentatum* as an alpine variety; but, without having seen any intermediate specimens, I do not feel justified in combining two forms so different in habit and foliage, as well as in some minor characters.

5. **P. dentatum,** *Sm. in Rees, Cycl.* xxvii. A tall shrub with elongated branches, hoary when young with a minute stellate pubescence. Leaves linear, obtuse, mostly 1½ to 3 in. long, the margins recurved and often minutely and remotely glandular-toothed, rather coriaceous, glabrous and smooth above, hoary underneath with a stellate tomentum, the midrib prominent. Flowers in short umbel-like racemes, axillary and pedunculate, but always much shorter than the leaves. Pedicels almost glabrous, 2 to 3 lines long. Calyx very small. Petals about 2 lines long, valvate. Longer stamens slightly exserted; filaments filiform, glabrous. Disk very small. Ovary glabrous. Cocci nearly orbicular, shortly beaked. Seeds black and shining.—*P. salicifolium,* A. Juss. in Mem. Soc. Nat. Hist. Par. ii. 134, t. 12; *Eriostemon umbellatus,* Turcz. in Bull. Mosc. 1849, ii. 15; F. Muell. Fragm. i. 104.

**N. S. Wales.** Port Jackson, *R. Brown* and others.

6. **P. Ralstoni,** *Benth.* A tall shrub, perfectly glabrous, the young branches angular. Leaves narrow-oblong or linear, obtuse and notched or 2-lobed at the end, 1 to 1½ in. long, the margins recurved and entire, narrowed into a short petiole, of a rather firm consistence, pale underneath. Flowers green or reddish, 3 to 5 in a terminal shortly pedunculate reflexed umbel. Calyx small. Petals narrow, valvate, fully 3 lines long, less open than in any other species. Stamens much exserted; filaments subulate, glabrous. Ovary glabrous, on a very short broad disk. Cocci short and broad, with a very short obtuse beak. Seeds smooth.—*Eriostemon Ralstoni,* F. Muell. Fragm. ii. 101, t. 14.

**N. S. Wales.**  Yokawa river, near Twofold Bay, *F. Mueller;* foot of Castle Rock Mountain, *Leichhardt.*

7. **P. bilobum,** *Lindl. in Mitch. Three Exped.* ii. 178.  An elegant usually divaricately branched shrub, sometimes tall and erect in wet valleys, glabrous or the young branches minutely stellate-pubescent.  Leaves sessile or nearly so, oblong or lanceolate, sometimes all under ½ in., sometimes 1 in. long or even more, truncate or 2-lobed at the top, the margins often serrate and recurved or revolute, rounded, narrowed or rarely cordate at the base, smooth and often shining on both sides, the midrib prominent underneath. Flowers small, in terminal erect sessile umbels, often on short lateral branches, rarely apparently axillary by the abortion of the branch.  Pedicels slender, 1 to 3 lines long.  Stamens shortly exserted; filaments filiform.  Disk small. Ovary glabrous, of 2 or 3, rarely 4, carpels.  Cocci oval-oblong, beaked.— *P. truncatum,* Hook. f. Fl. Tasm. i. 64, t. 9 ; *Eriostemon serrulatus,* F. Muell. Fragm. i. 4 ; *E. Hildebrandi,* F. Muell. in Trans. Phil. Soc. Vict. i. 10, and Pl. Vict. i. 127 ; Dietr. Fl. Univ. N. Ser. ii. t. 2.

**Victoria.**  Mount William, *Mitchell;* Cataracts and rocky rivulets in the Victoria ranges and Grampians, *F. Mueller.*

**Tasmania.**  Flinders Island, Bass's Straits, and Schouten Island, E. coast, *Gunn;* Mount Gog, *Archer.*

**S. Australia.**  Mount Lofty, *Whittaker;* sources of the Gawler river, *F. Mueller.*

In Mitchell's specimens, the leaves are broad and cordate at the base; in others, from the same locality, they are rounded or narrowed at the base, as in the generality of the Tasmanian ones.  The Mount Lofty specimens are small, divaricate, with short cordate leaves, as figured by Dietrich.  The pistil is usually 3-merous in Victoria, more frequently 2-merous in Tasmania, but variable in both.

8. **P. lamprophyllum,** *Benth.*  A densely branched glabrous shrub. Leaves crowded, oblong-lanceolate, acute, under ½ in. long, entire, coriaceous and shining, flat or concave, contracted into a very short petiole.  Flowers few, in terminal sessile umbels, with a small but usually leafy bract at the base of each pedicel.  Calyx small.  Petals and stamens not seen.  Carpels 5, of which 2 or 3 only ripen, ovate, beaked, glabrous.  Seeds smooth and shining.—*Eriostemon lamprophyllus,* F. Muell. Pl. Vict. i. 126.

**Victoria.**  Summit of Mount Ligar, towards the sources of Macalister river, *F. Mueller.*

9. **P. elatius,** *Benth.*  A tall shrub, glabrous or the branches very minutely pubescent, and usually tuberculate with prominent glands.  Leaves linear-cuneate or oblong, obtuse, ½ to ¾ in. long, entire or crenulate, thinly coriaceous, smooth and shining, narrowed into a very short petiole.  Peduncles 2- or more-flowered, terminal or in the uppermost axils, forming short terminal leafy corymbs or ovate panicles.  Calyx very small.  Petals valvate, not 2 lines long.  Stamens exserted; filaments subulate, glabrous; anthers small.  Ovary glabrous, on a raised almost stalk-like disk.  Cocci obliquely obovate, very minutely beaked.—*Eriostemon elatior,* F. Muell. Fragm. i. 181.

**N. S. Wales.**  New England, near Tenterfield, *C. Stuart.*  The species is very closely allied to the New Zealand *P. nudum,* Hook., differing chiefly in much smaller flowers, the calyx-lobes less prominent, the inflorescence not so flat-topped, etc.

10. **P. Oldfieldii,** *F. Muell. Herb.*  A densely branched shrub, quite glabrous or the branches pubescent.  Leaves narrow-oblong or slightly

cuneate, very obtuse or retuse, rarely exceeding ½ in., entire, coriaceous, and often shining, the margins flat or slightly recurved, contracted into a very short petiole. Flowers few, in short sessile terminal umbels. Sepals small. Petals and stamens not seen. Carpels 5, glabrous, ovate when ripe, shortly beaked. Seeds not seen.—*Eriostemon Oldfieldii*, F. Muell. Fragm. i. 3, and Pl. Vict. i. 125.

**Tasmania.** At the base of Mount Lapeyrouse, *Oldfield* and *Stuart.* The foliage is, at first sight, so much like that of *Eriostemon virgatus*, that the specimens without flowers first received were mistaken for that plant (Hook. f. Fl. Tasm. ii. 358); but even the leaves may be known by their end much more obtuse or retuse, and never mucronate.

11. **P. rotundifolium,** *Benth.* An erect much-branched shrub, the young branches minutely pubescent. Leaves crowded, almost imbricate, small, obovate or orbicular, obtuse or minutely mucronate, mostly 2 to 3 lines long, flat or concave, coriaceous, glabrous, very shortly petiolate or almost sessile. Flowers several, in a terminal sessile umbel, almost contracted into a head in our specimens, which are not fully out. Sepals small. Petals valvate, glabrous. Filaments filiform, glabrous. Ovary glabrous, on a very short disk, the terminal appendages of the carpels very short.—*Eriostemon rotundifolius*, A. Cunn., Endl. in Hueg. Enum. 15.

**N. S. Wales.** Hunter's River, *A. Cunningham.*

12. **P. brachyphyllum,** *Benth.* A dwarf shrub, with a thick woody base and numerous branching stems of 2 to 4 in., glabrous or minutely pubescent. Leaves small, crowded, sessile or nearly so, very spreading, obovate or orbicular, very obtuse, rarely exceeding 2 lines, thick, coriaceous and nerveless, very convex. Flowers few (usually 3 to 5), in terminal clusters or short racemes. Pedicels short. Sepals small. Petals about 1½ lines long, glabrous, valvate. Filaments filiform. Ovary glabrous, on a distinct stalk-like disk; the terminal appendages of the carpels very short.

**S. Australia.** Encounter Bay and near Coffin Bay, *F. Mueller.*

13. **P. diosmeum,** *A. Juss. in Mem. Soc. Hist. Nat. Par.* ii. 135, *t.* 11. An erect heath-like shrub, the branches more or less hirsute. Leaves crowded, linear, obtuse, mostly under ½ in., the margins revolute, scabrous or sprinkled with a few hairs. Flowers yellow, numerous, in a sessile terminal head, intermixed with linear bracts shorter than the calyx. Sepals linear, erect, pubescent, about half as long as the petals. Petals about 3 lines long, glabrous, valvate. Stamens exserted; filaments subulate, glabrous; anthers didymous. Carpels very short, with the terminal appendages 4 times as long, glabrous or hairy; style glabrous. Ripe fruit not seen.—*P. phylicoides*, Sieb. in Spreng. Syst. Cur. Post. 164; *Chorilæna angustifolia*, F. Muell. in Trans. Phil. Soc. Vict. i. 10; *Eriostemon phylicoides*, F. Muell. Fragm. i. 107, and Pl. Vict. i. 131.

**N. S. Wales.** Port Jackson to the Blue Mountains, *Sieber, n.* 110, *Fraser, A. Cunningham,* and others.
**Victoria.** Sandy heaths near Mount Imlay, abundant, *F. Mueller.*

SECT. 2. EUPHEBALIUM.—The whole plant, or at least the inflorescence and calyx, often also the petals and ovary, more or less covered with scurfy peltate scales, often fringed at the edge, those of the ovary often closely imbricate in one mass. Petals laterally imbricate or rarely almost valvate in the bud, with inflexed valvate tips.

14. **P. ozothamnoides,** *F. Muell. in Trans. Vict. Inst.* i. 31. A rigid shrub, the branches brown with scurfy scales. Leaves obovate, very obtuse, under ½ in. long, the margins recurved, narrowed into a short petiole, thick, coriaceous, glabrous and shining above when full-grown, white underneath with scurfy scales mixed with stellate hairs which are also sprinkled on the upper surface of the young leaves. Flowers few, in small terminal sessile umbels, like those of *P. squamulosum* in size and structure as well as in the scurfy scales.—*Eriostemon ozothamnoides,* F. Muell. Fragm. i. 103.

**Victoria.** Mitta-Mitta, Cabongra, and Livingstone rivers, *F. Mueller.*

15. **P. obcordatum,** *A. Cunn. Herb.* A small densely-branched shrub, silvery-white or hoary with scurfy scales. Leaves distinctly petiolate, either broadly obcordate and about 1 line long, or in luxuriant specimens broadly cuneate and attaining 2 lines, very obtuse and emarginate, flat, rather thick, glabrous above with 2 to 4 large prominent glands, silvery underneath. Flowers much smaller than in the allied species, few on short pedicels at the ends of the branches and uppermost axils, forming short terminal leafy corymbs. Structure of the flowers as in *P. squamulosum.*

**N. S. Wales.** S.W. of St. George's Range, *A. Cunningham.*

16. **P. glandulosum,** *Hook. in Mitch. Trop. Austr.* 199. Very closely allied to some of the smaller much-branched forms of *P. squamulosum,* with the same scurfy indumentum, inflorescence, and flowers, and recently united with that species by F. Mueller (Pl. Vict. i. 130). It appears however to me to differ sufficiently in the leaves, which are narrowly linear-cuneate, emarginate or almost 2-lobed at the end, with revolute or recurved margins varying from 2 or 3 lines to ¾ in. in length. In the ordinary form also the branches and leaves are covered with large glandular tubercles.—*P. sediflorum,* F. Muell. in Trans. Vict. Inst. i. 30 ; *Eriostemon sediflorus,* F. Muell. Fragm. i. 102.

**Queensland.** On the Upper Maranoa, *Mitchell.*
**N. S. Wales.** Eurylean scrub, *A. Cunningham.*
**Victoria.** Snowy River, Pinch Mountains, and the N.W. desert of the colony, *F. Mueller.*
**S. Australia.** Extending to Lake Torrens, *F. Mueller.*

Var. (?) *Daviesi.* Leaves narrow-linear, broader and emarginate at the end as in the ordinary form, but the glandular tubercles few or none.—*P. Daviesi,* Hook. f. Fl. Tasm. ii. 358.

**Tasmania.** E. coast near St. Helen's Bay, *Davies.* The ovary, in the flowers I have examined, has the peltate scurfy scales of the allied species.

17. **P. squamulosum,** *Vent. Jard. Malm. t.* 102. An erect shrub, varying in height but never arborescent, the young branches brown with scurfy scales. Leaves shortly petiolate, oblong or linear, obtuse but often mucronulate, ½ to 1½ in. long, somewhat coriaceous, the margins flat or slightly recurved, smooth above or slightly glandular-tuberculate, covered underneath with scurfy peltate scales. Flowers yellow, in terminal sessile, simple or compound umbels or corymbs, not exceeding the last leaves, the pedicels, calyx, and petals covered with comparatively large scurfy scales.

Calyx very short, truncate, with minute or short and broad teeth. Petals barely 2 lines long, slightly imbricate with inflexed valvate tips. Stamens exserted (1 or 2 occasionally wanting); filaments glabrous; anthers tipped by a small gland. Ovary densely covered with white or brown scurfy ciliate scales. Cocci small, broad, obscurely beaked. Seeds scarcely shining.— DC. Prod. i. 720; A. Juss. in Mem. Soc. Hist. Nat. Par. ii. 132; *P. elæag-nifolium,* A. Juss. l. c. 132, t. 11; *P. aureum,* A. Cunn. in Field, N. S. Wales, 331, with a figure (the specimens not so stunted as represented in the plate); *Eriostemon lepidotus,* Spreng. Syst. ii. 322; F. Muell. Fragm. i. 104, and Pl. Vict. i. 130.

**N. S. Wales.** Port Jackson to the Blue Mountains, *R. Brown, Sieber, n.* 112 (misnamed *P. anceps*); Liverpool plains, *A. Cunningham;* Clarence river, *Beckler.*

**Victoria.** Genoa Peak and river, *F. Mueller.*

Var. *alpinum.* Diffuse, with crowded more coriaceous leaves, rarely exceeding ½ in.—*P. podocarpoides,* F. Muell. in Trans. Vict. Inst. i. 31; *Eriostemon alpinus,* F. Muell. Fragm. i. 103.—Summits of the Australian Alps at an elevation of 5000 to 6000 ft.

Var. (?) *stenophyllum.* A small shrub. Leaves small, narrow, with the margins of the leaves closely revolute so as to be often almost terete.—In the Grampian Mountains and desert of the Tattiara country towards the Murray river, *F. Mueller.*—This form appears to me so constantly distinct, as far as our specimens show, that I should have described it as a separate species, were it not that F. Mueller includes it without any hesitation in the *P. squamulosum,* and I might thus be adding a useless synonym.

18. **P. tuberculosum,** *Benth.* An erect shrub, with rigid rather slender branches, covered with minute scurfy scales and prominent glandular tubercles as in *P. glandulosum.* Leaves narrow-linear, obtuse, rarely above ½ in. long, the upper surface channelled, glabrous and tubercular, the under side whitish with scurfy scales, the midrib prominent and the margins sometimes recurved. Flowers few, in terminal umbels, scurfy-scaly as well as the pedicels. Calyx small, the lobes or teeth prominent and usually at least as long as the tube. Petals broad, nearly 2 lines long, slightly imbricate with inflexed valvate tips. Stamens exserted, glabrous; anthers without any conspicuous gland. Ovary scaly.—*Eriostemon tuberculosus,* F. Muell. Pl. Vict. i. 130.

**W. Australia,** *Drummond, n.* 63; Fitzgerald river, *Maxwell.* This and the three following western species, like *P. squamulosum* and its allies in the east, are chiefly distinguished from each other by the foliage, and, as a whole, the four western species scarcely differ in anything but the foliage from the four or five eastern ones, except that the teeth or lobes of the calyx, small as they are, are more prominent.

19. **P. microphyllum,** *Turcz. in Bull. Mosc.* 1852, ii. 159. A heath-like shrub, the branches covered with scurfy scales. Leaves petiolate or nearly sessile, oblong or oblong-linear, obtuse, 2 to 4 lines long, the margins revolute, coriaceous, glabrous and shining above, and sometimes slightly glandular, white with minute scurfy scales underneath. Flowers few, in sessile terminal umbels, scurfy-scaly outside as well as the pedicels. Calyx small, the triangular lobes at least as long as the tube. Petals rather smaller and not so broad as in *P. tuberculosum,* but otherwise the same. Cocci small, broad, obscurely beaked.

**W. Australia.** Between Swan River and King George's Sound, *Drummond, 5th Coll. n.* 208, and other unnumbered specimens.

20. **P. Drummondii,** *Benth.* A small, elegant, much-branched shrub,

the branches covered with scurfy scales mixed with a minute stellate pubescence. Leaves very shortly petiolate, oblong, obtuse, 2 to 3 lines long, flat, coriaceous, glabrous and smooth above, silvery-white underneath with scurfy scales often mixed with a minute pubescence, the midrib not prominent. Flowers yellow, in terminal sessile umbels shortly exceeding the leaves and of the size of those of *P. squamulosum*, scurfy-scaly outside as well as the pedicels. Calyx-lobes triangular or lanceolate, as long as or longer than the tube. Petals, stamens, and ovary of *P. squamulosum*.

**W. Australia,** *Drummond, n.* 13.

21. **P. filifolium,** *Turcz. in Bull. Mosc.* 1852, ii. 159. An erect virgately-branched shrub, hoary all over with minute scurfy scales, or the young branches rust-coloured. Leaves narrow-linear, almost terete, obtuse, ½ to 1 in. long, whitish and scurfy-scaly on both sides. Flowers few, on rather long terminal pedicels. Calyx-lobes broadly triangular, as long as the tube. Petals, stamens, and ovary of *P. squamulosum*. Cocci broad, marked with deep transverse wrinkles.

**W. Australia,** *Drummond, 4th Coll. n.* 178; *5th Coll. n.* 206; *J. S. Roe.*

22. **P. Billardieri,** *A. Juss. in Mem. Soc. Hist. Nat. Par.* ii. 134. An erect shrub or small tree, the branches angular and clothed with small brown scurfy scales. Leaves oblong, lanceolate or linear, obtuse or acute, rarely under ½ in. and often 3 in., or in very luxuriant specimens 4 or 5 in. long, entire, coriaceous, flat or with recurved margins, glabrous above, silvery-white underneath with minute scales. Flowers in axillary corymbs, shortly pedunculate, but always shorter than the leaves; peduncles and pedicels thick and scaly. Calyx small, lobed. Petals about 2 lines long, glabrous, slightly imbricate, with inflexed valvate tips. Stamens exserted; filaments often hairy in the lower portion. Ovary glabrous. Cocci small, broad, with a very short beak. Seeds shining.—Hook. f. Fl. Tasm. i. 63; *Eriostemon squameus,* Labill. Pl. Nov. Holl. i. 111, t. 141; F. Muell. Fragm. i. 104, and Pl. Vict. i. 129; *P. retusum,* Hook. Journ. Bot. i. 254, and Ic. Pl. t. 57; *P. elatum,* A. Cunn. in Field, N. S. Wales, 331; *P. elæagnoides,* Sieb. Pl. Exs.

**N. S. Wales.** Port Jackson to the Blue Mountains, *R. Brown, Sieber, n.* 111, and others; northward to Hastings river, *Beckler,* and Clarence river, *C. Moore;* southward to Illawara, *Backhouse.*

**Victoria.** Damp forest-valleys near Apollo Bay, towards Cape Otway, and near the sources of the Barwon river, *F. Mueller.*

**Tasmania.** Port Dalrymple, *R. Brown;* abundant throughout the colony in damp woods, *J. D. Hooker.*

23. **P. argenteum,** *Sm. in Rees' Cycl.* xxvii. A tall, stout, erect shrub, the younger branches angular and covered with white scurfy scales. Leaves lanceolate, acute or obtuse, 2 to 3 or sometimes 4 in. long, entire, flat, narrowed at the base, glabrous on both sides when full grown, sprinkled underneath when young with a few scurfy scales. Flowers larger than in most species, in small axillary or terminal simple or compound cymes, much shorter than the leaves, the whole inflorescence as well as the calyx and petals densely covered with silvery-scurfy scales. Calyx-lobes about as long as the tube. Petals 2½ to near 3 lines long, valvate. Stamens shorter than the petals, glabrous. Ovary densely scaly. Cocci truncate, with short divergent beaks.

—*P. anceps*, DC. Prod. i. 719; A. Juss. in Mem. Soc. Hist. Nat. Par. ii. 133, t. 12; Bartl. in Pl. Preiss. i. 171; *Eriostemon anceps*, Spreng. Syst. ii. 322; F. Muell. Fragm. i. 103.

**W. Australia.** King George's Sound, *Menzies, R. Brown,* and others; Port Leschenault and Princess Royal Harbour, *Preiss, n.* 2011; and various localities near the S. coast, *Drummond, Oldfield,* and others.

24. **P. ovatifolium,** *F. Muell. in Trans. Phil. Soc. Vict.* i. 99. A compact, much branched, bushy shrub, with much the aspect of the European Box, the young branches rusty or hoary with scurfy scales. Leaves shortly petiolate, broadly ovate, very obtuse, rarely exceeding ½ in. and often smaller, flat or with slightly recurved thickened margins, coriaceous, smooth and shining above, hoary or white underneath with scurfy scales. Peduncles axillary, 1-flowered, shorter than the leaves, bearing 2 or 3 small leafy bracts. Calyx-lobes triangular, with few scurfy scales. Petals nearly 3 lines long, without scales (only seen fully expanded). Stamens shorter than the petals, the filaments slightly dilated. Ovary densely covered with silvery scales. Cocci very minutely beaked.—*Eriostemon ovatifolius,* F. Muell. Fragm. i. 103; Pl. Vict. i. 131.

**Victoria.** Alpine regions of the Munyang mountains and among rocks between Mount Wellington and Hardinge range towards the sources of Macalister river, *F. Mueller.*

25. **P. rude,** *Bartl. in Pl. Preiss.* i. 172. A much-branched bushy shrub, the young branches white with scurfy scales. Leaves crowded, broadly cuneate, obcordate or obovate, very obtuse, truncate or shortly 2-lobed, ½ in. long, or less on the flowering branches, twice as long on luxuriant barren shoots, entire, narrowed at the base, flat, green on both sides or whitish with scurfy scales. Peduncles axillary, 1- or very rarely 2-flowered, shorter than the leaves, covered as well as the calyx and petals with silvery scales. Calyx small, truncate, with very small teeth. Petals 2 lines long or rather more, valvate. Stamens shorter than the petals; filaments glabrous, dilated at the base. Ovary scaly. Cocci with a conical beak.—*P. bilobum,* Bartl. in Pl. Preiss. i. 172, not Lindley; *Eriostemon bilobus,* F. Muell. Fragm. i. 102.

**W. Australia.** King George's Sound and islands on the S. coast, *R. Brown, A. Cunningham,* and others; Bald Head and Konkongerup hills, *Preiss, n.* 2038 and 2039, and other parts of the S. districts, *Drummond, 4th Coll. and 5th Coll. n.* 207, and others.

26. **P. amblycarpum,** *Benth.* Shrubby, the young branches white with scurfy scales. Leaves linear-cuneate, very obtuse, not exceeding ½ in., notched or sometimes 2-lobed at the top, but otherwise entire, narrowed at the base, thick, scurfy-scaly when young, green when full grown. Peduncles axillary, 1-flowered, shorter than the leaves, more or less covered as well as the calyx and petals with scurfy scales. Calyx-teeth very short and broad. Petals not 2 lines long, valvate or very slightly imbricate, with inflexed valvate tips. Stamens shorter than the petals; filaments glabrous. Ovary almost without scales. Cocci angular at the top, but scarcely beaked.—*Eriostemon amblycarpus,* F. Muell. Fragm. i. 102.

**W. Australia.** Fitzgerald river, *Maxwell.*

27. **P. Baxteri,** *Benth.* A rigid shrub, the young branches white with scurfy scales. Leaves crowded and clustered in the axils, linear-cuneate, ½ to

¾ in. long, much dilated at the summit, with 2 diverging or divaricate lobes, otherwise entire, rigid, the margins revolute, glandular-scabrous above, scurfy-scaly underneath. Flowers much larger than in any other *Phebalium*, on very short axillary pedicels with 2 or 3 leafy bracts. Calyx-lobes almost as long as the tube. Petals 4 to 5 lines long, densely scaly outside, lanceolate with small inflexed tips, but the bud not seen. Longer stamens almost equalling the petals, filaments flattened, glabrous; anthers minutely apiculate. Ovary bearing a few scales. Fruit not seen.

**W. Australia.** S. coast, *Baxter* (*Hb. R. Br.*).

### 7. MICROCYBE, Turcz.

Sepals 5, small, thin, free or slightly united. Petals 5, slightly imbricate in the bud. Disk none. Stamens 10, exserted; filaments filiform, glabrous or ciliate at the base; anthers tipped with a small gland. Carpels 2, distinct; styles inserted above the middle and immediately united into one filiform style, with a minute stigma. Ovules 2, collateral, pendulous. Cocci 2-valved, rounded at the top and not beaked, the endocarp cartilaginous and separating elastically. Seeds usually solitary.—Heath-like shrubs, glabrous except scurfy scales on the young branches and under side of the leaves. Leaves numerous, small. Flowers small, in dense terminal sessile heads, with small leafy bracts at the base of the outer ones.

The genus is limited to Australia, and might be considered as a section of *Phebalium*. A peculiar habit, however, accompanied by a marked difference in the ovary, has induced me to retain it as a separate genus.

Leaves very spreading, linear, smooth or rough, with small glandular tubercles, the upper ones usually exceeding the flower-heads . . . 1. *M. pauciflora.*
Leaves slightly spreading, linear, with few large prominent transparent glands shorter than the flower-heads . . . . . . . . . . 2. *M. multiflora.*
Leaves very small, ovate, convex, reflexed, shorter than the small flower-heads . . . . . . . . . . . . . . . . . . . . . 3. *M. albiflora.*

1. **M. pauciflora,** *Turcz. in Bull. Mosc.* 1852, ii. 167. Branches rigid, hoary or almost tomentose with peltate fringed scales or stellate hairs. Leaves spreading, linear, obtuse, 2 to 4 lines long, the margins revolute, so as to be almost terete, coriaceous, glabrous and smooth above, or rough with very smooth glandular tubercles, the under side scaly-tomentose but usually concealed. Flower-heads about 3 lines diameter, sessile amongst the upper leaves, which usually exceed them. Sepals linear-lanceolate, transparent, small, and easily overlooked. Petals scarcely 1½ lines long. Filaments glabrous or ciliate. Cocci small, rounded at the top, the valves coriaceous, pitted but not wrinkled, and usually without scales. Seeds tuberculate.—*Asterolasia chorilænoides*, F. Muell. Trans. Vict. Inst. i. 116; *Eriostemon capitatus*, F. Muell. Fragm. i. 106.

**S. Australia.** Seacoast near Lake Hamilton, *Wilhelmi*; Venus Bay, *Warburton.*
**W. Australia,** *Drummond, 5th Coll. n.* 209; King George's Sound, *A. Cunningham;* E. Mount Barren, *Herb. Mueller.*

2. **M. multiflora,** *Turcz. in Bull. Mosc.* 1852, ii. 166. Glabrous, or the young branches slightly scaly. Leaves linear, obtuse, rarely exceeding 2 lines, the margins revolute so as to conceal the under surface, coriaceous,

almost shining, with 6 to 8 large prominent glandular tubercles. Flower-heads rather larger than in *M. pauciflora.* Sepals linear-spathulate. Petals nearly 2 lines long. Filaments glabrous. Cocci rounded as in *M. pauciflora,* but reticulate, and often retaining the scales of the ovary. Seeds reticulate.

**W. Australia,** *Drummond, 5th Coll. n.* 211.

3. **M. albiflora,** *Turcz. in Bull. Mosc.* 1852, ii. 167. Smaller than the other two species; the young branches scaly. Leaves ovate, obtuse, seldom above 1 line long, reflexed, convex, coriaceous, marked with a few large prominent transparent glands, the upper ones shorter than the flowers. Flower-heads mostly of only 3 or 4 small flowers. Sepals lanceolate, trans-parent, united to the middle, according to Turczaninow, but free or nearly so in our specimens. Petals scarcely 1 line long, slightly scaly outside. Ovary less scaly than in the other species. Fruit not seen.

**W. Australia,** *Drummond, 5th Coll. n.* 210.

## 8. GELEZNOWIA, Turcz.

(Sandfordia, *Drumm.*)

Sepals 5, large, petal-like, imbricate, exceeding the petals. Petals 5, ob-long, imbricate in the bud. Disk inconspicuous. Stamens 10, shorter than the petals; filaments subulate, glabrous; anthers not apiculate. Carpels 5, distinct or nearly so; styles inserted near the summit, immediately united into one filiform style, with a peltate obscurely lobed stigma. Ovules 2, superposed. Cocci 2-valved, not beaked.—Rigid, usually glaucous shrubs. Leaves alternate, small, rigid, crowded or almost imbricate. Flowers 1 to 3 together, sessile at the ends of the branches, remarkable for the large, leafy or petal-like bracts and sepals, exceeding the leaves.

The genus is limited to Australia, and in common with several others united by F. Mueller with *Eriostemon,* but the peculiar habit, large calyx, and insertion of the styles appear to me sufficient to retain it as a genus.

Sepals oblong, not much exceeding the petals.
    Carpels of the fruit rounded at the top, not longer than broad . .   1. *G. verrucosa.*
    Carpels of the fruit narrowed at the top, fully twice as long as broad   2. *G. macrocarpa.*
Sepals broadly ovate or orbicular, the petals much shorter. Carpels
    of *G. verrucosa* . . . . . . . . . . . . . . . . . 3. *G. calycina.*

1. **G. verrucosa,** *Turcz. in Bull. Mosc.* 1849, ii. 13. A bushy, rigid, glabrous, often glaucous shrub. Leaves crowded, obovate-oblong, obtuse, rarely exceeding 2 lines, thick, flat or concave above, convex underneath, and tuberculate with large prominent glands, a few of the upper leaves passing into sepal-like bracts. Sepals not 4 lines long, narrower than in *G. calycina,* the petals nearly as long, and both more or less glandular-warted outside. Ovary covered with minutely ciliate wart-like scales. Style elongated. Cocci (not yet quite ripe) not half so long as the petals, as broad as long, rounded at the top.—*Eriostemon Geleznowii,* F. Muell. Fragm. i. 107.

**W. Australia,** *Drummond, n.* 8. Some specimens from Sharks Bay, *Denham,* and Dirk Hartog's Island, *Milne,* appear to belong to the same species, but they are not in flower.

2. **G. macrocarpa,** *Benth.* From the fragmentary specimens we pos-sess, this appears to be nearly allied to *G. verrucosa,* with similar small leaves,

except that they are not so thick. Flowers large, the sepals narrow as in *G. verrucosa,* but attaining ½ in. Petals nearly 5 lines. Cocci (not yet fully ripe) more than twice as long as broad, narrowed at the top, attaining about 3 lines, covered upwards with wart-like glands.

**W. Australia.** Murchison river, *Oldfield.*

3. **G. calycina,** *Benth.* Rigid and erect, glaucous, and often turning yellow in drying, glabrous, or with a few hairs under the flowers. Leaves crowded, obovate or oblong, obtuse, in some specimens 2 to 3 lines long, in others attaining ½ in., the uppermost passing into sepal-like bracts. Sepals broadly ovate or almost orbicular, attaining 4 or 5 lines. Petals very much shorter and narrower. Ovary covered with wart-like scales. Style rather short. Cocci (not yet quite ripe) not half so long as the petals, as broad as long, rounded at the top.—*Sandfordia calycina,* Drumm. in Hook. Kew Journ. vii. 54; *Eriostemon Sandfordii,* F. Muell. Fragm. i. 107.

**W. Australia.** Sand plains, Hill river, and S. of the Irwin, *Drummond;* Murchison river, *Oldfield.*

## 9. PHILOTHECA, Rudge.

Calyx 5-cleft. Petals 5, imbricate in the bud. Disk slightly lobed. Stamens 10, shorter than the petals; filaments united into a glabrous tube at the base, free upwards, and very hairy; anthers oblong, all perfect, minutely apiculate. Carpels 5, nearly distinct from the base; styles inserted below the middle, and immediately united in a single style, hirsute in the middle; stigma small. Ovules 2 in each carpel, superposed. Cocci truncate, 2-valved, the endocarp cartilaginous and separating elastically. — Erect, heath-like shrubs, glabrous, or nearly so. Leaves crowded, alternate, narrow-linear. Flowers terminal, nearly sessile, solitary or two or three together.

A genus entirely Australian, differing from *Eriostemon,* with which F. Mueller unites it, only in the monadelphous stamens.

Leaves obtuse, mostly under 3 lines long . . . . . . . . 1. *P. australis.*
Leaves acute, mostly above 3 lines long . . . . . . . . 2. *P. Reichenbachiana.*

1. **P. australis,** *Rudge, in Trans. Linn. Soc.* xi. 298, *t.* 21. Glabrous or sprinkled with a minute pubescence. Leaves numerous, linear, obtuse, rarely exceeding 3 lines, rather thick, flat or channelled above, very convex underneath, or almost terete. Flowers usually solitary, but sometimes 2 or 3 together. Sepals small, broadly triangular. Petals 3 or 4 lines long, broadly lanceolate, minutely hoary-pubescent on both sides, except a broad glabrous central line outside. Stamens rather shorter than the petals. Cocci shortly beaked.—*Eriostemon salsolifolius,* Sm. in Rees, Cycl. xiii.

**N. S. Wales.** Port Jackson, *R. Brown, Sieber, n.* 307, and others.

Var. *parviflora.* Leaves more ciliate. Flowers much smaller; the petals scarcely 2¼ lines long.—*P. ciliata,* Hook. in Mitch. Trop. Austr. 347.

**Queensland.** Near Mount Faraday, *Mitchell.*

2. **P. Reichenbachiana,** *Sieb.; Spreng. Syst. Cur. Post.* 253. Very near *P. australis,* with which F. Mueller proposes to unite it, but the leaves always appear to be acute and longer, although rarely exceeding ½ in., the point sometimes quite pungent. Flowers usually larger than in *P. australis,*

and the hairs of the upper part of the filaments so long and dense as completely to cover the anthers.—Reichb. Icon. Exot. t. 200 (incorrect as to carpological details); *P. longifolia*, Turcz. in Bull. Mosc. 1849, ii. 16.

**N. S. Wales.** Port Jackson, *R. Brown, Sieber, n.* 308, and others; in the interior to the northward of Bathurst, *A. Cunningham.*

*P. Gaudichaudi*, G. Don, Gen. Syst. i. 792, from N. S. Wales, is not described so as to be recognizable.

## 10. DRUMMONDITA, Harv.

Sepals 5, short. Petals 5, erect, concave, imbricate in the bud. Disk fleshy, 5-lobed. Stamens 10, the filaments united into a long hairy tube, free at the top, 5 longer ones without anthers, plumose with long hairs, 5 shorter ones bearing anthers bearded on the back, acute at the top. Carpels 5, glabrous, free from the base; styles inserted near their summit, and immediately united into one filiform style; stigma capitate. Fruit unknown.—Shrub with heath-like leaves, and solitary terminal yellowish flowers.

The genus is limited to a single species, and appears from the character to differ from *Philotheca* only in the abortion of half the anthers. The only specimen, however, which I have seen, is a mere fragment insufficient for proper examination, and I am therefore unwilling to make any change without further information.

1. **D. ericoides,** *Harv. in Hook. Kew Journ.* vii. 53. An erect, branching, heath-like shrub. Leaves crowded, linear, semiterete, channelled above, ciliolate, with a large terminal gland, and sprinkled with black glandular dots. Flowers terminal, solitary, erect, almost sessile. Petals yellowish, green at the extremity. Staminal tube longer than the petals, white-tomentose outside, purple above the middle, sparingly pubescent inside.

**W. Australia.** Near the summit of White Peak, *J. Drummond.*

## 11. ASTEROLASIA, F. Muell.

(Urocarpus, *Drumm.*)

Calyx very minute or obsolete. Petals 5, tomentose outside, valvate and usually induplicate in the bud. Disk none. Stamens 10 or more, free, filaments filiform, glabrous or very slightly ciliate, anthers not apiculate. Carpels 2 to 5, united to the middle, or nearly to the top, into a single shortly-lobed or truncate ovary of 2 to 5 cells. Style inserted between the lobes, filiform, with a large reflexed peltate or deeply-lobed stigma. Cocci tardily separating, truncate, and often beaked, 2-valved; endocarp cartilaginous, separating elastically.—Shrubs or undershrubs, more or less stellate-tomentose, or, in one species, the tomentum united into scurfy scales. Leaves alternate, simple. Flowers sessile or pedicellate, axillary or terminal, solitary or few together.

The genus is limited to Australia, and, with several of the preceding ones, has been recently united with *Eriostemon* by F. Mueller; but the union of the carpels, more complete than in the exceptional *Eriostemon trachyphyllus*, the large reflexed stigma, the great reduction or abortion of the calyx, and the æstivation of the petals, are accompanied by differences in habit, which seem fully to justify the maintenance of the genus. I have now added *Urocarpus*, Drumm., as a section, for, on a detailed examination of all the species, the differences are reduced to the number of carpels of the ovary, which is variable. The curious tendency to an increase in the usual number of stamens is observable in some species of both sections.

SECT. 1. **Euasterolasia.**—*Ovary* 5-*merous.*
Stigma reflexed-peltate, scarcely lobed.    Ovary with 5 erect lobes.
Flowers pedicellate.
   Leaves ovate to lanceolate, 1 to 2 in., glabrous and smooth
      above . . . . . . . . . . . . . . . 1. *A. correifolia.*
   Leaves obovate to narrow-oblong, rarely above 1 in., rough
      above with stellate hairs . . . . . . . . . 2. *A. Muelleri.*
Flowers sessile.  Leaves obovate, coriaceous, glabrous above . 3. *A. buxifolia.*
Stigma with 5 distinct reflexed lobes.  Ovary truncate, scarcely
   lobed, slightly depressed in the centre.
   Leaves flat, obovate-oblong or lanceolate, ¾ to 1¼ in., tomentose
      on both sides.  Flowers shortly pedicellate . . . . . . 4. *A. mollis.*
   Leaves under ½ in.  Flowers sessile.
     Leaves obovate or cuneate, flat or concave, tomentose on both
       sides . . . . . . . . . . . . . . . . 5. *A. pleurandroides.*
     Leaves ovate or oblong, the margins revolute, glabrous above 6. *A. trymalioides.*

SECT. 2. **Urocarpus.**—*Ovary* 2–3-*merous.*
Indumentum scaly.  Ovary divided to the middle.  Leaves oblong 7. *A. squamuligera.*
Indumentum of stellate hairs.  Leaves mostly ovate.
   Stamens 10 to 15.  Ovary usually 2-merous.
     Ovary with 2 erect lobes . . . . . . . . . . . 8. *A. pallida.*
     Ovary truncate, not lobed . . . . . . . . . 9. *A. phebalioides.*
   Stamens above 20.  Ovary usually 3-merous . . . . . 10. *A. grandiflora.*

SECTION 1. EUASTEROLASIA.—Ovary 5-merous.

1. **A. correifolia,** *Benth.*   A tall shrub, the branches densely tomentose.  Leaves petiolate, from ovate to lanceolate, obtuse, mostly 1 to 2 in.
long, flat, glabrous and smooth above, softly velvety-tomentose underneath.
Flowers (white?) on short pedicels, in axillary or terminal clusters.  Calyx
exceedingly minute, concealed under the stellate hairs.  Petals about 2½ lines
long, valvate and slightly induplicate, tomentose outside.  Stamens 10.
Ovary densely tomentose, with 5 short, erect lobes.  Stigma large, reflexed-
peltate, scarcely lobed.  Cocci small, truncate, with incurved beaks on their
outer angle.—*Phebalium correæfolium,* A. Juss. in Mem. Soc. Hist. Nat.
Par. ii. 130, t. 10; *P. ovatum,* Sieb. Pl. Exs.; *Eriostemon correifolius*
(partly), F. Muell. Fragm. i. 105; Pl. Vict. i. 132.

   **N. S. Wales.** Port Jackson, *R. Brown, Sieber, n.* 113, *A. Cunningham;* Paramatta,
*Woolls.*

2. **A. Muelleri,** *Benth.*   A low shrub, allied to *A. correifolia,* with
which F. Mueller now unites it, but from the specimens I have seen it appears to me better to consider it as a distinct species, as he originally proposed.  Leaves petiolate, from obovate to narrow-oblong, very obtuse, rarely
exceeding 1 in. when very luxuriant, and often much smaller, flat, narrowed
at the base, rough above with scattered stellate hairs, densely tomentose
underneath.  Flowers of *A. correifolia,* but the pedicels usually longer, and
the calyx rather more conspicuous.  Cocci truncate as in that species, but
the heads much more horizontally divaricate.—*Phebalium asteriscophorum,*
F. Muell. in Trans. Vict. Inst. i. 31; *Eriostemon correifolius* (partly), F.
Muell. Fragm. i. 105, and Pl. Vict. i. 132.

   **Victoria.** Ravines of Buffalo mountains, Buffalo river, and Mount Disappointment,
*F. Mueller.*

3. **A. buxifolia,** *Benth.* A rigid shrub of several feet, the young branches densely tomentose. Leaves petiolate, from obovate to oblong-cuneate, very obtuse, mostly about ⅓ in. long; the margins slightly recurved, narrowed at the base, coriaceous, glabrous and shining above, white underneath with a short dense tomentum. Flowers terminal or axillary, sessile within 3 or 4 ovate concave leafy bracts, assuming the appearance of sepals. Calyx entirely obsolete. Petals 2½ to nearly 3 lines long, tomentose outside. Stamens often 2 or 3 more than 10. Ovary glabrous, with 5 short erect lobes. Stigma large, reflexed-peltate, slightly lobed at the edge. Cocci glabrous, with shortly divaricate obtusely triangular beaks.—*Phebalium buxifolium,* A. Cunn. Herb.

**N. S. Wales.** Blue Mountains, *A. and R. Cunningham.*

4. **A. mollis,** *Benth.* An erect spreading shrub, softly tomentose, with stellate spreading hairs. Leaves petiolate, from obovate to oblong or lanceolate, obtuse, ¾ to 1½ in. long, flat, tomentose on both sides. Flowers shortly pedicellate, few together in terminal or rarely axillary clusters. Sepals small, lanceolate, closely appressed, so as to be almost concealed under the dense tomentum of the petals. Petals about 3 lines long. Ovary densely stellate-tomentose, rounded at the top, and slightly depressed in the centre, where the styles are inserted. Stigma large, reflexed, 5-lobed. Fruit not seen.— *Phebalium hexapetalum,* A. Juss. in Mem. Soc. Hist. Nat. Par. ii. 131, t. 11.

**N. S. Wales,** *Gaudichaud.* Arbuthnot's Range in the N.W. interior, *Fraser.* The flowers on Gaudichaud's specimen are very few, and one is certainly 5-merous; it is therefore probably by accident only that those examined by Jussieu were 6-merous.

5. **A. pleurandroides,** *F. Muell.* A low rigid shrub, densely tomentose or almost woolly. Leaves crowded, obcordate, spathulate or oblong-cuneate, very obtuse or truncate, rarely exceeding 4 lines, thick, flat or concave, stellate-hairy on both sides. Flowers yellow, closely sessile, solitary, terminal, although from the shortness of the branches they often appear axillary. Calyx none, unless it be represented by 3 or 4 upper smaller leaves, which appear to alternate with the petals. Petals induplicate-valvate, about 4 lines long, tomentose outside. Stamens 10. Ovary densely stellate-hirsute, truncate, scarcely depressed in the centre where the styles are attached. Stigma deeply divided into thick, linear, recurved, densely papillose lobes. Cocci tomentose, not beaked.—*A. phebalioides,* F. Muell. in Trans. Phil. Soc. Vict. i. 10; *Eriostemon pleurandroides,* F. Muell. Fragm. i. 106, and Pl. Vict. i. 133.

**Victoria.** Arid and stony slopes of the Serra and Victoria ranges, *F. Mueller.* I have adopted F. Mueller's change of the specific name from *phebalioides* to *pleurandroides,* as the latter is much more appropriate, and the former would clash with *Urocarpus phebalioides,* Drumm., now transferred to *Asterolasia.*

6. **A. trymalioides,** *F. Muell. in Trans. Phil. Soc. Vict.* i. 10. A low rigid shrub, the branches densely tomentose. Leaves ovate obovate or oblong, very obtuse, mostly 2 to 4 lines long, the margins much revolute, coriaceous, glabrous and shining above when full-grown, tomentose underneath. Flowers yellow, sessile, terminal, solitary or 2 or 3 together, with 2 small bracts at their base. Calyx very small, with thin almost transparent ovate lobes. Petals induplicate-valvate in the bud, spreading, and attaining about 3 lines. Stamens 10. Ovary tomentose, truncate, slightly depressed

in the centre where the styles are inserted. Stigma deeply divided into oblong, reflexed, densely papillose lobes. Cocci tomentose, truncate, not beaked. Seeds smooth and shining.—*Eriostemon trymalioides*, F. Muell. Fragm. i. 106, and Pl. Vict. i. 134.

**N. S. Wales.** Mount Kosciusko, *F. Mueller.*

**Victoria.** On the highest summits of the Australian Alps, not descending below 5000 ft. elevation, *F. Mueller.*

SECTION 2. UROCARPUS.—Ovary 2- or 3-merous.

7. **A. squamuligera,** *Benth.* A weak shrub or undershrub, the younger branches covered with minute scurfy scales, often fringed with short rigid hairs. Leaves oblong-lanceolate, obtuse, ½ to near 1 in. long, rather thick, nerveless, narrowed into a short petiole. Flowers few, in terminal umbels, surrounded by short coloured bracts, with occasionally 1 or 2 longer leafy ones. Pedicels slender, rarely exceeding ½ in. Calyx very minute. Petals narrow-ovate, 3 to 3½ lines long. Stamens 10. Ovary of 2 or rarely 3 carpels, forming erect lobes, narrowed upwards, covered with scurfy scales. Stigma divided into 2 or 3 large reflexed lobes. Cocci, when young, obtusely acuminate and erect, but not seen ripe.—*Phebalium squamuligerum*, Hook. Ic. Pl. t. 727; *Eriostemon Hookeri*, F. Muell. Fragm. i. 104.

**W. Australia.** Between Swan River and King George's Sound, *Drummond.*

8. **A. pallida,** *Benth.* Branches weak, almost herbaceous, clothed with stellate hairs, sometimes slightly united into scales. Leaves distinctly petiolate, ovate or orbicular, very obtuse, 3 to 5 lines long, flat, sprinkled above and more densely covered underneath with stellate hairs. Pedicels 1-flowered, axillary and solitary, or several together in terminal umbels, with small or leafy bracts at their base. Petals 2 to 2½ lines long, induplicate-valvate, the part exposed in the bud stellate-tomentose. Stamens 10 to 15. Ovary densely stellate-hairy, consisting of 2 carpels, with 2 short erect lobes, between which the styles are inserted. Cocci beaked, the conical beaks remaining erect for some time after the flowering is over, becoming somewhat lengthened and divaricate at the fruit ripens.

**W. Australia,** *Drummond, n. 42 and* 112.

9. **A. phebalioides,** *Benth.* Branches elongated, often appearing glabrous, but really clothed with a minute stellate pubescence. Leaves on rather long petioles, orbicular, ovate or oblong, obtuse, mostly under ½ in. long, rarely ¾ in. Pedicels slender, either in terminal umbels soon becoming lateral, or 2 or 3 together in the upper axils. Flowers as in *A. pallida*, at least when fully out. Ovary stellate-hairy, truncate and not lobed, the 2 carpels united at the top, and retaining the shape for some time after flowering, the outer angles at length growing out into long horizontally diverging beaks.—*Urocarpus phebalioides*, Drumm. in Hook. Kew Journ. vii. 55; *Eriostemon Drummondii*, F. Muell. Fragm. i. 105.

**W. Australia.** Mount Lesueur, *Drummond.*

10. **A. grandiflora,** *Benth.* Branches rather slender, clothed with short stellate hairs. Leaves shortly petiolate, ovate or oblong, obtuse, mostly under ½ in. long, the margins recurved, the midrib prominent underneath,

sprinkled above and more densely clothed underneath with short stellate hairs. Pedicels terminal, usually several together, with short ovate, coloured or leafy bracts at their base. Petals induplicate-valvate, tomentose outside, not large when first expanded, but attaining at length 5 or 6 lines. Stamens 20 to 25. Ovary densely stellate-hairy, with 3 short erect lobes. Fruit not seen.—*Phebalium grandiflorum*, Hook. Ic. Pl. t. 724; *Eriostemon grandiflorus*, F. Muell. Fragm. i. 105.

**W. Australia,** *Drummond.*

## 12. CORREA, Sm.

(Didymeria, *Lindl.*)

Calyx cup-shaped, truncate and 4- or 8-toothed, or 4-lobed. Petals 4, valvate, connate in a cylindrical or campanulate tube, sometimes separating as the flower expands, spreading at the top. Disk shortly lobed. Stamens 8, free; anthers without appendages. Ovary of 4 carpels nearly distinct from the base; styles inserted above the middle, and immediately united into one filiform style, with a small often shortly 4-lobed stigma; ovules 2 in each carpel, superposed. Cocci 4, truncate, 2-valved, the endocarp cartilaginous and separating elastically.—Shrubs or rarely small trees, stellate-tomentose or rarely glabrous. Leaves opposite, petiolate, simple. Flowers rather large and showy, red yellow white or green, usually pendulous, solitary or 2 or 3 together, axillary or terminal. Petals usually mealy-tomentose outside.

The genus is limited to Australia.

Petals free after the flower is expanded.
　Calyx with 4 lanceolate teeth as long as the tube. Filaments di
　　lated at the base . . . . . . . . . . . . . . . . . 1. *C. æmula.*
　Calyx truncate, with 4 minute teeth. Filaments filiform or scarcely
　　dilated . . . . . . . . . . . . . . . . . . . 2. *C. alba.*
Petals connate or cohering till they fall off.
　Calyx truncate, with 4 minute or very broad teeth.
　　Four of the filaments dilated below the middle . . . . . . 3. *C. speciosa.*
　　Filaments all equally filiform or scarcely dilated . . . . . 4. *C. Lawrenciana.*
　Calyx with 4 short broad and 4 longer filiform teeth . . . . 5. *C. decumbens.*

1. **C. æmula,** *F. Muell. Fragm.* i. 3, *and Pl. Vict.* i. 139, *t.* 7. A tall shrub, with spreading branches, hirsute or tomentose with stellate often stipitate hairs. Leaves shortly petiolate, orbicular, ovate or ovate-lanceolate, obtuse, rarely exceeding 1 in., except in luxuriant barren shoots, often slightly cordate, scabrous above, densely tomentose underneath. Pedicels axillary, 1-flowered, slender, bearing a pair of small orbicular leafy bracts near the base, and 2 smaller subulate ones higher up. Flowers pendulous, dull green or purple. Calyx sprinkled with stellate hairs, the lobes lanceolate acuminate, usually as long or longer than the tube. Petals linear, about 1 in. long, cohering when young, but separating as the flower expands. Filaments dilated and oblong near the base, filiform upwards. Ovary densely hirsute. Style glabrous.—*Didymeria æmula*, Lindl. in Mitch. Three Exped. ii. 198.

**Victoria.** Stony shady declivities of the Serra and Victoria ranges, *F. Mueller,* and previously gathered by *Mitchell* in the same district.

**S. Australia.** Rocky glens of the Barossa ranges and mountains near Encounter Bay, *F. Mueller.*

A. Cunningham's fruiting specimen, referred here by Lindley, appears to be rather the *C. speciosa*, with the calyx accidentally split up.

2. **C. alba,** *Andr. Bot. Rep. t.* 18. A compact much-branched shrub, rarely above 3 or 4 ft. high, and often much lower, the branches clothed with a hoary or rusty tomentum, either close or almost floccose. Leaves from orbicular to ovate obovate or elliptical, very obtuse, ½ to 1 in. long or rarely more, coriaceous, slightly tomentose or at length glabrous above, densely tomentose underneath. Pedicels terminal, very short, solitary or 2 or 3 together. Flowers white or pink. Calyx tomentose, truncate, with 4 very small teeth. Petals tomentose outside, not exceeding ½ in., free from their first opening, but connivent in a more bell-shaped and less elongated corolla than the other species. Filaments equally filiform or scarcely dilated.—Vent. Jard. Malm. t. 13; DC. Prod. i. 719; Bot. Reg. t. 515; F. Muell. Pl. Vict. i. 135; *C. cotinifolia*, Salisb. Parad. Lond. t. 100; *Mazeutoxeron rufum*, Labill. Voy. ii. 12, t. 17; *C. rufa*, Vent. Jard. Malm. in note to t. 13; Labill. Pl. Nov. Holl ii. 120; DC. Prod. i. 719; Hook. f. Fl. Tasm. i. 61.

**Victoria.** Frequent along the sandy or rocky seashore, *R. Brown, F. Mueller.*

**Tasmania,** *R. Brown;* abundant, especially near the coast, *J. D. Hooker.*

**S. Australia.** On the coast, extending to St. Vincent's Gulf, *F. Mueller;* Kangaroo Island, *Waterhouse.*

Var. *rotundifolia.* Densely hirsute. Leaves small and broad. Flowers sessile, terminal or in the forks of the upper branches.—*C. rotundifolia*, Lindl. in Mitch. Three Exped. ii. 219.—Near the Glenelg, *Mitchell;* apparently not uncommon along the coast of Victoria and S. Australia.

3. **C. speciosa,** *Ait. Epit. Hort. Kew.* 366. A shrub, variable in size and habit, usually rigid and low, and rarely exceeding 6 to 8 ft., the stellate tomentum very variable, usually loose and abundant on the branches or sometimes on the whole plant, dense and soft on the under side of the leaves, disappearing on the upper surface or sometimes on the whole plant, except the peduncles and flowers. Leaves very shortly petiolate, from broadly ovate or cordate to narrow-oblong or lanceolate, obtuse or retuse, usually from ¾ to 1¼ in. long, rarely all under 1 in., or the larger ones attaining 2 in. Flowers red, varying to white or yellowish-green, terminal, shortly pedicellate and pendulous, or a few rarely erect, solitary or 2 or 3 together. Calyx hoary or rusty-tomentose, truncate, with 4 minute teeth. Petals hoary-tomentose outside, united the greater part of their length into a cylindrical or slightly campanulate corolla of ¾ to 1½ in., with 4 spreading lobes. Stamens exserted, the filaments of those opposite the petals more or less dilated below the middle.—DC. Prod. i. 719; F. Muell. Pl. Vict. i. 136.

**N. S. Wales.** Port Jackson, *R. Brown, Sieber, n.* 238 *and* 239, and others; northward and southward to the limits of the colony, apparently not extending inland far beyond the Blue Mountains.

**Victoria.** Not rare in heathy and barren rocky localities, not ascending to alpine elevations; known to the colonists as *Native Fuchsia, F. Mueller.*

**Tasmania,** *R. Brown;* abundant throughout the colony, *J. D. Hooker.*

**S. Australia.** From the Great Australian Bight to Lake Torrens, *F. Mueller.*

**W. Australia.** King George's Sound, *Maclean.*

I follow F. Mueller in uniting under one name all *Correas* with a truncate 4-toothed calyx, united petals, and 4 of the filaments dilated. At the same time, although the following races may occasionally be found to pass one into another, yet they appear generally so distinct, that I feel some hesitation in refusing to recognize them as species.

*a. normalis.* Branches loosely and copiously tomentose, sometimes almost woolly or very hirsute. Leaves mostly cordate-ovate, rarely narrow, convex or bullate, with recurved and sometimes undulate or crisped margins, glabrous scabrous or loosely tomentose above, densely tomentose or woolly underneath. Flowers usually elongated, cylindrical.—*C. speciosa,* Andr. Bot. Rep. t. 653 ; Bot. Reg. t. 26 ; Bot. Mag. t. 1746 (flowers more erect than I have ever seen them) ; *C. rubra,* Sm. Exot. Bot. ii. 26 ; *Antomarchia rubra,* Colla, Hort. Ripul. App. ii. 345 ; *C. cordifolia,* Lindl. in Mitch. Three Exped. ii. 233 ; *C. virens,* Sm. Exot. Bot. ii. 25, t. 72 ; Bot. Reg. t. 3 ; Bot. Mag. t. 1901 ; *C. viridiflora,* Andr. Bot. Rep. t. 436 ; Bonpl. Jard. Malm. 33, t. 12 (the last 2 names referring to a green-flowered variety) ; *C. cardinalis,* F. Muell. ; Hook. Bot. Mag. t. 4912 (a narrow-leaved variety).—N. S. Wales, Victoria, and Tasmania.

*b. Backhousiana.* Branches rather closely tomentose. Leaves ovate or oblong, scarcely or not at all cordate, coriaceous, flat, glabrous above, closely but usually densely tomentose underneath. Flowers nearly cylindrical, above 1 in. long.—*C. Backhousiana,* Hook. Journ. Bot. i. 253, and Ic. Pl. t. 2 ; Hook. f. Fl. Tasm. i. 61 ; *Mazeutoxeron reflexum,* Labill. Voy. ii. 66, t. 19 ; *C. reflexa,* Labill. Pl. Nov. Holl. ii. 120.—N. coast of Tasmania and islands of Bass's Straits.

*c. leucoclada.* Branches closely and often minutely tomentose. Leaves small, ovate or oblong, not cordate, coriaceous, flat, glabrous above, closely and often minutely tomentose underneath. Flowers generally under 1 in., and more campanulate than in the preceding varieties.—*C. leucoclada,* Lindl. in Mitch. Three Exped. ii. 39.—N. S. Wales (Mount Aiton, *A. Cunningham*) and Victoria.

*d. glabra.* Leaves ovate or oblong, small, flat, glabrous on both sides as well as the branches. Flowers of the var. *leucoclada* or rather longer.—*C. glabra,* Lindl. in Mitch. Three Exped. ii. 48 ; *C. Schlechtendahlii,* Behr, in Linnæa, xx. 630.—Victoria and S. Australia, and the single W. Australian specimen. *C. pulchella,* Sw. Fl. Austral. t. 1, belongs probably to this variety ; the leaves were found to be sprinkled with stellate (fasciculate) hairs when young, glabrous with age.

The species, being highly ornamental, has long been cultivated in British gardens, and numerous garden varieties, hybrids and crosses, have been raised, amongst which the following have been figured as species :—*C. pulchella,* Mackay ; Bot. Reg. t. 1224 ; Bot. Mag. t. 4029 ; Maund, Botanist, t. 152 ; *C. longiflora,* Paxt. Mag. Bot. vii. 195 ; *C. Harrisii,* Paxt. Mag. Bot. vii. 79 ; *C. bicolor,* Paxt. Mag. Bot. ix. 267.

4. **C. Lawrenciana,** *Hook. Journ. Bot.* i. 254. A shrub, usually tall and rather slender, sometimes growing into a small tree ; branches more or less tomentose. Leaves petiolate, from ovate to oblong, obtuse, in some specimens ½ to 1 in., in others 1 to 2 in. long or even larger, flat, coriaceous, glabrous above, tomentose underneath. Flowers 1 to 3 together, axillary or terminal, shortly pedicellate and pendulous. Calyx tomentose, truncate with 4 small teeth. Petals tomentose outside, united the greater part of their length into a cylindrical corolla of ¾ to 1 in., the lobes usually shorter and more obtuse than in *C. speciosa.* Stamens exserted ; filaments all filiform from the base or equally and very slightly dilated.—Hook. f. Fl. Tasm. i. 61 ; F. Muell. Pl. Vict. i. 138 ; *C. ferruginea,* Backh. in Ross, Hobart. Alm. ; Hook. Comp. Bot. Mag. i. 276, and Ic. Pl. t. 3 ; Maund, Botanist, t. 124 (a large-leaved variety) ; *C. Latrobeana,* F. Muell. in Dietr. Fl. Univ. N. Ser. t. 11 (a still larger form).

**Victoria.** In subalpine situations, descending along rivulets and torrents to 1000 ft. elevation, *F. Mueller.*

**Tasmania.** Derwent river, *R. Brown ;* abundant throughout the colony, *J. D. Hooker.* In foliage this species can scarcely be distinguished from some forms of *C. speciosa,* var. *Backhousiana,* but it is always readily known by the filaments all similar and scarcely perceptibly dilated.

Var. *glabra.* Leaves narrow, oblong, lanceolate or almost linear, glabrous on both sides as well as the branches. Derwent river, *R. Brown,* and in some other Tasmanian collections.

2 A 2

5. **C. decumbens,** *F. Muell. in Trans. Phil. Soc. Vict.* i. 30, *and Pl. Vict.* i. 137. A decumbent shrub with ascending branches, densely stellate-tomentose. Leaves oblong, from almost ovate to linear, obtuse, mostly 1 to 1½ in. long, coriaceous, nearly glabrous above, densely tomentose underneath. Flowers terminal, solitary, shortly pedicellate, usually pendulous. Calyx tomentose, with 4 lobes opposite the petals, triangular or lanceolate, rather longer than the tube, and 4 lobes alternating with them, subulate and nearly twice as long. Petals tomentose outside, united the greater part of their length into a nearly cylindrical corolla of ¾ to 1 in. Stamens exserted, the filaments all slightly dilated below the middle.

**S. Australia.** Lofty Range and Onkaparinga river, *F. Mueller;* Kangaroo Island, *Waterhouse.*

### 13. NEMATOLEPIS, Turcz.

(Symphyopetalum, *Drumm.*)

Calyx small, 5-cleft. Petals 5, valvate, united the greater part of their length in a cylindrical tube, spreading at the top. Disk small, crenate. Stamens 10; filaments slightly dilated at the base into an adnate scale fringed with long hairs; anthers not apiculate. Ovary of 5 distinct carpels, the styles inserted below the middle, and immediately united into one filiform style with a minute stigma; ovules 2 in each carpel, superposed. Cocci truncate, 2-valved, the endocarp cartilaginous and separating elastically.—A shrub, clothed with peltate scurfy scales. Leaves simple, alternate. Flowers axillary.

The genus consists of a single species, limited to W. Australia, allied to *Correa* in the united petals, to *Chorilæna* in the stamens, and to *Phebalium* in habit and indumentum.

1. **N. phebalioides,** *Turcz. in Bull. Mosc.* 1852, ii. 158. An erect, rigid, bushy shrub, the young branches, under side of the leaves, and pedicels covered with silvery scurfy scales. Leaves ovate or oblong, very obtuse, mostly under ½ in., but occasionally ¾ in. long, coriaceous, glabrous above, with more or less prominent tubercular glands. Pedicels short, axillary, recurved, 1-flowered. Sepals short, orbicular, almost cordate, smooth or sprinkled with a few scales. Corolla glabrous, ½ to ¾ in. long. The scale of the filaments forms a slight prominence inside, terminating the dilated base, and fringed with long hairs.—*Symphyopetalum correoides,* Drumm. in Hook. Kew Journ. vii. 54.

**W. Australia.** Near Middle Mount Barren, *Drummond;* Point Henry, *Oldfield.*

### 14. CHORILÆNA, Endl.

Flowers collected in dense pendulous cymes or heads, surrounded by a few subulate bracts. Sepals 5. Petals 5, very narrow, valvate or nearly so. Disk small, shortly lobed. Stamens 10, much exserted; filaments dilated at the base into an adnate scale, fringed with long hairs; anthers not apiculate. Ovary of 5 distinct carpels; styles inserted below the middle, and immediately united into 1 filiform style, with a small obscurely 5-lobed stigma; ovules in each carpel 2, superposed. Cocci truncate; endocarp cartilaginous, separating elastically.—Shrubs, with the habit of some *Thomasias,* hispid or tomentose with stellate hairs. Leaves alternate, sinuate-lobed. Flower-cymes pedunculate, axillary.

A genus limited to W. Australia, approaching *Diplolæna* in inflorescence, *Nematolepis* in the bearded appendage of the filaments, and connected with *Phebalium* through *P. Ralstoni.*

Leaves at length glabrous above, densely and softly tomentose under-
　neath. Sepals lanceolate . . . . . . . . . . . . . 1. *C. quercifolia.*
Leaves loosely stellate-hirsute. Sepals filiform . . . . . . . . 2. *C. hirsuta.*

1. **C. quercifolia,** *Endl. in Hueg. Enum.* 17. A tall shrub, the branches densely clothed with a soft close or velvety tomentum, often assuming a golden colour. Leaves petiolate, ovate, very obtuse, mostly 1½ to 3 in. long, sinuately lobed or broadly pinnatifid, somewhat coriaceous, the upper surface sprinkled when young with a slight stellate pubescence, glabrous when full grown, the under side densely and softly velvety-tomentose. Peduncles re-curved, scarcely exceeding ½ in. Cymes often at least 1 in. diameter, of 7 to 14 flowers, the outer ones at the ends of the branches appearing pedicellate, the inner ones sessile. Bracts filiform, shorter than the calyx. Sepals lan-ceolate, tomentose outside. Petals rather longer, attaining 3 lines, tomentose outside. Stamens fully twice as long.—Bartl. in Pl. Preiss. i. 172.

**W. Australia.** King George's Sound, *Huegel, Fraser ;* Bald Head and Island, *Preiss, n.* 2043, *Oldfield, Maxwell.*

2. **C. hirsuta,** *Benth.* A tall erect shrub, the branches densely hirsute with stellate hairs. Leaves petiolate, ovate, obtuse, mostly 2 to 3 in. long, sinuately lobed or broadly and obtusely pinnatifid, thinner than in *C. querci-folia,* the upper surface rough with scattered stellate hairs, the under side pale, more copiously hirsute. Peduncles solitary or 2 or 3 together, recurved, rarely above ½ in. long. Cymes nearly as in *C. quercifolia.* Bracts linear-filiform or slightly cuneate, very hirsute, the outer ones often 5 lines long, much more numerous than in *C. quercifolia,* and passing gradually into the sepals, of which the innermost are often under 3 lines. Petals very narrow, hirsute outwards along the centre. Stamens fully twice as long. Cocci short, glabrous or sprinkled with a few stellate hairs. Seeds smooth and shining.

**W. Australia.** Swan River, *Drummond;* Flinders Bay, *Collie;* Wilson's Inlet, *Oldfield.*

## 15. DIPLOLÆNA, R. Br.

Flowers sessile, in dense heads, surrounded by an involucre of broad bracts, imbricate in 3 or 4 series, the inner ones larger and petal-like. Calyx none. Petals 5, small, narrow. Disk small. Stamens 10, much exserted; filaments filiform, bearded with long hairs above the base; anthers not apiculate. Ovary 5-lobed; styles united into a single elongated style, with a shortly 5-lobed stigma; ovules 2 in each cell, superposed. Cocci 2-valved, the endocarp cartilaginous and separating elastically.—Shrubs, clothed with stellate tomen-tum. Leaves alternate, petiolate, entire. Flower-heads terminal, shortly pe-dunculate or nearly sessile.

The genus is limited to W. Australia, and, like *Chorilæna,* is chiefly distinguished by the inflorescence. In other respects it only differs from *Phebalium* in the abortion of the calyx and the narrow petals. The 4 following species will be found perhaps, when better known, to run too much into one another to be otherwise separated than as marked varieties.

Leaves tomentose or hoary on both sides.
　Flower-heads and leaves large.　Outer bracts ovate, inner ones
　　broadly elliptical . . . . . . . . . . . . . . 1. *D. grandiflora.*
　Flower-heads and leaves small.　Bracts lanceolate . . . . . 2. *D. microcephala.*
Leaves green and glabrous above, tomentose underneath.
　Leaves oblong, flat.　Bracts broad . . . . . . . . . . 3. *D. Dampieri.*
　Leaves linear, the margins revolute.　Bracts narrow . . . . . 4. *D. angustifolia.*

1. **D. grandiflora,** *Desf. in Mem. Mus. Par.* iii. 451, *t.* 19.　A shrub of 5 or 6 ft., with rigid divaricate branches, hoary or rusty with a close tomentum.　Leaves ovate or broadly oblong, very obtuse, 1 to 2 in. long, hoary on both sides and especially underneath with a close tomentum.　Flower-heads very shortly pedunculate, attaining sometimes 1½ in. diameter.　Outer bracts 5, broadly ovate, herbaceous, tomentose, 4 or 5 lines long, united at the base.　Inner ones about 10, longer, narrower, and more petal-like, those of the first 1 or 2 series broadly elliptical, obtuse, pubescent, passing into a few (innermost) much narrower ones, sometimes linear and acute.　Petals linear, ciliate, quite concealed within the head.　Stamens much longer than the bracts.　Cocci 3 or 4 lines long, coriaceous, glabrous, smooth or transversely wrinkled.

**W. Australia.** Sharks Bay and Dirk Hartog's Island, *A. Cunningham, Milne.*

2. **D. microcephala,** *Bartl. in Pl. Preiss,* i. 173.　A shrub of 2 or 3 ft.　Leaves obovate or oblong, very obtuse, sometimes all under ½ in., and rarely exceeding 1 in., rather thick, hoary-tomentose above, and densely and softly tomentose underneath.　Flower-heads much smaller than in *D. grandiflora* or *D. Dampieri.*　Bracts lanceolate, the outer herbaceous ones not much shorter than the inner ones.　Filaments more densely hirsute than in other species with reddish hairs.

**W. Australia.** Stony barren mountains of Grantham district, *Preiss, n.* 2018 ; near Cape Riche, *Preiss, n.* 2019, *Oldfield ;* between Perth and King George's Sound, *Harvey ;* Darling Range, *Collie ;* Murchison river, *Oldfield.*
Var. *Drummondi.* Leaves oblong, ¾ to 1½ in. long ; tomentum looser and sometimes disappearing with age on the upper side, which however has not the smooth texture of *D. Dampieri.*—Swan River, *Drummond, Coll.* 1843, n. 91 ; Phillips river, *Maxwell.* To this variety, rather than to the true *D. Dampieri,* ought perhaps to be referred the *D. Dampieri,* Lindl. Bot. Reg. 1841, t. 64, figured with narrow-lanceolate bracts.

3. **D. Dampieri,** *Desf. in Mem. Mus. Par.* iii. 452, *t.* 20.　Nearly allied to *D. grandiflora,* and chiefly distinguished by the leaves, quite glabrous green and smooth on the upper side.　In the form originally described, they are oblong or somewhat cuneate, the flower-heads are rather smaller than in *D. grandiflora,* and the bracts not so broad ; but in the Murchison river specimens the leaves and bracts are nearly as broad as in that species.—Bot. Mag. t. 4059 ; Bartl. in Pl. Preiss. i. 173.

**W. Australia.** From Swan River, *Harvey, Oldfield.* and others, and Darling range, *Preiss, n.* 2042, to Champion Bay and Murchison river, *Oldfield.*

4. **D. angustifolia,** *Hook. Bot. Mag. under n.* 4059.　Branches hoary or rusty with a close tomentum.　Leaves linear or linear-cuneate, obtuse, ¾ to 2 in. long, the margins revolute, glabrous above, white with a close but

dense tomentum underneath. Flower-heads rather larger than in *D. micro-cephala* in Preiss's specimens, considerably larger in Drummond's, the bracts numerous and lanceolate, or the outer ones ovate-lanceolate.—*D. salicifolia,* Bartl. in Pl. Preiss. i. 173.

**W. Australia,** *Drummond, 1st Coll., Roe, Preiss, n.* 2020.

TRIBE II. ZANTHOXYLEÆ.—Trees or shrubs. Leaves pinnate or 3-folio-late with opposite leaflets or 1-foliolate (truly simple in *Geijera*), the leaflets usually large. Ovary lobed. Fruit separating into distinct 2-valved cocci. Endocarp persistent or separating elastically. Seeds in most genera albumi-nous; the cotyledons flattened and broader than the radicle, but in a few genera the albumen is wanting, and the cotyledons are thick and fleshy.—The tribe differs from *Boronieæ* more in habit than in any definite character.

## 16. BOSISTOA, F. Muell.

Flowers hermaphrodite? Calyx small, 5-toothed. Petals 5, valvate or slightly imbricate, with inflexed tips. Disk thick. Stamens 10. Ovary of 5 distinct carpels; styles almost terminal, united upwards, but soon sepa-rating; ovules 2 in each carpel, superposed. Cocci distinct, large, coriaceous, 2-valved; endocarp cartilaginous, separating. Seeds solitary; testa membra-nous; albumen none; cotyledons thick and fleshy, radicle small.—A tree. Leaves opposite, pinnate. Panicles terminal.

The genus is limited to a single Australian species, allied in some respects to *Melicope* and *Evodia,* but very different in habit as well as in the seeds, which have the structure of *Pilocarpus* and some other American genera.

1. **B. sapindiformis,** *F. Muell. Herb.* A tree with the habit of a *Cupania,* the young shoots, petioles and inflorescence minutely pubescent. Leaves pinnate; leaflets 7 to 11, opposite in pairs, the terminal odd one oc-casionally wanting, oblong-lanceolate, 4 to 8 in. long, more or less serrate-toothed, especially above the middle, narrowed at the base, on a short petio-lule or nearly sessile. Panicles terminal, trichotomous, shorter than the leaves. Buds globular. Calyx small, very shortly and unequally toothed. Petals about 2 lines long. Filaments dilated at the base, attenuated upwards, glabrous; anthers large. Carpels very hirsute, on a raised disk. Styles short. Cocci broadly and very obliquely ovate, about 1 in. long, hard, almost woody, tomentose and rugose outside.—*Evodia pentacocca,* F. Muell. Fragm. iii. 41.

**Queensland.** Ipswich, *Nernst.* (A single leaf and loose fruit from F. Muell.)
**N. S. Wales.** Richmond and Clarence rivers, *Beckler.* (Specimens in flower only.)

## 17. MELICOPE, Forst.

Flowers more or less unisexual. Sepals 4. Petals 4, valvate, or slightly imbricate, with inflexed tips. Disk thick, entire or lobed. Stamens 8. Ovary of 4 nearly distinct carpels; styles inserted above the middle, united immedi-ately or at the summit into one, with a capitate 4-lobed stigma; ovules 2 in each carpel, superposed or collateral. Cocci distinct, spreading, 2-valved; endocarp cartilaginous or horny, separating. Seeds usually solitary; testa crustaceous, shining; albumen fleshy, embryo straight or slightly curved, with oblong or ovate cotyledons.—Trees or shrubs. Leaves opposite, 3-foliolate, or (in species not Australian) 1-foliolate or simple. Flowers rather small, in terminal or axillary cymes or panicles.

Besides the Australian species, which are endemic, there are 2 from New Zealand and a few from the Pacific islands. F. Mueller proposes to unite *Melicope* with *Evodia*, but the double number of stamens is a more constant character than many others distinguishing the received genera of *Zanthoxyleæ*.

Petals thin. Styles lateral. Leaflets mostly under 4 in. Panicles
    terminal.
  Young branches pubescent. Petals and filaments glabrous. Ripe
    carpels erect   .   .   .   .   .   .   .   .   .   .   .   .   . 1. *M. neurococca.*
  Branches and leaves glabrous. Petals minutely pubescent. Fila-
    ments ciliate. Ripe carpels divaricate   .   .   .   .   .   .   . 2. *M. erythrococca.*
Petals and stamens rigid. Styles terminal. Leaflets more than 6 in.
    Panicles lateral .   .   .   .   .   .   .   .   .   .   .   .   .   . 3. *M. australasica.*
  The first 2 species are the nearest allied to the New Zealand *M. ternata*, Forst., the third is in some respects anomalous.

1. **M. neurococca,** *Benth.* A small tree, the young branches, petioles, and peduncles pubescent with simple spreading hairs. Leaves of each pair generally unequal, the larger one with a common petiole of 2 in. or more, the other with a much shorter petiole; leaflets 3, ovate-lanceolate or lanceolate, acuminate, mostly 3 to 4 in. long, glabrous above, sprinkled with a few hairs underneath. Panicles terminal, trichotomous, corymbose. Sepals small, orbicular, concave, ciliate. Petals about 2 lines long, glabrous, valvate or nearly so. Filaments glabrous, dilated to the middle. Ovary hirsute, the carpels almost distinct from the base. Styles inserted below the summit. Cocci distinct, nearly erect, broad, about 3 lines long, the valves coriaceous and transversely wrinkled.—*Evodia neurococca*, F. Muell. Fragm. i. 28, and ii. 103.

**Queensland.** Brisbane river, *W. Hill* and *F. Mueller;* Wide Bay and Archer's Creek, used by the natives to make their spades, *Leichhardt.*
**N. S. Wales.** Richmond, Hastings, and Clarence rivers, *Beckler.*

2. **M. erythrococca,** *Benth.* A moderate-sized tree, quite glabrous. Leaflets 3 or rarely 1 only, oblong-lanceolate, obtuse, 1½ to 3 in. long, coriaceous, entire or obscurely crenulate, on a common petiole of ¾ to 1½ in. Panicles terminal or in the upper axils, loose, scarcely longer than the leaves. Sepals small, triangular, slightly ciliate. Petals 1½ lines long, slightly imbricate, valvate at the tips, minutely pubescent outside. Disk obscurely lobed. Filaments dilated and ciliate to above the middle. Ovary slightly hirsute, the carpels almost distinct. Styles inserted above the middle. Cocci 4 (or very rarely 5), very spreading, ovate, about 2 lines long, wrinkled, of a reddish colour.—*Evodia erythrococca*, F. Muell. Fragm. i. 28.

**Queensland.** Wide Bay, *C. Moore;* Moreton Bay and Brisbane river, *W. Hill, F. Mueller.*
**N. S. Wales.** Clarence river, *Beckler, C. Moore.*

3. **M. australasica,** *F. Muell. Herb.* A handsome tree, glabrous in all its parts. Leaves digitately 3-foliolate, the common petiole several times shorter than the leaflets; leaflets oblong-elliptical, or rarely obovate-oblong,

obtuse or .shortly acuminate, 6 to 10 in. long, somewhat coriaceous, entire.: Panicles axillary, trichotomous, loose and many-flowered, but much shorter than the leaves. Pedicels short. Sepals ovate. Petals narrow, about 4 lines long, of a firm consistence, reflexed above the middle, minutely pubescent outside; æstivation not seen. Filaments slightly dilated, ciliate and rigid, especially the larger ones, subulate upwards; anthers small. Disk inconspicuous. Carpels nearly glabrous, but tapering into strictly terminal short pubescent styles united at the summit. Cocci erect, distinct, angular; acuminate, not 2 lines long. Seeds shining.—*Evodia octandra,* F. Muell. Fragm. ii. 102.

**N. S. Wales.** Clarence river, *Beckler.*

### 18. EVODIA, Forst.

Flowers more or less unisexual. Sepals 4 or 5, imbricate. Petals 4 or 5, valvate or very slightly imbricate. Disk sinuate. Stamens 4 or 5; filaments subulate or slightly dilated. Ovary of 4 or 5 carpels, usually distinct and style-like in the male flowers, more or less united in the females, styles attached below the middle, more or less united with a 4- or 5-lobed stigma. Ovules 2 in each carpel, collateral or superposed. Fruit separating more or less completely into coriaceous 2-valved cocci, the endocarp separating elastically. Seeds with a crustaceous testa, usually smooth and shining; albumen fleshy; embryo straight with ovate cotyledons.—Unarmed trees or shrubs. Leaves opposite, usually digitately 3-foliolate or pinnate, rarely 1-foliolate or simple; leaflets entire, often large. Cymes or panicles axillary or rarely terminal. Flowers small.

A considerable genus, spread over tropical Asia and the islands of the Pacific and of the Madagascar group; the only Australian one is endemic. The genus differs from *Melicope* chiefly in the stamens equal to, not double, the number of petals, from *Zanthoxylum* by the leaves all or mostly opposite, generally by the more valvate petals and more united styles, besides minor characters offering occasional exceptions.

1. **E. micrococca,** *F. Muell. Fragm.* i. 144, *and* ii. 180. A tree often of considerable size, quite glabrous. Leaves digitately 3-foliolate with long petioles; leaflets obovate-oblong, obtuse, mostly 1½ to 3 in. long, entire, narrowed at the base, the central one almost petiolulate. Flowers in dense cymes or trichotomous panicles on short lateral peduncles below the young shoots. Sepals 4, orbicular, small. Petals 4, about 2 lines long, glabrous, slightly imbricate, with inflexed valvate tips. Filaments slightly dilated, ciliate, the attenuate tips folded inwards in the bud, exserted in the open flower. Cocci not 2 lines long, not separating so completely as in the *Melicopes,* rugose-glandular outside. Seeds black and shining.

**Queensland.** Moreton Bay, *W. Hill.*
**N. S. Wales.** Near Richmond, *R. Brown;* Blue Mountains, *Miss Atkinson;* northward to Clarence and Hastings rivers, *Beckler;* and Tenterfield, *C. Stuart;* southward to Illawara, *Ralston.*

### 19. MEDICOSMA, Hook. f.

Sepals 4, broad, imbricate. Petals 4, broad, much imbricate in the bud, the tips erect or recurved. Disk lobed. Stamens 8, filaments dilated, almost cohering by their woolly margins; anthers oblong. Ovary slightly 4-lobed, 4-celled. Style almost terminal, filiform, with a small 4-lobed stigma; ovules 2 in each cell, collateral. Fruit separating into distinct, 2-valved cocci; endocarp separating elastically. Seeds with a crustaceous shining testa, albumen fleshy; embryo straight with broad cotyledons.—A tree. Leaves mostly opposite, 1-foliolate. Flowers large, in axillary panicles.

The genus is limited to a single species endemic in Australia. F. Mueller proposes to include it as well as *Melicope* (with which it agrees in the double number of stamens) under *Evodia*, but the habit, that of *Acronychia*, and the large, much-imbricate petals, appear to be a sufficient distinction, unless nearly the whole of *Zanthoxyleæ* be united into one genus.

1. **M. Cunninghamii,** *Hook. f. in Benth. and Hook. Gen. Pl.* 297. A small tree, glabrous, or the young shoots and inflorescence minutely pubescent. Leaves mostly opposite, consisting of a single leaflet obscurely articulate on a short petiole, oblong-elliptical or rarely obovate-oblong, obtuse or acuminate, 3 to 6 in. long. Panicles axillary, 3-chotomous, with few large flowers. Sepals orbicular, 2 to 3 lines long, with a prominent midrib. Petals nearly ¾ in. long, broadly ovate, minutely tomentose outside, with a prominent midrib. Disk thick and glabrous. Ovary hirsute; style slender. Cocci about 3 lines long, quite distinct, scarcely coriaceous, hirsute. Seeds black.—*Acronychia Cunninghamii*, Hook. Bot. Mag. t. 3994; F. Muell. Fragm. i. 27; *Evodia Cunninghamii*, F. Muell. Fragm. iii. 2.

**Queensland.** Brisbane river, Moreton Bay, *A. Cunningham, F. Mueller*, and others. **N. S. Wales.** Richmond and Clarence rivers, *Beckler*.

The subsucculent cocci, originally described in our ' Genera Plantarum,' are shown by subsequently received specimens to have been diseased.

## 20. ZANTHOXYLUM, Linn.

(Blackburnia, *Forst*.)

Flowers more or less unisexual. Calyx 3-, 4- or 5-lobed. Petals 3, 4 or 5, imbricate or rarely valvate or wanting. Disk small or obsolete. Stamens in the males 3, 4 or 5, the ovary rudimentary or conical, or of 3, 4 or 5 distinct style-like carpels. Female flowers without stamens or with scale-like staminodia. Ovary of 1 to 5 distinct carpels. Styles nearly terminal, distinct or united upwards; ovules 2 in each carpel, usually collateral. Fruit of 1 to 5 distinct cocci, dry or drupaceous, usually 2-valved; the endocarp separating or adherent. Seeds with a hard or crustaceous shining testa; albumen fleshy; embryo straight or curved, with broad flat cotyledons.—Shrubs or trees, often armed with scattered prickles, and sometimes climbing. Leaves alternate, usually pinnate. Flowers small, in axillary or terminal cymes or panicles.

A large genus, dispersed over the tropical and subtropical regions of the whole world. Of the following species, two are endemic in Australia, the third is also in Norfolk Island. All three belong to the section *Blackburnia*, characterized chiefly by solitary carpels, which are rare in the rest of the genus.

Stems and branches prickly. Panicles axillary. Flowers 2 to 3
    lines long . . . . . . . . . . . . . . . . 1. *Z. brachyacanthum.*
Unarmed or with very few minute distant prickles.
    Leaflets very oblique, coriaceous. Panicles axillary and termi-
        nal. Flowers 2 to 3 lines long . . . . . . . . . 2. *Z. Blackburnia.*

Leaflets scarcely oblique, not coriaceous.  Panicles terminal.
Flowers very numerous, under 1½ line . . . . . . . . 3. *Z. parviflorum.*

1. **Z. brachyacanthum,** *F. Muell. Pl. Vict.* i. 108.  A slender glabrous tree, the trunk and branches covered with short conical prickles.  Leaves pinnate, the common petiole 6 to 10 in. long ; leaflets usually 9 to 13, opposite in pairs, with or without a terminal odd one, petiolulate, from ovate to oblong-elliptical, shortly acuminate, 2 to 3 or rarely 4 in. long, equal or oblique at the base, coriaceous and shining.  Panicles axillary, much shorter than the leaves, irregularly 2-3-chotomous.  Flowers on very short pedicels, the males nearly 3 lines long, the females shorter.  Sepals 4, small and broad. Petals obtuse, much imbricate.  Ovary rudimentary in the male flowers ; in the females consisting of a single carpel with a large oblique stigma, nearly sessile or on a very short style, terminal but excentrical.  Fruit opening wide to the middle in 2 valves.

**Queensland.**  Moreton Bay, Upper Brisbane river, etc., *A. Cunningham, F. Mueller,* and others; Araucaria ranges on the Burnett river, *F. Mueller ;* Rockhampton, *Thozet.*

**N. S. Wales.**  Clarence river, *Herb. Mueller.*

2. **Z. Blackburnia,** *Benth.*  A shrub or small tree, glabrous and unarmed.  Leaves pinnate, with a common petiole of 4 to 8 in.  Leaflets 3 to 9, very obliquely ovate, shortly acuminate, usually 2 to 3 in. long, very unequal at the base and petiolulate.  Panicles axillary or terminal, loose, but shorter than the leaves.  Flowers rather smaller than in the last species. Petals imbricate in our specimens (induplicate-valvate, according to Endlicher). Ovary and fruit of *Z. brachyacanthum.*—*Blackburnia pinnata,* Forst. ; Endl. Prod. Fl. Norf. 88.

**N. S. Wales.**  Lord Howe's Island, *Milne.*  The specimen being in leaf only, its identity with the Norfolk Island plant, from which the above character is taken, is not certain, but the foliage corresponds so well, that I am unwilling to omit it, in order to give the Lord Howe's Island flora as complete as possible.

3. **Z. parviflorum,** *Benth.*  A small tree, glabrous and unarmed, or with very few minute distant prickles.  Leaves pinnate, with a common petiole of 4 to 6 in., angular but not winged; leaflets usually 9 to 11, opposite in pairs, the terminal odd one occasionally wanting, ovate-lanceolate, acuminate, rarely above 2 in. long, entire or slightly denticulate, usually oblique, the upper edge most rounded at the base, membranous or at length scarcely coriaceous. Panicles terminal, 3-chotomous, broad, with numerous small 4-merous flowers. Sepals small, triangular.  Petals scarcely 1½ lines long, slightly imbricate. Stamens in the males 4, about as long as the petals.  Ovary rudimentary, of 1 or 2 carpels.  Female flowers not seen.  Cocci solitary, 3 to 4 lines long, coriaceous, rugose outside, opening broadly to below the middle in 2 valves, endocarp persistent.  Seeds with a hard bony testa enveloped in a thin black shining epiderm.

**N. Australia.**  Goulburn Island, *A. Cunningham ;* Port Essington, *Armstrong ;* islands of the Gulf of Carpentaria, *R. Brown.*

## 21. GEIJERA, Schott.

(Coatesia, *F. Muell.*).

Flowers hermaphrodite.  Sepals 4 or 5.  Petals 4 or 5, valvate or imbricate. Disk thick and fleshy. Stamens 4 or 5 ; filaments subulate. Ovary depressed, partly immersed in the disk, 4- or 5-lobed ; styles terminal, immediately united into a single short style, with a capitate 4- or 5-lobed stigma. Fruit of 4 or 5 or sometimes fewer, distinct, 2-valved cocci, the endocarp adherent or partially separating. Seeds with a hard or crustaceous shining testa; albumen fleshy; embryo straight; cotyledons broad.—Trees or shrubs. Leaves alternate, simple, not articulate on the petiole. Flowers small, in terminal panicles. Sepals small.

The genus is limited to Australia, and differs from *Zanthoxylum* chiefly in the simple leaves and hermaphrodite flowers.

Panicles compact. Petals imbricate. Leaves broad . . . . . . 1. *G. Muelleri.*
Panicles loose. Petals valvate.
    Leaves from ovate to lanceolate . . . . . . . . . . . 2. *G. salicifolia.*
    Leaves linear . . . . . . . . . . . . . . . . . 3. *G. parviflora.*

1. **G. Muelleri,** *Benth.* A glabrous tree. Leaves ovate or obovate-oblong, 2 to 3 in. long, narrowed into a rather long petiole, coriaceous, with a prominent midrib, the lateral veins slender and rather distant. Panicle compact, scarcely equalling the last leaves. Flowers rather larger than in the other species. Petals nearly 1½ lines long, distinctly imbricate, obtuse, without inflexed tips. Cocci 2 to 3 lines long, distinctly but very shortly beaked, very spreading, but cohering at the base. Endocarp persistent.—*Coatesia paniculata*, F. Muell. Fragm. iii. 26.

**Queensland.** Cumberland islands, *R. Brown ;* Araucaria woods near Moreton Bay, *F. Mueller ;* Curtis Island, *Henne.* This species was generically distinguished by F. Mueller, on account of the imbricate æstivation of the petals, and a slight difference in the fruit, but the habit is that of the other species, and the genus is too closely allied to *Zanthoxylum,* which contains species with valvate as well as with imbricate æstivation, to admit of dividing it solely on that ground.

2. **G. salicifolia,** *Schott, Fragm. Rut. t. 4.* A moderately-sized tree, glabrous or with a minute hoary pubescence on the inflorescence, and sometimes on the under side of the leaves. Leaves from ovate to ovate-lanceolate or rarely oblong-lanceolate, obtuse or acuminate, mostly 3 to 4 in. long, entire, coriaceous, narrowed or rarely rounded at the base, with a rather long petiole. Panicles rather loose, broadly pyramidal, but much shorter than the last leaves, alternately branched, with numerous small white flowers. Petals about 1 line long, valvate. Cocci often reduced to 1 or 2, obovoid, not beaked, 2 to 3 lines long, the endocarp persistent or partially separating.—*G. latifolia,* Lindl. in Mitch. Trop. Austr. 236.

**Queensland.** Broad Sound, *R. Brown ;* Moreton Bay and Brisbane river, *A. Cunningham, F. Mueller,* and others ; Brigalow scrub on the Burdekin, and near Warwick, *F. Mueller ;* Wide Bay, *C. Moore ;* Port Denison, *Fitzalan ;* Rockhampton, *Thozet ;* Mantua Downs, *Mitchell.*

**N. S. Wales.** Clarence river, *C. Moore ;* near Paramatta, *Woolls.*

Schott's figure represents a remarkably narrow-leaved form, which I have only seen in Brown's specimens, and in those from Warwick and from Rockhampton. These, however, pass into the common broad-leaved form.

3. **G. parviflora,** *Lindl. in Mitch. Trop. Austr.* 102. A tall shrub or small tree, with slender, erect or pendulous branches, glabrous or the inflorescence and young parts slightly hoary. Leaves linear, acute or obtuse,

3 to 6 in. long, and rarely above 3 lines broad, coriaceous, narrowed into a rather short petiole, the midrib prominent underneath. Flowers and fruit of *G. salicifolia*, or the flowers sometimes, but not always, rather smaller.—*G. pendula*, Lindl. in Mitch. Trop. Austr. 251. Possibly a variety only of *G. salicifolia*.

**Queensland.** Broad Sound, *R. Brown;* Burdekin river, *F. Mueller;* Belyando river, *Mitchell.*

**N. S. Wales.** Liverpool plains, *A. Cunningham;* Narran river, *Mitchell;* between the Darling and Lachlan rivers, *Victorian Expedition.*

**Victoria.** Murray desert, *F. Mueller.*

Var. (?) *crassifolia.* Leaves 1 to 2 in. long, very obtuse or retuse, thick, with the midrib scarcely conspicuous. Perhaps a distinct species.—*Eriostemon linearifolium*, DC. Prod. i. 720; *Zanthoxylum australasicum*, A. Juss. in Mem. Mus. Par. xii. 503.

**S. Australia.** Near Adelaide, *Herb. Hooker;* Spencer's Gulf, *F. Mueller;* South coast, *R. Brown;* isles of St. Francis, *Herb. Mus. Par.*

**W. Australia.** King George's Sound, *Maclean.*

## 22. PENTACERAS, Hook. f.

Sepals 5. Petals 5, valvate. Torus thick. Stamens 10; filaments subulate, glabrous. Ovary of 5 nearly distinct carpels, each with a glandular terminal appendage. Styles inserted below the middle, and immediately united into one filiform style, with a small stigma; ovules 2 in each carpel, superposed. Fruit-carpels 5 or fewer, often solitary by abortion, indehiscent, expanded all round into a membranous wing, forming obovate or oval-oblong samaræ, the centre almost drupaceous, with a cartilaginous endocarp. Seeds usually solitary; testa thick; albumen not copious; embryo straight, with ovate cotyledons.—Tree. Leaves alternate, pinnate. Flowers numerous, small, paniculate.

The genus is limited to a single species, endemic in Australia. It differs from *Evodia* in its habit, alternate leaves, and in some measure in the ovary resembling that of several *Diosmeæ*, and from that and all other *Zanthoxyleæ* by the fruit, which, at first sight, is like that of an *Ailanthus;* but the dotted leaves and superposed ovules, which place it among *Rutaceæ*, besides the inflorescence and other minor characters, amply distinguish *Pentaceras* from *Ailanthus.*

1. **P. australis,** *Hook. f. in Benth. and Hook. Gen. Pl.* 298. A glabrous tree, small according to A. Cunningham, attaining 60 ft. according to W. Hill. Leaves pinnate, with a common petiole of from 4 or 5 in. to nearly 1 ft.; leaflets usually 7 to 11, opposite in pairs, with a terminal odd one, ovate to lanceolate, obtuse or acuminate, 2 to 4 in. long, entire or obscurely crenate, the lateral ones more or less oblique and decurrent on the petiolule on the lower side, like those of a *Clausena*. Panicles large, terminal, spreading, loose, with numerous white flowers, pedicellate along the ultimate branches. Petals about $1\frac{1}{2}$ lines long. Stamens nearly as long as the petals. Ovary glabrous. Ripe samaræ 1 to $1\frac{1}{2}$ in. or rather more in length, $\frac{1}{2}$ to $\frac{3}{4}$ in. broad.—*Cookia australis*, F. Muell. Fragm. i. 25, and iii. 27; *Ailanthus punctata*, F. Muell. Fragm. iii. 42.

**Queensland.** Brisbane river, *A. Cunningham;* Moreton Bay district, "White Cedar" of the colonists, *W. Hill, F. Mueller;* M'Connell's Brush, *Leichhardt.*

**N. S. Wales.** Richmond river, *C. Moore.*

266

TRIBE III. TODDALIEÆ.—Trees or shrubs. Leaves pinnate or 3-foliolate with opposite leaflets, or 1-foliolate, the leaflets usually large. Ovary not lobed. Fruit several-celled, indehiscent or rarely dehiscent. Seeds albuminous (in the Australian genus). The tribe has the habit of *Zanthoxyleæ*, with the ovary and nearly the fruit of *Aurantieæ*.

## 23. ACRONYCHIA, Forst.

(Cyminosma, *Gærtn.*)

Flowers polygamous. Calyx 4-lobed. Petals 4, valvate. Torus thick. Stamens 8; filaments subulate. Ovary 4-celled; style terminal; stigma entire or obscurely 4-lobed, ovules 2 in each cell, superposed. Fruit 4-celled, usually succulent, with a coriaceous or hard endocarp, opening loculicidally, or drupaceous and indehiscent. Seeds usually solitary in each cell, with a crustaceous black testa; albumen fleshy; embryo straight; cotyledons oblong. —Trees or shrubs. Leaves opposite or alternate, 1-foliolate. Flowers white or yellowish, in axillary or rarely terminal small panicles or loose cymes.

The genus extends over tropical Asia and the islands of the S. Pacific, to New Caledonia and New Zealand. Of the Australian species, one is also found in New Caledonia, the two others are endemic.

Flowers minutely tomentose, in short oblong panicles. Petals ovate . 1. *A. Baueri.*
Flowers glabrous, in axillary 3-chotomous cymes. Petals narrow.
  Leaves thin and scarcely coriaceous. Fruits 4-angled, depressed on
    the summit . . . . . . . . . . . . . . . . 2. *A. lævis.*
  Leaves very coriaceous. Fruits obovoid-globular . . . . . . 3. *A. imperforata.*

1. **A. Baueri,** *Schott, Fragm. Rut. t. 3.* A moderate-sized tree, glabrous or the young shoots and inflorescence minutely hoary-tomentose. Leaves opposite, of a single leaflet, on a rather long petiole, ovate, elliptical or obovate, obtuse or very shortly and obtusely acuminate, narrowed at the base, 3 to 4 or very rarely 5 in. long, thinly coriaceous. Panicles axillary, oblong, the side branches and pedicels very short, sometimes reduced to a small spike. Flowers small, not numerous. Sepals very broad, short, ciliate. Petals ovate, valvate with inflexed tips, minutely pubescent outside, 1 to 1½ lines long. Filaments thin, dilated, and ciliate to above the middle. Ovary pubescent; style pubescent, short, with a rather large stigma. Fruit nearly globular or 4-angled, obtuse or shortly acuminate, ½ in. diameter or rather smaller, not very succulent. Testa of the seeds hard and bony.—*A. Hillii,* F. Muell. Fragm. i. 26.

**Queensland.** Northumberland Islands and Richmond district, *R. Brown;* Moreton Bay and Brisbane river, *A. Cunningham, F. Mueller,* and others; Five Islands, *A. Cunningham.*

**N. S. Wales.** Macleay and Clarence river, *Beckler;* Port Stephens, *Harvey;* Illawarra, *Herb. Mueller;* Ash Island, *Miss Scott.* Some specimens from Hastings river resemble rather more in foliage the Norfolk Island *A. Endlicheri,* Schott, but the flowers are diseased, and they cannot be determined.

2. **A. lævis,** *Forst. Char. Gen. 53, t. 27.* A tree, attaining 60 ft., glabrous except the stamens. Leaves irregularly opposite or alternate, of a single leaflet, obovate-oblong to oblong-elliptical, obtuse, 1½ to 3 or rarely nearly 4 in. long, coriaceous when old. Cymes 2- or 3-chotomous, usually shortly

pedunculate and few-flowered. Sepals very short, rounded, glabrous. Petals narrow, induplicate-valvate, with inflexed tips, 2 to 2½ lines long, glabrous. Filaments rather thick, dilated and ciliate towards the base, subulate and inflexed at the top. Ovary hirsute round the base of the style, otherwise glabrous ; style rather long, the stigma not thickened, obscurely 4-lobed. Fruit succulent, with a crustaceous 4-celled endocarp, obtusely 4-angled, truncate at the top, and depressed in the centre, ½ in. diameter or rather smaller.— *Lawsonia Acronychia*, Linn. f. ; Labill. Sert. Austr. Caled. 66, t. 65 ; *Cyminosma oblongifolium*, A. Cunn. in Bot. Mag. 3222 ; *Acronychia laurina*, F. Muell. Fragm. i. 27.

**Queensland.** Keppel Bay, *R. Brown ;* Moreton Bay and Brisbane river, *A. Cunningham, F. Mueller*, and others ; Rockhampton, *Thozet.*

**N. S. Wales.** Port Jackson to the Blue Mountains, *R. Brown, A. Cunningham,* and others; northward to Clarence and Hastings rivers, *Beckler ;* southward to Yowaka river and Lake King, *F. Mueller.*

According to F. Mueller, the leaves are occasionally 3-foliolate, but I have never seen them so.

3. **A. imperforata,** *F. Muell. Fragm.* i. 26. A moderate-sized tree, very nearly allied to *A. lævis.* Leaves of the same shape and size, but on much shorter petioles, and much more coriaceous, the minute pellucid dots only visible before a strong light. Inflorescence and flowers as in *A. lævis,* except that the peduncles are much shorter and the flowers rather larger. Filaments much ciliate. Fruit somewhat obovoid and obscurely or not at all angular, and not depressed at the top.

**Queensland.** N.E. coast, *R. Brown ;* Brisbane river, *W. Hill, F. Mueller.*

TRIBE IV. AURANTIEÆ.—Trees or shrubs. Leaves pinnate with alternate leaflets or 1-foliolate or simple. Stamens twice as many as petals or more. Ovary not lobed. Fruit indehiscent. Seeds without albumen.

## 24. GLYCOSMIS, Corr.

Calyx 5-cleft, the lobes broadly imbricate. Petals 5, imbricate in the bud. Stamens 10, filaments dilated at the base, anthers often tipped with a small gland. Ovary 3- to 5- or rarely 2-celled ; style very short, thick and persistent, the stigma scarcely broader, ovules solitary in each cell. Berry succulent or almost dry, usually 1-seeded. Seeds with a membranous testa, without albumen ; cotyledons fleshy.—Unarmed trees or shrubs. Leaves alternate, pinnate, with few alternate leaflets, or 1-foliolate. Flowers small, in axillary or terminal panicles.

A genus of very few species, dispersed over tropical Asia and the Eastern Archipelago, the Australian one being the most widely spread over the whole region.

1. **G. pentaphylla,** *Corr. ; Oliv. in Journ. Linn. Soc.* v. *Suppl.* 37. A tall shrub or small tree, quite glabrous. Leaves occasionally 1-foliolate, on short petioles, but more generally pinnate, with 2 or 3 leaflets, from ovate-elliptical or ovate-lanceolate to oblong-lanceolate, obtuse or acuminate, 2 to 4 or rarely 5 in. long. Panicles dense, shorter, or scarcely longer than the petiole of the pinnate leaves. Petals about 2 lines long. Ovary 5- or some-

times 4-celled, contracted into a very short, thick style. Berry globular, ½ in. in diameter, or smaller.

**Queensland.** Northumberland islands, *R. Brown;* islands of Torres Straits, *F. Mueller;* scrub near Rockhampton, *Thozet.*

The species has a very wide range in tropical Asia and is very variable in the size of the leaves and flowers, full details of which and of the consequently extended synonymy of the species will be found in Oliver's paper above quoted. The character given above has special reference to the Australian variety, which is almost identical with the Chinese and Eastern form, usually distinguished as *G. citrifolia,* Lindl. ; Benth. in Fl. Hongk. 51, and figured as *Limonia parvifolia,* Hook. Bot. Mag. t. 2416.

## 25. MICROMELUM, Blume.

Calyx 5-toothed or entire. Petals 5, valvate in the bud, or nearly so. Stamens 10 ; filaments linear-subulate. Ovary 2- to 6- usually 5-celled, the dissepiments spirally twisted after the flowering ; style deciduous with a small capitate stigma ; ovules 2 in each cell, superposed. Fruit a dry berry. Seeds usually 1 or 2 ; testa membranous ; albumen none ; cotyledons leafy, very much folded.—Unarmed trees. Leaves alternate, pinnate, with alternate oblique leaflets. Flowers small, in terminal corymbose panicles.

Besides the Australian species, which is widely dispersed over tropical Asia and the Eastern Archipelago, only 2 are known from Penang or the Philippine Islands.

1. **M. pubescens,** *Blume; Oliv. in Journ. Linn. Soc. v. Suppl.* 40. Young branches and leaves more or less pubescent. Leaflets 9 to 15, or sometimes more, from ovate to broadly lanceolate, 1 to 3 in. long, obtuse or shortly acuminate, oblique at the base, often becoming glabrous above, pubescent underneath. Corymbs nearly sessile above the last leaves, many-flowered. Calyx more or less 5-toothed. Petals about 2 lines long, more or less pubescent. Ovary usually hairy. Berry small, ovoid, glabrous or pubescent.

**N. Australia.** S. Goulburn Island and Port Essington, *A. Cunningham;* islands of the Gulf of Carpentaria, *R. Brown.*

**Queensland.** Albany and Cairncross Islands and from the Burdekin to Moreton Bay, *F. Mueller;* Cape Upstart and Barnard Isles, *M'Gillivray;* Wide Bay, *Bidwill;* Rockhampton, *Thozet.*

The various forms assumed by this species and the consequent synonymy are given in detail by Oliver in the above-quoted paper. The Australian specimens belong to the small-flowered variety, with rather broad leaflets, common in the S. Pacific islands, which I formerly described as *M. glabrescens,* in Hook. Lond. Journ. ii. 212.

## 26. MURRAYA, Linn.

Calyx 5-cleft. Petals 5, narrow, imbricate in the bud. Stamens 10, free ; filaments subulate ; anthers small. Ovary 2- to 5-celled. Style elongated, at length deciduous, stigma capitate. Ovules solitary, or 2 in each cell, superposed, or nearly collateral. Berry 1- or 2-seeded. Testa glabrous or woolly ; albumen none ; cotyledons equal, not folded.—Unarmed trees or shrubs. Leaves pinnate, leaflets alternate, usually oblique at the base. Flowers often rather large, in terminal corymbs, or few together in the upper axils.

The genus comprises few species, dispersed over tropical Asia and the Eastern Archipelago ; neither of the Australian ones are endemic.

Ovary 2-celled.   Flowers nearly ¼ in. long  . . . . . . . . . 1. *M. exotica.*
Ovary 5-celled.   Flowers numerous, not 3 lines long  . . . . . . 2. *M. crenulata.*

1. **M. exotica**, *Linn.; Oliv. in Journ. Linn. Soc.* v. *Suppl.* 28.   A shrub or small tree, glabrous, or the young branches and petioles pubescent. Leaflets usually 5 to 7, from ovate, cuneate-obovate, or almost rhomboidal to ovate-lanceolate, ¾ to 2 in. long, coriaceous and shining when full-grown. Flowers white, very fragrant, in compact, terminal, sessile corymbs, or few together in the common varieties.   Petals nearly ½ in. long, erect at the base, spreading in the upper half.   Ovary 2-celled.   Berry globular or almost ovoid, usually 2-seeded.—Wight, Ic. t. 96.

**N. Australia.**  Islands of the Gulf of Carpentaria, *R. Brown.*
**Queensland.**  Scrub near Rockhampton, *Thozet.*  These specimens are past flower and have only a few young fruits, which are more ovoid than they generally are in the species, but in other respects they appear to belong as well as Brown's to the few-flowered var. β of Oliver, or *M. paniculata,* Jack.   The species is common from N.W. India to the New Hebrides.

2. **M. crenulata**, *Oliv. in Journ. Linn. Soc.* v. *Suppl.* 29 ?   A glabrous shrub or tree.   Leaflets usually 7 to 11, very oblique, from oval-oblong to oblong-elliptical, obtuse or shortly acuminate, 2 to 3 in. long, entire or obscurely crenulate.   Flowers (in the Philippine specimens) in terminal corymbs, much more numerous and much smaller than those of *M. exotica.*   Petals 2½ to nearly 3 lines long.   Fruit depressed-globular, 5 or 6 lines diameter, 5-celled, but with 3 or 4 cells abortive.   Seeds 1 or 2; cotyledons plano-convex, thick and fleshy.—*Glycosmis crenulata,* Turcz. in Bull. Mosc. 1858, i. 250.

**Queensland.**  Eastern subtropical Australia, *Herb. Mueller.*  The specimens are in fruit only, but the foliage, the inflorescence, and calyx are so precisely those of the Philippine Island ones that there is little doubt that they belong to the same species.   The structure of the fruit is quite that of *Murraya;* the cotyledons of the seed very readily distinguish it from *Micromelum,* which in many respects has a similar habit and inflorescence.

## 27. CLAUSENA, Burm.

Calyx 4- or 5-cleft.   Petals 4 or 5, broad, imbricate in the bud.   Stamens 8 or 10; filaments dilated at the base or in the middle; anthers short. Ovary 4- or 5-celled, or rarely 2- or 3-celled; style deciduous, with an entire or lobed stigma; ovules 2 in each cell, collateral or superposed.   Berry ovoid oblong or globular.   Seeds with a membranous testa; no albumen; cotyledons plano-convex.—Unarmed trees or shrubs.   Leaves pinnate, with alternate, usually oblique leaflets.   Flowers small, usually clustered in terminal or axillary panicles or racemes.   Berries small.

The genus, although not large, comprises more species than any other one of the tribe *Aurantieæ,* and extends over tropical Asia and Africa; the only Australian species known is endemic.

1. **C. brevistyla**, *Oliv. in Journ. Linn. Soc.* v. *Suppl.* 31.   Apparently a shrub, glabrous, or the young branches and petioles slightly pubescent. Leaflets 10 to 15, very obliquely ovate or somewhat rhomboidal, shortly and obtusely acuminate and emarginate, mostly 2 to 4 in. long, membranous, often obscurely sinuate-dentate, on petiolules of about 2 lines.   Flowers

2 B

4-merous or 5-merous, in terminal, loose, oblong or pyramidal panicles. Petals about 2 lines long. Filaments thick and dilated at the base, arched. Ovary glabrous or nearly so, narrowed at the base, 4- or 5-celled. Style very short. Fruit not seen.

**Queensland.** Hope Islands, *M'Gillivray*. The species is allied to *C. heptaphylla*, W. and Arn., from E. India, but the leaflets are much more oblique, the style much shorter, besides minor differences.

## 28. ATALANTIA, Corr.

Calyx 3- to 5-cleft. Petals 3 to 5, imbricate in the bud. Stamens twice as many or rarely more, free or irregularly united at the base; anthers ovate or oblong. Ovary 2- to 5-celled; style deciduous, with a capitate stigma; ovules solitary or 2 in each cell, collateral or rarely superposed. Berry globular, with a thickened rind, 1- to 5-seeded. Seeds obovoid or oblong, testa membranous; albumen none; cotyledons flat or convex, more or less fleshy.—Shrubs or small trees, unarmed or thorny. Leaves simple, coriaceous. Flowers in axillary clusters or short racemes or small cymose panicles, occasionally solitary. Fruits usually larger than in the preceding genera.

The genus is dispersed over tropical Asia. The Australian species are both endemic; one however is in some measure doubtful, the flowers being unknown, and the other is slightly anomalous in character though congener in essential points and habit. The genus, in the increased number of stamens of two species, and in the inflorescence, fruit, and seeds, connects the anomalous *Citrus* with the rest of the tribe.

Leaves narrow. Spines straight or incurved. Pedicels clustered in the
  axils of the leaves . . . . . . . . . . . . . . . . . . 1. *A. glauca.*
Leaves ovate. Spines mostly recurved. Racemes short, axillary or terminal . . . . . . . . . . . . . . . . . . . . . . . 2. *A. recurva.*

1. **A. glauca,** *Hook. f., in Benth. and Hook. Gen. Pl.* 305. A rigid glaucous shrub of 2 or 3 ft., often armed with straight or incurved axillary spines of ½ in. or under, the young shoots whitish with a very minute pubescence. Leaves oblong-linear or slightly cuneate, very obtuse or emarginate, mostly 1 to 1½ in. long, thick, rigid, veinless, narrowed into a short petiole; those on the barren shoots sometimes marked with a few coarse crenatures. Flowers usually 2 or 3 together in the axils, on pedicels of 1 to 2 lines. Sepals 3 or 4, short and broad. Petals 3 or more frequently 4, obovate or broadly oblong, 2 to 2½ lines long, thin, concave, much imbricate. Stamens 8 to 12, or sometimes more, the filaments often slightly united at the base. Disk thick, annular. Ovary 4- or 5-celled, with 1, or occasionally 2, superposed ovules in each cell. Style rather thick. Berry globular, about ½ in. diameter. Seeds 3 or 4, obovoid, slightly compressed; cotyledons slightly fleshy, but not thick.—*Triphasia glauca*, Lindl. in Mitch. Trop. Austr. 353; Oliv. in Journ. Linn. Soc. v. Suppl. 26.

**Queensland.** Broad Sound, *R. Brown;* Maranoa river, *Mitchell;* Suttor and Burdekin rivers, *F. Mueller;* Port Denison, *Fitzalan.* The species, although anomalous in some respects, has the foliage and inflorescence of *Atalantia,* and is allied in several respects to *A. Hindsii,* Oliv., approaching like that species to *Citrus* in the increased number of stamens.

2. **A. (?) recurva,** *Benth.* Glabrous, armed with axillary spines, very spreading or recurved. Leaves broadly ovate, obovate or elliptical, mostly

very obtuse, 1½ to 2½ in. long, coriaceous, on petioles of 1. to 3 lines.
Racemes axillary, sometimes 2 together, ⅓ to 1 in. long, or terminal and
slightly branched. Pedicels very short. Calyx minute, 3- or rarely 4-lobed.
Petals and stamens not seen. Berries globular, either 1-seeded and 3 or 4
lines diameter, or 2-seeded and larger.

**N. Australia.** Careening Bay,N.W. coast, *A.Cunningham ;* islands of the Gulf of Carpentaria, *R. Brown* (*Hb. R. Br.*). The flowers are wanting, to determine absolutely the affinities
of this species. R. Brown's specimens are however in very good fruit. A. Cunningham's
are in leaf only, with some remains of the inflorescence and calyx.

## 29. CITRUS, Linn.

Calyx 3- to 5-lobed. Petals 4 to 8, thick, imbricate in the bud. Stamens
indefinite, usually numerous, filaments flattened at the base and variously
connate, anthers oblong. Disk large, cupular or annular. Ovary of 6 or
more cells ; style deciduous, with a capitate lobed stigma ; ovules 4 to 8 in
each cell, in 2 rows. Berry globular or oblong, with a thickened rind, several-
celled, with thin dissepiments, the cells more or less filled with transverse
pulpy cellules. Seeds with a coriaceous testa ; albumen none ; embryos often
more than one ; cotyledons fleshy, plano-convex.—Trees or shrubs, often
armed with axillary spines. Leaves 1-foliolate, the petiole often winged.
Flowers white, axillary, solitary clustered or shortly paniculate.

The really wild species are few, chiefly from tropical Asia, but long culture in most hot
countries has produced numerous permanent varieties. The Australian ones differ from the
others in the short petiole not at all winged.

Fruit globular. (Stamens about 10 ?) . . . . . . . . . . . 1. *C. australis.*
Fruit oblong. Stamens above 20 . . . . . . . . . . . 2. *C. australasica.*

**1. C. australis,** *Planch. in Hort. Donat.* 18 (*partly*). A tree of 30 ft.
or more, quite glabrous, with axillary straight thorns of about ½ in. Leaves
ovate, obovate, or almost rhomboidal, 1 to 2 in. long, obtuse or emarginate,
the petiole not exceeding 3 lines, and not winged. Flowers wanting in our
specimens, but according to A. Cunningham, he found a single one which had
10 free stamens. Fruit in the specimens which I have seen globular, from 1
to 1½ in. diameter, with a hard rind ; cells 6 to 8, more or less pulpy, with
usually 3 or 4 seeds in each.—*Limonia australis*, A. Cunn. in Sweet. Cat.

**Queensland.** Brisbane river, *A. Cunningham, Fraser ;* Moreton Bay, *Leichhardt.*
Cunningham's specimens of this the "Native Orange" are in leaf with fruits attached ;
Leichhardt's are only loose fruits. All our specimens in flower have much narrower leaves,
and I therefore refer them to the following species, to which also probably belongs the polyandrous flowering specimen described by Planchon.

**2. C. australasica,** *F. Muell. Fragm.* i. 26. A rigid shrub (according to A. Cunningham), quite glabrous, with axillary straight slender spines
of ½ in. or less. Leaves from obovate-oblong to oblong-cuneate or lanceolate,
very obtuse and emarginate, 1 to 1½ or rarely 2 in. long, coriaceous, the
petiole usually very short, and not winged. Flowers solitary or rarely 2
together, on very short pedicels. Sepals 5, small, spreading, concave, minutely
ciliate. Petals oblong, nearly 4 lines long. Stamens 20 to 25, free. Ovary
in the flowers examined 6-celled. Style very short, with a thickened, obtuse,
furrowed stigma. Ovules 4 in each cell. Fruit oblong, almost cylindrical,

2 or 3 times as long as broad, the largest seen about 2 in. long, with usually 2 or 3 seeds in each cell.

**Queensland.** Brisbane river, *A. Cunningham, F. Mueller*, and others; Pine river, *Fitzalan.*

**N. S. Wales.** Clarence river, *Beckler ;* Richmond river, *Herb. Mueller.*

The specimens are very unsatisfactory; several with the narrowest leaves are in leaf only, others with rather broader leaves are in flower. None have the fruit attached; the loose fruits are deposited in F. Mueller's herbarium as belonging to one of the narrow-leaved specimens. The evidence, therefore, which has induced me to refer the flowering specimens with numerous stamens to the oblong rather than to the globular fruits, is far from conclusive, and the question cannot be determined until undoubted flowers of the globular-fruited tree shall have been more fully examined.

## Order XXIX. SIMARUBEÆ.

Flowers regular, diœcious or polygamous, more rarely hermaphrodite. Calyx usually small, 3- to 5-lobed, or divided into as many distinct sepals. Petals 3 to 5, hypogynous or slightly perigynous, imbricate or valvate in the bud, rarely wanting. Stamens either equal in number to the petals, and alternating with them, or double the number, anthers usually versatile, with 2 parallel cells opening longitudinally. Disk annular, cupular or elongated within the stamens, under or round the ovary, or rarely none. Gynœcium of 3 to 5, rarely more or fewer carpels, quite distinct, or more or less united into a single lobed or rarely entire ovary, with one cell to each carpel. Styles as many as carpels, united from the base or by the stigmas only, or entirely distinct. Ovules solitary in each cell, or very rarely 2, the micropyle superior. Fruit-carpels either distinct, dry or drupaceous, usually indehiscent, or united in a single drupe or capsule. Seeds usually solitary in each carpel or cell, pendulous; testa membranous; albumen abundant, or little, or none. Embryo straight or curved; cotyledons flat or convex, rarely twisted; radicle superior.—Shrubs or trees, with a bitter bark. Indumentum of simple not stellate hairs. Leaves alternate or rarely opposite, pinnate or simple, usually without glandular dots. Stipules none, except in *Cadellia.* Flowers usually small, in axillary or rarely terminal panicles or racemes.

The Order consists of a considerable number of small genera, chiefly tropical, dispersed over the New as well as the Old World. Of the 6 Australian genera, 3 belong to tropical Asia, one of which extends also into Africa, 2 are endemic, and the sixth is on the seacoasts of all tropical countries. The Order as a whole is somewhat heterogeneous, and especially has no peculiar habit. In technical characters it is closely allied to *Rutaceæ*, from which it differs chiefly in the bitter bark, the want of pellucid dots to the leaves, and in the solitary ovules, but each of these characters has some exceptions.

Tribe I. **Simarubeæ.**—*Ovary lobed or carpels distinct.*
Leaves pinnate.
   Stamens twice as many as petals. Fruit-carpels winged and samara-
     like . . . . . . . . . . . . . . . . . . . . . 1. AILANTHUS.
   Stamens equal in number to the petals. Fruit-carpels drupaceous . 2. BRUCEA.
Leaves simple. Stamens twice as many as petals.
   Calyx very small. Styles connate . . . . . . . . . 3. HYPTIANDRA.
   Sepals nearly or quite as long as the petals. Styles free.
     Sepals spreading under the fruit. Leaves thin . . . . . . . 4. CADELLIA.
     Sepals connivent over the fruit. Leaves almost fleshy . . . . 5. SURIANA.

Tribe II. **Picramnieæ.**—*Ovary entire.*
Leaves 3-foliolate . . . . . . . . . . . . . . . . . 6. Harrisonia.

Tribe I. Simarubeæ.—Ovary deeply divided, the carpels or lobes entirely distinct or connected by the styles or stigmas.

## 1. AILANTHUS, Desf.

Flowers polygamous. Calyx small, 5-lobed. Petals 5, valvate in the bud. Disk 10-lobed. Stamens 10, fewer or none in the female flowers; filaments without scales. Ovary 2- to 5-lobed; styles connate, with plumose stigmas; ovules solitary in each cell. Fruit of 1 to 5, oblong, membranous samaræ, thickened in the centre round the seed. Seed flattened; testa membranous; albumen scanty; cotyledons leafy, nearly orbicular.—Trees. Leaves alternate, pinnate; leaflets oblique. Flowers small, in terminal panicles.

Besides the Australian species, which is endemic, the genus comprises three others, natives of the warmer regions of Asia, one of them much planted in various parts of the globe.

1. **A. imberbiflora,** *F. Muell. Fragm.* iii. 42. A tree, quite glabrous in all its parts. Leaflets about 15 to 17, shortly petiolulate, apparently obliquely ovate-lanceolate and 2 or 3 in. long, but much broken in the only specimens seen. Panicles not much branched. Male flowers on short pedicels, in little clusters along the upper part of the branches. Calyx very small. Petals about 1¼ lines long, quite glabrous, valvate, not induplicate, and the points scarcely inflexed. Stamens exserted. Female flowers not seen. Samaræ in our specimens attaining at least 2 in. in length and ½ in. in breadth.

**Queensland.** Rockhampton, *Thozet.* Evidently, as suggested by F. Mueller, very nearly allied to the E. Indian *A. malabarica,* DC. Prod. ii. 89, Wight, Ic. t. 1604, which indeed seems only to differ in a slight pubescence on the panicle and in rather larger flowers and fruits.

*A. rhodoptera,* F. Muell. Fragm. iii. 43, mentioned as cultivated in New England, is the commonly planted *A. glandulosa,* Desf., DC. Prod. ii. 89. *A. punctata,* F. Muell. l. c., is *Pentaceras australis,* Hook. f., of which the fruit closely resembles that of an *Ailanthus* in outward form, although the inner structure as well as the flower are very different.

## 2. BRUCEA, Mill.

Flowers polygamous. Calyx small, 4-cleft. Petals 4, minute, linear, imbricate in the bud. Disk 4-lobed. Stamens 4. Ovary 4-lobed or of 4 distinct carpels, the styles free or connate at the base, the stigmas entire, spreading; ovules solitary in each cell. Drupes 4, ovoid, scarcely fleshy, the putamen rugose. Seed with a membranous testa; albumen copious; embryo straight, radicle superior.—Trees. Leaves alternate, pinnate; leaflets oblique. Flowers very small, in small cymes, in simple slender axillary spikes.

The genus comprises a very few species, spread over tropical Asia and Africa, extending into northern India. The Australian species is one of the commonest Asiatic ones.

1. **B. Sumatrana,** *Roxb. Fl. Ind.* i. 449. A shrub or tree, the young branches and petioles softly tomentose. Leaves 1 to 1½ ft. long or even more; leaflets 5 to 11, ovate-lanceolate, acuminate, about 3 in. long, coarsely toothed, usually oblique at the base, softly pubescent or tomentose-villous,

especially underneath. Flowers very small, purple, in little cymes or clusters along the peduncle, forming interrupted spikes or racemes of 6 to 10 in. in the males, much shorter in the females. Drupes about 3 lines long.

**N. Australia.** Arnhem's Bay, *R. Brown; Victoria river, F. Mueller.* The latter specimen has the leaflets very densely and softly velvety on both sides; in R. Brown's specimens they are not more so than in the majority of Indian specimens. (*Herb. R. Br.* and *F. Muell.*)

## 3. HYPTIANDRA, Hook. f.

Flowers hermaphrodite. Calyx small, of 4 or 5 distinct sepals. Petals 4 or 5, imbricate in the bud. Disk thick. Stamens 8 or 10; filaments flattened, densely villous. Ovary of 4 or 5 distinct carpels, connected upwards by a short style; stigma inconspicuous. Ovules solitary in each cell or accompanied by a second smaller abortive one. Fruit unknown.—A shrub or tree, pubescent with simple hairs. Leaves alternate, simple. Flowers axillary.

The genus is limited to a single species, endemic in Australia. We had, in our 'Genera Plantarum,' placed it doubtfully amongst *Rutaceæ-Boroniæ*, with which it is closely connected by the flowers, but, on further consideration, the want of glandular dots, the bitter bark, and simple hairs have induced us to remove it to *Simarubeæ*.

1. **H. Bidwilli,** *Hook. f. in Benth. and Hook. Gen. Pl.* 294. Probably a tall erect shrub or tree, the young shoots silky-pubescent with appressed simple hairs. Leaves petiolate, lanceolate, narrowed at each end, but usually obtuse, 3 to 4 in. long, entire, coriaceous, glabrous on both sides, or with a few small appressed hairs on the veins underneath, not dotted. Flowers small, shortly pedicellate, in axillary clusters, with a few appressed strigose hairs on the pedicels and petals. Petals ovate, much imbricate, rather more than 1 line long. Filaments dilated to above the middle and fringed, especially inside, with long hairs. Ovary hirsute.

**Queensland.** Wide Bay, *Bidwill.*

## 4. CADELLIA, F. Muell.

Flowers hermaphrodite. Sepals usually 5, nearly as long as the petals, enlarged and stellately spreading under the fruit, imbricate in the bud. Petals 5, imbricate in the bud. Stamens 10; filaments filiform. Disk none. Carpels 1 or 5, free; styles distinct, inserted on the inner angle above or below the middle; stigmas dilated or capitellate; ovules 2 in each carpel, collateral, pendulous or ascending. Fruit-carpels coriaceous, small, indehiscent or obscurely 2-valved. Seeds solitary, without albumen; testa membranous; embryo curved.—A tree. Leaves alternate, simple, with small, often deciduous stipules. Flowers in short loose axillary racemes.

The genus is limited to Australia. It only differs from *Suriana* in the arborescent habit and thinner spreading calyx.

Carpels 5. Leaves mostly obtuse. Racemes very loose . . . . . 1. *C. pentastylis.*
Carpels solitary. Leaves mostly acute or acuminate. Racemes short . 2. *C. monostylis.*

1. **C. pentastylis,** *F. Muell. Fragm.* ii. 25, *t.* 12. A tree, attaining 40 ft., the smaller branches very slender and minutely pubescent. Leaves from obovate-oblong to elliptical or lanceolate, obtuse, about 1½ to 2 in. long,

entire, narrowed into a short petiole, occasionally bearing a gland on one side, glabrous, penninerved and reticulate, not dotted. Peduncles in the upper axils slender, bearing a short raceme of 2 to 4 flowers. Sepals nearly 3 lines long at the time of flowering, enlarged to 5 or 6 lines, and stellately spreading under the fruit. Petals white, slightly exceeding the sepals. Carpels 5, the styles inserted above the middle. Ovules pendulous. Drupes about 1½ lines long, nearly globular, with an inner angle, somewhat coriaceous, with a crustaceous endocarp. Embryo much curved or circinate like that of *Suriana*; cotyledons much broader than in that plant, variously folded according to F. Mueller, in the seed I opened flat, except following the general curvature of the embryo.

**N. S. Wales.** Rocks at the falls of the Severn in New England, near Tenterfield, *C. Stuart.*

2. **C. monostylis,** *Benth.* A glabrous slender tree (or shrub?). Leaves petiolate, from ovate-lanceolate to elliptical-oblong, shortly acuminate, mostly 3 to 4 in. long, narrowed at the base, membranous or thinly coriaceous. Racemes, in the few specimens seen, very short, slender, 2- to 4-flowered. Pedicels about 2 lines long, in the axils of minute bracts. Sepals nearly 2 lines long, shortly united at the base, membranous, persistent, and spreading after flowering. Petals (1 only seen) about twice as long as the sepals. Stamens 10, but in some of the flowers 1 or 2 are semiabortive (or already withered away?). Carpels in all the flowers seen solitary, with the style quite basal as in *Suriana.* Ovules as in *C. pentastylis,* collateral, but horizontal or slightly ascending.

**N. S. Wales.** Clarence river, *Beckler.* The specimens seen are very few with very few flowers, the petals already almost all fallen away.

## 5. SURIANA, Linn.

Flowers hermaphrodite. Sepals 5, as long as the petals, persistent and closing over the fruit, imbricate in the bud. Petals 5, imbricate in the bud. Stamens 10, filaments filiform. Disk none. Carpels 5, free; styles distinct, filiform, inserted near the base of the carpels; stigmas capitellate; ovules 2 in each carpel, ascending. Fruit-carpels coriaceous, indehiscent. Seeds solitary, ascending, without albumen; testa membranous; embryo curved.—A maritime shrub. Leaves alternate, simple. Peduncles in the upper axils 1- or few-flowered.

The genus is limited to a single species widely spread over the seacoasts of most tropical countries. It is in many respects anomalous in the structure of the flowers, but is certainly allied to *Cneorum* and *Castela,* and, with them, appears to be better placed among *Simarubeæ* than in any other Order to which it has been referred, although it is deprived of the bitter principle of the majority of *Simarubeæ.*

1. **S. maritima,** *Linn.; W. and Arn. Prod.* 361. A rigid, much-branched shrub, more or less hoary· or tomentose with simple, often capitate hairs. Leaves crowded, linear-spathulate, obtuse, 1 to 1½ in. long, narrowed at the base, quite entire, rather thick, scarcely veined. Peduncles short in the upper axils, bearing 1 or very few flowers, often forming short leafy terminal corymbs. Sepals rather thick, acute or acuminate, 3 to 4 lines long, slightly enlarging and closing over the fruit. Petals yellow, scarcely as long as the

sepals. Nuts or drupes about half as long as the calyx, minutely pubescent, with a thin epicarp and crustaceous endocarp. Embryo in the seeds examined as much curved as in *Cadellia*, but the cotyledons narrower.

**Queensland.** Islands off the N.E. coast, *R. Brown, F. Mueller,* and others.

TRIBE II. PICRAMNIEÆ.—Ovary 2- to 5-celled, entire or rarely shortly lobed.

## 6. HARRISONIA, R. Br.

Flowers hermaphrodite. Calyx small, 4- or 5-cleft. Petals 4 or 5, almost valvate. Disk hemispherical or cupular. Stamens 8 or 10, with a small 2-cleft scale at the base of the filaments. Ovary globular, entire or shortly lobed, 4- or 5-celled. Styles connate or distinct at the base; stigma furrowed. Ovules solitary in each cell, pendulous. Drupe small, globular, with 2 to 5 pyrenes or nuts. Seeds solitary, nearly globular; testa rather thick; albumen scanty; cotyledons folded towards the middle.—Trees, usually armed with prickles. Leaves alternate, compound. Flowers small, in pedunculate axillary cymes.

The genus comprises only two species, natives of the Indian Archipelago, one of them extending to Australia.

1. **H. Brownii,** *A. Juss. in Mem. Mus. Par.* xii. 540, *t.* 28. A shrub. Branches glabrous, often armed with short conical prickles, usually in pairs, one on each side of the leaf, but probably not really stipulary. Leaflets 3, ovate, acuminate, 1½ to 3 in. long, the lateral ones petiolulate and oblique at the base, the terminal one narrowed at the base ; all glabrous or sprinkled with a few hairs underneath. Flowers small, few together in axillary cymes, on slender peduncles, shorter than the leaves. Calyx and petals quite glabrous. Filaments hairy at the base. Drupe small, depressed, globular, furrowed between the nuts.

**N. Australia.** Islands of the Gulf of Carpentaria, *R. Brown* (*Herb. R. Br.*). We have it also from Timor and from the Philippine Islands, and it probably extends over other intervening islands.

## ORDER XXX. BURSERACEÆ.

Flowers regular, hermaphrodite or polygamous. Calyx usually small, 3- to 5-lobed or divided into as many distinct sepals. Petals 3 to 5, hypogynous or perigynous, imbricate or valvate in the bud. Stamens twice as many as petals, or rarely of the same number, inserted on or around the disk ; anthers versatile, with 2 parallel cells opening longitudinally. Disk usually annular or cupular, often adnate to the base of the calyx. Ovary free, 2- to 5-celled, tapering into a single style, with an entire or lobed stigma. Ovules 2 in each cell or rarely solitary, usually pendulous, the micropyle superior. Fruit a drupe, either indehiscent or the epicarp opening in 2 valves, pyrenes 2 to 5, bony or chartaceous, distinct or united. Seeds solitary in each pyrene, pendulous; testa membranous; albumen none. Cotyledons usually membranous, folded or rarely thick and fleshy.—Shrubs or trees, often yielding a balsamic fluid. Leaves usually alternate, pinnate, or in genera not Australian 3-foliolate, without or rarely with stipules. Flowers small, in racemes or panicles.

The Order is spread over most tropical regions. The two Australian genera are both widely dispersed over tropical Asia, one is also in Africa, and the other in tropical America.

Calyx 5-lobed, the disk lining the tube, with the stamens on the margin . 1. GARUGA.
Calyx 3-lobed, the disk free, with the stamens outside or on the margin . 2. CANARIUM.

## 1. GARUGA, Roxb.

Flowers polygamous. Calyx campanulate, 5-lobed, valvate. Petals 5, inserted above the middle of the calyx-tube, induplicate-valvate. Disk thin, lining the calyx-tube. Stamens 10, inserted with the petals. Ovary 4- or 5-celled; styles elongated; ovules 2 in each cell. Drupe indehiscent, with 5 or fewer bony nuts, rugose outside. Seeds solitary in each nut; cotyledons folded.—Trees. Leaves pinnate. Flowers rather large for the Order, in terminal panicles.

The genus is dispersed over tropical Asia and America; the Australian species extends at least to Timor, and is perhaps a variety of a common Asiatic one.

1. **G. floribunda,** *Dcne. Herb. Tim. Descr.* 149. Branches thick, marked with the broad scars of the fallen leaves. Leaves crowded at the ends of the branches; leaflets 7 or 8 pairs, very shortly petiolulate, very obliquely ovate-lanceolate, acuminate, 2 to 3 in. long, crenate especially on the outer edge, glabrous when full grown, the common petiole 8 in. to 1 ft. long, slightly pubescent or at length glabrous. Panicles broad and dense, terminating leafless branches. Flowers numerous, much smaller than in the common Indian *G. pinnata*, Roxb., arranged in cymes along the last ramifications, the pedicels and flowers hoary with a minute tomentum. Calyx about 2 lines long. Petals linear-oblong, twice as long as the calyx-lobes. Fruit not seen.

**N. Australia.** Port Nelson, N.W. coast, *A. Cunningham.* I have followed Planchon (in Herb. Hook.) in referring this to the Timor species described by Decaisne, although I have seen no specimens from that island. It differs from some forms of *G. pinnata*, Roxb., in little besides the much smaller flowers in a more compound panicle.

## 2. CANARIUM, Linn.

Flowers hermaphrodite or polygamous. Calyx campanulate, usually 3-lobed, valvate. Petals usually 3, valvate, or slightly imbricate in the bud. Disk annular, rather thick. Stamens twice as many as petals, inserted on the margin of or outside the disk. Ovary usually 3-celled; stigma sessile, capitate, 3-lobed; ovules 2 in each cell. Drupe ovoid or ellipsoid, often 3-angled, the putamen 1-celled by abortion. Seed solitary; testa membranous; cotyledons folded.—Trees, with large pinnate leaves. Flowers small, in axillary panicles.

The largest genus of the Order, dispersed over tropical Asia and especially the Indian Archipelago, with a few African species. The Australian one is endemic.

1. **C. australasicum,** *F. Muell. Fragm.* iii. 15. Branches thick, marked with the broad scars of fallen leaves, the young ones minutely hoary. Leaflets 5 to 9, petiolulate, ovate or oval-oblong, or the lower ones nearly

orbicular, very obtuse, or rarely shortly acuminate, 2 to 4 in. long, glabrous, coriaceous, with parallel pinnate veins, and smaller reticulations conspicuous on both sides. Stipules linear-subulate, deciduous. Panicles raceme-like in the upper axils, shorter than the leaves, the cymes shortly pedunculate along the simple rhachis. Bracts and bracteoles small, deciduous. Flowering calyx 1 line long, tomentose. Petals about 2 lines, glabrous. Stamens 6, the filaments shortly united in a cup at the base. Drupes ellipsoid, the woody nut nearly 1 in. long, smooth, usually 1-celled, rarely with 2 cells and seeds. Cotyledons much folded and crumpled.

**N. Australia.** Careening Bay, N.W. coast, *A. Cunningham;* Port Essington, *Armstrong;* islands of the Gulf of Carpentaria, *R. Brown, Henne.*

**Queensland.** Estuary of the Burdekin, *Fitzalan.* The species does not come very near to any other one known to me.

## Order XXXI. MELIACEÆ.

Flowers regular, usually hermaphrodite. Calyx small, 4- or 5-lobed, or divided into as many distinct sepals. Petals 4 or 5, rarely more, or 3 only, free or adnate to the staminal tube, imbricate or rarely valvate. Stamens as many, or more frequently twice as many, as petals; the filaments, in *Meliaceæ* proper, united in a tube; anthers sessile or shortly stipitate within, or at the summit of the tube; in *Cedreleæ*, filaments free. Disk various, often annular or tubular, free within the staminal tube. Ovary free, entire, 3- to 5-celled; style simple; stigma thick, disk-shaped or pyramidal. Ovules in each cell 2, or (in *Carapa* and the *Cedreleæ*) 4 or more, the micropyle superior. Fruit a capsule, berry, or rarely a drupe, indehiscent, or septicidally or loculicidally dehiscent. Seeds 1, rarely 2, or in *Cedreleæ* few in each cell, with a ventral hilum; albumen fleshy or none, embryo flat or nearly so, radicle superior.—Trees or shrubs, the wood often coloured and sometimes fragrant, the bark rarely bitter. Leaves alternate or very rarely opposite, simple, or more frequently pinnate, the petiole often continuing long to grow out and produce fresh leaflets; leaflets without dots, except in *Flindersia.* Flowers paniculate, often small.

The Order is found abundantly in the tropical or warm regions of Asia and America, more rarely in Africa. Of the 10 Australian genera, 3 are endemic, 3 are common to the tropical regions both of the New and the Old World, the remaining 4 are Asiatic, one of them extending also into Africa.

*Meliaceæ* proper are at once known among the allied Orders by their staminal tube. *Cedreleæ*, with free stamens, are in that respect anomalous, and might technically be referred to some of the preceding Orders containing pinnate-leaved trees; but the habit, the large disk-like stigma, and some minor characters, have referred them with common consent to *Meliaceæ* as a tribe. *Flindersia*, however, with its pellucid-dotted leaves, is really as nearly connected with *Rutaceæ-Zanthoxyleæ* as with *Meliaceæ*, but retained among the latter on account of its fruit and seeds so nearly those of *Cedrela.*

Tribe I. **Melieæ.**—*Stamens united in a tube. Ovules 2 in each cell. Seeds not winged, albuminous.*

Leaves simple. Petals very long and narrow. . . . . . . . . 1. Turræa.
Leaves bipinnate . . . . . . . . . . . . . . . . . . . . . 2. Melia.

Tribe II. **Trichilieæ.**—*Stamens united in a tube. Ovules 2, rarely 1, or (in Carapa) more than 2 in each cell. Seeds not winged, without albumen. Leaves pinnate.*

Disk tubular or cup-shaped, enclosing the ovary . . . . . . . . 3. Dysoxylon.

Disk annular, or undistinguishable from the thickened base of the ovary.
Stamens equal in number to or not twice as many as petals. Flowers
  very small, globular . . . . . . . . . . . . . .     4. AGLAIA.
Stamens twice as many as petals.
  Staminal tube truncate or scarcely crenulate, the anthers included
    or scarcely protruding. Capsule hard.
    Ovules 1 (rarely 2 superposed) in each cell . . . . . . .   5. AMOORA.
    Ovules 2, parallel, attached to a pendulous placenta, which in
      the fruit is a thick arillus between the two seeds . . . .  6. SYNOUM.
  Staminal tube toothed, with the anthers protruding between the
    teeth. Ovules solitary. Drupe globular, with a woody or
    stony putamen . . . . . . . . . . . . . . . . .    7. OWENIA.
  Staminal tube truncate or crenate. Ovules more than 2 in each cell.
    Leaflets reticulate . . . . . . . . . . . . . .    8. CARAPA.
TRIBE III. **Cedreleæ.**—*Stamens free. Ovules more than 2 in each cell. Seeds
winged. Leaves pinnate or rarely simple.*
Petals erect. Disk thick. Capsule smooth. Leaves not dotted  . .  9. CEDRELA.
Petals spreading. Disk broadly cupular. Capsule muricate. Leaves
  pellucid-dotted . . . . . . . . . . . . . . . . . . . 10. FLINDERSIA.

TRIBE I. MELIEÆ.—Stamens united in a tube. Ovules 2 in each cell.
Seeds not winged, albuminous. Leaves various.

## 1. TURRÆA, Linn.

Calyx 4- or 5-toothed or lobed. Petals 4 or 5, elongated, free. Staminal tube
cylindrical, toothed at the summit, anthers 8 or 10, within the summit of the
tube. Disk annular or none. Ovary 5-, 10- or 20-celled ; style filiform, with
a disk-like stigma ; ovules 2 in each cell, superposed. Capsule 5- or several-
celled, opening loculicidally in as many coriaceous valves. Seeds oblong, with
a broad ventral hilum, sometimes winged; albumen fleshy, cotyledons leaf-
like.—Trees or shrubs. Leaves simple. Peduncles axillary, bearing few, white
flowers.

The genus extends over tropical Asia and Africa; the Australian species is found also in
the Indian Archipelago.

1. **T. pubescens,** *Hellen.; Willd. Spec. Pl.* ii. 555. A shrub or small
tree. Leaves at the time of flowering small, from obovate and emarginate to
ovate-lanceolate and acuminate, pubescent as well as the young shoots; when
full-grown ovate, shortly acuminate, 2 to 3, or even 4 in. long, somewhat
coriaceous, quite glabrous or slightly pubescent underneath. Flowers white,
sweet-scented, in axillary clusters or short racemes of 3 to 6. Petals narrow,
linear-spathulate, 1 to 1½ in. long. Staminal tube rather shorter, with 10
short teeth, each one more or less divided into 2 to 4 lobes, or rarely entire.
Style exserted. Fruit nearly globular, 5-celled, furrowed opposite the dissepiments, 3 to 4 lines diameter in some specimens, ½ in. in others, opening
loculicidally in 5 valves, leaving the greater part of the membranous dissepiments attached to the axis. Seeds not winged.—*T. Billardieri,* A. Juss. in
Mem. Mus. Par. xix. 218; Benn. Pl. Jav. Rar. 181 (from the character
given) ; *T. concinna,* Benn. Pl. Jav. Rar. 182.

**Queensland.** Broad Sound, Keppel Bay, etc., *R. Brown ;* Cape York, *M'Gillivray ;*
Sunday Island, N.E. coast, *A. Cunningham ;* Burdekin and Pine rivers, *Fitzalan ;* Tarama
hills, *Leichhardt ;* Rockhampton, *Thozet ;* Mount Lindsay, *W. Hill.*

The species appears to be generally dispersed over the Indian Archipelago; the lobes of the teeth of the staminal tube, upon which the distinction of *T. pubescens, T. Billardieri*, and *T. concinna* is chiefly founded, are very variable, even on the same specimen.

## 2. MELIA, Linn.

Calyx 5- or 6-cleft. Petals 5 or 6, linear-spathulate, spreading. Staminal tube 10- or 12-toothed; anthers 10 or 12, within the summit. Disk annular. Ovary 3- to 6-celled; style slender, with a capitate lobed stigma; ovules 2 in each cell, superposed. Drupe succulent, with a bony 1- to 5-celled putamen. Seeds solitary in each cell; testa crustaceous; albumen fleshy, sometimes scanty or none, cotyledons leaf-like.—Trees. Leaves usually twice or thrice pinnate, with petiolulate toothed· leaflets. Flowers paniculate.

The genus comprises but very few species, natives of tropical Asia, one of them generally planted in many parts of the globe. The Australian species is one of the Asiatic ones.

1. **M. composita,** *Willd.; W. and Arn. Prod.* 117. An elegant tree, the young leaves, shoots, and inflorescence sprinkled with .a mealy stellate tomentum which disappears with age. Leaves twice or rarely thrice pinnate; leaflets petiolulate, opposite with a terminal odd one, ovate to almost lanceolate, acuminate, 1 to 2 in. long, entire, coarsely toothed or sometimes lobed. Panicles loose, shorter than the leaves, retaining the mealy tomentum late, especially on the calyx and petals. Sepals small, ovate. Petals 4 to 5 lines long. Staminal tube hirsute inside behind the anthers, the teeth alternately entire and 2-cleft; anthers glabrous or slightly hirsute. Ovary 5-celled. Drupe ovoid, $\frac{1}{2}$ to $\frac{3}{4}$ in. long.—*M. australasica*, A. Juss. in Mem. Mus. Par. xix. 257.

**N. Australia.** Albert river, *Henne.*

**Queensland.** Burdekin river, *F. Mueller;* Broad Sound, *R. Brown;* Rockhampton, *Thozet.*

**N. S. Wales.** Macleay, Hastings, and Clarence rivers, *Beckler;* Newcastle, *Leichhardt.*

The Australian tree appears to me identical with the *M. composita* of East India and the Archipelago, and scarcely differs from the more common *M. Azedarach*, except in the more abundant mealy tomentum, especially on the inflorescence and flowers. The drupe is also usually larger and more ovoid.

## 3. DYSOXYLON, Blume.

### (Hartighsea, *A. Juss.*)

Calyx small, 4- or 5-toothed, or divided into 4 or 5 sepals. Petals 4 or 5, free or adnate to the staminal tube, spreading at the top. Staminal tube truncate or 8- or 10-toothed; anthers 8 or 10, within the summit. Disk tubular, as long as or usually much longer than the ovary. Ovary 3- to 5-celled; style elongated; stigma disk-like; ovules 2 in each cell, or rarely solitary. Capsule globular or pear-shaped, 1- to 5-celled, opening loculicidally in 2 to 5 thickly coriaceous valves. Seeds with or without an arillus, oblong, with a broad ventral hilum; testa coriaceous; albumen none; cotyledons large.—Trees, often fœtid. Leaves pinnate, leaflets opposite or alternate in the same species, entire, often oblique. Panicles axillary, loose, but often small. Flowers not very small.

A considerable genus, spread over tropical Asia and the Indian Archipelago, extending also to New Zealand. The Australian species are all endemic. The genus is readily known by the tubular disk enclosing the ovary within the staminal tube.

Calyx cupular, shortly toothed.   Petals free.   Flowers 4-merous.
 Ovary-cells 2, 2-ovulate . . . . . . . . . . . . . 1. *D. latifolium.*
Calyx cupular, shortly toothed or lobed.   Petals adnate to the sta-
 minal tube.
 Flowers 4-merous.   Ovary-cells 3, 2-ovulate.   Leaflets 5 to 9.
  Panicles small, loose.   Tubular disk short and broad . . . . 2. *D. Fraseranum.*
 Flowers 4-merous.   Ovary-cells 4, 1-ovulate.   Leaflets 11 to 21.
  Panicles large.   Staminal tube hirsute.   Tubular disk long and
  slender . . . . . . . . . . . . . . . . 3. *D. Muelleri.*
 Flowers 4- or 5-merous.   Ovary-cells 4 or 5, 1-ovulate.   Leaflets
  4 to 6.   Panicles loose, few-flowered.   Staminal tube glabrous . 4. *D. Lessertianum.*
Calyx of 5 distinct sepals.   Petals adnate to the staminal tube.
 Flowers 5-merous.   Ovary-cells 5, 2-ovulate . . . . . . . 5. *D. rufum.*

**1. D. latifolium,** *Benth.*   Leaves glabrous; leaflets in our specimens 4 or 5, ovate or broadly oval-oblong, shortly acuminate, 3 to 4 in. long, oblique at the base, somewhat coriaceous.   Flowers in sessile or shortly pedunculate clusters, along a simple, axillary, nearly glabrous peduncle of 4 to 5 in.   Pedicels short, slightly pubescent.   Calyx cupular, not 1 line long, with 4 very short broad teeth.   Petals 4, pubescent outside, about 3 lines long, valvate in the bud, free from the staminal tube.   Staminal tube truncate, and shortly and irregularly 8-toothed.   Disk broadly tubular, sprinkled with a few minute hairs.   Ovary, in the flowers examined, 2-celled, with 2 ovules in each cell, pubescent, tapering into an elongated style; stigma disk-like. Fruit not seen.

**Queensland.**   Frankland Islands, *M'Gillivray.*

**2. D. Fraseranum,** *Benth.*   A tree of 80 to 130 ft., the young leaves and shoots slightly pubescent, glabrous when full-grown.   Leaflets 5 to 9, oblong-lanceolate or elliptical, acuminate, 3 to 6 in. long, narrowed and equal at the base, bearing occasionally tufts of hairs in the axils of the principal veins underneath.   Panicles in the upper axils short, loose, divaricately branched, slightly pubescent.   Calyx cupular, about 1 line long, shortly and broadly 4-lobed.   Petals 4, about 3 lines long, nearly glabrous, adnate to the staminal tube to about half their length.   Staminal tube 8-toothed, glabrous outside. Disk broadly tubular, rather longer than the ovary.   Ovary hirsute, 3-celled, with 2 ovules in each cell.   Fruit not seen.—*Hartighsea Fraserana,* A. Juss. in Mem. Mus. Par. xix. 262, t. 15.

**N. S. Wales.**   Hastings river, *Fraser ;* Woods of Paris Exhibition, n. 238, *M'Arthur.*

**3. D. Muelleri,** *Benth.*   A tree of 60 ft. or more, glabrous or nearly so, except the very young shoots and inflorescence.   Leaves 1 to 2 ft. long; leaflets 11 to 21, from ovate to almost lanceolate, shortly acuminate, 3 to 6 in. long, very oblique at the base, one side rounded, the other truncate and shorter, almost coriaceous.   Panicles pyramidal, ¾ to 1 ft. long, much-branched and many-flowered.   Calyx cupular, ½ to ¾ line long, pubescent, 4-lobed.   Petals 4, nearly glabrous, about 5 lines long, adhering to the stami-nal tube to about two-thirds their length.   Staminal tube truncate and mi-nutely crenulate, hirsute outside.   Disk narrow-tubular, nearly half as long

as the staminal tube.    Ovary hirsute, 4-celled, with 1 ovule in each cell.
Fruit only seen very young, soon becoming glabrous.

**Queensland.**  Brisbane river, Moreton Bay, *W. Hill, F. Mueller.*
**N. S. Wales.**  Clarence river, *Beckler.*

4. **D. Lessertianum,** *Benth.*  Quite glabrous, or the young shoots and
panicles minutely pubescent.    Leaflets 4 to 10, usually without any terminal
odd one, elliptical or lanceolate, shortly and obtusely acuminate, 4 to 5 in.
long.    Panicles loose, extra-axillary, 3 to 4 in. long.    Calyx short, cupular,
entire or irregularly crenulate.    Petals 4 or 5, glabrous, more or less adherent
to the staminal tube at their base, rarely at length free.    Staminal tube glabrous, 8- or 10-toothed.    Tubular disk broad, scarcely longer than the ovary.
Ovary hirsute, 4- or 5-celled, with 1 ovule in each cell.    Fruit hard, obovoid,
about $\frac{1}{2}$ in. long in the specimens seen.    Arillus of the seeds thin.—*Hartighsea Lessertiana,* A. Juss. in Mem. Mus. Par. xix. 264.

**N. S. Wales.**  Williams River, *R. Brown;* Clarence river, *Wilcox, Beckler.*
Var. *pubescens.*  Young shoots, petioles, under side of the leaflets, and inflorescence softly
pubescent.  Clarence river, *Beckler (Hb. F. Muell.).*

5. **D. rufum,** *Benth.*  A slender tree of 30 to 40 ft., the young branches,
petioles, and under side of the leaves clothed with a soft often rust-coloured
pubescence.    Leaves 1$\frac{1}{2}$ to 2 ft. long; leaflets numerous, very shortly petiolulate, ovate-lanceolate or lanceolate, acuminate, 3 to 6 in. long, very oblique
at the base, glabrous on the upper side.    Panicles axillary or lateral, not
much branched, pubescent.    Flowers sessile.    Sepals 5, almost free, orbicular, imbricate, about 1 line long.    Petals 5, pubescent, $\frac{1}{2}$ in. long, adhering
to the staminal tube to about the middle.    Staminal tube truncate, with 10
retuse short lobes or teeth; anthers tipped with a short point.    Disk broadly
tubular, very hairy.    Ovary hirsute, 5-celled, with 2 ovules in each cell.
Fruit depressed-globular, 1 in. diameter, densely hirsute with short, rigid,
almost golden hairs.    Seeds arillate.—*Hartighsea rufa,* A. Rich, Sert. Astrol.
29, t. 11.

**Queensland.**  Moreton Bay, *A. Cunningham, W. Hill, F. Mueller.*
**N. S. Wales.**  Port Macquarie, *A. Cunningham;* Hastings river, *Fraser;* Clarence
river, *C. Moore.*  The wood, known to the colonists as *Bastard Cedar-pencil wood,* is soft
and easily worked, used in house-building.
Var. (?) *glabrescens.*  Leaves quite glabrous.  Fruit tomentose, with very short golden
hairs.—Rockhampton, *Thozet.*

4. **AGLAIA,** Lour.

(Milnea, *Roxb.;* Nemedra, *A. Juss.*)

Flowers polygamous.    Calyx 4- or 5-toothed or cleft.    Petals 4 or 5,
short, connivent, imbricate in the bud.    Staminal tube globular or urceolate,
entire or shortly toothed; anthers as many as petals or rarely more, within
the summit of the tube.    Disk none, or not distinct from the base of the
ovary.    Ovary 2- or 3-celled, with a short thick style and disk-like stigma;
ovules 1 or 2 in each cell.    Fruit coriaceous or almost succulent, indehiscent.
Seeds 1 or 2, enveloped in a mealy pulp, without any arillus.—Trees, either
glabrous or clothed with small scurfy scales or rarely with stellate tomentum.

Leaves pinnate, with entire leaflets. Flowers very small, nearly globular, in axillary panicles.

The genus is dispersed over tropical Asia and the islands of the Indian Archipelago and the Pacific. The only Australian species is also a native of New Caledonia and New Guinea.

1. **A. elæagnoidea,** *Benth.* A tree of 20 to 30 ft., the young branches, inflorescence, and under side of the leaves covered with silky or rust-coloured scurfy scales, often fringed at the edges. Leaflets 3 or rarely 5, petiolulate, ovate-oblong, or the terminal one obovate, acuminate, rarely ovate-lanceolate, 2 to 3 in. long or rarely more, coriaceous, glabrous above when full-grown. Flowers globular, about 1 line diameter, numerous in loose panicles which rarely exceed the leaves. Calyx shortly 5-, rarely 4-lobed. Petals 5, rarely 4, much imbricate, sprinkled as well as the ovary with the scurfy scales that cover the calyx and inflorescence. Anthers usually 5, but in some flowers 6, 7, or even more, within the short urceolate tube, which is thickened into raised filaments below the anthers. Ovary 3-celled, with 1 (or sometimes 2 ?) ovules in each cell. Fruit obovoid, about 1 in. long, covered with minute rust-coloured scurfy scales. Seeds 1 or 2, enveloped in a mealy pulp.—*Nemedra elæagnoidea*, A. Juss. in Mem. Mus. Par. xix. 259, t. 14 ; *Aglaia odoratissima*, Benth. in Hook. Lond. Journ. ii. 213, but probably not of Blume.

**N. Australia.** Islands of the Gulf of Carpentaria, *R. Brown* (specimens in fruit and flower) ; Entrance Island, Endeavour Straits, *Leichhardt.* Found also in New Caledonia, the New Hebrides, and in New Guinea. The station, King George's Sound, given by A. de Jussieu on the authority of the Paris Herbarium, is evidently one of those errors of locality which occurs in many of the early collections of Australian plants deposited there. A. de Jussieu having found as many as 10 stamens, gives that as the typical number, although he observes at the same time that there are sometimes fewer. We, therefore, not having then any Australian specimens, failed to recognize his plant, and from the technical characters referred it in our ' Genera Plantarum ' to *Amoora.* Having since, however, examined Leichhardt's and R. Brown's Australian specimens, and also some flowers from A. de Jussieu's specimens, kindly transmitted to me by M. Brongniart, I have been able satisfactorily to identify the species, which, notwithstanding an occasional increase in the number of stamens, belongs undoubtedly to *Aglaia*, a very natural genus if extended so as to include *Milnea.* In the majority of specimens examined I find almost always 5 stamens, and only now and then 6. Out of three unexpanded flowers from A. de Jussieu's plant, I found 7 stamens in two of them, and only 5 in the third.

## 5. AMOORA, Roxb.

Flowers polygamous. Calyx 3- to 5-toothed or lobed. Petals 3 to 5, imbricate in the bud, free from the staminal tube. Staminal tube urceolate or nearly globular, truncate or crenate ; anthers within the tube, twice as many as petals. Disk none, besides the thickened base of the ovary. Ovary 3 to 5-celled or rarely 2-celled, with 1 or 2 superposed ovules in each cell ; style short or long with a disk-like stigma. Capsule obovoid or globular, coriaceous or hard, opening loculicidally in 3 to 5 valves (or sometimes indehiscent ?). Seeds solitary in each cell, enclosed in a fleshy arillus (or sometimes without an arillus ?).—Trees. Leaves pinnate, with entire leaflets. Flowers small, but usually larger than in *Aglaia.*

The genus is spread over tropical Asia and the Indian Archipelago ; the Australian species is endemic.

1. **A. nitidula,** *Benth.* A tall tree, quite glabrous. Leaflets 2 or 4,

opposite, without any terminal odd one, elliptical-oblong, 3 to 4 in. long or
sometimes more, obtuse or shortly and obtusely acuminate, somewhat coriaceous and shining, narrowed at the base, the common petiole often slightly
dilated towards the end. Panicles axillary, loose, but shorter than the leaves.
Calyx very short, with 5 short teeth or lobes. Petals 5, about 2 lines long,
glabrous or minutely ciliate. Staminal tube broadly urceolate; anthers 10;
the tips slightly protruding. Ovary 2- or 3-celled, with 1 ovule in each cell.
Fruit obovoid, hard and almost woody, narrowed almost into a stipes at the
base, 2- or 3-celled. Seeds nearly globular, laterally attached near the top,
apparently without any arillus.

**Queensland.** Moreton Bay, *W. Hill.*
**N. S. Wales.** Richmond and Clarence rivers, *Beckler.*

The species has much of the habit of some *Dysoxyla*, but the want of any free disk and
the form of the staminal tube agree better with *Amoora.*

## 6. SYNOUM, A. Juss.

Calyx 4- rarely 5-cleft. Petals 4, rarely 5, valvate or slightly imbricate in
the bud. Staminal tube cylindrical, slightly crenulate; anthers twice as many
as petals, within the summit of the tube. Disk continuous with the thickened
base of the ovary. Ovary 3-celled; style short, with a disk-like stigma;
ovules 2 in each cell, attached collaterally to a thickish placenta pendulous
from the apex of the cavity. Capsule 3-celled, opening loculicidally in 3
valves, or reduced by abortion to 2 valves and cells. Seeds 2 in each cell,
attached by a broad lateral hilum, and half embedded collaterally in a fleshy
arillus formed by the enlarged placenta.—A tree. Leaves pinnate, with entire leaflets.

The genus consists of a single species, limited to Australia.

1. **S. glandulosum,** *A. Juss. in Mem. Mus. Par.* xix. 227, *t.* 15. A
moderate-sized tree, glabrous or the young leaves and shoots slightly silky-
tomentose. Leaflets 5 to 9, elliptical-lanceolate, acuminate, mostly 2 to 3
in. long, narrowed at the base, somewhat coriaceous, the lateral veins few and
scarcely prominent. Flowers in short dense axillary panicles, rarely exceeding
1 in. Sepals small, orbicular, spreading. Petals about 2¼ lines long. Staminal tube broad, slightly crenulate, glabrous or with a few hairs inside; anthers sometimes slightly protruding. Ovary villous. Capsule depressed-
globular, glabrous, about ¾ in. diameter, furrowed opposite the dissepiments
so as to be almost 3-lobed.—*Trichilia glandulosa,* Sm. in Rees' Cycl. xxxvi.

**Queensland.** Moreton Bay, *W. Hill.*
**N. S. Wales.** Sandy shores about Port Jackson, *R. Brown* and others; to the southward, *A. Cunningham;* inland to the Blue Mountains, *Miss Atkinson;* northward to Hastings river, *Beckler.* "Native Rosewood" of some colonists. It has the general habit of
some *Dysoxyla*, but, besides the want of any free disk and the curious insertion of the ovules
and seeds, it is easily recognized by its very short inflorescence.

## 7. OWENIA, F. Muell.

Sepals 5, short, orbicular, much imbricate. Petals 5, imbricate in the bud.
Staminal tube short or long, with 10 entire or 2-lobed teeth; anthers protruding between the teeth. Disk small, annular or not distinct from the

ovary. Ovary 3- or 4-celled, or in one species 12-celled, with 1 ovule in each cell; style rather thick; stigma globular or conical, entire or lobed, on a disk-like expansion of the summit of the style. Drupe globular, the epicarp more or less succulent, putamen thick, woody or bony, rugose outside, 2- to 4-celled, or in one species 12-celled. Seeds solitary in each cell, the outer coating spongy, the hilum broad lateral; cotyledons oblong, thick.—Trees, with the juice often (perhaps always) milky, the young shoots often viscous or gummy. Leaves pinnate. Flowers small, in axillary panicles. Fruits rather acid, eaten by the aborigines.

The genus is endemic in Australia, and differs from all other known *Trichiliæ* in its globular drupaceous fruit.

Leaflets numerous, lanceolate, acute.
 Leaflets 1-nerved. Panicles narrow. Flowers 2½ lines long . . . 1. *O. acidula.*
 Leaflets with the lateral veins conspicuous. Panicles divaricate. Flowers
  very numerous, about 1 line long . . . . . . . . . . . 2. *O. vernicosa.*
Leaflets 2 to 4 pairs, obtuse, penninerved or reticulate.
 Leaflets oblong or broadly lanceolate, narrowed at the base, quite
  glabrous. Fruit 4-celled . . . . . . . . . . . . . 3. *O. venosa.*
 Leaflets pubescent. Fruit 12-celled . . . . . . . . . . . 4. *O. cerasifera.*
 Leaflets large, ovate or ovate-lanceolate, broad and sessile at the base,
  very prominently reticulate underneath . . . . . . . . 5. *O. reticulata.*

1. **O. acidula,** *F. Muell. in Hook. Kew Journ.* ix. 304, *and Fragm.* iii. 14. A small or moderate-sized tree, glabrous, with the young shoots glutinous. Leaves crowded at the ends of the often pendulous branches; leaflets from 9 to nearly 30, linear-lanceolate, acute or mucronate, 1 to 1½ in. long, oblique, the midrib prominent underneath, but otherwise almost nerveless, the common petiole 3 to 6 in. long. Panicles narrow, shorter than the leaves. Flowers nearly sessile, in clusters or on short branches of the panicle. Sepals about 1 line long. Petals about 2 lines. Teeth of the staminal tube subulate, but more or less connected by an undulate crenate or almost fringed membrane. Disk small, annular. Ovary 3-celled. Drupe ¾ to 1 in. or rather more in diameter, said to resemble a russet apple, the epicarp pulpy, of a rich crimson; putamen very hard.

**Queensland.** Desert of the Suttor and Burdekin, *F. Mueller.*
**N. S. Wales.** Arbuthnot's Range, *Fraser;* near the Gwydir river, *Mitchell* (figured in Mitch. Three Exped. i. 82, without any name); Darling Desert, *Victorian Expedition;* Castlereagh river, *Herb. F. Mueller.*

2. **O. vernicosa,** *F. Muell. Fragm.* iii. 15. Quite glabrous. Branches thick, marked with the broad scars of the fallen leaves, the young shoots glutinous. Leaves larger than in *O. acidula,* the common petiole slightly flattened; leaflets 15 to nearly 30, lanceolate, acuminate, often above 2 in. long, oblique, with a prominent midrib and transverse reticulations. Panicles 3 or 4 in. long, with divaricate branches and numerous flowers, much smaller than in *O. acidula.* Sepals about ½ line long, slightly ciliate. Petals little more than 1 line. Staminal tube short, with 10 subulate teeth. Fruit the size of that of *O. acidula,* the stony endocarp thicker and harder, usually 3-celled.

**N. Australia.** Cambridge Gulf, *A. Cunningham;* mouth of the Victoria river, *F. Mueller.*

Var. (?) *pubescens.* Young shoots and inflorescence softly pubescent; flowers still smaller and more numerous.—Mouth of the Victoria river, *F. Mueller (Hb..F. Muell.).*

3. **O. venosa,** *F. Muell. in Hook. Kew Journ.* ix. 304. A tall arborescent shrub, quite glabrous, the young shoots slightly glutinous. Leaflets 6 or 8, obliquely oblong or ovate-lanceolate, obtuse or emarginate, 2 to 3 or rarely 4 in. long, coriaceous, prominently penninerved, slightly reticulate underneath, the petiole angular or sometimes broadly winged. Panicles narrow, 3 to 5 in. long, glabrous. Flowers not yet open in our specimen, but apparently like those of *O. acidula,* except that the staminal tube is exceedingly short, but possibly it may grow out as the bud advances. Sepals orbicular, about 1 line diameter.

**Queensland.** Between the Dawson and Burnett rivers, *F. Mueller ;* Rockhampton, *Thozet.*

4 ? **O. cerasifera,** *F. Muell. in Hook. Kew Journ.* ix. 305. A small tree. Leaflets 6 to 10, obliquely oval-oblong, obtuse, 1½ to 3 in. long, narrowed into a very short petiolule, glabrous ·above, pubescent underneath as well as the common petiole. Flowers not seen. Drupe globular, 1 to 1½ in. diameter, black, with a red sarcocarp. Putamen hard, rugose outside, 12-celled, with 1 seed in each cell..

**Queensland.** Burdekin river, *F. Mueller.* Until the flowers have been seen, this species must remain in some measure doubtful.

5. **O. reticulata,** *F. Muell. in Hook. Kew Journ.* ix. 305. A small tree, quite glabrous. Leaves often above a foot long, the common petiole angular or slightly dilated, terminating in a short point. Leaflets 4, 6, or 8, sessile, ovate or broadly ovate-lanceolate, obtuse, 4 to 8 in. long, oblique at the base, coriaceous, smooth above, with very prominent pinnate veins and numerous raised reticulations underneath. Panicles loose, very divaricate, the branches often 6 in. long or more. Flowers sessile, clustered. Sepals above 1 line long, orbicular. Petals twice as long. Staminal tube often divided to near the middle into 10 flat 2-lobed teeth or lobes. Ovary 2- or 3-celled. Fruit 1½ in. diameter, the epicarp fleshy but not thick. Putamen hard and very rugose.—*O. xerocarpa,* F. Muell. Fragm. iii. 13.

**N. Australia.** Near Nichol Bay, *Walcott ;* islands of the Gulf of Carpentaria, *R. Brown, F. Mueller, Henne.*

## 8. CARAPA, Aubl.

(Xylocarpus, *Kœn.*)

Calyx small, 4- or 5-lobed. Petals 4 or 5, free, imbricate in the bud. Staminal tube urceolate, crenate or lobed ; anthers 8 or 10, within the summit. Disk thick, surrounding the ovary. Ovary 4- to 5-celled, with 2 to 6 ovules in each cell ; style short, with a large disk-like stigma. Capsule globular or ovoid, fleshy or woody, the dissepiments often disappearing. Seeds several in a compact mass round the remains of the central axis, large, thick, with a ventral hilum ; testa spongy ; cotyledons superposed, often united ; radicle dorsal.—Maritime trees. Leaves pinnate with entire leaflets. Panicles axillary.

The species are few, ranging over the tropical seacoasts either of America and Africa or of Africa and Asia. The Australian one belongs to the latter category.

1. **C. moluccensis,** *Lam.; DC. Prod.* i. 626. A tree, glabrous in all its parts. Leaflets 4, rarely 2 or 6, opposite, ovate, obtuse, shortly acuminate or rarely acute, 2 to 3 or rarely 4 in. long, somewhat coriaceous, more reticulate than in any of the preceding genera. Panicles short, loose, and few-flowered, sometimes reduced to simple racemes or with few divaricate branches. Calyx small, irregularly lobed. Petals 4 or rarely 5, 2½ to 3 lines long. Staminal tube crenate or splitting into short lobes. Ovary very small, in the centre of a large thick depressed disk. Ovules 2, 3, or 4 in each cell, excessively minute. Fruit often 3 or 4 in. diameter, irregularly globular. Seeds usually 4 to 6, large, irregularly shaped, closely packed; testa very thick, of a hard spongy consistence.—*Xylocarpus Granatum*, Kœn.; Willd. Spec. Pl. ii. 328.

**N. Australia.** Saltwater Creek, near Macadam Range, *F. Mueller;* islands of the Gulf of Carpentaria, *Henne.*

**Queensland.** N.E. coast, *A. Cunningham;* islands of Howick's group, *F. Mueller;* Port Denison, *Fitzalan* (in leaf only, with loose fruits).

Common on the seacoasts of tropical Asia, extending westward to E. Africa and eastward to the Moluccas. It varies considerably in the more compact or looser inflorescence, in the size of the flowers, and in the teeth of the staminal tube.

Tribe III. CEDRELEÆ.—Stamens free. Ovules more than 2 in each cell. Seeds winged. Leaves pinnate or rarely simple.

## 9. CEDRELA, Linn.

Calyx small, 5-cleft. Petals 5, imbricate. Disk thick or raised. Stamens 4 to 6, inserted on the summit of the disk, alternating sometimes with as many staminodia, filaments subulate, anthers versatile. Ovary 5-celled, style filiform, with a disk-like stigma; ovules 8 to 12 in each cell, in 2 rows. Capsule membranous or coriaceous, 5-celled, opening in 5 valves, leaving the dissepiments attached to the persistent axis. Seeds flattened, winged; albumen scanty; cotyledons flat; radicle short, superior.—Tall trees, with coloured wood. Leaves pinnate. Flowers small, in large panicles.

The genus is spread over tropical America and Asia. The Australian species is a common Asiatic one.

1. **C. Toona,** *Roxb. Pl. Corom.* iii. 33, *t.* 238. A tall, handsome tree, quite glabrous or the young shoots minutely pubescent. Leaves large, deciduous; leaflets 11 to 17, opposite or irregularly alternate, ovate-lanceolate, acuminate, 3 to 5 in. long, oblique at the base, petiolulate, membranous. Panicles large, pyramidal, many-flowered, glabrous. Pedicels short. Sepals orbicular, ciliate, very small. Petals nearly 3 lines long. Stamens 5, as long as the petals, inserted in cavities on the outside of the very thick pubescent disk. Ovary half immersed in the disk. Capsule glabrous, oblong, 1 to 1½ in. long.—Wight, Ic. t. 161; *C. australis*, F. Muell. Fragm. i. 4.

**Queensland.** Moreton Bay, *Herb. F. Mueller;* Mackenzie's Station, *Leichhardt.*
**N. S. Wales.** Illawarra, *Herb. F. Mueller.* "Red Cedar" of the colonists.
Var. *parviflora.* Petals scarcely 2 lines long.—Clarence river, *Wilcox.*

## 10. FLINDERSIA, R. Br.

(Oxleya, *A. Cunn.*; Strzeleckia, *F. Muell.*)

Calyx small, 5-lobed. Petals 5, imbricate in the bud, spreading. Disk broad, concave. Stamens 5, inserted on the outside of the disk, with as many or fewer staminodia alternating with them, sometimes wanting; filaments subulate; anthers versatile. Ovary 5-celled, 5-lobed; style short, thick, inserted between the lobes; stigma capitate; ovules 4 to 6 in each cell. Capsule oblong, hard, tuberculate or muricate, opening septicidally in 5 boat-shaped válves or cocci, without any persistent axis. Seeds flat, winged, 2 or 3 on each side of a flat placenta, which almost divides each cell into two; albumen none; cotyledons flat, radicle very short.—Trees. Leaves alternate or more frequently opposite, pinnate or rarely simple, marked with pellucid dots. Flowers in terminal panicles.

The species are all endemic in Australia. The genus, although allied to *Cedrela* and therefore placed by common consent in *Meliaceæ*, is nevertheless, as observed by R. Brown very closely connected with *Rutaceæ-Zanthoxyleæ*, and might be very well placed there next to *Geijera*, with which it is connected, especially through *F. maculosa*.

Leaves alternate (on different branches from the flowers). Petals tomentose outside. Seeds winged at one end only . . . . . .   1. *F. australis.*
Leaves opposite (on the flowering branches). Petals glabrous outside
  or nearly so.
  Leaflets mostly 3 to 6 pairs, very oblique, slightly coriaceous.
    Leaflets almost sessile, broad at the base. Petals slightly hairy
      inside  . . . . . . . . . . . . . . . . . . .   2. *F. Schottiana.*
    Leaflets narrowed into a distinct petiolule. Petals quite glabrous.
      Seeds winged at both ends . . . . . . . . . . .   3. *F. Oxleyana.*
  Leaflets 3 or 5, short, oblique, very coriaceous. Seeds winged at
    one end only . . . . . . . . . . . . . . . . .   4. *F. Bennettiana.*
Leaves simple or leaflets 3 to 5, narrow, with the petiole broadly
  winged. Fruit small. Seeds winged at both ends . . . . .   5. *F. maculosa.*

1. **F. australis,** *R. Br. in Flind. Voy.* ii. 595, *t.* 1. A tree of moderate size, with a rugged bark. Leaves alternate, crowded at the end of short barren branches, glabrous; leaflets 3 to 6, broadly lanceolate or oblong-elliptical, obtuse or scarcely acuminate, 2 to 4 in. long, scarcely oblique. Panicles much branched, terminating short branches without any leaves except a few scale-like bracts, sprinkled with a stellate tomentum. Flowers numerous. Calyx open, tomentose, with 5 short broad obtuse lobes. Petals about 2 lines long, tomentose outside, except a narrow border, slightly pubescent inside. Fruit almost woody, 2 or 3 in. long. Seeds (according to the plate quoted) winged at the upper end only.

**Queensland.** Scrub near Upper Head, Broad Sound, *R. Brown* (*Hb. R. Br.*).

2. **F. Schottiana,** *F. Muell. Fragm.* iii. 25. A tree of moderate size, or sometimes tall. Leaves opposite, crowded under the panicle; leaflets 8 to 12, with or without a terminal odd one, ovate-lanceolate, obtuse or acuminate, 4 to 5 in. long, more or less falcate, sessile, with a broad very oblique base, somewhat coriaceous, glabrous on both sides or softly pubescent underneath when young. Panicles ample and many-flowered, but not exceeding the leaves.

Petals about 2 lines long, glabrous outside, sprinkled on the inside as well as the anthers with a few hairs. Fruit not seen.

**Queensland.** Wide Bay, *Bidwill;* Cumberland Islands, *Herb. F. Mueller ;* Brisbane river, *A. Cunningham.*

**N. S. Wales.** Hastings river, *Thozet ;* Clarence river, *Beckler.*

3. **F. Oxleyana,** *F. Muell. Fragm.* i. 65; iii. 25. A tall, much-branched tree, attaining often 100 ft. Leaves opposite, crowded under the panicles ; leaflets 4 to 10, with or without a terminal odd one, broadly lanceolate, obtuse or shortly acuminate, 2 to 4 in. long, oblique and almost falcate, narrowed into a distinct petiolule, glabrous or sprinkled underneath with minute stellate hairs, thinly coriaceous, rather sparingly glandular-dotted. Panicles loose and many-flowered, but shorter than the leaves. Sepals very small. Petals about 2 lines long, obovate-oblong, glabrous or nearly so. Fruit woody, 3 to 4 in. long, muricate. Seeds winged at both ends.—*Oxleya xanthoxyla,* A. Cunn. in Hook. Bot. Misc. i. 246, t. 54.

**Queensland.** Brisbane river, *Fraser, A. Cunningham, F. Mueller.* "Yellow Wood" of the colonists.

4. **F. Bennetiana,** *F. Muell. Herb.* A large tree. Leaves opposite, crowded under the panicles ; leaflets 3 or 5, from ovate to ovate-lanceolate or oblong-elliptical, obtuse or scarcely acuminate, 2 to 3 in. long in some specimens, 4 to 5 in. in others, glabrous, very coriaceous, not oblique, and scarcely petiolulate, the common petiole angular. Panicles ample, sometimes short, sometimes exceeding the leaves, minutely stellate-pubescent. Petals about 2 lines long, rather broader than in *F. Oxleyana,* glabrous or nearly so. Fruit 2 or 3 in. long, muricate. Seeds winged at the upper end only, or some with a very small wing also at the lower end, but only seen in one capsule.— *F. australis,* F. Muell. Fragm. iii. 26, not of R. Brown.

**Queensland.** Wide Bay, *Bidwill;* Brisbane river, Moreton Bay, *A. Cunningham, Fraser, W. Hill.*

**N. S. Wales.** Clarence river, *Beckler.*

5. **F. maculosa,** *F. Muell. in Journ. Pharm. Soc. Vict.* ii. 44. A small tree, the trunk remarkably spotted by the falling off of the outer bark in patches. Leaves opposite or nearly so, glabrous, coriaceous, the glandular dots often only visible on the young ones, in some specimens all simple, linear-oblong or lanceolate, obtuse or emarginate and mucronate, 1 to 2 in. long or rather more ; in other specimens a few of the leaves break out into 2 or 3 narrow continuous lobes, in others, again, all are pinnate, with 3 or 5 leaflets, like the simple leaves, but smaller, and a winged common petiole. Panicles terminal, rather dense, usually shorter than the leaves. Sepals scarcely 1 line long. Petals about 2 lines long, glabrous. Capsule oblong and muricate, like those of the other species, but much smaller, often not more than 1 in. long when fully ripe. Seeds winged at both ends and along the back.—*Elæodendron maculosum,* Lindl. in Mitch. Trop. Austr. 384 ; *Strzeleckya dissosperma,* F. Muell. in Hook. Kew Journ. ix. 308 ; *Flindersia Strzeleckiana,* F. Muell. Fragm. i. 65.

**Queensland.** Scrub on the Burdekin and Burnett rivers, *F. Mueller ;* St. George's Bridge on the Balonne river, *Mitchell ;* Port Bowen and Broad Sound, *Herb. F. Mueller.* " Spotted Tree " of the colonists.

**N. S. Wales.** Between the Darling and Lachlan rivers, *Victorian Expedition.*
The simple-leaved specimens which are the most frequent in N. S. Wales have much the
habit of *Geijera*, to which in fact the genus is very nearly allied; the pinnate-leaved specimens are chiefly tropical, but not exclusively so.

## Order XXXII. OLACINEÆ.

Flowers regular, hermaphrodite or rarely unisexual.  Calyx small, 4- or 5-,
rarely 6-toothed, free or adnate to the disk (in *Cansjera* scarcely distinguishable from the corolla).  Petals 4, 5, or rarely 6, free or united in a campanulate or tubular corolla, valvate in the bud (except *Villaresia*).  Stamens as
many or twice as many as petals or rarely fewer, adnate to the base of the
petals, or free and hypogynous; anthers 2-celled, versatile, or rarely adnate.
Disk free, or adnate to the ovary or to the calyx, or divided into scale-like
glands.  Ovary free or immersed in the disk, 1-celled or imperfectly 2- or 3-
celled; style simple; stigma entire or lobed.  Ovules 2, 3, or rarely 1, pendulous from a central placenta into the imperfect cells, or from the side or
apex of the cavity.  Fruit usually an indehiscent drupe, either superior or inferior by the growth over it of the disk and tube of the calyx.  Seed solitary,
pendulous, or sometimes, owing to the adnate nerve-like remains of the placenta, apparently erect; testa very thinly membranous; embryo very small in
the apex of a fleshy albumen, or larger and axile; or, in a genus not Australian,
occupying the whole seed without albumen; cotyledons flat or terete; radicle
superior.—Trees, shrubs, or climbers.  Leaves usually alternate, entire, penninerved, without stipules.  Flowers few and axillary, or rarely in terminal panicles, usually small.

The Order is widely dispersed over the tropical and subtropical regions of the globe.  The
six Australian genera are none of them endemic, one extending to New Zealand, one to tropical Asia, two to tropical Asia and Africa, one to tropical Asia and America, and one is
common to Asia, Africa, and America.  The Order is more nearly allied to *Loranthaceæ*
among *Calycifloræ*, and especially to *Santalaceæ* among *Monochlamydeæ*, than to any (except *Ilicineæ*) of the *Discifloræ*, amongst which it is technically placed.

Tribe I. **Olaceæ.**—*Stamens twice as many as petals or fewer, or if the same number
as petals, opposite to them.  Ovary often 2- or 3-celled at the base, 1-celled at least at
the top; placenta central, with 2 or 3 pendulous ovules.*

Calyx not enlarged after flowering.  Stamens twice as many as petals;
   anthers oblong or linear  . . . . . . . . . . . . . . . . . . 1. Ximenia.
Calyx enlarged enclosing the fruit.  Stamens 3; staminodia (in
   the Australian species) 5; anthers short  . . . . . . . . . 2. Olax.

Tribe II. **Opilieæ.**—*Stamens as many as petals and opposite to them.  Ovary 1-
celled, with 1 ovule.*

Perianth apparently simple, shortly 4-lobed.  Stamens 4, included,
   alternating with 4 glands or scales  . . . . . . . . . . . 3. Cansjera.
Calyx minute.  Petals 5, free.  Stamens 5, exserted, alternating with
   5 scales . . . . . . . . . . . . . . . . . . . . . . . 4. Opilia.

Tribe III. **Icacineæ.**—*Stamens as many as petals and alternate with them.  Ovary
1-celled, with 1 or 2 pendulous ovules.*

Petals strictly valvate.  Ovule 1, the placenta not prominent.  Flowers
   in a much-branched corymbose panicle . . . . . . . . . . 5. Pennantia.

Petals slightly imbricate. Ovules 2, the placenta forming a half-dissepiment on one side of the cavity. Flowers in a narrow raceme-like panicle . . . . . . . . . . . . . . . . . . . . . . . 6. VILLARESIA.

TRIBE I. OLACEÆ.—Stamens twice as many as petals or fewer, or if the same number as petals, opposite to them. Ovary often 2- or 3-celled at the base, 1-celled at least at the top; placenta central, with 2 or 3 pendulous ovules.

## 1. XIMENIA, Linn.

Calyx minutely 4- or 5-toothed, not enlarged after flowering. Petals 4 or 5, bearded inside, valvate in the bud. Stamens twice as many as petals, free; filaments filiform; anthers linear, erect. Ovary 3-celled at the base; stigma capitate; ovules 3, descending into the incomplete cells from a central placenta. Drupe ovoid or globular, with a thick sarcocarp. Seed spuriously erect; embryo minute.—Shrubs or trees, often thorny. Flowers white, rather large for the Order, in small axillary cymes or solitary.

The Australian species is spread over almost all tropical countries, the few other species are American or African.

1. **X. americana,** *Linn.; DC. Prod.* i. 533. A glabrous shrub, or sometimes a small tree, with spreading branches, often armed with axillary spines (abortive peduncles). Leaves petiolate, ovate, obtuse, or scarcely acute, 1 to 2 in. long, entire, the veins inconspicuous, except the midrib. Peduncles short, bearing little cymes of 3 to 7 yellowish sweet-scented flowers, rarely reduced to a single one. Petals 3 to 4 lines long, densely bearded inside with long white hairs. Drupe attaining 1 in. diameter or rather more. —*X. elliptica,* Forst.; Labill. Sert. Austr. Caled. 34, t. 37; *X. laurina,* Delile, in Ann. Sc. Nat. ser. 2, xx. 89; *X. exarmata,* F. Muell. in Trans. Phil. Inst. Vict. iii. 22.

**N. Australia.** Ranges of the Suttor and Mackenzie rivers, *F. Mueller.* The species is widely spread over the tropical regions of both the New and the Old World, varying in most places with or without thorns. The Pacific and New Caledonian *X. elliptica* has been distinguished from the common form as having a globular, not elliptical fruit; but some of Gardner's specimens from Brazil have certainly also the fruit globular. F. Mueller's Australian specimens, like the majority of those in our herbaria, are without fruit; they are unarmed, or have only small nascent spines in the axils of some of the young leaves.

## 2. OLAX, Linn.

(Spermaxyrum, *Labill.*)

Calyx small, cup-shaped, truncate, enlarged after flowering and enclosing the fruit. Petals 5 or 6, free, or slightly cohering, valvate in the bud. Stamens usually 3, alternate with the petals, the filaments adnate to the petals and connecting them in pairs; staminodia as many as petals and opposite to them, filiform or flat, entire or 2-cleft. Ovary free, 1-celled, or very shortly 3-celled at the base; stigma entire or slightly 3-lobed; ovules 3, pendulous from a central placenta. Drupe globular or oblong, enclosed in the enlarged calyx, but free from it, the sarcocarp thin. Seed spuriously erect; embryo very small in the apex of a fleshy albumen.—Trees, shrubs, or undershrubs, rarely half climbing, the Australian species all erect shrubs, with small alternate, entire, distichous leaves, the veins inconspicuous, except the midrib. Flowers axillary, solitary in the Australian species, several in short racemes or spikes in some others.

The genus is confined to the Old World, extending over tropical Asia and Africa. The Australian species are all endemic, and differ from all except the E. Indian *O. nana*, Wall., in their solitary axillary flowers and small leaves. They have all 5 petals, 3 stamens, and 5 staminodia.

Staminodia undivided.
  Leaves oval or broadly oblong, retuse. Flowers glabrous inside.
    Staminodia subulate . . . . . . . . . . . . . . . 1. *O. phyllanthi.*
  Leaves narrow-oblong, mucronate. Staminodia linear, bearded at
    the base . . . . . . . . . . . . . . . . . . 3. *O. stricta.*
  Leaves reduced to minute scales. Flowers densely bearded inside.
    Staminodia linear . . . . . . . . . . . . . . 5. *O. aphylla.*
Staminodia 2-cleft to the middle.
  Leaves rather thin, narrow, retuse (Eastern species) . . . . . 2. *O. retusa.*
  Leaves rather thick, from linear to obovate or obcordate (Western
    species) . . . . . . . . . . . . . . . . . 4. *O. Benthamiana.*

1. **O. phyllanthi,** *R. Br. Prod.* 358. A shrub of 4 or 5 ft., the leafy branches, when dry, having much the aspect of those of a *Phyllanthus*. Leaves oval or broadly oblong, truncate or emarginate, from ⅓ to 1 in. long, sessile, with a broad base, thin, glabrous, and somewhat glaucous. Pedicels very short, slender. Petals nearly 1½ lines long, glabrous. Filaments flattened below the middle; staminodia glabrous, undivided, subulate, shorter than in the other species. Fruit ovoid-globular, about 2 lines long.—*Spermaxyrum phyllanthi*, Labill. Pl. Nov. Holl. ii. 84, t. 233 (the figure incorrect as to the shape of the petals and anthers); *Lopadocalyx phyllanthoides*, Klotzsch, in Pl. Preiss. i. 178, corrected to *O. phyllanthi*, l. c. ii. 230.

**W. Australia.** King George's Sound, *Labillardière, R. Brown*, and others; rocky places near Albany, *Preiss, n.* 1211.

2. **O. retusa,** *F. Muell. Herb.* (as a var. of *O. stricta*). A glabrous shrub, with the slender virgate branches of *O. stricta*. Leaves linear-cuneate or narrow-oblong, truncate and emarginate, or almost 2-lobed, minutely mucronate, rarely exceeding ½ in. and smaller on the lateral branches, rounded at the base. Pedicels very short. Flowers about 2 lines long. Filaments glabrous, dilated at the base; staminodia bearded below the middle, glabrous above and divided into 2 linear lobes. Fruit ovoid-oblong, not exceeding 3 lines in the specimens seen.

**Queensland.** Moreton Island, *M'Gillivray, F. Mueller*. This is believed by F. Mueller to be a variety of *O. stricta;* but besides the shape of the leaves, which is nearer to that of *O. phyllanthi*, I have found, in the few flowers I have been able to examine, the staminodia always 2-cleft, as in *O. Benthamiana* and in the Indian species.

3. **O. stricta,** *R. Br. Prod.* 358. An erect, glabrous shrub, of 2 or 3 ft., with slender virgate branches. Leaves narrow-oblong or linear, acute or obtuse, but always mucronate, ¼ to ½ in. or rarely ¾ in. long, flat, with a prominent midrib, narrowed or rarely rounded at the base. Pedicels scarcely 1 line long. Petals varying from 2 to 3 lines. Filaments flattened to very near the anthers, glabrous; staminodia linear, entire, more or less bearded below the middle. Fruit obovoid-oblong, often 4 lines long or rather more.

**Queensland.** Edges of lagoons, Moreton Island, *F. Mueller.*
**N. S. Wales.** Port Jackson, *R. Brown, Sieber, n.* 130, and others; Blue Mountains, *Miss Atkinson;* Port Macquarie, *Backhouse;* barren brushes, N.W. interior, *Fraser.*

4. **O. Benthamiana,** *Miq. in Pl. Preiss.* i. 228. A glabrous shrub of about 2 ft., usually much-branched and more rigid than *O. stricta,* and not drying so black. Leaves in the ordinary form linear or narrow-oblong in the lower part of the branches, about ½ in. long, terminating in a recurved point, narrowed at the base, rather thick, convex underneath, with the midrib less prominent than in the preceding species, the upper leaves, especially the floral ones, passing into a short broadly obovate form; in a few luxuriant specimens, all the leaves are obovate-oblong, 1 in. long or rather more; in others, all are broadly obovate, cuneate, or obcordate, ¼ to ½ in. long, and not mucronate. Flowers 2 to 3 lines long as in *O. stricta,* but the staminodia are pubescent only, or slightly bearded, and divided to the middle into 2 linear, oblong, or spathulate lobes, nearly as long as the petals. Fruit globular, attaining 4 or 5 lines diameter.

**N. Australia.** Bay of Rest, N.W. coast, *A. Cunningham.* (A single specimen with small obovate leaves.)
**S. Australia.** Port Lincoln, *Wilhelmi.* (Specimens with obovate leaves, not seen in flower and therefore doubtful, although precisely resembling some W. Australian ones.)
**W. Australia.** Swan River, *Drummond, Preiss, n.* 2095, *Oldfield,* etc. (leaves mostly narrow and pointed); Murchison river, *Oldfield* (leaves all obovate or oblong); Gardiner and Kalgan rivers, *Oldfield* (leaves cuneate, emarginate, or obcordate); Swan River, *Drummond, n.* 729 (leaves, especially the floral ones, small and broad, flowers small, the lobes of the staminodia oblong-spathulate and petaloid).

*Lopadocalyx uliginosus,* Kl. in Pl. Preiss. i. 178, corrected to *Olax uliginosa,* Kl. l. c. ii. 230, from swampy places in the plains between Mounts Melville and Elphinstone, *Preiss, n.* 1210, which I have not seen, would appear, from the very imperfect description given, to be the ordinary narrow-leaved form of *O. Benthamiana.*

5. **O. aphylla,** *R. Br. Prod.* 358. A shrub of several feet, with numerous, wiry, virgate, slightly pubescent branches. Leaves all reduced to minute scales. Flowers very small, almost sessile in the axils of orbicular ciliate bracts rather longer than the calyx, towards the ends of the branches. Petals scarcely more than 1 line long, densely bearded inside about the middle. Staminodia linear and entire, or slightly spathulate and emarginate at the top. Fruit ovoid, about 2 lines long.

**N. Australia.** N. coast, *R. Brown;* barren stony ridges on the Fitzmaurice river, *F. Mueller;* Arnhem's Land, *Leichhardt.*

Tribe II. OPILIEÆ.—Stamens as many as petals or corolla-lobes and opposite to them, usually alternating with as many hypogynous glands or scales. Ovary 1-celled, with a single ovule, erect or suspended from an erect central placenta. Seed spuriously or sometimes perhaps really erect; radicle superior.

### 3. CANSJERA, Juss.

Perianth apparently simple, the calyx very minute and often not distinguishable, at the base of the tubular or urceolate 4-lobed corolla. Stamens 4, opposite to the petals or corolla-lobes, and more or less adherent at the base; filaments filiform; anthers small. Hypogynous scales (or lobes of the disk) 4, alternating with the stamens. Ovary small, fleshy; ovule 1, apparently erect or suspended from a short placenta in the centre of the minute cavity. Drupe with a thin sarcocarp. Seed erect; embryo small or sometimes elongated.—Weak or climbing shrubs. Leaves alternate, entire. Flowers small, in short axillary spikes.

Besides the Australian species, which is also in New Ireland, the genus comprises 2 or perhaps 3 from tropical Asia.

1. **C. leptostachya,** *Benth. in Hook. Lond. Journ.* ii. 231. A climbing shrub, glabrous or the young shoots very minutely tomentose. Leaves ovate-lanceolate, long-acuminate, 2 to 3 in. long, membranous, glabrous. Spikes 1 or 2 together in the axils, rarely exceeding $\frac{1}{2}$ in. Flowers in the young bud strigose-pubescent, sessile in the axils of narrow minute bracts which soon fall off, when fully open about 1 line long, nearly globular and glabrous, the lobes very short and spreading. Filaments slender, but shorter than the perianth. Hypogynous scales short, broad, entire or rarely 3-toothed. Fruit not seen.—Meisn. in DC. Prod. xiv. 519.

**Queensland.** Cape York and islands off the N.E. coast, *A. Cunningham, M'Gillivray.* The species is also in New Ireland. The flowers are about half the size of those of the common *C. Rheedii*, Gmel., and I have not succeeded in detaching the calyx from the corolla, as I have readily done in Malacca specimens of *C. Rheedii* or of an allied species.

## 4. OPILIA, Roxb.

Calyx minute, 5- or rarely 4-toothed. Petals 5, rarely 4, hypogynous, valvate in the bud. Stamens as many, alternating with the petals, free; filaments filiform; anthers ovate. Disk of 5, rarely 4 scales, alternating with the stamens. Ovary 1-celled, tapering into a short thick truncate style; ovule solitary, suspended from a central filiform placenta very early adnate to it. Drupe with a thin sarcocarp and crustaceous endocarp. Seed spuriously erect; embryo linear, short, or nearly as long as the albumen.—Shrubs or small trees, sometimes climbing. Leaves alternate, entire. Flowers in axillary racemes; pedicels 3 together in the axils of peltate bracts, which are imbricate at an early stage but fall off before the flowers expand.

A genus of 2 or perhaps 3 species, natives of tropical Asia and Africa, the Australian species one of the widest dispersed.

1. **O. amentacea,** *Roxb. Pl. Corom.* ii. 31, *t.* 158. A scrambling half-climbing shrub or small weak tree, glabrous, or the young leaves and shoots minutely tomentose-pubescent. Leaves petiolate, ovate, ovate-lanceolate, or almost oblong, acute or acuminate, 2 to 3 or even 4 in. long, or rarely shorter and very obtuse, entire, thinly coriaceous, the veins usually prominent though fine. Racemes before flowering resembling little cylindrical cones of $\frac{1}{2}$ in., the peltate imbricate but almost squarrose bracts alone visible, when in flower slender, about 1 in. long, without bracts. Flowers very small, on filiform pedicels of about 1 line. Petals about $\frac{1}{2}$ line long, very deciduous. Drupe ovoid or globular, $\frac{1}{2}$ to $\frac{3}{4}$ in. long. Embryo linear, nearly as long as the albumen.—Wight, Illustr. t. 40; *O. javanica*, Miq. Fl. Ind. Bat. i. part i. 784.

**N. Australia.** York Sound, N.W. coast, *A. Cunningham;* Victoria river, *Bynoe, F. Mueller;* Port Essington, *Armstrong;* Point Pearce, *F. Mueller.* Also in the Indian Peninsula, in Ceylon and in Java. *O. pentitdis*, Blume, Mus. Bot. i. 246, from New Guinea, is also probably, as he himself suggests, the same species. The fruit is on some

Indian specimens globular, as described by Roxburgh. Wight figures it as ovoid, and so it appears to be on Horsfield's Javanese specimens, and certainly on F. Mueller's from Victoria river. All our other specimens from India as well as from Australia are in flower only or with young fruit.

TRIBE III. ICACINEÆ.—Stamens as many as petals or corolla-lobes, and alternate with them. Ovary 1-celled, with 2, rarely 1 ovule, pendulous from one side or the apex of the cavity. Seed pendulous.

### 5. PENNANTIA, Forst.

Flowers diœcious or polygamous. Calyx minute. Petals 5, hypogynous, glabrous, valvate in the bud. Stamens 5, alternating with the petals ; anthers oblong-sagittate. Ovary 1-celled; stigma nearly sessile, entire or 3-lobed; ovule solitary, suspended from the apex of the cavity. Drupe with a hard putamen, or almost baccate with a slightly coriaceous endocarp. Seed pendulous; embryo small within the apex of the fleshy albumen.—Trees. Leaves thinly coriaceous, entire or (in New Zealand species) coarsely toothed. Flowers in terminal corymbose panicles.

Besides the Australian species, which is endemic, there is one from Norfolk Island and another from New Zealand.

1. **P. Cunninghamii,** *Miers, in Ann. Nat. Hist. ser.* 2, ix. 491, *and Contrib.* 80, *t.* 12. A glabrous, suberect, tall shrub. Leaves ovate or broadly elliptical, acuminate, 4 to 6 in. long, entire, coriaceous and shining when old, narrowed into a petiole of ½ in. or more. Flowers numerous, in broad rather dense panicles, either terminal or in the upper axils, the males only known. Calyx scarcely prominent. Petals nearly 1½ lines long. Filaments bent in below the summit in the bud ; anthers oblong, sagittate. Rudimentary ovary narrow, with 2 or 3 erect style-like lobes, and occasionally containing an imperfect pendulous ovule. Drupes or berries ovoid, about ½ in. long, the endocarp scarcely hardened. Seed pendulous; testa thinly membranous ; embryo much shorter than the albumen.

**N. S. Wales.** Illawarra district, *A. Cunningham, M‘Arthur, Shepherd;* Kiama, *Harvey;* Clarence river, *Moore.* The ovaries described by Miers appear to me to have been imperfect, at least I find none but male flowers in the specimen he examined, nor in any others I have seen. It is probable that the female flowers, as in the New Zealand species, are smaller, and have therefore not attracted the notice of collectors.

### 6. VILLARESIA, Ruiz and Pav.

(Pleuropetalum, *Blume ;* Chariessa, *Miq.*)

Flowers hermaphrodite or polygamous. Sepals 5, distinct, broad, imbricate. Petals 5, with the midrib prominent inside, imbricate or almost valvate in the bud. Stamens 5, alternating with the petals ; anthers cordate. Ovary 1-celled, the cavity marked on one side with a raised ridge half dividing it ; style short, thick ; ovules 2, suspended from the summit of the raised ridge. Drupe ovoid or globular, the endocarp forming a prominent half-dissepiment which penetrates into a deep vertical furrow in the seed. Embryo small, in the apex of the albumen.—Lofty trees (or tall woody climbers ?). Leaves alternate, coriaceous, entire or toothed. Flowers in small cymes, along the simple rhachis of a raceme-like panicle.

Besides the Australian species, which may be endemic, there is one (perhaps not really different) from the Indian Archipelago, one from the S. Pacific islands, and several from S. America. The genus is exceptional in *Olacineæ* by the more or less imbricate petals. I have not seen the 2 cells to the ovary which Miers met with in one species, possibly in accidentally abnormal flowers.

1. **V. Moorei,** *F. Muell. Herb.* A lofty handsome tree, glabrous except the inflorescence. Leaves ovate-lanceolate or oblong, acuminate, 3 to 4 in. long, entire, narrowed into a short petiole, coriaceous and shining, but not so thick as in the American species. Raceme-like panicles irregularly lateral or axillary, 2 to 4 in. long, hoary with a minute pubescence. Cymes numerous, few-flowered, on short peduncles along the rhachis. Flowers almost sessile in the cymes, those seen all males. Petals 1 line long, very slightly imbricate. Drupes globular, the putamen hard, about ½ in. diameter, rugose outside, the half-dissepiment projecting quite to the centre of the cavity and there slightly thickened, forming a column, up the centre of which the placenta appears to pass, as if the endocarp had grown over it as in the New Zealand *Pennantia*. Seed quite enclosing the half-dissepiment, its transverse section being horseshoe-shaped.

**N. S. Wales.** Clarence river, *Moore.* The Javanese *V. suaveolens* (*Pleuropetalum suaveolens*, Blume) is unknown to me, but must, from the character given, be nearly allied to this species. *V. Samoensis* (*Pleuropetalum Samoense*, A. Gr.) which we have also from the Fiji islands, appears to be quite distinct.

## Order XXXIII. ILICINEÆ.

Flowers regular, hermaphrodite or unisexual. Calyx of 4 or 5, rarely 3 or more than 5 sepals, imbricate, usually persistent. Petals 4 or 5 or rarely more, hypogynous, imbricate in the bud, sometimes united in a lobed corolla. Stamens of the same number as petals, hypogynous, free or adhering to the corolla at the base; anthers 2-celled, opening inwards. Disk none, except the thickened base of the ovary. Ovary free, 3- to 5-celled, rarely many-celled; stigma broad or capitate, sessile or supported on a distinct style. Ovules 1 or 2 in each cell, pendulous, with a superior micropyle. Fruit a drupe, with as many one-seeded pyrenes as cells. Seeds pendulous; testa membranous; embryo very small in the apex of a fleshy albumen.—Trees or shrubs. Leaves alternate, simple, without stipules. Flowers small, in axillary umbels or cymes, rarely solitary or terminal. Fruits small.

The Order, limited to the large genus *Ilex*, and two small ones separated from it, is dispersed over the greater part of the world, but most abundant in America, very rare however in Africa, absent from New Zealand, and represented by one species only in Australia.

### 1. BYRONIA, Endl.

Petals and stamens 5 or more. Ovary-cells and pyrenes of the fruit 10 or more. Other characters and habit those of the Order.

Besides the Australian species, which is endemic, the genus only comprises two others, from the islands of the Pacific.

1. **B. Arnhemensis,** *F. Muell. Fragm.* ii. 119.   A shrub or tree, perfectly glabrous.   Leaves elliptical, obtuse or obtusely acuminate, 3 to 5 in. long, entire, coriaceous, shining above, narrowed into a petiole of ¼ to ½ in. Umbels few-flowered, on axillary or lateral peduncles of about ½ in., sometimes several in a short axillary leafless branch.   Flowers not seen.   Fruiting pedicels 3 or 4 lines long.   Fruit (not quite ripe) small, nearly globular, umbonate, the persistent calyx small, of 5 to 7 sepals.   Pyrenes about 12.

    **N. Australia.**   Valleys near Providence Hill, Arnhem's Land, *F. Mueller.*

## ORDER XXXIV. CELASTRINEÆ.

Flowers regular, hermaphrodite or polygamous.   Calyx small, persistent, 4- or 5-cleft, rarely 3- or 6- cleft.   Petals as many as calyx-segments, spreading, imbricate or rarely valvate in the bud.   Stamens as many as petals and alternate with them, inserted round the base or on the margin of the disk, or upon the disk itself; filaments usually short, incurved; anthers short, 2-celled, the cells in a few genera confluent into one.   Disk usually conspicuous, more or less fleshy, flat or broadly cup-shaped, or thick and conical, nearly free, or adnate to the base of the calyx or confluent with the ovary.   Ovary sessile on the disk, 2- to 5-celled, tapering to a short style with an entire or lobed stigma; ovules usually 2 in each cell, ascending with a ventral raphe, occasionally several, rarely 1 only, or pendulous with a dorsal raphe.   Fruit a capsule, berry, drupe, or samara, rarely divided into distinct carpels.   Seeds usually enveloped in an arillus, sometimes winged; albumen fleshy or almost horny or none; embryo usually rather large, with flat cotyledons and a short radicle next to the hilum.—Trees or shrubs, occasionally thorny, or woody climbers.   Leaves opposite or alternate, entire or toothed.   Stipules minute and very deciduous or none.   Flowers small, white or greenish, in axillary cymes or small racemes or in terminal panicles.

A considerable Order, dispersed over the greater part of the globe, more abundantly within the tropics than in temperate regions.   Of the six Australian genera one only is endemic, the others are all Asiatic, one extends to Africa and S. Europe but is not American, one is also tropical American but not hitherto found in Africa, and two are both in America and Africa.   The peculiar disk readily characterizes the greater number of genera, where that is wanting the insertion of the ovules and inferior radicle are the chief points separating *Celastrineæ* from *Ilicineæ*; from *Rhamneæ*, with which the real affinity is much closer, the stamens alternating with the petals is a constant distinctive mark.   The majority of *Celastrineæ* assume also when dry a peculiar pale-green colour, very rare in allied Orders.

    TRIBE I. **Celastreæ.**—*Stamens the same number as petals, inserted round the disk or on its margin.   Seeds albuminous.*

Leaves alternate.   Ovules 2 in each cell.   Capsule loculicidal, coriaceous.

    Flowers in racemes or panicles.   Stamens on the margin of the disk . . . . . . . . . . . . . . . . . .   1. CELASTRUS.

    Flowers in cymes.   Stamens under the disk . . . . . .   2. GYMNOSPORIA.

Leaves alternate.   Ovules 3 or more in each cell.   Capsule loculicidal, woody or bony.   Flowers in cymes.   Stamens on the margins of the disk . . . . . . . . . . . . . . . . . . . . .   3. DENHAMIA.

Leaves mostly opposite.   Ovules 2 in each cell.   Drupe indehiscent, 2- or 3-celled . . . . . . . . . . . . . . . . . . . .   4. ELÆODENDRON.

Leaves alternate. Ovules numerous in separate cells. Drupe indehiscent, with numerous pyrenes . . . . . . . . . . . 5. SIPHONODON.

TRIBE II. **Hippocrateæ.**—*Stamens usually 3, with a 5-merous calyx and corolla,
inserted on the disk ; filaments usually recurved at the top. Albumen none.*
Leaves opposite. Ovules 2 or several in each cell. Carpels distinct,
flat, 2-valved. Seeds winged . . . . . . . . . . . 6. HIPPOCRATEA.

TRIBE I. CELASTREÆ.—Stamens the same number as petals, inserted
round its disk or on its margin, the filaments usually incurved. Seeds albuminous.

## 1. CELASTRUS, Linn.

Flowers polygamous. Calyx 5-cleft. Petals 5, spreading. Disk broad,
concave. Stamens 5, inserted on the margin of the disk; filaments subulate,
flattened at the base ; anthers ovoid or oblong. Ovary not immersed in the
disk, 2- to 4-celled ; style usually short, the stigma lobed, spreading ; ovules
2, collateral, erect, the funicle cup-shaped. Capsule globular oblong or obovoid, coriaceous, 2- to 4-celled, opening loculicidally. Seeds 1 or 2 in each
cell, usually enveloped partially or wholly in a fleshy arillus, sometimes connecting the seeds in a mass, sometimes nearly or quite wanting ; testa membranous or almost crustaceous ; albumen fleshy ; cotyledons leafy.—Trees or
shrubs, often climbing, unarmed. Leaves alternate, petiolate, entire, or serrate. Stipules minute and deciduous, or none. Flowers small, in terminal
or axillary oblong panicles or racemes. Pedicels articulate. Bracts very
small.

The genus extends chiefly over tropical and eastern extratropical Asia, with 1 Mascarene
and a few N. American species. The Australian species are all endemic, although one is
nearly allied to a common Indian one.

Tall climber. Panicles terminal. Ovary 3-celled . . . . . . 1. *C. australis.*
Trees or tall shrubs. Racemes or pedicels lateral or axillary. Ovary
  2-celled.
  Leaves ovate or elliptical.
    Leaves quite entire, much narrowed into a long petiole.
      Flowers 5-merous . . . . . . . . . . . . . . . 2. *C. Muelleri.*
      Flowers 4-merous . . . . . . . . . . . . . . . 3. *C. dispermus.*
    Leaves entire or toothed, petiole short. Flowers 5-merous . 4. *C. bilocularis.*
    Leaves linear or narrow-lanceolate, entire . . . . . . . 5. *C. Cunninghamii.*

1. **C. australis,** *Harv. and Muell. in Trans. Phil. Soc. Vict.* i. 41. A
tall, woody, glabrous climber. Leaves from ovate-lanceolate to oblong-elliptical or lanceolate, acuminate, 2 to 4 in. long, entire or minutely and usually
remotely serrate, narrowed into a petiole of 1 to 3 lines. Panicles terminal,
or rarely in the upper axils, narrow, loose, rarely above 2 in. long. Flowers
white. Calyx-lobes broad, rounded, ciliate. Petals twice as long, attaining
a little more than 1 line, broadly ovate or orbicular. Disk almost free from
the calyx. Ovary 3-celled ; style short, with 3 spreading stigmatic lobes.
Capsule nearly globular, rarely exceeding 3 lines diameter. Seeds enveloped
in a fleshy arillus.—Reissek, in Linnæa, xxix. 265 ; F. Muell. Fragm. iii. 94.

**Queensland.** Burnett and Dawson rivers and Moreton Bay, *F. Mueller.*
**N. S. Wales.** Port Jackson, *R. Brown* ; northward to Clarence river, *Beckler,*

*Wilcox;* New England, *C. Stuart;* southward to Illawarra, *A. Cunningham, Backhouse, M'Arthur.*

**Victoria.** Moist forests on the Snowy and Buchan rivers, *F. Mueller.*

The species differs slightly from the E. Indian *C. paniculatus,* Willd., in the narrower and more acuminate, not obovate leaves, usually more coriaceous, and in the rather smaller flowers and fruits.

2. **C. Muelleri,** *Benth.* Probably a tree, quite glabrous, flowering before the leaves are fully out. Branches apparently weak and slender. Leaves in our specimens still young, elliptical or broadly lanceolate, acutely acuminate, quite entire, narrowed into a rather long petiole. Flowers small, white, in simple lateral racemes of about ½ in., occasionally growing out into leafy branches. Pedicels 1 to 2 lines long, articulate about the middle, thickened under the flower. Calyx-lobes 5, ovate, half as long as the petals. Petals 5, oblong, about 1½ lines long. Disk broad, adnate to the calyx at the base only. Ovary 2-celled, tapering into a very short style, with 2 scarcely prominent stigmatic lobes. Adult leaves and fruits not seen.

**N. Australia.** Near Macadam Range, *F. Mueller.* I had at first thought that this might have been the flowering state of *C. dispermus,* but the flowers are constantly 5-merous.

3. **C. dispermus,** *F. Muell. in Trans. Phil. Inst. Vict.* iii. 31. A small glabrous tree. Leaves elliptical, obovate-oblong, or rarely broadly lanceolate, obtuse or slightly acuminate, 2 to 3 in. long, quite entire, much narrowed into a rather long petiole. Racemes axillary or lateral, not seen in flower, when in fruit 1 to 1½ in. long, the pedicels 1 to 2 lines. Persistent calyx very small, with 4 triangular lobes. Capsule obovoid or obcordate, slightly compressed, 3 to 4 lines long, 2-celled and 2-valved, with usually 2 seeds, covered at the base, according to F. Mueller, with a thick arillus, but I find no remains of it on our specimens; very rarely the capsule is 3-angled and 3-celled.

**Queensland.** Araucaria forests near Moreton Bay, *F. Mueller;* Port Denison, *Fitzalan.* Until the flowers have been seen, some doubts must remain as to the affinities of this species.

4. **C. bilocularis,** *F. Muell. in Trans. Phil. Inst. Vict.* iii. 31. A small much-branched glabrous tree. Leaves ovate, oblong, or broadly lanceolate, obtuse or slightly acuminate, 1½ to 2½ or very rarely 3 in. long, entire sinuate or bordered by acute teeth, rounded or cuneate at the base, on a short petiole. Racemes axillary or lateral, rarely 1 in. long. Pedicels 1 to 2 lines. Calyx-lobes 5, broad and short. Petals 5, ovate, about 1 line long. Ovary 2-celled; style exceedingly short, with 2 broad short spreading stigmatic lobes. Capsule 2-valved, coriaceous, pear-shaped or nearly globular, under 3 lines diameter. Seeds enclosed in a thin arillus.

**Queensland.** Dawson and Burnett rivers, *F. Mueller;* Brisbane and Logan rivers, *Fraser* (all with entire or slightly toothed leaves); Warwick, *Beckler* (with sharply toothed leaves).

5. **C. Cunninghamii,** *F. Muell. in Trans. Phil. Inst. Vict.* iii. 30. A tall shrub or small tree, quite glabrous and often somewhat glaucous. Leaves linear or narrow-lanceolate, mucronate, 2 to 3 in. long in some specimens, all under 1 in. in others, entire, rigid, the midrib alone prominent underneath.

Flowers small, in short loose axillary or lateral racemes, occasionally growing out into leafy-branches. Pedicels slender, 2 to 3 lines long. Calyx-lobes 5, orbicular, not ciliate. Petals broadly ovate, about 1 line long. Disk rather thick, but less so than in *Gymnosporia.* Ovary 2-celled, with a short style and 2 short spreading stigmatic lobes. Capsule globular or ovoid, 2 lines diameter, or rather more, 2-valved, 1- or 2-seeded. Seeds enclosed in a pulpy arillus.—*Catha Cunninghamii*, Hook. in Mitch. Trop. Austr. 387.

**N. Australia.** Victoria river, *F. Mueller;* islands of the Gulf of Carpentaria, *R. Brown.*

**Queensland.** Broad Sound, *R. Brown;* Moreton Bay ?, *A. Cunningham;* Rockhampton, *Thozet;* Warwick, *Beckler;* St. George's Bridge, *Mitchell.*

**N. S. Wales.** Port Jackson and Hunter's River, *R. Brown;* Hastings, Clarence, and Macleay rivers, *Beckler;* New England, *C. Stuart;* Blue Mountains, *Miss Atkinson;* Penrith and St. Aubyn's, *Backhouse;* Paramatta, *Woolls;* Lachlan river, *A. Cunningham.*

This and the three preceding species appear to have the erect habit but not the cymose inflorescence nor the thick disk of *Gymnosporia,* and the stamens always proceed from the margin of the disk.

## 2. GYMNOSPORIA, W. and Arn.

Calyx 4- or 5-cleft. Petals 4 or 5, spreading. Stamens 4 or 5, inserted under the disk; filaments subulate; anthers short. Disk broad, sinuate or lobed. Ovary attached by a broad base or partially immersed in the disk, 2- or 3-celled; style short; stigma 2- or 3-lobed; ovules 2 in each cell. Capsule obovoid or nearly globular, 2- or 3-celled, opening loculicidally. Seeds 1 or 2 in each cell, the arillus complete or imperfect, or sometimes wanting; testa coriaceous; albumen fleshy; cotyledons leafy.—Shrubs or small trees, the small branches often thorny. Leaves alternate, entire or serrate, without stipules. Flowers small, in dichotomous cymes, either axillary or on the old nodes.

The genus is widely diffused over the warmer regions of the Old World, one species being found as far north as Spain, and a few extending to the Pacific islands. The Australian species is an Indian and African one.

**1. G. montana,** *W. and Arn. Prod.* 159 (under *Celastrus*). A tall glabrous shrub or small tree, the smaller branches occasionally terminating in stout thorns. Leaves obovate, very obtuse, 1½ to 2½ or rarely 3 in. long, entire or minutely crenulate, narrowed into a petiole of 2 or 3 lines, membranous or thinly coriaceous, of a pale-green. Cymes 2 or 3 together in the axils or on the old nodes, rarely above 1 in. long, with slender dichotomous branches. Calyx-lobes 5, very short, broad, ciliate. Petals 5, obovate, about 1 line long. Ovary 3-celled; style very short, with 3 spreading stigmatic lobes. Capsule flat at the top, obtusely 3-angled, about 3 lines diameter in the Australian specimens, usually smaller in India. Arillus of the seeds cup-shaped.—*Celastrus montanus,* Roxb.; W. and Arn. l. c., with all the synonyms quoted; Wight, Ic. Pl. t. 382.

**Queensland.** Cape York, *M'Gillivray.* Common in the Indian Peninsula, and apparently the same as the tropical African *Celastrus senegalensis,* Lam.; I have seen no specimens from the Indian Archipelago. The Australian specimens are unarmed, but that is frequently the case with Indian ones, with which they agree in every respect except the larger capsules.

## 3. DENHAMIA, Meisn.

(Leucocarpon, *A. Rich.*)

Calyx 5-cleft. Petals 5. Stamens 5, inserted on the margin of the disk; filaments subulate; anthers ovate. Disk broadly cupular, rather thick. Ovary 1-celled, with 3, or rarely 4 or 5 parietal placentas, or completely divided into as many cells; style short, with as many stigmatic lobes as cells or placentas. Ovules 3 to 8 to each cell or placenta. Capsule ovoid or globular, opening in thick woody or bony valves, bearing the placentas or dissepiments in their centre. Seeds enclosed in a fleshy arillus; albumen fleshy; cotyledons flat. —Shrubs or small trees, glabrous and more or less glaucous. Leaves alternate, rigid, entire, or toothed. Flowers small, in few-flowered cymes or racemes.

The genus is exclusively Australian, and, on account of the parietal placentation of two species, has been by some referred to *Bixineæ*; but the disk, stamens, general habit, etc., are those peculiarly characteristic of *Celastrineæ*.

Ovary 1-celled; placentas (4- to 8-ovulate) not meeting in the axis.
  Veins of the leaves not very prominent.
    Flowers racemose. Style distinct . . . . . . . . . . 1. *D. oleaster.*
    Flowers in cymes or narrow panicles. Style very short, branched  2. *D. obscura.*
Ovary 3-celled, placentas (3- or 4-ovulate) united in the axis. Leaves
  prominently veined . . . . . . . . . . . . . . . 3. *D. pittosporoides.*

1. **D. oleaster,** *F. Muell. in Trans. Phil. Inst. Vict.* iii. 29. A tall shrub with slender branches. Leaves lanceolate, acute, or rarely obtuse, 2 to 3 in. long, entire or remotely toothed, narrowed into a very short petiole, coriaceous, the veins scarcely conspicuous. Flowers in short, simple, axillary or terminal racemes, the pedicels very rarely bearing 2 flowers. Calyx-segments broadly ovate or orbicular. Petals nearly 2 lines long. Disk thicker, and filaments longer than in the other two species. Ovary 1-celled, tapering into a style of at least $\frac{1}{2}$ line, the stigmatic lobes very short. Placentas 3, with 4 to 6 ovules to each. Fruit not seen.—*Melicytus (?) oleaster,* Lindl. in Mitch. Trop. Austr. 383.

**Queensland.** St. George's Bridge, Balonne river, *Mitchell.*

2. **D. obscura,** *Meisn. in Walp. Rep.* i. 203. A tall shrub or small tree, the young branches generally pendulous. Leaves mostly oblong-lanceolate, acuminate, 2 to 3 in. long, entire, with often wavy margins, narrowed into a rather long petiole, coriaceous, finely but not prominently veined; on barren branches the leaves are sometimes broadly ovate and bordered by coarse prickly teeth like those of a Holly. Flowers in small pedunculate cymes in the upper axils, or forming a short oblong terminal panicle. Calyx-segments ovate. Petals rather broad, $1\frac{1}{2}$ lines long. Ovary 1-celled, with 3 to 5 placentas; style very short, with 3 to 5 oblong-linear stigmatic branches. Ovules 4 to 8 to each placenta. Capsule ovoid or globular, attaining about 1 in., of a pale-whitish hue when dry, the thick valves bearing slightly projecting placentas along their centre.—*Leucocarpon obscurum,* A. Rich. Sert. Astrol. 46, t. 18; *Denhamia xanthosperma,* F. Muell. Trans. Phil. Inst. iii. 28, and *D. heterophylla,* F. Muell. l. c. 29.

**N. Australia.** York Sound, N.W. coast, *A. Cunningham;* Melville Island (not

402 XXXIV. CELASTRINEÆ. [*Denhamia.*

Moreton Bay), *Fraser ;* Victoria river and Arnhem's Land, *F. Mueller ;* Port Essington, *Armstrong.*

**Queensland,** *Mitchell ;* Broad Sound, *R. Brown ;* Newcastle range, between Gilbert and Burdekin rivers, *F. Mueller.*

3. **D. pittosporoides,** *F. Muell. in Trans. Phil. Inst. Vict.* iii. 30. A tree, the trunk, according to Thozet, beautifully striated. Leaves lanceolate or rarely ovate-lanceolate, obtuse, 2 to 3 or rarely 4 in. long, obtusely serrate, narrowed into a petiole, coriaceous, with very prominent pinnate and reticulate veins, not so glaucous as in the other two species. Cymes pedunculate, few-flowered, on short leafless branches on the old wood or at the base of young leafy branches. Calyx-segments broadly orbicular. Petals ovate, about 1 line long, rather thick at the base. Ovary fleshy, completely 3-celled, with 3 or 4 ovules in each cell. Capsule globular, attaining in our specimens ½ in. or rather more, but many of them opening when not half that size, the thick woody valves bearing the dissepiments on their centre.

**Queensland.** Wide Bay, *Bidwill ;* sources of the Burnett river, *C. Moore ;* Rockhampton, *Thozet ;* Warwick, *Beckler ;* Keppel Bay and Fitzroy river, *Herb. F. Mueller.*

4. **ELÆODENDRON,** Jacq. f.

Flowers often polygamous. Calyx 4- or 5-cleft, rarely 3-cleft. Petals as many as calyx-segments, spreading. Disk thick. Stamens as many as petals, inserted under the edge of the disk; filaments short; anthers nearly globular. Ovary continuous with the disk, conical, 3-celled, rarely 2- or 4- or 5-celled; style very short; ovules 2 in each cell. Drupe succulent or nearly dry, the putamen hard, 1–2- or 3-celled. Seeds usually solitary, without any arillus; testa membranous or spongy; albumen scanty or copious, cotyledons flat.— Shrubs or small trees, usually quite glabrous. Leaves opposite or alternate, entire or crenate. Flowers small, in dichotomous cymes, usually axillary or lateral, often clustered.

The species are numerous in East India and southern Africa, with a very few in tropical America; none are known from tropical Africa. The two Australian ones are endemic.

Ovary 2-celled. Drupe red. Veins of the leaves scarcely conspicuous above . . . . . . . . . . . . . . . . . 1. *E. australe.*
Ovary 3-celled. Drupe black. Veins of the leaves conspicuous on both sides . . . . . . . . . . . . . . . . . 2. *E. melanocarpum.*

1. **E. australe,** *Vent. Jard. Malm. t.* 117. A glabrous, small or middle-sized tree. Leaves opposite, or here and there alternate, ovate, obovate, elliptical, or oblong-lanceolate, obtuse or obtusely acuminate, 2 to 4 in. long, entire or broadly crenate, narrowed into a very short petiole, coriaceous, the reticulate veins slightly prominent underneath and scarcely conspicuous above. Flowers 4-merous, in slender cymes, much shorter than the leaves. Calyx-segments broadly ovate. Petals from a little more than 1 line to nearly 2 lines long, ovate, often broadly and shortly 3-lobed, Ovary confluent with the disk in a conical mass, 2-celled; style either very short or attaining ¾ line. Drupe ovoid or globular, rarely above ½ in. long, of a bright-red colour, which it often retains in the dried specimens. Putamen hard and woody, usually 1-seeded, but showing the traces of the abortive cell.

Albumen copious.—F. Muell. Fragm. iii. 61 ; *Portenschlagia australis*, Tratt. Arch. t. 250.

**Queensland.**   Wide Bay and Moreton Bay, *C. Moore ;* Ipswich, *Nernst.*

**N. S. Wales.**   Hunter's River, *R. Brown;* Hastings, Macleay, and Clarence rivers, *Beckler ;* Illawarra, *A. Cunningham* and others; Kiama, *Harvey.*

Var. *angustifolia.*   Leaves lanceolate or narrow-oblong, entire or nearly so ; fruit more ellipsoid.—*Portenschlagia integrifolia,* Tratt. Arch. t. 284 ; *Elæodendron integrifolium,* G. Don, Gen. Syst. ii. 12.—Burnett, Dawson, and Pine rivers, in Queensland, *F. Mueller ;* Warwick, *Beckler.*

According to F. Mueller, the fruit in *E. australe* is occasionally 3-celled ; but this must be rarely the case, as I have never found more than 2 cells to the ovary in any of the numerous specimens I have examined.   The above references to Trattinick's Archiv are quoted after G. Don ; I do not find the second volume of that work in any of our libraries.

2. **E. melanocarpum,** *F. Muell. Fragm.* iii. 62.   A glabrous tree. Leaves opposite, obovate or oval-elliptical, broadly crenate, scarcely to be distinguished from those of *E. australe,* except that the veins are more conspicuous on the upper as well as the lower side.   Flowers smaller than in *E. australe,* the males more numerous, in slender cymes like those of the small-flowered Indian *Hippocrateas,* usually 3-merous.   Female flowers in less-branched cymes and often 4-merous.   Ovary 3-celled, but very imperfect in the flowers examined.   Drupe ovoid or globular, shining-black, rather larger than in *E. australe,* the hard putamen always 3-celled, or showing the traces of a second or third cell when reduced to one.   Albumen copious.

**N Australia.**   Arnhem North Bay, *R. Brown.*

**Queensland.**   Keppel Bay, *R. Brown ;* Port Bowen, *A. Cunningham ;* Fitzroy and Lizard Islands, *M'Gillivray ;* Port Denison, *Fitzalan ;* Rockhampton, *Thozet.*


## 5. SIPHONODON, Griff.

Calyx 5-cleft.   Petals 5, spreading.   Disk not distinct from the base of the calyx.   Stamens 5, connivent round the pistil, the filaments flattened. Ovary half immersed in the disk or base of the calyx, conical, the summit hollowed and stigmatic in the cavity round a central style-like column ; cells numerous, in 2 to 4 series ; ovules solitary in each cell, alternately ascending and pendulous.   Drupe globular, hard-fleshy, with numerous 1-seeded bony pyrenes superposed in rings of about 10 round the central axis.   Testa of the seed membranous ; albumen almost horny ; cotyledons large, flat ; radicle short.—Glabrous trees.   Leaves alternate, entire or crenate.   Stipules minute, deciduous.   Peduncles short, axillary, few-flowered.

Besides the Australian species, which is endemic (and referred to this genus from the fruit), it comprises only one from the Indian Archipelago, from which the floral characters are taken.

1. **S. (?) australe,** *Benth.*   A tree of 40 ft. or more.   Leaves obovate or broadly oblong, obtuse, 2 to 3 in. long, entire or slightly sinuate, coriaceous, drying of the pale colour so frequent in *Celastrineæ.*   Flowers unknown. Peduncles very short, bearing 1 or 2 fruits on pedicels of $\frac{1}{4}$ to $\frac{1}{2}$ in., as in *S. celastrineus,* Griff.   Drupe globular, $\frac{3}{4}$ to 1 in. diameter, the flesh hard and dry, with the stigmatic scar at the top, and the scar of the calyx at the base, as in *S. celastrineus.*   Nuts numerous, appearing to have been arranged in 2 rows in each of 5 cells, irregularly ovoid, somewhat compressed, 3 to 4

lines long.   Testa of the seed brown; albumen not very thick; cotyledons broadly ovate.

**Queensland.**   Brisbane river, *A. Cunningham.*
**N. S. Wales.**   Clarence river, *Beckler.*

Until the flowers have been seen, this plant must remain in some measure doubtful, but the habit and fruit are so nearly those of *S. celastrineus,* that I have little hesitation in referring it to that genus.   The ovary must probably be considered as 5-celled with many ovules in each cell, separated by spurious transverse dissepiments.

TRIBE II.   HIPPOCRATEÆ.—Stamens usually 3 only, with a 5-merous calyx and corolla, inserted on the disk itself; filaments usually incurved at the base but recurved under the anther, which thus opens outwards.   Seeds without albumen.

## 6.  HIPPOCRATEA, Linn.

Calyx small, 5-cleft.   Petals 5, valvate or imbricate.   Stamens usually 3, the filaments thick at the base, connivent round the ovary, recurved at the top; anthers at first divided into 2 or 4 cells, at length confluent into 1 transverse cell.   Disk conical or broad.   Ovary 3-celled, style short, stigma 3-lobed; ovules 2 or more in each cell.   Fruit of 3 distinct, flat, coriaceous carpels, opening along the middle in 2 boat-shaped valves.   Seed compressed, usually produced at the base into a wing adnate to the raphe; albumen none; embryo in the upper end of the seed; cotyledons flat, connate; radicle inferior.—Small trees or woody climbers.   Leaves opposite, entire or serrate. Stipules very small and deciduous.   Flowers in axillary cymes or panicles.

A large genus, widely distributed over tropical Asia, Africa, and America, the Australian species being one of the common Asiatic ones.   It belongs to the section with comparatively large flowers and valvate petals.   The other section common in India, including *H. indica,* with minute globular flowers and imbricate petals, has not yet been observed in Australia.

1. **H. obtusifolia,** *Roxb.; W. and Arn. Prod.* 104, var. *barbata.*   A tall, woody, glabrous climber.   Leaves ovate, obovate, or oblong, obtuse or obtusely acuminate, 2 to 4 in. long, entire, coriaceous, somewhat shining. Flowers in short, loose, axillary cymes, the upper ones forming sometimes large leafy terminal panicles.   Petals fully 2 lines long, lanceolate, rather thick, valvate in the bud, and in the Australian specimens bearded inside above the middle, the disk and ovary also occasionally villous or pubescent.   Ovules 6 to 10 in each cell of the ovary.   Carpels about 2 in. long, either broadly oblong and entire or broader and emarginate at the top.—*H. macrantha,* Korth. Verhand. Nat. Gesch. Bot. 187, t. 39; *H. barbata,* F. Muell. in Trans. Phil. Inst. Vict. iii. 23.

**Queensland.**   Moreton Bay, *W. Hill, F. Mueller.*
**N. S. Wales.**   Clarence river, *Beckler.*   The species is widely distributed over tropical Asia.   The common Indian form, figured in Wight, Ic. t. 963, has glabrous petals, but the variety with bearded petals as described by Korthals from Borneo, and of which we have specimens from Ceylon, is the same as the Australian one; and the amount of hairiness both on the petals and ovary appears to be variable.

ORDER XXXV.  **STACKHOUSIEÆ.**

Flowers regular, hermaphrodite.   Calyx small, 5-lobed or 5-cleft.   Petals

5, perigynous, with elongated claws, usually free at the base, but united upwards in a tubular corolla, with spreading lobes, imbricate in the bud. Disk thin, lining the calyx-tube. Stamens 5, inserted on the margin of the disk; filaments free, slender; anthers oblong. Ovary free, 2- to 5-lobed, 2- to 5-celled; style single, with 2 to 5 lobes, stigmatic along the inner side. Ovules solitary in each cell, erect, anatropous. Fruit of 2 to 5 globular, angular, or winged indehiscent cocci, at length seceding from the axis. Seeds solitary, erect; testa membranous; albumen fleshy; embryo straight; cotyledons short; radicle inferior.—Herbs, usually forming a perennial stock, with erect, little branched, virgate stems, often assuming a yellowish colour, rarely dwarf and tufted. Leaves alternate, narrow, entire, often somewhat fleshy. Stipules none or very minute. Flowers in terminal spikes, rarely solitary, with 3 minute or linear bracts (1 bract and 2 bracteoles) at their base. Stamens included in the corolla-tube, of very unequal lengths. Pistil almost always 3-merous.

The Order is limited to a single genus, almost endemic in Australia, one species extending to the Philippine Islands, and another represented by a closely allied species in New Zealand.

## 1. STACKHOUSIA, Sm.

(Tripterococcus, *Endl.*; Plokiostigma, *Schuch.*)

Characters and distribution those of the Order.

Corolla-lobes oblong, obtuse.
Flowers solitary, terminal, sessile among the leaves of dwarf tufted stems . . . . . . . . . . . . . . . . . 1. *S. pulvinaris.*
Stems elongated. Spikes terminal.
Cocci acutely angled or winged. Leaves obovate or obovate-oblong 2. *S. spathulata.*
Cocci obovoid or globular, reticulate. Leaves lanceolate, linear or filiform.
Spikes dense at the top, usually interrupted as the flowering advances. Flowers 4 to 6 lines long.
Leaves flat, lanceolate or linear or rarely terete. Bracts small 3. *S. monogyna.*
Leaves very narrow or terete. Bracts filiform.
Spikes or the whole plant pubescent. . . . . . . . 4. *S. pubescens.*
Glabrous except sometimes the cocci . . . . . . . 5. *S. Huegelii.*
Spikes short, dense. Flowers about 3 lines long . . . . . 6. *S. flava.*
Spikes filiform. Flowers distant, not 3 lines long. Leaves narrow, often very few . , . . . . . . . . . 7. *S. muricata.*
Corolla lobes acute or acuminate.
Cocci obovoid or globular, reticulate. Corolla 3 lines or less.
Spikes short, dense. Leaves linear . . . . . . . . 6. *S. flava.*
Spikes long and slender. Flowers or clusters of flowers distant.
Leaves oblong or linear, sometimes few or very small . . . . 8. *S. viminea.*
Flowers few, solitary along the broom-like branches. Leaves all reduced to minute scales . . . . . . . . . . . . 9. *S. scoparia.*
Cocci broadly winged. Corolla more than 4 lines, with filiform points to the lobes . . . . . . . . . . . . . . . . 10. *S. Brunonis.*

1. **S. pulvinaris,** *F. Muell. in Trans. Phil. Soc. Vict.* i. 101; *Fragm.* ii. 359, iii. 88; *and Pl. Vict.* ii. *t.* 14. A dwarf, glabrous, much branched, and densely tufted or prostrate herb. Leaves crowded, linear-oblong, obtuse, rather thick, usually 3 or 4 lines long. Flowers solitary and almost sessile amongst the last leaves, and but little exceeding them. Bracts very small, obtuse. Calyx-lobes ovate. Corolla about 3 lines long, with oblong obtuse

lobes, a little shorter than the tube.  Anthers glabrous.  Cocci rather large in proportion to the plant, smooth or obscurely reticulate.—Hook. f. Fl. Tasm. ii. 359.

**Victoria.**  Summits of the higher mountains of Gipps' Land, at an elevation of 6000 to 7000 ft., *F. Mueller.*

**Tasmania.**  Western mountains, *Archer.*

*S. minima*, Hook. f., from New Zealand, differs very slightly in the acute lobes of the corolla and pubescent anthers.

2. **S. spathulata,** *Sieb. in Spreng. Syst. Cur. Post.* 124.  Glabrous, usually much branched at the base, with stout decumbent or ascending branches of about ½ ft., but sometimes lengthening to 1 ft. or more.  Leaves from obovate to oblong, usually very obtuse, rather thick, and ½ to ¾ in. long, but in luxuriant stems lengthening out to 1 in. or more and almost acute. Spikes dense, with the flowers almost of *S. monogyna.*  Corolla-tube 3 to 4 lines long, lobes much shorter, oblong, obtuse.  Cocci fully 2 lines long, with 3 prominent vertical acute angles or narrow wings.—F. Muell. Fragm. iii. 86 ; *S. maculata*, Sieb. in Hook. Journ. Bot. ii. 421 ; Hook. f. Fl. Tasm. i. 79 (the name originating in a clerical error in Sieber's label) ; *Tripterococcus spathulatus*, F. Muell. in Hook. Kew Journ. viii. 208 ; Schuch. in Linnæa, xxvi. 20 ; F. Muell. Fragm. iii. 86 ; *S. monogyna*, Labill. Pl. Nov. Holl. i. 77, t. 104 (as to the fruit).

**Queensland.**  Sandy Cape, Hervey Bay, *R. Brown ;*  Moreton Island, *M'Gillivray, F. Mueller.*

**N. S. Wales.**  Southward of Botany Bay, *R. Brown ;*  Port Jackson, *Sieber, n.* 246, and others ; frequent on the seashore, *A. Cunningham ;* and on all the grass-lands of the interior, *Fraser* (but probably confounded with *S. monogyna*) ; Hastings river, *Beckler.*

**Victoria.**  Seacoast, Wilson's Promontory, Portland Bay, etc., *F. Mueller.*

**Tasmania.**  Islands of Bass's Straits, *Gunn, Bynoe.*  A specimen not in fruit from Recherche Bay, *C. Stuart*, is also probably the same.

**S. Australia.**  Mouth of the Glenelg and Rivoli Bay, *Allitt.*

3. **S. monogyna,** *Labill. Pl. Nov. Holl.* i. 77, *t.* 104 (partly).  Glabrous, with a perennial base, and erect, simple or slightly branched, stout or slender stems, usually 1 to 1½ ft., but sometimes twice that height.  Leaves linear or lanceolate, acute or obtuse, crowded or few and distant, usually ½ to 1 in. long, or when very luxuriant 2 in.  Racemes at first dense, but often lengthening out to 4 or 5 in., the lower bracts sometimes leaf-like, passing into the very small lanceolate upper ones, and often all very small.  Calyx-lobes narrow.  Corolla-tube 3 to 4 lines long ; lobes much shorter, oblong, obtuse.  Cocci obovoid, prominently reticulate, not angled.—Lindl. Bot. Reg. t. 1917 ; Hook. f. Fl. Tasm. i. 79 ; *S. obtusa*, Lindl. in Bot. Reg., under n. 1917 ; *S. linariæfolia*, A. Cunn. in Field. N. S. Wales, 356 ; F. Muell. Fragm. iii. 87 ; *S. Gunnii*, Hook. f. Fl. Tasm. i. 79 ; Schlecht. Linnæa, xx. 642 ; *S. aspericocca*, Schuch. in Linnæa, xxvi. 12 ; *S. Muelleri*, Schuch. l. c. 16 ; *S. Gunniana*, Schlecht. in Schuch. l. c. 18.

**Queensland.**  Keppel Bay, Broad Sound, *R. Brown ;*  Port Curtis, *M'Gillivray ;* Dawson and Bowen rivers, *F. Mueller.*

**N. S. Wales.**  Richmond and Grose river, *R. Brown ;*  Blue Mountains, and plains and country about Bathurst, also southward of Port Jackson, *A. Cunningham* and others ; Twofold Bay, *F. Mueller.*

**Victoria.** Common in fertile as well as in sterile soils, ascending in the Alps to 4500 ft., *F. Mueller.*

**Tasmania.** Derwent river, *R. Brown ;* abundant throughout the island, *J. D. Hooker.*

**S. Australia.** From the Murray to Spencer's Gulf, and in the interior to Lake Torrens, *F. Mueller.*

Although Labillardière confounded this species with *S. spathulata,* and represented and described the fruit of the latter species, yet the common one, of which he described the flowering specimens, has been so universally known under his name, that it would only increase the confusion to adopt a later name for that species. Among its numerous forms, the luxuriant specimens with more conical spikes which commonly pass for the true *S. monogyna,* and the smaller ones with fewer flowers and the young spike more obtuse, published by Lindley as *S. obtusa,* pass into each other by innumerable gradations. It is to the former that Schlechtendal gave the name of *H. Gunnii,* whilst Hooker's variety of that name is nearer to *H. obtusa.* A rather more distinct variety, with elongated slender stems, narrow and more distant leaves, sometimes very few and small, and rather smaller flowers, with smaller and smoother cocci, is amongst the more common Victorian and S. Australian forms, and is more especially the *S. linariæfolia,* A. Cunn., or *S. Muelleri,* Schuch. It has sometimes the almost terete leaves of *S. Huegelii,* from which it then differs in its very short bracts. The calyx in this variety is often strongly ribbed after flowering, but still more so in a slender northern variety, which has larger almost muricate cocci. A few Queensland specimens (Port Denison, *Fitzalan*), very slender, with small flowers in short dense spikes, seem almost to connect this with *S. muricata.* Indeed, different as are the extreme forms, the numerous specimens I have had before me show scarcely any definite limits between *S. monogyna, pubescens, Huegelii, flava, muricata,* and *viminea.*

4. **S. pubescens,** *A. Rich, Sert. Astrol.* 89, *t.* 33. Stems usually erect, nearly simple, 1 to 1½ ft. high, glabrous or pubescent. Leaves very narrow-linear, often 1 in. long in the lower part of the plant, glabrous or pubescent. Spike at first dense and conical, elongating to 2 or 3 in., always pubescent. Bracts linear, subulate-acuminate, usually exceeding the young buds. Calyx-lobes acuminate, usually strongly ciliate. Corolla of the size and shape of that of *S. monogyna,* with oblong obtuse lobes. Cocci strongly reticulate, usually pubescent.—Bunge, in Pl. Preiss. i. 180; Schuch. in Linnæa, xxvi. 10; *Plokiostigma Lehmanni,* Schuch. l. c. 40 (young buds, with the style not yet grown out).

**W. Australia.** King George's Sound, *R. Brown, Lesson, Oldfield ;* Swan River, *Drummond, Preiss, n.* 1972, and others; Rottenest Island, *Preiss, n.* 1364.

5. **S. Huegelii,** *Endl. in Hueg. Enum.* 17. Glabrous, with erect nearly simple stems of ½ to 1½ ft., with a terminal spike at first dense, afterwards elongated as in *C. monogyna,* and the flowers about the same size, with oblong, obtuse corolla-lobes ; but the leaves are very narrow-linear, often almost terete, and the bracts and calyx-lobes also very narrow, as in *S. pubescens,* from which this species differs slightly in the want of any pubescence, excepting sometimes in the cocci.—Schuch. in Linnæa, xxvi. 14.

**W. Australia.** Swan River, and northward to Murchison river, *Drummond, Oldfield,* and others ; King George's Sound, *R. Brown ;* Kalgan river, *Oldfield ;* Stirling ranges, *Maxwell.* This ought perhaps to be considered as a variety only of *S. pubescens.*

6. **S. flava,** *Hook. Ic. Pl. t.* 269. Glabrous. Stems numerous, branching at the base, decumbent or ascending to from 6 in. to 1 ft. in height. Leaves linear, flat, rarely above ½ in. long, rather thick, those of the short sterile branches sometimes broader and oblong. Flowers yellow, much smaller than in *S. monogyna,* clustered in short, dense, terminal spikes, the pedicels

often ½ line long.  Bracts very short, broad and obtuse.  Calyx small, with ovate lobes.  Corolla about 3 lines long, with oblong-lanceolate, rather acute lobes.  Cocci not seen.—Hook. f. Fl. Tasm. i. 80; Schuch. in Linnæa, xxvi. 26.

**Tasmania.**  Woolnorth, in poor sandy soil, *Gunn.*
**W. Australia.**  Flinders Bay, *Collie* (with the spike rather more elongated).

7. **S. muricata,** *Lindl. in Bot. Reg. under n.* 1917.  Glabrous.  Stems slender, simple or branched, often above 1½ ft. long.  Leaves narrow-linear, sometimes almost filiform, ½ to 1½ in. long.  Spikes long, very slender, with distant clusters of 2, 3, or more small flowers, usually under 3 lines and sometimes not 2 lines long.  Calyx-lobes small, obtuse.  Corolla-lobes narrow but obtuse, sometimes as long as the tube, sometimes not half so long.  Cocci strongly reticulate, sometimes almost muricate.—Schuch. in Linnæa, xxvi. 25.

**N. Australia.**  Sturt's Creek, *F. Mueller.*
**Queensland.**  Port Essington, *Armstrong ;* Port Curtis and Dunk Island, *M'Gillivray ;* Brigalow scrub in the interior, *Mitchell ;* Peak Downs, *F. Mueller.*
**N. S. Wales.**  St. George's river, *R. Brown ;* Peel's Range on the Lachlan, *A. Cunningham.*

This species, which we have also from the Philippine Islands, varies considerably and sometimes approaches *S. viminea,* but the leaves are never so broad, and the corolla-lobes obtuse.  The Sturt's Creek specimens belong to a more branched and compact form, with very small flowers more frequently solitary, and the leaves few, small, and distant.  Some smaller specimens, like those from the Philippine Islands, are less branched and perhaps sometimes annual.

8. **S. viminea,** *Sm. in Rees' Cycl.* xxxiii.  Glabrous.  Stems erect or ascending, slender, often 1 to 1½ ft. high.  Leaves on the barren shoots often rather broad, oblong, obtuse, ½ to 1 in. long, narrowed at the base, on the flowering-stems fewer, often small and narrow-linear, and sometimes scarcely any.  Spike slender, elongated, with distant clusters of small flowers, sometimes numerous in the clusters, sometimes solitary or nearly so.  Calyx small, with acute lobes.  Corolla rarely exceeding 3 lines and often not above 2 lines long, slender, with narrow acuminate or acute lobes.  Cocci small, strongly reticulate or muricate.—Schuch. in Linnæa, xxvi. 22 ; *S. nuda,* Lindl. in Bot. Reg. under n. 1917; Schuch. l. c. 22 ; *S. monogyna,* Sieb. Pl. Exs. ; *S. dorypetala,* Schuch. l. c. 24.

**N. Australia.**  Islands of the Bay of Carpentaria, *R. Brown ;* Goulburn Island, *A. Cunningham.*
**Queensland.**  Warwick, *Beckler.*
**N. S. Wales.**  Port Jackson and to the southward, *R. Brown, A. Cunningham, Sieber, n.* 245 and 591, and others; Blue Mountains, *Miss Atkinson ;* New England, *C. Stuart ;* Macleay and Clarence rivers, *Beckler.*
**W. Australia.**  Swan River, *Drummond, n.* 92; Phillips river, *Maxwell ;* between Moore and Murchison rivers, *Drummond, n.* 81.

Var. *elata.*  Branches numerous and more erect, attaining 5 ft. according to Maxwell, but several of Drummond's are under 1 ft.; leaves all narrow; the whole plant drying more yellow than usual in the eastern variety, although some specimens of the latter are also yellow.—*S. elata,* F. Muell. Fragm. iii. 86.  To this variety belong Maxwell's specimens above mentioned and Drummond's n. 92.  A few Port Jackson ones can scarcely be distinguished from them.

Var. *micrantha.*  Small, slender, and much-branched; flowers small, as in *S. muricata,* but the acuminate lobes as well as the narrow leaves are those of *S. viminea.*—To this are referrible Drummond's specimens, n. 81, and R. Brown's and Cunningham's from the N. coast.

The distinction between this species and *S. muricata*, and the value of the character derived from the acute or obtuse corolla-lobes, requires further investigation on the living plant.

9. **S. scoparia,** *Benth.* Glabrous, erect, with numerous stout, rigid, broom-like, apparently leafless branches, 8 to 10 in. high in our specimens. Leaves all reduced to minute distant scales. Flowers small, solitary and distant along the ends of the branches, shortly pedicellate, with minute bracts. Calyx-lobes narrow and acute. Corolla about 2½ lines long, with narrow acuminate lobes about as long as the tube. Cocci not seen.

**W. Australia.** Between Swan River and King George's Sound, *Drummond.*

10. **S. Brunonis,** *Benth.* Glabrous. Stems erect, simple or branched, attaining 1 to 2 ft. or even more. Leaves narrow-linear or almost terete, usually free and small, except at the base of some of the stems, rarely more generally scattered and attaining ½ to 1 in. Spikes sometimes short and crowded, but more frequently elongated, with rather distant shortly pedicellate flowers. Bracts subulate, very variable in length. Calyx-lobes narrow-linear or acuminate. Corolla-tube slender, usually about 3 lines long, but varying from 2½ to 3½ lines; lobes narrow, acuminate, often almost subulate, as long as the tube or much shorter. Cocci with 3 longitudinal scarious wings, marked with transverse veins, the 2 marginal ones from 1 to 2 lines broad, the dorsal one much narrower,.but all remarkably variable in width even on the same specimen.—*Tripterococcus Brunonis,* Endl. in Hueg. Enum. 18; Schuch. in Linnæa, xxvi. 31; *T. simplex,* Bunge, in Pl. Preiss. i. 181; Schuch. l. c. 35; *T. junceus,* Bunge, l. c. 181; Schuch. l. c. 37; *T. brachystigma,* Schuch. l. c. 33.

**N. Australia.** Regent river, N.W. coast, *A. Cunningham.*

**W. Australia.** King George's Sound, *R. Brown, Fraser,* and others, to Swan River and Murchison river, *Drummond, Oldfield,* and others; *Preiss, n.* 1971 and 1973.

## ORDER XXXVI. RHAMNEÆ.

Flowers regular, hermaphrodite, or rarely polygamous. Calyx campanulate, urceolate, or cylindrical, the tube persistent and often adnate to the ovary or disk; lobes 4 or 5, valvate, usually with a raised longitudinal line inside and deciduous. Petals 4 or 5, concave or hood-shaped, inserted at the base of the calyx-lobes, alternating with and rarely exceeding them, or none. Stamens 4 or 5, alternating with the calyx-lobes, inserted with the petals and opposite to them when present; filaments short, filiform; anthers small, often enclosed in the petals, rarely oblong or exserted. Disk rarely wanting, usually filling the calyx-tube or lining it, or annular round the ovary when inferior, rarely cup-shaped and free. Ovary sessile on the disk or immersed in it, or more or less inferior, 3-celled, or rarely 2- or 4-celled; style short, entire, or with as many lobes or branches as ovary-cells; stigmas terminal, capitate or club-shaped. Ovules solitary in each cell, erect, anatropous, with a dorsal or rarely lateral raphe. Fruit a drupe or capsule, the border of the adnate base of the calyx forming a ring at the base or round the fruit or at the summit;

epicarp thin and dry or fleshy; endocarp separating into as many membranous coriaceous or hard cocci as cells, or woody or bony, divided into cells. Seeds solitary, erect, usually ovate and somewhat compressed, often arillate; testa coriaceous or crustaceous and shining or rarely membranous; albumen fleshy or almost horny, often scanty, rarely wanting; embryo usually straight, with flat rather thick cotyledons and a short inferior radicle.—Shrubs or trees, very rarely, in genera not Australian, herbs, erect or climbing. Leaves alternate or rarely opposite, undivided, entire, or toothed. Stipules usually present but very deciduous, rarely spinous and persistent. Flowers small, usually green or yellowish, in cymes or umbel-like clusters, either solitary or forming axillary or terminal compound cymes, racemes or panicles.

A considerable Order, ranging over the tropical and temperate regions of both the New and the Old World. Of the 12 Australian genera, 3 are widely spread tropical or northern genera, and 1 tropical Asiatic, all represented in Australia by single or very few species, a fifth is South American, with one Australian and one New Zealand species, the remaining 7, several of them numerous in species, are endemic or nearly so; *Alphitonia* extending to the Pacific islands, and *Pomaderris* to New Zealand. The Order is a well-marked one, the floral characters separating it very readily from all except *Ampelideæ*, from which it is distinguished by the habit, by the drupaceous or capsular, not baccate fruit, and by the seeds; but most of the genera, even the most natural ones, are difficult to characterize. The differences in their flowers and fruits are very trifling; they often pass into each other by the finest gradations, and habit, foliage, and inflorescence must often be relied upon for fixing generic limits.

Calyx spreading. Disk broad, concave or filling the calyx-tube. Ovary free or immersed in the disk. Leaves usually alternate, rather large, often serrate. Fruit above 2 lines long or broad, succulent or dry.
Leaves 3- or 5-nerved.
Drupe succulent, the putamen woody or bony, 1- to 4-celled. Stipules usually spinescent . . . . . . . . . . .  2. ZIZYPHUS.
Drupe with a thin epicarp, covering membranous or crustaceous cocci. Unarmed . . . . . . . . . .  4. COLUBRINA.
Leaves penninerved.
Panicle branches elongated and raceme-like. Nut 1-seeded, produced into a long wing-like appendage . . . . .  1. VENTILAGO.
Panicle or cyme 2–3-chotomous. Endocarp separating into cocci.
Ovary immersed in the disk. Epicarp thick. Leaves white or rusty underneath . . . . . . . . . .  5. ALPHITONIA.
Ovary sessile on the disk. Epicarp thin. Leaves green on both sides . . . . . . . . . . . . . .  6. EMMENOSPERMUM.
Flowers in axillary clusters. Ovary sessile on the disk. Epicarp succulent . . . . . . . . . . . . .  3. RHAMNUS.
Calyx campanulate or tubular. Disk none, or annular, or lining the calyx-tube. Ovary partially or wholly inferior. Leaves alternate, usually small and entire (except a few *Pomaderrises*). Fruit under 2 lines diameter.
Calyx-tube entirely adnate, or lined by the disk up to the lobes.
Petals none, or concave, not enclosing the anthers, which are either oblong or on long filaments. Flowers usually pedicellate. Bracts very deciduous . . . . . . . .  7. POMADERRIS.
Petals enclosing the small anthers. Flowers pedicellate. Bracts very deciduous . . . . . . . . . . . .  8. TRYMALIUM.
Petals enclosing the small anthers. Flowers sessile, surrounded by small, imbricate, persistent, brown bracts . . . . .  9. SPYRIDIUM.

Calyx-tube produced above the ovary and disk.
  Flowers sessile or nearly so, in cymes, often contracted into
    heads surrounded by imbricate brown bracts . . . . . 10. STENANTHEMUM.
  Flowers solitary or in leafy spikes, sometimes contracted into
    heads, or pedicellate, individually surrounded by brown
    bracts . . . . . . . . . . . . . . . . . . . 11. CRYPTANDRA.
Calyx campanulate or tubular, the tube produced above the ovary
  and annular disk. Spines and small leaves opposite . . . 12. DISCARIA.

## 1. VENTILAGO, Gærtn.

Calyx 5-lobed, spreading. Petals hood-shaped or none. Stamens 5, scarcely exceeding the petals when present. Disk flat or concave, filling the short calyx-tube. Ovary more or less immersed in the disk, 2-celled; style short, with 2 short erect stigmatic lobes. Nut globular at the base, produced into an oblong or linear coriaceous wing, 1-celled and 1-seeded, indehiscent. Seed globular; testa membranous; albumen none; cotyledons thick and fleshy.—Climbing shrubs or trees. Leaves alternate, penninerved. Flowers small, clustered along the branches of axillary or terminal panicles.

The genus is dispersed over the tropical regions of the Old World. The Australian species is endemic, differing from the others in habit and foliage as well as in the absence of petals.

1. **V. viminalis,** *Hook. in Mitch. Trop. Austr.* 369. A small glabrous tree. Leaves narrow-lanceolate, 2 to 4 or even 5 in. long, entire, narrowed into a petiole, coriaceous, the pinnate veins very oblique and sometimes almost parallel with the midrib, without the elegant transverse venation of the rest of the genus. Panicles not much branched, or almost reduced to simple racemes, shorter than the leaves, solitary or clustered in the axils. Calyx about 1 line long. Petals none. Disk entirely adnate to the short broad calyx-tube. Ovary slightly immersed in the disk. Fruit glabrous, about 1 in. long, including the wing, the turbinate adnate base of the calyx not attaining above a quarter the length of the globular nut.

**N. Australia.** Nicholson river, Gulf of Carpentaria, *F. Mueller.*
**Queensland.** High sandy ridges on the Maranoa, *Mitchell.*
**N. S. Wales.** Tributaries of the Upper Darling river, *Bowman.*

## 2. ZIZYPHUS, Juss.

Calyx 5-lobed, spreading. Petals hood-shaped or rarely none. Stamens 5, included in the petals or scarcely exceeding them, when present. Disk flat, filling the short calyx-tube. Ovary immersed in the disk, 2-, rarely 3- or 4-celled; style shortly branched or styles distinct; stigmas small. Drupe ovoid or globular, putamen woody or bony, 1- to 4-celled, 1- to 4-seeded. Seeds with a smooth fragile testa; albumen none or scanty; cotyledons thick. —Trees or shrubs, usually armed with stipular prickles. Leaves alternate, 3- or 5-nerved, often distichous and very oblique. Flowers small, greenish, in axillary cymes. Fruit often edible.

The genus ranges over the tropical and subtropical regions of the New and the Old World. Two of the Australian species are also common Asiatic ones, the third is endemic.

Leaves green on both sides, softly pubescent or villous, or at length
  glabrous. Drupe small, 2-celled . . . . . . . . . . 1. *Z. Œnoplia.*

Leaves white or rusty underneath, with a close tomentum.
Ovary and drupes 2-celled . . . . . . . . . . . . 2. *Z. jujuba.*
Ovary and drupes 4-celled . . . . . . . . . . . . 3. *Z. quadrilocularis.*
(*Z. melastomoides*, A. Cunn. Herb. and Steud. Nom., is a *Celtis.*)

1. **Z. Œnoplia,** *Mill.; W. and Arn. Prod.* 163 (with the synonyms adduced, except *Z. Napeca*). A shrub of several feet, with very divaricate branches, the young ones rusty-pubescent or villous. Stipular spines short, in pairs, one straight and deciduous, the other hooked or recurved and more persistent. Leaves very obliquely ovate, obtuse or slightly acuminate, 1 to 2 in. long, entire or crenulate, 3- or 5-nerved, membranous, green on both sides, softly pubescent or villous, especially underneath, or sometimes glabrous when full grown. Cymes small, compact, few-flowered, and almost sessile. Ovary 2-celled, style short, the stigma scarcely divided. Drupe globular, 2 or 3 lines diameter, 2-celled or 1-celled by abortion.—*Z. celtidifolia*, DC. Prod. ii. 20 (from the character given); Fenzl, in Hueg. Enum. 20 ; *Z. rufula*, Miq. Fl. Ind. Bat. i. part 1, 643.

**N. Australia.** Islands of the Gulf of Carpentaria and Arnhem S. Bay, *R. Brown.* Common in East India and the Archipelago, but apparently not in Africa. Of the two Linnæan *Rhamni* doubtfully referred here by Wight and Arnott, *R. Œnoplia* is quite correct; *R. Napeca* however is *Zizyphus lucida*, Moon ; Thw. Enum. Pl. Ceyl. 74. The Linnæan herbarium has very good authentically named specimens of both.

2. **Z. jujuba,** *Lam.; W. and Arn. Prod.* 162 (with the synonyms adduced). A tall shrub or small tree, with short stipular prickles, occasionally wanting. Leaves ovate or nearly orbicular, usually very obtuse, 1 to 3 in. long, entire or toothed, 3-nerved, glabrous above, covered underneath, as well as the petioles and branches, with a close white or rusty tomentum. Cymes small, compact, and nearly sessile. Ovary 2-celled, tapering into a short 2-lobed style. Drupe globular, usually about $\frac{1}{2}$ to nearly $\frac{3}{4}$ in. diameter, 2-celled or 1-celled by abortion.

**Queensland.** Torres Straits, *Dubouzet.* Very common, both wild and cultivated, throughout tropical Asia, extending also to tropical Africa.

3. **Z. quadrilocularis,** *F. Muell. Fragm.* iii. 57. A tall shrub or small tree. Stipules lanceolate, appressed, very rigid and pointed, but not so spinous and more deciduous than in the other species. Leaves ovate, shortly acuminate, or rarely obtuse, 2 to 3 in. long, entire or scarcely crenulate, very oblique at the base, 3-nerved, glabrous above, rusty or hoary-tomentose underneath, as well as the young branches. Cymes small, dense, very shortly pedunculate. Ovary 4-celled, with a short 4-lobed style. Drupe globular, of the size of that of *Z. jujuba*, but the thick bony putamen 4-celled and 4-seeded.

**N. Australia.** Upper Victoria river, *F. Mueller.*

## 3. RHAMNUS, Linn.

Calyx 4- or 5-lobed, broadly campanulate or spreading. Petals hood-shaped, involute or nearly flat, or rarely none. Stamens 4 or 5, scarcely exceeding the petals when present. Disk broadly concave or lining the calyx-tube, with a free margin. Ovary free, sessile on the disk (not immersed), 2-celled in the Australian species, 3- or 4-celled in most others, tapering into a style, with as many short stigmatic lobes as ovary-cells. Drupe succulent, globular or oblong, containing 2 to 4 bony or cartilaginous pyrenes, indehiscent or scarcely dehiscent. Seeds with a smooth testa; albumen fleshy; cotyledons flat or recurved.—Shrubs or trees. Leaves alternate, petiolate, penninerved, entire or toothed, usually green on both sides. Stipules small, deciduous. Flowers in clusters, either axillary and solitary or in axillary or terminal racemes.

The genus is widely dispersed over the northern hemisphere, rare in tropical regions. The Australian species, which is in some measure doubtful, extends to the Fiji Islands.

1. **R (?) vitiensis,** *Benth.* Quite glabrous, the branches slender. Leaves ovate or oval-oblong, shortly acuminate, 2 to 3 in. long, entire or serrate-crenate, green on both sides, thin and apparently deciduous. Flowers in axillary sessile clusters, on slender pedicels of 3 or 4 lines. Calyx about 2 lines long, the tube broadly hemispherical, the lobes triangular, rather thin. Petals involute, enclosing the stamens. Disk concave, broadly cup-shaped, the margin free. Ovary broadly sessile, 2-celled, tapering into a short style. Fruit not seen.—*Colubrina vitiensis,* Seem. Syst. List Vit. Pl. 4.

**Queensland.** Cape York, *M'Gillivray.* Until the fruit is known, the genus of this plant cannot be free from doubt. The inflorescence and disk, however, are those of *Rhamnus*, and the species seems to differ from *R. javanica,* Miq., chiefly in its thinner leaves. Apparently the same species was gathered in the Fiji Islands by Seemann, and his specimens have young fruits, of an obovoid-oblong shape, which, as far as they go, agree with those of *Rhamnus.*

## 4. COLUBRINA, L. C. Rich.

Calyx 5-lobed, spreading. Petals hood-shaped. Stamens 5, included in the petals. Disk thick, filling the calyx-tube. Ovary immersed in the disk, 3- or rarely 4-celled, tapering into a 3-, rarely 4-cleft style, with obtuse stigmas. Drupe nearly globular, obscurely lobed, the epicarp thin or succulent, the endocarp separating into 3, rarely 4 membranous or crustaceous cocci, opening inwards by a longitudinal slit. Seeds without any arillus; testa smooth, shining, coriaceous; albumen fleshy but thin; cotyledons flat or incurved, thin or rather thick.—Erect or half-climbing shrubs or trees. Leaves alternate, 3-nerved at the base or penninerved in species not Australian. Stipules small, deciduous. Flowers small, in axillary cymes or clusters.

The species are nearly all American, tropical or subtropical, with one from tropical Asia, extending also into Australia.

1. **C. asiatica,** *Brongn.; W. and Arn. Prod.* 166 (with the synonyms adduced). A large shrub or small tree, unarmed, and quite glabrous, with long, slender, often flexuose branches. Leaves petiolate, ovate or broadly cordate, acuminate, 2 to 3 in. long, crenate-serrate, 3-nerved and penninerved, smooth and shining, but scarcely coriaceous. Cymes shortly pedunculate, rarely exceeding the petioles. Flowers greenish, about 2 lines diameter. Fruit about 4 lines diameter, depressed at the top, furrowed opposite the dissepiments, the endocarp separating more or less perfectly into 3 or rarely 4 membranous cocci.

**Queensland.** Cape York, *M'Gillivray;* Cape Grafton and Rodd's Bay, *A. Cunningham;* Howick's Group, *F. Mueller;* Shoalwater passage, *R. Brown;* Port Denison, *Fitzalan.* The species is common in tropical Asia, extending to the Pacific islands.

## 5. ALPHITONIA, Reissek.

Calyx 5-lobed, spreading. Petals involute. Stamens 5, included in the petals. Disk thick, filling the calyx-tube. Ovary immersed in the disk, 2- or rarely 3-celled, tapering into a shortly lobed style. Drupe globular or broadly ovoid, the epicarp of a dry, mealy or somewhat corky substance; endocarp of 2 or 3 hard coriaceous nuts or cocci, opening inwards by a longitudinal slit. Seeds with a shining hard testa, completely enclosed in a membranous brown shining arillus, open at the top, but with the edges folded over; albumen cartilaginous or horny; cotyledons flat.—Tree. Leaves alternate, penninerved. Cymes dichotomous, many-flowered. Seeds often persisting on the torus after the pericarp has fallen off.

The genus is probably limited to a single species, ranging from Australia to the Pacific islands.

1. **A. excelsa,** *Reissek, in Endl. Gen.* 1098. A tall hard-wooded timber-tree, the young branches, petioles, and inflorescence hoary or rusty with a close tomentum. Leaves petiolate, varying from broadly ovate or almost orbicular and very obtuse, to ovate or lanceolate and acute or acuminate, usually 3 to 6 in. long, entire, coriaceous, glabrous or slightly hoary above, white, or rarely rust-coloured underneath with a close tomentum, the parallel pinnate veins very prominent. Flowers 2 to 3 lines diameter, in little umbel-like cymes, arranged in dichotomous cymes in the upper axils or in a terminal corymbose panicle. Calyx tomentose. Disk broad and nearly flat. Fruit 3 or 4 lines diameter, or sometimes rather larger.—*Colubrina excelsa*, Fenzl, in Hueg. Enum. 20.

**N. Australia.** Islands of the Gulf of Carpentaria (Cape Van Diemen), *R. Brown;* Sweers Island, *Henne;* Arnhem's Land, *F. Mueller.*

**Queensland.** Curtis Island, *Henne;* Rockhampton, *Thozet;* Port Denison, *Fitzalan;* Brisbane river, Moreton Bay, *A. Cunningham, Fraser, F. Mueller,* and others.

**N. S. Wales.** Hunter's, Paterson's, and Williams rivers, *R. Brown;* Hastings and Clarence rivers, *Beckler* and others; Blue Mountains, *Miss Atkinson;* Illawarra, *M'Arthur.*

The Carpentaria island specimens belong to a variety with remarkably large obtuse leaves, the flowers rather larger than usual, and the tomentum somewhat rusty. To this belongs *Zizyphus pomaderroides*, Fenzl, in Hueg. Enum. 20, judging from R. Brown's specimens corresponding to Bauer's. *Alphitonia zizyphoides*, A. Gray, Bot. Amer. Expl. Exped. i. 278, t. 20 (*Rhamnus zizyphoides*, Soland.), which extends from Borneo and New Caledonia to the Pacific islands, does not appear to differ at all from some of the eastern Australian specimens; whilst *A. franguloides.* A. Gray, l. c. 280, is very like some of the more tomentose N. Australian specimens.

## 6. EMMENOSPERMUM, F. Muell.

Calyx 5-lobed, the tube campanulate. Petals hood-shaped, inserted with the stamens on the margin of the disk. Stamens 5, enclosed in the petals. Disk thin, lining the calyx-tube. Ovary inserted on the disk in the bottom of the calyx-tube, but not immersed, 2-celled or rarely 3-celled, tapering into a shortly-cleft style. Fruit almost capsular, with a very thin almost dry epicarp, the endocarp separating into 2 or rarely 3 cartilaginous almost crustaceous cocci, opening along the inner face in two valves. Seeds inserted on a turbinate or slightly cup-shaped funicle, without any arillus; testa hard and shining; albumen cartilaginous; cotyledons flat.—Trees. Leaves opposite or alternate, penninerved. Cymes or panicles trichotomous, many-flowered. Seeds often persisting on the torus after the pericarp has fallen off.

The genus is endemic in Australia. It is closely allied in technical characters to the S. African *Noltia*, but with a different habit.

Leaves opposite or nearly so . . . . . . . . . . . . . 1. *E. alphitonioides.*
Leaves alternate . . . . . . . . . . . . . . . . . 2. *E. Cunninghamii.*

1. **E. alphitonioides,** *F. Muell. Fragm.* iii. 63. A tall hard-wooded timber-tree, quite glabrous. Leaves opposite or nearly so, petiolate, ovate, acuminate, 2 to 3 in. long, entire, coriaceous, shining above, green on both sides. Flowers numerous, in little dense umbel-like cymes, arranged in trichotomous cymes or corymbose panicles in the upper axils or terminal. Calyx-lobes almost petal-like, nearly 1 line long. Fruits apparently about 3 lines long, but either unripe or already open in our specimen. Seeds persistent, like those of *Alphitonia*, but without the peculiar arillus of that species.

**Queensland.** Brush of Brisbane river, *M'Arthur;* Peri creek, *Leichhardt.*

**N. S. Wales.** Clarence river, *C. Moore, Wilcox;* Illawarra, known under the name of "Dogwood," *M'Arthur, Backhouse, Ralston.*

2. **E. (?) Cunninghamii,** *Benth.* Leaves alternate, similar to those of *E. alphitonioides*, except that the petioles are longer. Flowers not seen. Umbel-like cymes apparently not numerous, in a terminal corymbose panicle. Fruits rather larger than in *E. alphitonioides*, 3- or 4-celled; epicarp scarcely any; cocci 2-valved. Seeds red and shining as in that species, but not persistent on the torus, and the funicle very small.

**N. Australia.** Port Warrender, N.W. coast, *A. Cunningham.* The specimens are very imperfect; they were referred to *Croton* by Cunningham, but the seeds are erect and present all the characters of *Rhamneæ*, as already observed by Planchon in Herb. Hook.

## 7. POMADERRIS, Labill.

Calyx-tube entirely adnate to the ovary, the limb divided to the base into 5 lobes, usually deciduous or reflexed. Petals either concave or nearly flat, not enclosing the anthers, or none. Stamens 5, the filaments long and usually suddenly inflected and attenuate near the top; anthers oblong or ovoid. Disk annular, surrounding the ovary at the base of the calyx-lobes, often scarcely conspicuous, and never very prominent. Ovary half-inferior or rarely almost entirely inferior. Style 3-cleft, or rarely almost entire. Capsule protruding above the border of the calyx-tube, septicidally 3-valved, the endocarp separating into 3 crustaceous or membranous cocci, opening by a broad operculum at the base of the inner face, or by the separation of the whole inner face, or rarely by a longitudinal slit. Seed inserted on a short, thickened, turbinate or cup-shaped funiculus.—Shrubs, with the young branches and under side of the leaves white, hoary or rusty with a close stellate tomentum, often mixed with or concealed by longer, simple, soft, often silky hairs. Leaves alternate, penninerved. Stipules brown and scarious, usually very deciduous. Flowers pedicellate, in small umbel-like cymes, usually forming terminal panicles or corymbs, or rarely solitary in the axils of the leaves. Bracts brown and scarious, but so deciduous as to be seldom visible at the time of flowering.

The genus is confined to Australia and New Zealand; the Australian species are all endemic and from the eastern and southern districts, with the exception of two which are also found in New Zealand.

Flowers with petals.
  Calyx-tube turbinate, at least half as long as the lobes.  Cocci
    opening by an operculum below the middle.
    Leaves mostly ovate-lanceolate, 2 to 3 in. long.  Panicles many-
      flowered.
      Leaves hoary or tomentose above, softly tomentose underneath.
        Calyx about 2 lines long, very villous . . . . . . .  1. *P. lanigera.*
      Leaves glabrous or sparingly scabrous-pubescent above, densely
        ferruginous, tomentose underneath.  Calyx 1 to 1½ lines
        long, softly hairy . . , . . . . . . . . . .  2. *P. ferruginea.*
      Leaves somewhat coriaceous, glabrous above, very white un-
        derneath.  Calyx 1½ lines long, silky-hairy . . . . .  3. *P. grandis.*
    Leaves ovate, and obtuse or oblong-elliptical, often above 2 in.
      long, glabrous above, white underneath.  Panicles many-
      flowered,  Calyx 1 to 1½ lines, closely tomentose or hairy .  4. *P. elliptica.*
    Leaves firm, rarely above 1 in. long.  Panicles small and com-
      pact.  Calyx of *P. elliptica* . . . . . . . . . .  5. *P. phillyreoides.*
  Calyx-tube exceedingly short.  Cocci opening by their whole inner
    face.  Leaves small.  Panicles compact.
    Leaves broadly ovate or orbicular.  Calyx hoary.  Petals broad .  6. *P. vacciniifolia.*
    Leaves obovate or broadly oblong.  Calyx silky.  Petals very
      narrow . . . . . . . . . . . . . . . . . . .  7. *P. myrtilloides.*
    Leaves narrow-oblong.  Calyx silky.  Petals narrow . . . .  8. *P. ledifolia.*
Flowers without petals.
  Cymes rather loose, numerous in much-branched panicles.
    Calyx stellate-tomentose or hoary, with a very short tube.
      Leaves 2 to 4 in. long, irregularly crenate and rugose . . .  9. *P. apetala.*
      Leaves 1 to 2 in. long, ashy-white, not rugose . . . . . 10. *P. cinerea.*
    Calyx softly hairy, with a turbinate tube.
      Leaves mostly obtuse, scabrous above, often crenulate and
        rugose . . . . . . . . . . . . . . . . . . 11. *P. prunifolia.*
      Leaves mostly acute, smooth above, quite entire . . . . 12. *P. ligustrina.*
  Cymes condensed into heads, in oblong panicles.  Calyx-tube very
    short . . . . . . . . . . . . . . . . . . . . 13. *P. betulina.*
  Cymes loose, few, in close corymbs.  Leaves obcordate or bifid.
    Calyx-tube turbinate . . . . . . . . . . . . . . 14. *P. obcordata.*
  Cymes loose, usually few-flowered, axillary, or in narrow, oblong,
    or raceme-like panicles.  Calyx-tube very short.
    Leaves ovate, obovate, or broadly oblong, flat.
      Leaves thick, ½ to 1 in. long, white or cottony underneath . 15. *P. racemosa.*
      Leaves ½ to 1 in. long, loosely pubescent and scarcely white
        underneath . . . . . . . . . . . . . . . . 16. *P. subrepanda.*
      Leaves under ¼ iu., obovate, white underneath . . . . . 17. *P. elachophylla.*
    Leaves linear or oblong, the margins revolute.  Flowers very
      small and numerous . . . . . . . . . . . . . 18. *P. phylicifolia.*

1. **P. lanigera,** Sims, *Bot. Mag. t.* 1823.  An erect branching shrub, nearly allied to *P. elliptica,* with which it is united by F. Mueller, differing chiefly in the leaves softly though minutely tomentose on the upper side, and the larger more villous flowers.  Leaves oblong or ovate-lanceolate, the under side as well as the young branches clothed with a soft velvety tomentum often rust-coloured.  Panicles often larger and less corymbose than in *P. elliptica.*  Calyx about 2 lines long, very densely and softly hairy, the

turbinate tube about half as long as the lobes.  Petals ovate, concave, on slender claws.  Fruit as in *P. elliptica*, but larger and more hairy.—DC. Prod. ii. 33, excluding the var. β; *Ceanothus laniger*, Andr. Bot. Rep. i. 569; *P. obscura*, Sieb. Pl. Exs.

**N. S. Wales.**  Port Jackson, *R. Brown, Sieber, n.* 216; rocky gullies near King's Fall, *A. Cunningham*; New England, *C. Stuart*; Hastings river, *Beckler*.

2. **P. ferruginea,** *Sieb. ; Fenzl, in Hueg. Enum.* 21.  Very near *P. elliptica*, and united with it by F. Mueller, having the leaves glabrous above, and the small flowers of that species, but the leaves are usually rather longer for their breadth and more acute, and the down of the under side is much more dense, velvety and usually ferruginous.  The flowers are more numerous, the calyx more softly and densely hairy, and the petals usually narrower. The fruits are the same.—Hook. f. Fl. Tasm. i. 76; *P. lanigera, var.* β, DC. Prod. ii. 33; *P. viridirufa*, Sieb. Pl. Exs.; *Ceanothus Wendlandianus*, Rœm. and Schult. Syst. v. 299 (from the character given); *Pomaderris Wendlandiana*, G. Don, Gen. Syst. ii. 39.

**N. S. Wales.**  Port Jackson, *R. Brown, Sieber, n.* 209 and 214, and *Fl. Mixt. n.* 545; Paramatta, *A. Cunningham, Woolls*; Blue Mountains, *Miss Atkinson*.
**Victoria.**  Macalister river, Gipps' Land, *F. Mueller*.
**Tasmania.**  Flinders Island, Bass's Straits, *Gunn*.
Var. *pubescens*.  Leaves pubescent above with short scattered hairs, but green; flowers small, as in the normal form.—*P. hirta*, Reissek, in Endl. Nov. Stirp. Dec. 31 (from the description).—Illawarra, Twofold Bay, and Genoa river, *F. Mueller*; and other localities in southern N. S. Wales and eastern Victoria.
Var. *canescens*.  Leaves 3 to 4 in. long, white and less ferruginous underneath.  Intermediate almost between *P. ferruginea* and *P. elliptica*.—Percy Island, *A. Cunningham*.

3. **P. grandis,** *F. Muell. Fragm.* iii. 68. · Very nearly allied to *P. ferruginea*, and differing chiefly in the silvery whiteness of the tomentum. Leaves ovate-lanceolate or oblong-elliptical, rather acute, 2 to 3 in. long, glabrous above, silvery-white underneath, with a soft silky tomentum.  Panicles many-flowered, corymbose, as in *P. ferruginea* and *P. elliptica*, and flowers about the same size.  Calyx with a turbinate adnate tube, densely clothed with soft white silky hairs.  Petals broad.  Style-branches exceedingly short, but not shorter than in some N. S. Wales specimens of *P. elliptica*.

**W. Australia.**  Mount Manypeak river, *Maxwell*.  From the single specimen upon which this species is founded, it does not appear to me to differ more from *P. elliptica* than *P. ferruginea* and *P. phillyreoides*, and, if these are joined to it as varieties, *P. grandis* must surely follow, notwithstanding the distant habitat.

4. **P. elliptica,** *Labill. Pl. Nov. Holl.* i. 61, *t.* 86.  A tall shrub or small tree, the young branches rusty with a very close stellate down, intermixed occasionally with a few longer hairs.  Leaves petiolate, ovate, oblong or ovate-lanceolate, obtuse or rarely almost acute, usually 2 to 3 in. long and ¾ to 1¼ in. broad, entire or the margins slightly waved, glabrous above and smooth or scarcely scabrous, white underneath with a very close tomentum, the prominent midrib and principal parallel veins often rust-coloured. Cymes numerous, in dichotomous panicles, usually more or less corymbose. Stipules lanceolate, brown and scarious as well as the broad concave bracts, but all falling off in a very early stage so as to be rarely seen at the time of flowering.  Calyx about 1½ lines long, white with a minute stellate tomen-

tum, often intermixed with longer simple hairs, especially on the turbinate tube. Petals usually broadly cordate or nearly orbicular, concave, on slender claws, but often much narrower, sometimes deeply toothed and occasionally abortive. Style-branches short, with capitate stigmas. Capsule about 1½ lines diameter, slightly hairy, the free part rather shorter than the adnate portion, the cocci opening in a round valve or operculum below the middle.—Bot. Mag. t. 1510; DC. Prod. ii. 33; Hook. f. Fl. Tasm. i. 76; F. Muell. Fragm. iii. 69.

**N. S. Wales.** Port Jackson to the Blue Mountains, *R. Brown* and others; northward to New England, *C. Stuart,* and southward to Twofold Bay, *F. Mueller.*

**Victoria.** Monkey Creek, Gipps' Land, *F. Mueller.*

**Tasmania.** Common, especially in the northern portion of the island, *J. D. Hooker.* Also in the northern island, New Zealand.

Two species are usually distinguished, *P. elliptica,* with broader more obtuse leaves and without any silky hairs mixed with the stellate tomentum of the calyx, and *P. discolor,* DC. Prod. ii. 33, Sweet, Fl. Aust. t. 41, with the calyx, at least the tube, more or less silky-hairy and the leaves often less obtuse. Labillardière's specimens belong to the former, but his description agrees better with the latter; and in many instances the two forms pass one into the other. Sieber's specimens, n. 208 (*P. malifolia,* Sieb.; *P. multiflora,* Fenzl, in Hueg. Enum. 21), are very broad-leaved, with the tomentose calyx of the first form; n. 213 (*P. discolor*) belongs to the second; n. 210 (*P. intermedia,* Sieb.; DC. Prod. ii. 33) has the leaves narrower than usual and the indumentum of the calyx variable. *Ceanothus discolor,* Vent. Jard. Malm. t. 58, has the more acute leaves of the second form with the close tomentum of the first. *P. acuminata,* Link. Enum. Hort. Berol. 235, is probably established on the same garden-plant as Ventenat's.

F. Mueller considers *P. lanigera, ferruginea,* and *phillyreoides* as varieties only of this species, and it is certainly sometimes difficult to draw precise limits between them in the dried state. If they are united, the species should surely include also *P. grandis.*

5. **P. phillyreoides,** *Sieb. in DC. Prod.* ii. 33. A shrub, said to be of much smaller stature than *P. elliptica.* Down of the young branches sometimes very close and white or rusty, sometimes loose and more rusty, almost as in *P. ferruginea.* Leaves much smaller than in any of the preceding species, seldom attaining 1½ in. and usually much shorter, oblong or oval, obtuse or acute, entire, of a firm consistence, glabrous or minutely hoary above, soft underneath with a white or rusty down. Flowers rather larger than in *P. elliptica,* but variable in size, the cymes compact, in small terminal panicles. Calyx softly silky-hairy, the turbinate tube shorter than the lobes. Petals nearly of *P. elliptica,* but usually narrower. Styles more deeply cleft, the branches club-shaped at the top, with somewhat decurrent stigmas. Capsule of *P. elliptica.*—*P. andromedæfolia,* A. Cunn. in Field, N. S. Wales, 351; Bot. Mag. t. 3219; *P. phillyreæfolia,* Fenzl, in Hueg. Enum. 22 (from the character given).

**N. S. Wales.** Port Jackson, *Sieber, n.* 215; rocks in the Blue Mountains and stony barren hilly districts, *A. Cunningham* and others. I have failed in identifying in R. Brown's herbarium the plant described by Fenzl, but have little doubt of its belonging to this species, which F. Mueller unites with *P. elliptica.*

Var. *nitidula.* Leaves more coriaceous, usually acute; tomentum closer, very white on the under side of the leaves.—New England, *C. Stuart ;* Mount Lindsay, *W. Hill.*

6. **P. vacciniifolia,** *Reissek and Muell. in Linnæa,* xxix. 266. A shrub, with slender divaricate branches. Leaves ovate or nearly orbicular, very obtuse, seldom above ½ in. long, glabrous above, white underneath. Cymes small, in ovoid terminal panicles of about 1 in. Buds nearly globular, about

1¼ lines diameter, hoary with a very close stellate tomentum, without silky hairs, the calyx-tube exceedingly short. Petals broad. Summit of the ovary remarkably prominent, and birsute with white hairs. Style-branches short, with capitate stigmas. Fruit nearly 1½ lines long, the free part much longer than the adnate base; cocci thin, opening by the separation of the whole inner face, which often splits along the centre.—F. Muell. Fragm. iii. 71.

**Victoria.** Watts river, *F. Mueller.*

7. **P. myrtilloides,** *Fenzl, in Hueg. Enum.* 22. Apparently a low, erect, dichotomous shrub, the tomentum of the younger branches and under side of the leaves very close but dense, and having a silky appearance on the younger leaves. Leaves from obovate to obovate-oblong, very obtuse or almost acute, slightly emarginate, mostly about ½ in. long, in the original specimens narrowed at the base, glabrous above and quite entire. Cymes few, loose, forming small terminal corymbs, shorter or but little longer than the last leaves. Buds ovoid, or at length nearly globular. Calyx 1½ lines long, very silky with short hairs, the tube very short. Petals narrow-linear. Style almost entire. Fruit not seen.

**W. Australia.** Goose Island Bay, S. coast, *R. Brown.*

Var. *major.* Leaves larger, often 1 in. long; flowers larger.—*P. stenopetala*, F. Muell. Fragm. iii. 69. Point Henry, *Oldfield.*

8. **P. ledifolia,** *A. Cunn. in Field, N.S. Wales,* 351. A slender and apparently a low shrub, the tomentum of the younger branches white and very close, and soon disappearing. Leaves narrow-oblong, obtuse, mostly about ½ in. long, coriaceous, quite entire, glabrous above, the margins slightly recurved, white underneath, with the midrib alone prominent. Flowers few, in little loose shortly pedunculate cymes in the upper axils. Buds ovoid, about 1 line long, silky-hairy. Calyx-tube exceedingly short. Petals narrow, slightly concave. Styles rather short, free almost to the base. Ovary very hairy. Capsule obovoid, nearly glabrous, fully 1 line long, the free part much longer than the adnate tube, very obtuse and depressed or umbilicate at the top. Cocci opening by the separation of the whole inner face, which often splits also along the centre.—*Trymalium helianthemifolium*, Reissek, in Linnæa, xxix. 271.

**N. S. Wales.** Rocky hills near Cox's river, *A. Cunningham.*

**Victoria.** Avon river, Gipps' Land, *F. Mueller* (only seen in fruit).

Var. (?) *angustifolia.* Leaves narrower, sprinkled on the upper side with stellate hairs.— Macalister river, *F. Mueller.* The foliage in some measure comes near to that of *P. phylicifolia*, but the capsule is that of *P. ledifolia.* Flowers not seen.

9. **P. apetala,** *Labill. Pl. Nov. Holl.* i. 62, *t.* 87. A shrub of 3 to 6 feet, the stellate tomentum of the young branches and under side of the leaves usually dense, but close, sometimes however loose and floccose. Leaves petiolate, ovate-lanceolate or broadly oblong, obtuse or rarely acute, 2 to 4 in. long, irregularly crenulate, glabrous, but rough and much wrinkled on the upper side, the principal veins very prominent underneath. Flowers small and very numerous, in loose oblong thyrsoid panicles, leafy at the base. Buds ovoid or nearly globular. Calyx 1¼ lines long, with stellate hairs, the tube very short. Petals none. Anthers tipped by a small gland. Styles divided to the middle, with club-shaped almost capitate stigmas. Capsule obtuse,

2 E 2

with a few stellate hairs; cocci opening with a short valve, as in *P. elliptica.*
—Hook. f. Fl. Tasm. i. 77; F. Muell. Fragm. iii. 73; *P. aspera*, Sieb. in
DC. Prod. ii. 33; A. DC. Pl. Rar. Jard. Gen. 5ᵉ Not. 18, t. 4.

**N. S. Wales.** Nepean river, *R. Brown;* Port Jackson, *Sieber, n.* 211, and others;
abundant in open forest-lands south of the colony, *A. Cunningham;* Twofold Bay, *F.
Mueller.*

**Victoria.** King's Island and Port Phillip, *R. Brown;* extending over the southern and
eastern districts of the colony, *F. Mueller.*

**Tasmania.** Abundant throughout the island, *J. D. Hooker.*

**S. Australia.** Kangaroo Island, *Waterhouse;* specimens in leaf only, and therefore
doubtful.

The species varies much in the quantity of stellate tomentum, and also in the size of the
flowers, but does not appear to be separable into distinct varieties.

10. **P. cinerea,** *Benth.* A tall shrub, with numerous slender branches,
hoary with a minute tomentum. Stipules filiform. Leaves ovate or elliptical, obtuse or scarcely acute, 1 to nearly 2 in. long, quite entire, hoary above
and white underneath with a close minute tomentum, the primary veins prominent underneath, but not impressed above. Cymes loose, many-flowered,
in terminal leafy panicles. Bracts narrow, falling off very early, as in the
rest of the genus. Buds small, globular, white-tomentose, not yet quite open
in the specimens seen. Calyx-tube exceedingly short. Petals none.

**N. S. Wales.** Mount Imlay, Twofold Bay, *F. Mueller.*

11. **P. prunifolia,** *A. Cunn.; Fenzl, in Hueg. Enum.* 22. Stellate
tomentum of the branches and under side of the leaves dense and white, or
sometimes ferruginous. Leaves ovate or oblong, obtuse or mucronate, seldom
above 1½ in. long, wrinkled, and often scabrous above, with short, simple or
stellate hairs. Flowers small and numerous, in many-flowered compact cymes,
arranged in thyrsoid terminal panicles as in *P. ligustrina.* Calyx obovoid,
about 1 line long, the tube turbinate, the stellate tomentum usually concealed
by long silky hairs. Petals none. Styles cleft nearly to the base. Capsule
about 1 line diameter, hirsute, obtuse, only slightly protruding from the adnate tube of the calyx.—F. Muell. Fragm. iii. 75.

**N. S. Wales.** Near Liverpool, *A. Cunningham;* Paramatta, *Woolls.* In some herbaria Cunningham's labels of this and *P. betulina* are interchanged.

**Victoria.** Genoa river and coast near Snowy River, *F. Mueller.* (Leaves almost smooth
above. Capsule rather more prominent.)

12. **P. ligustrina,** *Sieb. in DC. Prod.* ii. 34. Branches slender, the
tomentum soft and rust-coloured. Leaves lanceolate or ovate-lanceolate, 1
to 2 in. long, glabrous above, quite entire, rusty-tomentose or almost woolly
underneath. Flowers small and numerous, in rather loose thyrsoid terminal
panicles. Calyx obovoid, scarcely above 1 line long, softly silky-hairy. Petals
none. Styles usually divided to the middle, with club-shaped stigmas. Capsule about 1 line diameter, hirsute, rather obtuse, the exserted part about as
long as the adnate tube; operculum of the cocci about half their length.—
F. Muell. Fragm. iii. 71.

**N. S. Wales.** Port Jackson, *Sieb. n.* 212, and *Fl. Mixt. n.* 544, and others; Blue
Mountains, *A. Cunningham;* northward to Hastings river, *Beckler;* southward to Twofold
Bay, *F. Mueller.*

13. **P. betulina,** *A. Cunn. in Bot. Mag. t.* 3212. A slender shrub or small tree, with elongated branches. Tomentum of the young branches and under side of the leaves often rust-coloured and usually close. Leaves oblong or obovate, obtuse, seldom above 1 in. long. Flowers nearly sessile, in dense globular heads, either solitary or more frequently two or three together, on short axillary or terminal peduncles. Bracts more persistent than in most species. Buds obovoid-globular. Calyx about 1 line long, densely clothed with long silky hairs. Petals none. Style cleft to the middle with club-shaped branches, stigmatic some way down.—F. Muell. Fragm. iii. 76.

**N. S. Wales.** In a water-gully at the base of the Pine Ridge, Macquarie river, *A. Cunningham.*

**Victoria.** Gravelly rocky banks of the Upper Genoa river, *F. Mueller.*

The foliage of this species is not unlike that of *P. prunifolia,* but the inflorescence is very different.

14. **P. obcordata,** *Fenzl, in Hueg. Enum.* 23. A low much-branched shrub, the young branches hoary with a minute tomentum. Leaves cuneate, obcordate, or broadly 2-lobed at the top, with rounded entire or crenate lobes, rarely above $\frac{1}{2}$ in. long, and often much less, much contracted at the base, the margins usually recurved, pale-coloured, but glabrous above, much whiter underneath with a minute close tomentum. Flowers in loose cymes, forming small terminal corymbs, of about $\frac{1}{2}$ in. diameter or rather more. Bracts rather large, but very deciduous, as in other species. Calyx fully 1 line long, slightly hoary. Petals none, in our specimens. Stamens long, with oblong anthers. Disk slightly prominent. Style 3-cleft to the middle. Fruit obovoid, nearly 2 lines long, the exserted part stellate-tomentose and rather longer than the adnate base. Cocci slightly wrinkled on the inner face, indehiscent or opening by the whole inner face, or sometimes in two valves.—*Trymalium bilobatum,* F. Muell., Reissek, in Linnæa, xxix. 279; *T. biauritum,* Reissek and Muell. l. c. 281; *Pomaderris biaurita,* F. Muell. Fragm. iii. 73, and Pl. Vict. ii. t. 22.

**S. Australia.** Memory Cove, *R. Brown;* dry hills on the Glenelg and thence to Guichen Bay, *F. Mueller;* Port Lincoln, *Wilhelmi;* Spencer's Gulf, *Warburton.*

**W. Australia.** King George's Sound, *M'Lean.*

This species in some measure connects *Pomaderris* with *Trymalium,* but both the inflorescence and flowers are much more those of the former genus than of *Trymalium,* especially if they are really apetalous, as I find them in all the specimens I have examined, although Reissek describes broadly hood-shaped petals with slender claws.

15. **P. racemosa,** *Hook. Journ. Bot.* i. 256. A small much-branched shrub, the stems and under side of the leaves covered with stellate tomentum, sometimes short and close, but often copious or loose and floccose, white or of a deep rust-colour. Leaves small, seldom exceeding an inch, and often not above $\frac{1}{2}$ in., from broadly ovate to oblong or obovate, obtuse, entire or irregularly crenate. Flowers on very short pedicels, and generally few in each cyme, of which 3 to 6 form short compound racemes in the upper axils, and sometimes the whole inflorescence reduced to 5 or 6 flowers. Buds globular. Calyx 1 to $1\frac{1}{2}$ lines long, with stellate hairs. Petals none. Style cleft to the middle, with club-shaped branches stigmatic some way down.—Hook. f. Fl. Tasm. i. 77; F. Muell. Fragm. iii. 75.

**N. S. Wales.** Desert of the Darling and Murray, *F. Mueller.* (I have not seen these specimens.)

**Victoria.** Port Phillip, *R. Brown;* on the coast from Wilson's Promontory to the Murray, Buchan river in Gipps' Land and in the Murray desert, *F. Mueller.*

**Tasmania.** N. coast about the mouth of the Tamar, *Lawrence, Gunn, C. Stuart.*

**S. Australia.** Memory Cove, *R. Brown;* from the Murray river to Spencer's Gulf and inland to Lake Torrens, *F. Mueller.*

The species is very variable, the following being the three principal forms observed:—

*a.* Leaves very scabrous on the upper surface and rather large; flowers rather large and numerous.—*P. oraria,* F. Muell. and Reissek, in Linnæa, xxix. 268.

*b.* Leaves quite glabrous above; flowers rather large and few.

*c.* Leaves slightly stellate-downy above; flowers small and usually numerous.—*P. paniculosa,* F. Muell. and Reissek, in Linnæa, xxix. 269.

16. **P. subrepanda,** *F. Muell., Reissek, in Linnæa,* xxix. 267. Branches slender, the tomentum of the young ones and under side of the leaves close, stellate, and white or rust-coloured. Leaves oval or oblong, seldom 1 in. long and usually ½ to ¾ in., entire or slightly and irregularly toothed, glabrous above with impressed veins. Cymes few-flowered, often reduced to 1 or 2 flowers, in short loose thyrsoid compound racemes in the upper axils, forming oblong leafy terminal panicles. Buds globular. Calyx stellate-tomentose, about 1 line long, the tube very short. Petals none. Ovary very villous. Styles short, with almost capitate stigmas. Capsule ovoid, scarcely 1 line long, the free part longer than the adherent base. Cocci membranous, opening by a longitudinal slit, or at length by nearly the whole inner face.— F. Muell. Fragm. iii. 74.

**Victoria.** Yarra Yarra river and Forest Creek, *F. Mueller.* The foliage is very nearly that of some specimens of *P. prunifolia,* but the flowers and fruit are quite distinct.

17. **P. elachophylla,** *F. Muell. Fragm.* ii. 131. A tall shrub, with numerous slender divaricate branches, rather loosely stellate-tomentose. Leaves broadly obovate, very obtuse, rarely ¼ in. long, and often under 2 lines, entire, glabrous above or sprinkled with a few minute stellate hairs, white-tomentose underneath. Cymes few-flowered or reduced to 1 or 2 flowers in the upper axils of the smaller branches, forming loose leafy racemes or narrow thyrsoid panicles. Buds globular. Calyx stellate-tomentose, about ¾ line long, the tube very short. Petals none. Styles short, club-shaped. Young capsule hairy, the free part much longer than the adnate calyx-tube.

**Victoria.** On the river Tyers, an affluent of the Latrobe river, *F. Mueller;* Upper Yarra river, *E. B. Heyne.*

18. **P. phylicifolia,** *Lodd. Bot. Cab. t.* 120. A heath-like shrub with numerous erect branches, densely villous or rarely only stellate-downy. Leaves linear or narrow-oblong, nearly sessile, seldom above ½ in. long, the margins usually much revolute so as often to conceal the under surface; which bears a close white tomentum, whilst the upper side is more or less scabrous with short simple or stellate hairs; more rarely the leaves are broader and nearly flat. Flowers small and few, in little loose cymes in the upper axils, scarcely longer than the leaves, but very abundant along the smaller branches, and the upper ones forming thyrsoid leafy panicles. Calyx globular, densely pubescent or villous, scarcely 1 line diameter. Petals none. Capsule ovoid, hirsute, about 1 line long, scarcely obtuse, the free part longer than the adnate base. Cocci membranous, opening by the whole inner face. —DC. Prod. ii. 34; *P. ericifolia,* Hook. Journ. Bot. i. 257; Hook. f. Fl.

Tasm. i. 78 ; Reissek, in Linnæa, xxix. 270 ; *P. polifolia*, Reissek, in Linnæa, xxix. 269.

**Victoria.** Banks of subalpine streams under the Australian Alps, descending into the plains of Gipps' Land on the Hume and Murray rivers, *F. Mueller.*

**Tasmania.** Mersey river, *Gunn ;* St. Paul's river, *C. Stuart.*

Found also abundantly in the northern island of New Zealand. Some specimens of *P. ledifolia* come near to this species in habit, but they may be readily known when in flower by the petals, and in fruit by the very truncate or depressed apex of the capsule.

Var. *latifolia.* Leaves oblong, ½ to 1 in. long, the margins scarcely revolute.—Genoa river in Victoria, *F. Mueller.*

## 8. TRYMALIUM, Fenzl.

Calyx-tube entirely adnate to the ovary, the limb divided to the base into 5 lobes, usually deciduous or spreading. Petals 5, hood-shaped, entire or 3-lobed, but not usually enclosing the anthers. Stamens 5, the filaments rather short, incurved, with small, ovoid anthers. Disk annular or divided into 5 glands, surrounding the ovary at the base of the calyx-lobes. Ovary half-inferior or almost entirely inferior, 3- or rarely 2-celled. Style 3-cleft or rarely 2-cleft at the top or to the middle. Capsule protruding above the adnate calyx-tube or rarely on a level with it, the endocarp separating into crustaceous or rarely membranous cocci, indehiscent or open internally in 2 valves. Seeds of *Pomaderris.*—Shrubs, with the habit and deciduous stipules and bracts of *Pomaderris*, but with smaller flowers and a more slender inflorescence, the panicles usually narrow, or the cymes few-flowered. Flowers always pedicellate.

The species are all confined to West Australia.

Panicles or racemes elongated, terminal, or longer than the leaves.
  Leaves ovate or broadly oblong, flat.
    Leaves hoary on the upper side with a minute tomentum . . . 1. *T. albicans.*
    Leaves glabrous above, or hirsute . . . . . . . . . . 2. *T. Billardieri.*
    Leaves linear-oblong or linear, the margins revolute . . . . . 3. *T. ledifolium.*
  Cymes few-flowered, shorter than or scarcely exceeding the leaves.
    Leaves linear. Ovary 3-celled.
      Petals entire. . . . . . . . . . . . . . . . . 3. *T. ledifolium.*
      Petals 3-lobed . . . . . . . . . . . . . . . . 4. *T. angustifolium.*
    Leaves cuneate, hoary on both sides. Ovary 2-celled . . . . 5. *T. Wichuræ.*

1. **T. albicans,** *Reissek, in Pl. Preiss.* ii. 280. Apparently a tall shrub, the branches white or hoary with a close stellate tomentum. Leaves broadly ovate or obovate, very obtuse, 1 to 2 in. long, soft and more or less hoary on the upper side, white underneath with a minute down. Flowers in thyrsoid terminal panicles, larger and fewer than in *T. Billardieri.* Calyx fully 1 line long, white with a close tomentum. Capsule very obtuse, 1½ lines in diameter, the broad stellately pubescent exserted portion as long as the turbinate adnate base ; cocci crustaceous, muricate or wrinkled on the inner face, apparently indehiscent.—*Pomaderris albicans*, Steud. in Pl. Preiss. i. 184.

**W. Australia,** Swan River, *Drummond, 5th Coll. n.* 229 ; sides of Mount Eliza, *Preiss, n.* 1689.

2. **T. Billardieri,** *Fenzl, in Hueg. Enum.* 25. A tall shrub, the young branches hoary with stellate hairs and often villous with simple ones. Leaves

sometimes broadly ovate or obovate, very obtuse, 1 to 2 in. long, sometimes ovate or ovate-lanceolate, more or less acuminate, 2 to 3 in. long, entire or with a few coarse crenatures, glabrous or pubescent above, white or hoary, or, in the hirsute variety, villous underneath. Flowers numerous, in loose narrow terminal panicles, sometimes almost racemiform and 2 to 3 in. long, more frequently forming compound leafy panicles of ½ ft. or more. Bracts very small. Pedicels very slender. Calyx less than 1 line long, the tube very short and densely pubescent. Capsule very obtuse, stellate-pubescent, the broad exserted portion longer than the adnate tube; cocci indehiscent, the inner face very rugose.—Reissek, in Pl. Preiss. ii. 282; *Ceanothus spathulatus*, Labill. Pl. Nov. Holl. i. 60, t. 84; *Pomaderris spathulata*, G. Don, Gen. Syst. ii. 38; *T. floribundum*, Steud. in Pl. Preiss. i. 185.

**W. Australia.** Swan River, *Drummond;* in stony rocky places, *Preiss, n.* 1680; King George's Sound, *R. Brown* and others; Harvey and Blackwood rivers, *Oldfield;* Mount Manypeak river, *Maxwell.*

Var. *hirsutum*, Reissek, in Pl. Preiss. ii. 282. Branches, and often the leaves also, hirsute and scarcely white underneath. Some specimens have so different an aspect from the typical form that they seem to indicate a distinct species, but the two are connected by numerous intermediates.—*T. expansum*, Steud. in Pl. Preiss. i. 185. King George's Sound, *Brown;* Kalgan river, *Oldfield;* Todyay valley, Victoria district, *Preiss, n.* 1683 (*Hb. R. Brown, Sonder, F. Muell.*).

3. **T. ledifolium,** *Fenzl, in Hueg. Enum.* 24. A low shrub, with slender branches, with a slight stellate tomentum. Leaves linear or sometimes linear-lanceolate or oblong, from ½ to 1 in. long, the margins more or less revolute, glabrous above, hoary or sometimes very white underneath, with a very prominent midrib. Panicles slender and raceme-like, usually 1 to 2 in. long and terminal, but sometimes scarcely longer than the leaves and on short lateral shoots so as to appear lateral, the rhachis slightly tomentose. Bracts small and very deciduous. Buds globular. Calyx little more than ½ line long, usually very tomentose or pubescent, especially the tube, and the ovary and disk pubescent, but sometimes· the whole flower quite glabrous. Style short. Capsule ovoid, truncate at the top, in the normal form not projecting beyond the adnate calyx-tube, and usually crowned by the persistent calyx-lobes. Cocci crustaceous, much wrinkled on the inner face.—Reissek, in Pl. Preiss. ii. 282.

**W. Australia.** King George's Sound, *R. Brown ;* Swan River, *Drummond, 1st Coll., Oldfield;* Blackwood and Vasse rivers and Darling range, *Oldfield.*

Var. *rosmarinifolium.* Leaves usually narrow and much revolute; capsule protruding considerably beyond the adnate calyx-tube.—*Pomaderris rosmarinifolia*, Steud. in Pl. Preiss. i. 184; *Cryptandra floribunda*, Steud. l. c. 186; *C. glaucophylla*, Steud. l. c. i. 187; *Trymalium rosmarinifolium*, Reissek, in Pl. Preiss. ii. 283.—Swan River, *Drummond, Preiss, n.* 1674, 1675, and 1684.

Var. *daphnifolium.* Leaves rather short, oblong, the margins less revolute than in the normal form; capsule protruding considerably beyond the adnate calyx-tube.—*T. daphnifolium*, Reissek, in Pl. Preiss. ii. 283.—Swan River, *Drummond, 5th Coll. n.* 237; between Perth and King George's Sound, *Harvey.*

Var. (?) *obovatum.* Leaves obovate or obovate-oblong, flat.—Rocks at Todyay, *Oldfield.* The specimens are small and in bud only, the petals appear to be broader than usual. *Cryptandra anomala*, Steud. in Pl. Preiss. i. 187, appears also to be a variety of *T. ledifolium.*

4. **T. angustifolium,** *Reissek, in Pl. Preiss.* ii. 284. An apparently

low heath-like shrub, with erect twiggy branches, hoary with short stiff hairs. Leaves linear, mostly 3 to 4 lines long, the margins much revolute, hispid with stiff hairs, hoary or silky underneath. Flowers very small, in axillary cymes, forming short, dense, terminal, raceme-like leafy panicles of $\frac{1}{2}$ to 1 in. Bracts minute. Pedicels short. Calyx-tube very hairy. Petals rather shorter than the calyx-lobes, with a lateral concave lobe on each side almost as large as the central one, and contracted below the lobes into a short claw. Disk annular. Capsule $1\frac{1}{4}$ to $1\frac{1}{2}$ lines diameter, globular, very hispid and acuminate with the persistent base of the style. Cocci almost membranous, apparently indehiscent.

**W. Australia.** Swan River, *Drummond, 1st Coll.*

5. **T. Wichuræ,** *Nees; Reissek, in Pl. Preiss.* ii. 281. A muchbranched slender shrub, the young branches and both sides of the leaves hoary with a minute close tomentum. Leaves obovate-cuneate or spathulate, very obtuse or rarely emarginate, 2 to 4 lines long, much contracted at the base. Flowers very small, 2 to 4 together in little terminal cymes. Calyx about $\frac{3}{4}$ line long, minutely hoary. Disk prominent. Petals small, hood-shaped, entire. Ovary 2-celled. Style minutely 2-lobed at the top. Capsule obovoid, $1\frac{1}{2}$ lines long, the exserted portion very obtuse and shorter than the adnate tube, splitting to the base into 2 valves, the 2 cocci opening in 2 valves.

**W. Australia,** Swan River, *Drummond;* between Perth and King George's Sound, *Harvey;* King George's Sound, Wilson's River, and Hay Inlet, *Maxwell.*

## 9. SPYRIDIUM, Fenzl.

(Stenodiscus, *Reissek.*)

Calyx-tube entirely adnate or shortly free above the ovary, but not above the disk, the limb divided to the disk into 5 usually persistent lobes. Petals 5, hood-shaped, usually enclosing the anthers. Stamens 5; filaments short; anthers small, ovoid. Disk annular or divided into 5 glands, either close round the ovary and filling the calyx-tube, or lining the calyx-tube when produced above the ovary. Ovary wholly inferior, 3-celled. Style entire or minutely 3-toothed. Capsule enclosed in the calyx-tube and crowned by the persistent lobes, 3-valved at the top, the endocarp separating into 3, sometimes reduced to 2 or 1, membranous or rarely crustaceous cocci, either indehiscent or opening inwards by a longitudinal slit. Seeds of *Pomaderris.*—Shrubs, with the indumentum of *Pomaderris.* Leaves usually small. Stipules scarious, brown, lanceolate, usually connate and persistent. Flowers sessile in heads or rarely solitary, surrounded by small, persistent, imbricate, brown scarious bracts, the heads small, sessile, usually several together in a compound head or in corymbose cymes, the outer heads in each having often a floral leaf, either like the stem-leaves, or smaller and broader, on a longer petiole and whiter, the head having the appearance of being inserted on the petiole.

The genus is entirely Australian and extratropical. It differs from *Trymalium* chiefly in inflorescence and habit, from *Stenanthemum* and *Cryptandra* in the calyx-tube not produced above the disk.

§1. *Heads very small and few-flowered, sessile along the branches, with very minute bracts. Leaves obcordate.*

Leaves 2 to 5 lines long, hoary on both sides . . . . . . 1. *S. tridentatum.*
Leaves 1 to 2 lines long, glabrous above, white underneath . . 2. *S. divaricatum.*

§ 2. *Heads several-flowered in cymes or compound heads, usually with one or more floral leaves. Leaves obovate, obcordate-ovate, or broadly oblong.*
Disk annular, or of 5 glands close upon the ovary or nearly so.
 Flower-heads in cymes, except in some of the last species, where they are in compound heads.
 Leaves herbaceous, pubescent or glabrous above.
  Leaves obovate, obcordate, or cuneate, mostly 2 to 3 lines
   long, glabrous above, the veins not impressed . . . 3. *S. serpyllaceum.*
  Leaves ovate, 3 to 6 lines or sometimes above 1 in. long,
   hoary or softly pubescent, or rarely glabrous above.
   Disk very prominent, almost closing over the ovary . . 4. *S. parvifolium.*
   Disk slightly prominent, of 5 distinct glands . . . . 5. *S. spadiceum.*
 Leaves coriaceous, glabrous and smooth above when full-grown.
  Leaves mostly 1 to 1½ in. long, ovate, on rather long pe-
   tioles. Heads numerous, in the cyme. Floral leaves
   rare. Plant generally canescent . . . . . . . 6. *S. globulosum.*
  Leaves mostly ½ to ¾ in., on short petioles. Cymes
   small, with 3 or 4 floral leaves. Tomentum rusty
   or very white.
   Leaves obovate or oblong, contracted at the base . . 7. *S. obovatum.*
   Leaves ovate, obtuse at the base . . . . . . . 8. *S. Gunnii.*
  Leaves rarely attaining 1 in.
   Leaves cuneate-obovate or spathulate, silky underneath . 9. *S. spathulatum.*
   Leaves small, broad, much revolute, smooth above,
    woolly underneath.
    Leaves orbicular or obovate. Flower-heads very
     villous . . . . . . . . . . . . . . 10. *S. Lawrencii.*
    Leaves broadly cordate. Flowers glabrous at the top . 11. *S. cordatum.*
   Leaves ovate or obovate, with raised reticulations above,
    silky underneath . . . . . . . . . . . 12. *S. phlebophyllum.*
Disk lining the calyx-tube above the ovary, with a thickened
 annular margin under the lobes. Flower-heads in compound
 compact heads. Leaves under ½ in.
 Leaves obtuse at the base, often emarginate . . . . . . 13. *S. coactilifolium.*
 Leaves obovate, narrowed at the base, folded lengthwise.
  Flower-heads globular . . . . . . . . . . . . 14. *S. complicatum.*

§ 3. *Heads several-flowered, in cymes or compound heads, usually with one or more floral leaves. Leaves linear, linear-oblong, narrow-cuneate or 2-lobed, the margins usually revolute.*
Flower-heads small, in cymes. Disk of 5 distinct glands.
 Leaves entire.
  Leaves glabrous above, silky underneath. Branches tomentose.
   Cymes little branched . . . . . . . . . . . 15. *S. westringiæfolium.*
  Tomentum hoary, mixed with long hairs. Cymes much
   branched . . . . . . . . . . . . . . . 16. *S. villosum.*
  Tomentum hoary or white. Cymes small, few-headed . . 17. *S. pauciflorum.*
Flower-heads united into one dense compound head.
 Leaves shortly 2-lobed. Disk dividing into distinct glands.
  Leaves cuneate, very pubescent . . . . . . . . . 18. *S. halmaturinum.*
  Leaves linear, minutely tomentose or glabrous . . . . 19. *S. bifidum.*
 Leaves entire. Disk undulate or entire.
  Calyx glabrous, very small . . . . . . . . . . 24. *S. microcephalum.*

Calyx tomentose or hirsute, at least the tube.
Stipules on the young shoots large. Calyx 1 to 1½ lines
long. Disk prominent, annular, close to the ovary.
Leaves tomentose or hoary on both sides, rarely glabrous above when old . . . . . . . . . . 20. *S. subochreatum.*
Leaves glabrous above, very narrow . . . . . . 21. *S. oligocephalum.*
Stipules small. Calyx under 1 line. Leaves usually glabrous above.
Floral leaves usually ovate, more petiolate than the stem-
leaves. Disk annular, close on the ovary . . . 22. *S. vexilliferum.*
Floral leaves like the stem-leaves. Disk lining the
calyx-tube with a thickened annular margin raised
above the ovary . . . . . . . . . . . 23. *S. eriocephalum.*

§ 4. *Flowers solitary or 3 together, each with separate bracts. Disk lining the calyx-tube, the thickened annular margin under the calyx-lobes far above the ovary. Leaves linear, the margins revolute* (**Stenodiscus**, Reissek) . . . 25. *S. ulicinum.*

1. **S. tridentatum,** *Benth.* Branches slender, wiry, slightly pubescent.
Leaves obovate, obcordate, or triangular, truncate or 3-toothed at the top,
narrowed at the base, 2 to 4 lines long, the margins not recurved, but the
leaf sometimes conduplicate as in *S. complicatum* or in *Stenanthemum,* usually
hoary on both sides with a minute close tomentum, or clothed with longer
appressed hairs underneath. Flowers very small, in small lateral heads, sessile among a few floral leaves, the brown bracts narrow and much smaller
than in any other species. Calyx not 1 line long, hoary-tomentose. Disk
annular, close round the ovary. Capsule ovoid, nearly 1½ lines long, crowned
by the calyx-lobes. Cocci almost crustaceous, opening inwards in 2 valves.
—*Cryptandra tridentata,* Steud. in Pl. Preiss. i. 186 ; Reissek, in Pl. Preiss.
ii. 289 ; *Stenanthemum tridentatum,* Reissek, in Linnæa, xxix. 295.

**W. Australia.** Swan River, *Preiss, n.* 1216 and 2421, Between Perth and King
George's Sound, *Harvey ;* Murchison river and Champion Bay, *Oldfield.*
This species was placed by Reissek in *Stenanthemum,* but the calyx has not the slender
tube produced above the disk and ovary which characterizes that genus.

2. **S. divaricatum,** *Benth.* A low, divaricately-branched, often spinescent shrub, the branches nearly glabrous, slender but rigid. Leaves in little
clusters along the branches, 1 to 2 lines long, obcordate or obtusely 2-lobed,
narrowed at the base, the margins revolute, glabrous and smooth above, white·
underneath. Flowers very minute, 2 or 3 together in the clusters of leaves, with
small imbricate acuminate bracts. Calyx little more than ½ line long, the short
tube pubescent, the lobes glabrous. Disk annular, close round the ovary.

**W. Australia.** Dirk Hartog's Island, *Milne ;* Murchison river, *Oldfield.*

3. **S. serpyllaceum,** *F. Muell. Fragm.* iii. 80. Branches numerous,
prostrate, slender and wiry, the young ones minutely tomentose, but soon
glabrous. Leaves obovate or obcordate, very obtuse, 2 to 3 or rarely 4 to 5
lines long, the margins recurved, glabrous or slightly tomentose above, with
the veins slightly impressed, hoary or white underneath. Flowers in small
very compact heads, forming small leafy cymes, the imbricate brown bracts
almost as long as the calyx. Calyx about 1 line long, densely tomentose.
Disk slightly raised above the ovary, lining the short tube and forming a ring
under the lobes. Cocci membranous.—*Cryptandra obcordata,* Hook. f. Fl.
Tasm. i. 71 ; *Trymalium serpyllaceum,* Reissek, in Linnæa, xxix. 280.

**Victoria.** Entrance of the Genoa river, *F. Mueller.*
**Tasmania.** Trap hills on the banks of the Tamar, and abundant on the Asbestos hills, *Gunn, J. D. Hooker.*

4. **S. parvifolium,** *F. Muell. Fragm.* iii. 79. Much-branched and rather slender, with a dense close tomentum or with a loose and more spreading pubescence, varying from hoary to a more or less rusty tint. Leaves obovate or orbicular, very obtuse or emarginate, seldom in the ordinary form above ½ in. and often not above 3 lines long, the margins usually recurved, soft and often hoary on the upper side, with the primary veins much impressed, softly hoary underneath, with the veins prominent. Flowers closely sessile in little heads, forming small dense terminal leafy cymes, and closely surrounded by the short brown imbricate bracts. Calyx very hirsute, about 1 line long. Disk very prominent over the ovary, almost concealing it. Capsule wholly inferior. Cocci crustaceous, slightly rugose on the inner face, indehiscent or opening tardily in 2 valves.—*Pomaderris parvifolia*, Hook. Journ. Bot. i. 257; Schlecht. Linnæa, xx. 636; *Cryptandra parvifolia*, Hook. f. Fl. Tasm. i. 73; *Trymalium parvifolium* and *T. hermannioides*, Reissek, in Linnæa, xxix. 275.

**N. S. Wales.** Twofold Bay, *F. Mueller.*
**Victoria.** Frequent in rocky, stony, and scrubby places, *F. Mueller.* In Mitchell's collections under the name of *T. majoranæfolium*, Lindl., but not Fenzl's species of that name.
**Tasmania.** N. coast, banks of the Tamar, and islands of Bass's Straits, *Gunn* and others.
**S. Australia.** Mouth of the Glenelg, *Allitt;* extending to Barossa ranges and St. Vincent's Gulf, *F. Mueller.*
Var. *molle.*—Softly hairy all over.—*Cryptandra mollis*, Hook. f. Fl. Tasm. i. 73. Flinders Island and Cape Barren Island, *Gunn.*
Var. *hirsutissimum*, very hispid all over.—In the Grampians, *Wilhelmi.*
Var. *grande*, F. Muell. Luxuriant, the leaves often above 1 in. long, and cymes loose and many-headed, thus assuming the aspect of *S. spadiceum*, but with the prominent disk of *S. parvifolium.*—*Trymalium eupatorioides*, Reissek, in Linnæa, xxix. 270; Dandenong in Victoria, *F. Mueller.*

5. **S. spadiceum,** *Benth.* Branches clothed with a soft but close often rusty tomentum, with more or less of soft spreading hairs. Leaves in the original form from narrow-oblong to nearly oval, obtuse, 1 to 1½ in. long, or ½ in. on the lateral branches, softly and minutely pubescent above, white underneath or the veins rusty. Flower-heads crowded in compact broad cymes, usually shorter than the leaves. Brown bracts broad and numerous. Calyx scarcely 1 line long, the tube very hairy. Petal-claws slender. Style short. Disk of distinct glands, alternating with the stamens and very slightly raised above the ovary. Capsule nearly 1½ lines long, crowned by the calyx-lobes. Cocci rather coriaceous, opening inside in 2 valves.—*Trymalium spadiceum*, Fenzl, in Hueg. Enum. 26; Reissek, in Pl. Preiss. ii. 280; *Pomaderris hirsuta*, Steud. in Pl. Preiss. i. 184; *Trymalium thomasioides*, Turcz. in Bull. Mosc. 1858, i. 459.

**W. Australia.** King George's Sound, *Huegel;* southern districts, *Drummond, n.* 231; rocky places at the back of Mount Clarence, *Preiss, n.* 1673 a, *Oldfield.*
Var. *majoranæfolium.* A smaller plant. Leaves usually under ½ in. long, rather more coriaceous than in the ordinary form, hoary on both sides with a close soft tomentum. Flower-heads small, in small compact cymes. Disk separating into 5 glands close to the

ovary.   Cocci membranous.—*Trymalium majoranæfolium*, Fenzl, in Hueg. Enum. 21;
Reissek, in Pl. Preiss. ii. 281; *Pomaderris commixta*, Steud. in Pl. Preiss. i. 184.   King
George's Sound, *R. Brown*, and others; Mount Clarence, *Preiss, n.* 1673 *b.*   Usually a
very marked form, but some specimens seem to pass into the larger variety.

Var. (?) *calvescens*, Reissek, in Pl. Preiss. ii. 28.   Leaves glabrous above, or nearly so,
usually small, of a firmer consistence, almost like those of *S. obovatum* and *S. Gunnii*, but
the flowers are much smaller and the disk different.—*Pomaderris subretusa*, Steud. in Pl.
Preiss. i. 183.—King George's Sound, *R. Brown;* Mount Baldhead, *Preiss, n.* 1687;
Princess Royal Harbour, *Maxwell.*

The species, although sometimes approaching *S. parvifolium* in habit, is readily known
by the disk.

6. **S. globulosum,** *Benth.*   A tall shrub, with larger leaves and more
of the appearance of a *Pomaderris* than most *Spyridia*, generally hoary with
a minute very close tomentum.   Leaves ovate, obovate or oblong, very obtuse,
1 to 1½ or rarely 2 in. long, almost coriaceous, glabrous above, white or
hoary underneath, or rarely slightly rusty.   Flower-heads nearly globular,
numerous in dense corymbose cymes in the axils of the leaves and not much
exceeding them.   Brown bracts pubescent, shorter than the calyx.   Calyx
pubescent or silky-villous, about 1 line long, broadly campanulate.   Disk of
5 distinct glands, close round the ovary.   Capsule scarcely 1½ lines long, the
pubescent convex summit slightly protruding from the calyx-tube, but covered
by the persistent segments.   Cocci membranous.—*Ceanothus globulosus*, Labill.
Pl. Nov. Holl. i. 61, t. 85; *Pomaderris globulosa*, G. Don, in Loud. Hort.
Brit. 84, and Gen. Syst. ii. 38; *Trymalium globulosum*, Fenzl, in Hueg. Enum.
25; Reissek, in Pl. Preiss. ii. 279; *Pomaderris polyantha* and *P. æmula*,
Steud. in Pl. Preiss. i. 182; *P. phillyreæfolia* and *P. pyrrhophylla*, Steud.
l. c. 183.

**W. Australia.**   Common about King George's Sound, *Labillardière, R. Brown*, and
others, and thence along the coast to Vasse river and Swan River, *Drummond, Oldfield,
Preiss, n.* 1676, 1677, 1678, 1679, 1681, 1690, and others.

7. **S. obovatum,** *Benth.*   Apparently a low and much-branched shrub,
the stellate tomentum usually somewhat rust-coloured.   Leaves obovate or
oblong, very obtuse or slightly emarginate, seldom exceeding ½ in., the margin recurved, firm and coriaceous, usually smooth and shining above, with
the primary veins impressed, softly but closely tomentose underneath.
Flower-heads small, in terminal cymes, with 1 to 3 floral leaves.   Bracts orbicular.   Calyx 1 line long, the tube hairy, the lobes glabrous or rarely hirsute.   Petal-claws slender.   Disk prominent, undulate, close round the ovary.
—*Pomaderris obovata*, Hook. Comp. Bot. Mag. i. 277; *Cryptandra obovata*,
Hook. f. Fl. Tasm. i. 74; *Trymalium obovatum*, Reissek, in Linnæa, xxix.
278.

**Tasmania.**   Common on the east coast, *Gunn* and others.   Some S. Australian broad-
leaved forms of *S. vexilliferum* appear to come very near to this species.

Var. *velutinum.*   Leaves minutely and softly tomentose on the upper side.—*Trymalium
velutinum*, Reissek, in Linnæa, xxix. 276.—Tasmania, *C. Stuart.*

8. **S. Gunnii,** *Benth.*   Very near *S. obovatum*, and the leaves have the
same coriaceous texture, but they are rather larger, mostly above ½ in. long and
more ovate or oval than obovate, glabrous or rarely tomentose above, densely
tomentose underneath.   Cymes more developed, with 2, 3, or more floral

leaves. Flowers larger, the calyx usually 1½ lines long, tomentose outside and the disk scarcely prominent. Cocci coriaceous.—*Cryptandra Gunnii*, Hook. f. Fl. Tasm. i. 73.

**Tasmania.** Banks of the Franklin river, near Macquarie Harbour, *Gunn.* Referred by F. Mueller to *S. parvifolium*, from which, however, it appears to me to differ considerably in flowers as well as in foliage.

9. **S. spathulatum,** *F. Muell. Herb.* Very much-branched, the stellate tomentum close and often assuming a yellowish-golden tint. Leaves cuneate-obovate, 3 to 5 lines long, the margins thickened but scarcely recurved, coriaceous, nearly glabrous above, the under surface hoary or yellowish with a more or less silky and shining pubescence consisting of appressed hairs. Flowers very minute, in little dense heads with a leafy bract at their base, forming short terminal cymes sometimes passing into racemes. Brown bracts minute. Calyx scarcely ½ line long. Disk prominent, undulate, close above the ovary. Capsule near 2 lines long, the persistent bracts much enlarged. Cocci membranous or chartaceous, apparently indehiscent.—*Trymalium spathulatum*, F. Muell. in Trans. Vict. Inst. 1855, 122 ; *T. daphnoides*, Reissek, in Linnæa, xxix. 278.

**S. Australia.** South coast, *R. Brown;* Lofty Ranges, *F. Mueller;* foot of the Marble range, *Wilhelmi;* Kangaroo Island, *Waterhouse.*

**W. Australia?** *Herb. Hooker,* specimens believed to be from Drummond.

Var. *microphyllum.* Leaves 2 to 3 lines long, usually silvery-white, branches slender, corymbose.—Kangaroo Island, *Waterhouse.*

10. **S. Lawrencii,** *Benth.* Low, much-branched, and prostrate or suberect, the tomentum hoary or rusty on the young branches. Leaves nearly orbicular, cordate, ovate or obcordate, very obtuse or emarginate, rarely above 2 lines long and often not more than 1 line, thickly coriaceous, the margins much recurved, glabrous or nearly so above, densely tomentose or woolly underneath. Cymes more or less leafy, very dense and hairy, the brown bracts pubescent outside. Calyx scarcely 1 line long, very hairy. Petals nearly sessile. Disk slightly prominent, immediately above the ovary. Cocci crustaceous.—*Cryptandra Lawrencii*, Hook. f. Fl. Tasm. i. 72 ; *Trymalium microphyllum*, Reissek, in Linnæa, xxix. 273.

**Tasmania.** E. coast, Great Swan Port, *Backhouse;* St. Paul's river, *Gunn, C. Stuart.*

11. **S. cordatum,** *Benth.* Apparently low and procumbent, much resembling *S. Lawrencii.* Leaves on rather long petioles, broadly cordate, very obtuse or emarginate, 2 to 3 lines long, coriaceous, tomentose above when young, at length glabrous, smooth and shining, the margins much recurved, white or rusty-tomentose underneath. Flower-heads in very compact compound heads, 3 to 4 lines broad, with 2 to 4 floral leaves. Calyx scarcely ¾ line long, the tube loosely villous, the lobes nearly glabrous. Disk little prominent, and almost concealed by the hairs of the top of the ovary, although in fact inserted at a small distance above it.—*Cryptandra cordata*, Turcz. in Bull. Mosc. 1858, i. 459.

**W. Australia.** *Drummond,* 5th Coll., *n.* 230.

12. **S. phlebophyllum,** *F. Muell. Herb.* Low, tortuous, and much-branched, with a dense, close, somewhat rusty tomentum. Leaves ovate or

nearly orbicular, very obtuse or emarginate, 3 to 4 lines long or rarely more, the margins thick and recurved, thickly coriaceous, glabrous above with raised reticulations, which distinguish this species from all others as yet known, silky-tomentose underneath with short appressed hairs. Flower-heads very small, in little dense cymes, usually with a small floral leaf. Brown bracts pubescent. Calyx rarely above ½ line long, hairy. Disk annular, undulate, slightly prominent, close above the ovary. Cocci coriaceous.—*Trymalium phlebophyllum,* F. Muell., Reissek, in Linnæa, xxix. 272.

**S. Australia.** Elders range, near Lake Torrens, *F. Mueller.*

13. **S. coactilifolium,** *Reissek, in Linnæa,* xxix. 291. Young branches rusty with a stellate tomentum mixed with spreading hairs. Leaves distinctly petiolate, ovate or obovate, very obtuse or emarginate, mostly 3 to 5 lines long, broad and obtuse at the base, flat on the edges, softly and densely pubescent on both sides, the upper ones often white and almost woolly. Flower-heads combined into very compact compound heads, like those of *S. Lawrencii,* with several white woolly floral leaves. Calyx slender, scarcely 1 line long, very hispid. Disk like that of *S. Lawrencii,* but the annular margin further removed above the ovary.

**S. Australia.** Encounter Bay, *Whitaker, F. Mueller.*

Var. *integrifolium.* Rather less tomentose, and the leaves not emarginate.—*S. thymifolium,* Reissek, in Linnæa, xxix. 289, and *S. Stuartii,* Reissek, l. c .290. The brown or black stipules and bracts are present in all these, as well as in the original form, but are smaller and less conspicuous in the more scrubby and woolly specimens than in the more luxuriant and elongated ones. F. Mueller unites both these forms with *S. vexilliferum,* but both the foliage and the disk appear to me to be quite different.

14. **S. complicatum,** *F. Muell. Fragm.* iii. 78. A rigid, divaricately-branched shrub, allied to *S. coactilifolium* in the indumentum and structure of the flowers, with nearly the foliage of *S. tridentatum,* and of some *Stenanthema.* Leaves nearly sessile, obovate or broadly cuneate, emarginate, with a short recurved point, ¼ to ¾ in. long, narrowed into a petiole, mostly folded lengthwise, rather thick, softly tomentose on both sides, especially underneath, or nearly glabrous above. Flower-heads compound, nearly globular, sessile, very dense, 3 to 6 lines diameter. Brown bracts very short. Calyx very hirsute, about 1 line long. Disk annular, lining the calyx-tube to a considerable distance above the ovary. Capsule globular or ovoid, 1½ lines long; cocci rather hard, opening in 2 valves.

**W. Australia.** Dirk Hartog's Island, *A. Cunningham;* Murchison river and Champion Bay, *Oldfield.*

15. **S. westringiæfolium,** *Benth.* Stellate tomentum of the young branches often mixed with short simple pubescence. Leaves narrowly cuneate-oblong, or almost oblong-linear, obtuse, above ½ in. long, much narrowed at the base, the margins recurved, glabrous or nearly so above, densely silky-tomentose with almost appressed hairs underneath. Flower-heads small, in short leafy scarcely branched cymes, often with 1 or 2 floral leaves to each head. Brown bracts ovate-acuminate or lanceolate, often pubescent. Disk of 5 distinct glands close above the ovary.—*Pomaderris westringiæfolia,* Steud. in Pl. Preiss. i. 185 ; *Trymalium westringiæfolium,* Reissek, in Pl. Preiss. ii. 284.

**W. Australia.** Limestone plains, Arthur's Head, *Preiss, n.* 1686. The specimen have seen is small and imperfect, but appears very distinct from any other species. The disk is that of *S. spadiceum,* but the foliage and indumentum are very different, and its affinity is more probable with the following species.

16. **S. villosum,** *Benth.* Tomentum of the young branches hoary or rusty, mixed with stiff spreading hairs. Leaves linear or linear-oblong, $\frac{1}{4}$ to $\frac{3}{4}$ in. long, or shorter on the side branches, mostly with a short recurved point, the margins much recurved, hoary with a minute tomentum or glabrous above, more densely tomentose underneath, and hispid with a few spreading hairs on the midrib and margins. Flower-heads very dense, in shortly pedunculate cymes, with one or two floral leaves. Brown bracts broad. Calyx about $\frac{3}{4}$ line long, tomentose. Disk prominent, divided into distinct glands immediately above the ovary. Petals rather long.—*Cryptandra villosa,* Turcz. in Bull. Mosc. 1858, i. 458.

**W. Australia,** *Drummond,* 5*th Coll. n.* 232.

17. **S. pauciflorum,** *Benth.* Young branches rusty-tomentose. Leaves narrow-oblong, obtuse, mostly about $\frac{1}{2}$ in. long, the margins much recurved, glabrous or minutely tomentose and hoary above, white underneath with a close stellate tomentum mixed with minute simple hairs. Cymes very small, consisting almost of single heads, usually with a floral leaf. Calyx scarcely $\frac{1}{2}$ line long, tomentose. Disk of 5 minute distinct glands close above the ovary. Capsule nearly 1 line long. Cocci membranous, opening inwards in 2 valves. —*Cryptandra pauciflora,* Turcz. in Bull. Mosc. 1858, i. 458.

**W. Australia.** Swan River, *Drummond,* 5*th Coll. n.* 233. Evidently allied to the last two species and may possibly prove to be a variety of one of them, but the specimens I have seen appear too distinct to justify their union without further materials.

18. **S. halmaturinum,** *F. Muell. Herb.* Low and erect, densely tomentose-villous with short spreading hairs, mixed with the closer stellate tomentum. Leaves cuneate-oblong, about $\frac{1}{2}$ in. long, divided at the top into 2 short obtuse spreading lobes, the margins much recurved, green and villous on both sides, or the under one more hoary. Flower-heads very dense, in compact terminal cymes. Brown bracts tomentose outside. Calyx nearly 1 line long, very hirsute. Petal-claws slender. Disk of 5 distinct prominent glands close above the ovary.—*Trymalium halmaturinum,* F. Muell., Reissek, in Linnæa, xxix. 283.

**S. Australia.** Sandy scrub, Kangaroo Island, *E. G. Sealy, Waterhouse.*

19. **S. bifidum,** *F. Muell. Herb.* A low heath-like shrub, the tomentum close and stellate. Leaves linear-cuneate, forked at the top, with 2 short obtuse or hooked lobes, $\frac{1}{4}$ to $\frac{1}{2}$ in. long, the margins much revolute, glabrous above, tomentose underneath. Flower-heads in compact terminal compound heads, usually with 2 or 3 prominent and very tomentose floral leaves. Calyx about line long, hirsute with white hairs. Petals clawed. Disk annular, prominent, close above the ovary, at length separating into distinct glands.— *Trymalium bifidum,* F. Muell., Reissek, in Linnæa, xxix. 282 ; *T. stenophyllum,* Reissek, l. c.

**S. Australia.** Boston Point and Marble Range, *Wilhelmi.*

20. **S. subochreatum,** *Reissek, in Linnæa,* xxix. 287. A much-

branched heath-like shrub, the tomentum very close, stellate and hoary.
Leaves linear or linear-oblong, obtuse, $\frac{1}{4}$ to nearly $\frac{1}{2}$ in. long, the margins
much revolute, stellate-tomentose on both sides, or becoming at length gla-
brous above, occasionally appearing perfectly so from the under side being
concealed by the revolute margins. Stipules large and conspicuous, especially
at the base of the young shoots, where they are often above 2 lines long.
Flower-heads in dense compound terminal heads, of $\frac{1}{4}$ to $\frac{1}{2}$ in. diameter,
sessile amongst the last leaves, the floral leaves not very prominent. Flowers
considerably larger than in *S. vexilliferum.* Calyx 1 to $1\frac{1}{2}$ lines long, hirsute
or tomentose. Petal-claws short. Disk annular, undulate, very prominent,
but close above the ovary. Capsule usually ripening a single membranous
coccus.—F. Muell. Fragm. iii. 82 ; *Trymalium subochreatum,* F. Muell. in
Trans. Vict. Inst. i. 122 ; *T. Behrii,* Reissek, in Linnæa, xxix. 274 ; *T. polycephalum,* Turcz. in Bull. Mosc. 1858, i. 460.

**N. S. Wales.** Desert of the Darling and Murray, *F. Mueller.*
**Victoria.** Murray scrub, *F. Mueller.*
**S. Australia.** S. coast, *R. Brown ;* Boston Point, *F. Mueller.*
**W. Australia,** *Drummond, 5th Coll. Suppl. n.* 91 (the same number affixed also to
*Stenanthemum humile*). Phillips river and E. Mount Barren, *Maxwell.* These western
specimens are rather coarser and more tomentose, with larger leaves and flowers.
*Trymalium leucopogon,* F. Muell. ; Reissek, in Linnæa, xxix. 274, from the Murray de-
sert, appears to be a slight variety, with smaller, more glabrous leaves, and the hairs of the
calyx very white.

21. **S. oligocephalum,** *Benth.* Very near *S. subochreatum* and may
be only a variety, differing from it chiefly in the leaves, like those of *S.
vexilliferum,* perfectly glabrous above, or only slightly hoary when very
young, and usually much longer and narrower. Stipules remarkably large.
Flowers in dense terminal compound heads, sessile amongst the last leaves.
Calyx about 1 line long, densely tomentose-hirsute. Disk annular, undulate, more prominent in some flowers than in others, but always less so than
in *S. subochreatum.*—*Trymalium oligocephalum,* Turcz. in Bull. Mosc. 1858,
i. 460.

**W. Australia.** Cape Riche, *Drummond, 5th Coll. n.* 236.

22. **S. vexilliferum,** *Reissek, in Linnæa,* xxix. 285. A low, straggling,
heath-like shrub, with prostrate or suberect branches, not above a foot high,
the close stellate tomentum rusty or hoary, the young shoots often somewhat
glutinous. Leaves linear linear-oblong or lanceolate, obtuse, mostly $\frac{1}{2}$ in.
long, or in some specimens shorter, the margins much revolute, glabrous or
nearly so above, except the floral ones, tomentose underneath. Flower-heads
compound, very compact, 2 to 3 lines diameter, usually pedunculate, with 1
or 2 petiolate ovate floral leaves, very white and tomentose on both sides, or
rarely more like the cauline ones. Brown bracts very numerous. Calyx hispid,
scarcely 1 line long. Disk annular, close to the ovary. Petal-claws very
short. Fruiting calyx 2 lines long, with membranous cocci.—*Cryptandra
vexillifera,* Hook. Journ. Bot. i. 257 ; Hook. f. Fl. Tasm. i. 71 ; *Spyridium
phylicoides,* Reissek, in Linnæa, xxix. 286 ; *S. diffusum,* Reissek, l. c. 288.

**Victoria.** Deserts of the Murray and Murrumbidgee, *F. Mueller ;* in the Grampians,
*Wilhelmi.*
**Tasmania.** Port Dalrymple, *R. Brown ;* northern districts, *Gunn.*

434        XXXVI. RHAMNEÆ.        [Spyridium.

**S. Australia.** S. coast, *R. Brown*; from the mouth of the Murray to St. Vincent's Gulf, *F. Mueller.*

Var. *latifolium.* More slender and apparently procumbent. Leaves oblong, sometimes rather broadly so, the margins much less recurved. Flower-heads small.—Victoria and S. Australia. Some specimens seem almost to connect this form with *S. obovatum.* F. Mueller proposes to consider *S. vexilliferum* itself as a variety only of *S. eriocephalum*; but, besides the floral leaves, in all the flowers I have examined I have found the disk much closer upon the ovary. It is possible, however, that this character may not be so constant as it has appeared to be.

23. **S. eriocephalum,** *Fenzl, in Hueg. Enum.* 24. An erect, spreading, or prostrate heath-like shrub, with the young branches stellate-tomentose. Leaves linear, rigid, mostly with a short callous or often pungent point, about ¼ or rarely near ½ in. long, the margins closely revolute, glabrous above, the under side usually quite concealed. Flower-heads compound, 2 to 3 lines diameter, sessile or shortly pedunculate, usually with 1 or 2 floral leaves like the cauline ones, but broader. Calyx scarcely 1 line long, hispid with white hairs. Disk lining the calyx-tube and forming a ring at some distance above the ovary. Petals clawed. Capsule usually with only one perfect membranous coccus.—*Cryptandra eriocephala,* Hook. f. Fl. Tasm. i. 72; *Spyridium prostratum,* Reissek, in Linnæa, xxix. 284; *S. uncinatum,* Reissek, l. c. 289 (with the leaves more frequently pungent).

**N. S. Wales.** Eurylean scrub, *A. Cunningham.*
**Victoria.** Desert of the Murray, *F. Mueller.*
**Tasmania.** Derwent river, *R. Brown;* dry places above Hobarton and South Esk, *Gunn.*
**S. Australia.** Arid places from the mouth of the Murray to Spencer's Gulf, *F. Mueller.*

24. **S. microcephalum,** *Benth.* Apparently procumbent, much-branched, and heath-like, the young branches slender, with a minute rusty tomentum. Leaves linear, obtuse, with a minute callous point, mostly 2 to 3 lines long, the margins closely revolute, glabrous above, the tomentose under side quite concealed. Flower-heads compound, compact, seldom above 3 lines diameter, terminal or lateral, often with 1 or 2 prominent tomentose floral leaves. Calyx less than 1 line long, glabrous. Disk undulate, close above the ovary.—*Cryptandra microcephala,* Turcz. in Bull. Mosc. 1858, i. 458.

**W. Australia,** *Drummond, 5th Coll. n. 234.*

25. **S. ulicinum,** *Benth.* Tall, much-branched, and heath-like, the tomentum hoary or rusty. Leaves crowded, linear or linear-oblong, obtuse, emarginate or shortly bifid, mostly about ½ in. long, the margins revolute, glabrous above or hoary when young, the under side hoary with a very close tomentum. Flowers 1 to 3 together, closely sessile amongst the last leaves of short lateral branches, the central one enclosed in 3 or 4 brown imbricate bracts, the lateral ones with 2 each. Calyx about 2½ lines long, silky-hairy, the lobes nearly as long as the free part of the tube. Petals and stamens at the base of the calyx-lobes, in the sinus of the disk, which lines the calyx-tube and forms a thick undulating ring round the throat, at a considerable distance above the ovary.—*Cryptandra ulicina,* Hook. Journ. Bot. i. 257; Hook. f. Fl. Tasm. i. 72; *Stenodiscus ulicinus,* Reissek, in Linnæa, xxix. 296.

**Tasmania.** Common on the banks of the Derwent above New Norfolk and Launceston, also on the summit of Mount Wellington, *Gunn.* Although this differs from other *Spyridia,* and approaches *Stenanthemum* and *Cryptandra* in the greater length of the calyx-tube and almost separate flowers, yet the disk is as in *Spyridium,* and it appears better to consider it an extreme form of that genus than a monotypic genus as proposed by Reissek.

## 10. STENANTHEMUM, Reissek.

Flowers sessile in heads, surrounded by small, persistent, imbricate brown bracts. Calyx-tube adherent at the base, free, slender, and often deciduous above the ovary and disk, 5-lobed at the top. Petals 5, hood-shaped, enclosing the anthers and inserted with the stamens at the top of the calyx-tube. Disk scarcely prominent, round the top of the ovary at the base of the calyx-tube. Ovary wholly. inferior, 3-celled. Style entire or minutely 3-toothed. Capsule enclosed in the base of the calyx-tube, which is often contracted over it or deciduous ; the endocarp separating into 3 membranous or crustaceous cocci opening in 2 valves. Seeds of *Pomaderris.*—Shrubs, with the habit of *Spyridium.* Flowers sessile, in heads, or in one species in a cyme, surrounded by small, persistent, imbricate brown bracts, and sometimes with 1 or 2 floral leaves, as in *Spyridium.*

The. genus is confined to Australia. The floral characters are those of *Cryptandra,* with the inflorescence of *Spyridium.*

Leaves obovate, obcordate, or broadly oblong.
  Leaves rusty-tomentose underneath . . . . . . . . . . 1. *S. pomaderroides.*
  Leaves densely silky underneath. Erect or ascending and much
    branched . . . . . . . . . . . . . . . . . . . 2. *S. leucophractum.*
  Leaves closely white-tomentose underneath. Stems prostrate . . 3. *S. pimeleoides.*
Leaves linear-cuneate, emarginate or toothed . . . . . . . . 4. *S. coronatum.*
Leaves narrow-linear, the margins closely revolute.
  Flower-heads dense. Calyx tube narrow, very densely hirsute . . 5. *S. humile.*
  Flower-cymes loose. Calyx tube short, loosely hirsute . . . . 6. *S. Waterhousii.*

1. **S. pomaderroides,** *Reissek, in Linnæa,* xxix. 295. Branches wiry, elongated, above 1 ft. long in our specimens. Leaves distant, obovate or oblong, obtuse or with a recurved point, often ½ in. long or even more, narrowed into a petiole, folded lengthwise or concave, hoary or at length nearly glabrous above, rusty or white-tomentose underneath. Flower-heads 3 to 5 lines diameter, surrounded by 2 or 3 floral leaves. Brown bracts numerous, ovate or oblong, nearly as long as the flowers. Calyx 2½ lines long, silkytomentose outside, tubular but not very slender. Anthers obtuse.—*Cryptandra pomaderroides,* Reissek, in Endl. Nov. Stirp. Dec. 29, and Pl. Preiss. ii. 288 (from the description) ; *Cryptandra tridentata, β tomentosa,* Reissek, l. c. 289, and therefore included in *Stenanthemum tridentatum,* Reissek, in Linnæa, xxix. 295.

**W. Australia,** *Drummond, n.* 212; Murchison river, *Oldfield.*

2. **S. leucophractum,** *Reissek, in Linnæa,* xxix. 295. A low, erect or ascending, very much branched shrub, sometimes only a few inches, sometimes several feet high, the young branches rusty-tomentose. Leaves obovate or obcordate, with a recurved point, about ¼ in. long, folded lengthwise and narrowed into a distinct petiole, the upper surface white with a close soft to-

mentum, the under softly pubescent or densely villous with appressed whitish or rust-coloured hairs. Flower-heads rarely above 3 lines diameter, surrounded by 2 or more floral leaves and several brown bracts. Flowers usually few. Calyx fully 2 lines long, slender and silky-hairy precisely as in *S. pimeleoides*, but it does not appear to be so constricted nor to break off so readily above the ovary.—*Cryptandra leucophracta*, Schlecht. Linnæa, xx. 640.

**Victoria.** Murray desert, *F. Mueller.*

**S. Australia.** Sandy deserts and arid hills, from the Murray to Spencer's Gulf, *F. Mueller;* Kangaroo Island, *Waterhouse.*

3. **S. pimeleoides,** *Benth.* Low, prostrate, and much-branched, the young branches loosely pubescent-tomentose. Leaves obovate or obcordate, mostly 2 to 3 lines long, flat or folded upwards, often undulate and the edges very slightly recurved, glabrous or the upper ones hoary-tomentose on the upper side, white underneath with a close stellate tomentum, usually mixed with a few longer appressed sometimes silky hairs. Flower-heads very dense, ¼ to ½ in. diameter, with numerous imbricate brown bracts and often 2 or 3 tomentose floral leaves. Calyx fully 2 lines long, very slender, hirsute outside with white hairs, after flowering constricted above the ovary and often breaking off when the fruit ripens. Anther-cells rather acute at the lower end.—*Cryptandra (Stenocodon) pimeleoides*, Hook. f. Fl. Tasm. i. 75, t. 12.

**Tasmania.** East coast, at Great Swan Port, *Backhouse, C. Stuart;* Spring Bay, *Gunn,* F. Mueller (Fragm. iii. 77) refers this to *S. leucophractum*, to which it is certainly very nearly allied, but it must be considered at least as a well-marked variety in its prostrate habit and the much closer tomentum, the adult leaves (except the floral ones) nearly glabrous.

4. **S. coronatum,** *Reissek, in Linnæa,* xxix. 295. Small and apparently prostrate, the branches pubescent with scattered stellate hairs. Leaves cuneate, emarginate or 3-toothed, 3 to 4 lines long, usually folded lengthwise and softly tomentose on both sides. Flower-heads 3 to 4 lines diameter, sessile amongst 2 to 4 floral leaves, the brown bracts very small and narrow. Calyx nearly 2 lines long, not so slender as in the other species, tomentose outside. Anther-cells obtuse.—*Cryptandra coronata*, Reissek, in Pl. Preiss. ii. 288.

**W. Australia,** *Drummond, 2nd Coll. n.* 722.

5. **S. humile,** *Benth.* Stems 2 or 3 in. high, bare below, the flower-heads and leaves closely crowded in the upper part. Leaves narrow-linear, seldom ½ in. long, the margins closely revolute, nearly glabrous above, tomentose and with a few long woolly hairs underneath. Heads few-flowered, almost sessile amongst the leaves. Brown bracts very broad, obtuse or the midrib ending in a fine point. Calyx slender, 2 lines long, densely hispid with long white woolly hairs.

**W. Australia.** Between Moore and Murchison rivers, *Drummond, n.* 91 (the same number as *Spyridium polycephalum*, but probably from a different set).

6. **S. Waterhousii,** *Benth.* An erect somewhat viscid shrub, the branches slightly tomentose. Stipules linear-lanceolate. Leaves linear, obtuse or with a recurved point, ½ to ¾ in. long, the margins closely revolute, glabrous above, slightly tomentose underneath. Flowers not numerous, in rather loose leafy terminal cymes, and not so closely sessile as in the other

species, the floral leaves like those of the stem, or broader, flatter, and more tomentose.   Brown bracts 3 under each flower, lanceolate or ovate-lanceolate.
Calyx above 1 line long, the tube hirsute with spreading hairs, narrow-turbi-
nate, produced above the disk as in other *Stenanthema*, but not so slender.
Disk undulate-lobed, shortly adnate to the calyx-tube, but at a considerable
distance from the lobes and the petals.   Fruiting-calyx 2 lines long.   Cocci
coriaceous, indehiscent.—*Spyridium Waterhousii*, F. Muell. Fragm. iii. 83.

**S. Australia.**   Kangaroo Island, *Waterhouse.*

## 11. CRYPTANDRA, Sm.

(Wichurea, *Nees.*)

Calyx-tube adherent at the base, free, campanulate or tubular and persistent above the ovary and disk, 5-lobed at the top or to the middle.   Petals
5, hood-shaped, enclosing the anthers and inserted with the stamens at the
top of the calyx-tube.  · Disk annular, or often scarcely prominent round the
top of the ovary, at the base of the calyx-tube.   Ovary wholly inferior, or
slightly prominent in the calyx-tube, 3-celled.   Style entire or minutely 3-
toothed.   Capsule enclosed in the base of the persistent calyx-tube but often
partially free within it, the endocarp or the whole capsule separating into 3
crustaceous or rarely membranous cocci usually opening inwards in 2 valves.
Seeds of *Pomaderris.*—Shrubs, mostly heath-like or thorny.   Leaves small,
narrow, often clustered, rarely ovate and flat, often nearly cylindrical, the
under surface usually tomentose and whitish, but often concealed by the
closely revolute margins.   Flowers sessile or shortly pedicellate, mostly surrounded by persistent imbricate brown bracts, either distinct along the smaller
branches or clustered in terminal spikes or heads intermixed with leaves, never
in cymes.

A genus confined to Australia.   Like the majority of *Rhamneæ*, it is chiefly distinguished
by habit.   The floral characters of the first section are nearly those of *Stenanthemum*, of
the second scarcely distinct from *Discaria*, those of *C. glabriflora* almost as in *Spyridium*.

SECT. 1. **Cryptandra.**—*Disk usually pubescent, continuous with the summit of the
ovary, either undistinguishable from it or forming a slightly prominent ring round it.*

Flowers pubescent or hairy, closely sessile in terminal or lateral heads.
  Brown bracts acuminate.
  Calyx tubular.
    Heads many-flowered.   Calyx narrow.   Ovary almost entirely
      inferior . . . . . . . . . . . . . . . . .   1. *C. ericifolia.*
    Heads few-flowered.   Calyx rather broad.   Free part of the
      ovary longer than the adnate base . . . . . . . . .   2. *C. hispidula.*
    Calyx very small, broadly campanulate.   Flower-heads densely
      globular.
      Flower-heads terminal . . . . . . . . . . . .   3. *C. spyridioides.*
      Flower-heads lateral . . . . . . . . . . . . .   4. *C. scoparia.*
Flowers pubescent or hairy (except *C. glabriflora*), sessile in spikes
  or short heads, or not crowded.   Brown bracts obtuse, very much
  shorter than the calyx-tube.
  Calyx 1 line long or more, the tube longer than the lobes.
    Calyx narrow, glabrous outside at the base, tomentose above.
      Adnate base of the ovary longer than the free top . . . .   5. *C. spinescens.*
    Calyx broadly campanulate or urceolate, tomentose all over.
      Free part of the ovary longer than the adnate base . . . .   6. *C. amara.*

Calyx urceolate-globular, densely covered with white wool . . 7. *C. lanosiflora.*
Calyx-lobes as long as the tube or longer.
Calyx campanulate, usually 1 line long or more, and glabrous
    outside at the base . . . . . . . . . . . . . . . . 8. *C. tomentosa.*
Calyx very open, under 1 line, tomentose all over . . . . . 9. *C. nutans.*
Calyx glabrous, divided almost to the base . . . . . . . . 10. *C. glabriflora.*
Flowers often large, pubescent or hairy (except *C. glabriflora*), sessile.
Brown bracts broad, imbricate, covering the whole or a great
    portion of the calyx-tube.
Leaves broadly ovate, flat, mostly ¼ in. long . . . . . . . 15. *C. buxifolia.*
Leaves narrow and heath-like or minute, the margins revolute.
Stems slender, prostrate.  Calyx-lobes shorter than the tube . 11. *C. alpina.*
Stems rigid, divaricate.  Calyx small, glabrous, divided almost
    to the base . . . . . . . . . . . . . . . . . 10. *C. glabriflora.*
Stems rigid, divaricate.  Calyx silky-hairy, usually above 2 lines,
    the lobes narrow, about equalling the tube.
Leaves slender, about 1 line long . . . . . . . . . 12. *C. leucopogon.*
Leaves slender, mostly 2 to 3 lines long . . . . . . . 13. *C. propinqua.*
Leaves minute, obovoid, mostly ½ line long . . . . . . 14. *C. parvifolia.*
Flowers very small, pedicellate within the minute bracts.
Flowers pubescent . . . . . . . . . . . . . . . . 16. *C. pungens.*
Flowers glabrous . . . . . . . . . . . . . . . . . 17. *C. mutila.*

SECT. 2. **Wichurea.**—*Disk glabrous or villous, distinct from the ovary, usually annular.  Calyx glabrous or very slightly tomentose.*

Leaves linear, with revolute margins.
Calyx campanulate, deeply lobed.  Disk and ovary glabrous . . 18. *C. longistaminea.*
Calyx ovoid, not 1 line long.  Disk glabrous.  Summit of the
    ovary villous . . . . . . . . . . . . . . . . . 20. *C. miliaris.*
Calyx tubular, about 2 lines long.  Disk villous.  Summit of the
    ovary glabrous . . . . . . . . . . . . . . . . . 19. *C. arbutiflora.*
Leaves spathulate or linear-cuneate.  Calyx tubular, about 2 lines
    long.  Disk and ovary glabrous . . . . . . . . . . 21. *C. nudiflora.*

(*C. australis,* a name attributed to Smith by Roem. and Schult. Syst. iv. 372, is imaginary, made up of a part of Smith's generic character with the generic habitat.  *C. spinosa,* A. Cunn., quoted by Don under *Solenantha* (*Hymenanthera*), is also imaginary ; Cunningham, in the place referred to, Field, N. S. Wales, 352, gives no name to the plant.)

SECTION 1.  CRYPTANDRA.—Disk usually pubescent, continuous with the summit of the ovary, either undistinguishable from it or forming a slightly prominent ring round it.  Some of the first species pass almost into *Stenanthemum.*

1. **C. ericifolia,** *Sm. in Trans. Linn. Soc.* x. 294, *t.* 18, *f.* 1.  Branches elongated and twiggy, with few smaller branchlets, always unarmed, more or less pubescent with simple appressed hairs.  Leaves linear-terete or with a slightly prominent midrib, 2 to 4 lines long, often clustered or crowded, glabrous or pubescent with simple appressed hairs.  Flowers crowded in little terminal heads surrounded by leafy bracts, and each flower by several imbricate, acuminate, and ciliate brown bracts, often half as long as the calyx.  Calyx narrow-campanulate, about 2 lines long, silky-hairy outside, the lobes short and spreading.  Ovary very small, slightly projecting above the very short adnate part.  Style pubescent at the base.  Disk inconspicuous.  Cocci opening in 2 valves.—*C. capitata,* Sieb. Pl. Exs.

**N. S. Wales.**  Moist heaths near Sydney, *R. Brown, A. Cunningham, Sieber, n.* 66⁾ and others, but apparently not very common.

2. **C. hispidula,** *Reissek, in Linnæa,* xxix. 294. Very near *C. erici-folia,* but the leaves are smaller and more frequently pubescent, the flowers fewer, more silky, the calyx rather broader, 1½ to 2 lines long when fully out, and the free part of the ovary within the calyx is much longer than the adnate portion.

**S. Australia.** Encounter Bay and St. Vincent's Gulf, *F. Mueller, Whittaker.*

3. **C. spyridioides,** *F. Muell. Fragm.* iii. 68. A low, much-branched' divaricate shrub, rarely spinescent, the young branches minutely hoary. Leaves oblong-linear, obtuse, 2 to 3 lines long, the margins revolute, green and usually glabrous above, often hoary or whitish underneath with a minute tomentum. Flowers very small, in dense terminal globular heads. Brown bracts fringed or ciliate, not half so long as the calyx. Calyx silky-pubescent, about 1 line long, the adnate base narrow, the tube campanulate above the ovary, the lobes rather shorter than the tube. Summit of the ovary much depressed, thickened round the edge into an obscure disk.

**W. Australia.** Murchison river, *Oldfield.* Very closely allied to *C. scoparia.*

4. **C. scoparia,** *Reissek, in Pl. Preiss.* ii. 283. A rigid shrub, the branches in the original specimens virgate, heath-like, and seldom spinous, in others divaricately branched and frequently spinescent, very slightly hoary when young. Leaves linear, obtuse, 2 to 3 lines long, or in luxuriant specimens rather acute and attaining 3 or 4 lines, the margins revolute so as to be almost terete, usually glabrous. Flowers in dense globular clusters, almost sessile along the principal branches, and surrounded by a few short floral leaves, or borne on very short leafy branches, often above 3 lines diameter when fully out. Each flower sessile within 3 or 4 broad, brown, scarious, ciliate or fringed, shortly pointed bracts, about half as long as the calyx. Calyx when first open about ½ line long and silky-pubescent, when fully out about 1 line long and nearly glabrous, or with a tuft of long hairs on each lobe, broadly campanulate, the lobes longer than the tube. Summit of the ovary much depressed, thickened round the edge into an obscure disk.

**W. Australia.** Swan River, *Drummond;* sandy woods near Perth, *Preiss, n.* 1215. Var. *microcephala.* More branched with numerous slender spines. Flowers and heads small. Murchison river, *Oldfield.*

5. **C. spinescens,** *Sieb. in DC. Prod.* ii. 38. Nearly allied to *C. amara,* and with nearly the same foliage, but the branches are usually more twiggy and the spinous branchlets more densely crowded. Leaves usually linear or linear-oblong, 2 or rarely 3 lines long, but occasionally small and obovate. Flowers smaller than in *C. amara,* and more distinctly although very shortly pedicellate. Calyx 1½ to 2 lines long, narrow-campanulate, the adnate base glabrous and suddenly contracted into a little stipes about the length of the imbricate brown bracts, the free part white-tomentose outside. Ovary almost entirely inferior, the pubescent summit slightly prominent above the adnate part and obscurely grooved opposite the stamens, but without any distinct disk. Capsule oblong, 1½ to 2 lines long, almost included in the glabrous, elongated, adnate base of the calyx-tube, shortly free in the upper part. Cocci thinly crustaceous.—*C. pyramidalis,* R. Br., Brongn. in Ann. Sc. Nat. x. 373.

**N. S. Wales.** About Port Jackson and on the Nepean river, *R. Brown, Sieber, n.* 68, and *Fl.' Mixt. n.* 691; N. of Bathurst, *A. Cunningham ;* Cabramatta, *Woolls.* This is considered by F. Mueller (Fragm. iii. 67) as an abnormal state of *C. amara,* but I find the characters constant in numerous specimens from various collectors, both in flower and fruit.

6. **C. amara,** *Sm. in Trans. Linn. Soc.* x. 295, *t.* 18, *f.* 2. A rigid, wiry, decumbent or suberect, much-branched shrub, the young branches minutely hoary with a close stellate down, the smaller ones often ending in a fine thorn. Leaves solitary or clustered, linear or linear-oblong, usually 1 to 2 and rarely 3 lines long, obtuse or acute, rigid, glabrous or nearly so, the margins usually recurved. Flowers almost sessile, solitary within the bracts, but usually several together, forming short leafy spikes or racemes on the smaller branches. Calyx at the time of flowering, 1 to 1½ lines long, campanulate, white outside with a close minute down, very shortly adnate by its obtuse base, the lobes usually shorter than the tube, the brown imbricate bracts not exceeding the adnate base and very obtuse. Ovary densely pubescent, included in the tube, but adnate only below the middle, the disk not distinct. Fruiting calyx often 3 lines long, enclosing the capsule, which remains adherent at the base only or below the middle. Cocci crustaceous.—DC. Prod. ii. 38 ; F. Muell. Fragm. iii. 66 ; *C. Sieberi,* Fenzl, in Hueg. Enum. 23 ; Hook. f. Fl. Tasm. i. 74 ; *C. campanulata,* Schlecht. Linnæa, xx. 639 ; F. Muell. Fragm. iii. 67, partly ; *C. nervata,* Reissek, in Linnæa, xxix. 291 ; *C. largiflora,* F. Muell., Reissek, in Linnæa, xxix. 292.

**Queensland.** Kent's Lagoons, *Leichhardt ;* Mount Mitchell, *Beckler.*
**N. S. Wales.** Port Jackson, *R. Brown, Sieber, n.* 67, and *Fl. Mixt. n.* 492 ; northward to Clarence River, *Beckler,* and New England, *C. Stuart ;* in the interior to the Lachlan river, *Fraser ;* St. George's Range, *A. Cunningham ;* Darling and Murray desert, *Herb. F. Mueller.*
**Victoria.** Arid hills and stony tracts, ascending into the Alps, *F. Mueller.*
**Tasmania.** North Esk river, *Lawrence, Gunn,* and others.
**S. Australia.** Between the Murray and St. Vincent's Gulf, *Behr, F. Mueller.*
Independently of the diversity in the size of the flowers resulting from age, there appear to be two distinct varieties with large and small flowers, the calyx in the latter usually broader and more deeply lobed, both of them included among Sieber's specimens ; the southern ones belong chiefly to smaller-flowered varieties. These have usually the free part of the ovary less prominent, but in Cunningham and Fraser's specimens from the interior the ovary and capsule are very prominent, whilst the calyx is small and much more loosely pubescent than usual. Some specimens are remarkable for their short, almost ovate leaves.

7. **C. lanosiflora,** *F. Muell. Fragm.* iii. 65. A divaricately-branched shrub, of 1 to 2 ft., the young branches minutely hoary, not spinescent in our specimens. Leaves linear or linear-oblong, 1 to 3 lines long, the margins revolute, glabrous. Flowers almost sessile, few together at the ends of the branches, forming short, leafy, oblong or almost globular spikes. Calyx globular, 1½ to nearly 2 lines diameter, densely covered with a very white crisped wool, the lobes much shorter than the tube, the brown imbricate bracts very broad and obtuse, about half as long as the tube. Ovary very short, almost wholly inferior, the summit expanded into a pubescent slightly undulate disk. Capsule more than half superior.

**N. S. Wales.** Mountains of New England on the Severn, *C. Stuart ;* Mount Mitchell, towards the Clarence river, *Beckler.*

8. **C. tomentosa,** *Lindl. in Mitch. Three Exped.* ii.ꞏ178. Very much branched, but seldom thorny, the young branches tomentose. Leaves linear or oblong, obtuse or acute, 1 to 2 lines long or rarely more, the margins recurved and frequently hoary underneath. Flowers usually 5 to 8 together, clustered at the ends of the branches, in short spikes or almost heads. Calyx varying in size from about 1 to 1½ lines, rather urceolate than campanulate, the lobes usually at least as long as the tube, very spreading when fully out, but often connivent again after flowering, slightly tomentose outside, except at the base. Ovary and capsule nearly as in *C. amara*, from which this species may be generally distinguished by its smaller leaves, by the whole plant often minutely hoary pubescent, by the flowers more crowded in shorter heads, and by the deeper-lobed calyx, glabrous outside at the base, and only slightly silky-tomentose on the lobes.—*C. propinqua*, Schlecht. Linnæa, xx. 638, not A. Cunn.; *C. erubescens*, F. Muell., Reissek, in Linnæa, xxix. 293; *C. Behriana*, Reissek, l. c.; *C. campanulata*, F. Muell. Fragm. iii. 67, partly.

**N. S. Wales.** In the interior, *Fraser.*
**Victoria.** In the Grampians, *Mitchell, F. Mueller;* on the Murray and generally in the N.W. interior, *Herb. F. Mueller.*
**S. Australia.** From the Murray to Spencer's Gulf, *F. Mueller.*
*S. divaricata*, Reissek, in Pl. Preiss. ii. 286, from Mitchell's early expeditions, must probably also be referred to *C. tomentosa*. I have seen no authentically-named specimen, but the only one of Mitchell's collections answering to the character given scarcely differs from the common forms of *C. tomentosa*.

9. **C. nutans,** *Steud. in Pl. Preiss.* i. 186. In habit and foliage this species much resembles *C. tomentosa*, but the flowers are different. Leaves rarely above 2 lines long, pubescent or glabrous. Flowers small, crowded in short terminal spikes, or sometimes few and not so close. Brown bracts not one-third the length of the calyx, and often shortly acuminate. Calyx very broadly campanulate, about 1 line long or rather less, hoary or almost silky outside, the lobes deep and very spreading. Free part of the ovary broader and flatter than in *C. tomentosa*. Disk inconspicuous.—*C. tomentosa*, Reissek, in Pl. Preiss. ii. 286, not of Lindl.

**W. Australia.** Swan River, *Drummond, 1st Coll.* and *2nd Coll. n.* 246, *Roe*, etc.; sandy woods near the sea, *Preiss, n.* 2424; Champion Bay, *Oldfield.*
Var. (?) *micrantha.* Flowers about ¾ line long, or even less.—Swan River, *Drummond, Roe, Harvey ;* William river, *Oldfield.*

10. **C. glabriflora,** *Benth.* Branches numerous, rather rigid, divari-cate, often spinescent, glabrous or nearly so. Leaves linear or oblong, obtuse, 1 or rarely 2 lines long, the margins revolute, glabrous. Flowers sessile and clustered along the branches, usually quite glabrous. Brown bracts broad, imbricate, covering the very short tube. Calyx very broadly campanulate, 1 to 1¼ lines long, the lobes very spreading, reaching almost to the ovary. Ovary more than half inferior, thickened into a broad disk at the top.

**W. Australia.** Murchison river, *Oldfield.* The habit of this species is entirely that of *Cryptandra*, whilst the extreme shortness of the calyx-tube above the ovary or disk brings it almost into *Spyridium.*

11. **C. alpina,** *Hook. f. Fl. Tasm.* i. 75, t. 12. A small prostrate species, with numerous slender wiry branches, rarely extending above 6 in., with little heath-like glabrous leaves, seldom more than 1 line long. Flowers

mostly solitary at the ends of the branches.   Brown bracts broad, imbricate, obtuse or acute, the inner ones often nearly as long as the calyx-tube. Calyx broadly campanulate, tomentose outside, rather more than 2 lines long, with ovate-lanceolate lobes, rather shorter than the tube.   Disk undulate, villous, scarcely distinct from the summit of the ovary.

**Tasmania.**   On the summits of the Western Mountains, about 3800 ft. elevation, *Gunn, Archer.*

12. **C. leucopogon,** *Meisn., Reissek, in Pl. Preiss.* ii. 287.   Very nearly allied to *C. propinqua,* and may prove to be a variety only, the flowers and bracts being similar in shape and relative proportions, but the slender branches and small leaves are more like those of *C. alpina,* except that the stems are apparently erect, not prostrate.   The flowers are also rather smaller than in *C. propinqua,* and the calyx-lobes have longer silky hairs.

**W. Australia.**   Sandy plains of the Gordon river, *Preiss, n.* 752.   (*Herb. Sond.*)

13. **C. propinqua,** *A. Cunn., Fenzl, in Hueg. Enum.* 23.   A rigid, divaricate, heath-like shrub, nearly glabrous.   Leaves crowded or clustered on the smaller branches, linear-terete, mostly 2 to 3 lines long, and usually acute.   Flowers 3 to 8 together at the ends of the branches, and larger than in most species.   Calyx varying from $2\frac{1}{2}$ to $3\frac{1}{2}$ lines long, very silky-hairy outside, the tube enclosed within the broad, brown, ciliate, imbricate bracts, the lobes narrow-lanceolate, fully as long as the tube.   Disk round the ovary continuous with it, but prominent and often nearly glabrous.

**N. S. Wales.**   In the interior, *A. Cunningham, Mitchell;* between Bathurst Plains and Wellington Valley, *Fraser;* N.W. branch of Hunter's River, *A. Cunningham;* Paramatta, *Woolls;* New England, near Tenterfield, *C. Stuart.*

Var. *grandiflora.*   Flowers exceeding 3 lines in length.—*C. magniflora,* F. Muell. Fragm. iii. 65.—Sandy desert between the Darling and Murray, *Herb. F. Mueller.*   This variety is also amongst Cunningham's plants, who had given it the name of *C. speciosa,* and designated the smaller variety by that of *propinqua,* as being near the larger one.   Unfortunately this latter name was the only one in the Vienna herbarium, and was thus, although inappropriate, adopted by Fenzl for the species, and has given rise to the opinion that some variety of the common *C. amara* was intended by it.

14. **C. parvifolia,** *Turcz. in Bull. Mosc.* 1858, i. 459.   Branches very rigid, divaricate, the young ones hoary with a minute stellate down, and appearing at first sight deprived of all leaves except distant clusters of minute stipules, amongst which however will generally be found 2 or 3 minute obovate to linear leaves, thick, very obtuse or with a minute recurved point, seldom 1 line long, the margins revolute.   Flowers solitary or 2 to 6 together, closely sessile at the summits of the branches.   Calyx about 3 lines long, the tube closely covered with large, brown, obtuse, imbricate bracts, the lanceolate lobes silky outside and spreading.   Summit of the ovary broad and depressed, thickened round the margin into a pubescent disk.

**W. Australia,** *Drummond, 4th Coll. n.* 156.

15. **C. buxifolia,** *Fenzl, in Hueg. Enum.* 23.   Stems erect from a woody rhizome, but little branched, hoary with a minute stellate tomentum.   Leaves ovate, obtuse or pointed, mostly about $\frac{1}{2}$ in. long, glabrous above, white underneath, giving the plant a very different aspect from the rest of the genus.   Flowers sessile, in terminal leafy heads.   Calyx tubular-campanulate,

nearly 3 lines long, hoary-tomentose outside, the tube nearly covered by the brown imbricate bracts, the lobes short, narrow, and spreading.   Ovary scarcely prominent at the bottom of the tube, flat at the top, but without any distinct disk.

**N. S. Wales.**  Rocky hills on the meridian of Bathurst, on the parallel of 30° 50′; Mount Yongo on the route to Hunter's River, and Goulburn river, *A. Cunningham.*

16. **C. pungens,** *Steud. in Pl. Preiss.* i. 187.   Resembling in habit *C. spinescens,* the numerous short branches terminating in slender spines. Leaves mostly fasciculate, 2 to 3 lines long, obtuse or with a slightly recurved point.   Flowers small and numerous, on pedicels of $\frac{1}{2}$ to nearly 1 line long, with minute, imbricate, acuminate, brown bracts at their base, and not under the calyx.   Calyx about $\frac{3}{4}$ line long, broadly campanulate, the lobes fully as long as the tube, softly pubescent outside.   Free part of the ovary very broad and flat, and slightly thickened on the edge into a villous disk.   Fruiting calyx more turbinate, above 1 line long, the pubescent capsule nearly as long as the calyx-lobes.—*C. holostyla,* Steud. in Pl. Preiss. i. 188.

**W. Australia.**  Swan River, *Drummond ;* sandy woods and limestone hills near the sea, *Preiss, n.* 2422 and 2423; south-west coast, *Baxter.*

17. **C. mutila,** *Nees, Reissek, in Pl. Preiss.* ii. 289.   A low heath-like shrub, with slender virgate almost spinescent branches, hoary with minute stiff hairs.   Leaves linear, mostly $1\frac{1}{2}$ to 2 or scarcely 3 lines long, the margins much revolute, glabrous or nearly so. Flowers in little sessile clusters in the upper axils, forming short, dense, terminal or nearly terminal leafy racemes, each flower on a pedicel of 1 to $1\frac{1}{2}$ lines, within 3 or more minute brown bracts at the base of the pedicel.   Calyx about $\frac{3}{4}$ line long, glabrous outside, the lobes very spreading.   Free part of the ovary broad and flat, the edge thickened into a minute almost 5-lobed disk.   Fruit not seen.

**W. Australia.**  Swan River, *Drummond, 2nd Coll., n.* 723 ; Freemantle, *Collie, Oldfield*; limestone hills near the sea, *Preiss, n.* 1217 and 1229.

SECTION 2.  WICHUREA, *Nees* (as a genus).—Disk glabrous or villous, distinct from the ovary, usually annular and rather broad.   Flowers usually glabrous, except in *C. longistaminea,* where they are slightly tomentose.   The characters of this section are very nearly those of *Discaria,* especially in the flower.   It is however at once known by the habit, alternate leaves, and small fruits.

18. **C. longistaminea,** *F. Muell. Fragm.* iii. 64.   A much-branched unarmed shrub of 2 or 3 ft., the smaller branches minutely hoary-tomentose. Leaves ovate or oblong, obtuse, 1 to 2 lines long, the margins recurved or revolute, glabrous above, minutely silky-tomentose underneath or almost glabrous.   Flowers numerous, crowded on the smaller branches, but not quite sessile.   Brown bracts imbricate round the base of the calyx-tube.   Calyx about 2 lines long, minutely silky outside, divided below the middle into spreading lobes.   Petals on slender claws, at first enclosing the stamens, but reflexed after the calyx opens, leaving the stamens erect and apparently exserted.   Disk annular, glabrous or very minutely tomentose, quite distinct from the ovary.   Ovary sessile or slightly immersed in the disk.   Style very shortly 3-lobed.   Fruit not seen.

**N. S. Wales.** New England, *C. Stuart.*

19. **C. arbutiflora,** *Fenzl, in Hueg. Enum.* 26. Branches virgate, slightly pubescent, with numerous short branchlets occasionally spinous. Leaves narrow-linear, obtuse or with a minute recurved point, 1 to 3 lines long, with the margins much revolute so as to be almost terete. Flowers white, fragrant, sessile, or very shortly pedicellate on the smaller branches, not crowded, quite glabrous, the broad obtuse imbricate brown bracts forming a minute cup at their base. Calyx about 2 lines long, broadly tubular, with very short lobes. Disk undulate, villous, covering the small glabrous top of the ovary, which is almost entirely free from the calyx, but enclosed in the tube. Capsule filling the calyx-tube, glabrous, the disk remaining round its base. Cocci indehiscent or 2-valved.—*Wichurea arbutiflora,* Nees, in Pl. Preiss. ii. 290; *C. suavis,* Lindl. Bot. Reg. 1844, t. 56.

**W. Australia.** Swan River, *Drummond, 1st Coll.;* sandy woods near Guildford, *Preiss, n.* 465 and 472; King George's Sound, *Huegel.*
Var. *tubulosa.* More slender and spinous, resembling *C. spinescens* in aspect; branches almost or quite glabrous; calyx-tube very slender.—*C. tubulosa,* Fenzl, in Hueg. Enum. 26; *Wichurea tubulosa,* Nees, in Pl. Preiss. ii. 291.—Swan River, *Huegel, Drummond;* shady rocks on the N. side of Mount Clarence, *Preiss, n.* 473; Vasse river and Murchison river, *Oldfield.*

20. **C. miliaris,** *Reissek, in Pl. Preiss.* ii. 288. Branches long and virgate, with numerous short spinous branchlets, as in *C. spinescens.* Leaves nearly as in that species, narrow-linear, 2 to 3 lines long, the margins recurved or revolute, glabrous or pubescent. Flowers very small, not quite sessile, forming little loose leafy racemes or clusters on the side-branches. Calyx campanulate, less than 1 line long in our specimens, but not fully out, the very obtuse, imbricate, brown bracts nearly half as long as the calyx; lobes of the calyx as long as the tube. Disk glabrous, undulate, close round the pubescent ovary. Fruit not seen.—*C. lasiophylla* and *C. glabrata,* Steud. in Pl. Preiss. i. 188.

**W. Australia.** Sandy woods near Perth, *Preiss, n.* 2420.
*C. tenuiramea,* Steud. in Pl. Preiss. i. 189, from W. Australia, Preiss, n. 2419, very imperfectly described from a specimen not yet in flower, which I have not seen, may be this species, but it is utterly unrecognizable.

21. **C. nudiflora,** *F. Muell. Fragm.* iii. 64. Branches decumbent or divaricate, the short branchlets often rigid but scarcely spinescent in our specimens. Leaves linear-cuneate or spathulate, obtuse or truncate, 2 to 6 lines long, flat or conduplicate. Flowers pedicellate, clustered with small leaves along the branches, but not crowded, the acuminate brown bracts very small at the base of the pedicels. Calyx quite glabrous, about 2 lines long, broadly tubular, the lobes short. Disk annular, rather thick, undulate, glabrous as well as the ovary, but quite distinct from it. Ovary quite free, sessile on the centre of the disk. Fruit not seen.

**W. Australia.** Port Gregory and Murchison river, *Oldfield.* In floral characters this species is almost a *Discaria,* but the habit is quite that of *Cryptandra.*

## 12. DISCARIA, Hook.

(Tetrapasma, *G. Don.*)

Calyx campanulate or tubular above the ovary, shortly 4- or 5-lobed. Petals hood-shaped, inserted with the stamens at the base of the calyx-lobes or none. Stamens 4 or 5, with short filaments, included in the petals when present. Disk annular in the base of the calyx-tube, the margin shortly free. Ovary more or less immersed in the disk, 3-lobed, 3-celled; style slender, with a shortly 3-lobed stigma. Drupe or capsule coriaceous, 3-lobed, the endocarp separating into 3 2-valved crustaceous cocci. Seeds with a coriaceous testa; albumen fleshy; cotyledons orbicular.—Much-branched rigid shrubs, with opposite, often thorny branchlets. Leaves small, opposite, 1-nerved or penninerved. Stipules and bracts small. Flowers axillary.

The genus is chiefly S. American, extratropical or alpine, with one species endemic in Australia and another in New Zealand.

1. **D. australis,** *Hook. Bot. Misc.* i. 157, *t.* 45. A scrubby, much-branched, thorny shrub of 1 to 2 ft., usually glabrous. Branches green, terete, the smaller ones reduced to stout spines of 1 to 1½ in. Leaves often appearing clustered from the shortness of the shoots, oblong or cuneate, obtuse or emarginate, rarely exceeding ½ in. Pedicels solitary or clustered in the axils of small leaves, which soon fall off from the very short branches, the flowers then appearing densely clustered under the spines. Calyx-tube broadly campanulate above the disk, the limb spreading to about 2 lines diameter. Petals narrow, hood-shaped. Ovary deeply immersed in the disk, the short free part 3-lobed. Fruit 2 to 3 lines diameter.—Hook. f. Fl. Tasm. i. 69; Reissek, in Linnæa, xxix. 266; F. Muell. Fragm. iii. 83; *Colletia pubescens,* Brongn. in Ann. Sc. Nat. x. 366; *Tetrapasma juncea,* G. Don, Gen. Syst. ii. 40; *Colletia Cunninghamii,* Fenzl, in Hueg. Enum. 23.

**N. S. Wales.** Cox's, Macquarie's, and Hunter's rivers, *A. Cunningham;* Liverpool plains, *Woolls;* Ben Lomond, New England, *Beckler.*

**Victoria.** Grassy hills and banks, ascending the Lower Alps, Delatite river, between Loddon and Creswick rivers, Snowy River, etc., *F. Mueller.*

**Tasmania.** Derwent river, *R. Brown;* Launceston road and South Esk river, *Gunn;* Great Swan Port, *Backhouse;* Brown river, *Oldfield.*

## Order XXXVII. AMPELIDEÆ.

Flowers regular, hermaphrodite or unisexual. Calyx small, entire or 4- or 5-toothed. Petals 4 or 5, free or cohering, valvate in the bud. Stamens 4 or 5, opposite the petals, inserted on the outside of the disk at its base or between its lobes. Disk free or adnate to the ovary. Ovary usually immersed in or surrounded by the disk, more or less perfectly 2- to 6-celled; style short and conical or subulate, or none; stigma small, capitate or lobed. Ovules 2 in each cell where there are 2 cells, solitary where there are more cells, erect, anatropous, with a ventral raphe. Fruit a berry, the dissepiments frequently disappearing. Seeds 1 to 6; testa hard, the inner coating frequently penetrating into the fissures of the ruminate albumen. Embryo short, in the base of the albumen; cotyledons oval; radicle short, inferior.—

Woody climbers or rarely erect shrubs or small trees. Branches often articulate. Leaves alternate or the lower ones opposite, simple or compound, the petiole usually articulate with the stem and expanded into a membranous stipule. Flowers small, in little umbels, cymes, racemes, or spikes, arranged in leaf-opposed, cymose, thyrsoid, or elongated panicles.

The Order, almost or quite limited to the two following genera, is widely dispersed over the tropical and warm regions of the globe, more abundant in the Old World than in America, and the smaller genus confined to the Old World. It is very nearly allied to *Celastrineæ*, and especially to *Rhamneæ*, from which it differs in habit, in the more developed petals, in the baccate fruit and in the smallness of the embryo.

Stamens free. Ovary 2-celled with 2 ovules in each cell. Woody climbers, with
tendrils . . . . . . . . . . . . . . . . . . . . . . . . . . . . . . 1. VITIS.
Stamens and petals connate with the disk. Ovary 3- to 6-celled with 1 ovule in
each cell. Erect, without tendrils . . . . . . . . . . . . . . . . . 2. LEEA.

## 1. VITIS, Linn.

### (Cissus, *Linn.*)

Petals free or cohering at the tips, and falling off together. Stamens inserted round the base of the short, annular, or lobed disk. Ovary 2-celled (sometimes imperfectly so), with 2 ovules in each cell.—Woody climbers or rarely bushy shrubs, with leaf-opposed tendrils (abortive inflorescences). Leaves simple or compound, sometimes marked with pellucid dots. Panicles in the Australian species cymose or rarely reduced to solitary umbels. Petals very concave, almost hood-shaped, but without the dorsal appendages of some Asiatic species.

The genus comprises nearly the whole of the Order, extending over the whole of its geographical area. Of the 14 Australian species, 3 are widely distributed over tropical Asia, another extends to the Fiji Islands, the remaining 10 are endemic. The Australian species appear tolerably constant in the division of their leaves, but that character is not to be absolutely relied on, for the trifoliolate, digitate, and pedate forms will occasionally pass one into the other.

*Leaves simple.*
　Leaves ovate, penniveined, or 3-nerved at the base, rather fleshy.
　　Leaves shortly acuminate, mostly toothed. Berries globular. Tall,
　　　woody climbers . . . . . . . . . . . . . . . . . . . . 1. *V. antarctica.*
　　Leaves very obtuse, quite entire. Berries obovoid. Bushy tree . 2. *V. oblonga.*
　Leaves broad-cordate, 5-nerved, membranous.
　　Branches glaucous. Veinlets reticulate, not prominent. Flowers
　　　at least 1 line diameter . . . . . . . . . . . . . . 3. *V. cordata.*
　　Not glaucous. Veinlets transverse. Flowers not ½ line diameter 4. *V. adnata.*
*Leaflets 3.*
　Leaflets ovate, rather thick and firm, shining. Cymes nearly globular,
　　on very short peduncles. Stigma very broad . . . . . . . 5. *V. nitens.*
　Leaflets large, broadly ovate or cordate, membranous. Cymes loose,
　　divaricate.
　　Leaves glabrous, or nearly so. Flowers fully 1 line diameter, on
　　　stout pedicels . . . . . . . . . . . . . . . . . . . . 6. *V. saponaria.*
　　Leaves hairy on both sides. Flowers about ½ line diameter, on
　　　filiform pedicels . . . . . . . . . . . . . . . . . . 7. *V. acris.*
　Leaflets mostly under 2 in., rather thick, or almost fleshy, coarsely
　　toothed. Cymes loose, divaricate . . . . . . . . . . . . 8. *V. trifolia.*

*Leaflets* 5 *to* 9, *pedate.*
  Leaflets small, ovate, acuminate, deeply toothed. Disk very promi-
    nent. . . . . . . . . . . . . . . . . . . . . . . 9. *V. clematidea.*
  Leaflets 2 to 3 in. long, oblong or cuneate, minutely and remotely
    serrate or entire. Disk inconspicuous . . . . . . . . . 10. *V. acetosa.*
*Leaflets* 5, *rarely* 3, *digitate.*
  Leaflets obtuse at the base, on a distinct slender petiolule, coriaceous,
    and very reticulate . . . . . . . . . . . . . . . 11. *V. hypoglauca.*
  Leaflets narrowed into a very short petiolule or sessile.
    Leaflets very coriaceous. Berries ovoid . . . . . . . 12. *V. sterculifolia.*
    Leaflets membranous. Berries globular.
      Leaflets linear-cuneate to oblong or obovate. Cymes loose . . 13. *V. opaca.*
      Leaflets narrow-linear, rarely broad and acuminate. Cymes
        compact . . . . . . . . . . . . . . . . . . 14. *V. angustissima.*

1. **V. antarctica,** *Benth.* Young shoots more or less clothed with short rust-coloured hairs, rarely entirely glabrous. Leaves simple, petiolate, ovate or oblong, mostly acuminate and slightly cordate, 3 to 4 in. long and 1½ to 2 in. broad, entire, sinuate or irregularly toothed, rather firm or almost coriaceous, penniveined and obscurely 3-nerved, with glands on the under side in the axils of some of the principal veins. Cymes dense, broadly corymbose, shorter than the petioles. Flowers tomentose-pubescent, the buds nearly globular, under 1 line diameter. Petals 4, separately deciduous. Disk prominent, undulate, obscurely 4-lobed. Style shortly conical. Berry globular.—*Cissus antarctica,* Vent, Choix, t. 21; DC. Prod. ii. 629; Bot. Mag. t. 2488; *C. glandulosa,* Poir. Dict. Suppl. i. 105.

**Queensland.** Brisbane river, Moreton Bay, *F. Mueller.*
**N. S. Wales.** Port Jackson, *R. Brown,* and others; northward to Hastings and Macleay rivers, *Beckler;* New England, *C. Stuart;* southward to Illawarra, *A. Cunningham, Herb. Mueller.* The specific name, although inappropriate, is too generally sanctioned by use to be altered.

2. **V. oblonga,** *Benth.* A small bushy tree (according to Henne's notes, but R. Brown's specimens have tendrils), quite glabrous or the young shoots minutely rusty-tomentose, the branches rigid and flexuose. Leaves petiolate, broadly oblong or ovate-oblong, very obtuse, 1½ to 2½ in. long, quite entire, firm but thinner than in *C. antarctica,* very finely penniveined and obscurely 3-nerved, with 2 large glands underneath in the axils of the lateral nerves. Flowers not seen. Fruiting cymes on short peduncles, bearing few obovoid berries.

**Queensland.** E. coast, *R. Brown;* Curtis Island, *Henne.* On some cymes the berries are replaced by a monstrous growth of dichotomous branches covered with small, broad, leafy scales, forming dense globular tufts of 3 or 4 in. diameter, like those often observed on some *Mæsas.* Although I have seen no flowers, the inflorescence, fruits, and seeds, as well as the tendency to articulation of the smaller branches, leave no doubt of the species belonging to the present genus.

3. **V. cordata,** *Wall. Catal. n.* 6008 (partly). Very glabrous and often somewhat glaucous in all its parts, the young stems succulent and disarticulating in the dried specimens. Leaves on rather long petioles, broadly cordate, 2½ to nearly 4 in. long and nearly as broad, entire, except small, almost bristle-like distant teeth, 5-nerved, the smaller veins reticulate, very few or none, transverse, and faintly conspicuous. Flowers in corymbose trichotomous

cymes, the buds about 1 line diameter. Petals 4, usually cohering at the top and falling off together. Style subulate. Berries obovoid-globular.—Benth. Fl. Hongk. 54; *Cissus cordata*, Roxb. Fl. Ind. i. 407; *Vitis cardiophylla*, F. Muell. Fragm. ii. 73.

**N. Australia.** N. coast, *R. Brown.*

**Queensland.** Barnard Islands, *M'Gillivray;* Burdekin river, *F. Mueller;* Rockhampton, *Thozet.* Common in the Archipelago and Eastern India, extending northward to Sikkim and Hongkong.

4. **V. adnata,** *Wall.; Wight and Arn. Prod.* 126 (with the synonyms adduced). Young shoots and under side of the leaves more or less covered with a short tomentum, which sometimes disappears with age. Leaves petiolate, broadly cordate, almost orbicular, acuminate, 3 to 6 in. diameter, bordered with small bristle-like teeth, 5-nerved and penniveined, the primary veins connected by transverse veinlets. Flowers scarcely $\frac{1}{2}$ line diameter, numerous in corymbose cymes. Petals 4, cohering by the tips and falling off together. Style shortly subulate, at least in the fertile flowers. Fruit globular, small.—*Cissus adnata*, Roxb.; Wight, Ic. t. 144.

**N. Australia.** N. coast, *R. Brown;* Sea Range, very rare, *F. Mueller.* Common in East India.

5. **V. nitens,** *F. Muell. Fragm.* ii. 73. Quite glabrous. Leaflets 3, ovate or oval-oblong, acuminate, mostly 3 to 4 in. long, remotely toothed, narrowed at the base, the lateral ones scarcely oblique, on short petiolules, rather firm, smooth and shining above. Umbel-like cymes almost glabrous, dense and nearly globular, 2 or 3 together or solitary on a very short common peduncle, the pedicels very short. Flower-buds ovoid, rather more than 1 line long. Petals 4 or rarely 5, oblong, falling off separately. Disk inconspicuous. Style very short and thick, with a broad, flat, almost fringed, slightly 2-lobed stigma. Berry ovoid.

**Queensland.** E. coast, *R. Brown;* Dawson and Burnett rivers, *F. Mueller;* Brisbane river, *Fraser, F. Mueller.*

**N. S. Wales.** Clarence, Macleay, and Hastings rivers, *Beckler;* Hunter's River, *R. Brown, F. Mueller.*

6. **V. saponaria,** *Seem. Syst. List Vit. Pl.* 4. Young leaves and shoots and inflorescence minutely hoary-tomentose. Leaflets 3, very broadly ovate, acuminate, entire or crenate, attaining 4 to 6 in., thin and glabrous when full-grown, penniveined and more or less distinctly 5-nerved at the base, especially the lateral ones, with transverse veinlets, the central one rounded at the base, the lateral ones obliquely cordate. Cymes loose, divaricate, many-flowered, on long peduncles. Flowers nearly globular, above 1 line diameter. Petals 4, usually falling off together. Disk broad. Style conical. Berry depressed-globular.

**Queensland.** Torres Straits, *R. Brown;* Cape York and Piper's Island, *M'Gillivray.* Also in the Fiji Islands, where, according to Seemann, the stems are used in washing linen. A. Gray in Bot. Amer. Expl. Exped. i. 272, had referred this plant with doubt to *Cissus geniculata*, Bl., and perhaps correctly so, for although Blume describes the central leaflet as oblong-lanceolate, yet he mentions a broad-leaved variety, but with more pubescent leaves. All are closely allied to the common E. Indian *V. pedata*, Wall., and may be a 3-foliolate variety of that very variable species.

7. **V. acris,** *F. Muell. Fragm.* ii. 75.    Branches and leaves softly pubescent or hairy.    Leaflets 3, broadly ovate, acuminate, crenate, 3 to 4 in. long, thin, hairy on both sides, penniveined with transverse veinlets, the lateral leaflets oblique, obscurely cordate, and more or less 5-nerved at the base, on petiolules of ¼ to ½ in.    Cymes loose and divaricate, on long slender peduncles, the branches almost filiform and nearly glabrous.    Flowers nearly globular, about ¼ line diameter.    Petals 4, apparently distinct.    Disk very prominent. Style short, conical.

**Queensland.**   Between Burnett and Pine rivers, *F. Mueller.*
**N. S. Wales.**   Richmond and Clarence rivers, *Beckler.*
The foliage is that of *V. mollissima,* Wall., from the Archipelago, from which the species appears to differ chiefly in the very slender inflorescence and small flowers. These may, however, not be full-grown in the very few specimens seen.

8. **V. trifolia,** *Linn. Spec. Pl.* 293.    Softly hoary-pubescent all over, especially the young shoots, or sometimes nearly or quite glabrous.    Leaflets 3, ovate-acuminate, obovate or rhomboid, usually 1 to 2 in., rarely 3 in. long, coarsely and irregularly toothed or crenate, softly herbaceous, usually thick and sometimes almost fleshy, the lateral ones very oblique, on short petiolules. Cymes many-flowered, divaricate, on long peduncles, hoary or pubescent. Flowers nearly globular, about 1 line diameter.    Petals 4, distinct.    Disk very prominent.    Style in some specimens short with a broad peltate stigma, in others slender with a small stigma.    Berry small, depressed-globular. —*Cissus carnosa,* Lam. ; DC. Prod. i. 630 ; *C. cinerea,* Lam. ; DC. l. c. 631 ; *C. crenata,* Vahl ; DC. l. c. ; *Vitis carnosa,* W. and Arn. Prod. 127 ; Wight, Ic. t. 171 (a broad-leaved form) ; *V. psoralifolia,* F. Muell. Fragm. ii. 75.

**N. Australia.**   N. coast, *R. Brown ;* Victoria river, *F. Mueller ;* Albert river, *Henne.*
**Queensland.**   Cape York, *M'Gillivray.*
The species is very common in East India and the Archipelago, and is probably described under several names besides those above quoted.

9. **V. clematidea,** *F. Muell. Fragm.* ii. 74.    Minutely tomentose, pubescent, or glabrous.    Branches angular-striate.    Leaflets usually 5, pedate, petiolate, ovate, acuminate, coarsely toothed or lobed, usually 1 to 2 in. long, narrowed at the base, herbaceous, rather thick and pubescent or thin and glabrous.    Cymes divaricate, rather dense, on long peduncles, minutely hoary-tomentose.    Pedicels short.    Flowers globular, about 1 line diameter.    Petals apparently separating.    Disk very prominent, entire.    Style filiform.    Berries depressed-globular, small.

**Queensland.**   Brisbane river, *Fraser, F. Mueller.*
**N. S. Wales.**   Port Jackson, *R. Brown ;* northward to Clarence river, *Beckler ;* New England, *C. Stuart ;* Newcastle, *Leichhardt ;* southward to Kiama, *Harvey.*

10. **V. acetosa,** *F. Muell. Herb.*    Glabrous or the young shoots and inflorescence very slightly hoary-tomentose.    Leaflets 5 to 7, pedate, petiolulate or the central one nearly sessile, oblong or obovate-cuneate, obtuse or rarely shortly acuminate, 2 to 3 in. long or rarely longer, entire or bordered by small teeth or minute distant serratures, narrowed at the base, herbaceous, but rather firm, pale underneath.    Cymes pedunculate, dense, divaricate or almost thyrsoid, the flowers often shortly racemose along the branches, on short pedicels.    Flowers purple-red, ovoid-globular, about 1 line long,

glabrous. Petals separating. Disk indistinct. Style very shortly conical or scarcely any, with a truncate stigma. Berries ovoid-globose.—*Cissus acetosa,* F. Muell. Trans. Vict. Inst. iii. 24.

**N. Australia.** N. coast, *R. Brown;* Victoria and Fitzmaurice rivers, *F. Mueller;* Sweers Island, *Henne.* The specimens first described were, according to F. Mueller's notes, from tall herbaceous not climbing stems, but others are evidently climbing, with the usual tendrils.

11. **V. hypoglauca,** *F. Muell. Pl. Vict.* i. 94. Young shoots rusty-tomentose or villous, adult specimens usually quite glabrous. Leaflets 5, digitate, obovate, oval or oblong-elliptical, shortly and often acutely acuminate, 2 to 3 in. long, the lateral ones smaller than the central ones, entire or toothed towards the top, obtuse at the base, on rather long petiolules, coriaceous, penniveined and finely reticulate, pale or glaucous underneath. Cymes rather dense, shortly pedunculate. Flowers yellowish, glabrous, ovoid, fully 1 line long. Petals separating or slightly cohering. Disk 4-lobed, but not very prominent. Style conical. Berry nearly globular, rather small.—*Cissus hypoglauca,* A. Gray, Bot. Amer. Expl. Exped. i. 272 ; *C. australasica,* F. Muell. in Trans. Phil. Soc. Vict. i. 8.

**N. S. Wales.** Port Jackson, *R. Brown,* and others; northward to Clarence river, *Beckler;* New England, *C. Stuart;* southward to Kiama, *Harvey;* Twofold Bay, *F. Mueller.*

**Victoria.** Forest streams and rivulets in eastern Gipps' Land, *F. Mueller.*

12. **V. sterculifolia,** *F. Muell. Herb.* Fruiting specimens quite glabrous. Leaflets 5, digitate, elliptical-oblong or somewhat obovate, shortly and obtusely acuminate, 3 to 4 in. long, entire, narrowed into a very short petiolule, coriaceous, penniveined, the reticulate veinlets much less conspicuous than in *V. hypoglauca,* with glands or foveolæ in the axils of some of the primary veins underneath, Flowers not seen. Fruiting cymes on short peduncles. Berries ovoid, rather large.

**N. S. Wales.** Hastings river, *Beckler.* One specimen has a very young flower-cyme, which is slightly rusty-pubescent, but not far enough advanced to give the floral characters.

13. **V. opaca,** *F. Muell. Herb.* Quite glabrous. Leaflets 5, rarely 3 or 4, digitate, from linear-cuneate to elliptical-oblong, obovate or narrow rhomboidal, obtuse or acuminate, mostly 1 to 2 in. long, entire or slightly toothed, narrowed at the base into very short petiolules or almost sessile, rather firm but not coriaceous, smooth, obscurely penniveined, usually pale underneath. Cymes rather loose, but not large. Flowers glabrous, globular, about 1 line diameter. Petals 5 or rarely 4, separating. Disk prominent, entire or scarcely lobed. Style short, conical. Berries depressed-globular.— *Cissus opaca,* F. Muell. in Trans. Vict. Inst. iii. 23.

**Queensland.** Burdekin river, *F. Mueller;* Brisbane river, Moreton Bay, *Fraser, F. Mueller;* Rockhampton, *Thozet;* Port Denison, *Fitzalan;* E. coast, *R. Brown* (with the leaves mostly 3-foliolate).

14. **V. angustissima,** *F. Muell. Fragm.* i. 141. Glabrous and rather slender. Leaflets usually 5, digitate, narrow-linear, 1 to 3 in. long, entire, coarsely toothed or lobed, narrowed at the base; occasionally, however, the lower ones are slightly pedate or united into 3 cuneate and coarsely toothed leaflets, or into a single broad palmately-lobed leaf. Cymes compact and

many-flowered, ½ to 1 in. broad, on rather long peduncles. Flowers fully 1 line diameter. Petals 5, separating. Disk broad, undulate. Style short, conical, with a truncate stigma. Berries nearly globular.

**W. Australia,** *Drummond, n.* 43 and 218; Murchison river, *Oldfield.* At first sight this closely resembles the S. American *Cissus palmata,* Poir., but that species has more ovoid buds, 4 petals falling off together, and a smaller disk.

## 2. LEEA, Linn.

Petals united in a campanulate corolla with 5 spreading or recurved lobes. Disk (resembling a staminal tube) cup-shaped, conical, or nearly globular, 5-lobed, enclosing the ovary. Stamens inserted in grooves outside the disk, the filaments incurved at the top, with the anthers inside the disk in the bud. Ovary enclosed in the disk, 3 to 6-celled, with 1 ovule in each cell.—Shrubs or small trees, without tendrils. Leaves once, twice, or thrice pinnate, with large entire or toothed penniveined leaflets. Panicles or cymes leaf-opposed, corymbose. Flowers usually larger than in *Vitis.*

The genus is dispersed over tropical Asia and Africa, the only Australian species being the most common among the Asiatic ones.

1. **L. sambucina,** *Willd. Spec. Pl.* i. 1177. A tall, glabrous, coarse shrub, the young branches occasionally furrowed. Leaves mostly twice or thrice pinnate; leaflets few in each pinna, from ovate to oblong-elliptical or lanceolate, acuminate, usually 3 to 6 in. long and 1½ to 2 in. broad, but sometimes twice as long, irregularly crenate, the primary arcuate pinnate veins and transverse veinlets very prominent underneath. Cymes large, divaricate, trichotomous, on short peduncles. Flowers about 2 lines long, on very short pedicels. Ovary 5-celled. Berries small, depressed-globular, usually ripening 4 to 6 seeds.—DC. Prod. i. 635; *L. staphylea,* Roxb., W. and Arn. Prod. 132, with the synonyms adduced; Wight. Ill. t. 58 and Ic. Pl. t. 78.

**N. Australia.** Raffles Bay, Goulburn Island, and other points of the N. coast, *A. Cunningham.*
**Queensland.** Islands of Howick's group, *F. Mueller.*
The species is common in tropical Asia, and is, perhaps, the same as a common African one.

ORDER XXXVIII. **SAPINDACEÆ.**

Flowers usually polygamous. Sepals 4 or 5, free or united in a small toothed or lobed calyx, imbricate or rarely valvate in the bud. Petals as many as sepals, or 1 fewer, sometimes minute or wanting, frequently bearing a scale inside. Disk various, in some genera unilateral, rarely wanting. Stamens 8, rarely fewer or more, inserted round the ovary within the disk (except in a few genera not Australian), sometimes unilateral; anthers versatile or erect. Ovary entire or lobed, 1- to 4-celled, most frequently 3-celled. Style simple, with a single stigma, or more or less divided. Ovules 1, 2, or rarely more in each cell, ascending, or rarely horizontal, with the micropyle inferior. Fruit dry or succulent, dehiscent or indehiscent, entire or separating into cocci. Seeds with or without an arillus, without albumen (except in a few genera not Australian). Embryo usually thick, frequently folded or

2 G 2

spiral, the cotyledons usually unequal, collateral or superposed ; radicle short, turned downwards or reascending towards the hilum.—Trees, shrubs, or rarely almost herbaceous, often climbers (especially in genera not Australian). Leaves alternate (or in genera not Australian opposite), usually compound, pinnate with, or more frequently without, a terminal odd one, the leaflets often irregularly alternate, rarely decompound ; 3-foliolate or simple. Flowers usually small.

*Sapindaceæ* are abundant within the tropics, both in the New and in the Old World, more rare in the temperate regions of the northern hemisphere, and those, chiefly of the genera *Æsculus, Acer,* and their allies, unrepresented in Australia ; there are very few also in southern extratropical Africa or America. Of the 16 Australian genera, 6 small ones are endemic or only extend to Timor, and the most numerous, *Dodonæa,* is nearly so, with the exception of 1 or 2 ubiquitous tropical species. Five of the genera are common to the tropical regions of the New and the Old World ; the remaining 4 restricted to tropical Asia or extend only into Africa.

The majority of *Sapindaceæ* are readily known by the disk outside, not inside the stamens, and by the 8 stamens in a 5-merous flower, with a 3-merous gynœcium ; but all these characters have exceptions, which render the technical limitations of the Order difficult, although really doubtful genera are very few. The position of the micropyle appears to be constant, but often difficult to observe. The arboreous genera with pinnate leaves, often numerous in species, especially in tropical Asia, may require considerable modification as to their characters, and probably some reduction, when those proposed by Blume come to be better known, as well as to flower as fruit.

Flowers irregular, either 1 petal fewer then the sepals, or the stamens or disk unilateral, and ovary excentrical.
  One ovule in each cell of the ovary.
    Herbaceous or half-herbaceous climber with biternate leaflets.
      Capsule inflated, membranous . . . . . . . . . .  1. CARDIOSPERMUM.
    Trees with pinnate leaves. Petals 1 fewer than sepals.
      Calyx valvately 5-lobed. Capsule loculicidally 3-valved . .  2. DIPLOGLOTTIS.
      Sepals 5, broadly imbricate. Fruit deeply divided into oblong indehiscent lobes . . . . . . . . . . .  3. ERIOGLOSSUM.
    Shrubs or trees, with 1 or 3 digitate leaflets. Sepals 4, broadly imbricate. Petals 4 or none. Fruit of 1 or 2 indehiscent lobes. . . . . . . . . . . . . . .  4. SCHMIDELIA.
  Two ovules in each cell of the ovary. Low shrubs or undershrubs, with entire, lobed, or pinnately dissected leaves . .  5. DIPLOPELTIS.
Flowers regular. Disk annular or none. Stamens all round the ovary.
  One ovule in each cell of the ovary. Trees or tall shrubs. Leaves pinnate (except *Heterodendron* and sometimes in *Atalaya*).
    Capsule loculicidally 3-valved.
      Sepals distinct, broadly imbricate . . . . . . . . .  6. CUPANIA.
      Calyx small, toothed, or the lobes valvate or slightly imbricate . . . . . . . . . . . . . . . . .  7. RATONIA.
    Fruit separating into winged samaras . . . . . . .  8. ATALAYA.
    Fruit divided into indehiscent or 2-valved lobes or irregularly loculicidal, the valves not separating from the axis.
      Leaves pinnate.
        Sepals broadly imbricate in 2 rows. Petals usually exserted. Fruit-lobes smooth, indehiscent . . . . .  9. SAPINDUS.
        Calyx-teeth or lobes valvate or slightly imbricate. Petals very small or none. Fruit-lobes smooth (in Australia), indehiscent or 2-valved . . . . . . . . . .  10. NEPHELIUM.

Calyx-segments imbricate. Petals very small or none.

Fruit-lobes tuberculate or muricate, indehiscent . . . 11. EUPHORIA.

Leaves coriaceous, simple, entire or pinnatifid. Calyx entire
or minutely toothed . . . . . . . . . . . . 12. HETERODENDRON.

Two ovules in each cell of the ovary.

Trees with pinnate leaves. Petals 4 or 5.

Calyx deeply divided into imbricate segments. Disk inconspicuous . . . . . . . . . . . . . . . . . 13. HARPULLIA.

Calyx campanulate, shortly lobed. Disk broad . . . . 14. AKANIA.

Shrubs or rarely small trees. Leaves simple or pinnate with
small leaflets. Calyx cup-shaped. Petals none. Disk inconspicuous.

Stamens in the male flowers 10 or fewer, usually 8 . . . 15. DODONÆA.

Stamens in the male flowers more than 10 . . . . . . 16. DISTICHOSTEMON.

## 1. CARDIOSPERMUM, Linn.

Flowers polygamous. Sepals 4, broadly imbricate, the 2 outer ones small. Petals 4, 2 larger with a large scale, 2 smaller with a crested scale. Disk one-sided, almost reduced to 2 prominent glands opposite the lower petals. Stamens 8, oblique. Ovary excentrical, 3-celled, with 1 ovule in each cell; style very short, with 3 stigmatic lobes. Capsule vesicular, membranous, more or less 3-cornered, 3-celled, opening loculicidally. Seeds globose, with a thick funicle or small aril; testa crustaceous; cotyledons large, transversely folded.—Herbs or undershrubs, mostly climbing. Leaves dissected. Flowers few, small, on long axillary peduncles, which usually bear a tendril under the panicle.

A small genus, chiefly American, of which 2 species are also spread over the Old World within the tropics, and a third is perhaps confined to the Old World. The Australian species is one of those most widely diffused in both worlds.

1. **C. Halicacabum,** *Linn.; DC. Prod.* i. 601. A straggling or somewhat climbing annual or perhaps perennial, attaining several feet in length, glabrous or slightly pubescent. Leaf-segments usually twice ternate, ovate or ovate-lanceolate, coarsely toothed or lobed, the upper leaves smaller, narrower and less divided. Peduncles 2 to 3 in. long, bearing a double or treble short recurved tendril under the small panicle, which is often reduced to an umbel of few small white flowers. Capsules flat on the top, usually pubescent.—A. Gray, Gen. Ill. t. 181; Wight, Ic. t. 508.

**N. Australia.** Victoria river, Sea range, etc., *F. Mueller;* Albert river, *Henne.*

**Queensland.** N.E. coast, *R. Brown;* Rockhampton, *Thozet.*

The species is common in most tropical regions. The Australian specimens belong either to the variety with fruits scarcely ¾ in. diameter, often considered as a distinct species (*C. microcarpum,* H. B. and K.), or are intermediate between that and the typical form, with fruits above 1 in. diameter.

## 2. DIPLOGLOTTIS, Hook. f.

Calyx deeply 5-lobed, valvate. Petals 4, the place of the fifth vacant, the inner scale divided into two. Disk one-sided, crescent-shaped. Stamens 8, ascending, unequal. Ovary 3-celled, style short, incurved; stigma entire or obscurely 3-lobed. Ovules solitary in each cell. Capsule nearly globular, thick, somewhat fleshy, loculicidally 3-valved. Seeds enclosed in a pulpy arillus.—A tree, with large pinnate leaves, more or less villous-tomentose. Flowers not very small, in large axillary panicles.

The genus is limited to a single species, endemic in Australia.

1. **D. Cunninghamii,** *Hook. f. in Benth. and Hook. Gen. Pl.* 395. A tree of 30 to 40 ft., the young branches, petioles and inflorescence densely clothed with a soft rust-coloured tomentum. Leaves very large, sometimes exceeding 2 ft.; leaflets 8 to 12, opposite or irregularly alternate, oblong-elliptical to ovate-lanceolate, acute or obtuse, usually 6 to 8 in., but sometimes above 1 ft. long, glabrous above, pubescent underneath, with raised parallel pinnate veins. Flowers numerous, on pedicels of 1 to 2 lines, clustered along the branches of the ample panicle. Calyx about 1½ lines long, rusty-tomentose. Petals about as long as the calyx, orbicular, thin, ciliate, the two inner scales not united, about as long as the petal itself, but thicker, and very hairy. Stamens exserted in some specimens, shorter than the petals in others. Fruit about ½ in. diameter, tomentose.—*Cupania Cunninghamii,* Hook. Bot. Mag. t. 4470.

**Queensland.** Brisbane river, *A. Cunningham;* also in *Leichhardt's* collection.

**N. S. Wales.** Hunter's River, *R. Brown;* Hastings river, *A. Cunningham, Fraser, Beckler;* Clarence river, *Wilcox;* Illawarra, *Ralston.* With the habit and fruit of a *Cupania*, this plant has the flowers of a *Paullinia.*

## 3. ERIOGLOSSUM, Blume.

Flowers polygamous. Sepals 5, broadly imbricate, the two outer ones smaller. Petals 4, the place of the fifth vacant, the scale hirsute with a terminal lobed appendage. Disk one-sided, lobed. Stamens 8, turned to one side, unequal. Ovary 3-lobed, 3-celled; style slender, obscurely 3-lobed; ovules solitary in each cell. Fruit divided to the base into 3 oblong indehiscent lobes. Seeds without any arillus; testa membranous, embryo straight; cotyledons thick.—Trees with pinnate leaves, more or less tomentose. Flowers not very small, in cymes or clusters along the branches of terminal panicles.

The genus contains very few species, natives of tropical Asia and Africa; one of the most widely spread extending into Australia. It differs from *Sapindus,* as *Diploglottis* from *Cupania,* in the irregular flowers.

1. **E. edule,** *Blume, Bijdr. and Rumphia,* iii. 119, *t.* 166. A tall tree, the young shoots, petioles and inflorescence more or less hoary or rusty with a close tomentum. Leaflets 8 to 12, elliptical-oblong or rarely ovate-lanceolate, more or less acuminate, 3 to 4 or rarely 5 in. long, glabrous above, pubescent underneath, with prominent parallel pinnate veins. Flowers numerous. Sepals orbicular, rather thick, pubescent outside, the inner larger ones about 1½ lines diameter. Petals rather longer, the scale shorter than the petal, very hairy in the lower part, the terminal glabrous appendage expanded either into 2 lobes or in a broad fringed erect crest, but very variable. Fruit not seen in the Australian specimens.—*Sapindus rubiginosus,* Roxb. Pl. Corom. i. 44, t. 62; W. and Arn. Prod. 112, with the synonyms quoted.

**N. Australia.** Brunswick Bay, N.W. coast, *A. Cunningham.* The species is widely spread over tropical Asia and the Indian Archipelago.

## 4. SCHMIDELIA, Linn.

Flowers polygamous. Sepals 4, broadly imbricate, the outer ones smaller. Petals 4, small, or rarely none. Disk one-sided, usually lobed or divided into 4 glands. Stamens 8, more or less one-sided. Ovary excentrical, 2 or rarely 3-celled ; style 2- or 3-lobed ; ovules solitary in each cell. Fruit of 1 or rarely 2 small ovoid or globular indehiscent, fleshy or almost dry berries. Seeds with a short arillus; embryo curved, cotyledons folded.—Shrubs or trees. Leaves with 1 or 3 leaflets. Flowers very small, in simple or loosely paniculate axillary racemes.

The species are numerous in tropical America, with several African ones, and a few in tropical Asia and the Indian Archipelago, one of the common Asiatic ones extending to Australia. The genus is one of the most easily recognized in the Order, by its foliage as well as by its small flowers and fruits.

1. **S. serrata,** *DC. Prod.* i. 610. A tree, the young leaves and shoots pubescent-tomentose, often glabrous when full-grown. Leaflets 3, ovate or obovate-oblong, obtuse or slightly acuminate, 2 to 4 in. long, irregularly and coarsely toothed, or rarely quite entire, sessile or narrowed into a short petiolule, glabrous above, pale or pubescent underneath, often bearing hairy tufts in the axils of the principal veins. Racemes slender, simple or slightly branched. Flowers ½ to nearly 1 line diameter, on short pedicels, clustered along the pubescent rhachis. Petals cuneate, with a minute scale. Disk of 4 small lobes or glands. Stamens glabrous. Berries small, globular.—W. and Arn. Prod. 110 ; *Ornitrophe serrata*, Roxb. Pl. Corom. i. 44, t. 61 ; *S. timoriensis*, DC., Dene. Herb. Timor. 115.

**N. Australia.** N. coast, *R. Brown* ; Port Essington, *Armstrong.* The latter specimens are nearly glabrous, with the leaflets more sessile and narrowed at the base, as described in *S. timoriensis.* Some of R. Brown's are similar ; others are more pubescent, like the common form in India, where these characters are very variable ; and, as suggested by W. and Arn., these plants may all be varieties only of *S. Cobbe,* Linn., which would thus have a very wide range over tropical Asia, including the Archipelago.

## 5. DIPLOPELTIS, Endl.

Flowers polygamous. Sepals 5, persistent, imbricate in the bud. Petals 4, the place of the fifth vacant, clawed, without any scale inside. Disk very oblique, produced into a concave or apparently double scale. Stamens 8, within the disk, turned to one side. Ovary 2- or 3-lobed, 2- or 3-celled ; style ascending, usually twisted ; ovules 2 in each cell, superposed halfway up the inner angles. Capsule 2- or 3-celled, opening loculicidally in as many valves, or separating into cocci. Seeds usually solitary in each carpel ; testa crustaceous ; arillus small ; embryo spirally rolled.—Shrubs or undershrubs, more or less glandular-pubescent. Leaves alternate, entire or pinnatifid. Panicles terminal, with scorpioid racemes. Flowers white pink or violet, larger than in most *Sapindaceæ.*

The genus is limited to Australia.

Fruit separating into distinct indehiscent cocci.
     Leaves ovate or obovate, on distinct, rather long petioles . . . . 1. *D. petiolaris.*
     Leaves linear, oblong, cuneate, or pinnatifid, narrowed into very short
         petioles or sessile . . . . . . . . . . . . . . . . 2. *D. Huegelii.*

Capsule membranous, loculicidally 3-valved. Leaves linear or cuneate,
entire or 3-lobed . . . . . . . . . . . . . . . . . . . 3. *D. Stuartii.*

1. **D. petiolaris,** *F. Muell. Herb.* Nearly allied to *D. Huegelii,* of
which F. Mueller thinks it may be a variety. Branches, panicles, and both
sides of the leaves very glandular, and apparently viscid. Leaves crowded,
ovate or obovate, ¾ to 1½ in. long, irregularly crenate or lobed at the base,
on petioles of 3 or 4 lines. Panicle more crowded than in *D. Huegelii,* with
smaller flowers. Cocci separating, and similar to those of *D. Huegelii,*
except that they are much more glandular and less hairy.

**W. Australia.** Murchison river, *Oldfield.*

2. **D. Huegelii,** *Endl. in Hueg. Enum.* 13. A shrub of 2 or 3 ft.,
but flowering also as an undershrub of 1 to 1½ ft., the branches and foliage
hoary with a minute tomentum, or softly pubescent or hirsute. Leaves either
undivided and from oblong-linear to broadly cuneate, entire or coarsely
toothed, or more or less deeply pinnatifid, with short, oblong or cuneate,
entire or 2- or 3-toothed lobes or segments, always narrowed at the base but
scarcely petiolate. Flowers racemose along the simple branches of a terminal
panicle, with a few glandular-tipped hairs on the branches and sometimes on
the sepals and ovary; the males and females usually in the same raceme.
Sepals broadly ovate, about 1 line long. Petals spreading, on short slender
claws, the lamina orbicular, about 3 lines broad, those next the vacancy often
smaller than the others. Ovary hirsute with simple and glandular hairs.
Fruit separating into 3 rather hard ovoid cocci, about 2 lines long, rugose,
usually indehiscent.—Lindl. Bot. Reg. 1839, t. 69; F. Muell. Fragm. iii. 12,
Lehm. in Pl. Preiss. ii. 235; *D. Preissii,* Miq. in Pl. Preiss. i. 223 (with
pinnatifid leaves); *D. Lehmanni,* Miq. l. c. i. 224 (with entire leaves).

**W. Australia.** Swan River, *Drummond,* 1st *Coll., Preiss, n.* 1281 *and* 1282, and
others, and thence to Murchison river, *Drummond, n.* 95, *Oldfield.* I have seen no specimens from King George's Sound or any of the southern districts. The foliage is very
variable, and the disk also appears to vary in shape; the inner margin or lobe is, however,
generally shorter than the outer one.

Var. (?) *eriocarpa.* Apparently diffuse, softly pubescent or hirsute. Leaves deeply pinnatifid with several cuneate, entire or toothed segments. Ovary very hirsute. The young
fruit also very hirsute, and, apparently longer, more lobed and more membranous than in the
ordinary form, but not seen full-grown.

**N. Australia.** Nichol Bay, N.W. coast, *F. Gregory.*

3. **D. Stuartii,** *F. Muell. Fragm.* iii. 12. A shrub apparently diffuse,
the branches pubescent and glandular. Leaves linear or cuneate, entire or
3-lobed at the end, ½ to ¾ in. long, nearly glabrous above, hirsute underneath.
Racemes simple in one specimen, divided into two in the other (both mere fragments), glandular-pubescent and hirsute. Flowers rather smaller than in *D.
Huegelii.* Margins or lobes of the disk nearly equal. Ovary very hirsute.
Capsule 4 or 5 lines long (3-lobed?), membranous, opening loculicidally in
3 valves.

**N. Australia.** Between Mount Morphett and Bonny river, *M'Douall Stuart* (*Herb.
F. Muell.*)

## 6. CUPANIA, Linn.

Flowers regular, polygamous. Sepals 4 or 5, imbricate in the bud. Petals either as many as sepals, small, with or without scales inside, or none. Disk usually annular. Stamens usually 8 to 10, inserted inside the disk; filaments short, rarely as long as the calyx. Ovary 2- or 3-celled, rarely 4-celled, with 1 ovule in each cell. Capsule obovoid or rarely globular, coriaceous or hard, 2- or 3-, rarely 4-celled, often angled or lobed, opening loculicidally in as many valves as cells. Seeds usually more or less covered by an arillus ; testa crustaceous or coriaceous ; embryo curved ; cotyledons plano-convex.— Trees or rarely tall shrubs. Leaves alternate, pinnate ; leaflets alternate or opposite, with or without a terminal one. Flowers small, in small axillary or terminal panicles, sometimes almost reduced to simple racemes. Petals rarely as long as the sepals.

A large tropical genus, both in the New and the Old World, the precise limits of which are very difficult to fix, and are very differently viewed by different botanists. The Australian species are all endemic, as far as hitherto known.

Sepals orbicular, much imbricate.
  Sepals glabrous or ciliate only.
    Leaflets obtuse, pale or glaucous underneath. Capsule nearly
      sessile, deeply 3-lobed . . . . . . . . . . . . 1. *C. semiglauca.*
    Leaflets acuminate, very oblique, green on both sides. Capsule
      stipitate, 3-angled . . . . . . . . . . . . . . 2. *C. punctulata.*
    Leaflets coriaceous, obtuse. Capsule nearly sessile, slightly 3-
      lobed, very coriaceous . . . . . . . . . . . 3. *C. anacardioides.*
  Sepals tomentose.
    Leaflets glabrous, acutely serrate . . . . . . . . . . 4. *C. serrata.*
    Leaflets tomentose underneath, nearly entire . . . . . . 5. *C. tomentella.*
Sepals ovate, slightly imbricate. Capsule 3-angled or globular, the
  valves almost woody.
    Leaflets numerous, acuminate, serrate. Capsule very hirsute . . 6. *C. pseudorhus.*
  Leaflets few, entire or slightly toothed.
    Panicles little-branched or racemes simple. Petals very short
      and broad. Capsule woody, villous inside.
    Inflorescence often branched, upper male flowers sessile.
      Young shoots and under side of the leaves usually tomen-
      tose-pubescent . . . . . . . . . . . . . . . 7. *C. xylocarpa.*
    Racemes simple. Flowers all pedicellate. Leaves glabrous 8. *C. nervosa.*
  Panicles terminal, much branched (though short). Flowers all
    pedicellate. Petals oblong. Filaments rather long . . . 9. *C. Bidwilli.*

1. **C. semiglauca,** *F. Muell. Herb.* A middle-sized tree. Leaflets 2 to 4 or rarely 6, oblong-elliptical, or from almost obovate to nearly lanceolate, obtuse or rarely almost acute or mucronate, 2 to 3 or rarely nearly 4 in. long, entire, narrowed into a short petiolule, coriaceous, glabrous and somewhat shining above, more or less glaucous underneath. Panicles either small and axillary or terminal and much branched, but shorter than the leaves, glabrous or minutely pubescent. Pedicels short. Sepals orbicular, ciliate, otherwise glabrous, the larger inner ones about 1 line diameter. Petals shorter, with 2 cuneate hairy scales as long as the petal. Stamens exserted. Ovary glabrous, 3-lobed. Capsule 4 to 5 lines diameter, glabrous, very shortly attenuate at the base, with divaricate compressed lobes. Seeds smooth and shining,

with a thin arillus.—*Arytera semiglauca,* F. Muell. in Trans. Vict. Inst. iii. 25.

**Queensland.** Moreton Bay, *W. Hill, F. Mueller.*
**N. S. Wales.** Hastings and Clarence rivers, *Beckler ;* Paramatta, *Woolls ;* Blue Mountains, *Miss Atkinson ;* S. of the colony, rare, *A. Cunningham ;* Kiama, *Harvey.*

2. **C. punctulata,** *F. Muell. Fragm.* iii. 12. A tall shrub, quite glabrous. Leaflets usually 4 to 7, on a long slender common petiole, very obliquely ovate-lanceolate, acuminate, 3 to 4 in. long, quite entire, thinly coriaceous, smooth and shining, minutely pellucid-dotted, narrowed into a petiolule of ½ in. or more. Flowers not seen. Fruiting panicles short, slender, clustered in the axils or at the ends of the branches. Pedicels short. Sepals often persistent or reflexed, orbicular, about 1 line long, glabrous. Capsule glabrous, 3-angled, flat at the top with the remains of the style forming a point in the centre, about 4 lines broad, contracted into a short obconical stipes, half opening in 3 coriaceous valves. Seeds not seen.

**Queensland.** Cumberland Islands, *Fitzalan.*

3. **C. anacardioides,** *A. Rich. Sert. Astrol.* 33, *t.* 13. A slender tree, quite glabrous or with a minute hoariness on the inflorescence. Leaflets 6 to 10, usually 8, from broadly ovate or obovate to elliptical-oblong, very obtuse, 2½ to 4 in. long, rounded at the base and shortly petiolulate, quite entire, coriaceous. Flowers rather large for the genus, in pedunculate cymes along the branches of loose panicles. Sepals orbicular, the inner ones 2 lines broad, slightly ciliate. Petals small, orbicular, with 2 very short obovate hirsute scales at the base. Stamens 10 ; filaments short, hirsute ; anthers oblong. Ovary villous. Capsule glabrous, coriaceous, acutely and divaricately 3-lobed, 6 to 8 lines broad, very shortly attenuate at the base.

**N. Australia.** Port Essington, *Armstrong.*
**Queensland.** Brisbane river, Moreton Bay, *Fraser, A. Cunningham, F. Mueller ;* Burdekin river, *F. Mueller.*
**N. S. Wales.** Port Jackson, *R. Brown* and others; Hastings river, *Fraser, Beckler ;* Clarence river, *Wilcox.*

4. **C. serrata,** *F. Muell. Fragm.* iii. 43. A tree, but flowering when still shrubby, the young branches rusty with a close tomentum. Leaflets usually 6 to 10, ovate-lanceolate or lanceolate, acute or acuminate, 3 to 6 in. long, sharply and coarsely serrate, rounded at the base and nearly sessile, rigid but not thick, shining above, very prominently pinnately veined and reticulate underneath. Panicles in the upper axils, little branched or almost reduced to dense racemes of 2 or 3 in., softly tomentose or pubescent. Flowers rather large, on very short pedicels. Sepals orbicular, the innermost fully 2 lines long. Petals much shorter, broad with a short 2-cleft scale at the base. Anthers 8, oblong, on very short filaments. Ovary in the males rudimentary, villous. Female flowers and fruit not seen.

**Queensland.** Pine river, Moreton Bay, *W. Hill.*

5. **C. tomentella,** *F. Muell. Herb.* Possibly a variety of *C. serrata,* of which it has the flowers. Branches, petioles, and inflorescence softly tomentose, almost villous. Leaflets 5 to 8, oblong or obovate-oblong, obtuse, 2 to 3 in. long, minutely and remotely denticulate or nearly entire, on petiolules

often 2 lines long, thinly coriaceous, glabrous above, softly tomentose under-neath. Panicles not much branched. Bracts rather large, tomentose, de-ciduous. Flowers nearly sessile. Sepals orbicular, and petals small with a short scale as in *C. serrata.* Anthers oblong, slightly pubescent. Capsule 3-angled, thickly coriaceous, velvety-tomentose and rugose, ¾ in. broad.

**Queensland.** Moreton Bay, *W. Hill.*

6. **C. pseudorhus,** *A. Rich. Sert. Astrol.* 34, *t.* 14. A spreading tree of moderate size, the young branches and petioles densely rusty-tomen-tose. Leaves crowded under the panicles ; leaflets 13 to 21 or even more, lanceolate or ovate-lanceolate, acuminate, 1½ to 3 in. long or rarely more, very oblique or almost falcate, nearly glabrous and shining above when full-grown, more or less tomentose or pubescent underneath. Panicles usually much-branched and rather dense, rarely exceeding the leaves, tomentose. Flowers rather small, on very short pedicels. Sepals ovate, less imbricate than in the preceding species, the longest scarcely exceeding 1 line. Petals orbicular, rather exceeding the sepals, the inner scales hirsute, as long as the lamina. Stamens 8 or 9 ; anthers oblong. Ovary villous. Capsule glo-bular, slightly lobed, almost woody, densely hirsute with short velvety hairs, about ½ in. diameter. Arillus small.

**Queensland.** Keppel Bay, *R. Brown ;* Brisbane river, *Fraser, A. Cunningham, F. Mueller ;* Wide Bay, *Bidwill ;* Mackenzie Island, *Wilcox.*

**N. S. Wales.** Hastings river, *Fraser, Beckler ;* Clarence river, *Beckler.*

7. **C. xylocarpa,** *A. Cunn. Herb. ; F. Muell. Trans. Vict. Inst.* iii. 27. A moderate-sized tree, the young branches rusty-tomentose. Leaflets 3 to 6, rarely more or reduced to 2, ovate obovate or elliptical-oblong, obtuse or scarcely acuminate, 2 to 3 in. long or rarely more, slightly and irregularly sinuate-toothed or entire, glabrous and shining above, more or less pubescent underneath or rarely almost glabrous, with hairy tufts almost always con-spicuous in the axils of the raised primary veins. Panicles short and little branched, often reduced to simple racemes and rarely above two inches long, shortly tomentose. Flowers small, the upper male ones sessile, the lower hermaphrodite and pedicellate. Sepals ovate, tomentose, under 1 line long, unequal and slightly imbricate. Petals very small, with a minute scale at the base. Stamens 8 to 10 ; filaments oblong. Ovary tomentose, occasionally 4-merous. Capsule nearly globular, 3-angled, about ½ in. broad, woody, glabrous or minutely tomentose outside, the valves villous inside. Arillus small.

**Queensland.** Burnett river, *F. Mueller ;* Brisbane river, *A. Cunningham ;* Logan river, *Fraser ;* Curtis Island, *Henne.*

**N. S. Wales.** Clarence river, *Beckler.* The foliage of this species often closely re-sembles that of *Nephelium tomentosum.*

8. **C. nervosa,** *F. Muell. in Trans. Vict. Inst.* iii. 27. A moderate-sized tree, the young branches and inflorescence minutely hoary-tomentose, otherwise glabrous. Leaflets 3 to 6, rarely more or reduced to 2, lanceolate or rarely elliptical-oblong, mostly 3 to 6 in. long, sinuate-toothed or entire, glabrous, with very rarely small tufts underneath in the axils of the raised primary veins. Racemes usually simple, axillary, 1 to 2 in. long, the flowers

all pedicellate and larger than in *C. xylocarpa.* Sepals narrow-ovate, slightly imbricate, above 1 line long. Petals very small, with a very short scale. Anthers oblong, hirsute at first, but soon glabrous. Capsule nearly globular, 3-angled, about ½ in. broad, woody, glabrous or nearly so outside, the valves villous inside.

**Queensland.** Moreton Bay, *F. Mueller;* Rockhampton, *Thozet;* also in A. Cunningham's and Leichhardt's collections, without the precise station.

**N. S. Wales.** Richmond river, *C. Moore;* Clarence river, *Beckler.*

Cunningham's and Leichhardt's are the only specimens I have seen in flower, the others are in fruit only, and may possibly include some glabrous specimens of *C. xylocarpa,* to which this species is very nearly allied. It is also closely allied to, although not quite identical with, *C. falcata,* A. Gray, from the Fiji islands.

9 ? **C. Bidwilli,** *Benth.* A tree, the young shoots and inflorescence minutely tomentose. Leaves 2 to 4, ovate oblong or ovate-lanceolate, obtuse or scarcely acuminate, 3 to 6 in. long, entire or obscurely sinuate-toothed, glabrous on both sides, with few or no tufts in the axils of the raised primary veins underneath. Panicles terminal, much branched, but shorter than the leaves. Flowers small, all pedicellate. Sepals tomentose, narrow-ovate, slightly imbricate, about 1 line long. Petals rather shorter than the calyx, oblong, concave, with 2 minute hirsute auricle-like scales at the base of the lamina. Stamens about 8; filaments nearly as long as the calyx; anthers oblong. Ovary hirsute. Fruit not seen.

**Queensland.** Wide Bay, *Bidwill.* Although I have not seen the fruit, this species has all the appearance of a true *Cupania.* It has some general resemblance to a Philippine Island species, n. 1237 of Cuming, which is I believe as yet unpublished.

## 7. RATONIA, DC.

(Arytera, *Blume.*)

Flowers regular, polygamous. Calyx small, cup-shaped, 4- or 5-toothed or lobed, open, valvate, or slightly imbricate in the bud. Petals 4 or 5, small, with or without scales inside, or none. Disk usually annular. Stamens 7 to 10, inserted inside the disk; filaments filiform, longer than the calyx. Ovary 2- or 3-celled, with 1 ovule in each cell. Capsule either 2-celled and compressed, or 3-celled and 3-angled or 3-lobed, loculicidally 2- or 3-valved, rarely almost indehiscent. Seeds more or less covered by an arillus; testa crustaceous; cotyledons thick, often curved or folded.—Trees. Leaves alternate, pinnate; leaflets alternate or opposite, usually without a terminal one. Flowers small, in terminal or axillary panicles. Petals rarely as long as the calyx.

A large tropical genus, with the same range as *Cupania,* but especially numerous in America. The Australian species are all endemic. It is closely allied to *Cupania,* with which it is usually joined, but the gamosepalous calyx and long filaments appear to give it at least as great a value as several other generally admitted genera of *Sapindaceæ.*

Capsule distinctly stipitate, 3-angled or pear-shaped. Leaflets coriaceous.
　Leaflets large, very coriaceous, the veins scarcely prominent. Petals 5.
　　Capsule glabrous inside . . . . . . . . . . . . 1. *R. pyriformis.*
　Leaflets thinly coriaceous, much reticulate. Petals none. Capsule
　　densely woolly inside . . . . . . . . . . . . 2. *R. anodonta.*
　Leaflets oblong-lanceolate, very coriaceous, the margins thickened;
　　primary veins prominent. Capsule glabrous inside . . . . . 3. *R. stipitata.*

Capsule nearly sessile, flattened and 2-celled or rarely 3-lobed.  Leaflets
    scarcely coriaceous.
Filaments long, very woolly.   Styles united to the middle  .  .  .  .  **4.** *R. tenax.*
Filaments short, slightly hairy.   Styles distinct from the base  .  .  **5.** *R. distylis.*

1. **R. pyriformis,** *Benth.*   A tree of considerable size, but flowering
sometimes as a shrub, glabrous except a minute hoariness on the young
shoots and panicles.   Leaflets 3 to 6, ovate or ovate-lanceolate, shortly acuminate, 4 to 6 in. long, entire, very coriaceous, on petiolules of ½ in. or more.
Flowers very small, shortly pedicellate, singly or in little cymes of 2 or 3
along the raceme-like branches of the panicle.   Calyx nearly 1 line diameter,
shortly and broadly 5-lobed.   Petals 5, scarcely exceeding the calyx-lobes,
cuneate or spathulate, the inner scales lining and bordering the base of the
lamina.   Stamens in the male flower 8, much exserted, the filaments slightly
hirsute, in the females few, with short filaments.   Ovary stipitate, slightly
hirsute, style filiform, with 3 diverging stigmatic lobes.   Capsule globular-
pear-shaped, about 4 lines diameter, narrowed into a long stipes, glabrous,
with 3 raised ribs, appearing almost drupaceous and scarcely dehiscent.
Seeds often reduced to 2 or 1, enclosed in the arillus ; cotyledons much
folded.—*Schmidelia pyriformis,* F. Muell. Fragm. i. 2.

**Queensland.**  Brisbane river, Moreton Bay, *A. Cunningham, W. Hill, F. Mueller.*

2. **R. anodonta,** *Benth.*   A tree of considerable size, flowering also as
a shrub, quite glabrous.   Leaflets 2, 3, or rarely 4, ovate or ovate-lanceolate,
obtuse or obtusely acuminate, 2 to 4 in. long, coriaceous, but not thick, very much
reticulate, narrowed into a petiolule of ¼ to nearly ½ in.   Panicle glabrous, slender, not much branched.   Calyx glabrous, about ¾ line diameter.   Petals none.
Filaments exserted, glabrous.   Ovary stipitate, almost glabrous ; style shortly
subulate, with diverging stigmatic lobes.   Capsule pear-shaped, somewhat
3-angled, nearly ½ in. broad, the valves almost woody, densely villous inside.
Seeds often reduced to 2 or 1, enclosed in the arillus.   Embryo much curved ;
cotyledons folded, but less so than in *R. pyriformis.*—*Schmidelia anodonta,*
F. Muell. Fragm. i. 2 ; *Cupania anodonta,* F. Muell. Fragm. ii. 76.

**Queensland.**  Brisbane river, Moreton Bay, *A. Cunningham, W. Hill;* Mackenzie
river, *Leichhardt.*

3. **R. stipitata,** *Benth.*   A moderate-sized tree, glabrous except a minute tomentum on the young branches and inflorescence.   Leaflets 3 to 6,
oblong-lanceolate, acute, 2 to 3 in. long, narrowed into a petiolule of 3 or 4
lines, coriaceous, very rigid, shining above, the primary veins very prominent
underneath.   Panicles axillary and terminal, divaricately branched.   Flowers
not seen.   Fruiting pedicels 2 to 3 lines long.   Calyx persistent, very small,
acutely 4 or 5-lobed.   Capsule 3-angled, depressed at the top, ½ in. broad, narrowed into a short but distinct stipes, valves thickly coriaceous, almost
woody, glabrous and reddish inside.   Seeds shining, in a thin arillus.—*Cupania stipata,* F. Muell. Fragm. ii. 75 and 175.

**N. S. Wales.**  Clarence river, *Beckler.*  I have corrected the specific name to *stipitata,* from the stipitate capsules, *stipata* (encircled) having been probably a clerical error.

4. **R. tenax,** *Benth.*   A moderate-sized tree, quite glabrous except the
flowers.   Leaflets usually 3, but varying from 2 to 6, from obovate to oblong-

elliptical or lanceolate, obtuse, 1¼ to 2 or rarely 3 in. long, much narrowed at the base but scarcely petiolulate, thinly coriaceous, shining above, pale or sometimes slightly glaucous underneath. Panicles small, little branched. Calyx a little above 1 line broad, 5-lobed. Petals small, broad, the scale inside very hairy. Stamens about 8, the exserted filaments woolly-hairy. Ovary stipitate, 2- or rarely 3-celled. Style rather short, with spreading stigmatic lobes. Capsule usually flattened, 2-celled, about ½ in. broad, contracted into a very short stipes; valves thick, densely villous inside. Seeds apparently only half enveloped in the arillus, but much injured in the specimens examined.—*Cupania tenax,* A. Cunn. Herb.

**Queensland.** Brisbane river, *A. Cunningham, W. Hill, F. Mueller;* Port Curtis, *C. Moore.*

5. **R. distylis,** *F. Muell. Herb.* A tree of considerable height, glabrous, except the inflorescence, and sometimes the very young shoots. Leaflets 2, or sometimes reduced to 1, at the end of a short common petiole, from obovate-oblong to elliptical or lanceolate, obtuse or shortly acuminate, 2 to 3 in. long, narrowed into a short petiolule, thinly coriaceous, reticulate. Panicles small, pubescent, with minute appressed hairs, the females often reduced to simple racemes. Calyx small; broad, shortly 5-toothed. Petals minute, orbicular, with a hairy scale at the base. Filaments shorter than in the other species, especially in the females; anthers rather large, pubescent. Ovary broadly obcordate, strigose-pubescent. Styles divided to the base, revolute. Capsule flattish, 2-celled, about ¾ in. broad; the valves coriaceous, slightly hairy inside. Seeds not seen.

**Queensland.** Brisbane river, Moreton Bay, *W. Hill;* Port Denison, *Fitzalan;* Bunija Creek Brush, *Leichhardt.*

## 8. ATALAYA, Blume.

Flowers regular, polygamous. Sepals 5, much imbricate in the bud. Petals 5, exceeding the sepals, with an inner scale or tuft of hairs. Disk annular. Stamens 8, inserted inside the disk. Ovary 3-celled, with 1 ovule in each cell. Style short, undivided. Fruit separating into 3 distinct carpels or samaræ, 1-celled, 1-seeded and indehiscent at the base, terminating in a long wing. Seeds without any arillus, testa coriaceous; cotyledons thick, unequal.—Trees or shrubs. Leaves pinnate or rarely simple. Flowers usually larger than in *Cupania* and *Ratonia,* in axillary or terminal panicles.

The genus is endemic in Australia, with the exception of one species, which extends to Timor. The flowers are nearly those of *Sapindus,* with the fruit of *Thouinia* and *Acer.*

Flowers and fruit, as well as the whole plant, quite glabrous . . . . 1. *A. salicifolia.*
Flowers and fruit more or less pubescent or tomentose.
  Leaflets ovate or broadly oblong, the petiole not winged. Panicle
    pedunculate, many-flowered. Carpels divaricate . . . . . . 2. *A. multiflora.*
  Leaflets narrow-oblong or linear, or leaves undivided, the petiole
    often winged. Carpels diverging.
    Plant glabrous, except the flowers . . . . . . . . . 3. *A. hemiglauca.*
    Branches, young leaves, and panicles velvety-tomentose. Leaflets
      and petiole-wings much reticulate . . . . . . . . 4. *A. variifolia.*

1. **A. salicifolia,** *Blume, Rumphia,* iii. 186. A small tree, quite glabrous, green or somewhat glaucous. Leaflets in our specimens 2 to 5, oblong or oblong-lanceolate, 3 to 5 in. long, narrowed at the base, but not petiolulate, thinly coriaceous, with numerous pinnate veins, and more or less reticulate, the margins not thickened. Panicles loose, perfectly glabrous, as well as the flowers, except a few hairs on the filaments and petal-scales. Flowers otherwise those of *A. hemiglauca.* Samaræ about ¾ to 1 in. long, including the wing, and perfectly glabrous.—*Sapindus salicifolius,* DC. Prod. i. 608 ; *Cupania salicifolia,* Dcne. Herb. Tim. Descr. 115 ; *Thouinia australis,* A. Rich. Sert. Astrol. 31, t. 12.

**N. Australia.** Careening Bay, N.W. coast, *A. Cunningham ;* Melville Island (not Moreton Bay), *Fraser* and *A. Cunningham.* Also in Timor, the specimens precisely similar to the Australian ones.

2. **A. multiflora,** *Benth.* A tall shrub or small tree, glabrous except the inflorescence. Leaflets 2 to 6, ovate or oblong, very obtuse, 2 to 3 in. long or rarely more, distinctly petiolulate, coriaceous and strongly reticulate. Panicle pedunculate above the last leaves, oblong or pyramidal, minutely tomentose-pubescent. Flowers very numerous, the small scale-like bracts more conspicuous than in the other species. Flowers of *A. hemiglauca.* Ovary slightly pubescent. Samaræ 1 to 1½ in. long, including the straight or falcate wing, very divaricate, pubescent or nearly glabrous.

**Queensland.** Cape York and Trinity Island, *M'Gillivray ;* Brisbane river, *W. Hill, F. Mueller.*

3. **A. hemiglauca,** *F. Muell. Herb.* A tall shrub or small tree, quite glabrous except the flowers, and more or less glaucous. Leaves usually pinnate ; leaflets few, from narrow-oblong to linear, obtuse or scarcely acute, from 2 or 3 to 7 or 8 in. long, often somewhat falcate, narrowed at the base but rarely petiolulate, rigidly coriaceous, with numerous pinnate and reticulate veins and a somewhat thickened margin, the common petiole terete or nearly so ; sometimes, however, the petiole becomes winged, or the leaves are quite simple, oblong, or linear, or the leaflets are decurrent on the petiole forming a large 2- or 3-lobed leaf, or rarely the simple leaf is ovate-lanceolate, and 8 to 10 in. long. Panicles rather dense, the rhachis and branches glabrous or nearly so ; pedicels 1 to 2 lines long. Sepals orbicular, nearly glabrous, 1½ or the inner ones nearly 2 lines long. Petals pubescent, oblong, 3 to 4 lines long, with a hirsute scale at the base. Filaments pubescent. Ovary densely silky-pubescent. Samaræ pubescent, with minute appressed hairs, 1 to 1¼ in. long, including the wing, which is nearly as broad as long, the cavity hairy or nearly glabrous inside.—*Thouinia hemiglauca,* F. Muell. Fragm. i. 98.

**N. Australia.** N.W. coast, *Bynoe ;* Hammersley Range, Nichol Bay, *F. Gregory ;* Albert river, *Henne.*

**Queensland.** E. coast, *R. Brown ;* Oxley's Station, *Leichhardt ;* Rockhampton, *Thozet ;* Brisbane river, *A. Cunningham, Fraser ;* Mooni river, *Mitchell.*

**N. S. Wales.** Liverpool plains, *A. Cunningham ;* Bowen and Castlereagh rivers, *F. Mueller ;* desert of the Darling, and thence to Stokes range and Cooper's Creek, *Victorian Expedition* and others.

4. **A. variifolia,** *F. Muell. Herb.* A tall shrub or small tree, the young

branches and panicles softly velvety-tomentose. Leaves or leaflets from oblong to linear, apparently as variable as in *A. hemiglauca*, but longer, often above 8 in., very much more reticulate, the common petiole usually broadly winged, the wing also much reticulate. Panicle loose. Sepals silky-pubescent, about 1½ lines long. Petals twice as long. Filaments hairy. Samaræ softly tomentose, 2 in. long including the wing, which is fully twice as long as broad, the cavity pubescent inside.—*Thouinia variifolia*, F. Muell. Fragm. i. 46.

**N. Australia.** Sea range, Macadam range, and near Fitzmaurice river, *F. Mueller.*

### 9. SAPINDUS, Linn.

Flowers regular, polygamous. Sepals 4 or 5, much imbricate in the bud. Petals as many, usually exceeding the sepals, with 1 or 2 inner scales or without any. Disk annular. Stamens usually 8 to 10. Ovary 2- to 4-lobed, 2- to 4-celled, with 1 ovule in each cell. Style with 2 to 4 stigmatic lobes. Fruit fleshy or coriaceous, divided into 2 to 4 globular or ovoid indehiscent lobes, not muricate. Seeds without any arillus; embryo straight or curved; cotyledons thick.—Trees or shrubs, rarely climbing. Leaves pinnate, rarely 1-foliolate. Flowers in terminal or axillary panicles.

The genus is widely dispersed over tropical regions, but less numerous in America than in Asia. The Australian species is, as far as known, endemic; but, like many others of the genus, it must remain in some measure doubtful until the fruit has been seen.

1. **S. (?) australis,** *Benth.* Young branches, petioles, and panicles pale or hoary with a very minute tomentum. Leaflets, in our specimens, 4 or 6, broadly ovate, obtuse, 3 to 5 in. long, entire, often oblique, narrowed into a short petiolule, coriaceous, glabrous, much veined, of a pale, almost glaucous colour. Panicle loose, longer than the leaves. Flowers shortly pedicellate, in little loose cymes along the divaricate branches. Sepals in the male flowers, the only ones seen, hoary-tomentose, rather above 1 line long. Petals nearly 2 lines long, oval-oblong, narrowed into a short claw, pubescent outside, with a single short broad scale inside fringed with long hairs. Stamens usually 8, as long as the petals. Filaments hairy.

**Queensland.** Cape York, *M'Gillivray.* In the absence of female flowers and fruit, I have referred this plant to *Sapindus*, from its general resemblance in habit and male flowers to *S. emarginatus,* Roxb.

### 10. NEPHELIUM, Linn.

Flowers regular, polygamous. Calyx small, cup-shaped, with 4 or 5 rarely 6 teeth or lobes, valvate or slightly imbricate in the bud. Petals none, or as many as calyx-lobes, small, with a 2-cleft scale or 2 scales inside. Disk annular. Stamens 6 to 10, inserted within the disk; filaments in the Australian species short, in others elongated. Ovary 2- or 3-celled, usually lobed, with 1 ovule in each cell. Style with 2 or 3 stigmatic lobes. Fruit usually deeply 2- or 3-lobed, or rarely entire, 2- or 3-celled, or reduced to a single carpel, the lobes indehiscent or 2-valved, or opening irregularly, muricate, or in the Australian species smooth. Seeds usually wholly or partially enclosed in an arillus; testa coriaceous; cotyledons thick.—Trees, with the habit of *Cupania*. Leaves abruptly pinnate; leaflets opposite or alternate, the primary parallel pinnate veins prominent underneath in all the Australian species except *N. microphyllum.* Flowers small, in axillary or terminal panicles.

The genus extends over tropical Asia, especially the Archipelago. The Australian species are all endemic, and differ from the majority of the Asiatic ones in their smooth fruit and shorter filaments. The flowers are nearly those of *Ratonia ;* but the fruit does not open in septiferous valves, even when, as in *N. connatum,* it is scarcely lobed. It is also very nearly allied to *Euphoria,* differing chiefly in the smaller gamosepalous calyx. The distinctions, however, between *Cupania, Ratonia, Nephelium, Euphoria,* and several others, are very slight.

Carpels quite connate, the capsule not depressed in the centre between
them.
  Leaflets slightly hoary or pubescent. Panicle much-branched,
    many-flowered. Petals 5. Capsule scarcely coriaceous . . .   1. *N. connatum.*
  Leaflets rigid, glabrous, mostly toothed. Panicles scarcely
    branched. Petals none. Capsule very coriaceous . . . .   2. *N. subdentatum.*
Carpels globular, the capsule depressed in the centre and deeply
lobed.
  Fruit densely villous. Leaflets 4 or more, mostly toothed, tomen-
    tose-villous underneath . . . . . . . . . . . .   3. *N. tomentosum.*
  Fruit minutely hoary. Leaflets 2, entire, coriaceous, glaucous
    underneath . . . . . . . . . . . . . . . .   4. *N. coriaceum.*
  Fruit thickly coriaceous, nearly glabrous outside, very hairy inside.
    Panicle very tomentose. Leaflets glabrous . . . . . . .   5. *N. foveolatum.*
  Fruit thinly coriaceous, glabrous. Panicle nearly glabrous. Leaflets
    quite glabrous . . . . . . . . . . . . . . . .   6. *N. leiocarpum.*
Carpels ovoid, united only by their attenuated bases.
  Leaflets 4 or 6, with few, parallel, prominent veins (as in all the
    preceding species). Panicles loose, many-flowered.
    Calyx divided to the base into imbricate segments . . . .   7. *N. Beckleri.*
    Calyx divided to the middle into broad obtuse lobes . . . .   8. *N. divaricatum.*
  Leaflets 2, small, with numerous, scarcely prominent veins. Pa-
    nicles short . . . . . . . . . . . . . . . .   9. *N. microphyllum.*

1. **N. connatum,** *F. Muell. Herb.* A tree of 20 to 40 ft., the young shoots and inflorescence minutely hoary-tomentose. Leaflets 2 to 6, from obovate to oblong-lanceolate, obtuse, 2½ to 4 in. long, narrowed at the base, but scarcely petiolulate, quite entire or very obscurely sinuate, thinly coriaceous, glabrous and shining above, somewhat glaucous or minutely tomentose underneath. Flowers small and numerous, in pyramidal panicles rarely exceeding the leaves. Calyx 5-lobed, about 1 line diameter. Petals about ¼ line long, the inner scale as long as the lamina. Filaments short ; anthers exserted, oblong, pubescent. Ovary 3-celled ; style thickened at the base. Fruit 3-furrowed or 3-lobed, but not deeply so, mucronate, and not depressed in the centre, somewhat inflated, scarcely coriaceous, hoary, indehiscent or splitting irregularly. Seeds small, shining, black, in a bright red cupular arillus.—*Spanoghea connata,* F. Muell. in Trans. Vict. Inst. iii. 26.

**Queensland.** Keppel Bay, *R. Brown ;* Brisbane river, Moreton Bay, *A. Cunningham, W. Hill, F. Mueller ;* Port Denison, *Fitzalan.* This is certainly the *Sapindus cinereus,* A. Cunn., referred to by A. Gray, in Bot. Amer. Expl. Exped. i. 258 ; but the plant from Hunter's River, more especially described by A. Gray, with coarsely serrate leaves and glabrous bracts, is probably different.

2. **N. subdentatum,** *F. Muell.* (as a var. of *N. connatum*). A tall shrub or small tree, the young shoots and inflorescence slightly pubescent

with minute appressed hairs. Leaflets 2 to 6, ovate or ovate-lanceolate, obtuse or scarcely acute, irregularly sinuate-toothed or rarely almost entire, coriaceous, glabrous on both sides and shining above. Panicles short, little branched. Pedicels short. Calyx truncate or shortly and broadly lobed. Petals none. Filaments very short; anthers oblong, scarcely pubescent. Ovary tomentose, 2- or 3-celled; fruit truncate at the top, slightly hoary with a minute tomentum, the lobes, usually 2 only, compressed-globular, united to the top, hard and indehiscent.

**N. S. Wales.** Tenterfield, New England, *C. Stuart;* "Tarampa Hill," *Leichhardt.* F. Mueller thinks that this may be a glabrescent form of *N. connatum,* but there is a considerable difference in general aspect ; the calyx is more open and less lobed, I can find no petals, and the fruit is differently shaped.

3. **N. tomentosum,** *F. Muell. in Trans. Vict. Inst.* ii. 64. A tree of 20 to 30 ft., the young branches and petioles clothed with a soft rust-coloured velvety tomentum. Leaflets 4 to 8, from oval-oblong to oblong-lanceolate, acute, or rarely obtuse, 2 to 4 in. long, acutely toothed or rarely almost entire, thinly coriaceous, pubescent above or at length glabrous, tomentose-pubescent underneath. Flowers small, crowded, on short slightly-branched tomentose panicles, sometimes reduced to simple racemes. Pedicels very short. Calyx nearly 1 line long, the lobes rather deep and acute. Petals none. Filaments very short; anthers oblong, exserted, glabrous or slightly pubescent. Ovary tomentose, 2- or 3-lobed; style short, with spreading stigmas. Fruit softly tomentose-villous, depressed at the top, of 2 or rarely 3 globular slightly compressed lobes, united to the top, 4 or 5 lines diameter, rather hard, indehiscent. Seeds half immersed in a yellowish arillus.

**Queensland.** Bremer river, Moreton Bay, *A. Cunningham, W. Hill, F. Mueller.* **N. S. Wales.** Clarence river, *Wilcox, Beckler.*

4. **N. coriaceum,** *Benth.* Young branches slightly hoary with a very minute tomentum. Leaflets in our specimens always 2, obovate-oblong or elliptical, 2½ to 4 in. long, very obtuse, quite entire, coriaceous, glabrous and shining above, pale or glaucous underneath, rounded at the base, on a short petiolule. Flowers not seen. Fruiting panicle branched, shorter than the leaves. Calyx small, with rather acute lobes. Fruits hoary-tomentose, mostly 3-lobed, much depressed in the centre, the lobes nearly globular, coriaceous, indehiscent.

**Queensland.** Brisbane river, *Fraser.*

5. **N. foveolatum,** *F. Muell. Herb.* A tree of considerable size, the young branches and inflorescence rusty-tomentose. Leaflets 4 to 6, ovate-lanceolate, or almost ovate, obtuse or acuminate, 3 to 5 in. long, entire or sinuate-toothed, narrowed into a distinct petiolule of 1 to 3 lines, thinly coriaceous, glabrous or rarely slightly pubescent underneath, having frequently a cup-shaped cavity in the axils of the primary veins. Panicles in the upper axils broad and many-flowered but shorter than the leaves, the flowers in little clusters or cymes along the principal branches. Calyx tomentose, deeply divided into lanceolate lobes of nearly 1 line, valvate in the bud. Petals minute or rudimentary. Filaments nearly as long as the calyx; anthers oblong, pubescent. Fruit tomentose, deeply divided into 2, 3, or sometimes

4 ovoid lobes, attaining sometimes ½ in., opening in 2 thickly coriaceous valves. Seeds completely enveloped in the arillus.—*Arytera foveolata,* F. Muell. in Trans. Vict. Inst. iii. 24.

**Queensland.** Moreton Bay, *W. Hill, F. Mueller.*

6. **N. leiocarpum,** *F. Muell. Herb.* A tall tree, usually glabrous except a very slight pubescence on the young leaves and shoots, and sometimes on the panicles. Leaflets 2 to 6, mostly oblong-elliptical, ovate-lanceolate or lanceolate, acuminate or obtuse, 3 to 4 or even 5 in. long, but more variable in size and shape than in most species, entire or rarely with a few deep serratures, narrowed into a very short petiolule, not coriaceous. Panicles loose, not much branched, usually glabrous. Calyx about 1 line diameter, with very short broad teeth. Petals broad and short but variable, the scale usually nearly as long as the lamina. Filaments often exceeding the calyx ; anthers oblong, glabrous or nearly so. Fruit sessile or nearly so, glabrous, with distinct globular lobes of 4 to 5 lines diameter, coriaceous, indehiscent or opening irregularly in a longitudinal slit, or breaking off transversely. Seed deeply enclosed in the arillus.—*Spanoghea nephelioides,* F. Muell. in Trans. Vict. Inst. iii. 25.

**Queensland.** Brisbane river, *F. Mueller ;* Curtis Island, *Henne* (a var. with smaller more obtuse and more coriaceous leaflets).

**N. S. Wales.** Port Jackson, *R. Brown ;* northward to Hastings river, *Fraser, Beckler ;* Richmond river, *C. Moore ;* Macleay and Clarence rivers, *Beckler ;* southward to Illawarra, *A. Cunningham, Backhouse ;* Kiama, *Harvey ;* Twofold Bay, *F. Mueller.*

7. **N. Beckleri,** *Benth.* A tree of considerable size, the young shoots and inflorescence slightly hoary with a minute tomentum, otherwise glabrous. Leaflets 3 to 6, ovate-lanceolate or oblong, obtuse or obtusely acuminate, 2 to 4 in. or when luxuriant 6 in. long, entire, narrowed into a petiolule of 3 to 6 lines, thinly coriaceous, quite glabrous. Panicles much branched. Flowers numerous, shortly pedicellate. Calyx pubescent, deeply divided into 5 orbicular or broadly-ovate very obtuse segments about ¾ line long. Petals in the males short, with a very small scale, in the females longer with a more developed scale. Filaments very short ; anthers oblong, pubescent. Fruit distinctly stipitate, glabrous, with 2 or 3 horizontally divaricate ovoid lobes of about ½ in., either indehiscent or rarely opening in a short slit ; often reduced to a single perfect lobe, the two others forming short tubercles at its base.

**N. S. Wales.** Clarence river, *Beckler.* The calyx is more deeply cleft and more imbricate than in any other *Nephelium,* thus approaching that of *Euphoria ;* but the species is too closely allied in fruit and other characters to *N. divaricatum* to be generically separated from it.

8. **N. divaricatum,** *F. Muell. Herb.* A handsome tree of considerable height, the young shoots and panicles slightly hoary with a minute tomentum, otherwise glabrous. Leaflets 4 or rarely 2, oval-oblong, elliptical or oblong-lanceolate, obtuse or acuminate, 2 to 3 or rarely 4 in. long, entire, narrowed into a petiolule of 2 or 3 lines, thinly coriaceous. Panicles loose, with few divaricate branches, the flower-cymes shortly pedunculate. Calyx very open, about ½ line long, pubescent, divided to the middle into 5 or rarely 4 broad obtuse lobes. Petals small, the inner scale short or in some females nearly

2 H 2

as long as the lamina. Filaments short; anthers oblong, pubescent. Ovary tomentose. Fruit glabrous, sessile or nearly so, with 1, 2, or 3 ovoid or nearly globular lobes, indehiscent or splitting longitudinally, more or less villous inside. Seed nearly enveloped in the arillus.—*Arytera divaricata*, F. Muell. in Trans. Vict. Inst. iii. 25.

**Queensland.** Brisbane river, Moreton Bay, *A. Cunningham, W. Hill, F. Mueller;* Pine river, *Fitzalan.*

9. **N. microphyllum,** *Benth.* Glabrous or the young shoots minutely hoary. Leaflets 2 or rarely 1 only, ovate or obovate, obtuse, ½ to 1½ in. long, entire, narrowed at the base but not petiolulate, somewhat coriaceous, the primary veins numerous and fine, not distant and raised as in other species. Flowers not seen. Fruiting panicles short and rather dense. Calyx small, 5-lobed. Fruit glabrous, almost sessile, with 1, 2, or rarely 3 ovoid lobes, about 5 lines long, splitting irregularly like those of *N. divaricatum*, hirsute inside.

**Queensland.** Wide Bay, *Bidwill.*

There are in R. Brown's herbarium specimens in flowers only, from Hunter's River, of what appears to be a *Nephelium* or *Cupania*, different from any of those above described; but, in the absence of fruit, I am unable to satisfy myself as to which genus it should be referred to, and therefore refrain from publishing it.

## 11. EUPHORIA, Juss.

Flowers regular, polygamous. Sepals 5, distinct, imbricate or valvate in the bud. Petals none or as many as sepals, with or without a scale inside. Disk annular. Stamens 6 to 10, inserted within the disk; filaments short. Ovary 2- or 3-celled, usually lobed, with 1 ovule in each cell; style deeply 2- or 3-lobed, or divided to the base into distinct styles. Fruit deeply 2- or 3-lobed, or reduced to a single carpel, the lobes usually indehiscent, tuberculate. Seeds enclosed in a pulpy arillus; testa coriaceous; cotyledons thick. —Trees, with the young shoots usually pubescent. Leaves pinnate; leaflets, as in *Nephelium*, with the primary pinnate veins raised underneath. Flowers small, in terminal panicles.

The genus extends over tropical Asia, especially the Archipelago, with one Australian endemic species. It is very nearly allied to *Nephelium*, differing chiefly in the distinct sepals (in which respect *N. Beckleri* comes very near to *Euphoria*), and from the Australian *Nephelia* in the tuberculate fruit.

1. **E. Leichhardtii,** *Benth.* Young branches, petioles, and inflorescence rusty-tomentose. Leaflets about 6, from obovate-oblong to ovate-lanceolate, obtuse or acuminate, 2 to 3 in. long; entire, rather thin, glabrous or nearly so above, tomentose or pubescent underneath, narrowed into a short petiolule. Panicles terminal, sessile, rather large, the flowers in little dense cymes along its branches. Sepals about 1 line long, tomentose, imbricate. Petals rather shorter, without any scale, but hairy inside, glabrous outside in the typical form. Filaments longer than the calyx; anthers ovoid. Ovary 3-celled. Style rather thick, with 3 divergent lobes. Young fruit deeply divided into 3 globular lobes, very tomentose and tuberculate, but not seen fully formed.

**Queensland (?),** *Leichhardt (Herb. F. Muell.).*

Var. *hebepetala.* Calyx rather smaller. Petals pubescent outside. " Nurrum Nurrum," *Leichhardt.(Herb. F. Muell.).*

## 12. HETERODENDRON, Desf.

Flowers regular, usually hermaphrodite. Calyx broadly cup-shaped, very shortly and irregularly toothed. Petals none. Disk small. Stamens 6 to 15, inserted within or upon the disk; anthers nearly sessile, longer than the calyx. Ovary 2- to 4-lobed, 2- to 4-celled, with 1 ovule in each cell; style short, with an obtuse lobed stigma. Fruit of 1 or 2, rarely 3 or 4 coriaceous or hard lobes, indehiscent. Seed half immersed in an arillus; testa crustaceous; cotyledons thick, flexuose.—Shrubs. Leaves simple, entire or lobed. Flowers small, in short terminal, slightly-branched panicles, often reduced to simple racemes.

The genus is limited to Australia.

Leaves entire, coriaceous, linear, oblong or rarely obovate, usually
above 2 in. long . . . . . . . . . . . . . . . . . . 1. *H. oleæfolium.*
Leaves entire, mucronate, toothed or pinnatifid, scarcely coriaceous,
rarely 2 in. long . . . . . . . . . . . . . . . . . 2. *H. diversifolium.*

1. **H. oleæfolium,** *Desf. in Mem. Mus. Par.* iv. 8, *t.* 3. A tall shrub, the young shoots hoary or glaucous with a minute silky pubescence. Leaves linear, lanceolate or narrow-oblong, rarely almost obovate, acute or obtuse, 2 to 4 in. long, quite entire, narrowed into a very short petiole, coriaceous and sometimes very rigid. Panicles usually few-flowered and much shorter than the leaves. Calyx broadly cup-shaped, varying from 1½ to nearly 3 lines diameter. Ovary usually 3- or 4-celled, densely tomentose. Fruit of 1, 2, or very rarely 3 or 4 nearly globular lobes, 3 or 4 lines diameter.—DC. Prod. ii. 92; F. Muell. Pl. Vict. i. 90.

**N. Australia.** Hammersley range, near Nichol Bay, *F. Gregory's Expedition.*
**Queensland.** Burdekin river, *F. Mueller;* Bowen river and Connor's Creek, *Leichhardt.*
**N. S. Wales.** N.W. interior, *Strutt;* Mount Brogden, *A. Cunningham;* plains of the Gwydir, *Mitchell;* Macquarie river and desert of the Darling and Murray, *Herb. F. Mueller.*
**Victoria.** Mallee scrub, on the rivers Murray, Wimmera, and Avoca, *F. Mueller.*
**S. Australia.** Lake Torrens, Flinders Range, and Cooper's Creek, *F. Mueller.*
**W. Australia.** Dirk Hartog's Island, *A. Cunningham, Milne;* Murchison river, *Oldfield.*

The Queensland specimens have smaller and more glabrous flowers than the more southern ones, with the ovary usually 2-carpellary. The north-western and some of the western ones have much broader leaves and more abundant flowers than the eastern.

2. **H. diversifolium,** *F. Muell. Fragm.* i. 46. A shrub, the young branches tomentose, pubescent, or perfectly glabrous. Leaves from linear-cuneate to oblong-cuneate or almost obovate, rarely 2 in. long and often under 1 in., usually mucronate with an almost pungent point, either entire or with a few sharp teeth or lobes towards the end, or pinnatifid with the triangular pungent lobes rigid and sometimes coriaceous, but less so than *H. oleæfolium.* Flowers few, in short panicles, pubescent or glabrous. Ovary 2-celled. Fruit-lobes very divaricate, ovoid, glabrous or tomentose.

**Queensland.** Keppel Bay, *R. Brown;* thickets at the foot of the dividing range, *A. Cunningham;* Rockhampton, *Thozet;* Warwick, *Beckler;* Comet river, *Leichhardt.*

There are two forms, one perfectly glabrous, the other with the young shoots and flowers pubescent, the fruit densely pubescent or tomentose. The specimens I have seen, although rather numerous, are not good.

## 13. HARPULLIA, Roxb.

Flowers regular, polygamous. Sepals 4 or 5. Petals as many, without any scale, but sometimes with inflected auricles at the base of the lamina. Disk inconspicuous. Stamens 5 to 8. Ovary 2-celled, with 2 ovules in each cell; style short, or elongated and spirally twisted. Capsule coriaceous, somewhat compressed, with 2 turgid lobes opening loculicidally in 2 valves. Seeds 1 or 2 in each cell, with or without an arillus; cotyledons thick.— Trees. Leaves pinnate; leaflets usually large, the primary veins prominent underneath. Flowers in loose terminal little-branched panicles, sometimes reduced to simple racemes. Capsules usually large, red or orange-coloured.

Besides the Australian species, which are endemic, there or two or three others, natives of tropical Asia or Madagascar.

Calyx persistent. Petals not auriculate.
  Petiole winged. Leaflets coarsely toothed . . . . . . . . 1. *H. alata.*
  Petiole not winged. Leaflets entire.
    Leaflets coriaceous, very obtuse . . . . . . . . . . . 2. *H. Hillii.*
    Leaflets membranous, shortly acuminate . . . . . . . . 3. *H. Leichhardtii.*
Calyx deciduous. Petals with inflected auricles . . . . . . 4. *H. pendula.*

1. **H. alata,** *F. Muell. Fragm.* ii. 103. A tall tree, the young branches and panicles minutely tomentose, otherwise glabrous. Leaflets usually 6 to 10, oblong-elliptical or lanceolate, acutely acuminate and coarsely toothed, almost lobed, 3 to 6 in. long, or more in the large leaves of barren shoots, rather rigid, green and much veined on both sides, the common petiole broadly winged. Panicles short, loose. Flowers few, larger than in the other species, on short pedicels. Sepals persistent, about 3 lines long, shortly tomentose. Petals about 4 lines long, oblong-cuneate, narrowed at the base, and not auricled. Stamens 7 or 8, about as long as the sepals in the males, shorter in the females. Capsule 1 to 1½ in. broad, coriaceous, nearly glabrous inside. Seeds enveloped in a yellowish arillus.

**N. S. Wales.** Clarence river, *Beckler;* Richmond river, *C. Moore.*

2. **H. Hillii,** *F. Muell. in Trans. Vict. Inst.* iii. 26, and *Fragm.* ii. 104. A tree of 60 to 80 ft., the young branches and inflorescence rusty with a close tomentum, otherwise glabrous. Leaflets usually 5 to 11, broadly-oblong or oval-oblong, very obtuse, 3 to 5 in. long, or more in the large leaves of barren shoots, thinly coriaceous, shining, the common petiole not winged. Panicles loose, little branched, shorter than the leaves. Pedicels 2 to 3 lines long. Sepals persistent, broadly ovate, 2 to 3 lines long. Petals oblong, 3 to 4 lines long, without auricles. Male flowers not seen. Stamens in the females 5 or 6, with very short filaments and acute anthers, probably imperfect. Capsule 1¼ in. broad, slightly tomentose outside, the turgid lobes hirsute inside. Seeds in the young state showing no arillus, but, according to Beckler, of an orange-yellow when ripe and enclosed in a rich red membrane.

**Queensland.** Wide Bay, *Bidwill;* Moreton Bay, *W. Hill;* Mackenzie river, *Leichhardt.*

**N. S. Wales.** Richmond river, *Beckler;* Clarence river, *C. Moore.*

3. **H. Leichhardtii,** *F. Muell. Herb.* Young shoots and inflorescence

minutely hoary-tomentose, otherwise glabrous.  Leaflets in the single specimen seen 10, elliptical, 3 to 5 in. long, membranous as in *H. pendula.*  Panicles almost reduced to simple racemes.  Flowers all females, on pedicels of
3 to 5 lines.  Sepals persistent, tomentose, about 2 lines long.  Petals and
stamens already fallen away.  Ovary tomentose, already enlarged, but the
fruit not fully formed.

**N. Australia.**  Port Essington, *Leichhardt.*  Although the specimen is very incomplete, it is evidently a distinct species, with the foliage nearly of *H. pendula,* and the persistent calyx of *H. Hillii.*

4. **H. pendula,** *Planch.; F. Muell. in Trans. Vict. Inst.* iii. 26, *and
Fragm.* ii. 104.  A tall tree, glabrous or the young shoots and panicles
minutely hoary-tomentose.  Leaflets 3 to 6, or rarely more, from ovate to
elliptical-oblong, obtusely acuminate, 3 to 5 in. long, membranous.  Panicles
loose and slender.  Pedicels in flower 3 to 4 lines, in fruit $\frac{1}{2}$ to 1 in. long,
slender.  Sepals deciduous, about 2 lines long.  Petals ovate, nearly 3 lines
long, with inflected ciliate auricles at the base, representing the inner scales
of many other *Sapindaceæ.*  Stamens 5 to 7, much longer than the calyx, with
slender filaments in the males, small and short in the females.  Ovary tomentose, with a long style twisted at the top.  Capsule glabrous or slightly
pubescent, 1 to $1\frac{1}{2}$ in. broad, the lobes inflated.  Seeds apparently without
any arillus.

**Queensland.**  Moreton Bay, known as "Tulipwood," *Fraser, A. Cunningham;* Wide
Bay, *C. Moore;* Port Denison, *Fitzalan;* Broad Sound, *Thozet.*
**N. S. Wales.**  Clarence river, *Beckler;* Richmond river, *C. Moore.*

14. **AKANIA,** Hook. f.

Flowers regular, hermaphrodite (or polygamous?).  Calyx campanulate,
with 5 short lobes, imbricate in the bud.  Petals 5, without any inner scale,
Disk adnate to the base of the calyx.  Stamens 5 to 10, inserted within the
disk.  Ovary 3-celled, contracted into a thickish style, with a capitate stigma;
ovules 2 in each cell.  Fruit not seen.—Tree.  Leaves pinnate.  Panicles
loose, axillary or terminal.

The genus is limited to a single species, endemic in Australia, allied to *Harpullia,* but
very different in the calyx and disk.

1. **A. Hillii,** *Hook. f. in Benth. and Hook. Gen. Pl.* 409.  An elegant
tree of 30 to 40 ft., glabrous except the panicle.  Leaves often above 2 ft.
long; leaflets numerous, lanceolate, acutely acuminate, often above 8 in. long,
bordered with acute often pungent serratures, rounded at the base and shortly
petiolulate, coriaceous, light green, shining above, marked underneath (in the
dried state) within each areola of the smaller reticulations with 3 or 4
round ovate or reniform dots.  Panicles long, loose, and little branched.
Pedicels long and slender.  Calyx tomentose, about 2 lines long, the lobes
rounded, with thin edges.  Petals inserted near the base of the calyx outside
the disk.  Anthers oblong.—*Cupania lucens,* F. Muell. Fragm. iii. 44.

**Queensland.**  Moreton Bay, *Leichhardt;* Pine river, *W. Hill.*
**N. S. Wales.**  Clarence river, *Beckler;* Richmond river, *C. Moore.*

## 15. DODONÆA, Linn.

### (Empleurosma, *Bartl.*)

Flowers polygamous or unisexual, often diœcious. Sepals 5 or sometimes fewer, valvate in the bud. Petals none. Disk small or inconspicuous. Stamens usually 8, sometimes fewer, rarely 10; filaments very short, anthers ovoid or linear-oblong. Ovary 3- or 4-, rarely 5- or 6-celled, with 2 ovules in each cell; style short or, in some flowers, very long, shortly lobed at the end. Capsule membranous or coriaceous, opening septicidally in as many valves as cells, each valve with a dorsal angle often produced into a vertical wing, and in falling off leaving the dissepiment attached to the persistent axis, or rarely the dissepiment splitting and remaining attached to the valves, thus closing the carpels and leaving only the central filiform axis persistent. Seeds 1 or 2, nearly globular or more frequently compressed, with a thickened funicle, but not arillate; testa crustaceous; embryo spirally curled.—Shrubs, often tall, but scarcely truly arborescent; the young shoots usually viscid, and often the whole plant. Leaves simple or pinnate, with small leaflets, with or without a terminal odd one. Flowers terminal or axillary by the abortion of the flowering branches, solitary, clustered, or in short racemes or panicles.

With the exception of *D. viscosa*, which is widely dispersed over almost all hot countries, and possibly one distinct Sandwich Island species, one from S. Africa, and one or two from Mexico, the *Dodonæas* are all endemic in Australia, and very difficult to distinguish by positive characters. The form of the wings of the capsule, which has been much relied on, is as variable as that of the leaves, and the species, which at first sight appear the most distinct, often pass one into the other by the most insensible gradations. Even the exceptional dehiscence of the capsule, in those species where the dissepiments are carried off with the valves, appears sometimes to be not quite constant, and is at most a purely artificial character separating species in all other respects very closely allied. Several species have in some, occasionally in nearly all the female flowers, a remarkably long style, sometimes ½ to 1 in., whilst other female flowers on the same specimen, or on other specimens of the same species, have no style at all, the stigma or stigmatic surface sessile on the ovary.

SERIES 1. **Cyclopteræ.**—*Leaves entire, toothed. or rarely lobed. Wings of the capsule extending from the base to the style or nearly so, each carpel, including its wing, nearly orbicular or longer than broad.*

Leaves flat, elliptical, oblong-lanceolate or spathulate or, if linear, not
    filiform, entire or obscurely sinuate, usually above 2 in. long,
    rarely between 1 and 2 in.
  Young branches very angular. Seeds smooth and shining. Leaf-
    veins indistinct.
    Sepals minute. Anthers linear . . . . . . . . . . . . 1. *D. triquetra.*
    Sepals 1 to 1½ lines long, from half as long to as long as the
      anthers . . . . . . . . . . . . . . . . . 2. *D. lanceolata.*
  Young branches very angular. Seeds opaque. Leaves long and
    narrow, often serrate . . . . . . . . . . . . 12. *D. ptarmicifolia.*
  Young branches terete or slightly angular. Seeds opaque.
    Leaves oval-oblong, on a rather long petiole, rounded at the base 3. *D. petiolaris.*
    Leaves narrowed into the petiole, the lateral veins more or less
      conspicuous.
      Leaves elliptical-oblong, lanceolate or spathulate, rarely almost
        linear-cuneate . . . . . . . . . . . . . 4. *D. viscosa.*
      Leaves narrow, linear-cuneate or long and linear . . . . 5. *D. attenuata.*

Leaves flat, more or less cuneate, entire or toothed at the end, rarely
   exceeding 1½ in., and usually under 1 in.
  Much-branched, erect or divaricate shrubs.   Terminal flowers
   clustered or shortly racemose.
    Leaves broad-cuneate, rounded or truncate at the end . . . .   **6.** *D. cuneata.*
    Leaves narrow-cuneate, rather acute, acuminate or 3-toothed at
     the end . . . . . . . . . . . . . . . . .   **7.** *D. peduncularis.*
  Prostrate shrub.   Leaves rather narrow-cuneate, mostly toothed or
   lobed.   Flowers solitary . . . . . . . . . . . . .   **8.** *D. procumbens.*
Leaves linear-filiform, heath-like or pine-like.
  Leaves crowded, under 1 in. long . . . . . . . . . .   **9.** *D. ericifolia.*
  Leaves 1 to 3 in. long, not crowded . . . . . . . . .   **10.** *D. filifolia.*
Leaves linear or lanceolate, mostly serrate or pinnatifid.
  Branches terete or nearly so.   Leaves linear or linear-cuneate, ob-
   tuse, mostly under 1½ in. long . . . . . . . . . .   **11.** *D. lobulata.*
  Branches very angular.   Leaves linear-lanceolate, acute, mostly 2
   to 4 in. long . . . . . . . . . . . . . . .   **12.** *D. ptarmicifolia.*

SERIES II. **Platypteræ.**—*Leaves quite entire, flat. Wings of the capsule very divergent or divaricate, not reaching to the style nor to the base, each carpel, including its wing, broader than long, transversely ovate or oblong.*

Leaves linear or lanceolate.   Branches very angular.   Dissepiments
  persisting on the axis . . . . . . . . . . . .   **13.** *D. truncatiales.*
Dissepiments splitting and coming off with the valves.
  Leaves oblong-elliptical . . . . . . . . . . . .   **14.** *D. platyptera.*
  Leaves narrow-linear . . . . . . . . . . . . .   **15.** *D. stenophylla.*

SERIES III. **Cornutæ.**—*Leaves entire or toothed at the end, the margins revolute or rarely flat. Wings of the capsule reduced to erect or divergent, usually falcate, horn-like appendages at the upper outer angle of the carpels.*

Leaves narrow-linear or subulate . . . . . . . . . .   **16.** *D. pinifolia.*
Leaves oblong or oblong-cuneate, obtuse, ½ to 1 in. long . . . .   **17.** *D. ceratocarpa.*
Leaves linear or cuneate, acute or 3-toothed, 2 to 4 lines long . .   **18.** *D. divaricata.*
Leaves broadly ovate or orbicular, mostly toothed . . . . . .   **22.** *D. Baueri.*

SERIES IV. **Apteræ.**—*Leaves entire or toothed. Capsule without wings, or the angles slightly and irregularly dilated into very narrow wings.*

Leaves flat, cuneate or obovate, rigid.
  Sepals lanceolate.   Buds ovoid or globular.
   Dissepiments persisting on the axis of the fruit.
    Branches scarcely angled.   Leaves obovate, cuneate, or trian-
     gular, glabrous or pubescent.   Flowers mostly axillary.
     Sepals narrow, short . . . . . . . . . .   **19.** *D. triangularis.*
    Branches acutely angled.   Leaves obovate, glabrous.   Ra-
     cemes short, terminal.   Sepals broad-lanceolate . . . .   **20.** *D. aptera.*
   Dissepiments splitting and coming off with the valves.   Branches
    terete.   Leaves obovate or oblong, glabrous . . . . .   **21.** *D. bursarifolia.*
  Sepals broad-ovate.   Buds very angular.
   Erect divaricate shrub.   Leaves obovate or orbicular, usually
    toothed . . . . . . . . . . . . . . .   **22.** *D. Baueri.*
   Prostrate shrub.   Leaves oblong-cuneate, often 3-toothed . .   **23.** *D. humifusa.*
Leaves short, linear, with recurved or revolute margins.
  Quite glabrous.   Stamens usually 6 . . . . . . . . .   **24.** *D. hexandra.*
  Hoary-tomentose, at least the capsules, rarely almost glabrous.
   Stamens usually 8 . . . . . . . . . . . . .   **25.** *D. ericoides.*

SERIES V. **Pinnatæ.**—*Leaves all pinnate or very rarely a few simple ones at the base of the branches. Capsule of the* Cyclopteræ, *except in D. oxyptera and D. inæquifolia, where it approaches that of the* Platypteræ, *and in D. humilis, where it is apterous.*

Tall shrubs or small trees. Leaflets flat, oblong, lanceolate or obovate, not coriaceous. Racemes or panicles terminal, loose.

Leaflets usually numerous, lanceolate or oblong. Capsule not inflated, the wings broad.

Leaflets ½ in. or less; rhachis scarcely winged. Sepals 3 to 4 lines long . . . . . . . . . . . . . . . . . . 26. *D. polyzyga.*

Leaflets ½ to 1 in.; rhachis broadly winged. Sepals 1 to 1½ lines 27. *D. megazyga.*

Leaflets few, obovate or oblong. Capsule large and inflated . . 28. *D. physocarpa.*

Much-branched, leafy shrubs. Pedicels solitary or clustered (racemose in *D. multijuga* and *D. pinnata*).

Leaflets obovate, cuneate or oblong, often toothed, the margins usually recurved or revolute. Plant usually pubescent or villous (except *D. humilis*).

Capsule winged, hirsute at least when young.

Villous. Leaflets 7 to 20 or more; rhachis winged. Sepals acuminate. Capsule-wings rounded.

Pedicels long, clustered . . . . . . . . . . . . 29. *D. vestita.*

Pedicels very short. Raceme terminal . . . . . . 30. *D. pinnata.*

Pubescent. Leaflets 3 to 7; rhachis angular. Pedicels short.

Sepals obtuse. Capsule-wings acutangular . . . . . 31. *D. oxyptera.*

Capsule not winged, covered with long, glandular setæ. Leaflets deeply toothed, glabrous. Flowers in dense corymbose clusters 32. *D. humilis.*

Capsule winged, glabrous or very sparingly pubescent. Plant pubescent or rarely glabrous.

Leaflets usually under 11. Pedicels short, clustered . . . 33. *D. boroniæfolia.*

Leaflets usually above 15. Flowers racemose.

Racemes loose. Pedicels slender . . . . . . . . 34. *D. multijuga.*

Racemes dense. Pedicels very short . . . . . . . 30. *D. pinnata.*

Leaflets linear-terete or linear-oblong. Plant glabrous, viscid.

Leaflets linear-oblong, flat, numerous. Capsules broadly winged 35. *D. larræoides.*

Leaflets narrow-linear, convex underneath. Capsules small.

Capsule-wings very divaricate; dissepiments remaining on the axis. Leaflets above 15 . . . . . . . . . . 36. *D. inæquifolia.*

Capsule-wings rounded; dissepiments splitting and coming off with the valves. Leaflets under 15 . . . . . . . 37. *D. adenophora.*

Leaflets almost terete, not thicker than the common petiole. Capsules rather large, the wings rounded.

Leaflets few, distant. Pedicels solitary . . . . . . 38. *D. stenozyga.*

Leaflets several, crowded. Pedicels shortly racemose . . . 39. *D. concinna.*

(*D. heterophylla,* Colla, and *D. scabra,* Lodd., inserted in Steud. Nom. Bot. ed. 2, as Australian plants, are unknown to me, nor can I find any description of them. They are probably garden names given to some of the species here enumerated.)

SERIES I. CYCLOPTERÆ.—Leaves entire, toothed, or rarely lobed. Wings of the capsule extending from the base to the style, or nearly so; each carpel, including its wing, nearly orbicular, or longer than broad. Dissepiments persistent on the axis.

In the following 12 species, great as is the diversity in the size of the capsule and the precise shape of the wings, these differences afford no specific characters, and are often very difficult to class as varieties, even when perfectly ripe and well-formed capsules are obtained; and the shape of the wing often alters much during growth, or is apparently affected by the manner in which the capsule has ripened. The very shining seeds distinguish 2 species, but where they are usually opaque they sometimes are somewhat shining. There remains little but the very uncertain character sderived from foliage to separate all these species, which are yet much too constantly dissimilar to be united into one.

1. **D. triquetra,** *Andr. Bot. Rep. t.* 230. Erect, usually tall, glabrous,

not very viscid, the young branches flattened or angular. Leaves from oval-elliptical to oblong-lanceolate, acuminate, 2 to 3 or rarely 4 in. long, the pinnate and reticulate veinlets few and fine, usually scarcely conspicuous. Pedicels slender, in short, oblong, compact panicles or racemes. Sepals minute, rarely ½ line long. Anthers linear, often 1½ lines long. Styles, when long, attaining ¼ in. Capsule of *D. viscosa,* usually middle-sized. Seeds brown, very smooth and shining.—DC. Prod. i. 617 ; F. Muell. Fragm. i. 75, and Pl. Vict. i. 226.—*D. laurina,* Sieb. in Spreng. Syst. Cur. Post. 152.—*D. longipes,* G. Don, Gen. Syst. i. 674 (from the character given).

**Queensland.** Brisbane river, Moreton Bay, *Fraser, Fitzalan.*

**N. S. Wales.** Port Jackson to the Blue Mountains, *R. Brown, Sieber, n.* 271 *and* 272, and others ; northward to Clarence and Hastings rivers, *Beckler,* and New England, *C. Stuart ;* southward to Twofold Bay, *F. Mueller.*

**Victoria.** Barren declivities and granite rocks of Genoa Peak, and elsewhere in the vicinity of Genoa river, *F. Mueller.*

The Fiji Island plant referred by A. Gray and Seemann to *D. triquetra,* appears to me to be one of the common forms of *D. viscosa.*

2. **D. lanceolata,** *F. Muell. Fragm.* i. 73. Very closely allied to *D. triquetra,* with the same angular branches, smooth, almost veinless leaves, slender pedicels, and very shining seeds, and scarcely distinguishable except by the sepals, which are from 1 to 1½ lines long. The leaves are perhaps generally rather narrower, and the capsule-wings broader, but neither of these characters can be relied upon.

**N. Australia.** Capstan Island, N.W. coast, *A. Cunningham* (the specimens rather doubtful, not being in fruit) ; Victoria river and Sea range, *F. Mueller ;* islands of the Gulf of Carpentaria, *R. Brown.*

**Queensland.** Northumberland Islands, *R. Brown ;* Cape Cleveland, *A. Cunningham ;* Sunday Island, *M'Gillivray ;* Palm Island, *Henne ;* Port Denison, *Fitzalan.*

**N. S. Wales.** Clarence river, *Beckler.*

3? **D. petiolaris,** *F. Muell. Fragm.* iii. 13. The single fragment in F. Mueller's herbarium has a few small oval-oblong leaves, veined as in *D. viscosa,* but much less narrowed at the base, on petioles of 2 or 3 lines. The single capsule is not yet full-grown, but, in that state, does not appear at all different from the larger varieties of *D. viscosa,* of which this plant may probably prove to be a variety.

**N. S. Wales.** Desert on the Darling river, *Neilson* (*Hb. F. Muell.*).

4. **D. viscosa,** *Linn.; DC. Prod.* i. 616. A shrub, sometimes low and stunted, more frequently tall, glabrous, and usually more or less viscid, the young branches frequently compressed or somewhat triangular, but much less so than in *D. triquetra.* Leaves simple, varying from broadly oblong-lanceolate, acute or acuminate, and 3 or 4 in. long, to narrow-lanceolate, or oblong-cuneate and very obtuse or almost linear-cuneate, always narrowed into a more or less distinct petiole, entire or obscurely sinuate, or rarely almost 3-toothed at the end, the pinnate veins usually rather numerous and very divergent, sometimes scarcely conspicuous. Panicles or racemes usually short and terminal, or reduced to axillary clusters. Sepals ovate, usually as long as or rather longer than the oblong obtuse anthers. Style rarely lengthened out. Capsule very variable in size, the wings continued from the base to the

style, or nearly so, either equally rounded at the top and at the base or more contracted at the base. Seeds rather large, dark-coloured or black, opaque or scarcely shining.—Hook. f. Fl. Tasm. i. 55 ; F. Muell. Pl. Vict. i. 85.

**N. Australia.** Apparently rare, but some specimens from the N.W. coast, *Bynoe*, probably belong to this species.

**Queensland.** Cumberland Islands, *R. Brown;* Endeavour river, *Banks ;* Rodd's Bay and Rockingham Bay, *A. Cunningham ;* Cape Upstart and Port Curtis, *M'Gillivray ;* Rockhampton, *Thozet ;* Moreton Bay, *Fraser, A. Cunningham,* and others.

**N. S. Wales.** From the borders of Queensland, *Beckler, C. Stuart,* and others, to Twofold Bay, *F. Mueller.*

**Victoria.** Rocky, scrubby, stony, and sandy localities, widely and copiously distributed over the colony, *F. Mueller.*

**Tasmania.** Common in poor soil, especially near the coast, *J. D. Hooker.*

**S. Australia.** Apparently common, at least in the eastern parts of the colony, *Herb. Mueller,* and others.

**W. Australia.** Blackwood river, *Oldfield.*

The species is abundantly distributed over tropical America, Africa, and Asia, extending to the Pacific Islands, and southward, beyond the tropics, to S. Africa and New Zealand. It includes probably the whole of the extra-Australian described *Dodonæas,* except, perhaps, the *D. eriocarpa* from the Sandwich Islands, *D. Thunbergiana,* Eckl. and Zeyh., from S. Africa, and one or two Mexican ones, which, whether varieties or species, do not occur in Australia. The almost protean forms the species assumes in Australia, even after deducting *D. attenuata, D. cuneata,* and *D. megazyga,* which F. Mueller unites with it, are very difficult to distribute into definite varieties, although at least the three following are usually considered as species.

*a. vulgaris.* Usually tall. Leaves large, obovate-oblong, broadly lanceolate or lanceolate, acuminate or rarely obtuse, the pinnate veins usually numerous and prominent. Capsules large, with rather broad wings, much rounded above and at the base, the terminal sinus (between 2 opposite wings) narrow, each carpel, including its wing, longer than broad.—*D. viscosa,* Linn., and *D. Burmanniana,* DC. ; Griseb. Fl. Brit. W. Ind. 127, with the synonyms adduced ; A. Gray, Gen. Ill. t. 182 ; Wight, Illustr. t. 52.—The most common form in America and tropical Africa, extending in Asia as far north as Scinde and Affghanistan, also in the Pacific islands ; and to this form belong most of the tropical Australian species as well as some from Hastings river, *Beckler.* Some specimens from Endeavour river, both in the Banksian and in Cunningham's collections, are remarkable for their thick, obscurely veined leaves.

*b. angustifolia.* Leaves narrow-lanceolate, mostly long and acutely acuminate, much narrowed at the base, the veins usually conspicuous. Capsules small, with very broad wings, leaving the terminal sinus very open and sometimes narrowed at the base, each carpel, including its wing, orbicular or rather broader than long, although much less so than in the *Platypteræ.*—*D. angustifolia,* Swartz ; Griseb. Fl. Brit. W. Ind. 128, with the synonyms adduced ; Lam. Ill. t. 304, n. 2, and consequently *D. salicifolia,* DC. Prod. i. 617, supposed to be from New Holland ; *D. neriifolia,* A. Cunn. in A. Gray, Bot. Am. Expl. Exped. i. 262.—This variety has nearly the same range within the tropics as the large-fruited one, and occasionally is found to pass into it. In Australia it includes many Queensland specimens, and is the common form in N. S. Wales collections. It occurs also in W. Australia, but in Victoria, S. Australia, and Tasmania, as in N. Zealand, it tends rather to pass into the spathulate-leaved form. *D. umbellata* and *D. Kingii,* G. Don, Gen. Syst. i. 674, from the characters given, belong probably to this variety.

*c. spathulata.* Usually a more bushy and not so tall a shrub as the preceding varieties, often very viscid. Leaves shorter (although much longer than in *D. cuneata*), obovate-oblong, oblong-cuneate, spathulate, oblanceolate or broadly linear-cuneate, usually obtuse or sometimes truncate, the lateral veins usually conspicuous, but in some thick-leaved specimens scarcely more so than in *D. cuneata.* Capsules very variable, but generally intermediate between those of the var. *vulgaris* and *angustifolia,* but nearer to the former.—*D. spathulata,* Sm. in Rees, Cycl. xii. ; DC. Prod. i. 616 ; *D. conferta,* G. Don, Gen. Syst. i. 674 ; *D.*

*viscosa*, var. *asplenifolia*, Hook. f. Fl. Tasm. i. 55.—This is the commonest, perhaps the only form, in Victoria, Tasmania, and S. Australia, and I have seen N. S. Wales specimens from Port Jackson, and northward to New England, Mount Mitchell, and Mount Aiton. It is the prevalent form in New Zealand, and some of the Sandwich Island specimens can be precisely matched in Australia. *D. oblongifolia*, Link, as figured in Bot. Reg. t. 1051, appears to represent rather a short-leaved form of this variety than a long-leaved *D. cuneata*. *D. asplenifolia*, Rudge, in Trans. Soc. xi. 297, t. 20, DC. Prod. i. 617, judging from N. S. Wales specimens agreeing with the figure, although not authentically named, is an apparently rare form with linear-cuneate, 3-toothed leaves, resembling those of luxuriant drawn-up shoots of *D. cuneata*, but longer.

5. **D. attenuata,** *A. Cunn. in Field, N. S. Wales*, 353. A viscid shrub, closely resembling the narrowest-leaved forms of *D. viscosa* on the one hand, and almost passing into *D. lobulata* on the other. Leaves linear or narrowly linear-cuneate, obtuse, often slightly sinuate-toothed, rather thick and rigid, 1-nerved, the lateral veins inconspicuous, $1\frac{1}{2}$ to $2\frac{1}{2}$ in. long in the original form, but sometimes longer. Flowers and ovate sepals of *D. viscosa*, in short usually simple racemes. Capsule of *D. viscosa*, usually intermediate between the extremes of the varieties *a* and *b* of that species. Seeds opaque.—Bot. Mag. t. 2860; *D. Preissiana*, Miq. in Pl. Preiss. i. 226; F. Muell. Fragm. i. 72.

**N. S. Wales.** Blue Mountains, *A. Cunningham* and others, and apparently common westward in the Darling and Murray desert, Mutanie ranges, Mount Brown, etc., *Herb. F. Mueller.*

**Victoria.** In the Murray desert and N.W. interior, *F. Mueller.*

**S. Australia.** Towards Spencer's Gulf, *Warburton.*

**W. Australia.** Mount Hardy, near York, *Preiss, n.* 2437; between Swan River and King George's Sound, *Drummond, 4th Coll., n.* 257; in the interior, *Roe.* The latter specimens have narrower, more rigid wings to the capsule, and more coriaceous leaves.

Var. *linearis.* Leaves long, narrow-linear, mostly acute, rigid, the margins often recurved. Capsule (only seen in few specimens) rather small, but with the terminal sinus between the wings narrow.—New England, *C. Stuart*; Mitta-Mitta, Genoa and Buchan rivers in Victoria, *F. Mueller*; Kangaroo Island, *Waterhouse, Sealy*; Swan River, *Drummond, n.* 203. The foliage nearly resembles that of *D. stenophylla*, which has a very different capsule.

6. **D. cuneata,** *Rudge, in Trans. Linn. Soc.* xi. 296, *t.* 19. A much-branched bushy shrub, glabrous, and usually viscid. Leaves obovate or cuneate, usually $\frac{1}{2}$ to 1 in. long and rather broad, rarely narrow-cuneate, attaining $1\frac{1}{2}$ in., rounded, truncate, emarginate or 3-toothed at the end, otherwise entire or rarely obscurely toothed, gradually narrowed into a very short petiole, thin or coriaceous; the lateral veins rarely conspicuous. Racemes short, terminal, scarcely branched, with slender pedicels, or the flowers few in axillary clusters. Sepals ovate-oblong, and capsules of *D. viscosa*, the wings usually not very broad and rather rigid, with the terminal sinus open. —DC. Prod. i. 617.

**Queensland.** Burnet river and Moreton Bay, *F. Mueller.*

**N. S. Wales.** Port Jackson, *R. Brown* and others; Blue Mountains, *Miss Atkinson*; Darling and Murray desert, *Victorian Expedition.*

**Victoria.** In the Grampians and Buffalo ranges, Wimmera and Murray rivers, *F. Mueller*, including a var. *coriacea*, with small, obovate, coriaceous leaves and small capsules with broad wings, and a var. *rigida*, with small, rigid, mostly obovate leaves, short pedicels, and rather large capsules with narrow wings. Luxuriant narrow-leaved N. S. Wales specimens occasionally almost pass into some unusual forms of *D. viscosa spathulata*, and the smaller forms come very near to *D. peduncularis.*

7. **D. peduncularis,** *Lindl. in Mitch. Trop. Austr.* 361. A very much branched glabrous and viscid shrub, closely allied to *D. cuneata,* the smaller branches terete, slender but rigid.    Leaves from linear-cuneate, to broadly spathulate, either acute or very shortly acuminate or rounded or truncate at the end, and often 3-toothed, ¼ to ½ in., or very rarely (when narrow) 1 in. long, coriaceous and rigid, 1-nerved, the margins often thickened, the lateral veins inconspicuous.    Pedicels rather slender, mostly axillary, solitary or clustered, or in short terminal racemes.    Sepals ovate, thicker than in *D. cuneata.*    Capsule of *D. viscosa.—D. pubescens,* Lindl. in Mitch. Trop. Austr. 342 (the supposed pubescence apparently a mistake).

**Queensland.**    Near Lindley's Range and on the Maranoa, *Mitchell.*
**N. S. Wales.**    Eurylean scrub in the N.W. interior, Liverpool plains, Hastings river, etc., *Fraser, A. Cunningham.*

8. **D. procumbens,** *F. Muell. in Trans. Vict. Inst.* i. 8, *and Pl. Vict.* i. 86.    A low, diffuse or prostrate, much-branched shrub, glabrous and scarcely viscid.    Leaves crowded, linear-cuneate, spathulate or almost triangular, mostly acute and often coarsely 3- to 5-toothed or lobed, about ½ or rarely ¾ in. long, coriaceous, 1-nerved, the lateral veins usually inconspicuous. Flowers mostly solitary, on short terminal pedicels.    Sepals lanceolate.    Style much more frequently elongated than in other species, often attaining nearly 1 in.    Capsule oblong, the angles produced into wings rounded at the top and base as in *D. viscosa,* but much narrower and not so thin.    Seeds not seen.

**Victoria.**    Subalpine and boggy plains, at the base of Mount Sturgeon and Mount Abrupt, and stony barren ridges near Snowy River, *F. Mueller;* also in *Mitchell's 1st Coll.*
**S. Australia.**    Clayey banks, eighteen miles W. of Glenelg river, *Robertson.*
F. Mueller describes the capsules as wingless, probably considering the wings, on account of their thickness, as angles of the capsule ; but they appear to me in this respect very much like those of the rigid varieties of *D. cuneata.*    These wings are indeed the chief character, besides the narrower sepals, to separate this species from *D. humifusa.*

9. **D. ericifolia,** *G. Don, Gen. Syst.* i. 674.    A heath-like, low but erect shrub, with numerous virgate branches, glabrous and sometimes viscid. Leaves usually crowded, narrow-linear, rather obtuse, ½ to ¾ in. long, nerveless and sometimes almost filiform.    Flowers few, in very short racemes or clusters in the upper axils or terminating short branchlets.    Sepals lanceolate, shorter than the anthers.    Capsule of *D. viscosa,* with rather broad wings. Seeds opaque.—*D. salsolifolia,* A. Cunn. in Hook. Journ. Bot. i. 251 ; Hook. f. Fl. Tasm. i. 55.

**Tasmania.**    Port Dalrymple, *R. Brown;* banks of rivers, etc., Launceston, New Norfolk, etc., not uncommon, *J. D. Hooker.*    The station, Port Jackson, usually given on the authority of plants raised in Kew Gardens, is, I believe, erroneous ; the seeds were probably from Fraser, who gathered the plant on the S. Esk river in Tasmania.    *D. filiformis,* Link, DC. Prod. i. 617, a garden plant of unknown origin, may be the same species, but too imperfectly characterized to justify the taking up the name.

10. **D. filifolia,** *Hook. in Mitch. Trop. Austr.* 241.    Erect, glabrous, and slightly viscid ; branches slender, terete or scarcely angular.    Leaves narrow-linear, almost filiform, terete or slightly flattened, often incurved, ob-

tuse or scarcely mucronate, 1 to 3 in. long, quite entire. Racemes very few-flowered, the pedicels rather long. Sepals lanceolate, about as long as the anthers. Capsule of *D. viscosa.*—*D. acerosa,* Lindl. in Mitch. Trop. Austr. 273; F. Muell. Fragm. i. 71.

**Queensland.** Newcastle ranges, between the Suttor and Burdekin rivers, *F. Mueller;* stony gullies near Mount Mudge, *Mitchell.*

11. **D. lobulata,** *F. Muell. in Linnæa,* xxv. 372. Closely allied on the one hand to *D. attenuata* and on the other to *D. ptarmicifolia,* glabrous and viscid, the branchlets scarcely angular. Leaves linear or linear-cuneate, obtuse, mostly 1 to 2 in. long, obtusely serrate or pinnatifid with short obtuse callous lobes, coriaceous and rigid, the midrib scarcely conspicuous. Flowers few, in short racemes, the pedicels rather slender. Sepals thin, broadly ovate. Capsule of the smaller forms of *D. viscosa,* the wings not very broad; Seeds smooth and shining.

**N. S. Wales.** Lachlan river, *Fraser, A. Cunningham;* between the Lachlan and Darling rivers, *Burkitt;* Mutanie ranges and Mount Goginga, *Victorian Expedition.*

**S. Australia.** S. coast, *R. Brown;* Flinders and Elder's ranges, *F. Mueller.*

**W. Australia.** In the interior, *Roe.* There are also some specimens of Drummond's which may belong to this species, with several of the leaves deeply 2- or 3-lobed, but they are evidently abnormal, the flowers being also monstrous with deformed stamens.

12. **D. ptarmicifolia,** *Turcz. in Bull. Mosc.* 1852, ii. 155. A tall shrub, glabrous and sometimes very viscid, the young branches very angular. Leaves linear-lanceolate, acuminate, acute or with a callous tip, from 1½ to 2 in. long in some specimens, 4 to 5 in. in others, sinuate-toothed, serrate or sometimes entire, gradually narrowed into a petiole, 1-nerved, the lateral veins inconspicuous. Flowers usually rather numerous, in short terminal racemes or panicles. Sepals ovate, about as long as the obtuse anthers. Capsule as in the var. *angustifolia* of *D. viscosa,* rather small, with broad wings, the terminal sinus rather open. Seeds opaque.—*D. denticulata,* F. Muell. Fragm. i. 97.

**W. Australia,** *Drummond, 5th Coll. n.* 248, Gardner river, *Herb. F. Mueller* (with short, regularly serrate leaves); Kojonerup Valley, *Herb. F. Mueller* (with long sinuate-toothed leaves).

Var. (?) *subintegra.* Scarcely viscid. Leaves long, entire or slightly toothed.—W. Australia, *Drummond, n.* 204 *and* 205. These specimens are in flower only, and resemble narrow-leaved forms of *D. truncatiales.* The species is very near to *D. viscosa angustifolia,* but with narrower leaves and the angular branches of *D. truncatiales,* and differs from both in the leaves usually toothed.

SERIES II. PLATYPTERÆ.—Leaves quite entire, flat. Wings of the capsule very divergent or divaricate, not reaching to the style nor to the base, each carpel including its wing, broader than long, transversely ovate or oblong.

13. **D. truncatiales,** *F. Muell. Fragm.* ii. 143, *and Pl. Vict.* i. 226. A tall glabrous shrub, scarcely viscid, the younger branches acutely angular. Leaves narrow-lanceolate or linear, rather acute, 2 to 4 or even 5 in. long, narrowed into a short petiole, entire or obscurely sinuate-toothed, the lateral veins little conspicuous. Racemes and flowers of *D. viscosa.* Sepals ovate, usually broad and nearly as long as the anthers. Capsule 4- or rarely 3-

lobed, flat at the top, the wings oblong, very diverging, not extending to the base of the carpels. Dissepiments remaining attached to the axis as in all the preceding species, or occasionally deciduous, but not splitting as in the two following species.—*D. calycina,* A. Cunn. Herb.; A. Gray, Bot. Amer. Expl. Exped. i. 262.

**N. S. Wales.** Port Jackson, *R. Brown;* frequent in the Blue Mountains, Croker's Range, and to the southward, *Fraser, A. Cunningham,* and others; Towamba and Yowaka rivers, *F. Mueller.*

**Victoria.** Wooded banks of Genoa river, *F. Mueller.*

14. **D. platyptera,** *F. Muell. Fragm.* i. 73. A tall shrub with the habit of the larger forms of *D. viscosa,* glabrous and viscid. Leaves elliptical-oblong or broadly lanceolate, rather obtuse, 1½ to 2½ in. long, entire, almost coriaceous, the pinnate veins rather numerous, but very fine, narrowed into a short petiole. Petals few, slender, in short racemes. Sepals narrow-ovate. Capsule flat at the top, the wings very diverging, obovate-oblong, not reaching to the style nor to the base of the carpels; dissepiments splitting and falling off with the valves, leaving only the filiform axis persistent.

**N. Australia.** Cygnet Bay, N.W. coast, *A. Cunningham;* Fitzmaurice river, *F. Mueller.* The specimens are not satisfactory. F. Mueller's have no flowers and only a few fruits; in A. Cunningham's the flowers are mostly fallen off, and I found amongst the capsules only one far enough advanced to identify them.

15. **D. stenophylla,** *F. Muell. Fragm.* i. 72. Glabrous and viscid. Leaves narrow-linear, rigid, 2 to 3 in. long, the margins usually thickened and entire. Flowers of *D. viscosa,* in short loose racemes or almost cymose panicles. Sepals ovate. Capsule small, the wings broadly oblong or obovate, diverging, not reaching to the style nor to the base of the carpels; dissepiments splitting and falling off with the valves, leaving only the filiform axis persistent.

**Queensland.** Broad Sound, *R. Brown;* Burdekin river, *F. Mueller;* Comet river, *Leichhardt.* In flower, this species is scarcely to be distinguished from *D. attenuata,* var. *linearis;* but the fruit is very different.

SERIES III. CORNUTÆ.—Leaves entire or toothed at the end, the margins revolute or rarely flat. Wings of the capsule reduced to erect or divergent, usually falcate, horn-like appendages at the upper outer angle of the carpels.

16. **D. pinifolia,** *Miq. in Pl. Preiss.* i. 227. A low shrub, with numerous divaricate or dichotomous branches, slender but rigid, terete or slightly angular, viscid when young. Leaves sessile, narrow-linear, obtuse or scarcely acute, in some specimens all under ½ in., in others exceeding 1 in., the margins revolute, entire or with a few teeth or short lobes when luxuriant. Flowers solitary or rarely 2 together, the males sessile, the females often shortly pedicellate. Sepals lanceolate. Anthers 6 to 8. Capsules 3 to 4 lines long, obtusely angled, the angles usually produced on the upper outer edge into short, erect, horn-like wings.—*Empleurosma virgata,* Bartl. in Pl. Preiss. ii. 228.

**W. Australia.** Swan River, *Drummond, n.* 117; York district, *Preiss, n.* 2166*b,* and 2438; Gordon river and Murchison river, *Oldfield.*

Var. *submutica.* Branches more angular; capsules apparently almost without horns, but not perfect in our specimens, *Drummond, 4th Coll. n.* 255.

17. **D. ceratocarpa,** *Endl. in Hueg. Enum.* 13. An erect or divaricate rigid shrub, the smaller branches virgate, acutely angled or almost winged, glabrous and often viscid. Leaves narrow-obovate oblong or narrow-cuneate, obtuse or acute, ¾ to 1¼ in. long, entire or when luxuriant sometimes 2- or 3-toothed, narrowed into a very short petiole, rather coriaceous, a few lateral veins sometimes conspicuous underneath, the margins usually recurved. Flowers on very short pedicels, few together in very short terminal leafy racemes or axillary clusters. Sepals broad, thin and almost petal-like, above 1 line long, the buds very angular. Style occasionally elongated. Capsule 2 to 3 lines long, glabrous, 4-angled, the angles produced at the upper outer edge into erect horn-like lanceolate or falcate wings, 1 to 2 lines long.—*D. pterocaulis*, Miq. in Pl. Preiss. i. 225.

**W. Australia.** Bald Head and Goose Island Bay, *R. Brown;* King George's Sound and towards Cape Riche, *A. Cunningham, Drummond, n.* 102, *and 5th Coll. n.* 246, 247, *Preiss, n.* 2440, and others.

18. **D. divaricata,** *Benth.* A low shrub, with divaricate branches, the smaller ones slender but rigid and sometimes almost spinescent, terete, glabrous or minutely pubescent and viscid. Leaves linear or linear-cuneate, 2 to 4 lines or rarely ½ in. long, entire or 3-lobed, rigid, with revolute margins. Flowers not seen. Capsules sessile or nearly so, obtusely 3- or 4-angled, often hirsute on the back, the outer angles produced into long lanceolate or falcate horn-like wings.

**W. Australia.** Between Moore and Murchison rivers, *Drummond, n.* 96, *and 4th Coll. n.* 256.

SERIES IV. APTERÆ.—Leaves entire or toothed. Capsules without wings, or the angles slightly and irregularly dilated into very narrow wings. Dissepiments persistent on the axis, except in *D. bursarifolia.*

19. **D. triangularis,** *Lindl. in Mitch. Trop. Austr.* 219 (male plant). An erect shrub of 3 to 4 ft., glabrous, pubescent or softly villous. Leaves obovate cuneate or almost triangular, rounded truncate or 3-toothed at the end, or very rarely elliptical-oblong, ½ to 1 in. or rarely 1½ in. long, coriaceous, 1-nerved, the lateral veins quite inconspicuous. Flowers axillary, solitary or clustered, on short pedicels. Sepals narrow-lanceolate, rather thick. Anthers as in *D. triquetra*, narrow, acuminate, exceeding the calyx. Capsule glabrous or pubescent, 3- or 4-angled, the angles rarely dilated towards the top into very narrow wings; dissepiments remaining attached to the axis, or very rarely deciduous but not splitting.—*D. mollis*, Lindl. in Mitch. Trop. Austr. 212 (with pubescent capsules); *D. trigona*, Lindl. l. c. 236 (with glabrous capsules); *D. Lindleyana*, F. Muell. Pl. Vict. i. 88.

**Queensland.** Suttor river, *F. Mueller;* near Mount Owen, Mount Faraday, and Mantuan Downs, *Mitchell;* near Brisbane and Ironbark forest, *Leichhardt.*
**N. S. Wales.** W. branches of Hunter's River, *A. Cunningham.*

20. **D. aptera,** *Miq. in Pl. Preiss.* i. 225. A shrub of 2 to 5 ft., glabrous and slightly viscid, the young branches very prominently angled. Leaves obovate, very obtuse or obcordate, mostly 1 to 1½ or even 2 in. long, entire, narrowed into a petiole, coriaceous, 1-nerved, the lateral veins inconspicuous. Racemes terminal, short, few-flowered. Sepals broadly lanceolate,

1 to 1½ lines long, rather thick.  Anthers narrow-oblong.  Style often elongated.  Capsule slightly 3- or 4-angled, not winged, glabrous, 3 or 4 lines long, the persistent dissepiments broad.  Seeds ovoid, smooth, and rather shining.—*D. sororia,* Miq. in Pl. Preiss. i. 225.

**W. Australia.**  Swan River and Rottenest Island, *Preiss, n.* 2388 *and* 2439, *Drummond, Coll.* 1844, *n.* 231 *and* 232, and others; Bonache Island, *Fraser.*  I can perceive no difference between the two forms described by Miquel.  The fruit-pedicels vary from 3 to 8 lines.

21.  **D. bursarifolia,** *Behr and F. Muell. in Trans. Vict. Inst.* i. 8.  A glabrous much-branched shrub, scarcely viscid, the smaller branches slender, terete or scarcely angled.  Leaves from obovate to oblong-cuneate or oblong, usually obtuse, under ½ in. or rarely ¾ in. long, entire, coriaceous, the lateral veins inconspicuous.  Pedicels short, solitary or 2 or 3 together, axillary or terminal.  Sepals narrow-lanceolate.  Anthers oblong, usually exceeding the calyx.  Styles often elongated.  Capsule 4 to 5 lines long, 3- or 4-angled, either not winged or with very narrow wings; dissepiments splitting and falling off with the valves, leaving only the filiform axis persistent.—F. Muell. Pl. Vict. i. 87, t. 5.

**Victoria.**  Murray desert, *F. Mueller.*
**S. Australia.**  Pine Forest, near Salt Creek, *Behr;* barren ridges and dry scrubby plains, near St. Vincent's Gulf, *F. Mueller.*
**W. Australia.**  In the interior, *Drummond, n.* 14 *and* 187.
Var. (?) *major.*  Leaves rather longer and not so broad in proportion, very rarely coarsely toothed.  Fruit not seen, and therefore the species doubtful.—Sharks Bay and Dirk Hartog's Island, *Milne.*

22.  **D. Baueri,** *Endl. in Hueg. Enum.* 13.  A small or spreading shrub, with short slender but rigid branches, glabrous and more or less viscid.  Leaves broadly ovate, obovate or almost orbicular, obtuse or truncate, usually slightly sinuate-toothed, mostly 4 to 6 lines long, coriaceous, 1-nerved, the lateral veins inconspicuous.  Pedicels short, recurved, axillary and solitary or few in a short terminal raceme.  Sepals broadly ovate, rather thick.  Capsule small, 3- or 4-angled, the angles very rarely produced into very narrow wings at the upper outer edge.—*D. deflexa,* F. Muell. in Trans. Vict. Inst. i. 8, and Pl. Vict. i. 87.

**Victoria.**  In the Murray scrub, *F. Mueller.*
**S. Australia.**  S. coast, *R. Brown;* Flinders Range and Spencer's Gulf, *F. Mueller;* Venus Bay, *Warburton.*

23.  **D. humifusa,** *Miq. in Pl. Preiss.* i. 226.  A low, diffuse or prostrate, much-branched shrub, often rooting at the nodes, glabrous or the young branches slightly pubescent and scarcely viscid.  Leaves crowded, linear-cuneate, oblong-spathulate or rarely almost triangular, obtuse or rarely acute, mostly under ½ in. and rarely ¾ in. long, entire or deeply 3-toothed, coriaceous, 1-nerved, the lateral veins rarely conspicuous.  Flowers usually 2 or 3 together on rather long pedicels.  Sepals ovate or ovate-lanceolate, often 2 lines long in the males, smaller in the females.  Style often elongated.  Capsule about 3 lines long, 3- or 4-angled, the angles acute or expanded towards the top into very narrow wings.

**W. Australia.**  Clayey and gravelly plains, Hay district, *Preiss, n.* 2441; towards Cape Riche, *Drummond, 5th Coll. n.* 250 *and* 251; Tone river and Tulbrunup lake, *Old-*

*field.* The general aspect is that of *D. procumbens,* from which it is readily known by the large sepals or by the fruit.

Var. *hirtella.* Branches hirsute with short spreading hairs. Leaves mostly 3-lobed.— *Drummond, 5th Coll. n.* 249.

24. **D. hexandra,** *F. Muell. in Trans. Vict. Inst.* 1855, 117. A low shrub, closely resembling *D. pinifolia* in habit, foliage, and flowers. Leaves narrow-linear with revolute margins, almost terete or subulate, under 1 in. long. Flowers solitary or 2 together, on very short recurved pedicels. Sepals ovate or lanceolate. Anthers usually 6. Capsule nearly globular or obscurely 4-angled, about 2 or nearly 3 lines diameter, not horned, but sometimes bearing small tubercles at the upper outer edge of the angles.

**S. Australia.** S. coast, *R. Brown;* Port Lincoln, Mount Greenly, and Marble Range, *Wilhelmi.*

25. **D. ericoides,** *Miq. in Pl. Preiss.* i. 227. A low shrub, with a thick rootstock and erect rather slender branching stems, often under 1 ft. high but sometimes twice as much, glabrous as well as the leaves or hoary-pubescent. Leaves sessile, linear, obtuse, 2 to 3 lines or rarely ½ in. long, the margins closely revolute, entire or with 2 or 3 small teeth or lobes. Flowers terminal, solitary, on very short recurved leafy peduncles. Sepals broadly lanceolate, acuminate, often 2 lines long, more or less hoary-tomentose. Capsule hoary-pubescent, nearly globular, with obtuse angles, neither winged nor horned, 3- or 4- rarely 5-celled.

**W. Australia.** In the interior, rare, *Preiss, n.* 2435 ; *Drummond, Coll.* 1843, *n.* 726.

SERIES V. PINNATÆ.—Leaves all pinnate or very rarely a few simple ones at the base of the branches. Capsule of the *Cyclopteræ,* except in *D. oxyptera* and *D. inæquifolia,* where it approaches that of the *Platypteræ,* and in *D. humilis,* where it is apterous; dissepiments persistent on the axis in all except *D. inæquifolia.*

26. **D. polyzyga,** *F. Muell. Fragm.* i. 74. A tall shrub, the short flowering branches nearly terete and, as well as the leaves, sparingly pubescent and glandular-viscid. Leaves pinnate, the rhachis slightly dilated or nearly terete; leaflets numerous, often above 30, oblong, acute, rarely exceeding ½ in., entire, obliquely rounded at the base and almost petiolulate, flat, 1-nerved, rather rigid but not coriaceous. Flowers not seen. Fruiting racemes terminal, loose, but much shorter than the leaves ; pedicels recurved, ½ in. long. Sepals lanceolate, foliaceous, 3 to 4 lines long. Capsule like the larger ones of *D. viscosa,* the wings rather broad but variable in shape, the terminal sinus usually open.

**N. Australia.** Upper Victoria river, *F. Mueller.*

27. **D. megazyga,** *F. Muell. Herb.* A tall shrub, glabrous and slightly viscid, the young branches acutely angled. Leaves mostly pinnate, the rhachis conspicuously winged ; leaflets usually numerous, sometimes above 30, lanceolate, acute, ½ to 1 in. long; in some specimens the lower leaves of the branches reduced to very few leaflets or to a simple linear-lanceolate leaf. Flowers rather large, in short axillary racemes or terminal panicles, the pedicels slender. Sepals ovate. Capsules small, with broad obovate or orbicular diverging wings of 3 or 4 lines.

**N. S. Wales.** Hastings river, *Herb. Lindley, Beckler;* Dogwood Creek, *Leichhardt;* Paramatta, *Woolls.* F. Mueller, Pl. Vict. i. 86, refers this to *D. viscosa,* on the ground of a few simple leaves occurring on Leichhardt's and Woolls's specimens; but even then the foliage and angular stems appear to me to be much more those of *D. truncatiales,* and the shape of the fruit rather different from both. Woolls's Paramatta specimens have no fully-formed fruits. The simple leaves are rare, and appear to occur only at the base of the branches.

28. **D. physocarpa,** *F. Muell. Fragm.* i. 74. A tall shrub, the flowering branches short, nearly terete, and as well as the leaves slightly pubescent as in *D. polyzyga,* but much less viscid. Leaves pinnate, the rhachis angular but scarcely dilated; leaflets rarely more than 10 and often only 4 to 6, obovate or oblong, obtuse or mucronate, mostly 3 to 4 lines long, entire or rarely obscurely 2- or 3-toothed, flat, 1-nerved, sometimes rather thick but not coriaceous. Racemes terminal, short, loosely few-flowered. Sepals lanceolate, obtuse, nearly 2 lines long. Anthers short, obtuse. Style often elongated. Capsule large, somewhat inflated, often 5- or 6-celled, the axis above $\frac{1}{2}$ in. long; wings not very broad, rounded above and below, but much injured in our specimens. Seeds opaque.

**N. Australia.** Sea range, Victoria river, *F. Mueller.*

29. **D. vestita,** *Hook. in Mitch. Trop. Austr.* 265. A much-branched shrub, densely villous, hirsute or pubescent, the hairs sometimes long and almost golden. Leaves pinnate, the rhachis winged; leaflets varying from few broadly obovate-cuneate and 2 or 3 lines long, to above 20, narrow-oblong and 4 or 5 lines long, entire or rarely 2- or 3-toothed, the margins always much recurved. Pedicels usually in clusters of 3 or 4, about $\frac{1}{2}$ in. long. Sepals lanceolate, acute, attaining 3 lines. Anthers 8 to 10, linear, hirsute, spirally twisted as they fade. Capsule when young hirsute with long hairs, the wings broadly orbicular, when far advanced the hairs mostly disappear and the wings are much narrower in proportion to the carpels.— *D. paulliniæfolia,* A. Cunn. Herb.; Steud. Nom. Bot. ed. 2.

**Queensland.** Belyando river, *Mitchell* (very hirsute specimens, with few, small, broad leaflets, and broadly winged, very hirsute young fruits); Endeavour river, *Banks, A. Cunningham* (scarcely more than pubescent, with numerous narrow leaflets and narrow-winged, scarcely hirsute, old fruits); Castle Creek and head of Boyd river, *Leichhardt* (leaves and indumentum intermediate, and on one specimen the young fruit, like Mitchell's, on one branch, and an old capsule, like Cunningham's, on another branch).

30. **D. pinnata,** *Sm. in Rees, Cycl.* xii. Branches terete, softly hirsute as well as the leaves as in *D. vestita.* Leaves pinnate, the rhachis winged; leaflets from about 8 to above 30, from obovate to oblong-obtuse, 2 to 4 lines long, the margins recurved, hirsute on both sides and hoary-tomentose underneath, the upper leaves often much reduced. Male flowers in short terminal compact racemes exceeding the leaves; pedicels short. Sepals broadly lanceolate, rather more than 1 line long. Anthers obtuse, hirsute, about as long as the calyx. Female flowers and fruit not seen.

**N. S. Wales.** Port Jackson, *R. Brown* (*Hb. R. Br. and Smith*). Intermediate in foliage between *D. vestita* and *D. multijuga.* This differs from both in inflorescence, but its affinities must remain doubtful until the fruit has been seen.

31. **D. oxyptera,** *F. Muell. Fragm.* i. 74. A shrub of several ft., the

branches virgate, terete, pubescent as well as the leaves and more or less viscid. Leaves pinnate, the rhachis angular but scarcely dilated; leaflets usually 5 to 11, narrow-oblong or oblong-cuneate, obtuse, 2 to 4 lines or rarely ½ in. long, the margins recurved. Flowers small, sessile or very shortly pedicellate. Sepals broad, acute, about 1 line long. Anthers obtuse, not exceeding the calyx, often hirsute. Capsule small, slightly hairy, the axis 2 to 3 lines long, the wings rigid, divergent, almost triangular and acute.

**N. Australia.** Islands of the Gulf of Carpentaria, *R. Brown*; dry rocky hills, Fitz-maurice river, Arnhem's Land, *F. Mueller.* Several of R. Brown's specimens have numerous male flowers and fruits on the same individual.

32. **D. humilis,** *Endl. Nov. Stirp. Dec.* 26, *and Atakta, t.* 31. A much-branched glabrous shrub, often viscid. Leaves pinnate, the rhachis slightly dilated; leaflets 5 to 13 or rarely more, broadly obovate-cuneate, deeply toothed at the end, 2 to 4 lines or rarely ½ in. long, narrowed at the base, the margins slightly recurved, 1-nerved, rather rigid. Flowers in short, dense, terminal corymbs, on very short pedicels. Sepals ovate or oblong, about 1½ lines long, often glandular-ciliate. Filaments more conspicuous than in most species; anthers slightly exceeding the calyx, tipped by a stipitate gland, spirally twisted as they fade. Capsule nearly globular, about 4 lines diameter, not winged, beset with rigid glandular-tipped bristles, otherwise glabrous.

**S. Australia.** Memory Cove, *R. Brown*; Port Lincoln, *Wilhelmi*; Spencer's Gulf and Streaky Bay, *Warburton.*

33. **D. boroniæfolia,** *G. Don, Gen. Syst.* i. 674. A much-branched shrub, usually pubescent or shortly hirsute, rarely glabrous, often viscid. Leaves pinnate, the rhachis more or less dilated; leaflets 5 to 9 or rarely more, obovate or cuneate-oblong, obtuse or truncate, and usually toothed at the end, 2 to 3 lines long or rarely more, coriaceous, with recurved margins. Pedicels clustered on very short lateral branches, those of the males very short, of the females often 3 to 4 lines long. Sepals ovate-lanceolate, about 1 line long. Anthers short, obtuse. Capsule of *D. viscosa*, glabrous, usually rather small, the wings not very broad, rounded at the top and at the base.— *D. Caleyana*, G. Don, Gen. Syst. i. 674 (from the character given); *D. hirtella*, Miq. in Linnæa, xviii. 94; F. Muell. Pl. Vict. i. 89.

**Queensland.** On the Maranoa, *Mitchell*; Kent's Lagoon and Bokhara flats, *Leichhardt.*
**N. S. Wales.** Liverpool plains, near Bathurst, Lachlan river, etc., *A. Cunningham*; Gwydir river, *Leichhardt*; between the Darling and Cooper's Creek, *Neilson.*
**Victoria.** Granite rocks between the Goulburn and Ovens rivers, *F. Mueller.* Several of these specimens have larger, more toothed leaflets, conspicuously marked with black dots.

34. **D. multijuga,** *G. Don, Gen. Syst.* i. 674. Shrubby and not so compact as the preceding species, pubescent or nearly glabrous, and very viscid. Leaves pinnate, the rhachis slightly dilated; leaflets usually from 15 to above 30, obliquely obovate or oblong, obtuse, often toothed, 3 to 4 lines long, the margins recurved. Flowers on slender pedicels in loose racemes, mostly terminal. Sepals lanceolate, acute, 1½ to 2 lines long. Anthers linear-oblong, nearly as long as the sepals. Capsule of *D. viscosa*, but usually larger than in *D. boroniæfolia.*

**N. S. Wales.** Port Jackson, *R. Brown*; Blue Mountains, *Miss Atkinson*; Illawarra,

*A. Cunningham, Shepherd.* Besides the numerous leaflets, this appears to be sufficiently distinct from *D. boroniæfolia*, in the longer sepals and anthers, and in inflorescence.

35. **D. larræoides,** *Turcz. in Bull. Mosc.* 1858, i. 408. Shrubby, glabrous, and very viscid, the young branches slightly angular. Leaves pinnate, the rhachis scarcely dilated; leaflets usually from 15 to near 30, linear-oblong, 2 to 4 lines long, or occasionally shorter and broader, entire or rarely minutely toothed, keeled underneath, rather rigid, the margins not recurved. Flowers not seen. Fruiting pedicels slender, clustered or very shortly racemose. Capsule of *D. viscosa*, not very large, the wings rounded at the top and at the base.—*D. multijuga*, F. Muell. Fragm. i. 219, not of G. Don; and therefore altered to *D. foliolosa*, F. Muell. Fragm. ii. 182.

**W. Australia,** Drummond, *3rd Coll., n.* 213; stony places, Geraldine mines, Murchison river, *Oldfield.*

36. **D. inæquifolia,** *Turcz. in Bull. Mosc.* 1858, i. 408. Shrubby, rigid, glabrous and usually very viscid. Leaves pinnate, the rhachis scarcely dilated; leaflets usually above 15, from linear-terete and 2 to 4 lines, to oblong and scarcely 1 line long, obtuse and often callous at the end, channeled above, convex underneath. Pedicels rather slender, clustered, those of the males very short. Sepals ovate, 1 to 1½ lines long. Anthers short and very obtuse. Capsules small, the wings usually ovate or obovate and very divergent, narrowed at the top and the base almost as in the *Platypteræ*. Seeds smooth and shining.—*D. leptozyga*, F. Muell. Fragm. i. 219.

**W. Australia,** Drummond, *4th Coll., n.* 258; Sharks Bay, *Denham;* Dirk Hartog's Island, *Milne;* Murchison river, *Oldfield.*

37. **D. adenophora,** *Miq. in Linnæa,* xviii. 95. A rigid shrub, glabrous and usually very viscid, the young branches angular. Leaves pinnate, the rhachis scarcely dilated; leaflets 3 to 9 or rarely 11, linear or slightly cuneate, obtuse and often callous at the tips, 2 to 4 lines long, very rarely slightly toothed at the end, convex or keeled underneath, flat above, rather thick and rigid. Pedicels slender, clustered. Sepals ovate, acute, or very shortly racemose, rather more than 1 line long. Anthers short, very obtuse. Capsule small, the wings rather broad, rounded at the top and at the base; dissepiments splitting and coming off with the valves, leaving only the filiform axis persistent as in *D. platyptera, D. stenophylla,* and *D. bursarifolia.*—*Thouinia (?) adenophora*, Miq. in Pl. Preiss. i. 224.—*D. tenuifolia*, Lindl. in Mitch. Trop. Austr. 248 (the Queensland and N. S. Wales specimens).

**Queensland.** Condamine river, *Leichhardt;* Belyando river, *Mitchell.*
**N. S. Wales.** Rocky hills near Liverpool plains, *A. Cunningham.*
**W. Australia,** Drummond, *5th Coll., Suppl., n.* 38; Darling range, *Preiss, n.* 2442.
Leichhardt's specimens are in leaf only, and Mitchell's in flower only. Cunningham's are in flower and fruit, but the capsules are not quite ripe enough to be certain of the dehiscence; as far as they go, however, I can see no difference whatever between them and Drummond's excellent fruiting specimens, which again agree perfectly with the fruiting fragments I have seen of Preiss's. Should, however, the eastern plant prove to have the persistent dissepiments of *D. viscosa*, it will stand as a distinct species, under the name of *D. tenuifolia*, Lindl., differing from *D. stenozyga* in its flat, linear leaflets, and clustered or racemose pedicels.

38. **D. stenozyga,** *F. Muell. Fragm.* i. 98, *and Pl. Vict.* i. 88. An

erect, compact, very much branched shrub, glabrous and often viscid, the last slender branchlets not much thicker than the petioles and leaflets. Leaves mostly pinnate with few usually distant linear and almost terete leaflets rarely above ½ in. long, channelled above and convex underneath like the common petioles. Male flowers not seen. Female pedicels solitary, 2 to 6 lines long. Sepals oblong-lanceolate, about 1 line long. Capsules of *D. viscosa*, rather large, the wings rounded at the top and at the base, the terminal sinus open; persistent dissepiments rather broad.

**N. S. Wales.** Desert of the Darling, *Dallachy* and *Goodwin.*

**Victoria.** Desert near the confluence of the Loddon and the Murray, *F. Mueller.*

**S. Australia.** S. coast, *R. Brown* (leaflets rather more numerous, but inflorescence of *D. stenozyga*).

**W. Australia,** *Drummond, n.* 188 (specimens precisely similar to the Victorian ones).

39? **D. concinna,** *Benth.* Very near *D. stenozyga,* and perhaps a variety, but the small specimens seen have a very different aspect. Leaflets 5 to 11, crowded on short coriaceous petioles, linear, almost terete, channelled above, convex underneath, 2 to 4 lines long. Flowers not seen. Fruiting pedicels several, in a very short raceme. Capsule of *D. viscosa,* the wings rounded at the top and at the base, the dissepiments broad and persistent as in *D. stenozyga,* not splitting and deciduous as in the true *D. adenophora.*— *D. adenophora,* F. Muell. Fragm. i. 98, not of Miquel.

**W. Australia.** In the south-west, *Herb. F. Mueller.*

## 16. DISTICHOSTEMON, F. Muell.

Characters of *Dodonæa* except that the sepals vary from 5 to 8, and the stamens are indefinite, usually above 20, closely packed in 2 or more series.— Pubescent shrub. Leaves simple. Inflorescence more nearly an interrupted spike than in any *Dodonæas.*

The genus is limited to a single species, endemic in Australia, scarcely sufficiently distinct from *Dodonæa.*

1. **D. phyllopterus,** *F. Muell. in Hook. Kew Journ.* ix. 306. A tall shrub, softly tomentose-pubescent or villous in all its parts. Leaves very shortly petiolate, oblong or rarely obovate, very obtuse, 1 to 3 in. long, entire, soft and velvety on both sides, the veins prominent underneath. Flowers nearly sessile, in terminal leafless interrupted spikes or racemes of 1 to 3 in., rarely branching into oblong panicles. Sepals most frequently 6, but in some specimens almost all 5. Stamens although usually above 20, yet occasionally only 12 to 15, and often above 30; anthers oblong-linear, crowded, with very short filaments as in *Dodonæa.* Styles occasionally elongated as in some *Dodonæas.* Capsule more or less tomentose, obovoid-triquetrous, the angles more or less produced into herbaceous erect wings, usually ovate, very obtuse, and only on the upper outer half of the carpels, but occasionally, especially in the Banksian specimens, not so broad, and continued almost to the base. Seeds very shining, usually 2 in each cell.—*Dodonæa hispidula,* Endl. Atakt. t. 30.

**N. Australia.** N.W. coast, *Bynoe*; Goulbourn Island and Cape Pond, *A. Cunningham*; Victoria river, Point Pearce, and Roper River, *F. Mueller*; Port Essington, *Arm-*

*strong ;* islands of the Gulf of Carpentaria, *R. Brown, Henne;* from Arnhem's Land to the sources of Gilbert's River, not rare, *F. Mueller.*

*Alectryon (?) canescens,* DC. Prod. i. 617, from the E. coast, with oblong, obtuse, closely pubescent leaves, axillary racemes the length of the leaves, the fruit nearly of *Cameraria,* surrounded by a wing connate with the style, and thick, oblong seeds, is unknown to me. From the above very unsatisfactory description, it cannot be an *Alectryon,* and is most probably not *Sapindaceous.*

## Order XXXIX. ANACARDIACEÆ.

Flowers unisexual polygamous or hermaphrodite, usually regular. Calyx of 3 to 5 lobes or distinct sepals. Petals 3 to 7, rarely none. Disk usually annular or broad. Stamens of the same number or twice as many as petals, very rarely indefinite, inserted round the disk or rarely upon it ; filaments free ; anthers versatile. Ovary superior, usually 1-celled, with 1 to 3 styles, or in the *Spondieæ* 2- to 5-celled, or very rarely of 2 to 5 distinct carpels, or in male flowers reduced to 4 or 5 rudimentary style-like carpels. Ovules solitary in the ovary or in each of its cells, pendulous or broadly adnate to the side of the cavity, or suspended from a free funicle erect from the base of the cavity, with a dorsal raphe and inferior micropyle ; very rarely in genera not Australian erect, with a ventral raphe and inferior micropyle. Fruit superior or rarely half inferior, free or adnate at the base to the enlarged calyx-tube or disk, 1-celled or (in *Spondieæ*) several-celled, usually drupaceous and indehiscent. Seed erect horizontal or pendulous ; albumen none or very thin. Embryo straight or incurved, cotyledons usually fleshy ; radicle short, inferior or more frequently turned upwards or superior.—Trees or shrubs, the bark often exuding a caustic, balsamic or gummy juice. Leaves alternate or very rarely opposite, without real stipules, simple or ternately or pinnately compound, usually without glandular dots. Inflorescence various, usually paniculate, with small flowers. Flesh of the drupes usually oily or full of caustic juice.

The Order is abundantly distributed over the tropical regions of the New and the Old World, more rare in temperate climates. Of the five Australian genera, two are common to the New and the Old World, two are Asiatic, and the fifth is endemic.

Ovary 1-celled or carpels distinct.
  Leaves pinnate or 3-foliolate.
    Stamens 5 or 10. Ovule suspended from an erect funicle . . . 1. RHUS.
    Stamens 10. Ovules suspended from the top of the cavity . . . 3. EUROSCHINUS.
  Leaves simple.
    Stamens 10. Carpels 5 or 6. Ovules suspended from an erect
    funicle . . . . . . . . . . . . . . . . . 2. BUCHANANIA.
    Stamens 5. Ovary 1-celled. Ovule suspended from the top of the
    cavity . . . . . . . . . . . . . . . . . 4. SEMECARPUS.
Ovary 2- or more celled. Leaves pinnate. Stamens 8 or 10. Ovules
suspended from the top of the cavity . . . . . . . . . . 5. SPONDIAS.

### 1. RHUS, Linn.

Flowers polygamous. Calyx small, of 4 to 6, usually 5, imbricate sepals. Petals as many as sepals, imbricate in the bud. Disk broad, flat or annular.

Stamens as many as petals or rarely 10, inserted round the base of the disk. Ovary 1-celled; styles 3, free or connate, with simple or capitate stigmas; ovule suspended from an erect filiform funicle. Drupe globular or compressed, usually small. Seed inverted or transverse, the radicle turned upwards.— Trees or shrubs. Leaves pinnate, 3-foliolate, or in species not Australian simple. Flowers small, in terminal or axillary panicles.

The species are numerous in the warmer extratropical regions of both the northern and southern hemispheres, especially in S. Africa, more rare within the tropics. The Australian species are both endemic.

Leaves pinnate, glabrous. Flowers rather large. Stamens 10. Drupes
globular . . . . . . . . . . . . . . . . . . . . . . 1. *R. rhodanthema.*
Leaves digitately 3- or 5-foliolate, tomentose underneath. Flowers very
small. Stamens 5 . . . . . . . . . . . . . . . . . 2. *R. viticifolia.*

1. **R. rhodanthema,** *F. Muell. Herb.* A tree of 70 to 80 ft., quite glabrous except little tufts of hairs along the midrib of the leaflets underneath. Leaves pinnate, the common petiole terete; leaflets usually 7 or 9, oblong, obtusely acuminate, mostly 2 to $2\frac{1}{2}$ in. long, entire, shortly petiolulate, the pinnate veins prominent underneath. Panicles pyramidal or broadly thyrsoid, dense. Flowers diœcious, red, very shortly pedicellate, larger than in most species. Sepals broadly ovate, very obtuse, about 1 line long. Petals ovate, recurved, about $1\frac{1}{4}$ lines. Stamens 10. Ovary broad; styles 3, short, thick, diverging, with capitate stigmas; ovule nearly globular, suspended as in the rest of the genus from an erect funicle. Drupe globular, shining, about $\frac{1}{2}$ in. diameter, putamen thick and woody, striate outside, lined with a separable cartilaginous layer inside. Seeds orbicular, flat; testa membranous, but rather thick.

**Queensland.** Wide Bay, *C. Moore*; Brisbane river, Moreton Bay, *Fraser, A. Cunningham, W. Hill, F. Mueller.*
**N. S. Wales.** Clarence river, *Herb. F. Mueller.*
This species differs from the greater part of the genus in its large red flowers, 10 stamens, and larger globular drupes. *R. simarubæfolia,* A. Gray, from the Fiji islands, approaches it in general habit and in the size of the flowers, but they are white and pentandrous, and the leaflets are firmer and more obtuse.

2? **R. viticifolia,** *F. Muell. Herb.* Branches, petioles, and inflorescence hoary-pubescent. Leaves digitately compound; Leaflets 3 or (according to F. Mueller) rarely 5, ovate or elliptical, acute, 2 to 3 in. long, entire or sinuate-toothed, narrowed into a petiolule, glabrous above, white or hoary underneath with a close tomentum. Flowers very small, in a pyramidal or thyrsoid terminal panicle. Sepals lanceolate, hirsute, about $\frac{1}{2}$ line long. Petals oblong, nearly 1 line long, glabrous. Stamens 5. Female flowers and fruit not seen.

**Queensland** (?), *Leichhardt.* Evidently closely allied to the S. African *R. tomentosa,* Linn. The leaves appear to be less coriaceous, but otherwise the fragmentary specimens are insufficient to give diagnostic characters. Can it be the species imported?

## 2. BUCHANANIA, Roxb.

Flowers hermaphrodite. Calyx short, obtusely 3- to 5-toothed. Petals 5, imbricate in the bud. Disk orbicular, crenate. Stamens 10, inserted round the disk. Gynœcium of 5 or 6 distinct carpels, of which one only perfect, the others rudimentary and style-like; style of the perfect one short, with a truncate stigma; ovule suspended from an erect filiform funicle. Drupe small, the putamen crustaceous or bony, 2-valved. Seed with thick cotyledons and a superior radicle.—Trees. Leaves alternate, simple, entire, coriaceous. Flowers small, white, in terminal or axillary panicles.

The genus extends over tropical Asia and the islands of the Pacific, the Australian species having also a wide Asiatic range.

1. **B. angustifolia,** *Roxb. Pl. Corom.* iii. 68, *t. 262.* A tree, either quite glabrous or the young shoots and panicles slightly rusty-tomentose or pubescent. Leaves oblong or cuneate-oblong, obtuse and rounded at the end, 3 to 8 in. long, and 1 to 2 in. broad, gradually narrowed into a short petiole, rather rigid, of a pale colour, the pinnate veins and transverse reticulate veinlets prominent on both sides. Panicles rather loose, shorter than the leaves, several together at the ends of the branches, each in the axil of a floral leaf usually reduced to a small bract; occasionally the central bud grows out and the panicles are placed at the base of the new branch. Flowers glabrous; petals nearly 1½ lines long. Drupe more or less compressed, oblique, from broadly ovate to nearly oblong, rarely exceeding ½ in.—W. and Arn. Prod. 169, with the synonyms adduced; Wight, Ic. t. 101.

**N. Australia.** Victoria river, *Bynoe, F. Mueller;* Port Essington, *Armstrong;* islands of the Gulf of Carpentaria, *R. Brown.*

**Queensland.** Albany Island, *F. Mueller;* N.E. coast, *A. Cunningham.*

The species is widely distributed over East India and the Archipelago.

### 3. EUROSCHINUS, Hook. f.

Flowers polygamous or diœcious. Calyx small, 5-lobed. Petals 5, imbricate in the bud. Disk orbicular, deeply crenate. Stamens 10, inserted round the disk. Ovary 1-celled, with 3 thick short styles, or in the males of 3 or 4 linear style-like rudiments; ovule pendulous from the top of the cavity. Drupe small, more or less compressed, the putamen coriaceous. Seeds compressed, with flat cotyledons; the radicle turned upwards.—Tree. Leaves pinnate. Flowers rather small, in terminal or lateral panicles.

The genus is limited to a single species, endemic in Australia. It is closely allied to the American genus *Schinus*, but with a rather different habit, a gamosepalous calyx, and the putamen of the fruit does not appear to contain the oily receptacles so conspicuous in that genus.

1. **E. falcatus,** *Hook. f. in Benth. and Hook. Gen. Pl. 422.* A low tree, glabrous or the young shoots minutely hoary. Leaflets 4 to 8, very oblique or falcate, ovate to lanceolate, shortly acuminate, 2 to 3 in. long, all but the terminal one very unequal at the base, on petiolules of 1 to 3 lines, penninerved and reticulate, the common petiole terete. Panicles divaricate, many-flowered, not exceeding the leaves. Flowers almost sessile, clustered along the branches, about 1 line long and glabrous. Calyx-lobes obtuse, slightly imbricate. Petals twice as long, oblong, very spreading. Drupes at first broadly and obliquely ovate, but in some specimens where they are better ripened more oblong, and attaining almost ½ in. in length.

**Queensland.** Sources of the Burdekin, *F. Mueller ;* Sunday Island, *M'Gillivray.*
**N. S. Wales.** Hastings river, *Beckler ;* Clarence river, *C. Moore.*
Var. *angustifolius.* Leaves falcate-lanceolate, much acuminate. Flowers rather larger.
—Northumberland Islands, *R. Brown ;* Rockhampton, *Thozet.*

### 4. SEMECARPUS, Linn. f.

Flowers polygamous. Calyx small, 5-lobed. Petals 5, imbricate in the bud. Disk orbicular, slightly lobed or crenate. Stamens 5, inserted round the disk. Ovary 1-celled, with 3 styles, and somewhat club-shaped stigmas ; ovule suspended from the top of the cavity. Drupe or nut reniform, seated on the much-enlarged, thick, succulent, fleshy, cupular or turbinate base of the calyx ; pericarp thick, hard, filled with resinous cells. Seed pendulous, the testa coriaceous, somewhat fleshy inside ; embryo thick, with plano-convex cotyledons and a very short superior radicle.—Trees. Leaves alternate, Flowers small, in terminal or lateral panicles.

The genus ranges over tropical Asia, the species most numerous in Ceylon ; the Australian one extending over nearly the whole area.

1. **S. Anacardium,** *Linn. ; W. and Arn. Prod.* 168, *var. (?) parvifolia.* Leaves broadly obovate, very obtuse, 3 to 4 in. long, entire, rounded at the base, on very short petioles, glabrous above, hoary or white underneath but scarcely tomentose, the pinnate veins and reticulate veinlets conspicuous on both sides. Male panicles pyramidal, shorter than or as long as the leaves. Flowers very small, sessile and clustered. Calyx very short. Petals scarcely 1 line long. Ovary minute and rudimentary or reduced to a tuft of hair. Female flowers and fruit of the Australian variety not seen.

**N. Australia.** Port Essington, *Armstrong.* The species is widely distributed over E. India, and has usually leaves from ⅓ to 1 ft. long, but, as far as our specimens go, I can see no character, besides the smaller leaves, to distinguish the Australian form.

There is also in Armstrong's Port Essington collection, a single leaf, 2½ ft. long by about 7 in. broad, and acutely acuminate, of what may be *S. cassuvium,* Roxb. Fl. Ind. ii. 85, a Molucca species.

### 5. SPONDIAS, Linn.

(Evia, *Comm. ;* Cytherea, *W. and Arn.*)

Flowers polygamous. Calyx small, 4- or 5-lobed or divided to the base. Petals 4 or 5, spreading, almost valvate in the bud. Disk orbicular, crenate. Stamens twice as many as petals, inserted round the disk. Ovary 3- to 5- (or sometimes 10- to 15- ?) celled, with as many short, conical, connivent styles ; ovules solitary in each cell, pendulous. Drupe with a fleshy epicarp, the putamen hard and bony, the cells erect or vertically curved and diverging at the top, the putamen pierced with a foramen corresponding to the apex of each cell. Seeds solitary in each cell, pendulous ; testa membranous ; embryo straight or slightly curved with the seed ; cotyledons oblong, radicle superior. —Trees. Leaves crowded at the ends of the branches, pinnate. Flowers small, in terminal or axillary panicles.

The genus is widely spread over tropical conntries, and some species are also cultivated under the name of *Hog Plums.* It is often divided into two : *Spondias,* chiefly American, with erect cells in the drupe, and *Evia* or *Cytherea,* chiefly Asiatic, with the cells divergent at the top. The Australian species, which is endemic, belongs to the latter group.

1. **S. Solandri,** *Benth.*   A moderate-sized tree, the trunk occasionally acquiring a very great thickness, quite glabrous in all its parts.   Leaflets 7 or 9, obliquely ovate or oblong, obtuse, 2 to 3 in. long, entire, very unequal at the base, pale underneath, with fine pinnate veins and reticulate veinlets. Flowers sessile, densely clustered, in short axillary interrupted spikes or racemes, rarely branching into panicles.   Calyx-lobes separate almost to the base, ovate, obtuse, about $\frac{1}{2}$ line long.   Petals 5, spreading, obtuse, about $1\frac{1}{2}$ lines long.   Stamens 10, inserted in or under the crenatures of the disk ; filaments slender ; anthers small.   Ovary half immersed in the disk, with 4 or sometimes 3 short conical styles.—*Spondias acida,* Soland. in Herb. Banks, not of Blume.

**Queensland.**   Endeavour river, *Banks* and *Solander ;* Keppel Bay, Shoalwater Bay, Broad Sound, and Northumberland Islands, *R. Brown.*   The above description is taken from R. Brown's notes, and from two flowering specimens in the Banksian herbarium, and one in R. Brown's.   There is also in the Banksian collection a packet of drupes named as belonging to this species and described as such in R. Brown's notes ; but perhaps really those of some allied species, for they have from 10 to 15, usually about 12 cells, although in every other respect like those of the section *Evia* of *Spondias.*   They are of a depressed globular form, the putamen with as many angles as cells, exceedingly hard, nearly 1 in. diameter ; the cells diverging at the top as in other *Evias.*

# INDEX OF GENERA AND SPECIES.

*The synonyms and species incidentally mentioned are printed in italics.*

# 494

JOHN EDWARD TAYLOR, PRINTER, LITTLE QUEEN STREET, LINCOLN'S INN FIELDS.